Advances in Intelligent Systems and Computing

Volume 327

Series editor

Janusz Kacprzyk, Polish Academy of Sciences, Warsaw, Poland
e-mail: kacprzyk@ibspan.waw.pl

About this Series

The series "Advances in Intelligent Systems and Computing" contains publications on theory, applications, and design methods of Intelligent Systems and Intelligent Computing. Virtually all disciplines such as engineering, natural sciences, computer and information science, ICT, economics, business, e-commerce, environment, healthcare, life science are covered. The list of topics spans all the areas of modern intelligent systems and computing.

The publications within "Advances in Intelligent Systems and Computing" are primarily textbooks and proceedings of important conferences, symposia and congresses. They cover significant recent developments in the field, both of a foundational and applicable character. An important characteristic feature of the series is the short publication time and world-wide distribution. This permits a rapid and broad dissemination of research results.

Advisory Board

Chairman

Nikhil R. Pal, Indian Statistical Institute, Kolkata, India
e-mail: nikhil@isical.ac.in

Members

Rafael Bello, Universidad Central "Marta Abreu" de Las Villas, Santa Clara, Cuba
e-mail: rbellop@uclv.edu.cu

Emilio S. Corchado, University of Salamanca, Salamanca, Spain
e-mail: escorchado@usal.es

Hani Hagras, University of Essex, Colchester, UK
e-mail: hani@essex.ac.uk

László T. Kóczy, Széchenyi István University, Győr, Hungary
e-mail: koczy@sze.hu

Vladik Kreinovich, University of Texas at El Paso, El Paso, USA
e-mail: vladik@utep.edu

Chin-Teng Lin, National Chiao Tung University, Hsinchu, Taiwan
e-mail: ctlin@mail.nctu.edu.tw

Jie Lu, University of Technology, Sydney, Australia
e-mail: Jie.Lu@uts.edu.au

Patricia Melin, Tijuana Institute of Technology, Tijuana, Mexico
e-mail: epmelin@hafsamx.org

Nadia Nedjah, State University of Rio de Janeiro, Rio de Janeiro, Brazil
e-mail: nadia@eng.uerj.br

Ngoc Thanh Nguyen, Wroclaw University of Technology, Wroclaw, Poland
e-mail: Ngoc-Thanh.Nguyen@pwr.edu.pl

Jun Wang, The Chinese University of Hong Kong, Shatin, Hong Kong
e-mail: jwang@mae.cuhk.edu.hk

More information about this series at http://www.springer.com/series/11156

Suresh Chandra Satapathy · Bhabendra Narayan Biswal
Siba K. Udgata · J.K. Mandal
Editors

Proceedings of the 3rd International Conference on Frontiers of Intelligent Computing: Theory and Applications (FICTA) 2014

Volume 1

Springer

Editors

Suresh Chandra Satapathy
Department of Computer Science
    and Engineering
Anil Neerukonda Institute of Technology
    and Sciences
Vishakapatnam
India

Bhabendra Narayan Biswal
Bhubaneswar Engineering College
Bhubaneswar, Odisha
India

Siba K. Udgata
University of Hyderabad
Hyderabad, Andhra Pradesh
India

J.K. Mandal
Department of Computer Science
    and Engineering
Faculty of Engg., Tech. & Management
University of Kalyanai
Kalyanai, West Bengal
India

ISSN 2194-5357   ISSN 2194-5365   (electronic)
ISBN 978-3-319-11932-8   ISBN 978-3-319-11933-5   (eBook)
DOI 10.1007/978-3-319-11933-5

Library of Congress Control Number: 2014951353

Springer Cham Heidelberg New York Dordrecht London

Printed on acid-free paper

Springer is part of Springer Science+Business Media (www.springer.com)

Sireesha Rodda, India
Srinivas Sethi, India
Jitendra Agrawal, India

Suresh Limkar, India
Bapi Raju Surampudi, India
S. Mini, India and many more

Organizing Committee

Committee Members from CSI Students Branch ANITS

M. Kranthi Kiran, Asst. Professor, Dept. of CSE, ANITS
G.V. Gayatri, Asst. Professor, Dept. of CSE, ANITS
S.Y. Manikanta, ANITS
Aparna Patro, ANITS
T. Tarun Kumar Reddy, ANITS
A. Bhargav, ANITS
N. Shriram, ANITS
B.S.M. Dutt, ANITS
I. Priscilla Likitha Roy, ANITS
Ch. Geetha Manasa, ANITS
S.V.L. Rohith, ANITS
N. Gowtham, ANITS
B. Bhavya, ANITS and many more

Committee Members from BEC

Sangram Keshari Samal, HOD, Aeronautical Engineering
Manas Kumar Swain, Prof. Computer Science
A.K. Sutar, HOD, Electronics and Telecommunication Engineering
Pabitra Mohan Dash, HOD, Electrical & Electronics Engineering
Rashmita Behera, Asst. Prof. Electrical & Electronics Engineering
Utpala Sahu, Asst. Prof. Civil Engineering
Sonali Pattnaik, Asst. Prof. Civil Engineering
V.M. Behera, Asst. Prof. Mechanical Engineering
Gouri Sankar Behera, Asst. Prof. Electrical & Electronics Engineering
Debashis Panda, Prof. Department of Sc &H and many more

Contents

Section I: Data Warehousing and Mining, Machine Learning

Section II: Mobile and Ubiquitous Computing

Program Chairs

Dr. Suresh Chandra Satapathy   ANITS, Vishakapatnam, India
Dr. S.K. Udgata   University of Hyderabad, India
Dr. B.N. Biswal, Director (A &A)   BEC, Bhubaneswar, India
Dr. J.K. Mandal   KalayanI University, West Bengal, India

International Advisory Committee/Technical Committee

P.K. Patra, India
Sateesh Pradhan, India
J.V.R Murthy, India
T.R. Dash, Kambodia
Sangram Samal, India
K.K. Mohapatra, India
L. Perkin, USA
Sumanth Yenduri, USA
Carlos A. Coello Coello, Mexico
S.S. Pattanaik, India
S.G. Ponnambalam, Malaysia
Chilukuri K. Mohan, USA
M.K. Tiwari, India
A. Damodaram, India
Sachidananda Dehuri, India
P.S. Avadhani, India
G. Pradhan, India
Anupam Shukla, India
Dilip Pratihari, India
Amit Kumar, India
Srinivas Sethi, India
Lalitha Bhaskari, India
V. Suma, India
Pritee Parwekar, India
Pradipta Kumar Das, India
Deviprasad Das, India
J.R. Nayak, India
A.K. Daniel, India
Walid Barhoumi, Tunisia
Brojo Kishore Mishra, India
Meftah Boudjelal, Algeria
Sudipta Roy, India
Ravi Subban, India
Indrajit Pan, India
Prabhakar C.J., India
Prateek Agrawal, India

Igor Belykh, Russia
Nilanjan Dey, India
Srinivas Kota, Nebraska
Jitendra Virmani, India
Shabana Urooj, India
Chirag Arora, India
Mukul Misra, India
Kamlesh Mishra, India
Muneswaran, India
J. Suresh, India
Am,lan Chakraborthy, India
Arindam Sarkar, India
Arp Sarkar, India
Devadatta Sinha, India
Dipendra Nath, India
Indranil Sengupta, India
Madhumita Sengupta, India
Mihir N. Mohantry, India
B.B. Mishra, India
B.B. Pal, India
Tandra Pal, India
Utpal Nandi, India
S. Rup, India
B.N. Pattnaik, India
A. Kar, India
V.K. Gupta, India
Shyam lal, India
Koushik Majumder, India
Abhishek Basu, India
P.K. Dutta, India
Md. Abdur Rahaman Sardar, India
Sarika Sharma, India
V.K. Agarwal, India
Madhavi Pradhan, India
Rajani K. Mudi, India
Sabitha Ramakrishnan, India

Organization

Organizing Committee

Chief Patron

Er. Pravat Ranjan Mallick,
   Chairman   KGI, Bhubaneswar

Patron

Er. Alok Ranjan Mallick,   KGI, Bhubaneswar
   Vice-Chairman   Chairman, BEC, Bhubaneswar

Organizing Secretary

Prof. B.N. Biswal,   BEC, Bhubaneswar
   Director (A &A)

Honorary Chairs

Dr. P.N. Suganthan   NTU, Singapore
Dr. Swagatam Das   ISI, Kolkota

Steering Committee Chair

Dr. B.K. Panigrahi   IIT, Delhi, India

reviewers who not only produced excellent reviews but also did in short time frames. We would also like to thank Bhubaneswar Engineering College (BEC), Bhubaneswar having coming forward to support us to organize the third edition of this conference in the series. Our heartfelt thanks are due to Er. Pravat Ranjan Mallick, Chairman, KGI, Bhubaneswar for the unstinted support to make the conference a grand success. Er. Alok Ranjan Mallick, Vice-Chairman, KGI, Bhubaneswar and Chairman of BEC deserve our heartfelt thanks for continuing to support us from FICTA 2012 to FICTA 2014. A big thank to Sri V Thapovardhan, the Secretary and Correspondent of ANITS and Principal and Directors of ANITS for supporting us in co-hosting the event. CSI Students Branch of ANITS and its team members have contributed a lot to FICTA 2014. All members of CSI ANITS team deserve great applause.

We extend our heartfelt thanks to Prof. P.N. Suganthan, NTU Singapore and Dr. Swagatam Das, ISI Kolkota for guiding us. Dr. B.K. Panigrahi, IIT Delhi deserves special thanks for being with us from the beginning to the end of this conference. We would also like to thank the authors and participants of this conference, who have considered the conference above all hardships. Finally, we would like to thank all the volunteers who spent tireless efforts in meeting the deadlines and arranging every detail to make sure that the conference can run smoothly. All the efforts are worth and would please us all, if the readers of this proceedings and participants of this conference found the papers and conference inspiring and enjoyable.

Our sincere thanks to all press print & electronic media for their excellent coverage of this conference.

November 2014

Volume Editors
Suresh Chandra Satapathy
Siba K. Udgata
Bhabendra Narayan Biswal
J.K. Mandal

Preface

This AISC volume-I contains 95 papers presented at the Third International Conference on Frontiers in Intelligent Computing: Theory and Applications (FICTA-2014) held during 14–15 November 2014 jointly organized by Bhubaneswar Engineering College (BEC), Bhubaneswar, Odisa, India and CSI Student Branch, Anil Neerukonda Institute of Technology and Sciences, Vishakhapatnam, Andhra Pradesh, India. It once again proved to be a great platform for researchers from across the world to report, deliberate and review the latest progresses in the cutting-edge research pertaining to intelligent computing and its applications to various engineering fields. The response to FICTA 2014 has been overwhelming. It received a good number of submissions from the different areas relating to intelligent computing and its applications in main track and five special sessions and after a rigorous peer-review process with the help of our program committee members and external reviewers finally we accepted 182 submissions with an acceptance ratio of 0.43. We received submissions from eight overseas countries including India.

The conference featured many distinguished keynote addresses by eminent speakers like Dr. A. Govardhan, Director, SIT, JNTUH, Hyderabad, Dr. Bulusu Lakshmana Deekshatulu, Distinguished fellow, IDRBT, Hyderabad and Dr. Sanjay Sen Gupta, Principal scientist, NISC & IR, CSIR, New Delhi. The five special sessions were conducted during the two days of the conference.

Dr. Vipin Tyagi, Jaypee University of Engg and Tech, Guna, MP conducted a special session on "Cyber Security and Digital Forensics", Dr. A. Srinivasan, MNAJEC, Anna University, Chennai and Prof. Vikrant Bhateja, Sri Ramswaroop Memorial Group of Professional colleges, Lucknow conducted a special session on "Advanced research in 'Computer Vision, Image and Video Processing". Session on "Application of Software Engineering in Multidisciplinary Domains" was organized by Dr Suma V., Dayananda Sagar Institutions, Bangalore and Dr Subir Sarkar, Former Head, dept of ETE, Jadavpur University organized a special session on " Ad-hoc and Wireless Sensor Networks".

We take this opportunity to thank all Keynote Speakers and Special Session Chairs for their excellent support to make FICTA 2014 a grand success.

The quality of a referred volume depends mainly on the expertise and dedication of the reviewers. We are indebted to the program committee members and external

Section III: AI, E-commerce and Distributed Computing

Section IV: Soft Computing, Evolutionary Computing, Bio-inspired Computing and Applications

Approximation Neural Network for Phoneme Synthesis

Marius Crisan

Polytechnic University of Timisoara,
Department of Computer and Software Engineering,
Blvd. V. Parvan 2, 300223 Timisoara, Romania
marius.crisan@cs.upt.ro

Abstract. The paper presents a dynamic method for phoneme synthesis using an elemental-based concatenation approach. The vocal sound waveform can be decomposed into elemental patterns that have slight modifications of the shape as they chain one after another in time but keep the same dynamics which is specific to each phoneme. An approximation or RBF network is used to generate elementals in time with the possibility of controlling the characteristics of the sound signals. Based on this technique a quite realistic mimic of a natural sound was obtained.

Keywords: Neural networks, Time series modeling, Phoneme generation, Speech processing.

1   Introduction

Speech synthesis remains a challenging topic in artificial intelligence since the goal of obtaining natural human-like sounds is not yet fully reached. Among the different approaches of speech synthesis, the closest to human-like sounds are obtained by the concatenation of segments of recorded speech. However, one of the main disadvantages of this technique, besides the necessity of a large database, is the difficulty of reproducing the natural variations in speech [1]. There are some attempts towards expressive speech synthesis using concatenative technique [2] but the lack of an explicit speech model in these systems makes them applicable mainly in neutral spoken rendering of text. The other techniques of speech synthesis such as formant/parametric synthesis and articulatory synthesis [3], although they employ acoustic models and human vocal tract models, cannot surpass the results of a robotic-sounding speech. The main difficulty resides in dealing with the nonlinear character of natural language phenomenon. Therefore the need of developing new models able to encompass the dynamics of speech became prominent in the recent years. More research works were invested in nonlinear analysis of speech signals in order to derive valid dynamic models [4], [5], [6], [7], [8]. A promising direction is given by the neural networks dynamic models. There are classic approaches in speech synthesis (text-to-speech synthesis) that use neural networks, but not for generating directly the audio signal [9], [10], [11]. However, neural networks did prove successfully to have the potential of predicting and generating nonlinear time-series [12], [13], [14].

© Springer International Publishing Switzerland 2015
S.C. Satapathy et al. (eds.), *Proc. of the 3rd Int. Conf. on Front. of Intell. Comput. (FICTA) 2014*
– *Vol. 1*, Advances in Intelligent Systems and Computing 327, DOI: 10.1007/978-3-319-11933-5_1

Different topologies have been studied starting from a feed-forward neural network architecture and adding feedback connections to previous layers [15], [16]. Applications in speech and sound synthesis have also been proposed [17], [18]. In a recent work [19] we have studied the possibility of training a feedback topology of neural network for the generation of three new periods of elemental patterns in phonemes. The phoneme sound was finally generated in a repetitive loop with promising results. In the present work we are interested in extending the ideas of dynamic modeling of speech sounds with neural networks, this time using approximation nets. The approximation or interpolation networks, also known as radial basis function (RBF) networks, offer a series of advantages for time-series prediction due to the nature of the non-linear RBF activation function.

The remainder of the paper consists of the following sections: A nonlinear analysis of the phoneme signals, the RBF network model and the experimental results. Finally, the summary and future researches conclude the report..

2   Nonlinear Analysis of Phoneme Signals

The purpose of nonlinear analysis of phoneme time series is to characterize the observed dynamics in order to produce new time series exhibiting the same dynamics. According to Takens' embedding theorem [20], a discrete-time dynamical system can be reconstructed from scalar-valued partial measurements of internal states. If the measurement variable at time t is defined by $x(t)$, a k-dimensional embedding vector is defined by:

$$X(t_i) = [x(t_i), x(t_i + \tau),\ldots, x(t_i + (k - 1)\, \tau)], \tag{1}$$

where τ is the time delay, and $k = 1\ldots d$, where d denotes the embedding dimension. The reason was to have samples from the original sound signal $x(t)$ delayed by multiples of τ and obtain the reconstructed d–dimensional space. The conditions in the embedding theorem impose $d \geq 2D + I$, where D is the dimension of the compact manifold containing the attractor, and I is an integer. According to the embedding technique, from sampled time series of speech phonemes the dynamics of the unknown speech generating system could be uncovered, provided that the embedding dimension d was large enough. The difficult problem in practical applications is finding the optimal length of the time series and the optimal time delay. However, there are some methods to estimate these parameters [21], [22]. We used the false nearest neighbor method and the mutual information method, for establishing the optimal embedding dimension and the time lag (embedding delay), respectively [23]. The optimal choice depends on the specific dynamics of the studied process. Prediction requires sufficient points in the neighborhood of the current point. As the dimension increases the number of such points decreases. Apparently a high embedding seems advantageous because a sufficiently large value ensures that different states are accurately represented by distinct delay vectors. When the value becomes unnecessarily high the data become sparse and each embedding dimension introduces additional noise. A technique that may provide a suitable embedding for

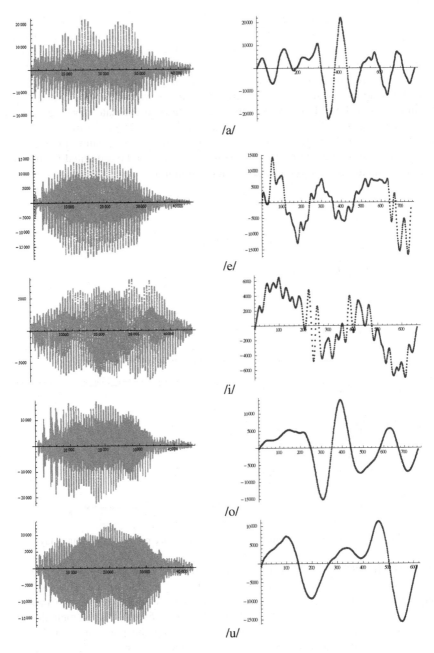

Fig. 1. The waveforms of phonemes /a/, /e/, /i/, /o/, /u/ along with a typical elemental pattern

most purposes is the false nearest neighbors (FNN) method. A provisional value for
d is first assumed. Then the nearest neighbor of each delay vector is located using the

d-dimensional metric. The same pairs of vectors are then extended by adding one more delay coordinate and are compared, this time using $d + 1$-dimensional metric. If they become far apart then we may consider that nearest neighbor to be false. Regarding the selection of τ, if the value is too large the successive components in a delay vector are completely unrelated. At contrary, if the value is too small the components are nearly identical and therefore adding new components does not bring new information. Out of several techniques available to estimate τ, we selected the minimum mutual information method. Mutual information is a measure of how much one knows about $x(t + \tau)$ if one knows $x(t)$. It is calculated as the sum of the two self-entropies minus the joint entropy. The optimal time lag can be estimated for the point where the mutual information reaches its first minimum.

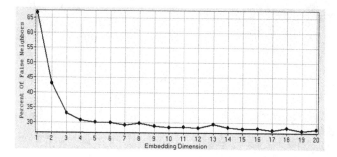

Fig. 2. The percentage of FNNs in dependence on the value of d

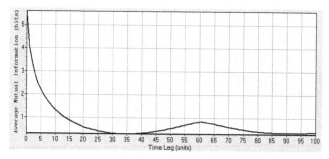

Fig. 3. The average mutual information in dependence on the embedding delay

For experimental purposes, we considered in this work the signals of the main vocal phonemes, /a/, /e/, /i/, /o/, and /u/ pronounced by a male person. The waveforms along with an instance of the corresponding elemental pattern are shown in Fig. 1. The vocal sound data were sampled at 96 kHz with 16 bits. The elemental pattern may be viewed as a basic component in constructing the phoneme signal. If this pattern is repeated in time (concatenated) the phoneme sound can be reconstructed. For these phonemes data we applied the FNN method and the mutual information method. As an example, the percentage of FNNs in dependence on the value of d can be seen in Fig. 2 for phoneme /a/. The percentage of FNN should drop to zero for the optimal global embedding dimension d. In Fig. 3 it is depicted the average mutual information in dependence on the embedding delay. The optimal time delays result in the point where the average mutual information reaches the first minimum. The results obtained for the phonemes under study are presented in Table 1.

Table 1. Phonemes nonlinear analysis results

Phonemes/ Vocals	Sample length	Optimal embedding dimension	Optimal time delay	LE $\approx$
/a/	43304	20	34	42
/e/	42984	14	59	16
/i/	42240	9	24	167
/o/	45120	15	48	53
/u/	42624	10	62	54

After the estimation of d and τ we could proceed with the embedding process. For a convenient exploration of the reconstructed phase-space we constructed the following three–dimensional map:

$$x = x(t)$$
$$y = x(t + k) \tag{2}$$
$$z = x(t + 2k),$$

and we selected the points along the z axis for $k = d$. The samples are depicted in Fig. 4 for phoneme /a/. These samples constituted the input vector to the RBF neural network as will be detailed in the next section.

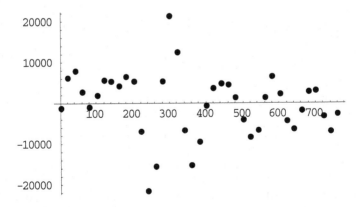

Fig. 4. Training samples taken out from the original time-series

3   The RBF Network Model and Experimental Results

By choosing the appropriate values for d and τ in the embedding process we may have a high degree of confidence that the intrinsic dynamics of the time-series was captured. If we apply the samples on a higher dimension as inputs to an interpolation neural network then we can have a good starting point in generating the phoneme sound. Using a RBF network in this case has several advantages. The network has three layers. The input and output layers have one neuron. The output of the network is a scalar function y of the input $(x_1, x_2, ...,x_n)$ and is given by

$$y(x_1, x_2, ..., x_n) = \sum_{i=1}^{S} w_i g_i(||x - c_i||), \qquad (3)$$

where S is the number of neurons in the hidden layer, w_i is the weight of the neuron i, c_i is the center vector for neuron i, and g_i is the activation function:

$$g_i(||x - c_i||) = exp^{-||x - c_i||^2}/_{2\sigma} \qquad (4)$$

The Gaussian function has the advantage of being controllable by the parameter σ. In this way, the Gaussian functions in the RBF networks centered on samples enable good interpolations. Small values of σ reflect little sample influence outside of local neighborhood, whereas larger σ values extend the influence of each sample, making that influence too global for extremely large values.

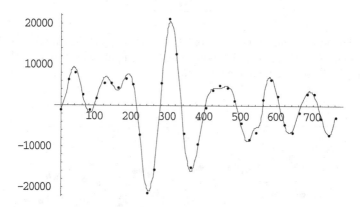

Fig. 5. The approximated elemental

In Fig. 5 it is shown the function approximation along with the data samples. The network consisted of 20 basis function of Gaussian type and was trained with 39 samples. The mean squared error decreased below 0.02 after only 5 iterations. We can observe a good match. The strength of this approach is given by the capability of controlling the width of the basis function and hence the final shape of the approximation. The dynamics of the elementals is still very well preserved if the number of nodes is not too low. In the next stage, a series of different elementals were generated by varying the width of the basis function according to a random source (for instance a quadratic map). Finally these elementals were concatenated to generate the sound signal. If the concatenation was performed with the same elemental the resulting sound created an artificial impression even if the original elemental was used. The impression of naturalness is not given by the phoneme elemental alone, but by the temporal perception of the slight variations of the elementals in succession.

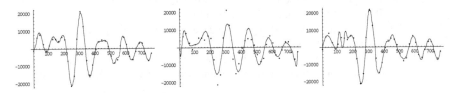

Fig. 6. Three elemental samples generated when σ was controlled by a random source

In Fig. 6, a series of three different elemental are depicted. Slight variations can be observed but with the preservation of the original dynamics as can be observed in the original signal. In conclusion, the method suggested proved to be simple and effective encouraging further researches.

4   Summary

A dynamic method for phoneme synthesis using an elemental-based concatenation technique was proposed. The phonemes' elementals could be generated by an approximation network and the signal parameters can be controlled, at every iteration, through the width of the Gaussian activation function. The resultant elementals were concatenated and finally assembled in the resultant phoneme sound. The sound impression was quite realistic. Future researches in this direction are encouraged by the positive results obtained in this work.

References

1. Edgington, M.: Investigating the limitations of concatenative synthesis. In: Proceedings of Eurospeech 1997, Rhodes/Athens, Greece, pp. 593–596 (1997)
2. Bulut, M., Narayanan, S.S., Syrdal, A.: Expressive speech synthesis using a concatenative synthesizer. In: Proceedings of InterSpeech, Denver, CO, pp. 1265–1268 (2002)
3. Balyan, A., Agrawal, S.S., Dev, A.: Speech Synthesis: A Review. International Journal of Engineering Research & Technology (IJERT) 2(6), 57–75 (2013)
4. Banbrrok, M., McLaughlin, S., Mann, I.: Speech characterization and synthesis by nonlinear methods. IEEE Trans. Speech Audio Process 7(1), 1–17 (1999)
5. Pitsikalis, V., Kokkinos, I., Maragos, P.: Nonlinear analysis of speech signals: Generalized dimensions and Lyapunov exponents. In: Proc. European Conf. on Speech Communication and Technology-Eurospeech-03, pp. 817–820 (September 2003)

6. McLaughlin, S., Maragos, P.: Nonlinear methods for speech analysis and synthesis. In: Marshall, S., Sicuranza, G. (eds.) Advances in Nonlinear Signal and Image Processing, vol. 6, p. 103. Hindawi Publishing Corporation (2007)
7. Tao, C., Mu, J., Xu, X., Du, G.: Chaotic characteristic of speech signal and its LPC residual. Acoust. Sci. & Tech. 25(1), 50–53 (2004)
8. Koga, H., Nakagawa, M.: Chaotic and Fractal Properties of Vocal Sounds. Journal of the Korean Physical Society 40(6), 1027–1031 (2002)
9. Lo, W.K., Ching, P.C.: Phone-Based Speech Synthesis With Neural Network And Articulatory Control. In: Proceedings of Fourth International Conference on Spoken Language (ICSLP 1996), vol. 4, pp. 2227–2230 (1996)
10. Malcangi, M., Frontini, D.: A Language-Independent Neural Network-Based Speech Synthesizer. Neurocomputing 73(1-3), 87–96 (2009)
11. Raghavendra, E.V., Vijayaditya, P., Prahallad, K.: Speech synthesis using artificial neural networks. In: National Conference on Communications (NCC), Chennai, India, pp. 1–5 (2010)
12. Frank, R.J., Davey, N., Hunt, S.P.: Time Series Prediction and Neural Networks. Journal of Intelligent and Robotic Systems 31, 91–103 (2001)
13. Kinzel, W.: Predicting and generating time series by neural networks: An investigation using statistical physics. Computational Statistical Physics, 97–111 (2002)
14. Priel, A., Kanter, I.: Time series generation by recurrent neural networks. Annals of Mathematics and Artificial Intelligence 39, 315–332 (2003)
15. Medsker, L.R., Jain, L.C.: Recurrent Neural Networks: Design and Applications. CRC Press (2001)
16. Kalinli, A., Sagiroglu, S.: Elman Network with Embedded Memory for System Identification. Journal of Information Science and Engineering 22, 1555–1568 (2006)
17. Coca, A.E., Romero, R.A.F., Zhao, L.: Generation of composed musical structures through recurrent neural networks based on chaotic inspiration. In: The 2011 International Joint Conference on Neural Networks (IJCNN), pp. 3220–3226 (July 2011)
18. Röbel, A.: Morphing Sound Attractors. In: Proc. of the 3rd. World Multiconference on Systemics, Cybernetics and Informatics (SCI 1999) AES 31st International Conference (1999)
19. Crisan, M.: A Neural Network Model for Phoneme Generation. Applied Mechanics and Materials 367, 478–483 (2013)
20. Takens, F.: Detecting strange attractors in turbulence. Lecture Notes in Mathematics 898, 366–381 (1981)
21. Small, M.: Applied nonlinear time series analysis: applications in physics, physiology and finance. World Scientific Publishing Co., NJ (2005)
22. Sprott, J.C.: Chaos and Time-Series Analysis. Oxford University Press, NY (2003)
23. Kononov, E.: Visual Recurrence Analysis Software Package, Version 4.9 (accessed 2013), http://nonlinear.110mb.com/vra/

Supplier Selection Using Ranking Interval Type-2 Fuzzy Sets

Samarjit Kar[*] and Kajal Chatterjee

Department of Mathematics, National Institute of Technology,
Durgapur-713209, India
{dr.samarjitkar,chatterjeekajal7}@gmail.com

Abstract. In face of global competition, supplier management is emerging as crucial issue to any companies striving for business success. This paper develops a framework for selecting suitable outsourced suppliers of upstream supply chain in uncertain environment. Emerging supply risk arising from outsourcing are analyzed to reduce cost and increase the sustainability of supply chain network. The study applies ranking based interval type-2 fuzzy set exploring the risk factors and ranking supplier companies. The performance rating weights of risk criteria in supply chain are evaluated based on decision makers. Finally, an empirical study is conducted for Indian Oil Corporation Limited (IOCL) to demonstrate the applicability of the proposed algorithm to select the suitable crude oil supplier(s).

Keywords: Supplier Selection, Supply chain risk, Interval type-2 fuzzy sets.

1   Introduction

Today's modern globalized world is a mass-production industrialized society connecting business through cross-boundary supply chain having multi-suppliers and production sites, relying on outsourcing business policy [1]. Uncertainty from supplier side led to high exposure to supply risk in complex upstream supply chain network. Modern oil supply chains have crossed geographical barriers distributed across multi-countries operating in multi-sectors like electronics, petroleum product and automobiles [2]. Various supply chain are structurally similar resulted in steep competition between supply chain networks, fluctuation of demand, financial risk and managing up-and down- stream relation between suppliers and customers [3]. Economic instability and disaster crisis has forced managers in multi-national companies to identify, categories risky supplier partners controlling supply and operational risks [4]. Researchers have now focused on Fuzzy Multi-attribute decision making (FMADM) model, where decision makers considers various attributes and alternatives giving decisions in linguistic terms, due to vague human nature[5]. Recent researchers are applying Fuzzy Type-2 sets having fuzzy-based membership function, dealing with risk uncertainties in a better way. Type-2 fuzzy sets (T2FS) [6] are

---

[*] Corresponding author.

© Springer International Publishing Switzerland 2015   9
S.C. Satapathy et al. (eds.), *Proc. of the 3rd Int. Conf. on Front. of Intell. Comput. (FICTA) 2014*
– *Vol. 1*, Advances in Intelligent Systems and Computing 327, DOI: 10.1007/978-3-319-11933-5_2

extension of ordinary type-1 fuzzy sets where traditional Type-1 fuzzy sets fails to deal with uncertainties, due to crisp nature of their membership function.

Zadeh [7] first proposed the concept of Type-2 fuzzy set and detailed discussion and more flexible approach are given by Mendel and John [8]. The membership functions of T2 FSs are three dimensional and include a footprint of uncertainty, it is the new third dimension of T2 FSs and the footprint of uncertainty that provide additional degrees of freedom that make it possible to directly model and handle uncertainties. In recent years many researchers have applied the Interval Type2 fuzzy sets (IT2FS) for handling FMADM. Chen and Lee [9] present a new method for handling fuzzy multiple criteria hierarchical group decision-making problems based on arithmetic operations and fuzzy preference relations of interval type-2 fuzzy sets. Chen and Lee [10] also presented an interval type-2 TOPSIS method to handle fuzzy MCDM problem. Lee and Chen [11] suggested a method to handle MCDM problems based on ranking values and arithmetic operation of interval type-2 fuzzy set. Wang et al [12] proposed a method for MAGDM based on interval type-2 fuzzy environment. Yang et al [13] presented a new method to handle fuzzy MAGDM problems based on ranking interval type-2 fuzzy sets. Celik et al [14] proposed an integrated novel type-2 fuzzy MCDM method to improve customer satisfaction in public transportation in Istanbul. Wang and Lee [15] applied interval type-2 fuzzy set in fuzzy decision making.

Oil and gas supply chain involves some environmental parameters which are uncertain in nature due to political turmoil and high requirement of storage. In response to hike in global oil prices, stiff competition among oil companies along with environmental problem, oil companies are forming global supply chain among intercontinental suppliers [5]. In foundation of investigation of OICL, we focus our attention on assessing risk attributes and ranking of crude oil supplier alternatives under ranking based interval type-2 fuzzy set based on information provided by decision maker.

2   Preliminaries

In this section, some basic concepts of Type-2 fuzzy sets, interval type-2 fuzzy sets and ranking interval type-2 fuzzy sets are introduced along with their basic operational rules related to it.

Definition 1: [14] A type-2 Fuzzy set (T2 FS), denoted by $\tilde{A}$ on a universal set X, represented by type-2 membership function $\mu_{\tilde{A}}(x,u)$ and expressed as

$\tilde{A} = \{(x,u), \mu_{\tilde{A}}(x,u) | \forall x \in X, u \in J_x \subseteq [0,1]$. $\tilde{A}$ can also be expressed as

$\tilde{A} = \int_{x \in X} \int_{u \in J_x} \mu_{\tilde{A}}(x,u)/(x,u)$ where $J_x \subseteq [0,1]$ and $\iint$ denote union over all x and

u. For discrete universe $\int$ is replaced by $\sum$.

Definition 2: [8] An interval type-2 fuzzy set $\tilde{A}$ is a special type of type-2 fuzzy set where, $\mu_{\tilde{A}}(x,u) = 1$ denoted by $\tilde{A} = \displaystyle\int_{x \in X} \int_{u \in J_x} 1/(x,u)$ where $J_x \subseteq [0,1]$. (1)

Definition 3: [11] Let two trapezoidal interval type-2 fuzzy sets be given below

$$\tilde{A}_1 = (\tilde{A}_1^U, \tilde{A}_1^L) = ((a_{11}^U, a_{12}^U, a_{13}^U, a_{14}^U, H_1(\tilde{A}_1^U), H_2(\tilde{A}_1^U)), (a_{11}^L, a_{12}^L, a_{13}^L, a_{14}^L, H_1(\tilde{A}_1^L), H_2(\tilde{A}_1^L)))$$

$$\tilde{A}_2 = (\tilde{A}_2^U, \tilde{A}_2^L) = ((a_{21}^U, a_{22}^U, a_{23}^U, a_{24}^U, H_1(\tilde{A}_2^U), H_2(\tilde{A}_2^U)), (a_{21}^L, a_{22}^L, a_{23}^L, a_{24}^L, H_1(\tilde{A}_2^L), H_2(\tilde{A}_2^L)))$$

The arithmetic operation of two IT2FSs $\tilde{A}_1 \oplus \tilde{A}_2 = (\tilde{A}_1^U, \tilde{A}_1^L) \oplus (\tilde{A}_2^U, \tilde{A}_2^L)$

$$= ((a_{11}^U + a_{21}^U, a_{12}^U + a_{22}^U, a_{13}^U + a_{23}^U, a_{14}^U + a_{24}^U, \min(H_1(\tilde{A}_1^U), H_1(\tilde{A}_2^U)), \min(H_2(\tilde{A}_1^U), H_2(\tilde{A}_2^U))$$

$$(a_{11}^L + a_{21}^L, a_{12}^L + a_{22}^L, a_{13}^L + a_{23}^L, a_{14}^L + a_{24}^L, \min(H_1(\tilde{A}_1^L), H_1(\tilde{A}_2^L)), \min(H_2(\tilde{A}_1^L), H_2(\tilde{A}_2^L)))) \quad (2)$$

The multiplicative operation of two ITFSs $\tilde{A}_1 \otimes \tilde{A}_2 = (\tilde{A}_1^U, \tilde{A}_1^L) \otimes (\tilde{A}_2^U, \tilde{A}_2^L)$

$$= ((a_{11}^U \times a_{21}^U, a_{12}^U \times a_{22}^U, a_{13}^U \times a_{23}^U, a_{14}^U \times a_{24}^U, \min(H_1(\tilde{A}_1^U), H_1(\tilde{A}_2^U)), \min(H_2(\tilde{A}_1^U), H_2(\tilde{A}_2^U))$$

$$(a_{11}^L \times a_{21}^L, a_{12}^L \times a_{22}^L, a_{13}^L \times a_{23}^L, a_{14}^L \times a_{24}^L, \min(H_1(\tilde{A}_1^L), H_1(\tilde{A}_2^L)), \min(H_2(\tilde{A}_1^L), H_2(\tilde{A}_2^L)))) \quad (3)$$

Definition 4: [13] For any interval type-2 fuzzy set $\tilde{A}_i$ located in second quadrant,

$$\tilde{A}_i = (\tilde{A}_i^U, \tilde{A}_i^L) = ((a_{i1}^U, a_{i2}^U, a_{i3}^U, a_{i4}^U, H_1(\tilde{A}_i^U), H_2(\tilde{A}_i^U)), (a_{i1}^L, a_{i2}^L, a_{i3}^L, a_{i4}^L, H_1(\tilde{A}_i^L), H_2(\tilde{A}_i^L)))$$

where $-1 \le a_{i1}^U \le a_{i2}^U \le a_{i3}^U \le a_{i4}^U \le 0$, $-1 \le a_{i1}^L \le a_{i2}^L \le a_{i3}^L \le a_{i4}^L \le 0$ and $1 \le i \le n$.

Shifting right interval type-2 fuzzy set $\tilde{A}_i$ from second quadrant into the interval type-2 fuzzy set $\tilde{A}_i^*$ in first quadrant, as follows:

$$\tilde{A}_i^* = (\tilde{A}_i^{*U}, \tilde{A}_i^{*L}) = ((a_{i1}^U + K_i, a_{i2}^U + K_i, a_{i3}^U + K_i, a_{i4}^U + K_i, H_1(\tilde{A}_i^U), H_2(\tilde{A}_i^U)),$$

$$(a_{i1}^L + K_i, a_{i2}^L + K_i, a_{i3}^L + K_i, a_{i4}^L + K_i, H_1(\tilde{A}_i^L), H_2(\tilde{A}_i^L)))$$

where K_i refers to right-shift distance of interval type-2 fuzzy set $\tilde{A}_i^*$ from second quadrant to first quadrant and $K_i = |a_{i1}^U|$. For any interval type-2 fuzzy set $\tilde{A}_i$ located between first and second quadrant, we can right shift $\tilde{A}_i$ to $\tilde{A}_i^*$ locate at first quadrant. The ranking value $RV(\tilde{A}_i)$ of the trapezoidal interval type-2 fuzzy set $\tilde{A}_i$ is defined as:

$$RV(\tilde{A}_i) = \left[\frac{(a_{i1}^U + K_i) + (a_{i4}^U + K_i)}{2} + \frac{H_1(\tilde{A}_i^U) + H_2(\tilde{A}_i^U) + H_1(\tilde{A}_i^L) + H_2(\tilde{A}_i^L)}{4} \right]$$

$$\times \left[\frac{(a_{i1}^U + K_i) + (a_{i2}^U + K_i) + (a_{i3}^U + K_i) + (a_{i4}^U + K_i) + (a_{i1}^L + K_i) + (a_{i2}^L + K_i) + (a_{i3}^L + K_i) + (a_{i4}^L + K_i)}{8} \right] \quad (4)$$

where

$$K_i = \begin{cases} 0 & if \ \min(a_{11}^U, a_{11}^U, \ldots\ldots\ldots, a_{11}^U) \geq 0 \\ \left| \min(a_{11}^U, a_{11}^U, \ldots\ldots\ldots, a_{11}^U) \right| & if \ \min(a_{11}^U, a_{11}^U, \ldots\ldots\ldots, a_{11}^U) < 0 \end{cases} \tag{5}$$

$$\left[\frac{(a_{i1}^U + K_i) + (a_{i2}^U + K_i) + (a_{i3}^U + K_i) + (a_{i4}^U + K_i) + (a_{i1}^L + K_i) + (a_{i2}^L + K_i) + (a_{i3}^L + K_i) + (a_{i4}^L + K_i)}{8} \right] \text{ is}$$

the basic ranking value of interval type-2 fuzzy set $\tilde{A}_i$, and $H_1(\tilde{A}_i^U), H_2(\tilde{A}_i^U),$ and $H_1(\tilde{A}_i^L)$ and $H_2(\tilde{A}_i^L)$ denotes membership value of elements $a_{i1}^U, a_{i3}^U, a_{i2}^L$ and a_{i3}^L in $\tilde{A}_i^U$, $\tilde{A}_i^U$, $\tilde{A}_i^L$, $\tilde{A}_i^L$ respectively.

3   An Algorithmic Ranking Based Interval Type-2 Fuzzy Set Approach for Supplier Selection Mitigating Supply Risk

For various MCDM problem, let $X = \{x_1, x_2, \ldots\ldots\ldots, x_n\}$ be a set of alternatives and $F = \{f_1, f_2, \ldots\ldots, f_m\}$ be a set of attributes. We utilize the kth decision-maker D_p, for determining the weighing matrix, DW_p, of the attributes and the evaluating matrix, Y_p, of the alternatives, where $1 \leq p \leq k$. The proposed fuzzy multiple attribute group decision making problem is based on the ranking interval type-2 fuzzy sets, (Yang et al, [13]. The algorithmic steps are as follows:

Step 1: Based on linguistic values of attributes and alternatives, construct the weighted decision matrix S_p of the pth decision-maker, $1 \leq p \leq k$

$$S_p = DW_p * Y_p = \begin{matrix} f_1 \\ f_2 \\ \vdots \\ f_m \end{matrix} \begin{bmatrix} \tilde{w}_1^p \\ \tilde{w}_2^p \\ \vdots \\ \tilde{w}_m^p \end{bmatrix} \times \begin{bmatrix} \tilde{y}_{11}^p & \tilde{y}_{12}^p & & \tilde{y}_{1n}^p \\ \tilde{y}_{21}^p & \tilde{y}_{22}^p & & \tilde{y}_{2n}^p \\ \vdots & \vdots & \cdots\cdots & \vdots \\ \tilde{y}_{m1}^p & \tilde{y}_{m2}^p & & \tilde{y}_{mn}^p \end{bmatrix} = \begin{matrix} f_1 \\ f_2 \\ \vdots \\ f_m \end{matrix} \begin{bmatrix} \tilde{s}_{11}^p & \tilde{s}_{12}^p & & \tilde{s}_{1n}^p \\ \tilde{s}_{21}^p & \tilde{s}_{22}^p & & \tilde{s}_{2n}^p \\ \vdots & \vdots & \cdots\cdots & \vdots \\ \tilde{s}_{m1}^p & \tilde{s}_{m2}^p & & \tilde{s}_{mn}^p \end{bmatrix} \tag{6}$$

where $\tilde{s}_{ij}^p = \tilde{y}_{ij}^p \times w_i^p \quad 1 \leq i \leq m, 1 \leq j \leq n$.

Step 2: Based on Eq. (6), Average decision matrix $\overline{S}$, is constructed:

$$\overline{S} = \begin{bmatrix} \tilde{s}_{11} & \tilde{s}_{12} & & \tilde{s}_{1n} \\ \tilde{s}_{21} & \tilde{s}_{22} & & \tilde{s}_{2n} \\ \vdots & \vdots & \cdots\cdots & \vdots \\ \tilde{s}_{m1} & \tilde{s}_{m2} & & \tilde{s}_{mn} \end{bmatrix}, where \ \tilde{s}_{ij} = \left[\frac{\tilde{s}_{ij}^1 + \tilde{s}_{ij}^2 + \ldots\ldots + \tilde{s}_{ij}^k}{k} \right] \tag{7}$$

$\tilde{s}_{ij}^p$ and $\tilde{s}_{ij}$ are interval type-2 fuzzy sets, $1 \leq i \leq m, \quad 1 \leq j \leq n \ and \ 1 \leq p \leq k$.

Step 3: Based on Equation.(4) and (5), ranking matrix RS is derived

$$RS = \begin{bmatrix} RV(\tilde{s}_{11}) & RV(\tilde{s}_{12}) & RV(\tilde{s}_{1n}) \\ RV(\tilde{s}_{21}) & RV(\tilde{s}_{22}) & RV(\tilde{s}_{2n}) \\ : & : & \cdots \cdots & : \\ RV(\tilde{s}_{m1}) & RV(\tilde{s}_{m2}) & RV(\tilde{s}_{mn}) \end{bmatrix} \tag{8}$$

Step 4: Based on Eq. (8) average agreement degree AD is constructed,

$$AD = \begin{bmatrix} \frac{1}{N}\sum_{J=1}^{N} RV(\tilde{s}_{ij}) \\ \frac{1}{N}\sum_{J=1}^{N} RV(\tilde{s}_{2j}) \\ : \\ \frac{1}{N}\sum_{J=1}^{N} RV(\tilde{s}_{mj}) \end{bmatrix} = \begin{bmatrix} AD_1 \\ AD_2 \\ : \\ : \\ AD_m \end{bmatrix}, \text{ where } 1 \leq i \leq m, 1 \leq j \leq n. \tag{9}$$

Step 5: Calculate the ranking value $R(x_j)$ of alternative x_j , shown as follows,

$$R(x_j) = \sum_{i=1}^{m} \frac{RV(\tilde{s}_{ij})}{AD_i}, \quad \text{where } 1 \leq j \leq n. \tag{10}$$

4   Case Study

Volatility of crude oil market, subsidy burden by government policy and business has forced the company to go for outsourced crude oil suppliers by continuous diversifying crude oil procurement sources adding more supplier countries [2]. IOCL is expanding its crude oil suppliers list as alternative arrangements due to difficulty in remitting import of crude oil from Middle East countries UAE, Saudi Arabia, Iran and Kuwait. Due to political Instability and steep price rise in barrel per dollar, Indian Oil is looking for new outsourcing suppliers from Africa, Canada and Latin America. Three expert decision-makers in their field, D_1 (Petroleum Engineer), D_2 (Marketing Manager) and D_3 (Economic analyst) are engaged for ranking outsourcing supplier of crude among three given Oil exporting companies of gulf region namely Iranian Oil Participants Limited (x_1) ,Basrah Oil corporation (x_2) ,Saudi Aramco (x_3) with respect to five risk attributes, Information risk (f_1) Supply risk (f_2) , Logistics risk (f_3) ,Market risk (f_4) and Environmental risk (f_5) .

Information risk of oil supply chain is because of incorrect information flow, crashes/changing in planning. Market risk arises due to change in demand and price, change in economic policy, stock market crash and financial inflation. Logistics risk refers to cargo damage, operation quality, agreement term breach and types and controlling export-import. Supply risk includes delivery time, global sourcing, and bankruptcy of suppliers. Environmental risk of oil supply refers to socio-economic turmoil, terrorism act and man-made and natural disasters. The linguistic weights of the attributes are shown in Table (1).The linguistic weights of the attributes given by the decision makers are given in Table (2). The evaluating values of alternatives given by the decision makers with respect to the attributes are shown in Table (3).

Table 1. (Chen and lee, 2010b): Linguistic terms and their corresponding Interval type-2 fuzzy sets

Linguistic terms	Interval type-2 fuzzy sets
Very low (VL)	((0, 0, 0, 0.1; 1, 1), (0, 0, 0, 0.5; 0.9, 0.9))
Low (L)	((0, 0.1, 0.1, 0.3; 1, 1),(0.05,0.1,00.1,0.2;0.9,0.9))
Medium low (ML)	((0.1, 0.3, 0.3, 0.5; 1, 1), (0.2, 0.3, 0.3, 0.4; 0.9, 0.9))
Medium (M)	((0.3, 0.5, 0.5, 0.7; 1, 1), (0.4, 0.5, 0.5, 0.6; 0.9, 0.9))
Medium High (MH)	((0.5, 0.7, 0.7, 0.9; 1, 1), (0.6, 0.7, 0.7, 0.8; 0.9, 0.9))
High (H)	((0.7, 0.9, 0.9, 1; 1), (0.8, 0.9, 0.9, 0.95; 0.9, 0.9))
Very High(VH)	((0.9, 1, 1, 1; 1, 1), (0.95, 1, 1, 1; 0.9, 0.9))

Table 2. Weights of the attributes evaluated by the decision- makers

Attributes	Decision makers		
	D_1	D_2	D_3
Information risk	H	MH	H
Supply risk	M	VH	MH
Demand risk	MH	H	H
Logistics risk	M	MH	ML
Environmental risk	H	L	L

Table 3. Evaluating values of alternatives given by decision makers w.r.t the attributes

Attributes	Alternatives	Decision makers		
		D_1	D_2	D_3
Information risk	x_1	H	MH	H
	x_2	H	M	H
	x_3	H	MH	ML
Supply risk	x_1	H	MH	ML
	x_2	MH	H	VL
	x_3	MH	H	H
Demand risk	x_1	H	H	M
	x_2	MH	M	H
	x_3	H	MH	MH
Logistics risk	x_1	VH	MH	H
	x_2	H	H	M
	x_3	M	VH	MH
Environmental risk	x_1	ML	M	ML
	x_2	H	MH	H
	x_3	H	VH	MH

[Step 1] We construct the evaluating matrix DW_p of the attributes and the matrix Y_p of alternatives , both given by decision–makers, $D_p(1 \le p \le 3)$ based on Table (2) and (3) respectively, as shown below.

$$
DW_1 = \begin{bmatrix} H \\ M \\ MH \\ M \\ H \end{bmatrix}, \quad
DW_2 = \begin{bmatrix} MH \\ VM \\ H \\ MH \\ L \end{bmatrix}, \quad
DW_3 = \begin{bmatrix} H \\ MH \\ H \\ ML \\ M \end{bmatrix}
$$

$$
Y_1 = \begin{bmatrix} MH & H & MH \\ H & MH & MH \\ H & MH & H \\ VH & H & M \\ ML & H & H \end{bmatrix}, \quad
Y_2 = \begin{bmatrix} H & M & M \\ MH & H & H \\ H & M & MH \\ M & H & VH \\ M & MH & VH \end{bmatrix}, \quad
Y_3 = \begin{bmatrix} MH & H & ML \\ ML & VH & H \\ MH & M & H \\ H & M & MH \\ ML & H & MH \end{bmatrix}
$$

[Step 3] Based on Eq. (4), (5), (9) we construct the ranking matrix RS, the average agreement degree AD and Ranking values of the alternatives

$$
RS = \begin{bmatrix} 0.637 & 1.021 & 0.585 \\ 0.638 & 1.006 & 0.680 \\ 0.790 & 0.821 & 0.486 \\ 0.587 & 0.422 & 0.539 \\ 0.194 & 0.614 & 0.579 \end{bmatrix} \quad
AD = \begin{bmatrix} 0.7477 \\ 0.7747 \\ 0.6990 \\ 0.5160 \\ 0.4623 \end{bmatrix} \quad
\begin{bmatrix} R(x_1) \\ R(x_2) \\ R(x_3) \end{bmatrix} = \begin{bmatrix} 4.3629 \\ 5.9846 \\ 4.6524 \end{bmatrix}
$$

Ranking order of the given alternatives x_1, x_2, x_3 is:

$$x_2 \succ x_3 \succ x_1 \quad as \quad R(x_2) \succ R(x_3) \succ R(x_1)$$

5   Result Discussion

Our analysis is based on using algorithm in section 3 and linguistic data given in Table (1), Table (2) and Table (3). Based on the result given in Section 4, it follows that Basrah Oil corporation (x_2) based on Iraq is the most risky crude oil supplier country followed by Saudi based company, Saudi Aramco (x_3) among the three given suppliers. So, Indian Oil should rely mainly on, Iran based company namely, Iranian Oil Participants Limited (x_1) for crude oil product for its market production in India. Increase in country risk of Middle East due to extreme events of war may significantly change India's crude Oil importing optimal decision. On the contrary, the shares of other oil region like Central America, North Africa, and North America can be chosen as alternative suppliers due to relatively low importing cost and low country risk.

6   Conclusion

In this paper, we have presented case study of supplier selection in supply chain network based on ranking method of interval type-2 fuzzy sets. The Crude oil exporting countries are facing political crisis like Iran from US and European Union, stiff competition like Canada from USA, and not getting proper market like Africa, Russia and Latin American countries. Alignment of objectives between Outsourcing companies and their crude oil suppliers in supply chain helps to reduce risk, surface new opportunities and differentiate them from competition. India's oil-importing plan should be adjusted and share of Middle East should be reduced in case of wars. From perspective of energy security, the paper introduces risk criteria of import disruption into oil-importing supplier selection which managers should pay attention.

References

[1] Kumar, S., Chan, F.: Global supplier development considering risk factors using fuzzy extended AHP-based approach. Omega 35, 417–431 (2007)
[2] Li, J., Tang, L., Sun, X., Wu, D.: Oil-importing optimal decision considering country risk with extreme events: A multi-objective programming approach. Computers & Operations Research 42, 108–115 (2014)
[3] Gaonkar, R.S., Viswanadham, V.: Analytical Framework for the management of risk in supply chains. IEEE Transactions on Automation Science and Engineering 4(2), 265–269 (2007)
[4] Sawik, T.: Supplier selection in make-to-order environment with risks. Mathematical and Computer Modeling 53, 1670–1679 (2011)
[5] Vosooghi, M.A., Fazli, S., KianiMavi, R.: Crude oil supply chain risk management with fuzzy analytical Hierarchy process. American Journal of Scientific Research 46, 34–42 (2012)
[6] Mendel, J.M., John, R.: Type-2 fuzzy sets made simple. IEEE Transactions on Fuzzy Systems 10, 117–127 (2002)
[7] Zadeh, L.A.: The concept of a linguistic variable and its application to approximate reasoning. Information Sciences 8(9), 199–249 (1975)
[8] Mendel, J.M., John, R.I., Liu, F.L.: Interval type-2 fuzzy logic systems made simple. IEEE Transactions on Fuzzy Systems 14(6), 808–821 (2006)
[9] Chen, S.M., Lee, L.W.: Fuzzy Multiple Criteria Hierarchical Group Decision-Making based on Interval Type-2 Fuzzy Sets. IEEE Transaction on Systems, Man and Cybernetics-Part A: System and Human 40(5), 1120–1128 (2010)
[10] Chen, S.M., Lee, L.W.: Fuzzy multiple attributes group decision-making based on interval type-2 TOPSIS method. Expert System with Applications 37, 2790–2798 (2010b)
[11] Chen, S.M., Lee, L.W.: Fuzzy multiple attributes group decision making based on ranking values and the arithmetic operations of interval type-2 fuzzy sets. Expert Systems with Applications 37, 824–833 (2010a)
[12] Wang, W., Liu, X., Qin, Y.: Multi-attribute group decision making models under interval type-2 fuzzy environment. Knowledge Based Systems 30, 121–128 (2012)
[13] Chen, S.M., Yang, M., Lee, L., Yang, S.: Fuzzy multiple attributes group decision making based on ranking interval type-2 fuzzy sets. Expert Systems with Applications 39, 5295–5308 (2012)

[14] Celik, E., Ozge, B., Erdogan, M., Gumus, A., Baracli, H.: An integrated novel interval type-2 fuzzy MCDM method to improve customer satisfaction in public transportation for Istanbul. Transportation Research Part E 58, 28–51 (2013)
[15] Chen, S.M., Wang, C.Y.: Fuzzy decision making systems based on interval type-2 fuzzy sets. Information Science 242, 1–21 (2013)

Pattern-Matching for Speaker Verification: A Review

S.B. Dhonde[1] and S.M. Jagade[2]

[1] Department of Electronics Engineering,
AISSMS Institute of Information Technology, Pune, India
[2] Department of Electronics & Telecommunication Engineering,
TPCT College of Engineering, Osmanabad, India
dhondesomnath@gmail.com, smjagade@yahoo.co.in

Abstract. This paper presents a brief survey on Pattern matching for speaker verification. Pattern matching performed for Speaker Verification (SV), is the process of verifying the claimed identity of a registered speaker by using their voice characteristics and further subdivided into text-dependent and text-independent. Paper outlines various models performance levels and the work undertaken by the researchers. It has been viewed over as, among these models GMM comparatively performs better and this paper is beneficial for researchers to proceed over in their work.

Keywords: Pattern-Matching, Modelling, Speaker Verification (SV), Text dependent and Text independent verification systems.

1 Introduction

Pattern Matching is the task of speaker verification of computing a match score, which is a measure of the similarity between the input feature vectors and some database model [1]. The features that are extracted from the speech signal are needed to construct speaker models. To introduce users into the system, based on the extracted feature a model of the voice, is generated and stored. Then, to authenticate a user, the matching algorithm compares/scores the incoming speech signal with the model of the claimed user [2]. According to the pattern matching system of the present invention, it is possible to carry out the pattern matching with a high accuracy even when a consonant at the beginning or end part of the word drops out in the input speech pattern.

In general the paper outlines the field of speaker verification for text-independent systems and not going deeply into mathematics. The examples for classification and pattern matching were displayed as well as a basic conceptual scheme for speaker verification.

1.1 Modelling

There are two types of models: stochastic models and template models:-

S.C. Satapathy et al. (eds.), *Proc. of the 3rd Int. Conf. on Front. of Intell. Comput. (FICTA) 2014*
– *Vol. 1*, Advances in Intelligent Systems and Computing 327, DOI: 10.1007/978-3-319-11933-5_3

1.1.1 Template Model

In this case the pattern matching is deterministic. The template method can be dependent or independent of time [2]. Usually templates for entire words are constructed. This has the advantage that, errors due to segmentation or classification [1].

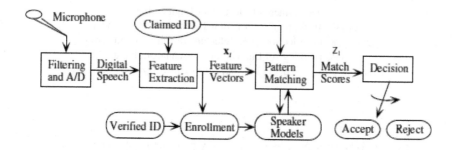

Fig. 1. Generic Speaker Verification System [2]

Template models include Vector Quantization (VQ) and Dynamic Time Warping (DTW) for text independent and text dependent respectively [3].

1.1.2 Stochastic Model

The pattern matching is probabilistic and results in a measure of the likelihood, or conditional probability, of the observation given the model, that is to deal with uncertain or incomplete information [2]. It includes Gaussian mixture model (GMM) and the Hidden Markov model (HMM) are the most popular models for text-independent and text-dependent recognition, respectively[3].A text dependent has a pre-defined text that is used for training and verification whereas text independent should be able to use any text [4].

2   Classification

2.1   Text Dependent Verification System

Text Dependent Speaker Verification (TDSV) is usually connected with the fact that a predefined utterance is used for training the system and for testing/using it. Major system types are text dependent with a predefined password, text dependent with a specific password for each customer, vocabulary dependent, the system or the user uses the text.

The two methods that are used for text dependent speaker verification system are Dynamic Time Warping (DTW) based methods and Hidden Markov Model (HMM) based methods [4].

2.1.1 Dynamic Time Warping [DTW]

To account for differences in speaking rates between speakers and utterances, dynamic time warping algorithms have been proposed around 1970 in the context of speech recognition,. The technique is useful, for example, when one is willing to find a low distance score between the sound signals corresponding to utterances [1] . DTW is a template model for measuring similarity between two sequences which may vary on the basis of time or speed. The most popular method to compensate for speaking-rate variability in template based systems is known as DTW. A well known application has been automatic speech recognition, to cope with different speaking speeds. In general, DTW is a method that allows a computer to find an optimal match between two given sequences on basis of time [1][2]. In order to prepare the reference template is the main problem in Dynamic time warping.

2.2.2 Hidden Markov Model [HMM]

It is the most popular stochastic approach that deals with probabilistic models.HMM enable easy integration of knowledge sources into a compiled architecture .The task is to find the most likely sequence of words 'W'. We need to find out all possible word sequences that maximize the probability [5]. The Strengths of HMM is its mathematical framework and its implementation structure. HMM method is fast in its initial training, and when a new voice is used in the training process to create a new HMM model [6].

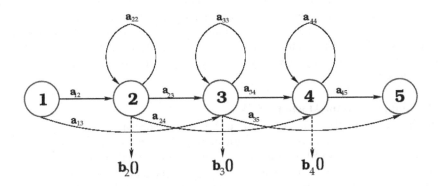

Fig. 2. Architecture Of HMM

In figure 2 oval shape represents random variable that can adopt a number of values. X (t) random variable whose value is hidden at variable time. In Figure 2 the numbers are the states. 1st is the start state and 5th is the end state. The b_i are the probability density function the a_{ij} are the probabilities.

A core advantage of the HMM algorithm is that it can work well even if several users perform the same gesture differently.

2.2   Text Independent Verification System

The users here are not restricted to any fixed phrases. It has no prior knowledge, what so ever about what the claimant will say during the verification session. It will therefore have to grant the claimant access on the basis of characteristics of his/her voice [4].

2.2.1 Vector Quantization (VQ)

It is often applied to ASR. It is useful for speech coders that is efficient data reduction [1]. The utterance feature vectors considered as 'X' and the reference vectors by 'R'. D(X;R) is the distance measure such as the Euclidean distance |X-R|. A smaller value indicates higher likelihood for X and R originating from same speaker. $D(X; R) \neq D(R;X)$ That is it not symmetric. The numbers of vectors are reduced by clustering method such as K map; this gives a reduced set of vectors known as codebook. The test speech is evaluated by all codebooks and chooses the word whose codebook yields the lowest distance measure. Thus, the spectral characteristics of each speaker can be modelled by one or more codebook entries that are representative of that speaker. In basic VQ, codebooks have no explicit time information that is phonetic segments in each word and their relative durations are ignored, since codebook entries are not ordered and can come from any part of the training words. The codebook entries are chosen to minimize average distance across all training frames [2][3].

2.2.2 Artificial Neural Networks

The artificial intelligence approach , it mechanize the recognition procedure according to the way a person applies intelligence in analyzing, visualizing, and characterizing speech based on a set of measured acoustic features[1].

The NN is trained to predict one true word or sentence at a time and whichever of these neurons gives the higher score wins[5].This approach has been mostly in the representation of knowledge and integration of knowledge sources. This method has not been widely used in commercial systems. Uncertainty is modelled not as likelihoods or probability density functions of a single unit, but by the pattern of activity in many units [9].

2.2.3 Support Vector Machine (SVM)

Support vectors are the data points that lie closest to the decision surface. They are the most difficult to classify. For data classification SVMs uses linear and nonlinear separating hyper-planes. An SVM is a kind of large-margin classifier: it is a vector space based machine learning method where the goal is to find a decision boundary between two classes that is maximally far from any point in the training data.SVMs can only classify fixed length data vectors, this method cannot be readily applied to task involving variable length data classification [1].

2.2.4 Gaussian Mixture Model (GMM)

The most successful likelihood function has been Gaussian mixture models for text-independent speaker recognition, as there is no prior knowledge of what the speaker will say, [7]. It is a stochastic model which has become the reference method in speaker recognition. The GMM can be considered as an extension of the VQ model, in which the clusters are overlapping. That is, a feature vector is not assigned to the nearest cluster as in, but it has a nonzero probability of originating from each cluster. The EM algorithm iteratively define the GMM parameters to monotonically increase the likelihood of the estimated model for the observed feature vectors, i.e., for iterations [1].

A universal background model (UBM) for alternative speaker representation. The GMM-UBM system is a likelihood ratio detector in which we compute the likelihood ratio for an unknown test utterance between a speaker independent acoustic distribution (UBM) and a speaker dependent acoustic distribution [3]. To allow maximum flexibility and performance for text-independent applications, both acoustic distributions are modeled by GMMs.

Sr. No.	Researchers	Proposed Work	Authors Contribution
1	Horng- Horng Lin et al 2011[11]	The paper discusses on properties of GMM and its capability of adaptation to background variations, Experiments show the proposed learning rate control scheme, gives better performance than conventional GMM approaches.	It's observed that robustness to quick variations in background as well as sensitivity to abnormal changes in foreground can be achieved simultaneously for several surveillance scenarios.
2	Selami Sadıc et al, 2011[12]	A new method which is a combination of CVA and GMM is proposed in this paper. The experimental studies indicate that GMM is superior to FLDA in terms of recognition rates.	GMM32 yields slightly better recognition rates than those obtained from GMM16. The processing time and memory requirement of GMM32 are much greater than those of GMM16.
3	R.H.Laskar et al, 2012[13]	A comparative analysis of artificial neural networks (ANNs) and Gaussian mixture models (GMMs) for design of voice conversion system using line spectral frequencies (LSFs) as feature vectors.	From the perceptual tests, it is observed that the voice conversion is more effective, if the source and target speakers belongs to different genders. Subjective evaluation also indicated that the developed voice conversion system using residual copying method performs better than residual selection method in terms of MOS
4	Nakamasa Inoue et al, 2012[14]	A fast maximum posteriori (MAP) adaptation method using a tree-structured GMM.	The calculation time of the MAP adaptation step is reduced by 76.2% compared with that of a conventional method. The total calculation time is reduced by 56.6% while keeping the same level of the accuracy.
5	Sami Keronen et al 2013[15]	The unreliable and noise-corrupted that is missing components are identified using a Gaussian mixture model (GMM) classifier based on a diverse range of acoustic features.	A missing data approach is used to compensate for challenging environ- mental noise containing both additive and convolutive components.

The UBM are trained using the EM algorithm. The GMM can be viewed as a hybrid between a parametric and nonparametric density model. Like a parametric model it has structure and parameters that control the behavior of the density in known ways, but without constraints that the data must be of a specific distribution type, such as Gaussian or Laplacian. Like a nonparametric model, the GMM has many degrees of freedom to allow arbitrary density modeling, without undue computation and storage demands [7].

A 2011 paper introduces a new volume exploration scheme with a unique ability to capture the data characteristics. This flexibility is helpful to inexperienced users because it provides an incremental GMM estimation scheme and it can automatically provide a suggestive volume classification using a greedy EM algorithm .But the limitations that is discussed over in this paper is regarding the number of the mixture components needs to be provided, because the user may not have a clear idea about how many features of interest are in the data set, which can be resolved [9].

A 2013 paper addresses robust speech recognition based on subspace Gaussian mixture models (SGMMs) using joint uncertainty decoding (JUD) for noise compensation [10].

3   Conclusions

The classification and pattern matching methods were briefly overviewed. Vector quantization method was outlined and an example of such classification was displayed. It is easier to understand it, since it is not statistical method. Later in the paper, we discussed statistical methods: HMM. It is obvious, that it is not so insignificant to implement those methods, since a deep knowledge of statistical mathematics is required. However, even more complicated methods known as Artificial Neural Network was shortly discussed. This methods is quite elegant for speaker verification, but hard to implement. Among these models GMM performs better as comparative to the other systems, this system is used for high-accuracy speaker verification. The GMM can be used as the likelihood function are that it is computationally inexpensive, is based on a well-understood statistical model and for text-independent tasks, is insensitive to the temporal aspects of the speech, modeling only the underlying distribution of acoustic observations from a speaker.

In order to obtain better results for processing power and memory management in GMM this can be improved by combination of CVA and GMM. This in order will result in better resolution rates and also improve its memory capacity.The HMM combination with Gaussian mixture model (GMM) with a joint distribution of provides both clean and noisy speech feature.

References

1. Anusuya, M.A., Katti, S.K.: Speech Recognition by Machine: A Review. International Journal of Computer Science and Information Security (IJCSIS) 6(3) (2009)
2. Campbell, J.: Speaker recognition: a tutorial. Proc. IEEE 85(9), 1437–1462 (1997)
3. Kinnunen, T., Li, H.: An overview of text-independent speaker recognition: features to supervectors. Science Direct, Speech Communication 52 (2010)

4. Rydin, S.: "Text dependent and Text independent speaker verification systems, Technology and applications" in a Centre for Speech Technology, Stockholm (December 2001)
5. Al-Alaoui, M.A., Al-Kanj, L., Azar, J., Yaacoub, E.: Speech Recognition using Artificial Neural Networks and Hidden Markov Models. IEEE Multidisciplinary Engineering Education Magazine 3(3) (September 2008)
6. Morgan, N.: Deep and Wide: Multiple Layers in Automatic Speech Recognition. IEEE Transactions on Audio, Speech, And Language Processing 20(1) (January 2012)
7. Reynolds, D.A., Quatieri, T.F., Dunn, R.B.: Speaker Verification Using Adapted Gaussian Mixture Models. Digital Signal Processing 10, 19–41 (2000)
8. Sturim, E., Reynolds, D.A., Dunn, R.B., Quatieri, T.F.: Speaker Verification using Text-Constrained Gaussian Mixture Models. In: IEEE 2002. MIT Lincoln Laboratory, Lexington (2002)
9. Wang, Y., Chen, W., Zhang, J., Dong, T., Shan, G., Chi, X.: Efficient Volume Exploration Using the Gaussian Mixture Model. IEEE Transactions on Visualization and Computer Graphics 17(11) (November 2011)
10. Lu, L., Chin, K.K., Ghoshal, A., Renals, S.: Joint Uncertainty Decoding for Noise Robust Subspace Gaussian Mixture Models. IEEE Transactions on Audio, Speech, And Language Processing 21(9) (September 2013)
11. Lin, H.-H., Chuang, J.-H., Liu, T.-L.: Regularized Background Adaptation: A Novel Learning Rate Control Scheme for Gaussian Mixture Modeling. IEEE Transactions on Image Processing 20(3) (March 2011)
12. Sadıc, S., Gulmezoğlu, M.B.: Common vector approach and its combination with GMM for text-independent speaker recognition. Expert Systems with Applications 38, Science Direct (2011)
13. Laskara, R.H., Chakrabarty, D., Talukdara, F.A., Sreenivasa Raoc, K., Bane, K.: Comparing ANN and GMM in a voice conversion framework. Applied Soft Computing, SciVerse ScienceDirect (2012)
14. Inoue, N., Shinoda, K.: A Fast and Accurate Video Semantic-Indexing System Using Fast MAP Adaptation and GMM Supervectors. IEEE Transactions on Multimedia 14(4) (2012)
15. Keronen, S., Kallasjoki, H., Remesa, U., Brownb, G.J., Gemmeke, J.F., Palomäki, K.J.: Mask estimation and imputation methods for missing data speech recognition in a multi-source reverberant environment
16. Wang, Y., Chen, W., Zhang, J., Dong, T., Shan, G., Chi, X.: Efficient Volume Exploration Using the Gaussian Mixture Model. IEEE Transactions On Visualization and Computer Graphics 17(11) (November 2011)
17. Halbe, Z., Bortman, M., Aladjem, M.: Regularized Mixture Density Estimation With an Analytical Setting of Shrinkage Intensities. IEEE Transactions on Neural Networks and Learning Systems 24(3) (March 2013)

Online Clustering Algorithm for Restructuring User Web Search Results

M. Pavani[1] and G. Ravi Teja[2]

[1] Department of Computer Science and Engineering
Lendi Institute of Science and Technology, Visakhapatnam, India
pavani.csebtech@gmail.com
[2] Department of Computer Science and Engineering
Satya Institute of Technology and Management, Visakhapatnam, India
teja.thesmile@gmail.com

Abstract. Now a day's web mining is very important area. When user issues a query on the search engine, it gives relevant and irrelevant information to the user. If the query issued by the user is ambiguous then different users may get different search results and they have different search goals. The analysis of user search results according to their user goals can be very useful in improving search engine experience, usage and relevance. In this paper, we propose to infer user search goals by clustering the proposed user search sessions. User search sessions are constructed from user search logs and these can efficiently reflects the information needed by the users. An Online Clustering algorithm is used for clustering the pseudo documents, and then we use another appoarch to generate pseudo documents for better representation of the user search sessions for clustering. Finally we are using Classified Average Precision to evaluate the performance of inferring user search goals.

Keywords: Query groups, user search sessions, search goals, click sequence.

1 Introduction

The web accessing in world drastically will increases for searching various purpose like managing the bank transactions, travel arrangements product purchases. Web mining is one of the important applications of data mining technique to discover the knowledge from the web. In web search applications, queries are submitted to search engines to represent the information needs of users. However, sometimes produced web results may not exactly represent users' specific information needs since many ambiguous queries users may want to get information on different aspects when they submit the same query. For example, when the query "Apple" is submitted to a search engine, some users want apple iphone home page to know the cost of the phone details, while some others want to know the information about eating apples. Therefore, it is necessary to capture different user search goals in knowledge extraction. User search goals can be considered as the clusters of information needs for a query. The inference and analysis of user search goals can have a lot of advantages like query recommendations [2] [3] [4], re-arranging the web search results [5] [6] [7] and re-ranking web search results are improving search engine relevance and user experience. .

In this paper, we aim at discovering the number of diverse user search goals for a query and depicting each goal with some keywords automatically. We first propose an approach to infer user search goals for a query by clustering user search sessions. The user search session is defined as the series of both clicked and unclicked URLs from user click-through logs. Then, we use optimization method to map user search sessions to pseudo-documents which can efficiently tells user information needs. At last, we cluster these pseudo-documents using Online Clustering algorithm to infer user search goals and with some keywords. Since the evaluation of clustering is also an important, we also use an evaluation criterion classified average precision (CAP) to evaluate the performance of the restructured web search results.

The remaining section of the paper organized as follows. Section 2 reviews the related work. Section 3 describes Proposed Work.. Section 5 presents evaluation results. Finally, we conclude paper in Section 5.

2   Related Works

In previous work [15] they used online clustering algorithm to cluster the user queries. After clustering the groups are created. By using one pass algorithm they discover the knowledge from the groups. My work is extension to this paper. After grouping we rearrange the user web results. After that we evaluate the results using criterion function.

Organization of user search results is very difficult for increasing usage and relevance of any search engine. Clustering search results is an efficient way to organize search results which allows a user to navigate into relevant documents quickly. Generally all existing work [7], [8] perform clustering on a set of top ranked results to partition results into general clusters, which may contain different subtopics of the general query term. However, this clustering method has two limitations so that it not always works well. First, created clustered groups do not give the necessary information needed by the user. Second, they give labels to the clusters which are more general and not informative to identify appropriate groups.

Wang and Zhai [9] proposed an approach to organize search results in user-oriented manner. They used search engines log to learn interesting aspects of similar queries and categorize search results into aspects learned. Cluster labels are generated from past query words entered by users.

R. Baeza-Yates, C. Hurtado, and M. Mendoza et.al [2] introduced a method that, when user submitted a query to a search engine, gives a list of related queries. The related queries are based in previously issued queries, and can be issued by the user to the search engine to redirect the search process. Query clustering method is proposed in which it cluster the similar user queries. The clustering process uses the content of user's history according to their preferences in the search log of the search engine. The method not only discovers the related queries, but also give ranks them according to a relevance criterion.

Deepali Agale, Beena Khade [10] used bisecting k-means algorithm to cluster the user web results. In this paper they are used two algorithms to cluster the user search goals. One is k-means clustering algorithm and another one is bisecting algorithm. In this initially apply k-means clustering algorithm on the documents that is retrieved based on the user queries and then on the result bisecting algorithm is applied. It is a time consuming process to compute the user queries using these two algorithms.

Doug Beeferman, Adam Berger [11] used agglomerative clustering algorithm to cluster the user search logs. in agglomerative clustering algorithm uses the bipirate graph on the search. In that graph vertices on one side representing user queries and another side vertices represents URLs. Apply clustering algorithm on that graph to find the similar queries and URLs. There are some disadvantage using this algorithm i.e. this algorithm is ignored the content of the queries and URLs. And find the similarity based on how they are co-occur within the click through data.

Zheng Lu, Zhengu Zhena, Weiyoo Lin [12] were using K-means clustering algorithm to cluster the user search logs. In that they described user needed information. For that initially they group the user queries and after grouping they evaluate the groups. But there is a limitation in that i.e. in K-means clustering we have to fix the k-value initially. But in web user queries are continuous so that after every iteration it is difficult to find the center and also difficult to find k-value.

R.Dhivya, R.Rajavignesh [13] used fuzzy c-means clustering algorithm to cluster the user queries. Fuzzy c-means clustering and similar algorithms have problems with high dimensional data sets and a large number of prototypes.

Wang, Zhai [14] clustered queries and learned aspects of similar queries. Limitations- This method does not work if we try to discover user search goals of any one single user query in the query cluster rather than a cluster of similar queries. Jones and Klinkner [9] predict goal and mission boundaries to hierarchically segment query logs. However, their method only identifies whether a pair of queries belongs to the same goal or mission and does not care what the goal is in detail.

3   Proposed Work

In this section, basic operations involved in proposed approach to discover user search goals/intents by clustering pseudo-documents are described. The flow of the proposed system design will be as shown in Fig. 1. Initially we first prepare user search sessions from the user search logs. After that pseudo documents are created. We apply clustering algorithm on pseudo documents. After creating the groups we analyze the groups and rearrange the web results based on the groups.

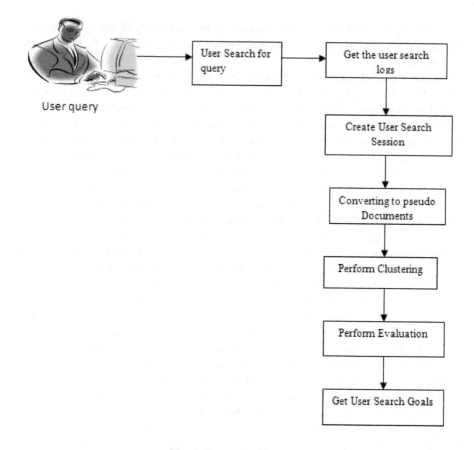

User query

Fig. 1. System Architecture

3.1   Representation of User Search Sessions

In this section first we describe the user search sessions. After that we explain how to convert these user sessions to pseudo documents [1].

3.2   User Search Sessions

User search sessions are enhanced user search logs. It contains both user clicked URLs, uncliked URLs and ends with the last URL that was clicked in a single session. In this all clicked and uncliked URLs are present and have been scanned and evaluated the URLs by the users. Therefore, besides the clicked URLs, the unclicked ones before the last click should be a part of the user feedbacks. Fig. 2 shows an example of a feedback session and a single session.

Search Results	Click sequence
www.smallseotools.com	2
https://paytm.com/	0
http://en.wikipedia.org/wiki/Cluster_analysis	1
http://en.wikipedia.org/wiki/Data_mining	1
https://www.apple.com/in/	2
http://www.forbes.com/companies/apple/	3
http://en.wikipedia.org/wiki/Apple	1
http://www.nutrition-and-you.com/apple-fruit.html	2
http://www.youtube.com/user/Apple	0
http://www.bloomberg.com/quote/AAPL:US	2

Fig. 2. User Search Sessions

The above example figure describes the user search queries on different search results. Left part contains the 10 search results for a user queries like apple fruit, apple ipod, paytm and data mining related concepts etc. Right part contains the clicked sequence whih described how many times user clicked particular url. Here zero (0) represents unclicked URL. In the user search sessions the clicked URLs tells that what user really want from the web and unclicked URLs represents what user relay do not care about.

3.3   Conversion of User Search Sessions into Pseudo Documents

3.3.1   Representing the URLs in the User Search Session

In this step, we first find the URLs in the user search history along with additional textual information by extracting the titles and snippets of the URL. In such way, each URL in the user search session is represented by a small text paragraph that consists of that URLs title and snippet. After that we apply textual process like stemming, removal of stop words and transforming all of letters to lowercases are implemented to those text paragraphs. Finally, each URL's title and snippet are represented by a Term Frequency-Inverse Document Frequency vector (TF-IDF), respectively, as in

$$T_i = [tw_1, tw_2 \ldots \ldots tw_n]^T \tag{1}$$

$$S_i = [sw_1, sw_2.........sw]^T \tag{2}$$

Where T_i and S_i are the TF-IDF vectors of the URL's title and snippet, respectively. $W_j= (1,2,3,.........n)$ is the jth term appearing in the enriched URLs. Here, a "term" is defined as a word or a number in the dictionary of document collections. twj and swj represent the TF-IDF value of the jth term in the URL's title and snippet, respectively. Considering that each URLs' titles and snippets have different significances, we represent the each enriched URL by the weighted sum of Tui and Sui, namely

$$R_i = wt.T_i + st.S_i$$
$$= [fw_1, fw_2.........fw_n]^T \tag{3}$$

Where Ri means the feature representation of the ith URL in the feedback session, and wt and st are the weights of the titles and the snippets, respectively.

3.3.2 Formation of Pseudo-Document

We propose an optimization method to combine clicked and unclicked URLs in the feedback session to obtain a feature representation. Let C be the feature representation of a feedback session, and UC be the value for the term w.
Let

$$C = (m = 1,2,3......M) \text{, and}$$

$$UC = (l = 1,2,3......L);$$

Let R be the feature representations of the clicked and unclicked URLs in this feedback session, respectively. Let C and UC be the values for the term w in the vectors. We want to obtain such a S that the sum of the distances between S and each C is minimized and the sum of the distances between S and each UC is maximized. Based on the assumption that the terms in the vectors are independent, we can perform optimization on each dimension independently, as shown in below equation.

$$S = [ff(w_1), ff(w_2).........ff(w_n)]^T \tag{4}$$

$$R_S(w) = \arg\min \sum_M (S(w) - C(w))^2 - \lambda \sum_L ((S(w) - UC(w))^2 \tag{5}$$

λ is a parameter balancing the importance of clicked and unclicked URLs. When λ in (4) is 0, unclicked URLs are not taken into account. On the other hand, if λ is too big, unclicked URLs will dominate the value of Uc. In this project, we set λ to be 0.5.

3.3.3 Clustering of Pseudo Documents

Online Clustering Algorithm

Input: The current singleton query group T_c containing the current query Q_c and set of clicks C_c

A set of existing query group $T = \{T_1, T_2 \ldots \ldots T_m\}$

A similarity threshold $R_s, 0 \leq R_s \leq 1$

Output:

The query group Q that best matches Q_c or a new one if necessary.

$Q = \phi$

$R_{Max} = R_s$

for $i = 1$ to m

if $S(T_c, T_i) > R_{Max}$

$Q = T_i$

$R_{Max} = S(T_c, T_i)$

If $Q = \phi$

$T = T \cup T_c$

$T = T_C$

Return S.

4   Evaluation Criterions

Average Precision

A possible evaluation criterion is the average precision (AP) which evaluates according to user implicit feedbacks. AP is the average of precisions which is computed at the point of each relevant document in the ranked sequence, shown in

$$AP = \frac{1}{N^+} \sum_{r=1}^{N} rel(r) \frac{R_r}{r} \tag{6}$$

Where N is the number of relevant (or clicked) documents in the retrieved ones, r is the rank, N is the total number of retrieved documents, rel(r) is a binary function on the relevance of a given rank, and Rr is the number of relevant retrieved documents of rank r or less.

Voted Average Precision (VAP)

It is calculated for purpose of restructuring of search results classes i.e. different clustered results classes. It is same as AP and calculated for class which having more clicks.

Risk

It is the AP of the class including more clicks? There should be a risk to avoid classifying search results into too many classes by error. So we propose the Risk.

Classified AP (CAP)

VAP is extended to CAP by introducing combination of VAP and Risk. Classified AP can be calculated by using the formula, as follows:

$$CAP = VAP(1 - Risk)^\gamma \qquad (7)$$

5   Experimental Results

In this experimental results section, we will show experiments of our proposed algorithm. The data set that we used is based on the user click logs from a commercial search engine collected over a period of one month, including totally 2,000 different queries, 2.0 million single sessions and 1.8 million clicks. We made comparisons on data set using three algorithms along with our algorithm. We calculate the classified average precision on the user queries (data set).

Method	Mean Average VAP	Mean Average Risk	Mean Average CAP
Online Clustering	0.78	0.255	0.732
K-means	0.755	0.224	0.623
Agglomerative	0.64	0.196	0.581

Fig. 3. The CAP comparison of three methods for 2,000 queries

The above figure 3 describes the comparative results of three methods on 2,000 user queries. We calculate the mean average VAP, Risk and CAP using miner tool. Our algorithm gives better results compared to other two algorithms. In k-means, agglomerative algorithms the Voted average precision and risk is less that means which represents the accuracy of user search results is less compared to online clustering.

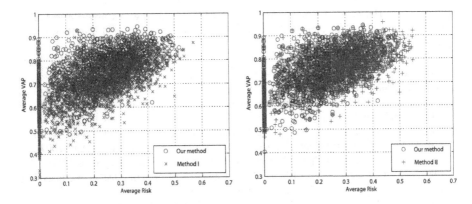

Fig. 4. Clustering of three methods. Each point represents the average risk and VAP of a query when evaluating the performance of restructuring search results.

6   Conclusion

The proposed system can be used to improve discovery of user search goals for a similar query by using online clustering algorithm for clustering user search sessions represented by pseudo-documents. By using proposed system, the inferred user search goals can be used to restructure web search results. So, users can find exact information quickly and very efficiently. The discovered clusters of query can also be used to assist users in web search.

References

1. Agale, D., Khade, B.: Inferring User Search goals Engine Using Bisecting Algorithm. International Journal of Scientific & Engineering Research 5(2) (2014) ISSN 2229-5518
2. Baeza-Yates, R., Hurtado, C., Mendoza, M.: Query Recommendation Using Query Logs in Search Engines. In: Lindner, W., Fischer, F., Türker, C., Tzitzikas, Y., Vakali, A.I. (eds.) EDBT 2004. LNCS, vol. 3268, pp. 588–596. Springer, Heidelberg (2004)
3. Cao, H., Jiang, D., Pei, J., He, Q., Liao, Z., Chen, E., Li, H.: Context-Aware Query Suggestion by Mining Click-Through. In: Proc. 14th ACM SIGKDD Int'l Conf. Knowledge Discovery and Data Mining (SIGKDD 2008), pp. 875–883 (2008)
4. Huang, C.-K., Chien, L.-F., Oyang, Y.-J.: Relevant Term Suggestion in Interactive Web Search Based on Contextual Information in Query Session Logs. J. Am. Soc. for Information Science and Technology 54(7), 638–649 (2003)
5. Chen, H., Dumais, S.: Bringing Order to the Web: Automatically Categorizing Search Results. In: Proc. SIGCHI Conf. Human Factors in Computing Systems (SIGCHI 2000), pp. 145–152 (2000)
6. Wang, X., Zhai, C.-X.: Learn from Web Search Logs to Organize Search Results. In: Proc. 30th Ann. Int'l ACM SIGIR Conf. Research and Development in Information Retrieval (SIGIR 2007), pp. 87–94 (2007)

7. Zeng, H.-J., He, Q.-C., Chen, Z., Ma, W.-Y., Ma, J.: Learning to Cluster Web Search Results. In: Proc. 27th Ann. Int'l ACM SIGIR Conf. Research and Development in Information Retrieval (SIGIR 2004), pp. 210–217 (2004)
8. Zamir, O., Etzioni, O.: Grouper: A dynamic clustering interface to web search results. Computer Networks 31(11-16), 1361–1374 (1999)
9. Wang, X., Zhai, C.-X.: Learn from Web Search Logs to Organize Search Results. In: Proc. 30th Ann. Int'l ACM SIGIR Conf. Research and Development in Information Retrieval (SIGIR 2007), pp. 87–94 (2007)
10. Baeza-Yates, R., Tiberi, A.: Extracting Semantic Relations from Query Logs. In: Proc. 13th ACM SIGKDD Int'l Conf. Knowledge Discovery and Data Mining (KDD) (2007)
11. Barbakh, W., Fyfe, C.: Online Clustering Algorith ms. Int'l J. Neural Systems 18(3), 185–194 (2008), Levenshtein, V.I.: Binary Codes Capable of Correcting Deletions, Insertions and Reversals. Soviet Physics Doklady 10, 707–710 (1966)
12. Sahami, M., Heilman, T.D.: A Web-based Kernel Fun ction for Measuring the Similarity of Short Text Snippets. In: Proc. the 15th Int'l Conf. World Wide Web (WWW 2006), pp. 377–386 (2006)
13. Chang, J.H., Lee, W.S., Zhou, A.: Findin g Recent Frequent Itemsets Adaptively over Online Data Streams. In: ACM SIGKDD In t'l Conf. on Knowledge Discovery and Data Mining (August 2003)
14. Charikar, M., Chen, K., Farach-Colton, M.: Finding Frequent Items in Data Streams. Theoretical Computer Science (January 2004)
15. Sunita, A., Yadwad, M.: Discovery of knowledge from the query groups using one pass algorithm. In: Satapathy, S.C., Udgata, S.K., Biswal, B.N. (eds.) FICTA 2013. AISC, vol. 247, pp. 493–499. Springer, Heidelberg (2014)

Haar Wavelet Transform Image Compression Using Various Run Length Encoding Schemes

Rashmita Sahoo[1], Sangita Roy[2], and Sheli Sinha Chaudhuri[3]

[1] ETCE Department, Balasore College of Engineering and Technology, Odisha, India
[2] ECE Department, Narula Institute of Technology, WBUT, Kolkata
[3] ETCE Department, Jadavpur University, Kolkata, India

Abstract. Image compression is a very important useful technique for efficient transmission as well as storage of images. The demand for communication of multimedia data through the telecommunication network and accessing the multimedia data through internet by utilizing less bandwidth for communication is growing explosively. Basically the image data comprise of significant portion of multimedia data and they occupy maximum portion of communication bandwidth for multimedia communication. Therefore the development of efficient image compression technique is quite necessary. The 2D Haar wavelet transform along with Hard Thresholding and Run Length Encoding is one of the efficient proposed image compression technique. JPEG2000 is a standard image compression method capable of producing very high quality compressed images. Conventional Run Length Encoding(CRLE),Optimized Run Length Encoding(ORLE),Enhanced Run Length Encoding(ERLE) are different types of RLES applied on both proposed method of compression and JPEG2000. Conventional Run Length Encoding produces efficient result for proposed method whereas Enhanced Run Length Encoding produces efficient result in JPEG2000 compression. This is the novel approach that the authors have proposed for compression of image using compression ratio (CR) without losing the PSNR, quality of image using lesser bandwidth.

Keywords: Compression ratio, Run Length Encoding (RLE), Haar wavelet Transform (HWT), Hard Thresholding (HT), Conventional Run Length Encoding (CRLE), Optimized Run Length Encoding(ORLE),Enhanced Run Length Encoding(ERLE).

1 Introduction

Image compression is an application of data compression. In image compression the original image is encoded with few bits [1]. The main objective of image compression coding refers to store original image into bit stream as compact as possible and to display the decoded image as similar to the original as possible as possible. The two fundamental principles behind image compression are redundancy and irrelevancy [2]. Redundancy removes repetition in the bit stream and irrelevancy omits the pixel values which are not noticeable by human eye. Image compression is applied in medical imaging, satellite imaging, artistic computer graphics design, HDTV (high

© Springer International Publishing Switzerland 2015
S.C. Satapathy et al. (eds.), *Proc. of the 3rd Int. Conf. on Front. of Intell. Comput. (FICTA) 2014*
– *Vol. 1*, Advances in Intelligent Systems and Computing 327, DOI: 10.1007/978-3-319-11933-5_5

definition television), digital storage of movies. The image is basically a 2D signal processed by human visual system. The signal representing the image is usually in analog form. But for image processing, storage, and transmission they are converted to digital form from analog form. Hence a digital image is a 2D array of numbers or matrix. Each element in the matrix is known as picture element or pixel. Basically there are two types of image compression. One is lossy compression and another is lossless compression. In Lossless compression scheme, the reconstructed image after compression is numerically identical to the original one. However in lossy compression some unneeded data which are not recognized by human eliminated. JPEG2000 compression technique basically uses DCT as a popular transform coding and Run Length Encoding as a lossless coding for efficient compression of image. Three different type of Run Length Encoding such as Conventional, Optimized, and Enhanced coding the Author has used both for existing and proposed methods. Conventional Run Length Encoding encodes the repetition of zeros in the image matrix. Optimized Run Length Encoding is almost same as CRLE, it represents single zero between nonzero characters with two digit sequence of (1, 0). In Enhanced Run Length Encoding the (1, 0) pair of optimized RLE is replaced by single zero. Wavelet Transform has been a recent addition in the field of image analysis both for time and frequency component. Initial Wavelet Transform is proposed by Mathematician Alfred Haar in 1909. Wavelet is a hierarchical decomposition function which improves image quality with higher compression ratio [3][4]. Haar Wavelet Transform (HWT) is the simplest form of 2D DWT. HWT is used as transform coding as DCT for proposed method. The remainder of the paper is organized as follow. Section 2 includes jpeg2000 image compression where as section 3 include different Run Length Encoding schemes. Section 4 includes proposed method with experimental results. Section 5 includes conclusion.

2   JPEG2000 Image Compression

JPEG2000 is basically a Lossy compression designed specifically to discard information that the human eye cannot see easily. In this technique the original image is divided into (8×8) blocks as shown in table 1. On each and every (8×8) block 2D DCT is computed. Then each and every (8×8) block is quantized by using a standard quantization table given in table 2. After quantization all blocks are combined and Conventional Run Length Encoding is applied on the whole image matrix. Cameraman.jpg, Rice.png, Moon.tif, and some other nonstandard images are taken for this compression. The encoding (zigzag scan, CRLE) steps achieve additional compression in lossless manner by encoding the coefficients after quantization. The encoding involves two steps. The first step converts the zigzag sequence of quantized coefficients in to intermediate sequence of symbols. The second step converts the symbols to data stream in which the symbols no longer have externally identifiable boundaries. This ordering helps to facilitate entropy coding by placing low frequency coefficients before high frequency coefficients. Then the sequence of AC coefficients of the whole image is encoded using three different types of Run Length Encoding to take the advantage of the long runs of zeros that normally results from the re ordering.

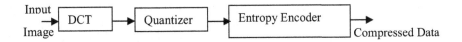

Fig. 1. Block Diagram of JPEG2000 Image Compression

Table 1. (8×8) Block of Original Image

11	32	2	36	68	10	20	23
86	4	9	34	63	11	19	22
82	20	21	43	70	11	19	22
61	19	17	37	67	12	20	23
39	14	8	28	64	12	20	23
33	19	10	29	66	12	20	23
29	14	5	26	66	12	20	23
20	13	5	28	69	12	20	23

Table 2. Quantization Matrix

16	11	10	16	24	4	51	61
12	12	14	19	26	5	60	55
14	13	16	24	40	5	69	56
14	17	22	29	51	8	80	62
18	22	37	56	68	1	103	77
24	35	55	64	81	1	113	92
49	64	78	87	10	1	120	10
72	92	95	98	11	1	103	99

Table 3. (8×8) Block after DCT and Quantization

93	-102	60	3	3	3	-1	1
4	10	5	2	3	1	0	0
0	0	2	1	1	0	0	0
0	0	1	0	0	0	0	0
0	0	0	0	0	0	0	0
1	0	0	0	0	0	0	0
0	0	0	0	0	0	0	0
0	0	0	0	0	0	0	0

3   Run Length Encoding

Run Length Encoding is a standard encoding technique basically used for block transform image/video compression. RLE has been classified into three parts, i) Conventional RLE (CRLE), ii) Optimized RLE (ORLE), iii) Enhanced RLE (ERLE).Run Length Encoding is applied for only AC coefficients. CRLE consists of two variables, out of which one is RUN and another is LEVEL. RUN represents the number of repeated zeros whereas LEVEL represents the non zero component present after the repetition of zero. CRLE is applied on table3 and the sequence is represented as [(0,-102) (0,4) (1,10) (0,60) (0,3) (0,5) (4,2) (0,2) (0,3) (0,3) (0,3) (0,1) (0,1) (1,1) (1,1) (2,1) (0,1) (0,-1) (0,1) (0,0)]. Conventional Run Length Encoding represents the 64 elements present in table3 to a sequence of 41 bits. Optimized RLE uses pair of (RUN, LEVEL) only when a pattern of consecutive zeros occurs at the input of the encoder. This method represents the single zero between two nonzero characters with a two digits sequence (1,0). The sequence obtained after applying Optimized RLE on table3 is denoted as [-102, 4, (1,0), 10, 60, 3, 5, (4,2), 2, 3, 3, 3, 1, 1, (1,0), 1, (1,0), 1, (2,1), 1, -1, 1, (35,0), (0,0)]. Enhanced Run Length Encoding(ERLE) is basically used for further minimization of Optimized RLE. In ERLE the (1,0) pair is simply replaced by a single 0 which reduces the number of bits required. After applying ERLE table 3 the obtained sequence is given as [-102, 4, 0, 10, 60, 3, 5, (4,2), 2, 3, 3, 3, 1, 1, 0, 1, 0, 1, (2,1), 1, -1, 1, (35,0), (0,0)] . The 64 elements table is reduced to a sequence of 29 elements in ERLE.

4   Proposed Method

In their previous work [5] authors had worked on Haar Wavelet Transform Image Compression using RLE. In the current work authors have worked on the same method with different RLE schemes, i.e., i) CRLE, ii) ORLE, and iii)ERLE. After that they compared with JPEG2000 and Haar Wavelet Transform with different RELs. In this method HWT is applied on each (8×8) block. Then to get high degree of compression hard thresholding is applied on each and every (8×8) block of image. Mathematically the hard thresholding is denoted as in equation 1. Finally different RLEs are applied.

$$T(x) = \begin{cases} 0 & if\ |x| \leq \epsilon \\ x & Otherwise \end{cases} \qquad (1)$$

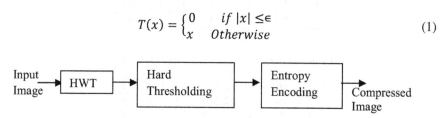

Fig. 2. Block Diagram of Proposed Method

Table 4. (8×8) Block of Original Image (same as table 1)

Table 5. (8×8) Block of Original Image after HWT and HT ($\varepsilon = 15$)

93	63	0	62	21	0	27	0
0	0	0	0	13	0	0	0
0	0	0	0	0	0	0	0
0	0	0	0	0	0	0	0
0	0	0	0	0	0	0	0
0	0	0	0	0	0	0	0
0	0	0	0	0	0	0	0
0	0	0	0	0	0	0	0

4.1   Experimental Results

The table 6 represents the comparison of various Run Length Encodings both proposed and existing methods. Compression ratio given in the table 6 is the ratio between the compressed data to uncompressed data. PSNR is denoted as peak signal to noise ratio which is computed for the best two methods, one from DCT and another from HWT and given in table 7.

Table 6. Different RLES Computation Table

Sl. No.		Algorithm used	Digits obtained in encoding sequence	Compression ratio
1	a.	DCT, Quantization, Entropy Encoding(zigzag scan and CRLE)	41	0.6406
	b.	DCT, Quantization, Entropy Encoding(zigzag scan and ORLE)	32	0.50
	c.	DCT, Quantization, Entropy Encoding(zigzag scan and ERLE)	29	0.4531
2	a.	HWT, HT, Entropy Encoding(zigzag scan, CRLE)	13	0.20313
	b.	HWT, HT, Entropy Encoding(zigzag scan, ORLE)	15	0.2343
	c.	HWT, HT, Entropy Encoding(zigzag scan, ERLE)	14	0.218

From the table 6 it is evident that HWT, HT along with different RLEs compressed the data most. Now authors experiment the PSNRs of lowest CRs of DCT with that of HWT.

Table 7. PSNR Computation Table

Sl. No	Algorithm used	Digits obtained in encoding sequence	Compression ratio	PSNR (dB)
1	DCT, Quantization, ERLE	29	0.4531	60.72
2	HWT, HT, CRLE	13	0.2031	65.59

The result shows that DCT with ERLE has the least CR value 0.4531 whereas HWT with CRLE has CR of 0.20313. The corresponding PSNRs are 60.72 and 65.59 respectively. This shows that PSNR with HWT and CRLE compressed the image significantly without degrading the image quality.

5   Conclusion

Authors first studied JPEG2000 image compression with different RLEs, and then they proposed a new scheme where they applied the same RLEs. They compared those RLEs. It is clear from the above discussion that proposed method with Run

Length Encodings are superior to than that of JPEG2000 for all three types of Run Length Encodings as shown in Table 6 .In the area of 3G and 4G LT the issue of band width is one of the main factors for better video communication. Since in this proposed method the PSNR increases appreciably than that of JPEG2000 .Therefore it can be concluded that the picture retrieved is of better quality and as the number of bits required to send the image is less, a lesser band width is utilized. Image can be send without compromising image quality.

References

1. Amritpal Singh, V.P.: Enhanced Run Length Coding for JPEG Image Compression. International Journal of Computer Application (0975-8887) 72(20) (June 2013)
2. Talukdar, K.H., Harada, K.: Haar Wavelet Based Approach for Image Compression and Quality Assessment of Image Compression. IAENG International Journal of Applied Mathmatics 36(1), IJAM_36_19 (2013)
3. Waic, C.C.: On Lossless and Lossy Compression of Step Size Matrics in JPEG Coding. IEEE (2013)
4. Nagarajan, A., Alagarsamy, K.: An Enhanced Approach in Run Length Encoding Scheme (ERLE). IJETT (July 2011)
5. Sahoo, R., Roy, S., Chaudhuri, S.S.: Haar Wavelet Transform Image Compression Using Run Length Coding. In: ICCSP 2014. IEEE-1-4799-3357-0 (2014)

An Integrating Approach by Using FP Tree and Genetic Algorithm for Inferring User Interesting Results

Md. Nishant Sharief and Sunita A. Yadwad

Department of Computer Science and Engineering,
Anil Neerukonda Institute of Technology and Sciences,
Sangivalasa, Bheemunipatnam [M], Visakhapatnam
{nishatsharief.cse,sunitayadwad}@gmail.com

Abstract. Search engine optimization is an interesting research issue in information retrieval for retrieving user interesting results. Satisfying the user search goal is a complex task while searching user specific query, because of billions of related and unrelated data available over the network. In this proposed work we are introducing an empirical model of search mechanism with FP Tree for finding frequent use of patterns (sequence of Urls) and genetic algorithm, which belongs to the larger class of evolutionary algorithms, which will be used for generating solutions for optimization problems using techniques such as mutation and crossover for optimal results with efficient feedback sessions, based on query clicks. From the time user starts clicking or visiting the urls the session is started and the session of feedback is generated reflecting the efficient user information needs.

Keywords: User Search goals, Feedback Sessions, FP Tree, Genetic Algorithm, Crossover, Mutation.

1 Introduction

Various approaches were released by various authors from years of research in the field of optimizing the search engines, every research work have its own pros and cons. Some of the mostly used search engines works based on relevance score, time stamps and query click graph .Latest technology of Search engines follows basic concepts of semantic comparison of keywords, localization and cache implementation for optimal performance.

There are several researches still going on for generating the user interesting results in web searches. The approaches like term based approach and log based approach[1] which matches keywords and synonyms from logs based on user clicks and retrieve the related documents based on frequency of keywords or terms. File relevance score is computed by using Term Frequency (TF) and Inverse Document Frequency (IDF)as parameters for finding frequency of keyword but not with the time stamp of documents and so there is no priority for recent updated documents. Time stamp based approaches works with the recent uploaded documents along with file relevance score. Clustering based techniques gives good results by combining both the above approaches [2].

© Springer International Publishing Switzerland 2015

S.C. Satapathy et al. (eds.), *Proc. of the 3rd Int. Conf. on Front. of Intell. Comput. (FICTA) 2014*
– *Vol. 1*, Advances in Intelligent Systems and Computing 327, DOI: 10.1007/978-3-319-11933-5_6

Many previous works has been done on user related results, from the search histories the user interesting measures are estimated and several papers on search histories were proposed[3][4].Clustering is the main functionality used in these approaches. *K means* is used for clustering the user queries[5]. Some of the search engines work with query clicks, based on previous user clicks or urls, server maintains the log of urls with respect to keywords and mines query oriented results for user interesting results. Query grouping mechanism to group the similar queries based on relevance and computes relevance with Query Fusion graph it is a combination of query reformulation(in this approach initially it compares the query whether is matches with previously accessed frequent queries) and Query click graph[3][4]. Further work on the user interesting results were implemented through saving the user session logs. Several representation methods were evaluated in the existing system like binary and pseudo documents[5].

In this paper we are proposing an integrating approach by eliminating pseudo documents, because in the previous approach weight is to be estimated for titles and snippets for feature representation for generating the Pseudo-documents. So we are introducing an integrated approach by using the Pattern mining and evolutionary algorithm for inferring the user interesting results.

The rest of the paper is organized as follows: The related work and the existing system is explained in section 2.The proposed system along with the framework of our approach is explained in section 3.The brief architecture description of each framework used in this paper is explained in Section 3.The section 4 concludes the paper.

2   Related Work

Many researches were done on analyzing the user queries information[6][7][8]. Many researches are still going on to produce an efficient search results for user. Work on improving the efficiencies and usability's of searches is being done [9][10].Automatically the searches are identified by past user clicks for identification of behavior of a user[11][15][16].R.B. Yates et al suggest that improving the search engine Results using Query Recommendation algorithm but using uncertain keywords is the problem[12].Some works are done for exploiting the query aspects by analyzing the search results from search engines [17].

In previous work, feedback sessions of query extracted from user click through logs along with respective pseudo documents. These documents can be clustered based on the similarity between the documents. Cosine similarity is the measure to compute the similarity between documents based on frequency of keywords in document. They proposed k means algorithm for clustering of documents based on the similarity between documents, the main drawback with k-means clustering is prior specification of number of clusters (K value) ,random selection of the centroid and not suitable for different density of objects[5].

We are proposing pattern mining approach instead of clustering approach for frequent use and relevant use of urls with respect to user query. Apriori algorithm is one of the simple frequent pattern mining algorithms, but the disadvantage of this algorithm is candidate set generation for every frequent item set generation and another major issue is multiple database scans[13]. FP Growth algorithm generates frequent item sets without candidate set generation.

In this paper we are proposing an efficient pattern based technique for identifying the interesting patterns of clicked urls with respect to user query. Fig 1 depicts the framework of our approach. Feedback session log maintains the user queries, session ids and Urls .Initially our approach searches the session oriented results for input query and find the frequent patterns with FP growth algorithm by constructing the FP tree, after the generation of the frequent patterns ,patterns can be forwarded to evolutionary approach for extraction of optimal patterns from FP tree generated patterns.

We are integrating an efficient evolutionary approach (Genetic algorithm) to the previous pattern mining approach for optimal results. In this approach we apply cross over (i.e. combines the existing patterns to generate a new pattern) and mutation (i.e. alter the genes (url) of the pattern) on mined patterns(sequence of urls) of user interesting results for optimal patterns in mined patterns [5].

3   Proposed Work

In this approach we are proposing an efficient mechanism to satisfy the user search goals based on the session based user search history based on the user click logs with respect to the user query. All the feedback sessions of a query are first extracted from user click-through logs. Table 1 shows the example of a feedback session of users in their session. It represents zeroes and non-zeroes for non-visited and visited urls respectively along with query, session id, url and sequence of clicks.

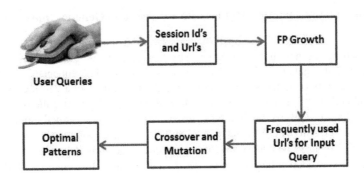

Fig. 1. Framework of our approach

Individual session based results involves a unique session id, query and respective urls, Consider each individual url as event and sequence of urls as patterns for mining of urls, which leads to the most frequent access by the users, for the mining of urls we are using FP tree algorithm for finding the frequent patterns of urls visited by the previous users. Table 2 is used as a Data set for FP growth for constructing FP Tree which gives frequent patterns of Url's

Table 1. Feedback Session of different users

Query	Session Id	URL's	Sequence of Clicks
Education	1	en.wikipedia.org/wiki/Education	0
Education	1	www.education.com/	1
Education	1	www.theguardian.com/education	2
Car	1	www.carwale.com	1
Education	2	www.theguardian.com/education	1
Tourism	2	www.tourism.gov.in/	1
Movies	2	www.ticketdada.com	1
Education	2	en.wikipedia.org/wiki/Education	2
Education	2	www.educationworld.com/	3
*	*	*	*
*	*	*	*
Education	4	www.education.com/	1

3.1   Optimal URL Pattern Mining with FP Tree and Genetic Approach

Initially we consider set of session based urls from feedback session log for an input query passed by the new user and patterns by the set of session based individual urls. From Table 2 the patterns are collected. Let us consider the sequence of URL's be represented as a, b, c, d, and so on. From Table 2 let's assume first cell i.e., first url be 'a', second cell be 'b' and so on. Again fifth cell will be 'a' because the url is already assigned with 'a' and so on...Patterns forwarded to the FP growth algorithm for generation of frequent patterns or set of frequently visited urls with respect to user query.

From the Table 1 the Query oriented results are generated based on the Query entered by User. Depending on that query the Session Id's and the URL's are collected. For Example the below Table shows the collected information after user enters Query.

Table 2. Url's visited by users in their particular sessions on the Query *"Education"*

Session Id	URL's	ID
1	www.education.com/,	a
	www.theguardian.com/education	b
2	www.theguardian.com/education,	b
	en.wikipedia.org/wiki/Education,	c
	www.educationworld.com/	d
3	*	*
*	*	*
10	www.theguardian.com/education	b

3.2 FP Growth Algorithm

The following pseudo code shows the FP growth algorithm. This pattern mining approach involves in two phases. One is FP tree construction; there it maintains the two passes over dataset and frequent item set generation done by traversing through FP tree. Sequential steps of FP growth algorithm is as follows,

1. It initially scans the data set and finds minimum threshold value for each item and discards infrequent items.
2. Frequent items can be sorted based on decreasing order of their support or threshold.
3. Every node in tree maintains a counter for items
4. Fixed order is used, so paths can overlap when transactions share items(when they have the same prefix).In this case counters are incremented
5. Pointers are maintained between nodes, single linked lists are created between the same items.
6. Frequent item sets extracted from the FP-tree.

3.3 Genetic Algorithm

In this we find negative and positive association rules. Negative rules are to consider not just all items within a transaction, but also all possible items absent from the transaction. Positive association rules are considering all present items from the transaction. In genetic algorithm the resultant chromosomes are encoded and Chromosome structure that is representation of association rule. Genetic algorithm [14] uses encoding, permutation encoding and binary encoding. Here we adopted Binary Coding. Binary Coding consists of two bits, 0 is representing an item is absent and 1 is representing an item is present.

In genetic algorithm there are two methods such as Crossover and Mutation. Crossover is is performed by selecting a random gene along the length of the chromosomes and swapping all the genes after that point. Mutation alters the new solutions so as to add stochasticity in the search for best solutions. This is the bit will be flipped within a chromosome (0 as 1, 1 to 0).

3.3.1 Representation of Chromosome

The resultant rules from FP-Growth algorithm is explained as follows:

For Example the items present in transaction is 'a, b, c, d, e' and the initial set is 00000. Consider a transaction is 'b c d'. This transaction is represented as 01110, which means if an item is present in the transaction that place is set to 1. In the above example 'b' is present so in second place bit is set to 1. This represented binary coded transaction is also called as chromosome in genetic algorithm.

3.3.2 Crossover and Mutation

Crossover applied as follows: For example consider two chromosomes 11001, 00111. In this crossover we take 2nd point crossover. First and second gene of the second chromosome is replaced with first and second gene of first chromosome which results another chromosome 11111. We explained about mutation that is flipping of the genes in chromosome example 11001 is 00110.

After completion of the crossover and mutation we have to calculate completeness, Confident factor, and fitness of the chromosomes. For finding these by Confidence Factor, $CF = TP / (TP + FN)$ We also introduce another factor completeness measure for computing the fitness function. $Comp=TP/ (TP+FP)$ $Fitness=CF *Comp$ The fitness function shows that how much we near to generate the rule.

TP = True Positives = Number of examples satisfying item set A and item set B
FN = False Positives = Number of examples satisfying item set A but not item set B
FP = False Negatives = Number of examples not satisfying item set A but satisfying item set B
TN = True Negatives = Number of examples not satisfying item set A nor item set B.

By finding these values we can calculate confidence factor and fitness function. If the fitness function value is greater than minimum confidence, that chromosome is optimized chromosome.

With more usage and development of data mining techniques and tools, much work has recently focused on finding negative patterns, which can provide valuable information. However, mining negative association rules is a difficult task, due to the fact that there are essential differences between positive and negative association rule mining. Mining association rules is not full of reward until it can be utilized to improve decision-making process of an organization. It is concerned with discovering positive and negative association rules. We present an FP-Growth algorithm that is able to find all valid positive and negative association rules in a support confidence framework. It finds all valid association rules quickly and overcome some limitations of the previous methods of mining. The complexity and large size of rules generated after mining have motivated researchers and practitioners to optimize the rule, for analysis purpose. This optimization can be done using Genetic Algorithm.

4 Conclusion

We are concluding our research work with efficient pattern mining and genetic approach for satisfying the user search goals.FP growth algorithm finds the frequent pattern of urls with respect to user query and the generated pattern forwarded to evolutionary approach for computation of fitness after cross over and mutation operation over chromosomes. In this approach we are introducing an efficient approach for generating the user wished or relevant URL's so that the restructured urls i.e., useful information or patterns are displayed first.

References

1. Pu, H.-T., Chiao, H.-C.: Web relevant term suggestion Using Log-based and Text-based Approaches (December 2006)
2. Dakka, W., Gravano, L., Ipeirotis, P.G.: Answering General Time-Sensitive Queries. IEEE Transactions on Knowledge and Data Engineering Member 24(2) (2012)
3. Hwang, H., Lauw, H.W., Getoor, L., Ntoulas, A.: Organizing User Search Histories. IEEE Transactions on Knowledge and Data Engineering (May 2012)
4. Wang, X., Zhai, C.-X.: Learn from Web Search Logs to Organize Search Results. In: Proc. 30th Ann. Int'l ACM SIGIR Conf. Research and Development in Information Retrieval (SIGIR 2007), pp. 87–94 (2007)
5. Lu, Z., Zha, H., Yang, X., Member, S., Lin, W., Zheng, Z.: A New Algorithm for Inferring User Search Goals with Feedback Sessions. IEEE Transaction on Knowledge and Data Engineering 25(3) (2013)
6. Joachims, T.: Optimizing Search Engines Using Click through Data. In: Proc. Eighth ACM SIGKDD Int'l Conf. Knowledge Discovery and Data Mining (SIGKDD 2002), pp. 133–142 (2002)
7. Wen, J.-R., Nie, J.-Y., Zhang, H.-J.: Clustering User Queries of Search Engine. In: Proc. Tenth Int'l Conf. World Wide Web (WWW 2001), pp. 162–168 (2001)
8. Beeferman, D., Berger, A.: Agglomerative Clustering of a Search Engine Query Log. In: Proc. Sixth ACM SIGKDD Int'l Conf. Knowledge Discovery and Data Mining (SIGKDD 2000), pp. 407–416 (2000)
9. Beitzel, S., Jensen, E., Chowdhury, A., Frieder, O.: Varying Approaches to Topical Web Query Classification. In: Proc. 30th International ACM SIGIR Conference of Research and Development (SIGIR 2007), pp. 783–784 (2007)
10. Cao, H., Jiang, D., Pei, J., He, Q., Liao, Z., Chen, E., Li, H.: Context-Aware Query Suggestion by Mining Click-Through. In: Proc. 14th ACM SIGKDD Int'l Conf. Knowledge Discovery and Data Mining (SIGKDD 2008), pp. 875–883 (2008)
11. Lee, U., Liu, Z., Cho, J.: Automatic Identification of User Goals in Web Search. In: Proc. 14th Int'l Conf. World Wide Web (WWW 2005), pp. 391–400 (2005)
12. Baeza-Yates, R., Hurtado, C.A., Mendoza, M.: Query Recommendation Using Query Logs in Search Engines. In: Lindner, W., Fischer, F., Türker, C., Tzitzikas, Y., Vakali, A.I. (eds.) EDBT 2004. LNCS, vol. 3268, pp. 588–596. Springer, Heidelberg (2004)
13. Han, J., Pei, J., Yin, Y.: Mining Frequent Patterns without Candidate Generation, http://Web.engr.illinois.edu/~hanj/pdf/sigmod00.pdf
14. Mitchell, T.: An introduction to Genetic Algorithms Melanie

15. Li, X., Wang, Y.-Y., Acero, A.: Learning Query Intent from Regularized Click Graphs. In: Proc. 31st Ann. Int'l ACM SIGIR Conf. Research and Development in Information Retrieval (SIGIR 2008), pp. 339–346 (2008)
16. Shen, J., Sun, Q.: Building Bridges for Web Query Classification. In: Proc. 29th Ann. Int'l ACM SIGIR Conf. Research and Development in Information Retrieval (SIGIR 2006), pp. 131–138 (2006)
17. Chen, H., Dumais, S.: Bringing Order to the Web: AutomaticallyCategorizing Search Results. In: Proc. SIGCHI Conf. HumanFactors in Computing Systems (SIGCHI 2000), pp. 145–152 (2000)

Improving Query Processing Performance Using Optimization among CPEL Factors

R. Kiran Kumar and K. Suresh

Department of Computer Science and Engineering,
Anil Neerukonda Institute of Technology and Sciences,
Sangivalasa, Bheemunipatnam [M], Visakhapatnam
{kirankumar.rajana38,kurumallasuresh}@gmail.com

Abstract. Query services in public servers are interesting factor due to its scalability and low cost. The owner of the data needs to check confidentiality and privacy before moving to server. The construction of cloud query services requires confidentiality, privacy, efficiency and low processing cost. In order to improve the efficiency of query processing, the system will have to compromise on computing cost parameter. So finding appropriate balance ratio among CPEL, is an optimization problem. The genetic algorithm can be the best technique to solve optimal balancing among CEPL (confidentiality, privacy, efficiency, and low cost). In this paper we propose a frame work to improve query processing performance with optimal confidentially and privacy. The fast KNN-R algorithm is designed to work with random space perturbation method to process range query and K-nearest neighbor queries. The simulation results show that the performance of fast-KNN-R algorithm is better than KNN-R algorithm.

Keywords: Query processing, Cloud computing, KNN query, Privacy, Range query.

1   Introduction

Cloud Computing is nothing but, both the applications which are distributed as services over the internet, the hardware and software in the data providing centers. The services are referred to as Software as a Service (SaaS). The data providing and processing center is called a Cloud. Building Data servers for query services in the data providing cloud, is remarkably popular because of the advantages in cost-saving and scalability. With the cloud infrastructures, the cloud service clients can conveniently scale down or up the service and pay for the period of using the servers. If cloud service providers (CSP) lose the security control over the data, data Confidentiality and query privacy will become the important problem. The requirements for constructing a query service in the real time cloud are based on PCLE factors: query Privacy, data Confidentiality, Low in-house processing cost and efficient query processing. In order to satisfy these requirements the complexity of constructing query services in the cloud will also increase. Random Space Perturbation (RASP) method is used to build practical range

© Springer International Publishing Switzerland 2015

S.C. Satapathy et al. (eds.), *Proc. of the 3rd Int. Conf. on Front. of Intell. Comput. (FICTA) 2014*
– Vol. 1, Advances in Intelligent Systems and Computing 327, DOI: 10.1007/978-3-319-11933-5_7

query and k-nearest-neighbor query services in the data clouds. This method addresses all the four aspects of the PCLE criteria and aim to have a good balance among them. The technique is used to transform the multidimensional datasets with a combination of order dimensionality expansion, preserving encryption, random project, and random noise injection. For processing range queries, this perturbation technique is designed to securely transform queried ranges into polyhedron in the perturbed data space and which can be efficiently processed by using indexing structures.

The Balancing among PCLE factors can be viewed as optimization problem. But it varies with application and dynamic load on data cloud. So there is a need for an approach that makes optimum balancing among privacy, confidentiality low cost and efficiency. This paper describes a frame work in which Genetic Algorithms can be used to solve the optimization problem. The main goal of this work is to discover the candidate chromosomes that can influence the balancing among PCLE.

Remaining sections of the paper describes as follows, section 2 describes the related works based on various subsections like existing works related to protecting cloud data, preserving query privacy and genetic algorithms. Section 3 describes the theoretical description related to random space perturbation. The proposed architecture and fast-KNN-R algorithm explained in section 4. Results and observations have reported in section 5 and finally conclude the paper in section 6.

2   Existing Work

The section is mainly divided into three sub sections described as follows

2.1   Background of Protecting Public Data

The following are the existing works for protecting data which is out sourced. Order Preserving Encryption (OPE) [1] maintains the dimensional value order after applying encryption. A well known attack is called prior knowledge attack on the original distributions of the attributes. If the attacker observes the original distributions and trying to identify the transformation mapping between the original attribute and its changed counterpart, a bucket based distribution can be performed to break the encryption for the attribute [2].

Crypto-Index mainly depends on column wise buckets. And a random ID number is assigned to each bucket; the values stored in the bucket are replaced with the ID of the bucket to generate the transformed data for indexing. To exploit the index for query processing, a normal range query condition should be transformed to a set-based query on the bucket IDs. A bucket-diffusion scheme [3] was proposed to protect the access pattern, which, however, has to down grade the precision of query results, thus it has to increase the client's cost of filtering the query result. Distance-Recoverable Encryption (DRE) [4] is the most perceptive method for protecting the nearest neighbor relationship.

2.2   Background Works for Preserving Query Privacy

Private information retrieval (PIR) [5] tries to fully shelter the privacy of access pattern, But PIR schemes has very in-house processing cost. To improve the PIR efficiency , in the work of Williams et al. [6] uses a pyramid hash index to maintain efficient privacy preserving data-block operations Hu et al. [7] addresses the privacy of query problem and it states that it needs the authorized query users, owner of the data, and the cloud to collaboratively process k-N-N queries.. Papadopoulos et al. [8] uses private information retrieval methods [2] to enhance location privacy.

2.3   Background of Genetic Algorithm

Genetic algorithms (GA) were first described by Holland et al [9][10][11]. Candidates of a population are represented as chromosomes, and genes in the chromosomes represent the solutions. The possible key chromosomes form a problem search space and are connected with a fitness function representing the value of solutions encoded as the chromosome. Search is carried out by evaluating the fitness of each of a population of chromosomes, and then point mutations and recombination is performed on the successful chromosomes.

3   Theoretical Description

RASP is one type of multiplicative perturbation, with the combination of OPE, random noise injection, dimension expansion, and random projection. Let's consider that the multidimensional data are numeric and in multidimensional vector space. The database has q searchable dimensions and r records, which makes a ds $\times$ r   matrix M. The *searchable* dimensions can be used in queries and thus should be indexed. Let y represent a ds-dimensional record, y $\in$ R^{ds}. Note that in the ds-dimensional vector space R, the range query conditions are represented as half-space functions and a range query is translated to find the point set in corresponding polyhedron area described by the half spaces. The RASP perturbation involves three steps. And the security of it is based on the existence of random invertible real-value matrix generator and random real value generator. For each q-dimensional input vector y,

1)   An order preserving encryption (OPE) scheme, EN_{ope} with keys Q, is applied to each dimension of y: ENope (y, Qope) $\in$ R^{ds} to change the dimensional distributions to normal distributions with each dimension's value order still preserved.

2)   The vector is then extended to ds + 2 dimensions as

$$H(y) = ((ENope(y))^{T}, 1, p)^{T},$$

where the (ds+1)th dimension is always a 1 and the (ds + 2)th dimension, p, is drawn from a random real number generator RNG that generates random values from a tailored normal distributions.

3)   The (ds + 2)-dimensional vector is finally transformed to

$$F(y, Q = \{A, Q \text{ ope}, RG\}) = A((ENope(y))^{T}, 1, p)^{T}$$

where A is a (ds+2)×(ds+2) randomly generated invertible matrix with A ∈ R such that there are at least two non-zero values in each row of A and the last column of A is also non-zero.

For example, consider a relation with attributes Student(sid,sname,marks, avg, division) as a original table and its corresponding table Student^S (Sid^S,Name^S,Marks^S,Avg^S,Division,Ds+1,Ds+2) stored at the Server side will be as follows:

Table 1. Original data of Student table

Sid	Name	Marks	Avg	Division
07	Kiran	500	80	Second
22	Raju	550	85	First
43	Pavan	480	70	Third
54	Navin	560	87	First
57	Nisath	530	83	Second

Table 2. Perturbated data of Student table

Sid^S	Name^S	Marks^S	Avg^S	Division^S	Ds+1	Ds+2
8@	Srllz	1444	@9	[nmzzq	1	0.827752665536363
:;	Zjt	1594	@>	Nrl~	1	0.735066808636797
<<	Xj lz	1384	?9	\qs}p	1	0.460988490125625
==	Vj tz	1624	@@	Nrl~	1	0.727846365295279
=@	Vr}l u	1534	@<	[nmzzq	1	0.569394528665298

4   Random Space Perturbation Using Genetic Algorithm

4.1   Architecture

The following figure represents the architecture of the proposed work. In which D, the original data set can be encrypted by using Order Preserving Encryption (OPE) function (F) with the key K, then it generates D', the perturbated data set. The D' now ready to host on the cloud in order to respond for the client query. And for the

Optimization part Genetic algorithm is used to balance among PCLE factors and based on that optimum balancing the strength of Encryption Function and key size will be adjusted. The original query is transformed into q' with knowledge of Encryption of D. The Q is a transformation function which takes q as an input and produce q' as output. Now the query is applied to D' in order to get results. Here the function H, takes D' and q' as input parameters and produces R' as a result. But R' is perturbated form, so the function G is used to get original results R.

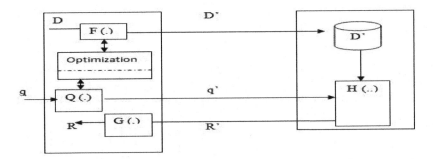

Fig. 1. Architecture

4.2   Range Query Processing

The RASP technique works with two types of query processing- range query and KNN query. In the range query the original space can be transformed to perturbated space. For example consider above table having [sid, name, marks, avg and division] as the attributes. The User then issue query: select count (*) from Student where avg > 75 and sid > 40. To process this range query, first it Encrypt all the records using Order Processing Encryption and the range will be transformed from one domain to another domain range but it preserves the order.

4.3   KNN-R Query Processing

The unique distance-based kNN query deals with finding out the nearby k points in the sphere-shaped range that is centered at the query point. The fundamental idea of our algorithm is to use rectangle ranges, instead of spherical ranges, to find the estimated kNN results, so that the RASP range query repair can be used. There are a number of key troubles to make this work steadily and competently. (1) How to efficiently find the lowest amount of square range that definitely contains the k outcomes, without many communications between the cloud and the client? (2) Will this solution preserve data confidentiality and query privacy? (3) Does the stand-in server's workload raise? To what extent?

4.4   Proposed Algorithm

The fast-KNN-R algorithm contains three rounds of interactions among server, client, and optimization modules. First client enquires optimality for current content, then it replies with key chromosome consisting of optimum balancing factors like key sizes, strength of transformation function. Second the consumer will send the original upper-bound range, which contains more than k points, and the original lower-bound range, which contains less than k points, to the server. The server finds the inner range and proceeds to the client. The client calculates the outer range based on the inner range and sends it reverse to the server. The server finds the records in the outer range and sends them to the client. The client decrypts the records and find the top k candidates as the final result.

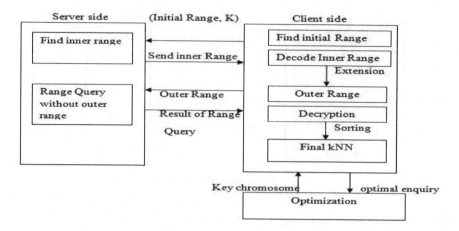

Fig. 2. Fast-KNN-R algorithm

4.5   Optimization among CPEL Factors

The main aim of this work is to optimize the balance among the factors;
the following table depicts the important genes for each and every factor.

Table 3. CPEL factors with their respective genes

S.no	Factor	Genes affecting factors
1	Confidentiality	a) key size b) No of rounds c) basic operation
2	Privacy	a)scope to co-task with confidentiality b)level of privacy c) need for privacy for data to be transferred
3	Efficiency	a) best choice of encryption key end rounds b)whether information to be exchange are public secrete
4	Low in house cost	a) remove unnecessary operations not required b) level of security

And the main task to be performed by fast KNN-R algorithm is to identify the key
set of genes called chromosomes. It works as follows [10].

Step1:- It encodes the genes in binary representation by varying its strengths.
Step2:- Choose initial composition factors.
Step3:- Apply the cross over
Step4:- perform the mutation
Step5:- Estimate the population selection
Step6:- Repeat the steps from 1 to 5 until satisfactory query performance arrives

5   Results and Observations

Experiments were conducted to observe the comparative study between frameworks using RASP method without genetic algorithm and with genetic algorithm. The results obtained were from naturally randomly generated data set. The system was simulated using visual studio 2010, ASP.NET and c# as coding language. Experiments were conducted on 3.4 GHz Intel Dual core with 2GB of primary memory and using windows7 as operating system.

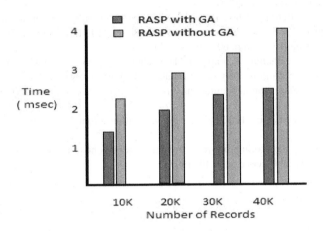

Fig. 3. Performance analysis

6   Conclusion

The requirements for constructing a query service in the real time cloud are based on PCLE factors: query Privacy, data Confidentiality, Low in-house processing cost and Efficient query processing, The low in-house workload is an important issue to fully appreciate the benefits of cloud computing. The efficient query processing is a measure for the quality, of query services. RASP perturbation is a composition of OPE, injection of random noise, expansion of dimensionality, random projection and genetic algorithm to provide unique security features. As a whole the paper presented an approach to solve optimization problem using Genetic Algorithms. And in the implementation of the random space perturbation (RASP) approach, the fast–KNN- R algorithm is designed to work with the RASP query algorithm to process KNN queries.

References

1. Agrawal, R., Kiernan, J., Srikant, R., Xu, Y.: Order preserving encryption for numeric data. In: Proceedings of ACM SIGMOD Conference (2004)
2. Chen, K., Kavuluru, R., Guo, S.: Rasp: Efficient multidimensional range query on attack-resilient encrypted databases. In: ACM Conference on Data and Application Security and Privacy, pp. 249–260 (2011)

3. Hore, B., Mehrotra, S., Tsudik, G.: A privacy-preserving index for range queries. In: Proceedings of Very Large Databases Conference, VLDB (2004)
4. Liu, K., Giannella, C.M., Kargupta, H.: An attacker's view of distance preserving maps for privacy preserving data mining. In: Fürnkranz, J., Scheffer, T., Spiliopoulou, M. (eds.) PKDD 2006. LNCS (LNAI), vol. 4213, pp. 297–308. Springer, Heidelberg (2006)
5. Chor, B., Kushilevitz, E., Goldreich, O., Sudan, M.: Private information retrieval. ACM Computer Survey 45(6), 965–981 (1998)
6. Williams, P., Sion, R., Carbunar, B.: Building castles out of mud: Practical access pattern privacy and correctness on untrusted storage. In: ACM Conference on Computer and Communications Security (2008)
7. Hu, H., Xu, J., Ren, C., Choi, B.: Processing private queries over untrusted data cloud through privacy homomorphism. In: Proceedings of IEEE International Conference on Data Engineering (ICDE), pp. 601–612 (2011)
8. Papadopoulos, S., Bakiras, S., Papadias, D.: Nearest neighbor search with strong location privacy. In: Proceedings of Very Large Databases Conference, VLDB (2010)
9. Holland, J.H.: Adaptation in Natural and Artificial Systems. University of Michigan Press, Ann Arbor (1975)
10. Haupt, R.L., Haupt, S.E.: Practical Genetic Algorithms. John Wiley & Sons, Inc., New Jersey (2004)
11. Tan, R.: Seeking the profitability-risk-competitiveness frontier using a genetic algorithm. Journal of Actuarial Practice 5(1), 49 (1997)

Sensitivity Analysis of MEMS Based Piezoresistive Sensor Using COMSOL Multiphysics

Ankit Mishra, Ishita Bahal, Jagrati Arya, Abhishek Pandey, and Shabana Urooj

Electrical Engineering Department, School of Engineering
Gautam Buddha University, Greater Noida 201312 U.P., India

Abstract. The present paper peruses MEMS based piezoresistive pressure sensor and its fabrication techniques. Simulation of the pressure sensor is done by using COMSOL Multiphysics software for P-type silicon piezoresistor. The deflection of N-type silicon diaphragm depends upon the Young's modulus of the material and varies with the amount of force applied to the diaphragm. The simulation result emphasizes that an appropriate selection of the piezoresistive material and the amount of force applied on the diaphragm impacts the sensor sensitivity levels upon low power consumption.

Keywords: MEMS, piezoresistivity, pressure sensor, diaphragm deflection.

1 Introduction

MEMS technology has become very important for microelectronics. This technology has originated from integrated circuit technologies; but it is evolving differently. The systems made from this technology are called Micro Electro Mechanical Systems (MEMS). These devices have the ability to sense, control and actuate on the micro scale and generate effects on the macro scale.[1] These systems are made of small components with size 1-100 micrometers; and device size is 0.02-1 millimetre. MEMS are not just about the miniaturisation of mechanical components or making things out of silicon. In fact, the term 'MEMS' is actually misleading as many micro machined devices are not mechanical in a strict sense. MEMS is a manufacturing technology; a paradigm for designing and creating complex integrated devices and systems using batch fabrication techniques similar to the technologies used in IC manufacturing extended into micro meter scales.[1] The MEMS market include applications in automotives, IT peripherals, telecommunication devices, consumer electronics & life style products, medical and life science applications, biomedical instruments, household appliances, industrial process control, aerospace, defence and homeland security.[1]

The MEMS concept has grown to encompass many other types of small things like thermal, magnetic, fluidic, and optical devices and systems with or without moving parts.

Common Features of MEMS technology are:

- It involves electronic and non-electronic elements.
- It can perform functions that include chemical/biochemical reactions and experiments.

© Springer International Publishing Switzerland 2015

S.C. Satapathy et al. (eds.), *Proc. of the 3rd Int. Conf. on Front. of Intell. Comput. (FICTA) 2014*
– *Vol. 1*, Advances in Intelligent Systems and Computing 327, DOI: 10.1007/978-3-319-11933-5_8

- Some MEMS involve large arrays of micro-fabricated elements such as un-cooled infrared imaging devices and both reflective and non reflective projection displays.

MEMS devices are made similarly to ICs, therefore standard IC technologies like Photolithography, oxidation, wet/dry etching and decomposition of standard materials can be used for MEMS.

Piezoresistive pressure sensor are some of the first MEMS devices to be commercialized compared to capacitive pressure sensor, as they are simple to integrate with electronics, their response is more linear and are shielded from RF noise.

MEMS have several distinct advantages as a manufacturing technology:

- The multifaceted nature of this technology and its micromachining techniques, as well as its diversity of applications, has resulted in an unparalleled range of devices across previously unrelated fields such as biology and microelectronics.
- MEMS, with its batch fabrication techniques, enables components and devices to be manufactured with increased performance and reliability, combined with the obvious advantages of reduced physical size, volume, weight and cost.
- MEMS provide the basis for the manufacture of products that cannot be made by other methods.

These factors make MEMS as pervasive technology as integrated circuit microchips.

In this paper the sensitivity of a square shaped pressure sensor is analysed. The pressure applied on diaphragm is causing a deflection in shape, thus changing the resistance in the sensor which can be read as change in current flow. Thus amount of current flow can be related to the magnitude of the applied pressure.

2   MEMS Fabrication Techniques

Most MEMS device use some form of lithography based micro fabrication borrowed from microelectronics industry enhanced with specialized techniques called micro machining [2].

2.1   Lithography

It is the process by which a pattern is transferred into a photosensitive material by selective exposure to a radiation source such as light.

A photosensitive material is a material that experiences a change in its physical properties when exposed to a radiation source. If a photosensitive material is selectively exposed to radiation the pattern of the radiation on the material is transferred to the material exposed. [3]

Photolithography is typically used with metal or other thin film deposition, wet and dry etching.

There are two types of photoresist: Positive and Negative Photoresist.

- For positive resists, the resist is exposed with UV light wherever the underlying material is to be removed. In these resists, exposure to the UV light changes the chemical structure of the resist so that it becomes more soluble in the developer.
- Negative resists behave in just the opposite manner. Exposure to the UV light causes the negative resist to become polymerized, and more difficult to dissolve.

2.2   Etching

Etching is the process of using strong acid (liquid or gaseous state) to cut into unprotected parts of metal surface to create a design in metal. There are two categories of etching process:

- **Wet Etching:** Wet chemical etching basically consists in selective removal of material by dipping a substrate into a solution that dissolves it. The chemical nature of this etching process provides a good selectivity, which means the etching rate of the target material is considerably higher than the mask material if selected carefully.[4]

- **Dry Etching:** Dry etching refers to the removal of material, typically a masked pattern of semiconductor material, by exposing the material to a bombardment of ions usually a plasma of reactive gases such as fluorocarbons, oxygen, chlorine, boron tri-chloride; sometimes with addition of nitrogen, argon, helium and other gases that dislodge portions of the material from the exposed surface.[5]

Table 1. Comparison between wet etching and dry etching

Wet Etching	Dry Etching
Highly selective	Easy to start and stop
No damage to substrate	Less sensitive to small changes in temp
Cheaper	More repeatable
	May have anisotropies
	Fewer particle in environment

- **Deep-Reactive Ion Etching:** Deep RIE is a highly anisotropic process for realizing, steep sided holes or trenches in silicon wafer with high aspect ratios. The Bosh Process was successful in producing a high aspect ratio (>100) with high etching selectivity to oxide and photo resist. The bosh process alternates between two modes: a standard, nearly isotropic plasma process and a deposition process of chemically inert passivation layer, it prevents etching of side wall of the trench.

3   Materials and Method

Nowadays, the finite element method (FEM) is widely used for thermal effect reduction, stress analysis and reliability enhancement of piezoresistive sensor. In this paper a structural model of sensor is built using this method using COMSOL Multiphysics v4.2 software to study structural stress and demonstrate sensor sensitivity. The FEM simulation of MEMS piezoresistive pressure sensor conducted in present study is significant advance towards device design optimization in MEMS prototyping.

COMSOL Multiphysics is a finite element analysis, solver and Simulation software / FEA Software package for various physics and engineering applications, especially coupled phenomena, or multi physics. The software also offers an extensive interface to MATLAB and its toolboxes for a large variety of programming, pre-processing and post-processing possibilities. COMSOL Multiphysics allows for entering coupled systems of partial differential equations (PDEs).

In the model N-type and P-type materials are used for the study of sensor. N-type silicon is used for sensor diaphragm whereas P-type Silicon has been taken as the piezoresistor material.

Table 2. Material Properties

Material Property	Diaphragm
Material	Silicon
Density	2330 [Kg/m^3]
Young Modulus	129 GPa
Poisson's Ratio	0.22 to 0.28
Dielectric	11.9
Thermal conduction	148 W/(m x k)
Electrical Resistivity	4.59 (ohm x cm)

4   Pressure Sensors

Pressure measurement is a key part of many systems, both commercial and industrial. Silicon has proved to be a good material from which small pressure sensors can be built. Pressure sensors constitute the largest market segment of mechanical MEMS device. MEMS based pressure sensor is based on piezoresistive effect [6]. Piezoresistivity is the change of resistance of material when submitted to stress. The effect was first discovered by Smith and it was proposed that the change in conductivity under stress in bulk n-type material and designed an experiment to measure the longitudinal as well as transverse piezoresistance coefficients. Kanda did a piezoresistance coefficient study about impurity concentration, orientations, and temperature [7]. Pfann designed several semiconductor stress gauge to determine shear piezoresistance effects. Lund studied temperature dependence of piezoresistance coefficient by four point bending experiment [8]. Pressure is measured by monitoring its effect on a specifically designed mechanical structure, referred to as sensing element. The application of pressure to sensing element causes a change in shape and resulting deflection (strain) in material can be used to determine magnitude of pressure.

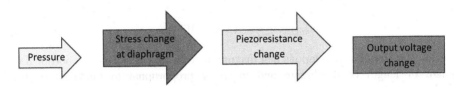

Fig. 1. Principle of piezoresistive pressure sensor

5   Concept and Sensor Design

The sensor design includes two basic elements: the thin elastic diaphragm and piezoresistive material. The diaphragm is made fixed around edges, with trace wire on the surface. The wire is made up of p-type piezoresistive material. When pressure is applied on the back of diaphragm, it deforms changing resistance of wire and thus pressure causing the deformation can be measured.

5.1   Mathematical Analysis

The analysis is done for square shape diaphragm deflection.

The Load-deflection relationships for square diaphragm with length L and thickness H are given below [9]:

$$\frac{P\,a^4}{EH^4} = \frac{4.2}{1-v^2}\left[\frac{w}{H}\right] + \frac{1.5}{1-v^2}\left[\frac{w}{H}\right]^3 \tag{1}$$

'P' is measured in Pascal (Pa); 'w' is center deflection of diaphragm, 'a' is half the side length, 'E' is Young's Modulus and 'v' is Poisson's Ratio. To keep deflection in range above formula is reduced to:

$$\frac{Pa^4}{EH^4} = \frac{4.2}{1-v^2}\left[\frac{w}{H}\right] \tag{2}$$

Maximum deflection at center of diaphragm is given by:

$$w_{max} = \frac{Pa^4}{4.2EH^4} \tag{3}$$

Maximum stress at center of each edge is given by:

$$\sigma_{max} = 0.308\left[\frac{L}{H}\right]^2 \tag{4}$$

$$\varepsilon = \frac{\sigma}{E} \tag{5}$$

Thus, the following relation can be established:

$$P = \frac{\sigma}{0.308} \left[\frac{H}{L}\right]^2 \tag{6}$$

$$w_{max} = \frac{1}{20.6976}(1 - v^2)\left[\frac{L^2}{H}\right]\varepsilon \tag{7}$$

It is clear from above relations that maximum deflection is directly proportional to square of length of diaphragm and inversely proportional to thickness of the diaphragm.

5.2   Simulation

In this design, a square membrane with side 1mm and thickness 20 μm is considered. Edges are 0.1mm wide to represent the remainder of the wafer. These edges are made to be fixed while the centre area of the membrane is left free for movement on application of pressure. Near to one edge of membrane an X-shaped piezoresistor is placed.

The piezoresistor is considered to be of a consistent p-type dopant density of 1.32×1019 cm–3 and a thickness of 400 nm. The diaphragm is made up of n-type silicon.

The edges of the die are aligned with the {110} orientation of the silicon with respect to the global X and Y axes. The piezoresistor is oriented to be at 45° to the die edge, and so lies in the {100} direction or orientation of the crystal.

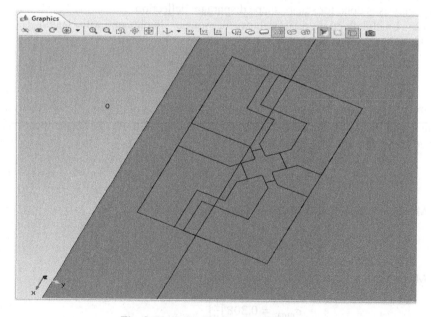

Fig. 2. Piezoresistor geometry 2D view

The piezoresistor is assumed to have a uniform p-type dopant density of 1.32×1019 cm−3 and a thickness of 400 nm. The interconnections are assumed to have the same thickness but a dopant density of 1.45×1020 cm−3. Only a part of the interconnections is included in the geometry, since their conductivity is sufficiently high that they do not contribute to the voltage output of the device.

6   Result and Discussion

Displacement of diaphragm as a result of 100 kPa pressure difference applied to membrane at its center is 1.2 µm. The result is in good agreement with theoretical and mathematical result. The RED colour shows the maximum displacement at the center of the diaphragm, similarly the displacement along the edges are zero as they are fixed, this is shown by BLUE colour.

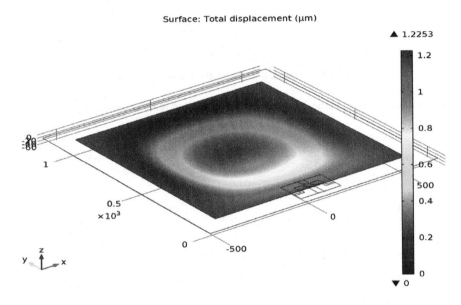

Fig. 3. Diaphragm displacement as result of 100kPa applied pressure

The stress along the edges shows a maximum magnitude of 38 MPa at centre of each of two edges along which plot is made. The stress is in negative direction along the edge having piezoresistor and the side geometrically opposite to it; while it is in positive direction along the remaining two edges.

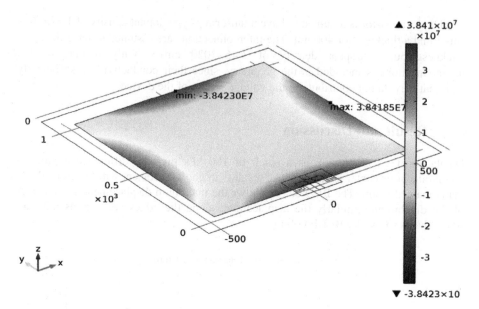

Fig. 4. Shear stress in the local co-ordinate system

Stress has its max value close to the piezoresistor with value of approximately -35Mpa.

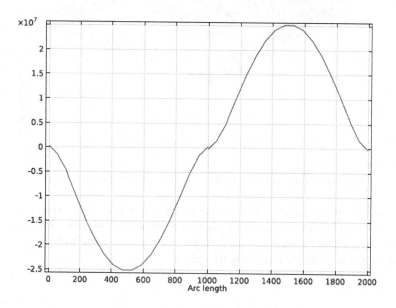

Fig. 5. Plot of local shear stress along two edges of diaphragm

The above graph shows the negative and positive magnitude of stress along the edges of the diaphragm. Here, the X-axis represents the arc length while the Y-axis depicts the stress experienced by the edges of the diaphragm.

With an applied bias of 3v a typical operating current of 5.9 mA is obtained. The model produces output voltage of 54 mV, similar to actual device output of 60mV.

References

1. van Heeren, H., Salomon, P.: Technology Watch, Electronics Enabled Products Knowledge Transfer Network, Wolfson School of Mechanical and Manufacturing Engineering, pp. 1,6,14,15. Loughborough University, Loughborough (2007)
2. Senturia, S.D.: Microsystems Design, pp. 29, 30. Kluwer Academic Publishers, New York (2002)
3. Senturia, S.D.: Microsystems Design, p. 50. Kluwer Academic Publishers, New York (2002)
4. Senturia, S.D.: Microsystems Design, p. 57. Kluwer Academic Publishers, New York (2002)
5. Senturia, S.D.: Microsystems Design, p. 58. Kluwer Academic Publishers, New York (2002)
6. Smith, C.: Piezoresistive effect in germanium and silicon. Physics Review 94, 42–49 (1954)
7. Kanda, Y.: A Graphical Representation of the Piezoresistance Coefficient in Silicon. IEEE Transactions on Electron Devices ED-29(1), 64–70 (1982)
8. Lund, E., Finstad, T.: Measurement of the Temperature Dependency of the Piezoresistance Coefficients in P-Type Silicon. In: Advances in Electronic Packaging-ASME, EEP, vol. 26(1), pp. 215–218 (1999)
9. Senturia, S.D.: Microsystems Design, p. 86. Kluwer Academic Publishers, New York (2002)

Optimal Computer Based Analysis
for Detecting Malarial Parasites

S.T. Khot[1] and R.K. Prasad[2]

[1] B.V.D.U.C.O.E., Pune, Maharashtra, India
[2] P.K. Technical Campus, Pune, Maharashtra, India
khotst@gmail.com

Abstract. Malaria poses a serious global health problem and it requires a rapid, accurate diagnosis to control the disease. An image processing algorithm for accurate and rapid automation in the diagnosis of malaria in blood images is developed in this research paper. The image classification system to identify the malarial parasites positively present in thin blood smears is designed, and differentiated into the various species and stages of malaria - falciparum and vivax prevalent in India. Method implemented presents a new approach to image processing in which the detection experiments employed the KNN rule, along with other algorithms such as ANN (Artificial Neural Networks), Zack's thresholding and Linear Programming and Template matching to find out the optimal classifier for detection and classification of malarial parasites with its stages.

Keywords: Image processing, KNN, malaria parasites, segmentation, ANN, Zack's thresholding, Linear Programming, Template Matching.

1 Introduction

Malaria is one of the predominant tropical diseases in the world causing wide spread sufferings and deaths in the developing countries [1]. The WHO reports 300 to 500 million clinical cases of malaria each year resulting in 1.5 to 2.7 million deaths [2]. About 40% of the world's population - about two billion people - are at risk in about 90 countries and territories [3][2]. A novel binary parasite detection scheme that is based on Template matching classifier which provides an adjustable sensitivity–specificity parasite detection is provided. The approach can be used to increase the total number of samples screened and for telemedicine application to enable pervasive healthcare at the base of the pyramid, along with other classifiers to get optimal results.

The Malarial Parasite

Malaria is transmitted by the infected female Anopheles mosquito which carry Plasmodium sporozoites in their salivary glands. The malarial parasites of the genus Plasmodium can be distinctly grouped into four species that can cause human infection: falciparum, vivax, ovale, and malaria[3][27]. During the life-cycle in

© Springer International Publishing Switzerland 2015
S.C. Satapathy et al. (eds.), *Proc. of the 3rd Int. Conf. on Front. of Intell. Comput. (FICTA) 2014*
– *Vol. 1*, Advances in Intelligent Systems and Computing 327, DOI: 10.1007/978-3-319-11933-5_9

peripheral blood, the different species may be observable in the four different life-cycle-stages which are generally morphologically distinguishable: ring, trophozoite, schizont, and gametocyte [4].

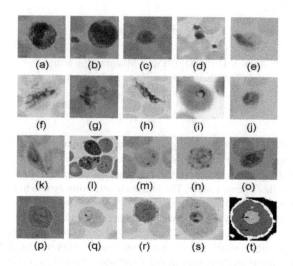

Fig. 1. Examples of stained objects: (a, b) white blood cells, (c, d) platelets, (e)–(h) artefacts, (i)–(l) P. falciparum ring, trophozoite, gametocyte, schizont, (m,n) P. malariae ringand schizont (o,p) P. ovale and P. vivax trophozoites, (q, r) P. vivax ring and gametocyte, (s) P. vivax ring, (t) extracted stained pixel group, green region(s) and the stained object

A total of four species of the parasite in four different life-cycle-stages (Fig.1) can be recognized as important and necessary in detection and diagnosis of malaria, and required for a detailed calculation of parasitemia [4]. The species differ in the changes of the shape of the infected (occupied) cell, presence of some characteristic dots and the morphology of the parasite in some of the life-cycle-stages .The life-cycle-stage of the parasite is defined by its morphology, size (i.e. maturity), and the presence or absence of malarial pigment (i.e. Haemozoin). However, because the parasite is a growing life-form it may appear in transient stages which may be difficult to classify into one of the adjacent stages [4].

With the aid of a microscope, visual detection and identification of the Plasmodium parasite is possible and highly efficient. This chemical process is called staining. Popular stains called as Giemsa and Leishman slightly color the red blood cells(RBCs) and highlights the parasites, white blood cells (WBC), platelets, and various artefacts [5]. For detecting the infection it is necessary to divide stained objects into two groups – parasite and non-parasite and differentiate the various stained cells between them. In order to specify the infection of the cells and to perform a detailed quantification on the cells, all four species of Plasmodium at four life-cycle-stages must be differentiated [5].

The existing research done by Timo R Bretschneider et al on thin blood smears obtained a reliable malaria count for substandard image quality smears. On an average an estimation error of less than 1% with respect to the average of manually obtained parasitemia counts was achieved [6]. On studies conducted by Raghuveer M. Rao & Vishnu V. Makkapati focused purely on segmentation of malarial parasites in blood smears by successful use of HSV color space [7][27]. Ross et al have incorporated a method of detecting the malarial parasites using thin smears by using neural networks to differentiate between the four species of human malaria, the positive predictive values were in the range of 28–81%, depending on the malaria species examined [8][14].A software was designed by Selena W.S. Sio et al called as MalariaCount, based on their research of malarial parasites it characterizes malaria infected RBC via edge based parameters. This software can readily be applied to many assays that are heavily dependent on parasitemia determination, and where large variations in values are expected [9].

2   Proposed System and Results

The algorithm can be proposed using the classifiers as mentioned, to achieve optimal accuracy and few of the results are mentioned to assess the performance of the classifiers.

i)   KNN

In image processing and pattern recognition, the **k-nearest neighbor algorithm** is a method used for classifying objects. This is achieved using the closest pixel value to the existing pixel in the training examples in the feature space. *KNN* is a type of instance-based learning, or local learning where the function is only approximated locally and all computation is deferred until classification [5][10]. The *k*-nearest neighbor algorithm is amongst the simplest of all machine learning algorithms: an object is classified by a majority vote of its neighbors with the object being assigned to the class most common amongst its *k* nearest neighbors [5][11]. The accuracy achieved using this classifier can be summed up in Table 1.

ii) Zack's Thresholding and Linear Programming

Triangle algorithm - This technique proposed by Zack is illustrated in Fig.2. A line is constructed between the maximum of the histogram at brightness b_{max} and the lowest value $b_{min} = (p=0)\%$ in the image. The distance d between the line and the histogram h[b] is computed for all values of b from b = b_{min} to b = b_{max}. The brightness value b_o where the distance between h[b_o] and the line is maximal is the threshold value, that is, $\Theta = b_o$. This technique is particularly effective when the object pixels produce a weak peak in the histogram [12].An automatic system for detecting and counting sister chromatid exchanges in human chromosomes had been developed by Zack. They used different digital image processing techniques and thresholding algorithms like Zack to separate chromosomes was estimated from size and shape measurements. Comparison of manual and computer estimates of S.C.E. (sister chromatid exchanges) frequency ([16] maximum or minimum) value of the objective functions, complies with all restrictions of the model. An algebraic representation of a generic formulation of linear programming model could be presented in equations (1) to(6).

Table 1. Accuracy of KNN

Sr. No	SPECIES	STAGE	ACCURACY in %
1	FALCIPARUM	1 (Ring Form)	83.33
2	FALCIPARUM	2 (Tropozoite)	92.45
3	FALCIPARUM	3 (Schizont)	87.50
4	FALCIPARUM	4 (Gametocyte)	91.67
5	VIVAX	1 (Ring Form)	91.67
6	VIVAX	2 (Tropozoite)	90.00
7	VIVAX	3 (Schizont)	91.25
8	VIVAX	4 (Gametocyte)	91.67

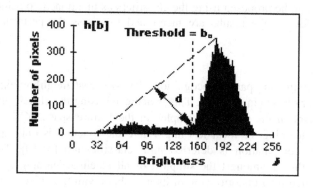

Fig. 2. Triangle algorithm as proposed by Zack

The objective function of Linear Programming:

$$Z = c1\ x1 + c2\ x2 \tag{1}$$

It is subject to restrictions:

$$a1\ x1 + b1x2 \leq \mu1 \tag{2}$$

$$a2x1 + b2x2 \leq \mu2 \tag{3}$$

$$a3\ x1 + b3x2 \leq \mu3 \tag{4}$$

$$a4x1 + b4x4 \leq \mu4 \tag{5}$$

$$\text{with } xi \geq 0\ (i = 1,2) \tag{6}$$

Where:-

(1) represents the mathematical function encoding the objective of the problem and is called objective function

(Z) in linear programming, this function must be linear.

(2)- (5) represent the linear mathematical function encoding the main restrictions identified, based on the

parasite species x1 and x2 values.

(6) non-negativity restriction, i.e. the number of infected red blood cells(x1) and non-infected red blood

cells(x2) may assume positive value or zero because negative value for these two categories is not possible.

"xj" corresponds to the decision-making variables that represent the number of infected red blood cells (j=1) and

number of non-infected red blood cells (j=2).

"ci" represents cost coefficients that each variable is able to generate or cost for parasite detection .

"µi" represents the species of parasite . We have four species of parasites described in the above sections.

"ai and bj" represent the quantity of resources each decision making variable consumes.

The accuracy obtained using the Zack's Thresholding with Linear Programming method is given by:

Table 2. Accuracy of Triangle Algorithm

Sr. No	SPECIES	STAGES	ACCURACY in %
1	FALCIPARUM	1 (Ring Form)	88.88
2	FALCIPARUM	2 (Tropozoite)	78.28
3	FALCIPARUM	3 (Schizont)	82.37
4	FALCIPARUM	4 (Gametocyte)	83.33
5	VIVAX	1 (Ring Form)	91.66
6	VIVAX	2 (Trophozoite)	76.65
7	VIVAX	3 (Schizont)	62.87
8	VIVAX	4 (Gametocyte)	91.66

iii) ANN

Back propagation networks and multi layered perceptrons, in general, are feed forward networks with distinct input, output, and hidden layers. The units function basically like perceptrons, except that the transition (output) rule and the weight update (learning) mechanism are more complex [13][14]. The ROC for ANN classifier can be given by Fig.4. It can be observed that saturation takes place at a value near 1 and the curve becomes linear at this value:

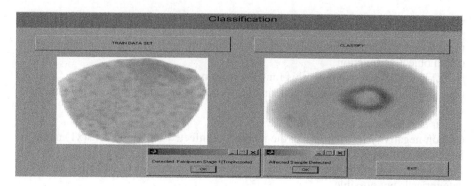

Fig. 3. Classification of an infected sample detected by Zack's Thresholding and linear programming module [23]

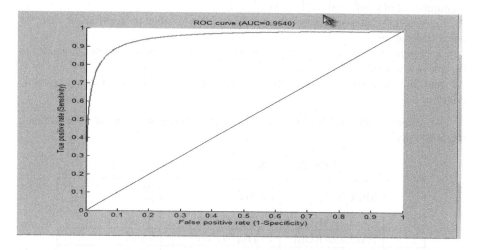

Fig. 4. ROC Curve for ANN classifier [23]

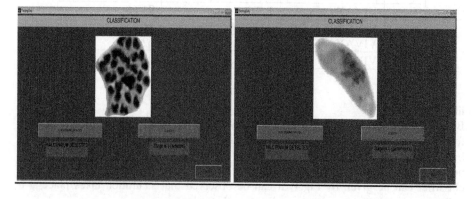

Fig. 5. a) Detection of Falciparum Stage 3 [19] b) Detection of Falciparum stage 4 [23]

Table 3. Accuracy of ANN Algorithm

Sr. No	SPECIES	STAGE	ACCURACY in %
1	FALCIPARUM	1	83.33
2	FALCIPARUM	2	92.34
3	FALCIPARUM	3	87.50
4	FALCIPARUM	4	89.50
5	VIVAX	1	91.67
6	VIVAX	2	90.00
7	VIVAX	3	91.87
8	VIVAX	4	91.67

iv) Template Matching

In template matching framework DTW(Dynamic Time Wrapping) distance is used, where a shape-based averaging algorithm is utilized to construct meaningful templates. It consists of two phases, i.e., training phase where templates are constructed, and a test phase where a query sequence is classified with the constructed templates. In the test phase, templates are retrieved and compared with the query sequence, and a class label of the nearest template will be the answer to the query [15]. Standard deviation concept is used for template matching. Using template matching for cell recognition template of normal cells and infected cells by malaria such as vivax, falciparum etc. are saved [16][17].The ROC for template matching given in Fig.6 shows the ROC curve along with the Area under the Curve (AUC) value shown, this differs ANN as the curve from 0.9 value to 1 value is sudden as compared to the gradual curve in Fig.4 which is evident in Fig.6:

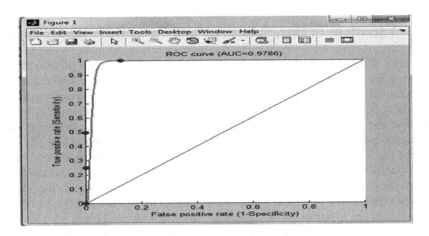

Fig. 6. ROC Curve for Template Matching [23]

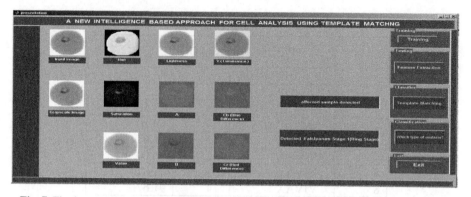

Fig. 7. Final output detecting the infected sample using Template Matching Algorithm [23]

Table 4. Accuracy of Template Matching

Sr. No	SPECIES	STAGE	ACCURACY in %
1	FALCIPARUM	1	90
2	FALCIPARUM	2	90
3	FALCIPARUM	3	90
4	FALCIPARUM	4	95
5	VIVAX	1	95
6	VIVAX	2	90
7	VIVAX	3	90
8	VIVAX	4	95

3   Variations in Hue, Saturation and Value (HSV)and YCbCr

The HSV scheme is used to signify the changes in the color of the sample after preprocessing stage to signify the level to which it has been infected by the malarial parasites [18]. The RGB and the HSV are related to each other where the geometric properties of the RGB presentation are changed and manipulated in order to be more intuitive and perceptually relevant than the Cartesian (Cube) representation. These values are then mapped to a cylinder resembling a traditional color wheel. The angle around the central vertical axis corresponds to "hue" and the distance from the axis corresponds to "saturation". These first two values give the two schemes the 'H' and 'S' in their names [19]. The height corresponds to a third value, the system's representation of the perceived luminance in relation to the saturation [22]. Fig. 8, 9, 10 indicates the variation in the values of hue, saturation and value also known as brightness of the image. 7 images have been tested for the variations in the HSV values, thus distinguishing the infected samples from non-infected ones. These variations are for the Zack's Thresholding and Linear Programming algorithm.

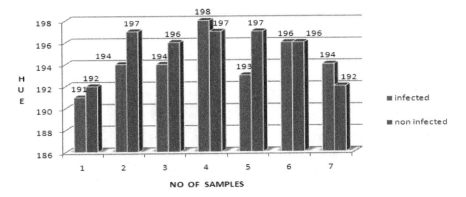

Fig. 8. Variations in Hue for infected and non-infected samples

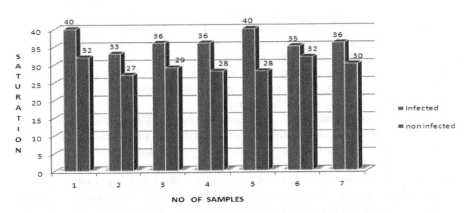

Fig. 9. Variation in saturation for infected and non-infected samples

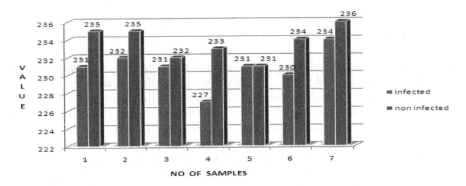

Fig. 10. Variation in value for infected and non-infected samples

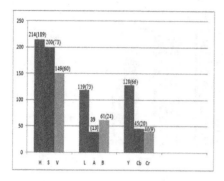 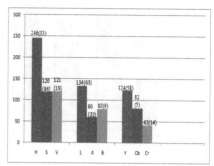

Fig. 11. HSV, LAB & YCbCr Results in template matching algorithm for a) Falciparum b) Vivax

A **Lab color space** is a color-opponent space with dimension **L** for lightness and **A** and **B** for the color-opponent dimensions. The L*A*B* color space includes all perceivable colors which means that its gamut exceeds those of the RGB and CMYK color models. One of the most important attributes of the L*A*B*-model is device independence. This means that the colors are defined independent of their nature of creation or the device they are displayed on [24].

Y' is the luma component and C_b and C_r are the blue-difference and red-difference chroma components. Y' (with prime) is distinguished from Y which is luminance meaning   light   intensity   is   nonlinearly   encoded   based   on gamma corrected RGB primaries. Y'CbCr is not an absolute color space rather, it is a way of encoding RGB information.   The   actual   color   displayed   depends   on   the actual RGB primaries used to display the signal [25]. Therefore a value expressed as Y'CbCr is predictable only if standard RGB primary chromaticities are used. As seen in Fig.11, template matching algorithm gives more efficient results in L*A*B* than YcbCr and HSV viz. subjected to more number of samples.

4   Conclusion

The system was tested with many classifiers and it can be easily concluded that the template matching method is giving the best results, though there are many researches using HSV [20][21] as base model. This result which differentiates and provides the basis for greater understanding of the algorithms and promising results for detection of Malarial parasites is evident from the accuracy table. By using 2 or more of these classifiers one can achieve even more accuracy and a system using the same is currently under testing. It is expected that this can provide an easy, accurate and instantaneous detection of malarial parasites along with their stages for efficient treatment where there is scarcity of other well established methods of diagnostics or help in supporting the diagnostics in places where such privileges already exist.

References

[1] Breman, J.G., et al.: Burden of malaria in India: retrospective and prospective view (2007)

[2] Phillips, R.S.: Current status of malaria and potential for control. Clinical Microbiology Reviews 14(1), 208–226 (2001)

[3] Sherman, I.W.: Biochemistry of Plasmodium (malarial parasites). Microbiological Reviews 43(4), 453 (1979)

[4] Tek, F.B., Dempster, A.G., Kale, I.: Computer vision for microscopy diagnosis of malaria. Malaria Journal 8(1), 153 (2009)

[5] Tek, F.B., Dempster, A.G., Kale, I.: Malaria Parasite Detection in Peripheral Blood Images. In: BMVC (2006)

[6] Le, M.-T., et al.: A novel semi-automatic image processing approach to determine Plasmodium falciparum parasitemia in Giemsa-stained thin blood smears. BMC Cell Biology 9(1), 15 (2008)

[7] Makkapati, V.V., Rao, R.M.: Segmentation of malaria parasites in peripheral blood smear images. In: IEEE International Conference on Acoustics, Speech and Signal Processing, ICASSP 2009. IEEE (2009)

[8] Ross, N.E., et al.: Automated image processing method for the diagnosis and classification of malaria on thin blood smears. Medical and Biological Engineering and Computing 44(5), 427–436 (2006)

[9] Sio, S.W., et al.: MalariaCount: an image analysis-based program for the accurate determination of parasitemia. Journal of Microbiological Methods 68(1), 11–18 (2007)

[10] Russ, J.C.: The image processing handbook. CRC Press (2010)

[11] Gonzalez, R.C., Woods, R.E., Eddins, S.L.: Digital image processing using MATLAB. Pearson Education India (2004)

[12] Zack, G.W., Rogers, W.E., Latt, S.A.: Automatic measurement of sister chromatid exchange frequency. Journal of Histochemistry & Cytochemistry 25(7), 741–753 (1977)

[13] Widrow, B., Lehr, M.A.: 30 years of adaptive neural networks: perceptron, madaline, and backpropagation. Proceedings of the IEEE 78(9), 1415–1442 (1990)

[14] Yegnanarayana, B.: Artificial neural networks. PHI Learning Pvt. Ltd. (2009)

[15] Lewis, J.P.L.: Fast template matching. Vision Interface 95(120123) (1995)

[16] Díaz, G., González, F.A., Romero, E.: A semi-automatic method for quantification and classification of erythrocytes infected with malaria parasites in microscopic images. Journal of Biomedical Informatics 42(2), 296–307 (2009)

[17] Halim, S., et al.: Estimating malaria parasitaemia from blood smear images. In: 9th International Conference on Control, Automation, Robotics and Vision, ICARCV 2006. IEEE (2006)

[18] Vardavoulia, M.I., Andreadis, I., Tsalides, P.: A new vector median filter for colour image processing. Pattern Recognition Letters 22(6), 675–689 (2001)

[19] Milligan, P.R., Morse, M.P., Rajagopalan, S.: Pixel map preparation using the HSV colour model. Exploration Geophysics 23(1/2), 219–224 (1992)

[20] Díaz, G., Gonzalez, F., Romero, E.: Infected cell identification in thin blood images based on color pixel classification: Comparison and analysis. In: Rueda, L., Mery, D., Kittler, J. (eds.) CIARP 2007. LNCS, vol. 4756, pp. 812–821. Springer, Heidelberg (2007)

[21] Mandal, S., et al.: Segmentation of blood smear images using normalized cuts for detection of malarial parasites. In: 2010 Annual IEEE India Conference (INDICON). IEEE (2010)

[22] Wikipedia contributors. HSL and HSV. Wikipedia, The Free Encyclopedia. Wikipedia, The Free Encyclopedia (April 2, 2014) (web April 14, 2014)

[23] MATLAB version R2011b. The Mathworks Inc., Pune (2011)
[24] Wikipedia contributors. Lab color space. Wikipedia, The Free Encyclopedia. Wikipedia, The Free Encyclopedia (April 2, 2014) (web April 17, 2014)
[25] Wikipedia contributors. YCbCr. Wikipedia, The Free Encyclopedia. Wikipedia, The Free Encyclopedia (February 8, 2014) (web April 17, 2014)
[26] Kareem, S., Kale, I., Morling, R.C.S.: Automated P. falciparum Detection System for Post- Treatment Malaria Diagnosis Using Modified Annular Ring Ratio Method. In: 2012 UKSim 14th International Conference on Computer Modelling and Simulation (UKSim). IEEE (2012)
[27] Ghosh, S., Ghosh, A.: Content based retrieval of malaria positive images fom a clinical database VIA recognition in RGB colour space. In: Satapathy, S.C., Avadahani, P.S., Udgata, S.K., Lakshminarayana, S. (eds.) ICT and Critical Infrastructure: Proceedings of the 48th Annual Convention of CSI - Volume II. AISC, vol. 249, pp. 1–8. Springer, Heidelberg (2014)

VLSI Implementation of Maximum Likelihood MIMO Detection Algorithm

Vakulabharanam Ramakrishna[1] and Tipparti Anil Kumar[2]

[1] Department of Electronics & Communication Engg.,
JNTU, Hyderabad, Telangana, India
ramakrishna.bharan@gmail.com
[2] Department of Electronics & Communication Engg.,
SR Engineering Collge, Warangal, Telangana, India
tvakumar2000@yahoo.co.in

Abstract. This paper presents FPGA implementation of low complexity Maximum Likelihood multiple input multiple output detection algorithm. FPGA Implementation of Maximum Likelihood detection for Multiple Input Multiple Output systems remains to be a great challenge. Devices which operate with battery power enforce severe silicon area and power limitations while trying for definite performance over a wide range of operating circumstances. First, multiple input multiple output system structure, mathematical model and Maximum Likelihood detection algorithm is presented. Performance of Maximum Likelihood Detection algorithm and its characteristics are studied. Next the developed system is implemented using MATLAB. Simulations are carried out and results are analyzed in terms of complexity and error performance. Finally, the low complexity Maximum Likelihood detector for multiple input multiple output is synthesized, and implemented on XC4VLX15SF363-12 device using Xilinx ISE tool. The results show that it is possible to implement Maximum Likelihood detector with a low power of 0.156W in a single FPGA chip.

Keywords: VLSI, MIMO, Detection Algorithms, ML, FPGA, BER, Low Complexity.

1   Introduction

Multiple Input Multiple Output (MIMO) is used to describe the multi-antenna wireless communication system, an abstract mathematical model takes the advantage of multiple transmitter antennas to transmit and multiple receiver antennas to receive and recover the original message. In MIMO communication system, channel capacity increases exponentially without increasing the transmission bandwidth [1]. MIMO technology is the future wireless communication systems to achieve high data rate transmission, to improve transmission quality (an important way to improve the system capacity). Maximum Likelihood (ML) detection algorithm is used to separate the spatially multiplexed data units at the receiver [2] and ML Detection also offers

© Springer International Publishing Switzerland 2015
S.C. Satapathy et al. (eds.), *Proc. of the 3rd Int. Conf. on Front. of Intell. Comput. (FICTA) 2014*
– *Vol. 1*, Advances in Intelligent Systems and Computing 327, DOI: 10.1007/978-3-319-11933-5_10

good error rate. The computational difficulty of a ML detector increases exponentially as the number of transmit antennas are increased. A number of effective suboptimal detection methods have been proposed or revised from the field of MIMO detection, but these methods are computationally complex and are often unable to exploit a great part of the diversity in comparison with the ML detector, and thus their performance be likely to be considerably inferior to that of ML detection[3].

The design tool should be chosen carefully as the signal processing applications enforce substantial limits on area, power dissipation, speed and cost. Digital signal processors (DSPs), Field programmable gate arrays (FPGAs) and application specific integrated circuits are the most widely used tools for the design of such application. The DSPs used for very complex math-intensive tasks but can't process great sampling rate applications due to its architecture. Application Specific Integrated Circuits (ASIC) faces lack of flexibility and need extensive design cycle. The shortcomings of DSP and ASIC are overcome by using single FPGA. Therefore FPGA has become the best choice for the signal processing system designs due to their greater flexibility and greater bandwidth, resulting from their parallel architecture [4]. Implementing the design using FPGA is very fast with lower development costs and takes less amount of time. VLSI implementation of the detection algorithms in the MIMO system will turn out to be a key methodology in the future wireless communication system.

This paper is arranged as follows: Section 2 gives brief overview of system model. Section 3 presents the Maximum likelihood detection algorithm. Section 4 shows the results of Bit Error Rates (BERs) and also hardware implementation results and section 5 is conclusion of the paper. This paper investigates the applicability of FPGA system for low complexity MIMO Detection algorithm in effective and economical way.

2   System Model

Below a MIMO is considered as that is having N_t sending antennas as well as N_r receiving antennas. A block diagram of the system is shown in figure 1.

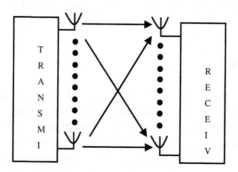

Fig. 1. Basic MIMO System Model

Received signal $\qquad\qquad\qquad Z_r = H_r X_r + n_r \qquad\qquad\qquad$ (1)

Where r=1, 2, 3 ... R, R is the number of sub carriers and received signal vector $Z_r = [Z_{r1}, Z_{r2},...Z_{rN_r}]^T$, Transmit Signal vector $X_r = [X_{r1}, X_{r2},...X_{rN_r}]^T$, and $n_r = [n_{r1}, n_{r2},...n_{rN_r}]^T$ is the additive noise vector where each noise component is usually demonstrated as independent identically distributed white Gaussian noise of variance σ^2. In general the transmitted signals are combined in the channel because of identical carrier frequency. At the receiver end, the received signal consists a linear grouping of all transmitted signals plus noise. Then the receiver can resolve for the transmitted signals by considering equation (1) as a scheme of linear equations. If channel is correlated, the method of linear equations will have increased number of unknowns than the equations. Correlation may occur due to two reasons. To avoid correlation due to spacing between antennas they are generally spaced at least half of the wavelength of the carrier frequency. Another reason correlation may takes place is due to absence of multipath components, so for this purpose rich multipath is desirable in MIMO systems. The multipath influence can be taken by each receive antenna being in a different channel H indicates the $N_r \times N_t$ channel matrix. Where

$$H = \begin{pmatrix} h_{1,1} & h_{1,2} & \cdots & h_{1,N_t} \\ h_{2,1} & h_{2,2} & \cdots & h_{2,N_t} \\ \vdots & \vdots & \vdots & \vdots \\ h_{N_r,1} & h_{N_r,2} & \cdots & h_{N_r,N_t} \end{pmatrix} \qquad\qquad (2)$$

Where every single $h_{i,j}$ indicates the attenuation and phase shift between the ith receiver and the jth transmitter. In order to identify the communicated data it would be best to use ML detector [3].

3   Maximum Likelihood (ML) Detection Algorithm

This is the theoretical optimum detection method and can provide full diversity gain [5]. In the most common case, the input symbol belongs to a code word covering space and time. Here the numbers of probable code words are finite and we denote CW as the set of all possible codewords. In the ML method, all possible codewords are verified and one that best fits the received signal according to the ML principle is selected as an estimation of the code word which is actually transmitted. The code word conveyed is a matrix denoted as χ with dimension $M_T \times N$. The columns of this matrix are the vectorial inputs of the MIMO system at time instant k to N and are denoted in this section as x_k, k = 1, . . . , N. The corresponding matrix output is denoted as $\tilde{y}$ and the columns of $\tilde{y}$ are denoted as y_k, k = 1, . . . , N where y_k is the received signal corresponding to the transmission of x_k. χ is the definite transmitted codeword. When testing all the probable input code words, a candidate codeword is Z where z_k, k = 1, . . . , N are the columns of X. The ML optimization principle finds the codeword with the minimal distance to the received signal

$$\min_{Z \in CW} \left\| y - Hz \right\|_F^2 \Leftrightarrow \min_{\{Z_1,...,Z_N\} \in CW} \sum_{k=1}^{N} \left\| y_k - H Z_k \right\|^2 \qquad (3)$$

The ML receiver can be simplified if the receiver proceeds first to estimating the input symbols and then to decoding. ML exploits the fact that a symbol belongs to a constellation and takes a finite number of values. As symbol x_i takes a finite number of values, so does the transmitted vectorial input $x=[x_1 \ldots x_{MT}]^T$. Here C denotes the set of all possible vectorial inputs. Calling z a candidate vectorial input in the set C, the ML criterion finds the value of z that best fits the received signal as

$$\min_{z \in C} \| y - Hz \|^2 \tag{4}$$

The decoder is error free if the value of z found through the ML optimization, is the actual transmit input x. ML receivers are computationally complex in general because one needs to test all the possible input vectorial inputs. Their number can become quite large especially as the number of antennas and constellation order increases. Hence simpler receivers are considered as more practical in general. This paper is focused on the 4 × 4 MIMO system simulation, constellation map of QPSK, at the receiving end using the maximum likelihood detection algorithm. ML detection algorithm is used to investigate channel estimation algorithm performance, so when testing using the transmitted signal space uses exhausted search approach to maximum likelihood detection [6], in order to fully demonstrate channel estimation algorithm BER performance.

Maximum likelihood detectors are the ideal detectors and have better BER performance in comparison with non-optimal detectors, but ML detectors are the most complex detectors [7]. The exhaustive detector complexity rises exponentially as the number of bits in constellation or the number of antennas increases. As a case in point, the number of probable transmitted symbol for a 4x4 QPSK is 28=256 and for a 4x4 16-QAM system this number grows to 216=65536.

4   Results and Discussion

A. Matlab Simulations

Figure2 shows the results of 1x1, 2x2 and 4x4 antenna bit error curves and from that it can be seen that the detection of the ML, when transmitting and receiving antennas are increased, error rate is reduced, so that the MIMO communication system can overcome the adverse multipath fading effects, to achieve the reliability of signal transmission, which increases the system capacity and hence improves the spectrum efficiency [8].

From the result which is shown in figure 3 it can be seen, the use of ML detection, the ideal channel and the estimated channel error rate overall trend is same, but the ideal channel bit error rate than the estimated channel bit error rate is low, indicating that based on the pilot channel estimation in MIMO communication system is feasible.

It can be seen from the result figure 4, when using the ideal channel, ML Detection algorithm error rate is small, and the same general trend, but at high SNR part MLD algorithm BER is obviously much lower, indicating that global search algorithm MLD is having excellent performance.

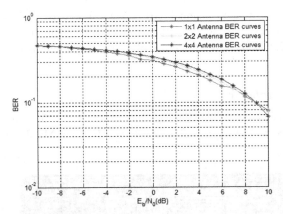

Fig. 2. ML Detection MIMO System 1x1, 2x2, 4x4 Antenna Performances

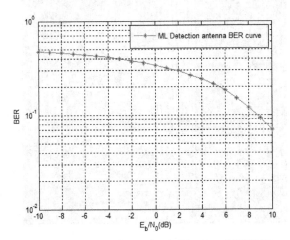

Fig. 3. Estimated channel, ML Testing under 4x4Antenna MIMO bit error rate curve

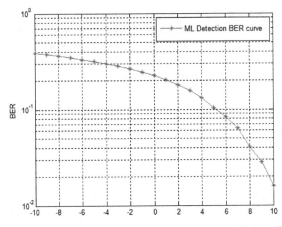

Fig. 4. Ideal channel, MIMO 4x4 antenna system ML detection

When using the estimated channel, MLD algorithm BER is less, where MLD algorithm is slightly lower SNR. From the computational point of view and run time, when the num = 256, loop = 1000, the use of a MLD algorithm is taking a total time of about 1200s.

B. FPGA Implementation

The design has been coded in Verilog and checked for functional and timing correctness with ISim. The Register Transfer Level (RTL) of ML detector is shown in below figure 5.

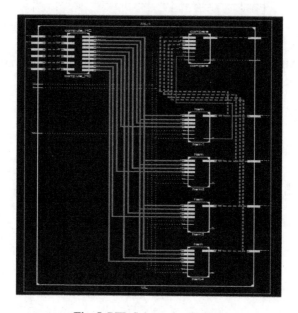

Fig. 5. RTL Schematic of MLD

The architecture was designed concentrated on great performance and low cost. The architecture was defined in Hardware Description Language and synthesized to Xilinx Virtex 4 FPGAs and the RTL schematic of maximum likelihood detector is shown in figure5. Table 1 is about device utilization summary for ML detection algorithm and the corresponding graph is shown in figure 6.

Table 1. XC4VLX15SF363-12 Device utilization summary

Logic Utilization	Used	Available	Utilization
Number of Slices	805	6144	13%
Number of Slice Flip Flops	405	12288	3%
Number of four input LUTs	1339	12288	10%
Number of IOs		138	
Number of bonded IOBs	138	240	57%
Number of GCLKs	1	32	3%

The synthesis results demonstrate that the architecture is capable to accomplish processing at faster rates and attains the real-time requirements. The architecture is consuming fewer resources when compared to related works [09, 10]. The proposed design implemented on VIRTEX 4 based FPGA. The total power consumption of the proposed scheme based on XC4VLX15SF363-12 FPGA device has been calculated and observed that the proposed design has consumed 156µW.

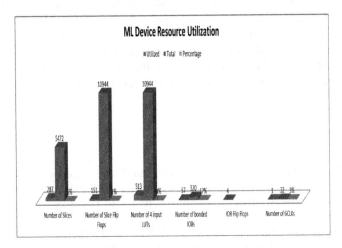

Fig. 6. ML Detection device utilization Summary

In Fig.7 first graph indicates the power by function i.e. the power consumed by each functional block. Next is power versus voltage and third graph is about the power variance in accordance with process and voltage. Finally power is plotted against junction temperature which is the temperature of the device in operation. Temperature grade is chosen while selecting the device and this grade defines a temperature range where the device will operate.

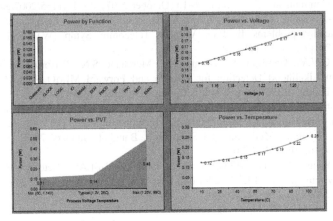

Fig. 7. Graphs for Power by function, Power vs Voltage, Power Variance and Power vs Temperature

5   Conclusion

This paper presents the design and simulation of efficient low complexity ML detector. The application of MATLAB on the system simulation is to analyze the Bit Error Rate. The proposed design has been implemented using XC4VLX15SF363-12 device. The results show that it is possible to implement ML detector with a low power in a single FPGA chip. . The architecture which has been used for ML detector delivers better area efficiency. The proposed design can provide great speed of operation by consuming significantly less resources in terms of slices and with less power to provide cost effective solution for high speed wireless communication applications. From the above method, we further understand that the outstanding advantages and performance of MIMO system laid the foundation for future 5G communications.

References

1. Goldsmith, A., Jafar, S.A., Jindal, N., Vishwanath, S.: Capacity Limits of MIMO Channels. Proc. IEEE J. Selected Areas in Comm. 21(5), 684–702 (2003)
2. Ahmadi, S.: Mobile WiMAX: A Systems Approach to Understanding IEEE 802.16m Radio Access, pp. 560–572. Academic Press (2011)
3. Su, Y., Tang, J., Mergen, G., Subrahmanya, P., Xiao, L., Sidi, J., Salvekar, A.: CQI algorithm design in MIMO systems with maximum likelihood detectors. Proc. Special Issue on Advances in MIMO-OFDM 4(4), 266–274 (2011)
4. Woods, R., McAllister, J., Yi, Y., Lightbody, G.: FPGA-based Implementation of Signal Processing Systems, pp. 111–125. John Wiley (2008)
5. Shin, M., Kwon, D.S., Lee, C.Y.: Performance Analysis of Maximum Likelihood Detection for MIMO Systems. In: Proc. IEEE 63 Veh. Technol. Conf., vol. 5, pp. 2154–2158 (May 2006)
6. Wong, K.-K., Paulraj, A.: Efficient Near Maximum Likelihood Detection For Underdetermined MIMO Antenna Systems Using A Geometrical Approach. EURASIP J. Wireless Commun. and Networking, 1–13 (October 2007), doi:10.1155/2007/84265
7. Shiu, D.-S., Kahn, J.M.: Near optimum maximum likelihood detector for structured communication problems. In: Proc. Wireless Telecomm. Symp. (WTS 2012), London, April 18-20, pp. 1–6 (2012)
8. Sharma, G.V.V., Ganwani, V., Desai, U.B., Merchant, S.N.: Performance Analysis of Maximum Likelihood Detection for Decode and Forward MIMO Relay Channels in Rayleigh Fading. IEEE Trans. Wireless Comm. 9(9), 2880–2889 (2010)
9. Ramakrishna, V., Kumar, T.A.: Low Power VLSI Implementation of Adaptive Noise Canceller Based on Least Mean Square Algorithm. In: Proc. 2013 4th Intl. Conf. on Intelligent Systems, Modelling and Simulation, Bangkok, January 29-31, pp. 276–279 (2013), doi:10.1109/ISMS.2013.84
10. Sun, L., Yang, W., Huang: FPGA Implementation of V-BLAST Detection Algorithm In MIMO System. In: Proc.IEEE Youth Conf. Information, Computing and Telecommunication (YC-ICT 2009), pp. 134–137 (2009)

E-FCM Algorithms for Collaborative Tagging System

Latha Banda and K.K. Bharadwaj

School of Computer and System Sciences
Jawaharlal Nehru University, Delhi
{latha.banda,kbharadwaj}@gmail.com

Abstract. Collaborative Tagging is the process of managing, classifying and describing tags where in user can easily describe or get an information about an item. As growth of information in social networking websites there is a possibility of increase of tags and it is very difficult to predict or recommend items to the user. In this paper we address the problem of sensitivity in the distribution of the tag data in clusters and optimization can be achieved by merging similar clusters in E-FCM and E-GK algorithm for collaborative Tagging Systems. Our experiments on each movie data set shows E-FCM and E-GK with sensitivity gives more accurate results as compared to baseline clustering approaches in collaborative tagging systems.

Keywords: Collaborative tagging System, sensitivity, E-FCM Algorithm, fuzzy cluster and E-GK algorithm.

1 Introduction

Folksonomy is a system of classification derived from the practice and method of collaboratively creating and translating tags to annotate and categorize content [8] [9]. This process is also known as collaborative tagging [12]. social classification, social indexing, and social tagging. CT is divided into broad and narrow tagging systems. In broad tagging system more number of users or multiple users tag particular item or different items with the variety of vocabularies. Where in narrow tagging systems few number of users tag the items with limited content of tags. while in both tagging systems search ability can be done but the narrow tagging system does not have same benefits as broad tagging systems [16] . Tagging, allows users to collectively classify and find information. In some websites tag clouds are included to visualize tags in a folksonomy [4]. However, tag clouds visualize only the vocabulary but not the structure of folksonomies [7].

Cluster cannot be precisely defined, which is one of the reasons why there are so many clustering algorithms [3]. Cluster is a group of objects and there so many algorithms and models are available. A "clustering" is essentially a set of such clusters, usually containing all objects in the data set. Clustering is of two types (i) hard clustering represents each object is in a cluster or not (ii) soft clustering is also known as fuzzy clustering [6] [17] represents an object is in cluster and its degree

© Springer International Publishing Switzerland 2015

S.C. Satapathy et al. (eds.), *Proc. of the 3rd Int. Conf. on Front. of Intell. Comput. (FICTA) 2014*
– *Vol. 1*, Advances in Intelligent Systems and Computing 327, DOI: 10.1007/978-3-319-11933-5_11

(likelihood of belonging to cluster). Our work in this paper is applying E-FCM (extended fuzzy clustering method) in collaborative tagging systems [15] with sensitivity [14] which gives accurate and efficient results.

2   Related Work

In this section, we briefly explain several major approaches we used in our research such as Collaborative Tagging, soft clustering and Time sensitiveness.

2.1   Collaborative Filtering Based on Tagging

Tagging [10][11] was popularized by websites associated with Web 2.0 and is an important feature of many Web 2.0 services. Tagging is having wide popularity due to the growth of social networking, photography sharing and bookmarking sites. These sites allow users to create and manage labels (or "tags") that categorize content using simple keywords. Tagging is a bottom-up classification process. Tagging can be categorized into two types: (i) Static tags or like/dislike tags in which a user post thumbs up for like and thumbs down for dislike. (ii) Dynamic tags in which a user give his opinion on item. These are also called as popular tags.

2.2   Soft Clustering

Soft clustering is also known as fuzzy clustering class of algorithms for cluster analysis in which the allocation of data points to clusters is soft. In hard clustering, data is divided into different clusters, where each object belongs to exactly one cluster. In fuzzy clustering [5], objects may belong to more than one cluster, and associated with each element is a set of membership levels. These indicate the strength of the association between that data element and a particular cluster. Fuzzy clustering is a process of assigning these membership levels, and then using them to assign data elements to one or more clusters. The fuzzy clusters are shown in figure 1. The first cluster represents items, second cluster represents tags and third one is user profile based cluster. One of the most algorithms are used in fuzzy clustering is Extended Fuzzy c-means (FCM) algorithm and Extended Gustafson and Kessel (E-GK).

The FCM algorithm is used to partition the number of elements $A = \{a_1, ..., a_n\}$ into a collection of c fuzzy clusters and the list of cluster centres $C = \{c_1, ..., c_c\}$ and partition matrix $W = w_{ij} \in [0, 1], i = 1, ..., n, j = 1, ..., c$ where each element w_{ij} is the degree to element A_i belongs to cluster C_i. FCM is used to minimize the objective function like K-means clustering and the standard function of FCM is:

$$W_k(x) = \frac{1}{\sum_j \left(\frac{d(center_k, x)}{d(center_j, x)} \right)^{2/(m-1)}} \tag{1}$$

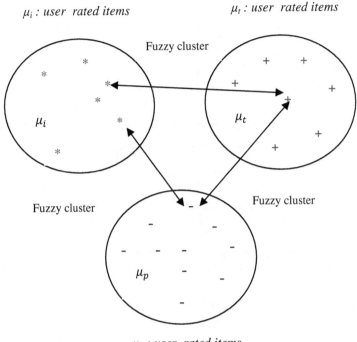

μ_i : *user rated items* μ_t : *user rated items*

Fuzzy cluster

μ_i

μ_t

Fuzzy cluster   Fuzzy cluster

μ_p

μ_p : *user rated items*

2.3   Time Sensitiveness

Time sensitiveness [10] means that once a tag clustering is available, it does not remain valid all the time. The tagging behavior of user changes over time, thus the tag clusters need to be updated. Ning et al. also suggest a real time update algorithm based on spectral clustering in graph, through which the system can insert and delete data points and change similarity between current items. When the dataset is updated the weights may change. If the similarity changes between two vertices and nodes are having less edge weight then the edges of these nodes are removed.

3   Proposed Work E-FCM Algorithm for Collaborative Tagging System

In this section we describe the extension of fuzzy clustering algorithms with volume prototypes and similarity based cluster merging [13].the proposed work shown in figure 2. A number of data points are close a cluster center may lie in the same cluster. This is especially the case when there are some clusters that are well separated from the others. It is then sensible to extend the core of a cluster from a single point to a region in the space.

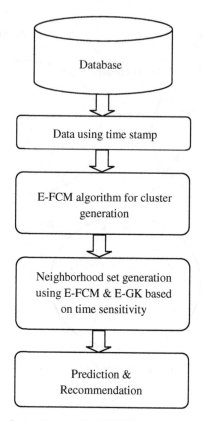

Fig. 2. Architecture for E-FCM-CFT and E-GK-CFT

A volume prototype $V \in R^n$ is a n-dimensional, convex and compact subspace of the clustering space.

The volume cluster prototypes v_i are hyper spheres with center v_i and radius r_i.

Distance measure is measured from a data point x_k to a cluster center v_i. Then, the distance d_{ik} to the volume prototype vi is determined by accounting for the radius r_i:

For the size of the cluster prototypes obtain in E-FCM [14] by using:

$$R_i = \sqrt{i}Q_i^T I Q_i \sqrt{i} = i \, , \tag{2}$$

For the size of the cluster prototypes obtain in E-GK [14] by using

$$R_i = \sqrt{i}Q_i^T |P_i|^{1/n} Q_{ii}^{-1} Q_i^T Q_i \sqrt{i} = |P_i|^{1/n} I, \tag{3}$$

Prototype radius in E-FCM is given by:

$$r_i = \sqrt{\prod_{j=1}^n \lambda_{ij}^{1/n}} = \sqrt{|P_i|^{1/n}}, \tag{4}$$

Prototype radius in E-GK is given by:

$$r_i = \sqrt{|P_i|^{1/n}}, \tag{5}$$

Extended Fuzzy c-Means and E-GK Algorithm

Given the data X, choose the initial number of clusters $1 < M^{(0)} < N$, the fuzziness parameter $m > 1$ and the termination criterion $\epsilon > 0$. Initialize $U^{(0)}$ (e.g random) and let $S_{i*j*}^{(0)} = 1$, $\beta^{(0)} = 1$.

Repeat for $l = 1, 2,...$

1. *Compute pointwise cluster prototypes*

$$V_i^{(l)} = \frac{\sum_{k=1}^{N}(u_{ik}^{(l-1)})^m x_k}{\sum_{k=1}^{N}(u_{ik}^{(l-1)})^m}, \quad 1 \le i \le M^{(l-1)}$$

2. *Compute radius of cluster prototypes from fuzzy covariance:*

$$P_i = \frac{\sum_{k=1}^{N}(u_{ik}^{(l-1)})^m (x_k - v_i^{(l)})(x_k - v_i^{(l)})^T}{\sum_{k=1}^{N}(u_{ik}^{(l-1)})^m}, \quad 1 \le i \le M^{(l-1)}$$

$$r_i = \beta^{(l-1)} \sqrt{|P_i|^{1/n} \Big/ M^{(l-1)}}, \quad 1 \le i \le M^{(l-1)}$$

3. a) *Compute the distance to the volume cluster prototype (for E-FCM)*

$$d_{ik} = \max\left(0, \sqrt{(x_k - v_i^{(l)})^T(x_k - v_i^{(l)})} - r_i\right),$$

$$1 \le i \le M^{(l-1)}, 1 \le k \le N.$$

b) *Compute the distance to the volume cluster prototypes (for E-GK):*

$$d_{ik} = \max\left(0, \sqrt{|p_i|^{1/n}(x_k - v_i^{(l)})^T P_i (x_k - v_i^{(l)})} - r_i\right),$$

$$1 \le i \le M^{(l-1)}, 1 \le k \le N.$$

4. *Update the partition matrix:*

$$for \ 1 \le k \le N, let \ \emptyset_k = \{i | d_{ik} = 0\}$$

$$if_k = \emptyset,$$

$$u_{ik}^l = \frac{1}{\sum_{j=1}^{M^{(l-1)}} \left(d_{ik} / d_{jk} \right)^{2/(m-1)}}, \quad 1 \le i \le M^{(l-1)},$$

Otherwise

$$u_{ik}^{(l)} = \begin{cases} 0 & if \ d_{ik} = 0 \\ 1/|k| & if \ d_{ik} = 0 \end{cases} \quad 1 \le i \le M^{(l-1)}.$$

5. *Select the most similar cluster pair:*

$$S_{ij}^{(l)} = \frac{\sum_{k=1}^N \min(u_{ik}^{(l)}, u_{jk}^{(l)})}{\sum_{k=1}^N u_{ik}^{(l)}}, \quad 1 \le i,j \le M^{(l-1)},$$

$$(i^*, j^*) = \arg \max_{\substack{(i,j) \\ i \ne j}} (S_{i,j}^{(l)}).$$

6. *Merge the most similar clusters:*

$$if \ \left| S_{i^*j^*}^{(l)} - S_{i^*j^*}^{(l-1)} \right| < \epsilon$$
$$let \ \alpha^{(l)} = 1 / (M^{(l-1)} - 1)$$
$$if \ S_{i^*j^*}^{(l)} > \alpha^{(l)}$$

$$u_{i^*k}^{(l)} := \left(u_{i^*k}^{(l)} + u_{j^*k}^{(l)} \right), \quad 1 \le k \le N,$$

remove row j^* from U,
else enlarge volume prototype

$$\beta^{(l)} = \min(M^{(l-1)}, \beta^{(l-1)}) + 1.$$

Until $\left\| U^{(l)} - U^{(l-1)} \right\| \le \epsilon.$

4   Experiments

We have conducted several experiments on different datasets amazon.com, Grouplens, Movielens and Flickr.com. To examine the effectiveness of our new scheme E-FCM-CTF and E-GK-CFT. In this we address the following issues.

(i)   The datasets have been rearranged using timestamp method
(ii)   We have conducted experiments on collaborative filtering based on tagging [2][11].
(iii)   Then we have applied E-FCM and E-GK algorithm for CFT.
(iv)   We have compared these approaches in terms of mean absolute error and number of predictions.

4.1   Datasets

For every dataset we have collected 5000 user data who have at least visited 40 items. For every dataset we have divided 1000 users split. Such a random separation was intended for the executions of one fold cross validation where all the experiments are repeated one time for every split of a user data. For each dataset we have tested set of 30% of all users.

To evaluate the effectiveness of the recommender systems [1], the mean absolute error (MAE) computes the difference between predictions generated by RS (Recommender Systems) and ratings of the user. The MAE is given by the following formula.

$$\text{MAE}\ (i) = \frac{1}{n_i}\sum_{j=1}^{n_i}\left|\text{pr}_{i,j} - \text{r}_{i,j}\right|\ . \tag{6}$$

4.2   Results

In this experiment we run the proposed E-FCM-CFT and E-GK-CFT compared its results with classical collaborative filtering based on collaborative tagging [5]. In this the active users are considered from 1 to 20. Based on these active users, we have computed the MAE and prediction percentage of both approaches and the results show that E-FCM-CFT and E-GK-CFT performs better than baseline methods and for some number of active users prediction percentage is more in E-FCM-CFT less in E-GK-CFT and vice versa. The results are shown in Table 1 and figure 3.

4.3   Analysis of Results

For E-GK-CFT and E-FCM-CFT, out of 10 runs for each active user, the run with the best weights was chosen and plotted the results from CF and CFT as shown in figure 3. The results summarized in Table.1 and show total average of MAE and for all the algorithms that E-GK-CFT and E-FCM-CFT outperforms rather than CFT. MAE for both algorithms was always smaller than the corresponding values for CFT.

Table 1. Total MAE for CFT and E-FCM-CFT

Split (100 users per split)	MAE(CFT)	MAE(E-FCM-CFT)	MAE(E-GK-CFT)
1	0.802	0.662	0.643
2	0.792	0.642	0.651
3	0.770	0.602	0.598

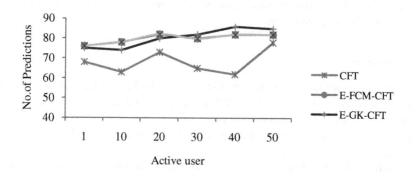

Fig. 3. Correct predictions percentage for active user

5   Conclusion

We have proposed a framework for collaborative Tagging using E-FCM and E-GK. In this approach we implemented E-FCM and E-GK algorithms and added time sensitivity to update data and to get more accurate predictions. Experimental results show that our proposed scheme can extensively improve the accuracy of predictions when we combine both E-FCM and E-GK algorithms.

References

1. Adomavicius, G., Tuzhilin, A.: Toward the next generation of recommender systems: A survey of the state-of-the-art and possible extensions. IEEE Transaction on Knowledge and Data Engineering 17(6), 734–749 (2005)
2. Banda, L., Bharadwaj, K.: Improving Scalability Issues in Collaborative Filtering based on Collaborative Tagging using Genre Interestingness Measure. In: ARTCom. LNEE, pp. 403–409. Springer, Berlin (2012)
3. Estivill-Castro, V.: Why so many clustering algorithms — A Position Paper. ACM SIGKDD Explorations Newsletter 4(1), 65–75 (2002)
4. Lamere, P.: Lamere, Paul. Social Tagging And Music Information Retrieval. Journal of New Music Research 37(2), 101–114 (2008)

5. Banda, L., Bharadwaj, K.K.: Evaluation of Collaborative Filtering Based on Tagging with Diffusion Similarity Using Gradual Decay Approach. 27, 421–428 (2014)

6. Han, L., Chen, G.: HFCT: A Hybrid Fuzzy Clustering Method for Collaborative Tagging. In: International Conference on Convergence Information Technology. IEEE (2007)

7. Lohmann, S., Diaz, P.: Representing and Visualizing Folksonomies as Graphs - A Reference Model. In: Proc. International International Working Conference on Advanced Visual Interfaces. ACM Press (2012)

8. Peters, I.: Folksonomies. Indexing and Retrieval in Web 2.0. De Gruyter Saur, Berlin (2009)

9. Pink, D.H.: Folksonomy.New York Times (2005)

10. Keller, P., Smadja, F.: Automated Tag Clustering: Improving search and exploration in the tag space (2006)

11. Kim, H.N., Ji, A.T., Ha, I., Jo, G.S.: Collaborative filtering based on collaborative tagging for enhancing the quality of recommendation 9(1), 73–83 (2010)

12. Golder, S.A., Huberman, B.A.: Usage patterns of collaborative tagging systems. Journal of Information Science 32(2), 198–208 (2006)

13. Setnes, M., Kaymak, U.: Extended fuzzy c-means with volume prototypes and cluster merging. In: Proceedings of Sixth European Congress on Intelligent Techniques and SoftComputing, vol. 2, pp. 1360–1364. ELITE (1998)

14. Kaymak, U., Hans, R., Van Natua, L.N.: A sensitivity analysis approach to introducing weight factors into decision functions in fuzzy multicriteria decision making. Fuzzy Sets and Systems 97(2), 169–182 (1998)

15. Kaymak, U., Setnes, M.: Extended Fuzzy Clustering Algorithms. ERIM Report Series Research in Managent (2000)

16. Thomas, V.W.: Explaining and Showing Broad and Narrow Folksonomies (2013)

17. Pedrycz, W.: Collaborative fuzzy clustering. Journal of Pattern Recognition Letters 23(14), 1675–1686 (2002)

An Efficient Approach of Multi-Relational Data Mining and Statistical Technique

Neelamadhab Padhy and Rasmita Panigrahi

Gandhi Institute of Engineering and Technology
Department of Computer Science, Gunupur -765022, Odisha, India
{neela.mbamtech,rasmi.mcamtech}@gmail.com

Abstract. The objective of data mining is to find the useful information from the huge amounts of data. Many researchers have been proposed the different algorithms to find the useful patterns but one of the most important drawbacks they have found that data mining techniques works for single data table. This technique is known as traditional data mining technique. In this era almost all data available in the form of relational database which have multiple tables and their relationships. The new data mining technique has emerged as an alternative for describing structured data such as relational data base, since they allow applying data mining in multiple tables directly, which is known as Multi Relational data mining. To avoid the more number joining operations as well as the semantic losses the researchers bound to use Multi Relational Data Mining approaches. In this paper MRDM focuses multi relational association rule, Multi relational decision tree construction, Inductive logic program (ILP) as well three statistical approaches. We emphasize each MR-Classification approach as well as their characteristics, comparisons as per the statistical values and finally found the most research challenging problems in MRDM.

Keywords: MRDM, Decision tree, ILP, Heavy Association rule mining Linear Regression Model.

1 Introduction

In data mining techniques the main aim is to extract the regularities from a huge amount of data .There are varieties of tools and algorithms are developed to handle the large database. The relational database is the best approach to deal with the structured database. To handle those kinds of data, different data mining tasks can be performed in the database [1] .These tasks are called as clusters or to recognize the individual piece of data termed as outlier which is not fitted with other set of data. An important task of multi relational data mining is able to directly the patterns that involve multiple relations from a relational database and have more advantages then propositional data mining approaches. The objective is to discover the knowledge directly from relational data. Knobbe et.al [2] which proposed the generalized frame work for MRDM which exploits SQL to gather the information needed for decision tree from Multi relational data. Mining. MRDM is entering a period of rapid expansion that will increasingly place it at the center of the KDD enterprise and its real-world impact [3][4][36].The number of data mining applications are increasing which involves the analysis of complex as well as structured types of data which requires the expressive pattern language. Some of these applications can't be solved by using the traditional rule mining algorithm which leads the MRDM.In MRDM where the relations are available in the form of parent and child .when we extract the data from different tables by using the primary and foreign key relationships then scalability problem will arises. Valencio et.al, *Human-centric computing and Information Science 2012 2:4[5]* was proposed the first time to avoid the scalability problem which occurred in the MRDM. To suppress this problem they proposed the new construction of tree called as MR-Radix tree. It is one type of the data structure which compresses the database memory.The MR-Radix algorithm provided the better efficiency when we compared with any traditional algorithms...Multi-relational data mining algorithms aims to extract of multiple relation patterns with efficiency without necessity to joining the data into the single table[6][7].

In this paper there are 6 sections .Section 1 describes the most feasible data mining methodology, section 2 describes the novel approaches of MRDM, section 3 describes the comparative study of MRDM approaches, section 4 describes novel comparative study on decision tree, section 5 emphasis on heavy association rule mining and section 6 presents linear regression and finally comparative study of statistical prediction technique.

In the last decade, a wide range of techniques for Structured Data Mining has been developed. These techniques are primarily into four categories, which can be characterised by the choice of representation of the structured individuals. These four categories are as below.

1.1   Graph Mining [8, 9, 10, 37, 38]

Graph mining is the techniques which will extract the required information from data represented in the form of graph structured form. A graph can be defined as the equation G={V,E},Where V ={v1,v2,v3,......................vn} is an ordered set of vertices in the graph and ={e1,e2,e3,........en}is the set of pair of edges. The term graph mining which can refer to discover the graph patterns.

1.2   Inductive Logic Programming (ILP) [11, 12, 13, 14]

If we deal with the logic based MRDM then we called as ILP (Inductive Logic Programming) which consists of the first order logic (Prolog) can be treated as to identify the subgroups. Through this technique to discover the required knowledge from the web based data. It is the dominant research areas of both machine learning and logic programming. The two important characteristics of the ILP, these are expressive representation formalism and find the exact required rules in the web intelligence. ILP can also be used for translating the relational databases to the first-order logic.

1.3   Semi-structured Data Mining

The web contains the huge numbers of data what we called as information resources.These information's in the web are represented in the form of XML structure. For mining the XML data we required to pre-processing or post-pressing so that the XML data is being converted into the relational structure. It is also used for integrating the databases. The XML form is useful to represent the semi-structured data further which can change the tabular data as well as the tree structure form when the database consists of XML documents which explain about the objects in the combination of structural and free-text information is called as the semi-structured data mining .[15,16].

1.4   Multi-Relational Data Mining (MRDM) [17, 18, 19, 20,39,40]

Multi relational data mining handles with the knowledge discovery from the relational database consisting of one or multiple tables. In MRDM, the crucial technique is ILP (Inductive Logic Program). MRDM can contribute to the foundation of data mining (FDM), the interesting perspective of "ILP-based MRDM for FDM" has not been investigated in the past.

2   MRDM Approaches

The different approaches are available in MRDM, these are as below:

- ➢ Selection Graph
- ➢ Multiview learning
- ➢ Probabilistic Approaches
- ➢ Tuple-ID Propagation

2.1   Selection Graph

In some of the database which contains the very large number of records and each relation may link with other relations. MRDM has a technique called the selection graph where multi relational patterns are expressed in terms of graphics form[21].It is generally used to represent the complex query in SQL which directly deal with the tables. There are number of algorithms are developed which converts directly selection graph into corresponding SQL query. Anna Atramentov et.al also focused the graphical representation of selection graph. [22]

2.2   TupleID Propagation

It is the process of transferring the information among difference relations by virtually joins them. It is observed that it requires less costly then physical join in both and time respect. It is the process of performing virtual joins among relations which less expansive then physical joins. It is necessary for searching the good predicates but the challenging task is to find the useful links as well as transferring information efficiently.

2.3   Multi View Learning

In this learning strategy each learner is assigned to a set of training data from which should be able to learn the target concept. In MVL problem with n views can be seen n interdependent relations .The MVC (Multi –Relational Classification) approach adopts the MVL framework to operate directly on MR-Database with the generalized data mining algorithms. [31].The multi view learning approach is more suitable for mining the relational database in the form of running time.

2.3.1   Analysis of MRDM (Multi-Relational Data Mining) Approaches on the Basis of SG, TIDP & MVC

When we compare with it shows some strong points of multi-view learning as compare to other approaches .The strong points of MVC are as below

2.3.2   First Approach

The relational database is able to keep its compact representation and normalized structure.

2.3.3   Second Approach

Require some of the framework which is able to directly incorporate any traditional single table data mining algorithm.

2.3.4   Third Approach

The MVL framework is highly efficient for mining relational databases in term of running time.

In the above table we demonstrated that the multi view learning is more powerful than other approaches of relational data mining. When we use very large class of data mining problem then it cannot be successfully approached using another relational learning without transformation

Tabular Representation for Multi Relational Data Mining Parameters

Table 1. Novel approaches for MRDM

Required Parameter	Some of the Relational database approaches		MVC(Multi-View Learning)
	(SG)Selection Graph	TIDP(TupleID Propagation)	
When advance techniques are integrated	No	NO	Yes
Incremental design support	Low	Low	High
Heterogeneous Learning classifier	No	NO	Yes
Single Table alg Incorporation	No	NO	Yes
Issues & Scalability	Low	Low	High
When structures are normalized	No	NO	Yes
When time will come to learn	Less	More	Less

2.3.5 Analysis of Structured Data Mining

In this section we have made a comparative study in between the different techniques available in SDM(Structured Data Mining) approach.

Table 2. for Structured Data Mining

Important Name of concept	Multi-Relational Data Ming (MRDM)	Inductive Logic Program(ILP)	Graph Mining (GM)	Semi Structured Data Mining (SSDM)
When used with graph/tree	Graph	Graph	Graph	Tree
When used with Numeric value	Yes	Yes	No	No
When directly used with the attributes	Yes	Yes	No	Yes
When we take the Intentional data	Yes	Yes	No	No
Order in structural parts	No	No	No	Yes

The above table derives as per the comparative study among MRDM, ILP, GM and SSDM

1. Attributes

Attribute are the field values of any relation (table).For example name field can be the attribute of a student relation from the above table we can know that all the technique support the attribute except the graph mining.

2. Numeric Values

Attributes with numeric values are supported by only the MRDM and ILP technique whereas GM and SSDM doesn't supports directly attribute with the numeric value.

3. Intentional Data

Intentional data are supported by both the technique MRDM and ILP, where as MRDM achieves it by means of view definitions and ILP treats it in a natural way.

4. Graph / Tree

Graphical representation is adopted by the entire structured data mining paradigms (MRDM, ILP, and GM) except the SSDM.

5.   Order in Structural Part

From the above example we can see that only SSDM can represent the parts in a structural way through tree except other three .for ex-A parent child relationship can be represented by tree properly rather graph.

6.   Structured Terms

We have studied a lot of papers related to MRDM and we are concluding that structured terms can be represented through both MRDM and ILP technique

Apart from the representational details we can make differentiate among the SDM approach techniques through the expressive power and the utilization of different language .This can be shown in the following tabular representation.

2.3.6   Expressive Power and Language Supported by SDM Approach

Table 3. For Expressive Power and Language supported by SDM approach

Names		Multi-Relational Data Mining (MRDM)	Inductive Logic Program(ILP)	Graph Mining (GM)	Semi Structured Data Mining (SSDM)
Language to be used	Database Language	RDBMS	Prolog (Representing for First Order Logic) Logic+Program	Creating Graphs	Extensive Markup Language (XNL)
	Bias Language	Object Oriented Language (UML,C++,Java)	Time to time different	-	DTD (Document Type Definition)
	Pattern Language	SQL(Structured Query Language)	Prolog (Representing for First Order Logic) Logic+Program	Creating Graphs	Creating Graphs
The concepts	Aggregate Functions	Yes	No	No	No
	Sub Graphs	No	No	Yes	No

In the above tabular representation we can get the result in the favor of MRDM that means MRDM is the best techniques among all the structured data mining approaches.

3   Decision Tree

This is the powerful and useful tools for classification and prediction. It is one type of classifier in the form of tree structure where each node is either a leaf nodes which indicates the value of target attribute. The decision node is the node where some decision to be taken on a single attribute-value It is used to classify an example from starting at the root of the tree to the decision tree can represents the decisions,.these decisions generates the rules for the classification of the data set by using the statistical methods like entropy, information gain, chi-square test, measurement error. There are different decision tree algorithms are available these are CART, CHAID, ID3, C4.5 and C5.0.By using the decision tree based analysis we can diagnosis the healthcare system.[24][25]. The decision tree will represent one form of a binary tree where a node represents some decision and edges emanating from a node represent the outcome of the decision. External nodes represent the ultimate decisions. In fact all the internal nodes represent the local decision if condition is true then decision 1 else decision 2 towards a global decision. Sometimes the decision tree classification is also not suitable when we deal with heterogeneous relational database. There are two important challenges are encountered these are as below [32] [33]. It is used to predict a pattern or to classify the class of a data. Suppose we have new unseen records of a person from the same location where the data sample was taken. The following data are called *test data* (in contrast to *training data)* because we would like to examine the classes of these data.

3.1   Motivational Example of Decision Tree

Table 4. For classification of data

Name of the person	Identification	How many cars have	Cost of travel	Status	Mode of the transportation
Nmp	Male	1	Statndard	High	?(unknown)
Gobind	Male	0	Cheap	Medium	?(Unknown)
Rasmi	Female	1	Cheap	High	?(Unknown)

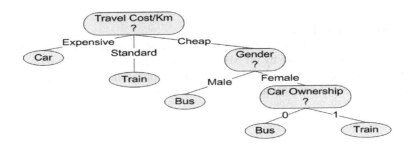

Fig. 1. In Section 3 for Decision Tree Construction

The question is what transportation mode would Nmp, Gobind and Rasmi use? Using the decision tree that we have generated.

We can start from the root node which contains an attribute of Travel cost per km. If the travel cost per km is expensive, the person uses car. If the travel cost per km is standard price, the person uses train. If the travel cost is cheap, the decision tree needs to ask next question about the gender of the person. If the person is a male, then he uses bus. If the gender is female, the decision tree needs to ask again on how many cars she own in her household. If she has no car, she uses bus, otherwise she uses train

3.2   Proposed Block Diagram of Multi Relational Data Mining

In this section we have proposed one methodology which will suitable for the researchers .It consists of four steps these are as below.

3.2.1 Information Gathering Stage

This is the stage where the researchers combined all the data which is to be required for next steps. This stage primarily used for the different types of inquires that can be carried our with the help of KDD process.

Information Gathering Stage
➤ Description of The Problem
➤ Input/ Output Requirement
Preprocessing Stage
➤ Data Selection
➤ Data Normalization
MRDM Techniques
1.   Algorithm Selection
➤   K-Mean and K-Medoid
➤   MRD-Tree
➤   Selecting Sub –Graph
2.   Tools Selection
Implementation Stage

Proposed Block Diagram

3.2.2 Preprocessing Stage

This is the proposed second stage of MRDM where the researcher Identifies the required data and then normalized them with different dimensions like (organizing, capturing, treating and preparing the data) for the KDD process.

3.2.3 MRDM Techniques

In this stage the researcher choose the MRDM based algorithms and implement those using suitable tools.

3.2.4 Implementation Stage

In this stage, the researcher can involves the simplification and presentation of knowledge models generated by the MRDM stage. In especially in this stage that the knowledge discovers specialist and the domain application specialist evaluate the results obtained and define the new alternative of data inquiry.

4   The Real Life Problem of MRDM

All most all data mining algorithms suffering some of the problems these are

4.1   Inability to Handle the Missing Attribute Values

In the case of gene localization task from KDD Cup 2001[36], 70% of CLASS, 50% of COMPLEX and 50% of MOTIF attribute values are missing. This problem can be solved effectively and efficiently by using the one of the multi relational data mining technique i.e. Multi relational decision tree learning construction (MRDTL). It is the approach of MRDM which handles the missing attribute values in the data set.

4.2   Slow Running Time

All most all the data mining algorithms stuck to handle huge amounts of data set like mutagenesis, thrombosis, KDD Cup, PKDD 2001 etc. MRDM provides the technique to reduce the slow running time and speeding up the running time

5   Justification of MRDM

Problem 1: Data stored in the relational database

Action
Construct one of the Multi-Relational data mining technique i.e. MRDTL (Multi Relational Decision Tree Construction for predicting target attribute in target table).

Problem 2
All structured data is available in the form of relational database and the representation of relational database based multi relational classification is also relational nature.

Action
Need not transform the structured data to any other form. The issue raised that how to directly use database operations to get TupleID propagation based classification and the scalability. To achieve some multi relational classification tasks than some of the tables may also be redundant .Reduction of redundancy among the table is also the major challenging task.

　　This challenging task can be solved by using the MRDM approaches like Inductive logic program (ILP), Selection graph based relational classification etc.

6   Heavy Association Rule Mining

We have studied the association rule mining[26] and proposed the new approach to improve the association rule mining in distributed database that is Heavy association rule mining[28] .In the age of information science the large organizations need useful knowledge or information. This knowledge need to be extracted from the multiple database of the organization over various branches. So a well known problem that restricts the use of association rule mining algorithm is generation of extremely large number of association rules from various sources of database of the organization. This large number of association rule makes the algorithm inefficient and increase the difficulty of the end users, to comprehend the generated rules. Thus, it is necessary to study data mining association rule in distributed databases. We will make the following contributions through my research work: Firstly, an extended model will be proposed for generating global patterns from local patterns in distributed databases. Secondly, the concept of heavy association rule in distributed databases needs to be analyzed along with the algorithm for generating such association rules in distributed database. Thirdly, the concept of exceptional association rule in distributed databases needs to be found out. Fourthly, an extension is made to the proposed algorithm to notify whether a heavy association rule is extremely-frequent or exceptional. We also make a comparative analysis of the existing algorithm.

7   Regression Model

A regression is a statistical analysis [34] assessing the association between two variables. It is used to find the relationship between two variables. It is basically used for forecasting purpose .we have studied different regression models like Linear Regression, Least Square, Simple Regression, logistic regression etc. we found linear regression is one of the best forecasting technique because time is used as the independent variable and headcount is the dependent variable as well as the different attributes are predicted . This model can use the least squares approach which means the overall solution (the line) minimize the sum of the distance between each observed point to the approximate line.

7.1   A Brief Comparison of Regression Algorithms

In this comparative analysis we found that both the model gives the good results but LMS(Least Median Square Regression) produce the better results but in terms of accuracy the linear regression was relatively equivalent to that of the lest median of squares algorithm .So the computational cost used by the LR is lower than that of LMS.

Table 5. For comparison of different regression algorithm

Used Algorithms	Model for Linear Regression	Model for Least Median Square Regression
The total time required to build the model	0.16s	10.84s
RAE(The Relative Absolute Error)	10.77%	10.01%
The Correlation Coefficient	0.9810	0.9803

8   Feature Direction for the Researcher

In this paper we have reviewed deeply Multi Relational Data Mining, Heavy Association Rule Mining and found these are the following challenges task in the 21st century.

    8.1   In the feature, how MRDM will handle the real world database with huge amount of data and intend to combine the fuzzy logic to predict the data.

    8.2   How the researcher can works efficiently to mine the data in a multi relational database which contains some information about the patients stroked down by the chronic hepatitis in MRDM domains.

    8.3   The concept of heavy association rule in distributed databases needs to be analyzed along with the algorithm for generating such association rules in distributed database.

    8.4   There will be the challenging task to identify exceptional association rule in the distributed database domains.

    8.5   The new model should be required to propose the identify the global patterns from the local patterns in the distributed database domains.

    8.6   From the web ,finding the useful links through the MRDM techniques is very difficult

9   Conclusion

Multi relational data mining is used to access the information across the different tables. In this paper we present the comparative study of Multi Relational Data Mining approaches as well as one of the statistical techniques. Multi relational data mining requires the more search space of the patterns which will be minimized more that creates the scaling up to large database. From the description we found that, the formulation of MRDM is in general and the data available in the form of trees, graphs.This paper presents the ideas about the multi relational data mining approach in terms of the real world scenario. The approaches which we have presented like (Decision tree induction, ILP, Heavy association rule mining) that can be overcome the problem of the realworld. In this study we found that MRDM not only hands the structured and propositional data mining but also represent the sequence of patterns with help of inductive logic program(ILP). We also found during this study that more accuracy, we should to avoids the multiple joining operations because there will be more chances for loss of data. MRDM never shows the semantic loses but in other case like (traditional algorithms) that can provide the errors during the joining operations in multi relational database. With the advent of liner regression, MRDM dominant other approaches because of accuracy and prediction. MRDM gives more flavor as comparisons to other approaches. We have studied different papers regarding statistical tetchiness and found Leaner regression is suitable for. Some of the feature research works presents in the feature direction section, which may begins of new era of MRDM in this 21st century.

References

1. Kantrardzic, M.: Data Mining: Concepts Models, and Algorithms. Wiley, New Jersyey (2003)
2. Garcia-Molina, H., Ulman, J.D.: Database Systems. The complete bopectok prentice Halll (2002)
3. Fayyad, U., Piatesky-Shapiaro, G., Padharic, S., et al.: From data mining to Knowledge Discovery, pp. 1–34. An overview The MIIT,Press (1996)
4. Hand, D., Mannila, H., Smyth, P.: Principles of Data Mining. MIT Press, Cambridge (1996)
5. Valencio, F.T., et al.: Human-Centric Computing and Information Science (2012), http://www.hcis-journals.com
6. Dzeroski, S., Raedt, L.D., Wrobel, S.: Multi-Relational data mining workshop report ACM SIGKDD, Exploration News letter 5(2), 200–202 (2003), doi:doi:1011415/980972981007
7. Domingos, P.: Prospect and challenges for Multi-Relational data mining. ACM SIGKDD Exploration News Letter 5(1) (2003)
8. Cook, D., Holder, L.: Graph-Based Data Mining. Intelligent Systems & their Applications 15(2), 32–41 (2000)
9. Inokuchi, A., Washio, T., Motoda, H.: An apriori-based algorithm for mining frequent substructures from graph data. In: Zighed, D.A., Komorowski, J., Żytkow, J.M. (eds.) PKDD 2000. LNCS (LNAI), vol. 1910, pp. 13–23. Springer, Heidelberg (2000)
10. Matsuda, T., Horiuchi, T., Motoda, H., Washio, T., et al.: Graph-bases induction for general graph structured data. In: Proceedings of DS 1999, pp. 340–342 (1999, 2000)
11. Kavuruchu, Y., Senkul, P., Totoslu, I.H.: A comparative study of ILP-based concept discovery system. Expert System with Applications 38, 11598–11607 (2011)
12. Mutlu, A., Senkul, P., Kavuruchu, Y.: Improving the scalability of ILP-based multi-relational concept discovery system through parallelization. Knowledge Based System 27, 352–368 (2012), http://www.elsevier.com/locate/knosys
13. Blockeel, H., De Raedt, L., Jacobs, N., Demoen, B.: Scaling up inductive logic programming by learning from interpretations. Data Mining and Knowledge Discovery 3(1), 59–93 (1999)
14. Dżeroski, S.: Inductive Logic Programming and Knowledge Discovery in Databases. AAAI Press (1996)
15. Asai, T., Abe, K., Kawasoe, S., Arimura, H., Sakamoto, H., Arikawa, S.: Efficient substructure discovery from large semi-structured data. In: Proceedings of DM 2002 (2002)
16. Miyahara, T., Shoudai, T., Uchida, T., Takahashi, K., Ueda, H.: Discovery of frequent tree structured patterns in semi structured web documents. In: Proceedings PAKDD 2001, pp. 47–52 (2001)
17. Dżeroski, S., Lavrač, N.: Relational Data Mining. Springer (2001)
18. Knobbe, A.J., de Haas, M., Siebes, A.: Propositionalisation and aggregates. In: Siebes, A., De Raedt, L. (eds.) PKDD 2001. LNCS (LNAI), vol. 2168, p. 277. Springer, Heidelberg (2001)
19. Knobbe, A.J., Siebes, A., Van Der Wallen, D.: Multi-relational decision tree induction. In: Żytkow, J.M., Rauch, J. (eds.) PKDD 1999. LNCS (LNAI), vol. 1704, pp. 378–383. Springer, Heidelberg (1999)
20. Leiva, H., Atramentov, A., Honavar, V.: Experiments with MRDTL – A Multi-relational Decision Tree Learning Algorithm. In: Proceedings of Workshop MRDM 2002 (2002)
21. Washio, T., Motoda, H.: State of the Art of Graph based Data Mining

22. Knobbe, A.J., Siebes, A., Van Der Wallen, D.: Multi-relational decision tree induction. In: Żytkow, J.M., Rauch, J. (eds.) PKDD 1999. LNCS (LNAI), vol. 1704, pp. 378–383. Springer, Heidelberg (1999)
23. Bickel, S., Scheffer, T.: Proceedings of the IEEE Conference on Data Mining (2004)
24. Cao, S.-Z.: 2012 International Conference on Page(s) of IEEE Xplore Systems and Informatics (ICSAI), Hangzhou, China, pp. 2251–2254. Healthcare Inf. Eng. Res. Center, Zhejiang Univ (2012)
25. Draper, W.R., Hawley, C.E., McMahon, B.T., Reid, C.A.: Journal of Occupational Rehabilitation published by Springer Link -(2013)
26. Grahne, G., Zhu, J.F.: Fast algorithms for frequent itemset mining using FPtrees. IEEE Transactions on Knowledge and Data Engineering 17(10), 1347–1362 (2005)
27. Chung, D.W.-L., Han, J., Ng, V.T.Y., Fu, A.W.-C., Fu, Y. (1996)
28. Yi, X., Zhang, Y.: Privacy-preserving distributed association rule mining via semi-trusted mixer. Data & Knowledge Engineering 63, 550–567 (2007)
29. Girish, K., Palshikar, A., Mandar, S., Kale, B., Apte, M.M.: Association rules mining using heavy itemsets. Elsevier (2006)
30. Liu, X., Zhai, K., Pedrycz, W.: An improved association rules mining method (2011)
31. Guo, O., Viktor, H.L.: Mining Relational Databases with Multi-view Learning
32. Yin, X., Han, J., Yang, J.: Efficient Classification from Multiple Database Relations: A cross Mine Approach. IEEE Transaction on Knowledge and Data Engineering 18, 770–783 (2006)
33. Zhang, S., Wu, X., Zhang, C.: Multi-database mining. IEEE Computational Intelligence Bulletin 2(1), 5–13 (2003)
34. De Marco, J.: Excel's data import tools. Pro Excel 2007 VBA, p. 43. Apress (2008) ISBN 1590599578
35. Domingos, P.: Prospects and Challenges for Multirelational Data Mining, SIGKDD Exploration, vol. 4(2)
36. Zhang, W.: Multi-Relational Data Mining Based on Higher-Order Inductive Logic Programming. IEEE Computer Society Global Congress on Intelligent Systems (2009)
37. Padhy, N., Panigrahi, R.: Multirelationa data mining approach: a data mining techniques. International Journal of Computer Applications 57(17), 975–8887 (2012)
38. Aggarwal, C.C., Wang, H. (eds.): Managing and Mining Graph Data, Advances in Database Systems. Springer Science Business Media, LLC 2010 (2010)
39. Liu, M., Guo, H.-F., Chen, Z.: On Multi-Relational Data Mining for Foundation of Data Mining, May 13-16, pp. 389–395. Nebraska Univ, Omaha (2007)
40. Spyropoulou, E., De Bie, T.: Interesting pattern mining in multi-relational data, Birlinghoven, 53754, Sankt Augustin, Germany. Springer (2014)

A Survey of Dynamic Program Analysis Techniques and Tools

Anjana Gosain and Ganga Sharma

University School of Information and Communication Technology,
Guru Gobind Singh Indraprastha University, New Delhi-110078, India
anjana_gosain@hotmail.com, ganga.negi@gmail.com

Abstract. Dynamic program analysis is a very popular technique for analysis of computer programs. It analyses the properties of a program while it is executing. Dynamic analysis has been found to be more precise than static analysis in handling run-time features like dynamic binding, polymorphism, threads etc. Therefore much emphasis is now being given on dynamic analysis of programs (instead of static analysis) involving the above mentioned features. Various techniques have been devised over the past several years for the dynamic analysis of programs. This paper provides an overview of the existing techniques and tools for the dynamic analysis of programs. Further, the paper compares these techniques for their merits and demerits and emphasizes the importance of each technique.

Keywords: dynamic analysis, static analysis, instrumentation, profiling, AOP.

1 Introduction

Analysing the dynamic behaviour of a software is invaluable for software developers because it helps in understanding the software well. Static analysis has long been used for analysing the dynamic behavior of programs because it is simple and does not require running the program [6], [24]. Dynamic analysis, on the other hand, is the analysis of the properties of a running program [1]. It involves the investigation of the properties of a program using information gathered at run-time. Deployment of software now-a-days as a collection of dynamically linked libraries is rendering static analysis imprecise [32]. Moreover, the widespread use of object oriented languages, especially Java, to write software has lead to the usage of run-time features like dynamic binding, polymorphism, threads etc. Static analysis is found to be ineffective in these kinds of dynamic environments. Whereas static analysis is restricted in analyzing a program effectively and efficiently, and may have trouble in discovering all dependencies present in the program, dynamic analysis has the benefit of examining the concrete domain of program execution [1]. Therefore dynamic analysis is gaining much importance for the analysis of programs.

Dynamic and static analysis are regarded as complementary approaches and have been compared for their merits and demerits[17], [18]. Some of the major differences of dynamic analysis with static analysis are listed in Table 1. The main advantage of

© Springer International Publishing Switzerland 2015
S.C. Satapathy et al. (eds.), *Proc. of the 3rd Int. Conf. on Front. of Intell. Comput. (FICTA) 2014*
– *Vol. 1*, Advances in Intelligent Systems and Computing 327, DOI: 10.1007/978-3-319-11933-5_13

dynamic analysis over static analysis is that it can examine the actual, exact run-time behaviour of the program and is very precise. On the contrary, the main disadvantage of dynamic analysis is that it depends on input stimuli and therefore cannot be generalized for all executions. Nevertheless, dynamic analysis techniques are proving useful for the analysis of programs and are being widely used. Efforts are also being made to combine dynamic and static analysis to get benefit from the best features of both. For example, static and dynamic analysis capability has been provided in a framework called CHORD [33], a dynamic analysis tool is used as an annotation assistant for a static tool in [34] etc.

Table 1. Comparison of Dynamic analysis with Static Analysis

Dynamic Analysis	Static Analysis
Requires program to be executed	Does not require program to be executed
More precise	Less precise
Holds for a particular execution	Holds for all the executions
Best suited to handle run-time programming language features like polymorphism, dynamic binding, threads etc.	Lacks in handling run-time programming language features.
Incurs large run-time overheads	Incurs less overheads

The remainder of this paper is structured as follows. Firstly, the main techniques of dynamic analysis are described followed by a comparison of each technique. Then, a description of the most widely used dynamic analysis tools is given. Finally, in the last section, we describe the main conclusions and the future work.

2   Dynamic Analysis Techniques

Dynamic analysis techniques reason over the run-time behavior of systems[17].In general, dynamic analysis involves recording of a program's dynamic state. This dynamic state is also called as profile/trace. A program profile measures occurrences of events during program execution[2]. The measured event is the execution of a local portion of program like lines of code, basic blocks, control edges, routines etc. Generally, a dynamic analysis technique involves the following phases: 1) program instrumentation and profile/trace generation, 2) analysis or monitoring [27]. Program instrumentation is the process of inserting additional statements into the program for the purpose of generating traces. These instrumented statements are executed at the same time when the program is running. The fundamental challenge for success of dynamic analysis lies in the creation of instrumentation and profiling infrastructures that enable the efficient collection of run-time information [32]. Depending upon the instrumentation provided, required information from one or more executions would be gathered for analysis purpose. The actual analysis or monitoring phase takes place on these traces. This phase is additionally augmented to handle violations of properties. The analysis or monitoring can be performed either offline or online.

An analysis is said to be online if the target system and the monitoring system are run in parallel. But it may increase the cost. Choosing between the two, one should consider whether the purpose of the analysis is to find error or to find and correct errors. In the first case, offline analysis should be performed; while the latter should go for the online analysis.

In the next sub-sections, we will study the main techniques for dynamic analysis of programs. We have omitted the dynamic analysis of a system based on models (like UML) because we are providing only those techniques which work on source code itself or some form of source code like binaries or bytecode. The section ends with a comparison of these techniques in Table 2.

2.1   Instrumentation Based

In this technique, a code instrumenter is used as a pre-processor to insert instrumentation code into the target program[27]. This instrumentation code can be added at any stage of compilation process. Basically it is done at three stages: source code, binary code and bytecode. Source code instrumentation adds instrumentation code before the program is compiled using source-to-source transformation. It can use some meta-programming frameworks like Proteus[47], DMS[3] etc. to insert the extra code automatically. These meta-programming frameworks provide a programming language that can be used to define context sensitive modifications to the source code. Transformation programs are compiled into applications that perform rewriting and instrumentation of source, which is given as input. Applying instrumentation to source code makes it easier to align trace functionality with higher-level, domain-specific abstractions, which minimizes instrumentation because the placement of additional code is limited to only what is necessary. The disadvantage is that it is target language dependent and can also become problematic when dealing with specific language characteristics, such as C/C++ preprocessing and syntactic variations.

Binary instrumentation adds instrumentation code by modifying or re-writing compiled code. This is done using specially designed tools either statically or dynamically. Static binary instrumentation involves the use of a set of libraries and APIs that enable users to quickly write applications that perform binary re-writing. Examples of such tools are EEL[28], ATOM[9] etc. Dynamic binary instrumentation (implemented as Just-In-Time compilers) is performed after the program has been loaded into memory and just prior to execution using tools like MDL[22], DynInst[43] etc. MDL is a specialized language that has two key roles. First, it specifies the code that will be inserted into the application program to calculate the value of performance metrics. This code includes simple control and data operations, plus the ability to instantiate and control real and virtual timers. Second, it specifies how the instrumentation code is inserted into the application program. This specification includes the points in the application program that are used to place the instrumentation code. DynInst[43] is an API that permits the insertion of code into a running program. Using this API, a program can attach to a running program, create a new bit of code and insert it into the program. The program being modified is able to continue execution and doesn't need to be re-compiled, re-linked, or even re-started[43]. Dynamic binary instrumentation has the advantage over its static counterpart in that profiling functionality can be selectively added or removed from the program without the need to recompile.

Bytecode instrumentation performs tracing within the compiled code. Again, it can be static or dynamic. Static instrumentation involves changing the compiled code offline before execution i.e., creating a copy of the instrumented intermediate code using high level bytecode engineering libraries like BCEL[11], ASM[8], Javaassist[10] etc. BCEL and ASM allow programmers to analyze, create, and manipulate Java class files by means of a low-level API. Javassist[10], on the other hand, enables structural reflection and can be used both at load-time or compile-time to transform Java classes. The disadvantage here is that dynamically generated or loaded code cannot be instrumented. Dynamic instrumentation, on the other hand, works when the application is already running. There exist various tools for dynamic bytecode instrumentation like BIT[29], IBM's Jikes Bytecode Toolkit[23] etc. There also exist tools which can provide both static and dynamic bytecode instrumentation. For example, FERRARI[4] instruments the core classes in a Java program statically, and then uses an instrumentation agent to dynamically instrument all other classes. This has the advantage that no class is left without instrumentation. Nevertheless, bytecode instrumentation is harder to implement, but gives unlimited freedom to record any event in the application.

2.2   VM Profiling Based

In this technique, dynamic analysis is carried out using the profiling and debugging mechanism provided by the particular virtual machine. Examples include Microsoft CLR Profiler[20] for .NET frameworks and JPDA for Java SDK. These profilers give an insight into the inner operations of a program, specifically related to memory and heap usage. One uses plug-ins (implemented as dynamic link libraries) to the VM to capture the profiling information. These plug-ins are called as profiling agents and are implemented in native code. These plug-ins access the profiling services of the virtual machine through an interface. For example, JVMTI[44] is the interface provided by JPDA. It is straightforward to develop profilers based on VM because profiler developers need only implement an interface provided by the VM and need not worry about the complications that can arise due to interfering with the running application. Benchmarks like SpecJVM [13] etc. are then used for actual run-time analysis. A benchmark acts like a black-box test for a program even if its source code is available to us. The process of benchmarking involves executing or simulating the behavior of the program while collecting data reflecting its performance. This technique has the advantage that it is simple and easier to master. One of the major drawbacks of this technique is that it incurs high run-time overheads[4], [39].

2.3   Aspect Oriented Programming

Aspect-oriented programming (AOP)[26] is a way of modularizing crosscutting concerns much like object-oriented programming is a way of modularizing common concerns. With AOP, there is no need to add instrumentation code as the instrumentation facility is provided within the programming language by the built-in constructs. AOP adds the following constructs to a program : aspects, join-point, point-cuts and advices. Aspects are like classes in C++ or Java. A join-point is any well defined point in a program flow. Point-cuts pick join-points and values at those points. An advice is a piece of code which is executed when a join-point is reached. Aspects specify point-cuts to intercept join-points. An aspect weaver is then used to modify the code of an application to execute advice at intercepted join-points [21]. The AOP paradigm makes it easier for developers to insert profiling to an existing application by defining a profiler aspect consisting of point-cuts and advice. Most popular languages like C++ and Java have their aspect oriented extensions namely AspectC++ [42] and AspectJ [25] respectively. There also exist frameworks that use AOP to support static, load-time, and dynamic (runtime) instrumentation of bytecode [7].

Some problems encountered by AOP approaches are the design and deployment overhead of using the framework[14]. AOP frameworks are generally extensive and contain a variety of configuration and deployment options, which may take time to master. Moreover, developers must also master another framework on top of the actual application, which may make it hard to use profiling extensively. Another potential drawback is that profiling can only occur at the join-points provided by the framework, which is often restricted to the methods of each class, i.e., before a method is called or after a method returns.

Table 2. Comparison of Dynamic Analysis Techniques

	Dynamic Analysis Technique			
	Instrumentation Based		VM Profiling Based	AOP Based
	Static	Dynamic		
Level of Abstraction	Instruction/Bytecode	Instruction/bytecode	Bytecode	Programming Language
Overhead	Runtime	Runtime	Runtime	Design and deployment
Implementation Complexity	Comparatively Low	High	High	Low
User Expertise	Low	High	Low	High
Re-compilation	Required	Not Required	Not Required	Required

Application-specific events occurring within a method call therefore cannot be profiled, which means that non-deterministic events cannot be captured by AOP profilers[21]. Still, AOP is getting popular to build dynamic analysis tools as it can be used to raise the abstraction level of code instrumentation and incurs less runtime overhead [18].

3   Dynamic Analysis Tools

Dynamic analysis tools have been widely used for   memory analysis [35], [38], [36], [40], [30], [37], [12], invariant detection [16], [19], deadlock and race detection[35], [40], [6] and metric computation[15], [41]. These tools are being used by companies for their benefits. For example, Pin[40] is a tool which provides the underlying infrastructure for commercial products like Intel Parallel Studio suite[49] of performance analysis tools. A summary of dynamic analysis tools is provided in Table 3.

Table 3. Dynamic Analysis Tools

Technique	Tool	Language	Cache Modelling	Heap Allocation	Buffer Overflow	Memory Leak	Deadlock Detection	Race Detection	Object LifeTime Analysis	Metric Computation	Invariant Detection
Instrumentation Based	Daikon	C, C++									✓
	Valgrind	C, C++				✓		✓			
	Rational Purify	C, C++, Java				✓					
	Parasoft Insure++	C, C++		✓		✓					
	Pin	C	✓								
	Javana	Java	✓						✓		
	DIDUCE	Java									✓
AOP Based	DJProf	Java		✓					✓		
	Racer	Java						✓			
VM Profiling Based	Caffeine	Java							✓		
	DynaMetrics	Java								✓	
	*J	Java								✓	
	JInsight	Java				✓	✓		✓		

Valgrind[35] is an instrumentation framework for building dynamic analysis tools. It can automatically detect many memory management and threading bugs, and profile a program in detail. Purify[38] and Insure++[36] have similar functionality as Valgrind. Whereas Valgrind and Purify instrument at the executables, Insure++ directly instruments the source code. Pin[40] is a tool for dynamic binary instrumentation of programs. Pin adds code dynamically while the executable is running. Pin provides an API to write customized instrumentation code (in C/C++), called Pintools. Pin can be used to observe low level events like memory references, instruction execution, and control flow as well as higher level abstractions such as procedure invocations, shared library loading, thread creation, and system call execution.

Javana [30] runs a dynamic binary instrumentation tool underneath the virtual machine. The virtual machine communicates with the instrumentation layer through an event handling mechanism for building a vertical map that links low-level native instruction pointers and memory addresses to high-level language concepts such as objects, methods, threads, lines of code, etc. The dynamic binary instrumentation tool then intercepts all memory accesses and instructions executed and provides the Javana end user with high-level language information for all memory accesses and natively executed instructions[30].

Daikon[16] and DIDUCE[19] are two most popular tools for invariant detection. The former is an offline tool while the latter is an online tool. The major difference between the two is that while Daikon generates all the invariants and then prunes them depending on a property; DIDUCE dynamically hypothesizes invariants at each program point and only presents those invariants which have been found to satisfy a property. Another major difference is that Daikon collects tracing information by modifying the program abstract syntax tree, while DIDUCE uses BCEL to instrument the class JAR files.

*J [15] and DynaMetrics[41] are tools for computing dynamic metrics for Java. While *J relies on JVMPI (predecessor of JVMTI but now rarely used) interface for metrics computation, DynaMetrics uses JVMTI interface. Another major difference between the two is that *J computes dynamic metrics specifically defined by its authors for Java whereas DynaMetrics computes major dynamic metrics from various dynamic metrics suites available in literature. JInsight [12] is used for exploring runtime behaviour of Java programs visually. It offers capabilities for managing the information overload typical of performance analysis. Through a combination of visualization, pattern extraction, interactive navigation, database techniques, and task-oriented tracing, vast amounts of execution information can be analyzed intensively. Caffeine [31] is a tool that helps a maintainer to check conjectures about Java programs, and to understand the correspondence between the static code of the programs and their behavior at runtime. Caffeine uses JPDA and generates and analyzes on-the-fly the trace of a Java program execution, according to a query written in Prolog. DJProf[37] is a profiler based on AOP which is used for the analysis of heap usage and object life-time analysis. Racer[6] is a data race detector tool for concurrent programs employing AOP. It has specific point-cuts for lock acquisition and lock release. These point-cuts allow programmers to monitor program events where locks are granted or handed back, and where values are accessed that may be shared amongst multiple Java threads.

4   Conclusion

Dynamic analysis has acquired a great importance in recent years because of its ability to determine run-time behavior of programs precisely. This paper provided the details of the techniques and tools of   dynamic analysis. An attempt is made to highlight the strength and weaknesses of each technique. Aspect oriented techniques have got an edge over other techniques and are being emphasized for dynamic analysis of programs.

References

1. Ball, T.: The concept of dynamic analysis. In: Wang, J., Lemoine, M. (eds.) ESEC 1999 and ESEC-FSE 1999. LNCS, vol. 1687, p. 216. Springer, Heidelberg (1999)
2. Ball, T., Larus, J.R.: Efficient path profiling. In: Proceedings of MICRO 1996, pp. 46–57 (1996)
3. Baxter, I.: DMS: Program Transformations for Practical Scalable Software Evolution. In: Proceedings of the 26th International Conference on Software Engineering, pp. 625–634 (2004)
4. Binder, W., Hulaas, J., Moret, P., Villazón, A.: Platform-independent profiling in a virtual execution environment. Software: Practice and Experience 39(1), 47–79 (2009)
5. Binkley, D.: Source Code Analysis: A Road Map. Future of Software Engineering (2007)
6. Bodden, E., Havelund, K.: Aspect-oriented Race Detection in Java. IEEE Transactions on Software Engineering 36(4), 509–527 (2010)
7. Boner, J.: AspectWerkz - Dynamic AOP for Java. In: Proceeding of the 3rd International Conference on Aspect- Oriented Development (AOSD 2004), Lancaster, UK (2004)
8. Bruneton, E., Lenglet, R., Coupaye, T.: ASM: A code manipulation tool to implement adaptable systems. In: Adaptable and Extensible Component Systems, Grenoble, France (2002)
9. Buck, B., Hollingsworth, J.K.: An API for Runtime Code Patching. International Journal of High Performance Computing Applications, 317–329 (2000)
10. Chiba, S.: Load-time structural reflection in java. In: Bertino, E. (ed.) ECOOP 2000. LNCS, vol. 1850, p. 313. Springer, Heidelberg (2000)
11. Dahm, M.: Byte code engineering. In: Java-Information-Tage(JIT 1999) (1999), http://jakarta.apache.org/bcel/
12. Zheng, C.-H., Jensen, E., Mitchell, N., Ng, T.-Y., Yang, J.: Visualizing the Execution of Java Programs. In: Diehl, S. (ed.) Dagstuhl Seminar 2001. LNCS, vol. 2269, pp. 151–162. Springer, Heidelberg (2002)
13. Dieckmann, S., Hölzle, U.: A study of the allocation behavior of the SPECjvm98 Java benchmarks. In: Guerraoui, R. (ed.) ECOOP 1999. LNCS, vol. 1628, pp. 92–115. Springer, Heidelberg (1999)
14. Dufour, B., Goard, C., Hendren, L., de Moor, O., Sittampalam, G., Verbrugge, C.: Measuring the dynamic behaviour of AspectJ programs. In: Proceedings of the ACM Conference on Object-Oriented Programming, Systems, Languages and Applications, pp. 150–169. ACM Press (2004)

15. Dufour, B., Hendren, L., Verbrugge, C.: *J: A tool for dynamic analysis of Java programs. In: Proc 18th Annual ACM SIGPLAN Conference on Object-Oriented Programming, Systems, Languages, and Applications, pp. 306–307. ACM Press (2003)

16. Ernst, M.D.: Dynamically Discovering Likely Program Invariants. (PhD Dissertation), University of Washington, Dept. of Comp. Sci. & Eng., Seattle, Washington (2000)

17. Ernst, M.D.: Static and Dynamic Analysis: Synergy and Duality. In: Proceedings of the 5th ACM SIGPLAN-SIGSOFT Workshop on Program Analysis for Software Tools and Engineering (2004)

18. Gupta, V., Chhabra, J.K.: Measurement of Dynamic Metrics using Dynamic Analysis of Programs. In: Applied Computing Conference, Turkey (2008)

19. Hangal, S., Lam, M.S.: Tracking Down Software Bugs using Automatic Anomaly Detection. In: ICSE 2002 (2002)

20. Hilyard, J.: No Code Can Hide from the Profiling API in the.NET Framework 2.0. MSDN Magazine   (January   2005), http://msdn.microsoft.com/msdnmag/issues/05/01/CLRProfiler/

21. Hilsdale, E., Hugunin, J.: Advice weaving in AspectJ. In: Lieberherr, K. (ed.) Aspect-oriented Software Development (AOSD 2004). ACM Press (2004)

22. Hollingsworth, J.K.: MDL: A language and compiler for dynamic instrumentation. In: Proceedings of International conference on Parallel architecture and compilation techniques (1997)

23. IBM Corporation. Jikes Bytecode Toolkit (2000), http://www-128.ibm.com/developerworks/opensource/

24. Jackson, D., Rinard, M.: Software Analysis: A Road Map. IEEE Transaction on Software Engineering (2000)

25. Kiczales, G., Hilsdale, E., Hugunin, J., Kersten, M., Palm, J., Griswold, W.G.: An Overview of AspectJ. In: Lindskov Knudsen, J. (ed.) ECOOP 2001. LNCS, vol. 2072, pp. 327–355. Springer, Heidelberg (2001)

26. Kiczales, G., et al.: Aspect-oriented programming. In: Proc of the 11th European Conference on Object-Oriented Programming, Finland. LNCS, vol. 1241, pp. 220–242. Springer (1997)

27. Larus, J.R., Ball, T.: Rewriting executable to measure program behavior. Software:Practice and Experience 24(2), 197–218 (1994)

28. Larus, J.R., Schnarr, E.: EEL: Machine independent executable editing. In: PLDI 1995 (1995)

29. Lee, H.B., Zorn, B.G.: BIT: A tool for instrumenting Java bytecode. In: Proceedings of USENIX Symposium on Internet Technologies and Systems, pp. 73–82 (1997)

30. Maebe, J., Buytaert, D., Eeckhout, L., De Bosschere, K.: Javana: A System for Building Customized Java Program Analysis Tools. In: OOPSLA 2006 (2006)

31. Mines de Nantes, E.: No Java without Caffeine. In: ASE 2002 (2002)

32. Mock, M.: Dynamic Analysis from the Bottom Up. In: 25th ICSE Workshop on Dynamic Analysis (2003)

33. Naik, M.C.: A static and dynamic program analysis framework for Java (2010), http://chord.stanford.edu/

34. Nimmer, J.W., Ernst, M.D.: Automatic generation and checking of program specifications. Technical Report 823, MIT Lab for Computer Science, USA (2001)

35. Nethercote, N., Seward, J.: Valgrind: A framework for heavyweight dynamic binary instrumentation. In: Proc. of ACM SIGPLAN Conference on Programming Language Design and Implementation, PLDI (2007)

36. Parasoft Inc. Automating C/C++ Runtime Error Detection With Parasoft Insure++. White paper (2006)
37. Pearce, D.J., Webster, M., Berry, R., Kelly, P.H.J.: Profiling with AspectJ. Software: Practice and Experience (2007)
38. Rastings, R., Joyce, B.: Purify: Fast Detection of Memory Leaks and Access Errors. Winter Usenix Conference (1992)
39. Reiss, S.P.: Efficient monitoring and display of thread state in java. In: Proceedings of IEEE International Workshop on Program Comprehension, St. Louis, MO, pp. 247–256 (2005)
40. Skatelsky, A., et al.: Dynamic Analysis of Microsoft Windows Applications. In: International Symposium on Performance Analysis of Software and System (2010)
41. Singh, P.: Design and validation of dynamic metrics for object-oriented software systems. (PhD Thesis), Guru Nanak Dev University, Amritsar, India (2009)
42. Spinczyk, O., Lohmann, D., Urban, M.: Aspect C++: an AOP Extension for C++. Software Developer's Journal, 68–76 (2005)
43. Srivastava, A., Eustace, A.: ATOM: A system for building customized program analysis tools. SIGPLAN Not 39(4), 528–539 (2004)
44. Sun Microsystems Inc. JVM Tool Interface, JVMTI (2004),
    http://java.sun.com/javase/6/docs/technotes/guides/jvmti/index.html
45. Sun Microsystems Inc. Java Virtual Machine Profiler Interface, JVMPI (2004),
    http://java.sun.com/j2se/1.4.2/docs/guide/jvmpi/
46. Sun Microsystem Inc. Java Platform debug Architecture (2004),
    http://java.sun.com/javase/6/docs/technotes/guides/jpda/
47. Waddington, D.G., Yao, B.: High Fidelity C++ Code Transformation. In: Proceedings of the 5th Workshop on Language Descriptions, Tools and Applications, Edinburgh, Scotland, UK (2005)
48. Waddington, D.G.: Dynamic Analysis and Profiling of Multi-threaded Systems. In: Taiko, P.F. (ed.) Designing Software-Intensive Systems: Methods and Principles. Information Science Reference Publishing (2008) ISBN 978-1-59904-699-0
49. https://software.intel.com/en-us/intel-parallel-studio-xe

DC (Drought Classifier): Forecasting and Classification of Drought Using Association Rules

B. Kavitha Rani[1] and A. Govardhan[2]

[1] Jyothishmathi Institute of Technology & Science,
Karimnagar, Andhra Pradesh, India
kavi_gdk1978@yahoo.co.in
[2] Department of CSE & Director, School of IT
JNTUH University, Hyderabad
govardhan_cse@yahoo.co.in

Abstract. Normally droughts are viewed as the nature disasters which show heavy economic impact in the affected regions. Indemnifying the information about the pattern, area, severity and timing of droughts effect, can be used for operational planning and decision making. In this work, combination of Artificial Neural Network (ANN) coupled with Fuzzy C-means and association rule mining are used to develop a model to identify the severity of drought by forecasting the climate conditions for upcoming season. A suitable Feed Forward Neural network (FFNN) is developed with forward selection to forecast the rainfalls for future years with the input dataset of several archived data. Later fuzzy c-means (FCM) clustering is used for partitioning the forecasted data in three groups like low, medium and high rainfall. Finally association rules are used to find associations among data belonging to the climate information using proposed rule based model. The low rain data group generated by FCM is used for classifying the drought effect from the predicted results.

Keywords: Drought, Prediction, Cluster, FCM, FFNN.

1 Introduction

Droughts are vicious cycle of impacts which show effect on agriculture unemployment and badly hit the economic conditions of the effected regions. Monsoons failure show disastrous consequences on the agriculture sector whereas most of the population in India is purely dependent on agriculture. This study focuses on Andhra Pradesh (AP) state [5], which was seriously affected by drought in many past seasons. Figure 1 show the rainfall levels and most drought affected regions in Andhra Pradesh state. The findings of this work mostly focus on identifying the regions which may possibly be affected by drought and water-scarce for the upcoming seasons for initiating suitable actions at early stages.

Since the issue is much focused on forecasting the climate for upcoming season, data mining techniques [10] such as classification and prediction are used for identifying the relationships in the data, to generate new knowledge.

© Springer International Publishing Switzerland 2015   123
S.C. Satapathy et al. (eds.), *Proc. of the 3rd Int. Conf. on Front. of Intell. Comput. (FICTA) 2014*
– *Vol. 1*, Advances in Intelligent Systems and Computing 327, DOI: 10.1007/978-3-319-11933-5_14

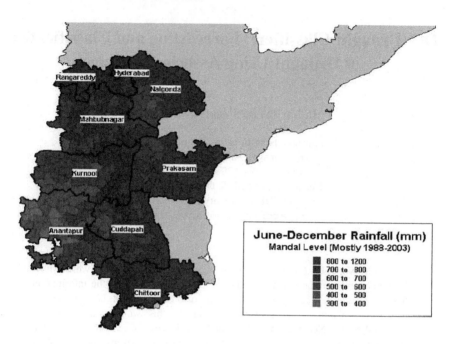

Fig. 1. Rainfall and drought prone regions in Andhra Pradesh (Source: Indian Agriculture Ministry)

This work is proposed to develop the prediction model that can forecast the climate and identify the regions of less rain fall to locate the probable drought prone areas under various conditions of changing climate. The steps used for developing this model are describes as:

- Analyzing and understanding the important patterns in past data to develop a suitable model for forecasting the rainfall for future seasons.
- Identifying the low, medium and high rainfall regions and further identifying the probable drought regions by using low rain fall data module.

2   Data Mining Techniques Used for Drought Classification

Since the huge collection of past data is required to analyze for understanding the relationships among different climate parameters for future rainfall prediction. Suitable management and quick processing of meteorological data becomes necessary in drought research. Data mining techniques [1] [2] facilitate handling huge data, analyzing the complex results belong to climate information by extracting patterns of drought. In our work, a suitable ANN model is developed for forecasting rainfalls for 5 years from the dataset of previous years. Later, Fuzzy C-means groups the sixty months forecasted rainfalls into three groups using a tri-angular fuzzy membership function which depicts low, medium and high rainfalls. Finally, association rule mining is used to develop rules for arriving at the decision of drought prediction.

2.1   FFNN Model for Predicting Monthly Rainfall in Coastal Andhra

This section discuss about the proposed Feed forward neural networks model for predicting monthly rainfall in Coastal Andhra region in AP State. Many researchers worked on prediction of rainfall using FFNN [6] and generated excellent prediction results. This work focuses on developing a suitable minimal model of FFNN with forward selection approach. The classical back propagation learning algorithm is implemented for arriving at a suitable model. The Root Mean Square Error (RMSE) determines the effectiveness of the model developed.

2.1.1   Dataset Description

The data of monthly rainfall (mm) in Coastal Andhra is collected from Indian Institute of Tropical Meteorology (IITM), Pune, India. The data set consists of 1692 monthly observations during years 1871 to 2011. In this study, to rescale the variables adjusted normalized technique is used. These adjusted values fall in range from -1 to +1. The used data is split into three different samples such as training sample, testing sample and hold out sample.

2.1.2 Structure of the FFNN

The model proposed uses three layer feed forward neural network [7] [8] with input layer, one hidden and one output layer. To develop a very simple FFNN a single hidden layer is chosen. The number of input neurons needed by model is two, each representing the values of lag_{12} (monthly rainfall of previous year) and Month (where month takes the values 1 to 12 for January to December respectively). The training dataset is grouped year and month wise. The input data set is a matrix with two columns and rows equal to the size of the training dataset. The predicted rainfall of a month is a function of the corresponding month of previous years available in the training dataset. For example predicted rainfall of 2011 Jan is the function of all previous years Jan rainfalls. The model uses only one output unit which indicates the forecast of monthly rainfall. Number of neurons in hidden layer is initially started with one neuron and based on the RMSE of the model the number of neurons are grown. After several simulations we find that the model provided fairly good results with seven neurons in the hidden layer. This demonstrates the use of Forward selection method to determine number of hidden neurons. The figure 2 of feed forward neural network gives clear idea about selected model for the given data.

To train the network back propagation algorithm is used. Total number of twenty one synaptic weights and eight bias values are tuned with the help of back propagation algorithm. The RMSE is computed from the values of actual rainfall and predicted rainfall to guide the learning of synaptic weights and arrive at the optimal model under investigation.

3   Forecasts of Monthly Rainfall in Coastal Andhra Region

We have forecasted the monthly rainfall (in Millimeters) for the out-of-sample set (years 2007-2011) and the out-of-sample forecasts of monthly rainfall in Coastal Andhra region using the model developed in the section 2.1.2 of the paper. The comparative graph between the actual rainfall and predicted rainfall is shown in the figure 3.

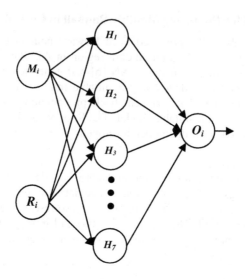

Fig. 2. FFNN for forecasting monthly rainfall in Coastal Andhra

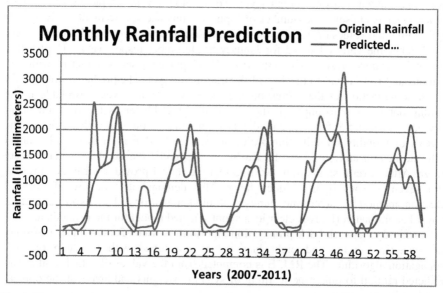

Fig. 3. Figure showing comparison between original and predicted rainfall of Coastal Andhra (2007-2011)

The model performance is shown in table 1 by calculating Root-Mean-Square Error (RMSE) and Mean Absolute Error (MAE) which are used to measure how close forecasts or predictions are to the eventual outcomes given in the data.

Table 1. Model performance (FFNN)

Data set	RMSE	MAE
In-Sample set	560.479	361.882
Out-of-Sample set	563.254	380.997

3.1   Forecasting Drought

To reduce the computational cost of the predicted rainfall of sixty months fuzzy c-means (FCM) clustering is used to split the data into three subsets such as low, medium and high-intensity rainfall.

A dataset of N vector Y_j, $j = 1,...,n$, can be partitioned into c fuzzy clusters using FCM clustering method [9]. In this every data point belongs to one cluster with a degree specified by its membership grade u_{ij} usually range from 0 to 1. Fuzzy partition matrix U contains the elements belong to membership u_{ij} where summation of the degrees that belongs to any data point is equal to 1, i.e.$\sum_{i=1}^{c} u_{ij} = 1$.

The goal of the FCM algorithm is to find c cluster centers so that the cost function of the dissimilarity measure is minimized. In our case C value is three as there are three groups like low rainfall, medium rainfall and high rainfall.   A tri-angular membership function is used for fuzzy C-means. As the problem focuses on finding the drought regions the data which belongs to low rainfall is used. Less rainfall does not indicate drought, since the drought data is hidden in that cluster the algorithm can accurately classify the drought regions as the search is restricted single cluster from huge data.

The major challenge of the drought classification is to avoid the chances of low rainfall and no drought situation. This is an interesting association which must be identified to improve the accuracy of prediction of drought. In our work, the main crux lies in this aspect to eliminate such cases or in other words find out fairly accurately the actual drought associated with low rainfalls. In our work we have developed and simulated an association rule mining approach for our purpose. Next section presents the approach and the procedures for the same.

3.1.1 Proposed Association Rule Mining
Association rule mining [3] quests for finding correlations or associations of interest among attributes in a specified dataset.

Algorithm: DCR-Drought Classification Rules
Input   : Rainfall Database, ms - minimum support, mc- minimum confidence, DCL-Drought Classification Labels, c- class label
Output: Drought Classification Rules
Method:
DCL= {Minor Drought, Moderate Drought, Extreme Drought}
F1= find frequent 1-itemset
for (K = 2; FK−1 ≠ Ø; K ++)
{
    CK= generate candidates by performing join operation on FK−1 and F1

FK = set of frequent itemsets
for each i∈ FK
* for each c ∈DCL*
*   if supp(i→c)≥ ms and conf(i→c) ≥ mc*
*   add i→c to DCR*
*   else discord the rule*
}

The above algorithm is standard Apriori like algorithm. It is used to classify the type of drought i.e., Severe Drought, Moderate Drought, Extreme Drought. It takes input as rainfall data and applies on the data. The value of mc (minimum confidence) varies on the application or data supplied to the system. Further support and confidence vales are to be calculated from the given data by using below given equations.

Support (I) = no of records containing I / total number of transactions in the database

and

Confidence(X→Y) = support (X U Y) / Support (X).

For performance testing we considered rainfall database which contains the data from years 1871 to 2011.

The Standardized Precipitation Index SPI [4] which is usually based on cumulative probability of certain rainfall occurrence at a region or place is used to categorize the drought regions. Previous data belonging to rainfall of the region or place is fitted to a gamma distribution, since gamma distribution is known to fit quite well for the precipitation distribution. This SPI values and ranges which are used for finding the drought regions are given in Table 2.

Table 2. SPI values for Drought

SPI values	Event
-1.25 to -0.5	Minor drought
-2.75 to -1.25	Moderate drought
Lesser than -2.75	Extreme drought

3.1.2 Rules Generated Using Drought Classifier (DC)

Further the classification rules are generated for identifying the drought prone regions. These rules are used for comparing region level rainfall data and identify the region with less rainfall which may be drought prone.

Due to space constraint we are unable to present all the rules generated by our algorithm. For the sake of discussion, here we are presenting some of the valuable rules generated by our algorithm and are as follows:

For the year 2007:   Region1 (Andhra) ^ Region2 (Rayalaseema) →Minor Drought
                     Region1 ^ Region3 (Telangana)→ Moderate Drought
                     Region1^Region2 ^ Region3→ Extreme Drought

For the year 2008:   Region2 ^ Region3 ➔Moderate Drought
                     Region1 ^ Region3 ➔ Minor Drought
                     Region1^Region2 ^ Region3➔ Extreme Drought

For the year 2009:   Region1 ^ Region2 ➔Minor Drought
                     Region1 ^ Region3 ➔ Minor Drought

For the year 2010:   Region1 ^ Region2 ➔Moderate Drought
                     Region1^ Region3 ➔ Moderate Drought
                     Region1^Region2 ^ Region3➔ Minor Drought

For the year 2011:   Region1^Region2 ^ Region3➔ Moderate Drought

4   Analysis of Results

Further the algorithm Drought Classifier is applied to find the drought regions in
forthcoming seasons by using the low rainfall data obtained by FCM from the
forecasted values. Table 3 shows the predicted value for each month and the drought
category assigned by the proposed Drought Classifier algorithm.

Table 3. Analysis of drought severity of coastal Andhra in year 2007

Month	Rainfall	Class label
1	0	Extreme Drought
2	25	Extreme Drought
3	140	Extreme Drought
4	385	Moderate Drought
5	730	Moderate Drought
6	530	Moderate Drought
7	1435	No Drought
8	590	Moderate Drought
9	2305	No Drought
10	614	Moderate Drought
11	16	Extreme Drought

5   Conclusion and Future Improvements

The objective of this work is to identify the drought regions for upcoming seasons.
Initially feed forward neural networks are used to predict the rainfall for future
seasons by using the existing huge data available. Once the forecasting of future
seasons rainfall is made available, fuzzy c-means clustering method is used to find the
low rain fall regions from the total predicted data. This process is used as to reduce
the complexity and work with less and reliable data. After obtaining the data belongs
to less rainfall regions, proposed algorithm Drought Classifier is used for identifying

the probable effect of drought from the data using rules generated and SPI. This result the areas which may expect to be hit by drought in upcoming days. As heavy rainfall data for the forthcoming seasons is also available in the cluster generated by FCM, in our future work we also propose to identify heavy rainfall regions which help of the administration to take suitable early measures for the safety of people in the effected regions. The overall predictions results obtained were satisfactory using RMSE and MSE. Further this model will be improved using the district wise rainfall data for accurate prediction of heavy and less rainfall areas. A more efficient FFNN may be evolved using intelligent approach for the prediction of rainfalls as a future research.

References

1. Rajput, A., et al.: Impact of Data Mining in Drought Monitoring. International Journal of Computer Science Issues (IJCSI) 8(6), 309–313 (2011)
2. Tadesse, T., et al.: Drought monitoring using data mining techniques: A case study for Nebraska. USA Natural Hazards 33(1), 137–159 (2004)
3. Harms, S.K., Deogun, J.S.: Sequential association rule mining with time lags. Journal of Intelligent Information Systems 22(1), 7–22 (2004)
4. Guttman, N.B.: Accepting The Standardized Precipitation Index: A Calculation Algorithm 1, 311–322 (1999)
5. Final Report, Volume 1, Drought in Andhra Pradesh: Long term impacts and adaptation strategies, South Asia Environment and Social Development Dept, World Bank (2005)
6. Sahai, A.K., Soman, M.K., Satyan, V.: All India summer monsoon rainfall prediction using an artificial neural network. Climate Dynamics 16(4), 291–302 (2000)
7. SrinivasRao, S.: Forecasting of Monthly Rainfall in Andhra Pradesh Using Neural Networks, Ph.D Thesis
8. Shukla, J., Mooley, D.A.: Empirical prediction of the summer monsoon rainfall over India. Monthly Weather Review 115(3), 695–704 (1987)
9. Bezdek, J.C., Ehrlich, R., Full, W.: FCM: The fuzzy<i> c</i>-means clustering algorithm. Computers & Geosciences 10(2), 191–203 (1984)
10. Kavitha, R.B., Govardhan, A.: Rainfall Prediction Using Data Mining Techniques-A Survey. In: The Second International Conference on Information Technology Convergence and Services (ITCSE), pp. 23–30 (2013)

Efficient Binary Classifier for Prediction of Diabetes Using Data Preprocessing and Support Vector Machine

Madhavi Pradhan[1] and G.R. Bamnote[2]

[1] Department of Computer Engineering, AISSMS College of Engineering,
University of Pune, Pune, Maharashtra, India
[2] Department of Computer Science and Engineering, PRMIT&R, SSGBAU,
Amravati, Maharashtra, India
{madhavipradhan,grbamnote}@rediffmail.com

Abstract. Diabetes offer a sea of opportunity to build classifier as wealth of patient data is available in public domain. It is a disease which affects the vast population and hence cost a great deal of money. It spreads over the years to the other organs in body thus make its impact lethal. Thus, the physicians are interested in early and accurate detection of diabetes. This paper presents an efficient binary classifier for detection of diabetes using data preprocessing and Support Vector Machine (SVM). In this study, attribute evaluator and the best first search is used for reducing the number of features. The dimension of the input feature is reduced from eight to three. The dataset used is Pima diabetic dataset from UCI repository. The substantial increase is noted in accuracy by using the data pre processing.

Keywords: SVM, Diabetes, Classifier, Preprocessing, Binary Classifier.

1   Introduction

Diabetes is a malfunctioning of the body caused due to lack of production of insulin in the body or resistance of the produced insulin by the body. Insulin is a hormone which regulates and controls the glucose in the blood. Lack of insulin leads to excessive content of glucose in the blood which can be toxic. 347 million people worldwide have diabetes[1] . Diabetes can lead to heart diseases which may increase the complications and can be fatal. Looking at all these instances and statistics, there is a need for early and accurate detection of diabetes.

Diagnosis of diabetes depends on many parameters and usually the doctors need to compare results of previous patients for correct diagnosis. So establishing a classifier system that can classify according to the previous decision made by experts and with minimal features can help in expediting this process. In the classification tasks on clinical datasets, researchers notice that it is common that a considerable number of features are not informative because they are either irrelevant or redundant with respect to the class concept. Ideally, we would like to use the features which have high predictive power while ignore or pay less attention to the rest. The predictive feature set can simplify the pattern representation and the classifier design. Also the resulting classifier will be more efficient.

S.C. Satapathy et al. (eds.), *Proc. of the 3rd Int. Conf. on Front. of Intell. Comput. (FICTA) 2014*
– *Vol. 1*, Advances in Intelligent Systems and Computing 327, DOI: 10.1007/978-3-319-11933-5_15

Feature reduction has been applied to several areas in medicine [2], [3]. Huang et al. [4] predicts type 2 diabetic patients by employing a feature selection technique as supervised model construction to rank the important attributes affecting diabetes control. K. Polat, S. Gunes[5] used PCA_ANFIS for diabetes detection and has got accuracy 89.47%. K. Polat, S. Gunes, A. Aslan[6] got 78.21% accuracy. They have used support vector machine for the diabetes detection. T. Hasan, Y. Nijat and T. Feyzullah [7] present a comparative study on Pima Indian diabetes diagnosis by using Multilayer Neural Network (MLNN) which was trained by Levenberg–Marquardt (LM) algorithm and Probabilistic Neural Network (PNN). MLNN have been successfully used in replacing conventional pattern recognition method for disease diagnosis system. LM used in this study provides generally faster convergence and better estimation results than other training algorithm. The classification accuracy of MLNN with LM obtained by this study using correct training was better than those obtained by other studies for the conventional validation method. K. Kayaer and T. Yildirim [8] have proposed three different neural network structures namely Multilayer Perceptron (MLP), Radial Basis Function (RBF) and General Regression Neural Network (GRNN) .These techniques were applied to the Pima Indians Diabetes (PID) medical data. The performance of RBF was worse than the MLP for all spread values tried. The performance of the MLP was tested for different types of back propagation training algorithms. The best result achieved on the test data is the one using the GRNN structure (80.21%). This is very close to one with the highest true classification result that was achieved by using the more complex structured ARTMAP-IC network (81%). P. Day and A. K. Nandi [9] introduce genetic algorithm based classifier. D. P. Muni, N. R. Pal and J. Das [10] proposed a Genetic Programming approach to design multiclass classifier.

Numerous classification models are proposed for prediction of diabetes but it is widely recognized that diabetes are extremely difficult to classify [11]. This paper presents an efficient binary classifier for detection of diabetes using data preprocessing and support vector machine. First we have applied feature selection methods to identify most predictive attributes and then used SVM algorithm for classification.

2   Proposed Classifier

Mathematically, a classifier can be represented as a function which takes a feature in p dimensional search space and assigns a label vector Lvc to it ;
$C: Sp \rightarrow Lvc$

Where,

      C is the classifier that maps search space to label vectors.
      Sp is the p dimensional search space.
      Lvc is the set of label vectors.

The objective here is to create 'C' using SVM. During the training phase of the classifier, samples of the form $\{x1, x2, x3, x4, x5, x6, x7, x8\} \in S^8$ and associated label vector Lvc={Diabetic(1), Non-diabetic(0)} are used to create the classifier. Classification of diabetes data set is a binary classification problem i.e. there are only 2 label vectors that can be assigned to each of the samples.

2.1   Preprocessing

The prediction accuracy of any classification algorithm is heavily based on the quantity and quality of the data. The clinical data is generally collected as a result of the patient care activity to benefit the individual patient. So the clinical datasets may contain data that is redundant, incomplete, imprecise and inconsistent.

Not all features used in describing data are predictive. The irrelevant or redundant features , noisy data etc affect the predictive accuracy. Hence the clinical data requires rigorous data preprocessing.

Data preprocessing is often a neglected but important step in the data mining process. Data preprocessing includes; [12]

Normalization
Aggregation
Sampling
Dimensionality Reduction
Feature subset selection
Feature creation
Discretization
Attribute Transformation

We have used Normalization, Discretization and feature selection.

2.1.1   Normalization
Normalization involves scaling all values for a given attribute so that they fall within a small specified range.[12] It is required to avoid giving the undue significance to the attributes having large range. SVM requires the normalization of numeric input.

2.1.2   Discretization
Discretization is the process of transforming continuous valued attributes to nominal. It is used as a preprocessing step for the correlation-based approach to feature selection. As we have used correlation-based approach we have used discretization.

2.1.3   Feature Selection
The purpose of the feature selection is to decide which of the features to include in the final subset. Feature selection techniques can be categorized according to a number of criteria [13]. One popular categorization is based on whether the target classification algorithm will be used during the process of feature evaluation. A feature selection method, that makes an independent assessment only based on general characteristics of the data, is named "filter" [14]; while, on the other hand, if a method evaluates features based on accuracy estimates provided by certain learning algorithm which

will ultimately be employed for classification, it will be named as "wrapper" [14, 15]. With wrapper methods, the performance of a feature subset is measured in terms of the learning algorithm's classification performance using just those features. For this reason, wrappers do not scale well to data sets containing many features [16]. Besides, wrappers have to be re-run when switching from one classification algorithm to another. In contrast to wrapper methods, filters operate independently of any learning algorithm and the features selected can be applied to any learning algorithm at the classification stage. Filters have been proven to be much faster than wrappers and hence, can be applied to data sets with many features [16]. Since the clinical data sets generally have many features we have use filter method.

Evaluation and Search Strategy
The focus of evaluation and search strategy is to simplify the data set by finding out the features which are significant and contribute the most in arriving at the diagnosis. Evaluators and search methods used by researchers are ;
Evaluators :
cfssubseteval
consistencysubseteval
PCA (principal component analysis)
Wrapper subseteval

Search Methods :
Best first search
Genetic search
Ranker
Exhaustive search
Forward selection

In this classifier design, we have selected cfssubseteval attribute evaluator and best first search for the feature selection. Correlation-based Feature Selector (CFS) algorithm is used in cfssubseteval attribute evaluator.

2.1.4 CFS Algorithm
The CFS algorithm takes the training dataset as input and generates a feature-class and feature-feature correlation matrix which is then used to search the feature subset space. The criterion used for search is the best first search. The search starts with an empty set of features and generates all possible single feature expansion [13][17]. The subset with highest evaluation is chosen and expanded in same manner by adding single feature[18]. If expanding a subset results into improvement, the search drop backs to next unexpanded subset and continues from there. CFS ranks feature subsets according to a correlation based heuristic evaluation function. The subsets that contain features that are highly correlated with the class and uncorrelated with each other are ranked as per evaluation function.

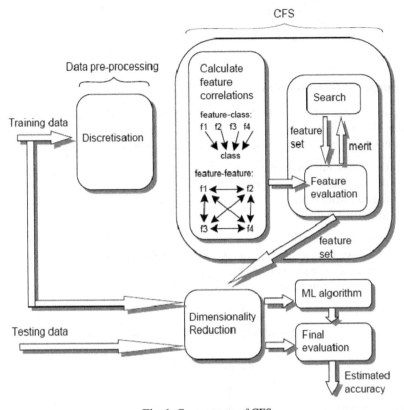

Fig. 1. Components of CFS

CFS is a simple filter algorithm that ranks feature subsets according to a correlation based heuristic evaluation function as per the equation shown below. Irrelevant features should be ignored because they will have low correlation with the class. Redundant features should be removed as they will be highly correlated with one or more of the features in the subset. The feature acceptance will depend on the extent to which it predicts classes in areas of the instance space not already predicted by other feature[19].

$$M_s = \frac{k\bar{r}_{cf}}{\sqrt{k + k(k-1)\bar{r}_{ff}}}$$

where M_s is the heuristic evaluation function "merit" of a feature subset S containing k features,

$\bar{r}_{cf}$ is the mean feature-class correlation (f ∈ S), and

$\bar{r}_{ff}$ is the average feature-feature inter correlation.

The numerator of the equation represents an indication of how predictive the feature - class correlation are and the denominator indicates the redundancy among the features.

Training and testing data is reduced to contain only the features selected by CFS. The dimensionally reduced data can then be passed to a machine learning algorithm for induction and prediction. In this classifier the Machine Learning (ML) algorithm used is SVM.

2.2   Support Vector Machine

A Support Vector Machine (SVM) is a discriminative classifier defined by a separating hyper plane. It uses, labelled training data, to output an optimal hyper plane which categorizes new examples. The basic concepts in SVM are:

1. the separating hyper plane,
2. the maximum-margin hyper plane,
3. the soft margin and
4. the kernel function.

Any training tuple that fall on hyper planes H1 or H2 are support vectors. Hyper planes with larger margin are less likely to over fit the training data. A hyper plane should separates the classes and has the largest margin.

A separating hyper plane can be written as

$$W \bullet X + b = 0$$

where $W = \{w1, w2, \ldots, wn\}$ is a weight vector and b scalar

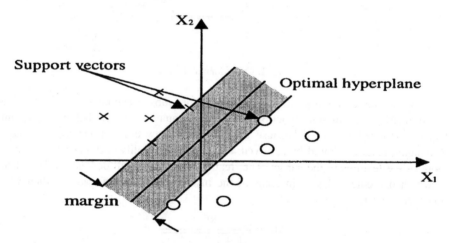

Fig. 2. SVM classification with a hyperplane that maximizes the separating margin between the two classes

The hyperplane shown in above Fig 2 defines the margin based on support vectors and can be mathematically written as :

H1: $w0 + w1\ x1 + w2\ x2 \geq 1$ for yi = +1, and
H2: $w0 + w1\ x1 + w2\ x2 \leq -1$ for yi = −1.

3   The Experimental Results

The dataset available for Pima Indian diabetes patients is taken as Training and testing dataset for the classifier design and experiments carried out on them are described below.

3.1   Diabetes Dataset

The dataset used is obtained from UCI Repository of Machine Learning Databases. This database is owned by National Institute of Diabetes and Digestive and Kidney Disease. All patients were Pima Indian females who were above 21 years of age.

There are eight attributes and one output variable which has either a value '1' or '0', where '1' means positive test of diabetes and '0' means negative test for diabetes.

1. Number of times pregnant
2. Plasma glucose concentration a 2 hours in an oral glucose tolerance test
3. Diastolic blood pressure (mm Hg)
4. Triceps skin fold thickness (mm)
5. 2-Hour serum insulin (mu U/ml)
6. Body mass index (weight in kg/(height in m)^2)
7. Diabetes pedigree function
8. Age (years)
9. Class variable (0 or 1)

The goal of the proposed classifier is to predict class variable using the remaining eight attributes. There are 268(34.9%) cases for class '1' and 500(65.1%) cases for class '0'. So if we assume that all are non diabetic then accuracy is 65.1%. Thus our classifier must do much better than this. We have used Weka tool for experimentation. It is freely available software. We have integrated libsvm with Weka.

Table 1. Statistical Analysis of the attributes

Attribute No	Mean	Standard Deviation	Min/Max
1	3.8	3.4	0/17
2	120.9	32.0	0/199
3	69.1	19.4	0/122
4	20.5	16.0	0/99
5	79.8	115.2	0/846
6	32.0	7.9	0/67.1
7	0.5	0.3	0.078/2.42
8	33.2	11.8	21/81

3.2   Result and Discussion

The SVM classifier is fed with the dataset in comma separated format. To avoid over fitting two-fold cross validation is used. Using Attribute Selection and best first search method we got three most predictive attributes Plasma glucose concentration , Body mass index and Age. The Table below shows the comparison of performance of classifier with and without pre processing.

Table 2. Performance of classifier

SVM CLASSIFIER	Correctly Classified Instances (%)	Incorrectly Classified Instances (%)	ROC(Area under ROC)
Without Pre-processing	65.10	34.89	0.49
After Normalisation	76.953	23.04	0.70
After Discretization	75.52	24.47	0.71
After attribute selection	86.46	13.54	0.83

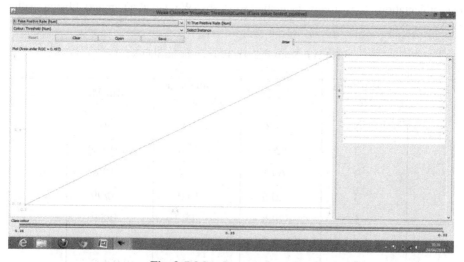

Fig. 3. ROC before preprocessing

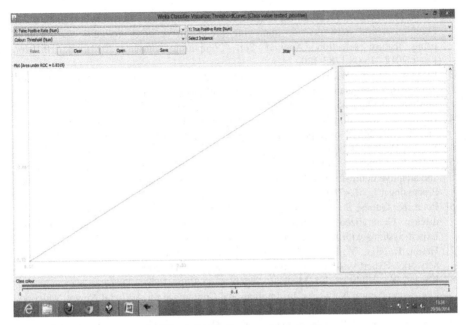

Fig. 4. ROC after preprocessing

4   Conclusion

In this paper, we have presented an approach for the binary classification problem using SVM. Very small set of non-redundant features are obtained while increasing the predictive accuracy. From the results we can see that because of pre processing the accuracy values we have got is much higher. There is improvement in performance of classification algorithm due to removal of non predictive features. Few features are selected which makes the classifiers more efficient in time and space. Less cost incurred for the collection, storage and processing of data. These results clearly indicate that the proposed method can be profitably used for diagnosis of diabetes.

Acknowledgement. This work is supported in part by University of Pune's research grant scheme fund from Board of College and University Development (BCUD) Ref No. # O.S.D./BCUD/330 .

References

[1] Danaei, G., Finucane, M.M., Lu, Y., Singh, G.M., Cowan, M.J., Paciorek, C.J., et al.: National, regional, and global trends in fasting plasma glucose and diabetes prevalence since 1980, systematic analysis of health examination surveys and epidemiological studies with 370 country-years and 2.7 million participants. Lancet 378(9785), 31–40 (2011)

[2] Fernández-Navarro, F., Riuz, R., Riquelme, J.C.: Evolutionary Generalized Radial Basis Function neural networks for improving prediction accuracy in gene classification using feature selection. Applied Soft Computing Journal (2012), doi:10.1016/j.asoc.2012.01.008,1787-1800

[3] Li, D.-C., Liu, C.-W., Hu, S.C.: A fuzzy-based data transformation for feature extraction to increase classification performance with small medical data sets. Artificial Intelligence in Medicine 52(1), 45–56 (2011)

[4] Huang, Y., McCullagh, P., Black, N., Harper, R.: Feature selection and classification model construction on type 2 diabetic patients' data. Artificial Intelligencte Medicine Journal 41, 251–262 (2007)

[5] Polat, K., Gunes, S.: An expert system approach based on principal component analysis and adaptive neuro-fuzzy inference system to diagnosis of diabetes disease. Digital Signal Processing 17, 702–710 (2007)

[6] Polat, K., Gunes, S., Aslan, A.: A cascade learning system for classification of diabetes disease: Generalized discriminant analysis and least square supoort vector machine. Expert Systems with Applications 34(1), 214–221 (2008)

[7] Hasan, T., Nijat, Y., Feyzullah, T.: A comparative study on diabetes disease using neural networks. Expert system with applications 36, 8610–8615 (2009)

[8] Kayaer, K., Yildirim, T.: Medical diagnosis on Pima Indian Diabetes using General Regression Neural Networks. In: International Conference on Artificial Neural Networks and Neural Information Processing (ICANN/ICONIP), pp. 181–184

[9] Day, P., Nandi, A.K.: Binary String Fitness Characterization and Comparative Partner Selection in Genetic Programming. IEEE Trans. on Evolutionary Computation 12(6), 724–735 (2008)

[10] Muni, D.P., Pal, N.R., Das, J.: A Novel Approach to Design Classifiers Using Genetic Programming. IEEE Trans. on Evolutionary Computation 8(2), 183–196 (2004)

[11] Khashei, M., Effekhari, S., Parvizian, J.: Diagnosing Diabetes Type II Using Soft Intelligent Binary Classification Model. Review of Bioinformatics and Biometrics (RBB) 1, 9–23 (2012)

[12] Han, J., Kamber, M.: Data Mining concepts and techniques, 2nd edn., pp. 61–77. Morgan Kaumann publication, An imprint of Elsevier (2006) ISBN: 978-81-312-0535-8

[13] Hall, M.A., Holmes, G.: Benchmarking attribute selection techniques for discrete class data mining. IEEE Transaction on Knowledge and Data Engineering 15(3), 1437–1447 (2003)

[14] Witten, H., Frank, E.: Data Mining: Practical Machine Learning Tools and Techniques with Java Implementation. Morgan Kaufmann, San Mateo (2000)

[15] Kohavi, R., Joh, G.H.: Wrappers for feature subset selection. Artificial Intelligence, pp. 273–324 (1997)

[16] Hall, M.A.: Correlation-based feature selection for discrete and numeric class machine learning. In: Proceedings of the Seventeenth International Conference on Machine Learning, pp. 359–360 (2000)

[17] Karagiannopoulos, M., Anyfantis, D., Kotsiantis, S.B., Pintelas, P.E.: Educational Software Development Laboratory. Department of Mathematics. University of Patras, Greece, http://math.upatras.gr

[18] Rai, S., Saini, P., Jain, A.K.: Model for Prediction of Dropout Student Using. ID Decision Tree Algorithm International Journal of Advanced Research in Computer Science & Technology (IJARCST 2014) 2(1), 142–149 (2014) ISSN : 2347 - 9817

[19] Bolón-Canedoa, V., Sánchez-Maroño, N., Alonso-Betanzos, A., Benítez, J.M., Herrera, B.F.: A review of microarray datasets and applied feature selection methods. Information Sciences 282, 111–135 (2014)

A Combined SVD-DWT Watermarking Scheme with Multi-level Compression Using Sampling and Quantization on DCT Followed by PCA

Sk. Ayesha, V.M. Manikandan, and V. Masilamani

Indian Institute of Information Technology Design and Manufacturing (IIITDM)
Kancheepuram, Chennai – 600127, India
ayeshanoormd@gmail.com,
{coe14d001,masila}@iiitdm.ac.in

Abstract. This paper proposes a watermarking scheme, which combine both singular value decomposition (SVD) and discrete wavelet transform (DWT) based watermarking techniques. In addition to this, the paper introduces multi-level compression on the watermark image to improve the visual quality of the watermarked image. This can be achieved in two levels. In first level, compression is done by applying sampling and quantization on discrete cosine transform (DCT) coefficients. Then in the second level compression is carried out by principal component analysis (PCA). The proposed method is compared with various existing watermarking schemes by using image quality measures like peak signal to noise ratio (PSNR), mean structural similarity index (MSSIM) and correlation coefficient. It is observed that the proposed approach survives unintentional linear attacks such as rescaling, rotation and some minor modifications.

Keywords: Watermarking, Multi-level Compression, SVD, DWT and PCA.

1   Introduction

For the past few decades the exchange of multimedia data such as images, videos and audio through internet is increased significantly. Unfortunately, this may lead to various issues such as illegal duplication, unauthorized modification, image forgery etc., Digital watermarking for images is the process of inserting data into an image in such a way that it can be used to make an assertion about the image. It is one of the most efficient content based image authentication techniques, which used as a tool for various purposes such as copy protection, usage monitoring, data authentication and forgery detection [1][14].

The process of watermarking can be roughly classified into visible watermarking and invisible watermarking according to the transparency of watermark on the original image. Visible watermarking is very prone to attack, because anyone can remove the watermark easily with the help any powerful image editing software such as Photoshop, Photo-Paint etc., After removing the visible watermarks, they can easily reuse or duplicate it without any legal constraints [2]. In case of invisible

© Springer International Publishing Switzerland 2015   141
S.C. Satapathy et al. (eds.), *Proc. of the 3rd Int. Conf. on Front. of Intell. Comput. (FICTA) 2014*
– Vol. 1, Advances in Intelligent Systems and Computing 327, DOI: 10.1007/978-3-319-11933-5_16

watermarking the watermark is not visible to the users; when the owner needs to prove his ownership or want to make sure that the image is not modified then they can extract the watermark and can ensure the authenticity.

Watermarking can be done either in spatial domain or in frequency domain. LSB watermarking technique is a traditional spatial domain watermarking method. Different variations and improvements on this method are also proposed. In frequency domain discrete cosine transform (DCT), discrete wavelet transform (DWT) based watermarking methods are excessively in use [3][14].

The proposed algorithm anticipated to produce a watermarked image which is very close to the original image without compromising the intend of watermarking. For this purpose, a multi-level compression on the watermark is applied. For compressing the watermark image, firstly DCT is applied on the watermark, then sampling and quantization is performed on DCT coefficients. Then the inverse DCT operation is performed on the resultant coefficients. These will be processed by PCA to increase the amount of compression on the watermark. Finally the compressed watermark is embedded in the original image by using combination of SVD and DWT based approach.

The novelty of proposed method is to make use of both DCT and PCA based compression technique for compressing the watermark. In addition two powerful watermarking algorithms SVD and DWT are combined to ensure the improvement in visual quality of the watermarked image.

The rest of the paper is organized as follows, Section 2 will discuss literature review and Section 3 gives details about proposed method. Section 4 presents the results obtained and its analysis, followed by Section 5 conclusion.

2   Literature Review

Basics of digital image watermarking and its importance are discussed in [14, 7]. Discrete cosine transform (DCT) based watermarking technique and its algorithm is clearly explained [14]. Singular value decomposition (SVD) based watermarking method is proposed by Ruizhen Liu *et al.* [4]. The authors decompose the original image into three matrices by using singular value decomposition method and the watermark image is inserted in the most significant part. A combination of SVD and DWT based watermarking scheme by incorporating biometrics is introduced in [9]. Benjamin Mathon et. al introduced a secure spread-spectrum based watermarking using distortion optimization. The elements of transportation theory is used to minimize global distortion [8]. A semi-fragile watermarking algorithm has been proposed for authenticating JPEG-2000 compressed images, which tolerates large amount of image data loss due to resolution or quality scaling in [10]. A robust watermarking method based on discrete wavelet transform (DWT) proposed by G. Kotteswara rao *et.al.* in [5]. Binary watermark is embedded in the detail sub-band coefficients of the cover image. LH2 and LH3 sub band coefficients of the cover image are used to form five element blocks. In each block, the first minimum and the second minimum are identified and one watermark bit is embedded in them. Principal component analysis (PCA) based watermarking scheme is proposed by Wang Shuozhong [6]. Here principal component analysis is applied into the original image after subdividing it into fixed sized blocks. On the PCA transformed image they have inserted the watermark to produce the watermarked image. A blind data hiding method in spread spectrum domain is introduced in [11]. Here the authors tried to extract unknown messages hidden in image hosts via multicarrier/signature spread-spectrum embedding. Neither the original host nor the embedding carriers are assumed available.

3   Proposed Watermarking Technique

In the proposed method, watermarking procedure can be mainly divided into two modules: first module is to compress the watermark and the second module is to insert the watermark into original image.

The watermark compression and watermarking is shown in Fig. 1 and Fig. 2 respectively.

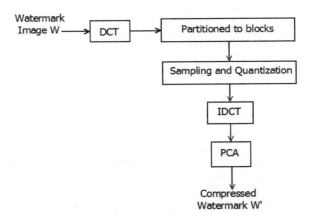

Fig. 1. Watermark Compression

Algorithm for watermark compression is as follows:

Input: Watermark image ($w_{p \times q}$) with p number of rows and q number of columns
Output: Compressed watermark ($w_{k \times q}$) with k number of rows and q number of columns

Step 1: Find the DCT of $w_{p \times q}$ say $D_{p \times q}$.
Step 2: Initially subdivide $D_{p \times q}$ into four equal sized blocks say B_{00}, B_{01}, B_{10} and B_{11}, select the left most top block B_{00} again divide it into four blocks of same size say BB_{00}, BB_{01}, BB_{10} and BB_{11}.
Step 3: Perform dense sampling in left most top blocks and sparse sampling in the remaining blocks, say result of sampling is $S_{p \times q}$.
Step 4: Apply quantization on S, such a way that the four most significant bits only used to represent the coefficients. Assume the result of quantization is in $SQ_{p \times q}$.
Step 5: Find the Inverse DCT from $SQ_{p \times q}$ say result is $SF_{p \times q}$, which will be similar to $w_{p \times q}$ with some amount of degradation in content.

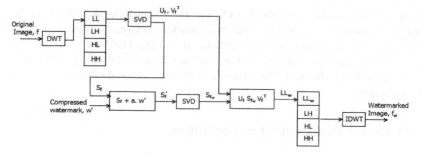

Fig. 2. Watermarking Module

Step 6: Consider each row of $SF_{p \times q}$ as samples in PCA, find the mean vector of these samples say $M_{p \times 1}$.

$$M_i = \frac{1}{q} \sum_{j=1}^{q} SF_{ij} \quad i = 1, 2, \dots p$$

Step 7: Find the covariance matrix $C_{p \times p}$ from $SF_{p \times q}$

$$C_{p \times p} = \frac{1}{q} \sum_{i=1}^{q} Y_i Y_i^T$$

$C_{p \times p}$ will be a square matrix of size p x p, Y_i denotes the variance of ith column. $Y_i = SF_{ji} - M_i$, i= 1, 2, 3..., p and j= 1, 2, ..., q.

Step 8: Find the eigen vectors and eigen values of covariance matrix.

Step 9: Find the PCA transform matrix $A_{p \times p}$ by rearranging the eigen vectors in such a way that, the eigen vector corresponding to highest eigen value should keep as the first row in A, and the eigen vector corresponding to second largest eigen value is on second row of A and so on.

Step 10: Select the first k number of rows from $A_{p \times q}$, say $A_{k \times q}$ (k<=p) which contain the most significant properties of the watermark.

Step 11: Find the compressed transformed watermark by

$$w'_{k \times q} = A_{k \times q} \cdot w_{p \times q}$$

Step 12: output the compressed watermark in transformed form $w'_{k \times q}$.

Algorithm for watermarking is as follows:

Input: Image to be watermarked ($f_{m \times n}$) and Compressed watermark ($w'_{k \times q}$)
Output: Watermarked Image ($fw_{m \times n}$)

Step 1: Find the one level DWT of original image $f_{m \times n}$ will produce LL, LH, HL and HH sub-bands.

Step 2: Select the first k number of rows from LL sub-band say LL.

Step 3: Apply SVD decomposition on LL_s and produce three matrices say U_f, S_f, and V_f.

Step 4: Define the scaling factor (a) and add the watermark, w ' into the S_1 will produce temp $S_f' = S_f + (a \times w')$

Step 5: Find the SVD of S_f', as U_{fw}, S_{fw} and V_{fw} then find the LL sub-band of watermarked image by $LL_n = U_f \cdot S_{fw} \cdot V_f^T$

Step 6: Find the LL sub-band of watermarked image (LL_w) by appending the left out (p-k) rows of LL that will get from step 1 to the watermarked portion (LL_w).

Step 7: Produce the watermarked image f_w by finding the Inverse DWT of LL_w, LH, HL and HH.

Watermark extraction technique is as shown in Fig. 3.

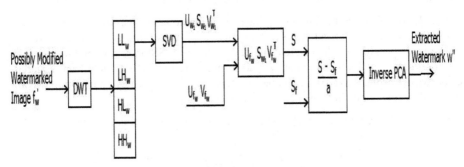

Fig. 3. Watermark Extraction

Algorithm for watermark extraction is as follows:

Input: Possibly attacked watermarked image $fw'_{m \times n}$
Output: Extracted watermark $w''_{m \times q}$
Step 1: Find the DWT of the fw' results in four sub-bands LL_w, LH_w, HL_w and HH_w.
Step 2: Select **k** number of rows from LL_w and say LL_s.
Step 3: Decompose LL_s by using singular value decomposition procedure as U_{w1}, S_{w1} and V_{w1}.
Step 4: Combine U_{fw}, S_{w1} and V_{fw}^T by multiplying them and say S
$$S= U_{fw} . S_{w1} . V_{fw}^T$$
Step 5: Extract the watermark by
$$w_e = (S-S_f) / a$$
w_e will be in PCA transformed form.
Step 6: Reconstruct the watermark by taking inverse PCA and the result will be w''.

4   Results and Analysis

The major objective of the proposed approach is to improve the visual perception of the watermarked image. The quality of the watermarked image can be analyzed by using peak signal to noise ratio (PSNR), mean structural similarity index (MSSIM) and correlation coefficient.

PSNR: Ratio between the maximum possible power of the signal and the power of corrupting noise that affects the fidelity of its representation. Because many signals have wide dynamic range, PSNR is usually expressed in terms of the logarithmic decibel (dB) scale.
$$PSNR= 10. \log (255^2 /E) \text{ dB}$$

Where E is mean square error (MSE), E can be defined as

$$E = \frac{1}{MN} \sum_{i=1}^{N} \sum_{j=1}^{M} [f(i,j) - f'(i,j)]$$

f(i, j) is the pixel value of the original image and f '(i, j) is the watermarked image.

MSSIM: Mean Structural Similarity Index is a method for measuring the similarity between two images. Structural Similarity Index is measuring based on a reference image here we will take the original image as reference. SSIM is designed to improve on traditional methods like peak signal-to-noise ratio (PSNR) and mean squared error (MSE), which have proven to be inconsistent with human eye perception. Mean of SSIM is considered here as measuring criteria [8].

Correlation Coefficient: Correlation coefficient between two images f and f ' can be calculated as

$$\frac{\sum_m \sum_n (f_{mn} - \bar{f})(f'_{mn} - \bar{f'})}{\sqrt{[\sum_m \sum_n (f_{mn} - \bar{f})^2][\sum_m \sum_n (f'_{mn} - \bar{f'})^2]}}$$

$\bar{f}$: Mean pixel value of image f
$\bar{f'}$: Mean pixel value of image f '

(a)   (b)   (c)

(d)   (e)

Fig. 4. (a) Original image   (b) Watermark image   (c) Watermarked image with 50% compression in the watermark (d) Extracted and reconstructed watermark from the watermarked image with 50% compression in the watermark (e) Extracted and reconstructed watermark from the watermarked image when there is no compression on watermark

In order to analyze the efficiency of proposed method, various images from standard image dataset [12, 13] is used as watermark images. Result obtained from the execution of the proposed scheme is shown in Fig. 4. Original image is shown in Fig. 4(a), watermark is shown in Fig. 4(b) and Fig. 4(c) shows the watermarked image Fig. 4(d) and Fig. 4(e) are the extracted and reconstructed watermarks images without compression and with 50% compression respectively.

The PSNR value between original image and watermarked image obtained for various amount of compression on watermark with different watermarking algorithms is shown in Fig. 5. From the Fig. 5 the PSNR of proposed algorithm is very high as compared to the other approaches.

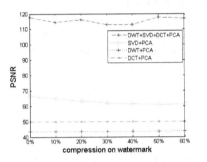
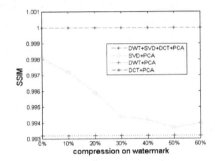

Fig. 5. PSNR vs Compression for various schemes

Fig. 6. MSSIM Vs Compression on various approaches

In the proposed approach, the correlation coefficient between the embedded watermark and the extracted watermark has been measured for authenticity determination and plotted with respect to the different compression on various approaches as shown in Fig. 7.

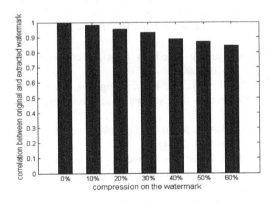

Fig. 7. Correlation between inserted watermark and extracted watermark

The correlation coefficient between watermarked image and original image of the proposed method is equal to 1. The correlation between inserted watermark and extracted watermark is also analyzed. The proposed watermarking scheme is tested with the various amounts of compressions on watermark, Fig. 7 shows that, if the compression on the watermark is more the correlation between inserted watermark and extracted watermark will be less.

5   Conclusion

A combined version of SVD and DWT watermarking scheme with multi-level compression on watermark has been implemented and the results are also analyzed. Watermarking algorithm insert this compressed watermark image into original image. We also compared the proposed method with the existing watermarking algorithms such as SVD, DWT, DCT watermarking techniques with and without PCA compression. The proposed system shows a great improvement on this watermarking area.

References

1. Potdar, V.M., Han, S., Chang, E.: A Survey of Digital Image Watermarking Techniques. In: IEEE International Conference on Industrial Informatics (INDIN-03), pp. 709–716 (2005)
2. Huang, C.-H., Wu, J.-L.: Attacking Visible Watermarking Schemes. IEEE Transactions on Multimedia 1, 16–30 (2004)
3. Hsieh, M.-S., Tseng, D.-C.: Hiding digital watermarks using multi-resolution wavelet transform. IEEE Transactions on Industrial Electronics 48, 875–882 (2001)
4. Ruizhen, L., Tan, T.: An SVD based watermarking scheme for protecting rightful ownership. IEEE Transactions on Multimedia 4, 121–128 (2002)
5. Koteswara Rao, G., Srinivaskausalyanandan, T.P., Prashanth, R.L., Anuragh, V.: A Novel Approach In Image Watermarking With Discrete Wavelet Transform. In: The Proceedings of International Conference on Electrical and Electronics Engineering (ICEEE), pp. 120–123 (2012)
6. Shuo-Zhong, W.: Watermarking based on Principal component analysis. Journal of Shanghai University 4, 22–26 (2000)
7. Wan, Y.H., Yuan, Q.L., Ji, S.M., He, L.M., Wang, Y.L.: A survey of the image copy detection. In: IEEE Conference on Cybernetics and Intelligent Systems, pp. 738–743 (2008)
8. Benjamin, M., Cayre, F., Bas, P., Macq, B.: Optimal transport for secure spread-spectrum watermarking of still images, p. 1 (2014)
9. Swanirbhar, M., Devi, K.J., Sarkar, S.K.: Singular value decomposition and wavelet-based iris biometric watermarking. IET on Biometrics 2, 21–27 (2013)
10. Angela, P., Safavi-Naini, R.: Scalable fragile watermarking for image authentication. IET on Information Security 4, 300–311 (2013)

11. Ming, L., Kulhandjian, M.K., Pados, D.A., Batalama, S.N., Medley, M.J.: Extracting Spread-Spectrum Hidden Data from Digital Media. IEEE Transactions on Information Forensics and Security 8, 1201–1210 (2013)

12. http://www.petitcolas.net/fabien/watermarking/image_database / (May 2014)

13. http://lear.inrialpes.fr/jegou/data.php#copydays (May 2014)

Diagnosis of Parkinson Disease Using Machine Learning and Data Mining Systems from Voice Dataset

Tarigoppula V.S. Sriram[1], M. Venkateswara Rao[2],
G.V. Satya Narayana[3], and D.S.V.G.K. Kaladhar[4]

[1] MCA, Raghu Engineering College, Visakhapatnam, India
[2] IT, GITAM University, Rushikonda, Visakhapatnam, India
[3] IT, Raghu Institute of Technology, Visakhapatnam, India
[4] Dept. of Bioinformatics, GITAM University, Visakhapatnam, India
{ramjeesis,dkaladhar}@gmail.com
mandapati_venkat@yahoo.co.in,
gv_satyanarayana@yahoo.com,

Abstract. Parkinson's disease is one of the most painful, dangerous and non curable diseases which occurs at older ages (mostly above 50 years) in humans. The data-set for the disease is retrieved from UCI repository. A relative study on feature relevance analysis and the accuracy using different classification methods was carried out on Parkinson data-set. Sieve multigram data and Survey graph provide the statistical analysis on the voice data so that the healthy and Parkinson patients would be correctly classified. KStar and NNge present good accuracy based classification methods. Sieve multigram shows the edges between the nodes such as Fhi, Flo, Jitter, JitterAb, RAP and PPQ. KStar and NNge have connections with Shimmer and ShimmerDB . ADTree shows 21 leaves with 31 leaves and SimpleCART shows 13 leaves and 7 leaves. Most of the clusters vary with DBScan and SimpleKMeans with 25% and 38% towards Parkinson disease.

1 Introduction

In a data warehouse, the user usually has a view of multiple data-sets collected from different data sources and understanding their relationships is an extremely important part of the KDD process. Data mining is a rapidly evolving advanced technology that used in bio-medical sciences and research in order to predict and analyze large volumes of medical data such as Parkinson's disease [4]. Data mining in neuro-degenerative diseases like Parkinson disease is an emerging field with enormous importance for deeper understanding of mechanisms relevant to disease, providing prognosis and complete treatment [1].

A relatively high prediction performance has been achieved with classification accuracy. Classification algorithms predict accuracy for discrete variables that are relevant on the other attributes present in the data-set. The continual increase of the number of features can significantly contribute with clustering-based feature

S.C. Satapathy et al. (eds.), *Proc. of the 3rd Int. Conf. on Front. of Intell. Comput. (FICTA) 2014*
– *Vol. 1*, Advances in Intelligent Systems and Computing 327, DOI: 10.1007/978-3-319-11933-5_17

analysis [2]. Clustering is a datamining technique that is mostly studied for predicting unstructured data to informal data. Clustering algorithms divide data into segments with clusters predicting similar branches and properties. Association algorithms find the correlations between different attributes in a data-set to conduct polynomial associations for better hypothesis.

The data mining intelligence systems can analyze voice fluctuations in patients suffering from Parkinson's disease (PD) by using data obtained as signals from bio-medical equipment like surface electromyographic (EMG) and accelerometer (ACC) [3]. Specific clinical applications requiring advanced analysis techniques, such as data mining and other techniques can be obtained from sensors to analyze large data sets. Medical data mining is a new emerging field that has great influence for exploring patterns from respective medical data sets in clinical research [5].

PD is the 2nd most common type of Neurodegenerative disorder that causes disturbances in speech and accurate voice [6]. No genetic component is evident that a disease begins after age 50 years [7]. Research and investigations on efficacy of intensive voice therapy are underway to improve the functional communication in the patients with idiopathic PD [8]. Parkinson's disease occurs between the ages of 50 and 95 years [9].

2   Methodology

The data-set for the PD was obtained from an online data repository UCI. Comparative classification studies on different data-sets in an entity have been applied for accuracy analysis and the time taken to execute the data-set in order to find the best classification rule. The data of healthy people and those with Parkinson can be correctly classified by using machine learning and data mining systems.

Bayesian theory is a mathematical model in calculus of degrees that predicts proposition of interest. BayesNet and NaiveBayes are the most practical learning methods that have a random sequence model within each class. System classification summarizes a sequence of classification methods using algorithms. Logistic function can be derived from simple classification problems, measuring from minus infinity to plus infinity. Simple logistic regression is used to explore associations between dichotomous outcome with continuous, ordinal, or categorical exposure variable.

Lazy classification scheme uses Hierarchical SVMs to select a subset of candidate classes for each test instance, in order to determine the overall best performer. K-Star is an Instance-Based learner. The class of a test instance is based upon the class of those that is determined by similarity function with similar training instances. It is different from the other instance-based machine learners using entropy-based distance function.

Meta-classification makes its binary decision by classifying synthesized feature vectors and   one meta-classifier for each class is built for each. Bagging (or Bootstrap aggregating) is an algorithm to improve machine learning of statistical classification and regression models in terms of stability and classification accuracy. A statistical method of distance based classification with that of best matching rule can be explained by NNGE (Non-Nested Generalized Exemplars).

The classification is used to follow the path dictated by the successive test placed along the tree until it finds a leaf containing the class to assign a new attribute. ADTree (An alternating decision tree), J48, Random Forest and Simple CART (A simple Classification And Regression Tree) are some of the trees being used in classification of Parkinson's disease.

A survey graph and sieve multigram is used to construct the statistical analysis on the multiple data-sets of the voice data-set from UCI.

3   Results and Discussions

The present experimentation is conducted to predict attributes that are associated and classified from voice data. Table 1 shows the classification of the voice data-sets, accuracy and time taken to execute the classified data-sets. KStar and NNge have presented good accuracy based analysis on the classification. The execution time taken is also less in providing the output for the submitted data-sets (<2 seconds). The classification methods i.e. Support Vector Machine method (SVM), can be used to distinguish people with Parkinson's disease from the healthy people [10]. UCI data set on PD was composed of a voice measurement from bio-medical instrumentation with 195 samples with 16 attributes. Two training algorithms like Scaled Conjugate Gradient (SCG) and Levenberg-Marquardt (LM) using Multilayer Perceptrons (MLPs) and Neural Network had performed with high accuracy. LM performed with an accuracy rate of 92.95% while SCG obtained 78.21% accuracy [11].

Table 1. Classification of voice data-sets

Classification type	Classification model	Correctly classified	Incorrectly classified	Time taken (in seconds)
Bayes	BayesNet	84.6	15.4	0.11
	NaiveBayes	70.26	29.74	0.02
Functions	Simple logistic	85.13	14.87	0.98
Lazy	KStar	100	0	0
Meta	Bagging	92.31	7.69	0.27
Rules	NNge	100	0	0.1
Trees	ADTree	2.05	97.95	0.22
	J48	98.97	1.03	0.47
	RandomForest	99.49	0.51	0.17
	SimpleCART	96.41	3.59	0.42

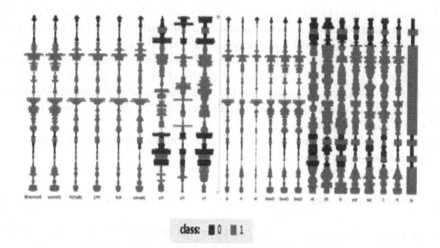

Fig. 1. Survey graph

Figure 1 and 2 presents the survey path and sieve multigram for the complete dataset. The survey graph shows high number of variances with healthy and diseased voice data-sets. Most of the healthy data-sets showed constant levels. Sieve multigram showed the edges between the nodes such as Fhi, Flo, Jitter, JitterAb, RAP and PPQ. Connections with edges are also present between Shimmer and ShimmerDB. The important observation by Ramani and Sivagami, 2011, shows the feature relevance analysis for better classification purpose and showed three important features like spread1, PPE and spread2 based on PD dataset [5].

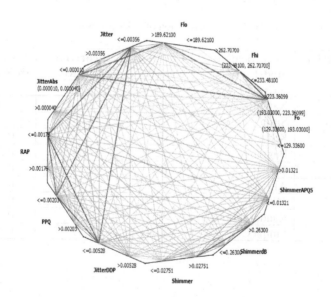

Fig. 2. Sieve multigram

ADTree and SimpleCART presented the relationships of nodes for the Parkinson data-sets. ADTree showed 21 leaves with 31 leaves and SimpleCART showed 13 leaves and 7 leaves. Fo data-sets can provide the focus in analyzing the frequencies for recognition of healthy and diseased individuals in Parkinson data-set (Figure 3).

Fig. 3. Decision Tree using ADTree and CART

Table 2 provided the clustering results for the complete data-set. All the algorithms that were analyzed have shown 2 clusters. Most of the clusters vary with DBScan and SimpleKMeans with 25% and 38% towards Parkinson disease. Clustered data by SimpleKMeans has been presented in Figure 4.

Table 2. Clustering

ClusterType	Number of clusters	Cluster report	
DBScan	2	0	144(75%)
		1	48(25%)
Hierarchial clustering	2	0	193(99%)
		1	2(1%)
SimpleKMeans	2	0	120(62%)
		1	75(38%)

```
kMeans
======
Cluster centroids:
                  Cluster#
Attribute              Full Data    0        1
                       (195)      (120)    (75)
===================================================
Fo                   154.2286  165.6164  136.0083
Fhi                  197.1049  211.9424  173.3649
Flo                  116.3246  125.6427  101.4157
Jitter                 0.0062    0.0042    0.0095
JitterAbs                   0         0    0.0001
RAP                    0.0033    0.0021    0.0052
PPQ                    0.0034    0.0022    0.0054
JitterDDP              0.0099    0.0064    0.0156
Shimmer                0.0297    0.0188    0.0472
ShimmerdB              0.2823    0.1731    0.4569
ShimmerAPQ3            0.0157    0.0099    0.0249
ShimmerAPQ5            0.0179    0.0111    0.0287
APQ                    0.0241    0.0149    0.0388
ShimmerDDA             0.047     0.0297    0.0746
NHR                    0.0248    0.0113    0.0466
HNR                   21.886    24.2069   18.1725
RPDE                   0.4985    0.4524    0.5724
DFA                    0.7181    0.7068    0.7361
spread1               -5.6844   -6.2804   -4.7308
spread2                0.2265    0.1934    0.2795
D2                     2.3818    2.2644    2.5696
PPE                    0.2066    0.1568    0.2862
class                       1         1         1
```

Fig. 4. Cluster centroids based on KMeans

Flo, Spread1 and APQ has shown best Attribute ranking based on RankSearch Method. Fhi, Flo, JitterDDP, APQ, NHR, spread1, spread2, D2 and PPE are the 9 locally selected attributes predicted based on CFS Subset evaluator.

4   Conclusions

Most of the work has been taken from the voice data-set. Diagnosing the voice data is useful in treating the disease by various voice exercises and also administering medicines at an early stage. Further research has to be done in these areas.

Acknowledgement. The author thanks the management and the staff of GITAM University and also RAGHU Institute of Technology, Visakhapatnam, India for the support extended in bringing out the above literature and experiment.

References

1. Joshi, S., Shenoy, D., Vibhudendra Simha, G.G., Rrashmi, P.L., Venugopal, K.R., Patnaik, L.M.: Classification of Alzheimer's Disease and Parkinson's Disease by Using Machine Learning and Neural Network Methods. In: Machine Learning and Computing (ICMLC), pp. 218–222 (2010)

2. Yang, M., Zheng, H., Wang, H., McClean, S.: Feature selection and construction for the discrimination of neurodegenerative diseases based on gait analysis. Pervasive Computing Technologies for Healthcare, 1–7 (2009)
3. Bonato, P., Sherrill, D.M., Standaert, D.G., Salles, S.S., Akay, M.: Data mining techniques to detect motor fluctuations in Parkinson's disease. In: Conf. Proc. IEEE Eng. Med. Biol. Soc., vol. 7, pp. 4766–4769 (2004)
4. Shianghau, W., Jiannjong, G.: A Data Mining Analysis of the Parkinson's Disease. IB 3(1), 71–75 (2011)
5. Geetha, R.R., Sivagami, G.: Parkinson Disease Classification using Data Mining Algorithms. International Journal of Computer Applications 32(9), 17–22 (2011)
6. Kaladhar, D.S.V.G.K., Nageswara, R.P.V., Ramesh, N.R.B.L.V.: Confusion matrix analysis for evaluation of speech on Parkinson disease using weka and matlab. International Journal of Engineering Science and Technology 2(7), 2734–2737 (2010)
7. Tanner, C.M., Ottman, R., Goldman, S.M., Ellenberg, J., Chan, P., Mayeux, R., William, L.J.: Parkinson Disease in Twins: An Etiologic Study. JAMA 281(4), 341–346 (1999)
8. Smith, M.E., Ramig, L.O., Dromey, C., Perez, K.S., Samandari, R.: Intensive voice treatment in parkinson disease: Laryngostroboscopic findings. Journal of Voice 9(4), 453–459 (1995)
9. Logemann, J. A., Gensler, G., Robbins, J., Lindblad, A. S., Brandt, D., Hind, J. A., Kosek, S., Dikeman, K., Kazandjian, M., Gramigna, G. D., Lundy, D., McGarvey-Toler, S., Miller, G. P. J.: A randomized study of three interventions for aspiration of thin liquids in patients with dementia or Parkinson's disease. J Speech Lang Hear Res. 51(1), 173–83 (2008).
10. Bhattacharya, I., Bhatia, M.P.S.: SVM classification to distinguish Parkinson disease patients. In: A2CWiC 2010, ACM, New York (2010)
11. Bakar, Z.A., Tahir, N.M., Yassin, I.M.: Classification of Parkinson's disease based on Multilayer Perceptrons Neural Network. Signal Processing and Its Applications (CSPA), 1–4 (2010)

Particle Swarm Optimization Based Adaptive PID Controller for pH-Neutralization Process

Shradhanand Verma and Rajani K. Mudi

Department of Instrumentation and Electronics Engineering
Jadavpur University, Sec-3, Block-LB/8, Salt-lake, Kolkata-700098
shradhanandei@gmail.com,
rkmudi@yahoo.com

Abstract. Though conventional PID controllers (CPID) are effectively used in a wide range of industrial processes, they usually fail to provide satisfactory performance for third order systems (pH-neutralization process) due to large overshoots and oscillation. Nonlinear and adaptive PID controllers (APID) are being developed towards achieving the desired control performance for such systems. In this study, we make an attempt to develop particle swarm optimization (PSO) based adaptive PID controller (PSO-APID) in order to attain adequate servo as well as regulatory performance. While designing our PSO-APID, first we formulate the structure of the adaptive PID controller followed by its optimal parameter estimation for a given system using PSO. Performances of PSO-APID for pH neutralization process are compared with those of CPID and APID reported in the leading literature. From detailed performance analysis PSO-APID is found to provide significantly improved performance over others.

1 Introduction

Conventional PID controllers (CPID) can provide satisfactory performance for first order and second order systems but their performance for third order systems (pH-neutralization process) usually not satisfactory due to associated intolerably large overshoot [1-3]. Several attempts have been made to overcome this limitation [4-6] by developing adaptive PID controllers (APID) through nonlinear parameterization. However, most of the APID parameters are selected based on trial and error or sometimes through heuristics [4-6]. Therefore, their performances may not be optimal. Keeping in mind this point, and excellent optimization power of particle swarm optimization (PSO) [7-9], in this study, we attempt to develop PSO based adaptive PID controller (PSO-APID) towards achieving optimum performance. Here PSO-APID involves two steps; first we define the structure of the adaptive PID then PSO is used to find its best set of parameters with respect to a given close-loop performance index or objective function. Performance of the developed PSO-APID for third order process with dead time (pH-neutralization process) is compared with other PID controllers. Considerably improved performance with respect to a large number of indices justifies the effectiveness of the developed PSO-APID.

2   The Proposed PID Controller

2.1   The Conventional PID Controller (CPID)

Block diagram of the discrete form of a conventional PID controller (CPID) is shown in Fig. 1. The three tunable gain parameters – proportional gain (K_p), integral gain (K_i), and derivative gain (K_d) play the key role in achieving the desired control performance.

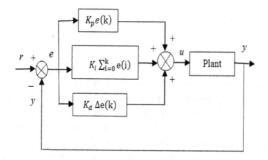

Fig. 1. Conventional PID controller

Output of the controller (CPID) can be expressed as:

$$u^c(k) = K_p \left[e(k) + \frac{\Delta t}{T_i} \sum_{i=0}^{k} e(i) + \frac{T_d}{\Delta t} \Delta e(k) \right]$$

or

$$u^c(k) = K_p\, e(k) + K_i \sum_{i=0}^{k} e(i) + K_d\, \Delta e(k) \tag{1}$$

Where K_p is the proportional gain, $K_i = K_p\left(\frac{\Delta t}{T_i}\right)$ is the integral gain, $K_d = K_p\left(\frac{T_d}{\Delta t}\right)$ is the derivative gain, T_i is the integral time, T_d is the derivative time, and Δt is the sampling period. We have used Ziegler-Nichols (ZN) continuous cycling method [3] for the initial settings of the PID parameters.

2.2   The Adaptive PID Controller (APID)

We have already mentioned that CPID usually fails to provide acceptable performance for third order systems (pH-neutralization process). In order to overcome such limitations several attempts have been made to develop adaptive PID (APID) [4]. In this study, we will concentrate on the APID presented in [4], where the three gain constants, *i.e.*, K_P, K_i, and K_d are continuously modified by an online updating factor alpha (α) with the following simple heuristic relations:

$$K_p^m(k) = K_p(1 + k_1 |\alpha(k)|) \tag{2}$$

$$K_i^m(k) = K_i(k_2 + k_3 \alpha(k)) \tag{3}$$

$$K_d^m(k) = K_d(1 + k_4|\alpha(k)|) \tag{4}$$

$$\text{here} \quad \alpha(k) = e_N(k) \times \Delta e_N(k) \tag{5}$$

with, $e(k) = r - y(k)$; $e_N(k) = \dfrac{e(k)}{|r|}$; and $\Delta e_N(k) = e_N(k) - e_N(k-1)$.

Now the APID can be redefined as:

$$u^m(k) = K_p^m(k)e(k) + K_i^m(k) \sum_{i=0}^{k} e(i) + K_d^m(k)\Delta e(k) \tag{6}$$

In Eqn. (6), $K_p^m(k)$, $K_i^m(k)$ and $K_d^m(k)$ are the modified proportional, integral and derivative gains respectively at k^{th} instant and $u^m(k)$ is the corresponding control action. k_1, k_2, k_3, and k_4 are the four additional positive constants. The objective behind such online gain adjustments as described by relations (2)–(4) is that, when the process is moving towards the set point, control action will be less aggressive to avoid possible large overshoots and/or undershoots, and when the process is moving away from the set point, control action will be more aggressive to make a rapid convergence of the system. Following this gain adaptive technique, in [4] a significantly improved performance of APID is found for higher order systems both in set-point and load disturbance responses.

However, out of the *seven* parameters of [4], i.e., K_P, K_i, K_d, k_1, k_2, k_3, and k_4, the first *three* constants, i.e., K_P, K_i, and K_d are selected based on ZN ultimate cycle rule, where as the remaining *four* constants, i.e., k_1, k_2, k_3, and k_4 are chosen by trial. Therefore, there is further scope to achieve more improved performance if we can find the most appropriate settings of these parameters.

Keeping in mind this objective and the PSO as a powerful optimization tool, we are motivated to develop the proposed PSO-APID. In the present work, *all* the seven parameters (i.e., K_p, K_i, K_d , k_1, k_2, k_3, and k_4) are selected through optimization using PSO. In the next section, we will describe the optimization process.

2.3 PSO Optimized APID (PSO-APID)

2.3.1 Particle Swarm Optimization (PSO)

Particle swarm optimization [7-9] is an artificial intelligence based technique used for maximization and minimization problems. It was inspired by social behavior of animals such as bird flocking and fish schooling. It starts with random set of solutions (known as particles). Each particle has positions (value of variables) and velocities. It improves the initial solutions by updating of velocities and positions.

2.3.2   Objective Function of PSO

The function to be optimized is known as objective function. Here, minimization of the *integral-absolute-error* (*IAE*) or *integral-time-absolute-error* (*ITAE*) or both (*IAE+ITAE*) is defined as the objective function (performance index or fitness function).

$$IAE = \int_0^\infty |e(t)| dt \quad \text{and} \quad ITAE = \int_0^\infty t|e(t)| dt$$

2.3.3   Different Terms, Variables and Its Range

Table 1.

Population	10 (it is the no of particle in a swarm, *i.e.*, set of solutions)
Variables	K_P, $K_{i,}$, K_d, k_1, k_2, k_3, and k_4
Range of variables	K_P, $K_{i,}$, K_d are ±20% of their respective CPID, k_1 [0,5], k_2 [0,1], k_3[0,5], and k_4[0,30].
Velocity range	[-l(span of the variable)l , l(span of the variable)l] Where span is the difference between largest and smallest value of that variable.
Local_best_particle	It is the best solution associated with minimum fitness or *IAE* of the current population.
Global_best_particle	It is the best solution associated with minimum fitness or IAE of the population found so far.

2.3.4   Different Operations Used in PSO

• **Population**

Here, the population size is chosen heuristically as 10. We have 7 variables, $K_p, K_i, K_d, k_1, k_2, k_3$, and k_4. We generate random set of solution by MATLAB-command, *random* (population size, variables). Since value of the variables is not in the defined range therefore we bring it in a defined range to get the real value of the optimization variable by following rule. Let *i* denotes the particles so *i* =1, 2, 3...10; and let *j* denotes the variables so *j* =1, 2, 3...7.

$$particle(i,j) = X_{min}(j) + \frac{X_{max}(j) - X_{min}(j)}{d_{max} - d_{min}} * (d(i,j) - d_{min})$$

Where, *particle (i,j)* = real value of variable; $X_{max}(j)$ = maximum value of *j*th variable of any set of solution; $X_{min}(j)$ = minimum value of *j*th variable of any set of solution; $d(i,j)$ = decimal value of variable; d_{max} = maximum value of decimal input; d_{min} = minimum value of decimal input.

- **Velocity , Velocity Update, and Position Update**

Each variable has velocity which is a vector quantity. Its vector nature helps the particles to traverse in all possible directions. Its initialization is similar as population.

We first update particle's velocity by

$$new\_velocity(i,j) = w*velocity(i,j) + c_1*r_1*(Local\_best\_particle(i,j) - particle(i,j)) + c_2*r_2*(Global\_best\_particle(i,j) - particle(i,j) \ ) \qquad (7)$$

Then we update particle's position by

$$particle(i,j) = particle(i,j) + new\_velocity(i,j) \qquad (8)$$

Where i =1, 2, 3…10 is the particles and j =1, 2, 3…7 is the variables. c_1 and c_2 are constant. $c_1 = 2$, is called the cognitive or personal or local weight. $c_2 = 2$, is called the social or global weight. Note that, as $c_1 = c_2$, here we give same weightage for both *Local_best_particle* and *Global_best_particle*. r_1 and r_2 are two random numbers both are in the range of (0, 1). w is the inertia weight, which is used to control the search. At the starting phase of search inertia weight is very high (almost equal to 0.99) and search is in exploration mode. At the end phase of search inertia weight is very low (almost equal to 0.01) and search is in exploitation mode. We make w as a time varying quantity (dynamic nature) to control the both modes, *i.e.*, w gradually decreases from 0.99 to 0.01 as the no of iteration increases. From the velocity update equation (7), if the *Global_best_particle* is too far from the *Local_best_particle*, then *Global_best_particle* has huge impact on the velocities and positions of the particles which are nearer to *Local_best_particle*. It means that for those particles higher change in velocities and positions occur. Therefore, particles finally move towards *Global_best_particle* .

2.3.5 Particle Swarm Optimization Algorithm
(1) Initialize the particles (solutions) and velocities.
(2) For each particle evaluate the fitness.
(3) Find out the particle associated with minimum fitness. Let it is ith particle then *Local_best_particle = particle(ith), Global_best_particle = particle(ith)*.
Start of PSO loop: repeat until slope of objective is almost zero or until maximum no of iteration is reached.

(a) For each particles, *i.e.*, i=1, 2, 3…10 and for each variables, i.e., j=1, 2, 3…7
    {Update the velocity by

$$new\_velocity(i,j) = w*velocity(i,j) + c_1*r_1*(Local\_best\_particle(i,j) - particle(i,j)) + c_2*r_2*(Global\_best\_particle(i,j) - particle(i,j))$$

   If $new\_velocity(i,j) > Vmax\ (j)$ then $new\_velocity(i,j) = Vmax(j)$
   Else if $new\_velocity(i,j) < Vmin(j)$ then $new\_velocity(i,j) = Vmin(j)$ }

(b) For each particles, *i.e.*, *i*=1, 2, 3...10 and for each variables, *i.e.*, *j*=1, 2, 3...7

 { Update the position by particle(i,j) = particle(i,j) + new_velocity(i,j)
 If   particle (i,j) > Xmax (j)   then   particle (i,j)=Xmax(j)
 Else if particle (i,j) < Xmin (j)   then   particle (i,j)=Xmin(j) }
(c) For each particle evaluate the fitness.
(d) Find out the particle associated with minimum fitness in the current population.
 Let for *i*th particle we are getting minimum fitness.
 Local_best_particle = particle(ith)
 If   f (Local_best_particle) < f (Global_best_particle)
       Global_best_particle = Local_best_particle
 End

Hence, *Global_best_particle* is our solution for which the fitness becomes minimum.

3   Results

For simulation study we consider the third order practical process, pH-neutralization with dead time (*L*) [4-6].

$$G_P(S) = \frac{e^{-LS}}{(1+S)(1+0.1S)^2} \tag{9}$$

Performance analysis of our PSO-APID is compared with conventional PID (CPID), PSO optimized conventional PID (PSO-CPID) and adaptive PID (APID) of [4]. For detailed comparison, in addition to the response characteristics, several performance indices, such as percentage overshoot (*%OS*), rise time (*t_r*), settling time (*t_s*), integral absolute error (*IAE*), and integral-time absolute error (*ITAE*) are calculated for each setting. Close-loop response curves (Fig. 2a and Fig. 2b) for different PID controllers are presented as: CPID (– - –),   PSO-CPID (– –), APID (- - -), and PSO-APID (—). Here, detailed performance analysis is provided for all the three objective functions, *i.e.*, minimization of *IAE, ITAE,* and *IAE+ITAE*. However, due to limitation of space, we provide response curves only for the objective function of *IAE* minimization.

Table 2 and Fig. 2 present the performance comparison of different controllers for process of (9) with *L* = 0.01s and *L* = 0.02s. Fig. 2a and Fig. 2b show the remarkably improved performance of our proposed PSO-APID during both set-point change and load disturbance applied at *t* =3.5s, and this fact is clearly established from the various indices of Table 2. Thus, our PSO optimized adaptive PID (PSO-APID) can significantly improve the performance over CPID, PSO-CPID, and APID, which justifies the usefulness of PSO for further enhancement of adaptive PID (APID).

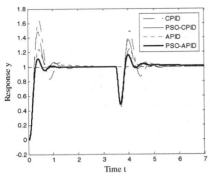

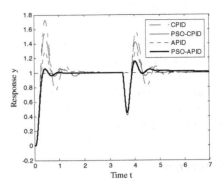

Fig. 2a. Responses of (9) with $L = 0.01\,s$ **Fig. 2b.** Responses of (9) with $L = 0.02\,s$

Table 2. Performance analysis for pH-neutralization process (9)

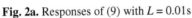

L	Controller	Obj. Fun	%OS	$t_r(s)$	$t_s(s)$	IAE	ITAE
	CPID		64.21	0.26	1.69	0.72	1.40
	PSO-CPID	IAE	48.42	0.25	1.23	0.52	0.94
	APID		23.94	0.30	1.77	0.53	1.13
	PSO-APID		11.38	0.30	0.79	0.39	0.82
	CPID		64.21	0.26	1.69	0.72	1.40
0.01s	PSO-CPID	ITAE	48.42	0.25	1.24	0.52	0.94
	APID		23.94	0.30	1.77	0.53	1.13
	PSO-APID		9.94	0.30	2.05	0.41	0.74
	CPID		64.21	0.26	1.69	0.72	1.40
	PSO-CPID	IAE	48.42	0.25	1.23	0.52	0.94
	APID	+	23.94	0.30	1.77	0.53	1.13
	PSO-APID	ITAE	9.75	0.31	0.81	0.39	0.79
	CPID		72.33	0.26	1.78	0.84	1.74
	PSO-CPID	IAE	53.95	0.26	1.27	0.59	1.08
	APID		28.49	0.30	1.84	0.56	1.18
	PSO-APID		5.74	0.35	0.80	0.42	0.88
	CPID		72.33	0.26	1.78	0.84	1.74
0.02s	PSO-CPID	ITAE	53.39	0.27	1.27	0.59	1.07
	APID		28.49	0.30	1.84	0.56	1.18
	PSO-APID		9.51	0.31	1.91	0.44	0.81
	CPID		72.33	0.26	1.78	0.84	1.74
	PSO-CPID	IAE	53.54	0.27	1.27	0.59	1.07
	APID	+	28.49	0.30	1.84	0.56	1.18
	PSO-APID	ITAE	0.0	3.96	0.87	0.48	0.96

4   Conclusion

In this work, we studied the performance of PSO optimized adaptive controller (PSO-APID) for pH-neutralization process with dead-time under both set-point change and load disturbance. Simulation results revealed that PSO-APID is capable of providing remarkably improved servo as well as regulatory performance compared to even PSO optimized conventional PID (PSO-CPID), and significantly overall improved performance in comparison with APID.

References

1. Shinsky, F.G.: Process control systems - application, design, and tuning. McGraw-Hill, New York (1998)
2. Astrom, K.J., Hang, C.C., Person, P., Ho, W.K.: Towards intelligent PID control. Automatica 28(1), 1–9 (1992)
3. Ang, K.H., Chong, G.C.Y., Li, Y.: PID control system analysis, design, and technology. IEEE Trans. Control Sys. Technology. 13(4), 559–576 (2005)
4. Dey, C., Mudi, R.K.: An improved auto-tuning scheme for PID controllers. ISA Transactions 48(4), 396–409 (2009)
5. Mudi, R.K., Dey, C., Lee, T.T.: An improved auto-tuning scheme for PI controllers. ISA Transactions 47(1), 45–52 (2008)
6. Mudi, R.K., Dey, C.: Performance improvement of PI controllers through dynamic set-point weighting. ISA Transactions 50(2), 220–230 (2011)
7. McCaffrey, J.,
   http://msdn.microsoft.com/en-us/magazine/hh335067.aspx
8. Eberhant, R.C., Knnedy, J.: A new optimizer using particle swarm theory. In: Proc. 6th Int. Symposium on Micro machine and Human Science, Nagoya, Japan, pp. 39–43 (1995)
9. del Valle, Y., Venayagamoorthy, G.K., Mohagheghi, S., Hernandez, J.C.: Particle Swarm Optimization: Basic Concepts, Variants and Applications in Power Systems. IEEE Trans. on Evolutionary Computation. 12(2), 171–195 (2008)

Extended Self Organizing Map with Probabilistic Neural Network for Pattern Classification Problems

Prasenjit Dey and Tandra Pal

Department of Computer Science and Engineering,
National Institute of Technology, Durgapur, India, 713209
{prasenjitdey13,tandra.pal}@gmail.com

Abstract. This paper presents a hybrid classifier based on extended Self Organizing Map with Probabilistic Neural Network. In this approach, at first we use feature extraction technique of Self Organizing Map to achieve topological ordering in the input data pattern. Then, with the use of Gaussian function, we obtain a better representation of the input dataset. After that, Probabilistic Neural Network is used to classify the input data. We have tested the proposed scheme on Iris, Glass, Breast Cancer Wisconsin, Wine, Ionosphere, Liver (BUPA), Sonar, Thyroid, and Vehicle data sets. The experimental results show better recognition accuracy of the proposed model than that of traditional Probabilistic Neural Network based classifier.

Keywords: Self Organizing Map (SOM), Probabilistic Neural Network (PNN), Feature Extraction, Radial Basis Function (RBF).

1   Introduction

Probabilistic Neural Network (PNN), a special type of Radial Basis Function Network (RBFN), is a well known classifier that has been extensively used in many applications of different domains. However, PNNs are prone to overfitting due to its less generalization capability, which in turn reduces the classification accuracy. Therefore, there still exists a need to find PNN based classifier having better capability in terms of accuracy by enhancing its generalization capability.

In the last few decades, researchers have put huge efforts to improve the classification accuracy of different classifiers. In [1], the authors have shown the pattern classification capability of PNN based classifier on several real life pattern classification problems, like - text independent speaker identification systems, classification of brain tissues in multiple sclerosis, speaker verification and EEG pattern classification. The author in [2] has proposed a new class of PNN for pattern classification in time varying environment. The applications of PNN for face recognition have been illustrated in [3], where wavelet decomposition and discrete cosine transform have been used to extract the features of the images. Then, the features are sent to back propagation neural network, linear vector quantization neural network, and PNN for image recognition. In [3], the authors have shown that PNN has better recognition accuracy compared to others. Schwenker et al. [4] have described various training methods of RBFN. Another application of RBFN is found in [5], where the authors have implemented RBFN for recognition of idiopathic pulmonary fibrosis in microscopic images. For the development of the RBFN classifiers, they have used fuzzy means clustering algorithm. A comparative analysis of the performance of RBFN with Gaussian Mixture Model (GMM) for voice recognition has been presented in [6]. They have trained GMM model with Expectation Maximization (EM) algorithm on a dataset containing 10 classes of vowels. The classification ability of Self Organizing Map (SOM) and its comparative study with other classification methods have been presented in [7]. Hybrid models of SOM have been used there for various real life applications. In [8], the authors have used a hybrid model of SOM and RBFN for induction machine fault detection. A multisite model of SOM and RBFN has been proposed in [9] for groundwater level prediction. In that work [9], at first the authors have found the number of hidden units of RBFN using SOM, which basically represents the number of clusters. Then, they have determined the position of the radial basis centres. A SOM-PNN Classifier for volume data classification and visualization has been proposed in [10], where the reference vectors from each of the classes of the trained SOM were used to estimate the probability distribution function. Hybrid SOM-RBFN classifiers have also been used for developing incremental intrusion detection system in [11].

In all of the above mentioned works, the researchers have tried to improve the classification accuracy on various types of applications. They have combined different neural network models to achieve better classification accuracy. In our proposed approach, the classification ability has been improved by using SOM and then a Gaussian function. The patterns have been topologically sorted by feature extraction of SOM and then the Gaussian functions map this input space into a better feature space. Subsequently, PNN performs pattern classification over the extracted feature space. Since, a well represented data pattern is less prone to overfitting, the proposed model of classifier shows better classification accuracy than the traditional PNN classifier.

The rest of this paper is organized as follows. Section 2 describes the preliminaries and the architecture of PNN and SOM required for the proposed model. Section 3 presents the proposed model for classification. Experimental results and analytical discussions are presented in section 4. Finally, we conclude in section 5.

2   Preliminaries

2.1   Feature Extraction

Many real life classification problems require supervised learning. In supervised learning, class probabilities and class conditional probabilities are unknown. Each instance is associated with a class label. In classification problems, some of the available features may be partially or completely irrelevant or redundant, i.e., relevant features are often unknown a *priori*. For the dataset whose dimension (number of features) is large and most of the features are irrelevant or redundant, learning may not work well if these unwanted features are not removed judicially. Reduction of number of irrelevant and redundant features may drastically improve the learning, as well as, it may reduce the learning time. This technique of reduction of irrelevant and redundant features by transforming the given large dimensional dataset into a lower dimensional space is called feature extraction.

2.2   Self Organizing Map (SOM)

Self Organizing Map Network (SOMN), a special type of artificial neural network (ANN), is trained using competitive unsupervised learning to map an input data from a high dimensional space to a low dimensional space of the training samples. SOMNs are different from other kind of ANNs in a sense that they use a neighborhood function to preserve the topological properties of the input space. It makes SOMN to provide low-dimensional views of the high-dimensional data [12]. This property of SOM, we have used in our model, for feature extraction.

SOMN, is used as m-dimensional feature map. There are two layers in SOMN, input layer and output map layer. The number of nodes in input layer is same as number of features of the input dataset. For m-dimensional feature map, the number of nodes in output map layer is m. All the connections of a node of the output map layer with all nodes of input layer are associated with weights, each of dimension same as the input. The output layer represents the m-dimensional feature map of the input data pattern. The architecture SOMN is shown in Fig 1 (a).

2.3   Probabilistic Neural Networks (PNNs)

PNN is defined as an implementation of statistical algorithm called Kernel discriminate analysis [1]. Parzen or a similar probability density function is used in PNN. Classifier based on PNN approaches to Bayes optimal classification, estimating the probability density function for each class based on the training samples. The training is performed by a single pass of each of the training vectors, rather than several passes used in Multilayer Perceptrons (MLPs). PNN is frequently used to solve variety of applications.

PNN is a multilayered feed-forward neural network with four layers: input layer, hidden layer, summation layer and output layer [13]. PNN is shown below in Fig 1(b). Number of nodes in the input layer is same as a number of features of the data set or application. All connections between input and hidden layers are associated with a weight value of 1, i.e., the input vector is directly passed to each hidden node. A hidden node corresponds to one training instance of the training data set. Each hidden node H_i has a centre point y_i associated with it, which represents i^{th} input instance. It also has a spread factor, σ_i, that determines the size of its respective field. A hidden node receives an input vector x and activates a Gaussian function, which returns a value of 1 if x and y_i are equal and drops continuously as the distance grows. Each hidden node in the network is connected to a single node in the summation layer. If the output class of ith instance is j, then H_i is connected to node C_j. Each node of summation layer computes the sum of the activations of the hidden nodes that are connected to it and passes this sum to a single decision node that belongs to the output layer. The decision node outputs the class with the highest summed activation in output layer.

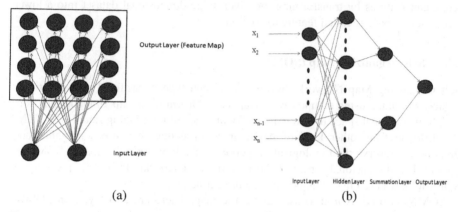

| (a) | (b) |

Fig. 1. (a) Architecture of a self organizing map network (SOMN) (b) Architecture of a probabilistic neural network (PNN)

3   Proposed Model

The proposed model for pattern classification using SOM and PNN is described here. At first, SOM is used to get the topological relationship in input patterns. Then, we use a Gaussian function to map the input space into a better feature space. After that, PNN is used for classification. The detailed method is as follows and Fig. 2 depicts the architecture of the proposed model.

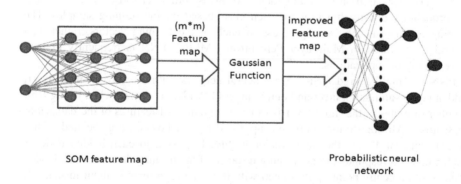

Fig. 2. A hybrid SOM (with Gaussian function) with PNN model

Let, the input layer of the SOM is an array of N neurons, denoted by

$$\mathbf{x} = [x_1, x_2, ..., x_N]^T \tag{1}$$

The output layer contains M neurons, denoted by u_j (j = 1, 2, . . ., M), which are organized in a 2D map. The weights of the connections between the neurons of input and output layers be w_{ij} (i = 1, 2, . . ., N; j = 1, 2, . . ., M). The weight vectors are denoted as follows in (2).

$$\mathbf{w_j} = [w_{1j}, w_{2j}, ..., w_{Nj}]^T , j = 1,2,....,M.$$ (2)

Before training, the weights are initialized randomly in the range [0, 1]. SOM computes a similarity measure d_j between the input vector $\mathbf{x}$ and the weight vector $\mathbf{w_j}$ of each neuron in output layer using Euclidean distance as defined in (3).

$$d_j = \left\| \mathbf{x} - \mathbf{w_j} \right\| = \sqrt{\sum_{i=1}^{M}(x_i - w_{ij})^2}$$ (3)

The output neuron with the weight vector having the smallest distance from the input vector is the winner. The weights of the winning neuron along with the weights of the neurons in topological neighborhood of the winning neuron are adjusted in the direction of the input vector. The influence of competition drops symmetrically from the location of the winning neuron location which is at the center of the topological neighborhood. In this work, the following function given in (4), has been used to find the neighborhood.

$$g_j = \exp\left(\frac{\left\| u_j - u_j^{"} \right\|}{2\sigma^2}\right)$$ (4)

Here, g_j is the topological neighbourhood function, σ is the width of the topological the neighbourhood function and u_j denotes the winning neuron. The change of the weight vector $\mathbf{w_j}$ is obtained as follows in (5).

$$d\mathbf{w_j} = \eta g_j (\mathbf{x} - \mathbf{w_j})$$ (5)

In (5), η is the learning-rate and updating weight vector $\mathbf{w_j}^{(t+1)}$ at time t+1 is

$$\mathbf{w_j}^{(t+1)} = \mathbf{w_j}^{(t)} + \eta^{(t)} g_j^{(t)} [\mathbf{x} - \mathbf{w_j}^{(t)}],$$ (6)

where $\eta^{(t)}$ and $g_j^{(t)}$ in (6) are respectively learning-rate and topological neighborhood at time t. The weight vectors tend to move toward the input pattern due to the neighborhood updating by repeated presentations of the training data. In other words, the adjustment makes the weight vectors to be similar to the input pattern. Hence, the winning neuron shows the topological location of the input pattern and the neighborhood of the winning neuron indicates the statistical distribution of the input pattern.

Then a special technique has been used to map the input space into a feature space using Gaussian function. The final value of output node u_j from SOM is obtained using a Gaussian function over the Euclidean distance between the input vector $\mathbf{x}$ and the weight vector $\mathbf{w_j}$ of each neuron u_j in output layer. The mapping function is given below in (7).

$$u_j = \exp\left(\frac{\|\mathbf{x} - \mathbf{w}_j\|}{2\sigma^2}\right) \tag{7}$$

Once this feature extraction process is complete, PNN classifier is used over this extracted feature space to perform the classification task. The extracted feature pattern vector $\mathbf{x} \in R^{m*m}$ is fed to PNN, where m*m is the SOM grid.

Let these patterns are assigned to one among k predefined classes. The conditional density of a particular class represents the uncertainty associated with the class attribute. Here, we have estimated the conditional density $p(\mathbf{x}|c_k)$ of each class C_k using Parzen window technique where $\mathbf{x}$ is an unknown test instance. In Parzen window technique, a sphere of influence $d(\mathbf{x}, \mathbf{s})$ around each training instance $\mathbf{s}$ is built and they are added up for each of the k classes, as given in (8). These estimates are combined with the Bayes' rule to get the posteriori class probabilities $p(c_k|\mathbf{x})$ that allow to make optimal decisions.

$$p(\mathbf{x}|c_k) = \sum_{\mathbf{x} \in c_k} d(\mathbf{x}, \mathbf{s}), \tag{8}$$

where, the basis function used as window is Gaussian Kernel is given by (9).

$$d(\mathbf{x}, \mathbf{s}) = \exp\left(-\frac{\|\mathbf{x} - \mathbf{s}\|^2}{2\sigma^2}\right) \tag{9}$$

In our approach, each hidden node computes the distance $d(\mathbf{x}, \mathbf{s})$ from the test vector $\mathbf{x}$ to each training instance $\mathbf{s}$ and produces output value according to (9). Each node of summation layer computes the sum of the activations of the hidden nodes those are connected to it and using (8) passes this sum to a single decision node in the output layer. The decision node outputs the class with the highest summed activation in output layer.

4   Experimental Results and Discussion

To assess the performance of the proposed model, we have used nine datasets, obtained from [14] to compare our classifier with traditional PNN based classifier. The summary of all the datasets is presented in Table 1. We have used Neural Network Toolbox of MATLAB (Version 8.1) for the simulation purpose. The spread value of PNN, for both the proposed model and the traditional PNN based classifier, is chosen to be 0.1 (default in MATLAB). For both the classifiers, we have first normalized the data and then performed 10-fold cross validation 10 times for each data set. The mean obtained from 10 runs of 10-fold cross validation by PNN and the proposed method is shown respectively in column 3 and column 4 of Table 2 for all the data sets. A statistical validation test, wilcoxon signed rank test [15] on the data sets has been performed to make a proper comparison of the proposed method with the traditional PNN. The result is shown in the last two columns of Table 2. We reject the null hypothesis at $\alpha = 0.05$ level of significance on the basis of the p-value obtained using Table 2. It implies that the proposed method is significantly better then the PNN.

Table 1. Data sets

Sl. No.	Data set	No of features	No of classes	Size of Data Set and Class Wise Distribution
1	SONAR	60	2	208 (97+111)
2	Ionosphere	34	2	351 (225+126)
3	Vehicle	18	4	846 (212+217+218+199)
4	Wine	13	3	178 (59+17+48)
5	Breast Cancer	9	2	699 (458+241)
6	Glass	9	6	214 (70+76+17+13+9+29)
7	Liver	6	2	345 (145+200)
8	Thyroid	5	3	215 (150+35+30)
9	Iris	4	3	150 (50+50+50)

Table 2. Classification rate (%) of PNN and the proposed model and Wilcoxon Signed-Ranks Test

Sl. No.	Dataset	PNN	Proposed Method	difference	S/R ="signed rank"
1	SONAR	62.02	67.04	-5.02	-9
2	Ionosphere	71.74	72.13	-0.39	-3
3	Vehicle	72.39	73.67	-1.28	-4
4	Wine	96.20	98.17	-1.97	-5
5	Breast Cancer	95.00	97.43	-2.43	-6
6	Glass	67.00	70.93	-3.93	-8
7	Liver	67.67	67.73	-0.06	-1
8	Thyroid	94.68	94.93	-0.25	-2
9	Iris	95.23	98.04	-2.81	-3

W=-45, z-score is -2.665570, p-value is 0.007686.

5   Conclusion

This paper proposes a new hybrid model for pattern classification problems by combining the features of SOM and PNN. At first, we use SOM to get the topological relationship of the input patterns. Then, we use a Gaussian function to map the topologically sorted input space into a better feature space. After this process, the features provide a better representation of the input space. Experimental results reveal that the proposed model possesses good generalization ability, measured in terms of accuracy, enhancing the capabilities of the traditional PNN based classifiers.

The proposed SOM-PNN hybrid model is more time consuming than the traditional PNN based classifier due to addition of the feature extraction procedure. Our future work will focus on overcoming this limitation.

References

1. Ibrahiem, M., El Emary, M., Ramakrishnan, S.: On the Application of various probabilistic neural networks in solving different classification problems, vol. 4(6), pp. 772–780. IDOSI Publications (2008)
2. Rutkowski, L.: Adaptive Probabilistic Neural Networks for Pattern Classification in Time-Varying Environment. IEEE Transactions on Neural Networks 15(4), 811–827 (2004)
3. Qiakai, N.I., Chao, G., Jing, Y.: Research of Face Image Recognition Based on Probabilistic Neural Networks. In: IEEE 24th Chinese Control and Decision Conference (2012)
4. Schwenker, F., Kestler, H.A., Palm, G.: Three learning phases for radial-basis-function networks. Neural Networks 14(4-5), 439–458 (2001)
5. Maglogiannis, I., Sarimveis, H., Kiranoudis, C.T., Chatziioannou, A.A., Oikonomou, N., Aidinis, V.: Radial Basis Function Neural Networks Classification for the Recognition of Idiopathic Pulmonary Fibrosis in Microscopic Images. IEEE Transactions on Information Technology in Biomedicine 12(1), 42–54 (2008)
6. Anifowose, F.A.: A Comparative Study of Gaussian Mixture Model and Radial Basis Function for Voice Recognition. International Journal of Advanced Computer Science and Applications 1(3) (2010)
7. Vasighil, M., Kompany-Zareh, M.: Classification Ability of Self Organizing Maps in Comparison with Other Classification Methods. Commun. Math. Comput. Chem. 70, 29–44 (2013)
8. Wu, S., Chow, T.W.S.: Induction Machine Fault Detection using SOM-Based RBF Neural Networks. IEEE Transactions on Industrial Electronics 51(1), 183–194 (2004)
9. Chen, L.-H., Ching-Tien, C., Yan-Gu, P.: Groundwater Level Prediction Using SOM-RBFN Multisite Model. Journal of Hydrologic Engineering 15, 624–631 (2010)
10. Ma, F., Wang, W., Tsang, W.W., Zesheng, T., Shaowei, X., Xin, T.: Probabilistic Segmentation of Volume Data for Visualization Using SOM-PNN Classifier. In: IEEE Symposium on Volume Visualization, pp. 71–78 (1998)
11. Li-Ye., T., Wei-Peng, L.: Incremental intrusion detecting method based on SOM/RBF. In: Proceedings of the Ninth International Conference on Machine Learning and Cybernetics, pp. 11–14 (2010)
12. https://www.princeton.edu/~achaney/tmve/wiki100k/docs/Self-organizing_map.html
13. Specht, D.F.: Probabilistic Neural Networks. Neural Networks 3(1), 109–118 (1990)
14. https://archive.ics.uci.edu/ml/datasets.html
15. http://scistatcalc.blogspot.in/2013/10/wilcoxon-signed-rank-test-calculator.html

Path Planning and Control of Autonomous Robotic Agent Using Mamdani Based Fuzzy Logic Controller and ARDUINO UNO Micro Controller

Pratap Kumar Panigrahi and Sampa Sahoo

Padmanava College of Engineering, Rourkela, Odisha, 769002, India
{pratapkumarpanigrahi,Sampaa2004}@gmail.com

Abstract. The autonomous mobile robots are used for various purposes like materials transportation, nuclear and military environments etc. In this paper fuzzy logic technique is used for controlling the mobile robot in an unknown environment. The main goal of robot is to reach the target point from a starting point with avoiding obstacles in the way. The Mamdani fuzzy logic controller is used to obtain collision free path where inputs are front obstacle distance (FOD), left obstacle distance (LOD), right obstacle distance (ROD), heading angle (HA) and the output corresponds to the steering angle (SA) of the mobile robot. The effectiveness of the controller is verified using Mamdani fuzzy inference system.

Keywords: Autonomous Mobile Robot, Collision-free Path, Mamdani Fuzzy Inference System, Target Seeking Behavior, Arduino Uno Microcontroller.

1   Introduction

The basic functions of mobile robot includes avoiding obstacles, finding best route, trajectory tracking and so on, which are research content for path planning for mobile robot. The usage of the general method becomes impossible and not cost-effective due to its additional requirements like total map of the unknown environment. Therefore different evolutionary methods like fuzzy systems or the neural networks are used to solve this problem. Fuzzy logic is used to handle the indistinct and inaccurate information. It employs the human knowledge for linear mapping between input sensory data and output control actions. The path planning of mobile robot includes a large amount of uncertainty and incomplete or the absence of priori knowledge of the real world. Fuzzy logic controller is much more suitable to solve such situations. Fuzzy logic framework make use of human reasoning and decision-making to formulate a set of simple and intuitive IF(antecedent)—THEN (consequent) rules, also known as fuzzy rule. These rules are written in easily understandable and natural linguistic representations. The fuzzy logic controller consists of fuzzy rules and membership functions (input and output) based on expert knowledge, modelling of process or learning. Fuzzy rules are usually based on Mamdani or Takagi–Sugeno–Kang (TSK) rule base system.

2   Related Work

The major challenge of autonomous robotics is to build robust control algorithms that reliably perform complex tasks in spite of environmental uncertainties and without human intervention. In the last decade, a great amount of research has been carried out to increase the autonomy of mobile robots. Several advanced control algorithms have been proposed to guarantee successful navigation in real-world applications. The authors [1], [6], [8], [11], [13] have designed various types of Intelligent controller based on fuzzy logic technique and optimized model of fuzzy logic controller for obstacle avoidance of mobile robot. They have verified that design of ordinal structure fuzzy logic is easier than conventional fuzzy logic controller. Their work also presents the comparison of incremental GA with direct GA, which is used for optimization of fuzzy logic controller. This paper [2] uses fuzzy-neural network for finding best route for mobile robot from starting point to ending point in dynamic environment. They put forward a novel membership function based on collision prediction. If the obstacles and the robot are closer and the expected collision time is shorter, then probability of collision is strong, otherwise collision is unlikely to happen. This paper [3] presents design of neural network based on radial basis function and reinforcement learning mechanism for tracking of a desired orientation profile of the mobile robot. The neural network is trained via reinforcement learning where neural controller is charged to enhance the control system by adding some degrees of award. The researchers [4] have used a hybrid technique which is a combination of fuzzy inference system and artificial neural network (MANFIS) for navigation of mobile robot. Fuzzy-reasoning radial basis function neural network (FRBFN) [5] is proposed in paper, which is a three-layer neural network. Here the training of rule-based fuzzy system is done through reinforcement learning. This paper [7] describes the use of multi-objective genetic algorithm (MOGA) for the path planning of autonomous mobile robot. This work shows that both MOGA and conventional GA are effective tools for solving point-to-point path planning problem. This paper [9] proposed a new fuzzy logic based navigation method for movement of mobile robot in an unknown environment. A fuzzy controller based on human sense and a fuzzy reinforcement algorithm is used to fine tune fuzzy rule base. In this paper [10] 3D –depth and color imaging is used .A remote server is used to perform target recognition and tracking of the robot and is done by neural network. Fuzzy logic is used to supply the control mechanism necessary to follow and navigate towards targets. This paper [12] describes the design, implementation and comparison of zero order Takagi-Sugeno and Mamdani fuzzy logic controllers for obstacle avoidance of mobile robot in indoor environment. The authors [14] have used Q-learning method for path planning of mobile robot in an unknown environment. The size of the Q-table is reduced by a new definition for the search space. The performance comparison of Q-learning method and potential field method shows that new approach takes less time to reach the target and it has high hit rate. The paper [15] behavior based robot controller is tested in maze like indoor environment. Simulation environment based 2D robot simulator is used for testing navigation algorithms and control logic for robots. For a given set of end points the simulation environment monitors the time elapsed between the start point and end point and their result shows that none of the target searches needed more than 200s.

3   Problem Statement

The mobile robot path planning problem is typically classified as follows: given a mobile robot and a description of an environment, mobile robot needs to plan a path autonomously between two specified locations, a start and goal point. The path should be collision free and shortest. In general the researchers apply different methods to solve the path planning problem according to two factors namely the environment type (i.e., static or dynamic) and the type of path planning algorithms (i.e., global or local). The static environment is an   environment which contains static/fixed obstacles of different shapes and sizes while the dynamic environment is an environment which contains dynamic/moving obstacles (i.e. human beings, moving robots and moving vehicles). Global path planning algorithm requires complete information about the environment (i.e. map, cells and grid etc.) so that a robot can perform navigation comfortably whereas in the local path planning the robot does not have any information about the environment at the same time robot has to sense the environment and autonomously decides the steering angle to move towards the target for obstacle avoidance.

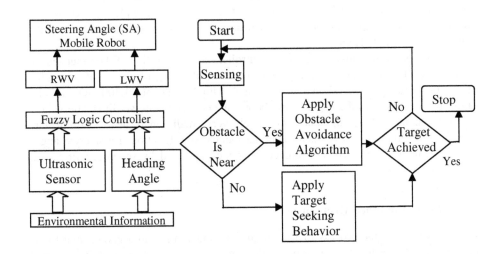

Fig. 1. Control Diagram of Wheel Mobile Robot

Fig. 2. Controller Flow Chart

In the research it is assumed that the robot moves in a flat ground and the robot moves without slipping (i.e. the robot is a non-holonomic mobile robot). The environment contains static obstacles of different shapes and sizes in disorganized order.

4   Fuzzy Logic Controller Algorithm

The main goal of the controller is divided into two parts; in the first part the mobile robot is to avoid the obstacles based on the information gained from the sensors located in the robot then in the second part, it should move towards the goal in the absence of the obstacles. While moving towards the goal point avoiding obstacles the robot changes its steering angle by changing left wheel and right wheel velocity as shown in the Fig.2.The fuzzy controller implemented in this work has four inputs (i.e. LOD, FOD, ROD and HA) and one output (SA).The left obstacle distance (LOD) is the distance of the robot from nearest obstacle in the left side of the environment. The front obstacle distance (FOD) is the distance of robot from nearest obstacle in the front side of the environment and so on. While the heading angle (HA) is the angle between the lines joining robot and goal point with respect to y-axis of the environment in clockwise direction. When robot is allowed to move in the environment the heading angle is always provided to the robot. If the robot finds no obstacle around it the target seeking algorithm works and when robot faces obstacles nearby in any direction then immediately obstacle avoidance algorithm is activated to avoid the obstacles. When obstacle avoidance algorithm works the robot makes left or right turn instead of straight towards the target until it completely avoids the obstacles. Once it avoids all the obstacles nearby it moves towards the target straight with target seeking behavior.

In the proposed Mamdani type fuzzy controller the fuzzification of the input sensor values of LOD, FOD and ROD are explained with the help of six set of triangular membership functions as shown in the Fig.4 with a scale of (0-300) unit as our ultrasonic senor senses the obstacle of about 3 meters. There are six linguistic values of all the input membership functions namely very small, small, near, medium, far, very far .The output has seventeen triangular membership functions for steering angle from -90^0 to 90^0 with 10^0 differences and are explained by seventeen linguistic values like X_1, X_2 etc.

4.1   Fuzzy Inference Mechanism

Fuzzy inference is the process of formulating the mapping from a given input to an output using fuzzy logic. The mapping then provides a basis from which decisions can be made, or patterns recognized. The used inference fuzzy mechanism is the MAMDANI fuzzy inference system.

4.2   Fuzzy Rules

The algorithm uses the lingual rules when robot faces different distances i.e. LOD, FOD and ROD within the environment while moving from one position to a target. The fuzzy rules are if- then statements that consist of a premise (antecedent) and consequent. Few rules are written below. In the proposed fuzzy controller 9 rule bases are prepared for controlling the steering angle (output) of mobile robot. Some of the rules are as follows:

Rule 1: If LOD is very far and FOD is very far and ROD is very far then SA is X_{14}.

Rule 2: If LOD is very small and FOD is very small and ROD is very small then SA is X_1.

Rule 3: If LOD is near and FOD is very small and ROD is very small then SA is X_{16}.

4.3   Defuzzification

The reverse process of fuzzification is called defuzzification. The rules of Mamdani FLC produce the required output in the form of linguistic variables (fuzzy variables). According to the real world requirements, linguistic variables have to be converted into crisp/real values. There are many defuzzification methods such as min-max method, center of gravity method and center of gravity of largest area method etc. In our problem the logical sum of inference result is steering angle (SA) of mobile robot which is obtained by center of gravity method. Corresponding crisp values of linguistic variables X_1, X_{10}, X_{13}, X_{14}, X_{16} and X_{17} used in output are -80^0, 10^0, 40^0, 50^0, 70^0 and 80^0 respectively

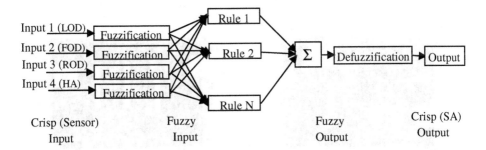

Fig. 3. Fuzzy Logic System

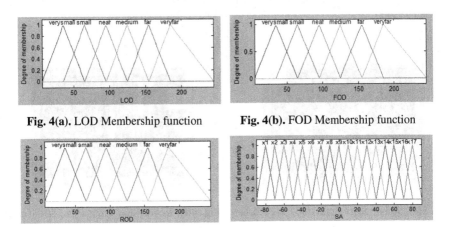

Fig. 4(a). LOD Membership function   **Fig. 4(b).** FOD Membership function

Fig. 4(c). ROD Membership function   **Fig. 4(d).** Output Membership function

5   Simulation Results and Discussion

The simulation of path planning of mobile robot is conducted using MATLAB in a 2-D work space (500×500) square unit using Mamdani based fuzzy inference system. The Fuzzy logic based obstacle avoidance and target seeking algorithm is verified

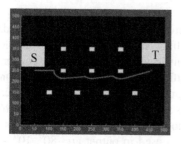

Fig. 5(a). Robot Start Point (50,250) Target (450,250)

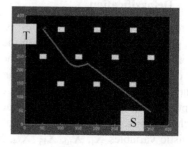

Fig. 5(b). Robot Start Point (350, 50) Target (50,350)

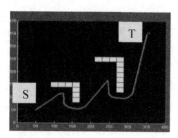

Fig. 5(c). Robot Start Position (50, 50), Target (400,400)

Fig. 5(d). Robot Start Position (450,250), Target (50,450)

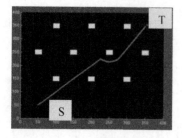

Fig. 5(e). Robot Start Point (50, 50), Target (350,350)

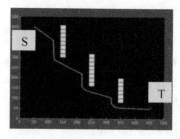

Fig. 5(f). Robot Start Point (50, 350), Target (350, 50)

keeping different type of static obstacles in the environment. Fig.5 (a) shows the path of the mobile robot while moving from (50,250) position to the target position (450,250) with 10 square size obstacles in the environment. It is evident that the mobile robot successfully reached the target without collision with obstacles. Similarly in Fig.5 (b) and 5(e) the robot avoids the rectangular shape obstacles and reached the target with different starting and target positions. In Fig.5 (c), (d) and (f) the mobile robot is allowed to navigate in presence of L and I-shaped obstacles environment from different start positions to target positions. It is observed that the autonomous mobile robot successfully reached the target making collision free path.

6   Experimental Set Up

In order to validate the theoretical results, experimental analysis can be carried out for four wheeled differential drive mobile robot. The robot is equipped with the following components

1. Microcontroller—Arduino UNO microcontroller board based on the ATmega2560. It has 14 digital input/output pins (of which 6 can be used as PWM outputs).
2. Obstacle sensor — Three HC-SR04 ultrasonic distance sensor are mounted in left , front and right directions of the body frame of mobile robot which detects the obstacles in the environment..
3. Position encoder—It is used for determining position and velocity of the wheel. It works on
4. Magnetic Compass—It is interfaced using I2C communication which determines the direction of the robot.
5. L2938—It is a motor driver which helps in controlling the speed of the two rear DC motors

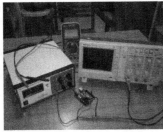 

Fig. 6(a). Experimental Set up   **Fig. 6(b).** Mobile Robot

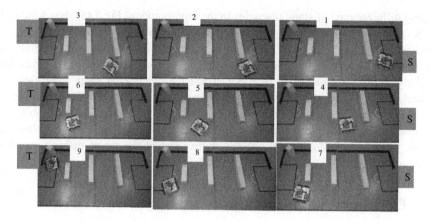

Fig. 7. Experimental Result of mobile robot with environment 5(d)

The performance of autonomous mobile robot in experimental mode is shown in Fig.7 with the environment of Fig.5 (d) in MATLAB simulation. During experiment the robot destination point is specified in the Arduino Uno microcontroller program. In this experiment robots different positions are shown while navigating from starting position to target position. It is evident that robot has successfully reached the target without collision with obstacles.

References

1. Samsudin, K., Ahmad, F.A., Mashohor, S.: A highly interpretable fuzzy rule base using ordinal structure for obstacle avoidance of mobile robot. International Journal of Applied soft Computing 11, 1631–1637 (2011)
2. Jiang, M., Yu, Y.X., Liu, F.Z., Hong, Q.: Fuzzy Neural Network Based Dynamic Path Planning. In: Proceedings of International Conference on Machine Learning and Cybernetics, Xian, pp. 15–17 (2012)
3. Bayar, G., Konukseven, E.I., Koku, A.B.: Control of a Differentially Driven Mobile Robot using Radial Basis Function Based Neural Networks. WSEAS Transactions on Systems and Control 3(12), 1002–1013 (2008)
4. Mohanty, P.K., Parhi, D.R.: A New Intelligent Motion Planning for Mobile Robot Navigation Using Multiple Adaptive Neuro-fuzzy Inference System. International Journal of Applied Mathematics and Information Sciences 8(5), 2527–2535 (2014)
5. Yang, Y.K., Lin, Y., Fang, W.L., Pan, J.K.: A Fuzzy-reasoning Radial Basis Function Neural Network with Reinforcement Learning Method. In: Proceedings of ICAI, pp. 1–5 (2012)
6. Eskandar, H., Salehi, P., Sabour, M.H.: Fuzzy Logic Tracking Control for a Three Wheel Circular Robot in Unknown Environment. World Applied Science Journal 11, 321–326 (2010)
7. Castillo, O., Trujillo, L., Melin, P.: Multiple Objective Genetic Algorithms for Path Planning Optimization in Autonomous Mobile Robots. International Journal of Computer Systems and Signals 6(1), 48–63 (2005)

8. Faisal, M., Hedjar, R., Suaiman, M.A., Mutib, K.A.: Fuzzy Logic Navigation and Obstacle Avoidance by a Mobile Robot in an Unknown Dynamic Environment. International Journal of Advanced Robotic Systems 10, 1–7 (2013)
9. Boubertakh, H., Tadjine, M., Glorennec, Y., Labiod, S.: A Simple Goal Seeking Navigation Method for a Mobile robot using Human Sense. Fuzzy Logic and Reinforcement learning, Journal of Automatic Control 18(1), 23–27 (2008)
10. Benavidez, P., Jamshidi, M.: Mobile Robot Navigation and Target Tracking System. In: Proceedings of IEEE, 6th International Conference on System of Systems Engineering, Albuquerque, December 27-30, pp. 299–304 (2011)
11. Pradhan, S.K., Parhi, D.R., Panda, A.K.: Fuzzy Logic Techniques for Navigation of Several Mobile Robots. International Journal of Applied Soft Computing 9, 290–304 (2009)
12. Farooq, U., Hasan, K.M., Abbas, G., Asad, M.U.: Comparative Analysis of Zero Order Sugeno and Mamdani Fuzzy Logic Controllers for Obstacle Avoidance of Mobile Robot Navigation. In: IEEE International Conference and Workshop on Current Trends in Information Technology (CTIT), Dubai, October 26-27, pp. 113–119 (2011)
13. Shayestegan, M., Marhaban, M.H.: Mobile Robot Safe Navigation in Unknown Environment. In: IEEE International Conference on Control System, Computing and Engineering, Penang, November 23-25, pp. 44–49 (2012)
14. Jaradat, M.A.K., Al-Rousan, M., Quadan, L.: Reinforcement Based Mobile Robot Navigation in Dynamic Environment. International Journal of Robotics and Computer Integrated Manufacturing 27, 135–149 (2011)
15. Rusu, P., Petriu, T.E., Cornell, W.A., Spoelder, H.J.W.: Behavior Based Neuro Fuzzy Controller for Mobile Robot Navigation. In: Proceedings of IEEE,19th Instrumentation and Measurement Conference, vol. 2(4), pp. 1617–1622 (2002)

Packet Hiding a New Model for Launching and Preventing Selective Jamming Attack over Wireless Network

Sonali Gavali[1], Ashwini Gavali[2],
Suresh Limkar[3], and Dnyanashwar Patil[4]

[1] Department of Computer Engineering, DYPCOE, Ambi, Pune, India
[2] Department of Computer Engineering, SBPCOE, Indapur, India
[3] Department of Computer Engineering, AISSMS IOIT, Pune, India
[4] Department of Computer Engineering, Shriram IET, Akluj, India
{sonygavali1991,dnyane.ash,sureshlimkar}@gmail.com

Abstract. In real world, wireless medium is widely used among various military agents and companies. It provides various challenging features such as

Its faster accessibility among wide set of users, but it is fragile by intentional interference attacks from internal adversary. These attacks can be easily launched by sending continuous radio signals or several short pulses. External adversaries can be easily found out. But internal adversaries are difficult to detect and prevent. To prevent these attacks various cryptographic schemes are implemented. In this paper, in order to overcome drawbacks of existing systems we propose a strong security scheme that prevents the safe transmission among communicated nodes although the jammer is present.

Keywords: Commitment scheme, overview, packet hiding methods, and selective jamming attacks.

1 Introduction

In wireless network various types of attacks such as jamming, squeeze or intentional interference attacks are invited because of its sharing medium. The adversaries with internal knowledge of network secrets take more effort on jamming the network or degrade the network performance [1, 2].Anyone which has transceiver can easily delivered these jamming attacks by emitting continuous signal or injecting dummy packets into the shared medium causing interference with existing communications or inject spurious messages or block the transmission of legitimate users. In the simplest form, the jammer classifies first few bytes of transmitted packet and corrupts them by creating proximity of the targeted receivers or FM modulated noise or electromagnetic interference such as magnetic radio waves.

In these schemes, jammer includes either continuous or random transmission of high interference signals [3,4], and cause several disadvantages as, first it has to spend its more amount of energy to jam frequency bands of interest and the second one, due to continuous presence of unusually high interference levels make these types of attacks easy to detect [1,2].The adversaries are active only for short period and targeting the message of high importance for example rout request, rout reply messages or TCP acknowledgement [5]. Before the wireless transmission completes, the very basic step of the jammer is to implement "Classify them jam" strategy [6].

Usually over the communication of nodes in wireless network, there may be a possibility of intentional interference in the communication. This intentional interference is also called as Jamming attacks. These attacks are much harder to detect and count.

Consider the communication of nodes as A and B. J is the intermediate (jamming) node within their communication range. Now A sends packet m to B, the target of J is classifies first few bytes of transmitted packet, then corrupt these few bytes and visualize to A as J is nothing but B (proximity of targeted receiver).The figure 1 gives the realization of jamming attack. He must have knowledge about each layer of the TCP protocol, is the required condition for the jammer [7, 8].

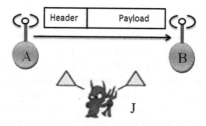

Fig. 1. Realization of selective jamming attack

The remaining paper is organized as follows. The existing system with its disadvantages and the cryptographic schemes are represented in the chapter 2. Proposed system and its advantages are studied in chapter 3. We are illustrating the implementation of working steps in chapter 4. At the last of paper we are studying the results in chapter 5 and chapter 6 we conclude.

2   Background Work

In related work, we are studying the reasons for jamming, how the real time packet classification is performed and strong hiding commitment scheme.

Because of jamming, the wireless network either stopped or disturbed. Noise, collision, interference these are various forms of jamming. Jamming may be performed intentionally or unintentionally, depending upon attack purpose or network load. There is no need of special hardware to execute these attacks.

Existing system more focused in case of external threat model but in case of internal threat model the compromise of single node is sufficient for getting useful information. Conventional anti-jamming techniques give probabilistic analysis and use matrices about collision or interference for attack detection [9]. They give prevention scheme but with less security. So there is need of strong security scheme with more focus on attack prevention.

A. Real Time Packet Classification

At first the packet m is encoded for creating narrow bandwidth analog signal then it is interleaved and then modulation is performed by using digital sequence [9]. Now, at the receiver side first demodulation then de-interleaving and at the last decoding is done to get original packet as shown in figure 2.

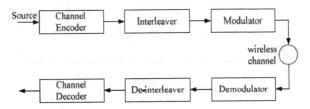

Fig. 2. Generic communication system diagram

B. Strong Hiding Commitment Scheme

Under this scheme, Symmetric encryption technique is used [10]. Sender S constructs commit message by using permutation key and key k of random length. At the receiver side any receiver R can computes by receiving d (de commit message) [11].

A message m with key k and initialization vector IV is set, message m is split into x blocks as m_1, m_2 . . . m_x, as m_i and each cipher text block c_i, is generated as

$$C_1 = IV, \; C_{i+1} = E_k \left(C_i \oplus m_i \right) \; where \; i = 1, 2, ..., x \tag{1}$$

Where $\oplus$ is exclusive OR function and $E_k(m)$ denotes the encryption of m with key k, the plaintext m_i is recovered by

$$m_i = C_i \oplus D_i \left(C_{i+1} \right) \; where \; i = 1, 2, ..., x \tag{2}$$

From equation 2 if k is known ($C_1 = IV$ is also known) the reception of C_{i+1} is sufficient for generating the original packet mi. Therefore, real time packet classification is still possible. If the key is compromised then it must have to update from time to time. But it will not be a proper solution if that key is generated from compromised node. One solution to the key compromise problem would be to update the static key time to time whenever it is compromised. The best solution for this key compromise problem is that identify a mechanism that find set of compromised nodes.

3   Overview of Proposed System

Proposed system uses Network Simulator 2.34 tool in which front end is tcl and back end is c++. Here two protocols are used. TCP protocol is used for establishing reliable connection and AODV routing protocol is used for finding routing path for

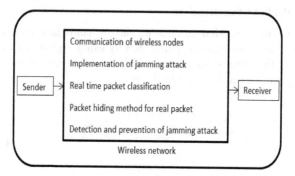

Fig. 3. Block diagram

data packets. The RTS/CTS mechanism enabled at MAC layer. The transmission rate is 11Mbps for every link. The continuous, random, targeted RREQ these jammers are kept between the communicated pairs [12]. But due to flooding feature of AODV the random jammer fails in disturbing route path. The above figure 3 gives the exacting carried out in the proposed system.

3.1   Problem Statement

Conventional anti-jamming system more focused in case of external threat model but in case of internal threat model the compromise of single node is sufficient for getting relevant information. By surveying some papers it is found that, they give only overview of jamming attacks and their types with small amount of information regarding prevention. These systems are more concentrated on attack detection for that various matrices are used for example per- packet reputation or credit and probabilistic analysis about collision or interference is also carried out for detection purpose [12, 13]. So as to overcome above there is use of our proposed system, for avoiding selective jamming attacks with prevention measures.

3.2   Scope

The proposed system provides the features as it works under an internal threat model. It gives detection of jamming attack with prevention mechanism. This system, allowed the jammer to jam the network by exploiting knowledge from compromised nodes easily. Selective jamming attack acts as DOS with very low effort on behalf of jammer. By using our proposed system we can achieve strong security protocols.

4   Actual Implementation

The proposed system works as follows. First nodes are communicated with each other. Then jamming attack is implemented, in the third step we are calculating throughput and packet classification. Our main goal is to hide real packets from the jammed packets using strong prevention scheme. We study each step in detail.

We are generating communication of 12 wireless nodes in the initial step. AODV routing protocol is used .For establishing the connection using NS2 tool .tcl file is created which gives sequence number and length of the packet, source and destination address and also contains additional fields.

In the second step we are implementing the jamming attack [14]. The ddos.o attack file is generated. After compiling the tcl file of the second step, the PSR and PDR these two graphs and one nam file is generated. The nam file shows how the packets are transmitted between intended nodes and how the jammer node is intruded between them. The PSR graph shows the packet sent ratio versus time as shown in figure 4.

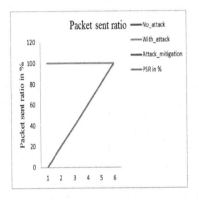

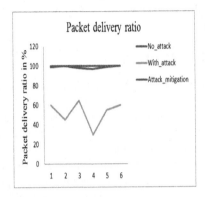

Fig. 4. Graph for packet sent ratio   **Fig. 5.** Graph for packet delivery ratio

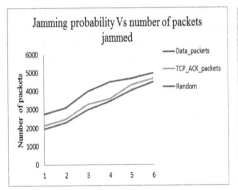

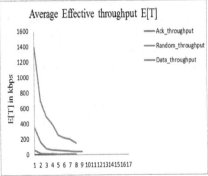

Fig. 6. Jamming probability Vs. No. of packets jammed   **Fig. 7.** Jamming probability vs. throughput

The PDR graph shows how the packets are delivered versus time, as jamming attack is implemented here so the PDR minimizes from 1 to 6 shown in figure 5.

In the third step, we are performing real time packet classification. Here we are calculating the jamming probability verses number of packets jammed and average throughput also. As jamming probability increases the number of packets jammed increases as shown in figure 6 and also throughput of the proposed system decreases shown in figure 7. So it is indicating that attack is implemented.

At the last step jamming attacks are detected by calculating the PDR it is in linear format indicating that attack is prevented and the packets are discarded that are transmitted from the jammer, this step must be performed carefully. By implementing further steps, the proposed system produces the required result so as to achieve great security.

5   Results and Comparative Study

In this chapter we are studying the results and the exact definition of PSR and PDR also.

5.1   Result Set

The results are generated in terms of various graphs. These graphs are calculated over different transmission rates and packet size. PSR graph is linear, now due to jamming attack is implemented as ddos.o in Makefile.in as configured. The PDR graph is non-linear. Now when we are performing classification of the real packets from jammed packet at that time we are calculating the number of packets jammed and the throughput of the proposed system which is decreases due to attack from the jammer. At the last step we are hiding the real packets by using cryptographic schemes [15] such as Hash based encryption and DES algorithm. In hash based encryption we are sending the packet or message in encrypted format with hash based value that is pre-shared before actual communication starts and at the receiver side the message is decrypted with the help of hashing value. In DES algorithm at the sender side input is given as plain text and 64 bit key from that the cipher text of 64 bit block is generated. Now at the receiver side DES decryption consists of the encryption algorithm with the same key but reversed key schedule [16]. In this way proposed system avoids the jamming attack over wireless network with great security.

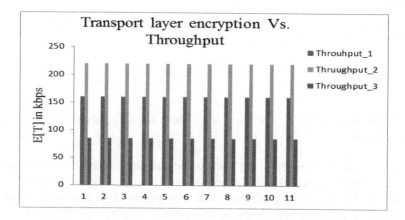

Fig. 8. Hash based encryption Vs throughput

The PSR is defined as ratio of packets that are successfully sent out by a trusted traffic source compared to the number of packets it wants to send out at the MAC layer. The PDR is defined as packets that are successfully delivered to a destination compared to the number of packets that have been sent out by the sender. This PSR and PDR are calculated at the back end using c++.Also encryption and decryption is done at the back end. The figure 8 shows the final result in which throughput of the proposed system is great because of achieving the strong security.

6   Conclusion and Future Work

This paper provides solution for jamming attack over wireless network. In this paper the internal threat model is considered, in which jammer is part of network and he is aware of network secrets and protocol specification. Jammer can perform classification in real time by decoding first few bytes of the transmitted packet. To prevent real time packet classification various schemes are developed. These schemes combine cryptographic primitives such as strong hiding commitment scheme with physical-layer characteristics so as to transform jammer to random one. We also measured how each jammer fared by their effect on the packet send ratio and packet delivery ratio. We analyze the security of our method and quantified their computational and communication overhead.

References

1. Xu, W., Wood, T., Trappe, W., Zhang, Y.: Channel surfing and spatial retreats: Defenses against Wireless denial of service. In: Proc. Third ACM Workshop Wireless Security, pp. 80–89 (2004)
2. Xu, W., Trappe, W., Zhang, Y., Wood, T.: The feasibility of launching and detecting jamming attack in wireless network. In: Proc. ACM Int'l Symposium. Mobile Ad-Hoc Networking and Computing (Mobi-Hoc), pp. 46–57 (2005)
3. Lazos, L., Liu, S., Krunz, M.: Mitigating Control-Channel Jmming Attacks in Multi-channel Wireless Ad Hoc Networks. In: 2nd ACM Conference on Wireless Network Security (WiSec 2009) (2009)
4. Simon, M.K., Omura, J.K., Schotlz, R.A., Levitt, B.K.: Spread spectrum communications Handbook. McGraw-Hill (2001)
5. Popper, C., Strasser, M., Čapkun, S.: Jamming-Resistant Broadcast communication without shared keys. In: Proc. USENIX Security Symposium (2009)
6. Proano, A., Lazos, L.: Packet-Hiding Methods for Preventing Selective Jamming Attacks. IEEE Transactions on Dependable and Secure Computing (TDSC) 9(1), 101–114 (2012)
7. Juels, A., Brainard, J.: Client Puzzles: A Cryptographic Countermeasure Against Connection Depletion Attacks. Springer, Heidelberg (1999)
8. Rivest, R.L.: All-or-nothing encryption and the package transform. In: Biham, E. (ed.) FSE 1997. LNCS, vol. 1267, pp. 210–218. Springer, Heidelberg (1997)
9. Desmedt, Y.: Broadcast Anti-jamming Systems. Computer Networks 35(2/3), 223–236 (2001)
10. Stinson, D.: Cryptography: Theory and practice. CRC Press (2006)

11. Manojkumar, K., Vinothkumar, M., Tholkappia Arasu, G.: An Analysis on Denial of Service attacks and packet defending methodologies in wireless sensor network. International Journal of Advanced Research in Computer Science and Software Engineering 2(11) (November 2012)
12. Dilip Kumar, D.P., Venugopal, H.: Avoiding selective jam attack by packet hiding method in wireless sensor network. International Journal of Science and Research (IJSR) 2(6) (June 2013)
13. Thamilarasu, G., Mishra, S., Shridhar, R.: Improving reliability of jamming attack detection in Ad hoc networks. IJCNIS 3(1) (April 2011)
14. Sodagari, S., Clancy, T.C.: Efficient jamming attack on MIMO channels. In: IEEE International Conference on Communications (ICC), VA,USA, pp. 1550–3607 (2012) ISSN :1550-3607
15. Damgård, I.B.: Commitment schemes and zero-knowledge protocols. In: Damgård, I.B. (ed.) Lectures on Data Security. LNCS, vol. 1561, pp. 63–86. Springer, Heidelberg (1999)
16. Noubir, G., Lin, G.: Low power DoS attacks in data wireless LANs and countermeasures. ACM SIGMOBILE Mobile Computing and Communications Review 7(3), 29–30 (2003)

Fuzzy Rule-Based Adaptive Proportional Derivative Controller

Ritu Rani De (Maity)[1] and Rajani K. Mudi[2]

[1] Dept. of Electronics & Instrumentation Engineering,
Dr. B.C. Roy Engineering College, Durgapur, India
ritu_maity_8@yahoo.co.in
[2] Dept. of Instrumentation & Electronics Engineering, Jadavpur University,
Salt-lake, Kolkata – 700 098, India
rkmudi@yahoo.com

Abstract. An improved fuzzy rule based auto-tuning PD controller is designed for integrating processes with dead-time. A large number of industrial processes are integrating in nature. It has been studied that for integrating systems, widely used Ziegler-Nichols (ZN) tuned PID controller (ZNPID) gives excessively large value of overshoot and settling time. Some improvement in overshoot and settling time has been possible for such systems by the development of dynamic set-point weighted PID (DSWPID) and augmented ZN tuned PID (AZNPID) controllers. Here, we propose a fuzzy auto-tuning PD (FAPD) controller where an updating factor 'α' continuously adjusts the derivative gain to provide an overall good performance during set point change and load disturbance. The value of α is updated online by 9 fuzzy if then rules defined on the value of error (e) and change of error (Δe) of the controlled variable. To study the effectiveness of the proposed controller, FAPD is tested and compared with other PID controllers for different integrating systems with varying dead-time.

Keywords: PID control, PD control, auto-tuning, integrating process.

1 Introduction

PID controllers are still now very popular in industrial close-loop control [1-4]. They are widely used for their simplicity and ease of implementation, but improper tuning sometimes results in a very poor performance. Though, ZNPID controllers give satisfactory performance for first order systems, they produce excessively large overshoot and settling time for higher-order and integrating systems. The overshoot may be considerably reduced by DSWPID [6, 7] but load regulation remains poor.

The integral action is likely to be the cause of oscillations for integrating plus dead-time process [2]. Properly designed PD controllers are capable of providing a satisfactory performance for integrating or zero-load process with delay. Though Derivative action of the controller are sensitive towards high frequency noise, still then PD controller [12] reduce the overshoot by introducing damping. There is a scope to design a good PD controller because unlike PID controller there are less running schemes for PD controller.

© Springer International Publishing Switzerland 2015
S.C. Satapathy et al. (eds.), *Proc. of the 3rd Int. Conf. on Front. of Intell. Comput. (FICTA) 2014*
– *Vol. 1*, Advances in Intelligent Systems and Computing 327, DOI: 10.1007/978-3-319-11933-5_22

Many Fuzzy controllers have been successfully designed with high performance than conventional controllers. Self-tuning concept is incorporated in fuzzy PD [11] controller with 49 fuzzy if–then rules on process error and change of error. However, the proposed controller FAPD is designed with only 9 fuzzy rules. The performance of FAPD is tested and compared with other PID controllers for integrating plus dead-time processes. The robustness of FAPD is proved with load disturbances.

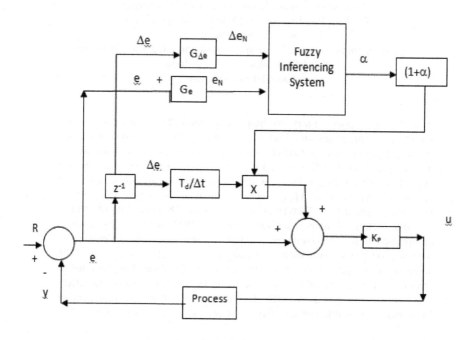

Fig. 1. Block diagram of the proposed FAPD

Here, Fig.1 shows the Block diagram of the proposed FAPD. The proposed controller is in positional form and the derivative gain K_d of the conventional controller (FAPD) is being online tuned by a gain updating factor 'α'. The value of α is determined using 9 fuzzy rules defined on two linguistic input variables, *i.e.*, the process error (e) and change of error (Δe). The next section presents the discrete form of the PD controller derived from the conventional ZN-tuned Proportional-Integral-Derivative controller termed as ZNPID [10], the concept of tuning strategy, and details of the proposed controller FAPD.

2   Design of FAPD Controller

The discrete form of the PD controller can be described as:

$$u(k) = K_p[e(k) + \frac{T_d}{\Delta t}\Delta e(k)] \tag{1}$$

or,

$$u(k) = [K_p e(k) + K_d \Delta e(k)] \tag{2}$$

where, $K_d = K_p(T_d/\Delta t)$. In (1), $e(k) = [R - y(k)]$ is the error, $\Delta e(k) = [e(k) - e(k-1)]$ is the change of error, R is the set value, $y(k)$ is the process output at the kth instant, and Δt is the sampling interval. Here, K_p, and T_d are the proportional gain, and derivative time, respectively. In (2), K_d is the derivative gain.

In case of our proposed controller (FAPD), the widely used Ziegler-Nichols tuned PID control without integral control is taken for setting the initial values of the controller parameters. The values of initial PD parameters are determined based on the following equations, according to the ZN ultimate cycle tuning rules [10]:

$$K_p = 0.6K_u , \tag{3}$$

$$T_d = 0.12T_u \tag{4}$$

Where, K_u and T_u are the ultimate gain and ultimate period, respectively determined by relay-feedback method [2]. In the proposed FAPD, the online fine tuning using fuzzy logic is done on the nominal value (ZN-tuned) of K_d.

2.1   Auto-tuning Using Fuzzy

The derivative gain, 'K^t_d' of the proposed FAPD is modified by α with a simple empirical relation :

$$K^t_d = K_d(1+\alpha) \tag{5}$$

Therefore, the effective value of the derivative gain of FAPD becomes K^t_d. The value of K^t_d does not remain constant while the controller is in operation, rather, K^t_d is continuously modified at each sampling time by 'α', depending on the trend of the controlled process. The following subsection describes the auto-tuning strategy of K^t_d.

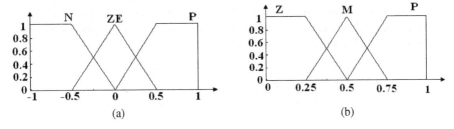

(a)   (b)

Fig. 2. Membership functions of (a) e and Δe, and (b) gain updating factor α

$\Delta e/e$	N	ZE	P
N	P	P	Z
ZE	M	Z	M
P	Z	P	P

Fig. 3. Fuzzy rules for computation of α

2.2   Membership Function, Scaling Factor, and Rule Base

Figure 2 shows the input (Fig. 2a) output (Fig. 2b) membership functions (MFs) of the fuzzy rules. The input MFs, *i.e.*, error and change of error (e, and Δe) are defined in [-1, 1] whereas the MF for α is defined within [0, 1].

Mapping of the actual values of input variables e and Δe into the common interval [-1 1] is done by the input scaling factors (SFs) G_e and $G_{\Delta e}$. Appropriate values of SF's are selected on the basis of knowledge about the process under control or by trial and error. Online computation of α is done on the basis of rule-base defined in terms of e and Δe. The relationship of SFs and input variables are as follows:

$$e_N = G_e.e \; ; \quad \Delta e_N = G_{\Delta e}.\Delta e.$$

In this study, we consider $G_e = 1$ and $G_{\Delta e} = 10$. The value of the K^t_d is continuously updated by α.

The gain updating factor is calculated online using the rules in Fig 3. The rules are in the form: *If e is E and Δe is ΔE then α is α*

2.3   Tuning Strategy

Observation on the close-loop operating cycle reveals the possibility of large over-shoot and large undershoot when the process is moving towards the set point (*i.e.*, columns A and C of Fig. 4). Considerable amount of damping can reduce such large overshoot and undershoot. Thus, when e and Δe are in opposite sign, the value of α is increased which subsequently increases the value of K^t_d.

Opposite of the above case, when the process is moving further away from the set-point (*i.e.*, columns B and D of Fig. 4), both e and Δe are of same sign. Here, the damping is increased more comparative to the previous condition. High value of K^t_d helps to give a more aggressive control action. The same situation is also valid under load disturbance. Thus, our proposed FAPD is expected to provide improved perfor-mance both under set-point change and load disturbance.

	A	B	C	D
e	+	-	-	+
Δe	-	-	+	+
α	*small*	*large*	*small*	*large*
K^t_d	Increased LESS	Increased MORE	Increased LESS	Increased MORE

Fig. 4. Tuning strategy for computation of K^t_d

3   Simulation Results

In order to study the performance and robustness of the proposed FAPD, itis tested on a number of integrating systems. The parameters included for performance measurement are the percentage overshoot (%OS), settling time (t_s), rise time (t_r), integral absolute error (IAE), and integral-time absolute error (ITAE).

3.1   Integrating Process with Delay(IPD)

$$G_P = \frac{Ke^{-\theta s}}{s} \qquad (6)$$

Here, we consider the well-known IPD process model with open loop gain $K =$ 0.0506 and dead-time $\theta = 6s$. Simulation study has been made with set-point change and load disturbance. Comparative performance analysis of the proposed FAPD is done with ZNPID, AZNPID [5], in Table 1 and Fig. 5. Table 1 shows that the overshoot is reduced almost 60% and the ITAE is also reduced to almost 50% compared to ZNPID. Though controller is tuned at θ =6, the performance for 25% increased dead-time ($\theta = 7.5s$) proves the robustness of the controller.

Table 1. Performance comparision of IPD (6)

θ	Controller	%OS	t_s	t_r	IAE	ITAE
6s	FAPD	-0.16	51.9	5.1	15.50	574.62
	ZNPID	68.22	79.7	10.9	26.06	1002.84
	AZNPID	21.44	100.4	11.6	21.4	927.35
7.5s	FAPD	21.40	95.1	5.1	22.71	1080.37
	ZNPID	115.72	89.5	12.4	36.38	1554.44
	AZNPID	50.21	96.9	13.1	25.33	1092.51

3.2   First-Order Integrating Process with Delay(FOPID)

$$G_p = \frac{Ke^{-\theta s}}{s(\tau s + 1)} \qquad (7)$$

Here, $K=1$, $\theta=4s$, $\tau=4s$ in (7). Performance analysis of (7) with FAPD and otherPID controllers are shown in Table 2 and in Fig. 6. The controller is tuned at dead-time 4s. Results (Table 2) show that FAPD has very low overshoot and less sensitive ondead-time variation.

$$G_p = \frac{Ke^{-\theta s}}{s(s + 1)} \qquad (8)$$

In (8), $K = 1$, $\theta = 0.1s$. Performance analysis of (8) under set-point change and load disturbance is presented in Table 3 and Fig. 7. The controller is tuned at $\theta = 0.1s$. The comparative study with the FAPD and AZNPID, and ZNPID in the Table 3 reveals that all the performance parameters have been minimized in the proposed controller. The high level of performance of FAPD for increased dead-time proves the usefulness of our proposed controller.

Table 2. Performance comparision of FOIPD (7)

θ	Controller	%OS	t_s	t_r	IAE	ITAE
4	FAPD	0.008	30.3	16.0	16.68	876.45
	ZNPID	68.45	81.9	12.5	35.36	2587.82
	AZNPID	23.82	116.1	13.6	29.4	2265.29
5	FAPD	2.23	41.1	7.7	19.31	1254.01
	ZNPID	90.0	81.3	13.2	43.33	3336.37
	AZNPID	37.63	115.4	14.1	32.38	2508.38

Table 3. Performance comparision of FOPID (8)

θ	Controller	%OS	t_s	t_r	IAE	ITAE
0.1	FAPD	8.21	4.2	1.3	1.96	30.01
	ZNPID	78.50	19.5	1.4	6.18	121.46
	AZNPID	34.98	10.3	1.7	3.22	56.81
0.2	FAPD	11.87	4.1	1.1	2.09	32.07
	ZNPID	94.27	31.9	1.4	9.59	211.02
	AZNPID	40.24	10.1	1.7	3.39	60.21

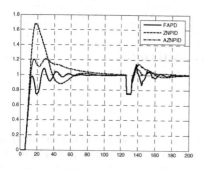

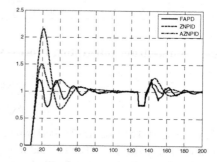

Fig. 5. a. Responses of (6) for FAPD(–), ZNPID(—), AZNPID(-.–) with $\theta = 6s$

Fig. 5. b. Responses of (6) for FAPD(–), ZNPID(—), AZNPID(-.–) with $\theta = 7.5s$

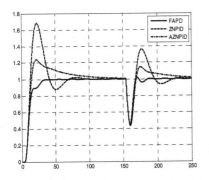

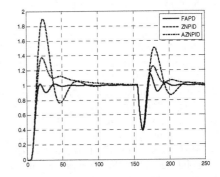

Fig. 6. a. Responses of (7) for FAPD(–), ZNPID(—), AZNPID(–.–) with $\theta = 4s$

Fig. 6. b. Responses of (7) for FAPD(–), ZNPID(—), AZNPID(–.–) with $\theta = 5s$

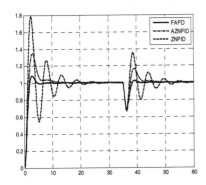

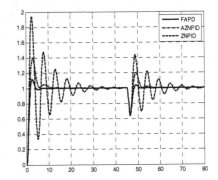

Fig. 7. a. Responses of (8) for FAPD(–), ZNPID(—), AZNPID(–.–) with $\theta = 0.1s$

Fig. 7. b. Responses of (8) for FAPD(–), ZNPID(—), AZNPID(–.–) with $\theta = 0.2s$

4   Conclusion

A fuzzy rule-based self-tuning scheme has been proposed for a PD controller. The derivative gain of the proposed controller has been continuously modified depending on the process trend. The proposed controller exhibited satisfactory performance under both set-point change and load disturbance. Performance analysis of the proposed FAPD with other PID controllers for different integrating system with dead-time established its effectiveness.

References

1. Seborg, D.E., Edgar, T.F.: Adaptive control strategies for process control: A survey. AICHE J. 32(6), 881–913 (1986)

2. Shinsky, F.G.: Process control systems — application, design, and tuning. McGraw-Hill, New York (1998)
3. Kristiansson, B., Lennartson, B.: Robust and optimal tuning of PI and PID controllers. IEE Proc. Control Theory Appl. 149(1), 17–25 (2002)
4. Astrom, K.J., Hang, C.C., Person, P., Ho, W.K.: Towards intelligent PID control. Automatica 28(1), 1–9 (1992)
5. Dey, C., Mudi, R.K.: An improved auto-tuning scheme for PID controllers. ISA Trans. 48(4), 396–409 (2009)
6. Dey, C., Mudi, R.K., Lee, T.T.: Dynamic set-point weighted PID controller. Control and Intelligent Systems 37(4), 212–219 (2009)
7. Mudi, R.K., Dey, C.: Performance improvement of PI controllers through dynamic set-point weighting. ISA Transactions 50, 220–230 (2011)
8. Ziegler, J.G., Nichols, N.B.: Optimum settings for automatic controllers. ASME Trans. 64, 759–768 (1942)
9. Mudi, R.K., Dey, C., Lee, T.T.: An improved auto-tuning scheme for PI controllers. ISA Transactions 47, 45–52 (2008)
10. Mudi, R.K., De Maity, R.R.: A noble fuzzy self-tuning scheme for conventional PI controller. In: Satapathy, S.C., Udgata, S.K., Biswal, B.N. (eds.) Proceedings of Int. Conf. on Front. of Intell. Comput. AISC, vol. 199, pp. 83–91. Springer, Heidelberg (2013)
11. Mudi, R.K., Pal, N.R.: A self-tuning fuzzy PD controller. IETE J. Res. 44(4-5), 177–189 (1998)
12. Dey, C., Mudi, R.K., Simhachalam, D.: A simple non linear PD controller for integrating process. ISA Trans. 53, 162–172 (2014)

Projected Clustering with LASSO
for High Dimensional Data Analysis

Lidiya Narayanan, Anoop S. Babu, and M.R. Kaimal

Dept. of Computer Science and Engineering, Amrita VishwaVidhyapeetham,
Amrita School of Engineering, Kollam 690525, Kerala, India
{lidiya483,mail2anoopsbabu}@gmail.com,
mrkaimal@am.amrita.edu

Abstract. It has always been a major challenge to cluster high dimensional data considering the inherent sparsity of data-points. Our model uses attribute selection and handles the sparse structure of the data effectively. We select the most informative attributes that do preserve cluster structure using LASSO (Least Absolute Selection and Shrinkage Operator). Though there are other methods for attribute selection, LASSO has distinctive properties that it selects the most correlated set of attributes of the data. This model also identifies dominant attributes of each cluster which retain their predictive power as well. The quality of the projected clusters formed, is also assured with the use of LASSO.

Keywords: sparsity problem, projected clustering, attribute selection, penalized regression, LASSO, attribute relevance index.

1 Introduction

High-dimensionality and large data set characterize many contemporary computational problems from genomics and neural science to finance and economics, which give statistics and machine learning opportunities with challenges. In response to the challenge of the complexity of data, new methods and algorithms started to flourish [2][4][10][22][28]. Methodology has responded dynamically to these challenges, and procedures have been developed to provide practical results. High - dimensional real problems often involve costly experimentations. The expensive experimental and computational costs make traditional computational procedures infeasible for high-dimensional data analysis and new techniques are needed to reduce the number of the experimental trials still guaranteeing satisfactory results [2][4][10][22][28].

Cluster analysis [1][4][22] is an unsupervised learning technique to assign observations into a number of clusters such that observations in the same cluster are similar to each other. The similarity is often quantified by some distance measures, such as the Euclidean distance and correlation. In spite of prevalence of a large number of clustering algorithms and their success in a number of different application domains, clustering remains a difficult problem[20]. Among the most discussed

© Springer International Publishing Switzerland 2015

S.C. Satapathy et al. (eds.), *Proc. of the 3rd Int. Conf. on Front. of Intell. Comput. (FICTA) 2014*
– *Vol. 1*, Advances in Intelligent Systems and Computing 327, DOI: 10.1007/978-3-319-11933-5_23

clustering algorithms, k- means clustering is one of the most popular clustering algorithms [22]. While the k-means clustering is conceptually simple andcomputationally efficient, its performance can be severely deteriorated when clustering high-dimensional data where the number of variables becomes large and many of them may contain no information about the clustering structure [11]. The *sparsity principle* [26] assumes that only a small number of predictors contribute to the response. Though there are very many methods for selecting the attributes, the LASSO (Least Absolute Shrinkage and Selection Operator) introduced by Tibshirani (1996) [1][3] is one of the most prominent and popular estimation methods for high-dimensional linear regression models. It shrinks some coefficients and sets others to 0. This modern learning regression technique thus provides an automatic method for the variable selection problem.

The paper discusses a method for finding projected clusters embedded in a high-dimensional space. To handle the sparseness of the data set effectively, attribute selection capability of LASSO [1][24][25], is utilized before applying the clustering algorithm. While estimating the coefficients, LASSO shrinks and selects the most informative attributes and sets redundant or irrelevant variables into zero. This supervised learning methodology chooses the most informative subspaces for all the data points. From this low-dimensional subspace, our model detects different subset of variables that are relevant for each cluster. Projection to this subspace that contains only relevant attributes is done by unsupervised learning strategy and the quality of projected clusters thus obtained is evaluated by the prediction capability of LASSO[3]. The clustering process makes use of the statistical properties of data points. The cluster structure that exists along every dimension is discovered and corresponding density information is binary coded. Based on this information, distance calculation during clustering phase is restricted to avoid unwanted computation.

The rest of the paper is organized as follows: Section 2 discusses the design of the model. Experimental results are given in the Section 3. Conclusion is discussed in the Section 4.

2   Selection of Attributes and Projected Clustering

2.1   Problem Formulation

Let D be a data set of m- dimensional points, where the set of attributes is denoted by $A = \{A_1, A_2 \ldots A_m\}$ and $X = \{X_1, X_2, \ldots X_N\}$be set of N data points, where $X_i = \{x_{i1}, x_{i2} \ldots x_{im}\}$. Each $x_i^j (i = 1 \ldots N; j = 1 \ldots , m)$ corresponds to the value of the data point $X_i$on the attributeA_j.It is assumed that each data point $x_i$belongs either to one projected cluster or to the set of outliers OUT.

2.2   Attribute Selection by LASSO

LASSO selects the attributes in the order by which it is correlated to the response vector, based on the shooting algorithm described in [26]. The most significant

d- attributes of the data set $D(\text{in}R^{NXm})$ is chosen to get a reduced data matrix $D\text{in}R^{NXd}$. For finding the LASSO solutions β is initialized to zero. The algorithm starts with the empty set of active variables, since $\beta_i^{(0)} = 0, i = 1,..p$ and active set contains no attributes, where the residual sum of square (RSS) is at its maximum. The LASSO can perform attribute selection while estimating the coefficients. It determines the significance of each attribute based on its correlation with the response and selects them accordingly.

2.3   Sparse Coding of Cluster Information

To distinguish between dense and sparse regions, a binary matrix is constructed from the data matrix D (which has only the selected attributes obtained by LASSO and QR factorization methods) based on k-nearest neighbour method. The goal is to identify the cluster structure by discovering dense regions and their locations in each attribute. The underlying assumptions for this phase is that, in the context of projected clustering, a cluster should have relevant dimensions in which the projection of each point of the cluster is close to a sufficient number of other projected points[4].

Sparseness degree [4] s_i^j of the 1D point x_i^j is a relative measure which gives an indication of variance of a data point x_i from its closest k- neighbours along the dimension A_j. As in [4] our model also takes $k = \sqrt{N}$. The value of sparseness degree s_i^j is computed as [4] :

$$s_i^j = \frac{\sum_{y \in knn_i^j(x_i^j)} \left(y - c_i^j\right)^2}{k + 1} \tag{1}$$

where $knn_i^j(x_{ij}) = \{ nn_k^j(x_i^j) \cup x_i^j \}$ and $nn_k^j (x_i^j)$ denotes the set of k-nearest neighbours of x_i^j in dimension D_j and term c_i^j in the equation 1 is the center (mean) of the set and is found out by equation 2 [4] :

$$c_i^j = \frac{\sum_{y \in knn_i^j(x_i^j)} y}{k + 1} \tag{2}$$

A large value of s_i^j means that x_i^j belongs to a sparse region, while a small one indicates that x_i^j belongs to a dense region. i.e. a large value of s_i^j indicates, the data point x_i is far away even from its closest neighours along the dimension A_j and does not belong to a cluster. The degree of sparseness gives a measure to distinguish between dense and sparse regions along each dimension. Since this measure is data dependent, we cannot simply choose a threshold value to discriminate dense region from sparse region. For that we found representatives from sparseness measure by dividing those values into k regions by traditional k-means clustering to get the median of each cluster. The median thus obtained is added to an array called REG. i.e. the set $\{REG\} = \{REG\} \cup median (region_i(A_j))$, where $region_i(A_j)$ implies the $region_i$ consists of similar sparseness values along the attribute A_j and REG_i indicates the location of $region_i$. [4] offers a selection algorithm to decide on the threshold value from these representatives for discriminating the dense or sparse regions based on minimum description length (MDL) principle.

2.4   Construction of Binary Matrix

The threshold value obtained from the MDL algorithm [4] is used to construct a binary matrix $Z_{(n\text{-}by\text{-}d)}$ in such a way that, if sparseness degree measure $s_i^j <$ threshold then set $z_i^j = 1$ and x_i^j belongs to a dense region; else (i.e. $s_i^j \geq$ threshold) then set $z_i^j = 0$ and x_i^j belongs to a sparse region.

This binary matrix gives an indication of outliers also. Outliers can be defined as a set of data points that are considerably dissimilar, exceptional, or inconsistent with respect to remaining data. It is obvious that outliers do not belong to any of the identified dense region; they are located in the sparse regions in $\mathcal{D}$. (I.e. the row contains $z_i^j = 0, \forall j, j = 1 \dots d$). The outliers are discarded from $\mathcal{D}$ and are stored in the set OUT, while their corresponding rows are eliminated from the matrix Z. This yields, a reduced data set with size $Nr = \mathcal{N}$ -$|OUT|$ and [4] a reduced data matrix $\mathcal{RD}$ is formed as: $\mathcal{RD} \leftarrow \mathcal{D} - OUT$. Based on $\mathcal{RD}$ and OUT extract T from Z. T is a binary matrix that contains entries from the Z matrix corresponding to the entire data set except the discarded outliers

2.5   Discovery of Projected Clusters

The binary matrix T is used to find the projected clusters. In order to accomplish identify projectedclusters, basic k-means is modified by using a distance function that considers contribution only from relevant dimensions when computing the distance between a data point and the cluster center [4]. The binary weights t_i^j ($i = 1, \dots , N_r$; $j=1,\dots,d$) in matrix T, which is a binary   matrix that contains no entries corresponding to outliers. This makes distance measure more effective because the computation of distance is restricted to subsets (i.e. projections) where the object values are dense [4].

Each projected cluster obtained will have a set of cluster members which are close to each other. The density information stored in binary matrix T is used to determine relevance index of each dimension. For each dimension, the sum of the binary weights of the data points belonging to the same cluster divided by total number of members in the cluster gives a meaningful measure of the relevance of the dimension. Based on this observation a relevance index measure can be defined as follows [4]:

$$RI_k^j = \frac{\sum_{ti \in Ck} t_i^j}{|C_k|} \tag{3}$$

RI_k^j represents the percentage of the points in the cluster k, which have the dimension j as a relevant dimension. As in [4], A dimension $\mathcal{D}_j$ is considered as δ-relevant for the cluster C_k if $RI_k^j > \delta$. The value of δ is a user defined parameter that controls the degree of relevance of the dimension $\mathcal{D}_j$ to the cluster C_k. It is clear that the larger the value of the relevance index, the more relevant the dimension to the cluster.

Algorithm 1: k-means Projected Clustering Algorithm

---

1. Choose the cluster centres $m_s^0 (s = 1, \ldots, k)$ randomly from $\mathcal{RD}$.
2. **Repeat**
3. Find the distance of data point to each of the cluster centroid.

$$d(X_i, m_s) = \sqrt{\sum_{j=1}^{d} t_i^j (x_i^j - m_s^j)^2}$$

4. Find the minimum distance d_{min}.
5.   **if** d_{min} is with cluster p **then**
6.   set $M_i^p = 1$
7.   **Else**
8.   set $M_i^p = 0$
9. Compute the cluster centres $m_s = \frac{\sum_{i=1}^{Nr} (M_i^s * t_i * x_i)}{\sum_{i=1}^{Nr} M_i^s}$

10. **until** no change in centroid coordinates
11. Based on equation 3 find the R matrix
12. Based on given δ determine the set SD_k.

---

3   Experimental Results

The details of data sets used for evaluation is given in Table 1. Data sets like Wisconsin Breast Cancer Data (WBCD), Parkinson's Disease Data, Cardiotocographic Data (CTG) and Ozone Level Detection Dataset are taken from UCI Machine Learning Repository[27][29].

Table 1. Datasets used for Testing and Analysis

Dataset Name	Description	No. of Attributes
Dataset 1A	Wisconsin Breast Cancer Data (WBCD)	9
Dataset 1B	Reduced Dataset of WBCD with selected attributes using LASSO	8
Dataset 2A	Original Parkinson's Disease Data	22
Dataset 2B	Reduced Dataset of Parkinson's Disease Data with selected attributes using LASSO	19
Dataset 3A	Original Cardiotocographic Data (CTG)	21
Dataset 3B	Reduced Dataset of CTG with selected attributes using LASSO	18
Dataset 4A	Ozone Level Detection Dataset	72
Dataset 4B	Reduced Dataset of Ozone Level Detection Data with selected attributes using LASSO	40

Projected clustering is performed on the reduced data set in low-dimensional space (that contains selected variables only) and the results thus obtained are compared with that of original data having the full dimensions. It is noticeable from the evaluation and analysis that accuracy of clustering is not affected by reduction in number of attributes; rather it improves the quality of learning by eliminating noisy or redundant values.

On applying the LASSO shooting algorithm [26] on CGT (Dataset 3A). Among the 21 attributes for the Dataset 3A, the LASSO chose 18 attributes and forms the reduced dataset (Dataset 3B). Because the β values of attributes 2, 9 and 12 are zero, which means these attributes least significant and hence these 3 attributed are removed. The LASSO solution of CGT is shown in Table 2.

Table 2. β coefficient values of Dataset 3A. The zero value indicates that the corresponding attribute is an irrelevant one. Hence attributes 2, 9 and 12 are dropped.

Attribute No	Attribute Name	Coefficient value (LASSO)	Attribute No	Attribute Name	Coefficient value (LASSO)
1	LB	5.75412939	12	Width	0
2	AC	0	13	Min	2.48064198
3	FM	-0.6277832	14	Max	1.58595446
4	UC	-2.8348201	15	Nmax	-0.0248683
5	DL	2.67121931	16	Nzeros	0.14159387
6	DS	1.98160764	17	Mode	-3.3885373
7	DP	10.3482585	18	Mean	-1.1957851
8	ASTV	6.73967537	19	Median	-4.6094432
9	MSTV	0	20	Variance	3.43303397
10	ALTV	9.74509782	21	Tendency	0.73685646
11	MLTV	1.0183781			

The projected clustering algorithm is applied on both the original (Dataset 3A) and reduced (Dataset 3B) datasets of CTG. Table 3, Table 4 and Table 5 are the relevance index matrices for cluster 1, cluster 2 and cluster 3 of Dataset 3A and Dataset 3B respectively. It is observed that the relative relevance of attributes which are retained after LASSO selection is the same for both Dataset 3A and Dataset 3B. It is also noted that the attributes that are discarded by LASSO shows relatively low relevance value.

Table 3. Relevance Index Matrix of cluster 1 formed from CTG data at $\delta = 0.75$

Att.Name	LB	AC	FM	UC	DL	DS	DP	ASTV	MSTV	ALTV	MLTV
Cluster1 Dataset 3A	0	0.418919	0.810811	0.148649	0.77027	1	1	0	0.148649	0.554054	0.027027
Cluster1 Dataset 3B	0		0.810811	0.148649	0.77027	1	1	0		0.554054	0.027027

Att.Name	Width	Min	Max	Nmax	Nzeros	Mode	Mean	Median	Variance	Tendency	
Cluster1 Dataset 3A	0	0	0	1	1	0	0	0	0.891892	1	
Cluster1 Dataset 3B		0	0	1	1	0	0	0	0.891892	1	

Table 4. Relevance Index Matrix of cluster 2 formed from CTG data at $\delta = 0.75$

Att.Name	LB	AC	FM	UC	DL	DS	DP	ASTV	MSTV	ALTV	MLTV
Cluster2 Dataset 3A	0.75578	0.397399	0.637283	0.127168	0.528902	0.992775	0.893064	0.251445	0.132948	0.630058	0.072254
Cluster2 Dataset 3B	0.75578		0.637283	0.127168	0.528902	0.992775	0.893064	0.251445		0.630058	0.072254

Att.Name	Width	Min	Max	Nmax	Nzeros	Mode	Mean	Median	Variance	Tendency	
Cluster2 Dataset 3A	0	0.16474	0.859827	0.981214	0.981214	0.406069	0.385838	0.459538	0.511561	1	
Cluster2 Dataset 3B		0.16474	0.859827	0.981214	0.981214	0.406069	0.385838	0.459538	0.511561	1	

Table 5. Relevance Index Matrix of cluster 3 formed from CTG data at $\delta = 0.75$

Att.Name	LB	AC	FM	UC	DL	DS	DP	ASTV	MSTV	ALTV	MLTV
Cluster3 Dataset 3A	0.847059	0.432353	0.595588	0.171324	0.594118	0.998529	0.923529	0.246324	0.159559	0.599265	0.0625
Cluster3 Dataset 3B	0.847059		0.595588	0.171324	0.594118	0.998529	0.923529	0.246324		0.599265	0.0625

Att.Name	Width	Min	Max	Ilmax	Nzeros	Mode	Mean	Median	Variance	Tendency	
Cluster3 Dataset 3A	0	0.060882	0.121324	0.958824	0.988971	0.610294	0.591176	0.613971	0.570588	1	
Cluster3 Dataset 3B		0.080882	0.121324	0.958824	0.988971	0.610294	0.591176	0.613971	0.570588	1	

The clustering accuracy is determined by comparing it with the domain knowledge. From the results it can be ensured that 53.34% of samples from Dataset 3A are clustered correctly. According to domain knowledge, Dataset 3A has 1655 are "*Normal*", 295 are "*Suspect*" and 176 are "*Pathologic*" samples. Clustering on Dataset 3A discovered that 1059 "*Normal*", 13 "*Suspect*" and 62 "*Pathologic*". This means 63% from "*Normal*" samples, 0.044% of "*Suspect*" and 35% of "*Pathologic*" are clustered accurately. The result is same from Dataset 3B. The data points are clustered in the similar way. This means that the reduction of attribute does not cause much of information loss and the attributes discarded by LASSO are truly irrelevant or redundant. Thus LASSO could be efficiently used to handle the clustering of data.

Similar results are obtained from the other three data sets Dataset 1A (WBCD), Dataset 2A (Parkinson's Disease Data) and Dataset 4A (Ozone Level Detection Data). For Dataset 1A, LASSO chose 8 attributes from 9 and formed Dataset 1B. For Dataset 2A, 19 attributes out of 22 are selected and formed Dataset 2B. For Dataset 4A, 40 attributes out of 72 are selected and formed Dataset 4B. For Dataset 1A and Dataset 1B, the clustering accuracy (96.5%) obtained is better than any of the existing clustering methods which are discussed in [4]. Clustering accuracy from Dataset 2A and Dataset 2B is 67.7%.

Complexity of Algorithm 1 is O ($d*Nr*k$). It is obvious that, normal projected clustering takes more computation time than the projected clustering enhanced with LASSO. Even if the LASSO attribute selection is performed, the computational time is increased as the number of instances and number of attributes of the dataset increases.

4   Conclusion

A projected cluster consists of similar data points that are projected to a low-dimensional space of attributes significant to that cluster. The main objective of our model is to identify projected clusters of a high-dimensional data. The quality of clustering is enhanced with LASSO that handles the sparse structure of the data effectively. From the experimental results shown in section 3 it is found that,

- LASSO selects the attributes that are most essential and informative without disturbing the cluster structures.
- LASSO is a reliable attribute selection tool that determines which variables are most essential and informative.
- Attributes show equivalent relevance index measure in the projected clusters formed for original and reduced datasets. This means cluster structure embedded in the data is not destroyed by attribute selection by LASSO and the attributes discarded by LASSO are actually redundant or irrelevant for the system under study.

Acknowledgments. We are grateful for the support and facilities provided by Amrita VishwaVidyapeetam University.

References

1. Hastie, T., et al.: Linear Methods for Regression in The Elements of Statistical Learning – Data Mining, Inference, and Prediction, 2nd cdn. Springer
2. Donoho, D.L.: High-Dimensional Data Analysis: The Curses and Blessings of Dimensionality
3. Tibshirani, R.: Regression Shrinkage and Selection via Lasso. Journal of the Royal Statistical Society 58(1), 267–288 (1996)
4. Bouguessa, M., Wang, S.: Mining Projected Clusters in High- Dimensional Spaces. IEEE Transactions on Knowledge and Data Engineering 21(4) (2009)
5. Yip, K.Y., et al.: HARP: A practical Projected Clustering Algorithm. IEEE Transactions on Knowledge and Data Engineering 16(11) (2004)
6. Agarwal, R., et al.: Automatic Subspace Clustering of High Dimensional Data. Data Mining and Knowledge Discovery 11(1), 5–33 (2005)
7. Yip, K.Y., et al.: Identifying Projected Clusters from Gene Expression Profiles. J. Biomedical Informatics 37(5), 345–357 (2004)
8. Aggarwal, C.C., Yu, P.S.: Redefining Clustering for High-Dimensional Applications. IEEE Transactions on Knowledge and Data Engineering 14(2) (2002)
9. Efron, B., et al.: Least Angle Regression. The Annals of Statistics 32(2), 407–499 (2004)
10. Johnstone, M., Titterington, D.M.: Statistical challenges of high-dimensional data. Phil. Trans. R. Soc. A 200(367) (2009)
11. Sun, W., et al.: Regularized k-means clustering of high-dimensional data and its asymptotic consistency. Electronic Journal of Statistics 6-148-167 (2012)
12. Lv, J., et al.: Prediction of Transient Stability Boundary Using the Lasso. IEEE Transaction on Power Systems 28(1) (2013)
13. Bondell, H.D., Reich, B.J.: Simultaneous Regression Shrinkage, Variable Selection and Supervised Clustering of Predictors with OSCAR. Biometrics 64, 115–123 (2008)
14. Zou, H., Hastie, T.: Regularization and variable selection via the elastic net. JRSSB 67(2), 301–320 (2005)
15. Zou, H.: The adaptive lasso and its oracle properties. JASA 101(476), 1418–1429 (2006)
16. Yuan, M., Lin, Y.: Model selection and estimation in regression with grouped variables. JRSSB 68, 49–67 (2006)
17. Fang, Y.: Asymptotic Equivalence between Cross – validation and Akaike Information Criteria in Mixed- Effects Models. Journal of Data Science, 15–21 (2011)

18. Aggrawal, C.C., et al.: Fast Algorithm for Projected Clustering. Proc. ACM SIGMOD 1999, 329–340 (2005)
19. Procopius, C.M., et al.: A Monte Carlo algorithm for fast projective clustering. In: Proc. ACM SIGMOID International Conference on Management of Data (2002)
20. Kriegel, H.P., et al.: Clustering High Dimensional Data: A Survey on Subspace Clustering, Pattern- Based Clustering, and Correlation Clustering. ACM Trans. Know. Discov. Data 3(1), Article 1 (2009)
21. Yip, K.Y., et al.: On discovery of Extremely Low- Dimensional Clusters using Semi-Supervised Projected Clustering. In: Proc. 21st International Conference on Management of Data Engineering (ICDE 2005), pp. 329–340 (2005)
22. Jain, A.K.: Data clustering: 50 years beyond K-means. Pattern Recognition Letters 31, 651–666 (2010)
23. She, Y.: Sparse Regression with exact clustering, Ph. D. Dissertation, Dept. Statistics. Stanford Univ (2008)
24. Fraley, C., Hesterberg, T.: Least Angle Regression and Lasso for large data sets, Technical Report, Insightful Corporation (2008)
25. Ma, S., et al.: Supervised group lasso with application to microarray data analysis. Technical Report, Department of Statistics and Actuarial Science (2007)
26. Fu, W.: Penalized Regressions: The Bridge Versus the Lasso (1998)
27. Little, M.A., et al.: Exploiting Nonlinear Recurrence and Fractal Scaling Properties for Voice Disorder Detection (2007)
28. Fan, J.: Selected Works of Peter J. Bickel. Springer (2013)
29. Bache, K., Lichman, M.: UCI Machine Learning Repository. University of California, School of Information and Computer Science, Irvine, CA (2013)

Probabilistic Classifier and Its Application to Face Recognition

S.G. Charan[1], G.L. Prashanth[1], K. Manikantan[1,*], and S. Ramachandran[2]

[1] Department of Electronics and Communication,
M.S. Ramaiah Institute of Technology, Bangalore, India
kmanikantan2009@gmail.com
[2] Department of Electronics and Communication,
S.J.B. Institute of Technology, Bangalore, India

Abstract. This paper proposes a method to classify different subjects from a large set of subjects. Taking correct decision in the process of classification of various subjects from the large set is an arduous task, since its probability is very low. This task is made *simple* by the proposed *Probabilistic Classifier* (PC). Maximum Likelihood Estimation (MLE) and Error Minimizing Algorithms (EMA) are the basis for the proposed classifier. Interpreting the EMA output in a probabilistic manner gives rise to PC. Concept of feedback is used in the classification process to enhance the decision rule. Experimental results obtained by applying the proposed classifier on various benchmark facial datasets, show its promising performance. Eventually, PC is found to be *independent of the datasets*.

Keywords: Probabilistic Classifier, Maximum Likelihood Estimation, Error Minimization, Correlation, Face Recognition.

1 Introduction

Classifier is the algorithm which separates out a particular entity from the colossal set of entities. The separation of the entities is done by using a rule known as decision rule. The key challenge is in framing a decision rule which works with all the algorithms. It is necessary for a classifier to have highest Classification Rate (CR) with least resource consumption. Ultimate goal of pattern recognition systems [1] is to improve the performance of classification.

Decision rule in classification is based on EMA[2]. There are several error minimization algorithms which perform the task of classification. Main criteria for choosing a classifier [3] is based on its accuracy of classification and resource consumption [4], [5]. Classifiers and decision rule will work more accurately when the classification is done from a small set of entities. If the amount of the entities increase, the CR goes low.

All the practical FR systems are affected by the variance in illumination, pose, expression and background [6]. There are algorithms which specifically

---

* Corresponding author.

© Springer International Publishing Switzerland 2015   211
S.C. Satapathy et al. (eds.), *Proc. of the 3rd Int. Conf. on Front. of Intell. Comput. (FICTA) 2014*
– Vol. 1, Advances in Intelligent Systems and Computing 327, DOI: 10.1007/978-3-319-11933-5_24

target few of the above mentioned variables [7], [8]. In recent days, open-universe identification in uncontrolled, web-scale scenarios is evaluated in Ref. [9].

Rest of the paper is organized as follows. In Section 2, we explain problem and proposed solution in brief. In Section 3 of the paper, we discuss detailed analysis of proposed PC and RF. Section 5 completely deals with the experiments and analysis conducted. Future work and conclusion is done in Section 6.

2   Problem Definition and Proposed Solution

Practical implementation of vision systems which gives a high CR is the key challenge. Resource usage of existing classifiers has to be minimized. *The large dataset on which the classification occurs has to be minimized and a compact classification efficient dataset must be generated.* Flexible classifier is required which can be implemented in any domain. Proposed solutions to the aforementioned problems have been described below.

2.1   Probabilistic Classifier for Reduction of Datasets

Proposed PC is simple and performs the operation on the output of EMA. *Large set of subjects is minimized to a smaller set of subjects. Using results of EMA probabilistically, decrease in the effective dataset size is achieved.* Classifiers working on this smaller set of subjects will result in high CR. Thus overhead on classifiers is drastically reduced and decision rule is made simple.

2.2   Robust Feedback for Test-Subject-Specific Variation Neutralization

System is made more intelligent by using a feedback rule at classification side. Test image has lot of variations in it and cannot be neutralized in a single algorithm. Use of feedback type algorithms help in *neutralizing the test subject specific variations.* In this paper we choose to neutralize pose of facial images.

3   Proposed Probabilistic Classifier

Input to EM and EMA are trained database and test subject. Analysis of the output of EMA and EM is done here and is interpreted in a probabilistic way. This interpretation of the result in a probabilistic way shows the solution to decrease the number of subjects. It creates a *custom minimized compact database* for that particular test image.

Metric chosen to show the importance of PC is FR. FR is basically recognition of the person whose certain images are trained in the database. Error minimization algorithms provide a resultant matrix. Subject which gave the highest match score for tested image was considered as probable subject to which tested image belongs. The highest match scored subject is the most *probable* result, which means it *may* be a result.

The error minimization (EMA) which we chose is the normal pixel distance between two galleries. PC does the pixel wise difference between training and testing image gallery. Resultant samples of subtraction are squared and added. This method is known as Least Square Error (LSE) calculation.

PC after performing EMA arranges large number of classes in the order of probability of recognition. Minimum value obtained from EMA of the above says that, the correlation between the testing and that training image is very high. In other words the minimum obtained from the EMA output corresponds to Maximum Probability of recognition. A training image corresponding to minima will refer to its class (subject). Similarly, we can say that probability of testing image belonging to that particular class is very high. Therefore, next minimum corresponding to different class corresponds to second Maximum Probability for recognition. LSE is carried out on all classes and this is the total output of EMA. All classes are arranged in the descending order of their probability of correct recognition. This final result is a probabilistic one and is termed as maximum probability matrix.

Maximum probability matrix gives a clear idea for user to choose proper class for correct recognition. PC gives the output with x-probabilities for recognition where 'x' is user defined. It just means we select 'x' classes from the probability matrix in their decreasing order of recognition. *'x' is the number of probable output classes chosen.* These 'x' classes are test image specific compact database. Fig. 1 a), b) explains proposed PC effects on decision device ('x' is chosen to be 2).

Fig. 1c) shows how PC works. Initially due to class C, the class separator is confused and will eliminate all dark gray features from class B. Class separator also removes minority light gray feature from class A. This leads to improper

---

Data: Training and testing Images
Result: Recognition of test image
Training stage;
while *(number of training images > 0)* **do**
| Apply pre-processing on training images to obtain the feature vectors.
end
Feature Selection;
Apply CBPSO on training dataset to produce feature vectors;
Testing stage;
if *(Training Dataset > 'x')* **then**
 Apply classification algorithm to recognize the test image;
 Choose 'x' best matching classes. This is new customized training dataset for this particular test image;
 Correct Class $\in$ (Chosen 'x' classes);
 Repeat Testing Procedure with chosen 'x' classes;
else
| Apply classification algorithm to recognize the test image;
end

Algorithm 1. Proposed Probabilistic Classifier applied to FR

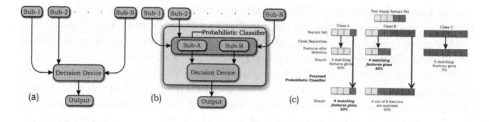

Fig. 1. (a) Normal load on decision device. (b) Reduced load on decision device due to PC. (c) Figure explaining the after effects of proposed PC. Initially test image is recognized as belonging to Class B (false statement). After application of PC, one can see that test image is recognized as belonging to Class A (true statement).

judgment leading to recognition as class B (4 features match, 20% for each feature). Now probabilistic classifier solves the above problem by eliminating redundant classes and selects top two matches. It is again fed to class separator where no features are removed from either of them. Class A retains light gray as the light gray feature is minority in class B. Class A gives 80% this time where as class B has 4 matching features out of 8 features contained by it. Therefore resulting in 40%. Thus correct class A is recognized after the application of the proposed PC. Therefore we can conclude that PC will improve selection of features thus aiding to recognition rate. Complete algorithm of the explained PC is shown in Algorithm 1.

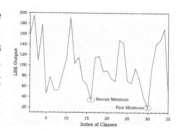

Fig. 2. The EMA resultant matrix where classes which gave first two minima are separated out by PC

Fig. 2 shows the output obtained of LSE for FERET database. Figure shows the 2 best matches, i.e., two minima. Extracting both first and second minima is the proposed process for reduction in the size of database. PC arranges EMA (it just takes 'x' maximum probable results). EMA gives the error output, thus 'x' minimum values (or highly matching classes) are chosen. Mathematical analysis is done in two parts. First the complexity analysis, second the likelihood score analysis.

Complexity Analysis

$$Existing\ complexity = {}^{N}C_1 = N \tag{1}$$
$$Proposed\ Combination = {}^{N}C_x \tag{2}$$
$$New\ Proposed\ Complexity = {}^{x}C_1 = x \tag{3}$$

N = Total Number Of Classes, x = Total probabilities chosen and $N \gg x$. Initial complexity of the machine is given by Eq. 1 (by permutations and combinations). The *selection* of a single subject from a large set of subjects is very complex as given in Eq. 1. *Combinations* of selecting a single class from 'N' is given by Eq. 1 and equals to 'N'. Reducing 'N' to one is the complexity here. The proposed PC does a simple, accurate job in reducing the Total Number of Classes (N) to 'x' probabilistically as explained above and is shown in Eq. 2.

Post classifiers which come after PC do a simple job of selecting one subject from a small set of subjects whose complexity is lower as in Eq. 3. *The combinations of selecting a single subject for these post classifiers is 'x' and $N \gg x$. Reducing 'x' to one is the task of the machine, which is far easy compared to initial (Eq. 1). These post classifiers now have to select one subject from a very small set and there is less confusion for the system here as the redundant subjects (N-x) are removed.*

Eq. 4 is the probability of choosing an output from large set of classes. This is the standard case where classifiers fail due to *low probability*.

Probability of choosing 'x' entities is more than the probability of choosing only one entity. Probability of PC is shown in Eq. 5. Eq. 5 yields higher value than Eq. 4. This shows that PC can be implemented easily. It also shows that PC is simple.

The probability of post classifier is given in the $P_{new}=1/x$. This gives *more value of probability of recognition* than Eq. 4 and Eq. 5. Thus complexity of post classifier is reduced drastically and accuracy is increased by using PC.

Close comparison of P_{new} and Eq. 4 shows an increase in the correctness of classification. *Initial rule of Eq. 4 is modified to new simple rule of P_{new} by PC which is more suited for practical purposes so as to achieve high CR.*

$$P_{standard} = \frac{1}{Total\ Number\ Of\ Classes} \tag{4}$$

$$P_{PC} = \frac{x}{Total\ Number\ Of\ Classes} \tag{5}$$

Maximum Likelihood Analysis. LSE gives those parameter values which provide *most accurate* description of the data. We use LSE [10] initially and obtain data where the parameter which gives the most accurate result is obtained. Using MLE on the data provided by LSE, we form the output of PC which enhances accuracy of correctness of the result.

MLE provides a solution saying that, parameter p attains maximum value at $\hat{A}$ (likelihood of the output is maximum for $\hat{A}$). Next near solution is $\hat{B}$. Just using $\hat{A}$ for the decision is the normal decision rule, this wouldn't provide a complete solution. We have to consider the parameter value $\hat{B}$ also in our decision. Considering $\hat{B}$ with $\hat{A}$ increases the chances of correctness. This method of choosing *certain nearest likelihood in addition to the maximum likelihood* brings down the problem from a large set to a simpler one. Now the problem is redefined as to provide a solution for only a minimal set of entities. $P = \sum_{i=1}^{N} P_i = 1$ shows a general form of probabilistic matrix. Total probability of success in recognition is denoted as P and other probabilities given by this classifier are named as $P_1, P_2....P_N$. P_1 corresponds to the maximum probability, P_2 corresponds to second maximum probability and so on. All these cannot aid to recognition at the same time, so these are mutually exclusive. $P = \sum_{i=1}^{4} P_i$ is the method that we have used from the PC output.

$P_1 \gg P_2 \gg P_3 \gg P_4 \gg .... \gg P_N$ is obtained as the preprocessing is done and the redundancy in the image is removed. The second or third probability when considered will increase the Recognition Rate (RR). RR is the metric and is the extension of CR in FR systems.

4   Proposed Robust Feedback Recognition

If testing image is recognized, it is fine, otherwise we send this testing image for further analysis. The failed testing image is again tested for recognition but, this time the pre-processing of the testing image is different. *We flip the image to neutralize the pose in the feedback loop.* This is like an adaptive feedback algorithm. There exists a chance that this testing image may be recognized second time. This process of RF is shown in Fig. 3.

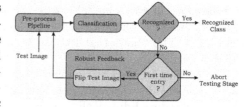

Fig. 3. Robust Feedback mechanism

The various methods to neutralize pose variations is reviewed in [7]. We here try to neutralize pose using a reflection or mirror method. We test an image normally using the basic algorithm which has the satisfactory recognition rate. If this algorithm fails, then we do the same pre-processing for the reflected image of the testing image. Suppose all the images that are trained are of only one profile, assume right pose, then in testing if we have the left pose image we can have a false recognition sometimes. This failure case is eliminated by sending the reflected image of testing image to the testing block.

The probability of success in the recognition $P_{standard}$ is given by Eq. 4. As one could see from Eq. 4, the success in recognition is very less, which also means the intricacy for the machine is very high. $PAL1$ and $PAL2$ are the probabilities of success for first algorithm and second algorithm respectively (Eq. 6). These two events are mutually exclusive as these cannot work or recognize simultaneously. This means the intersection between the above 2 events is a null matrix. Therefore we have $P_{standard} = PAL1 + PAL2$, this shows that the probability of success has increased (from only PAL1 to PAL1+PAL2) due to algorithm 2. The complexity of classifier is reduced to $P_{new} = (1/2)$, which gives more probability of classification value when compared to Eq. 7. Thus by using minimal number of the training images we can obtain high RR.

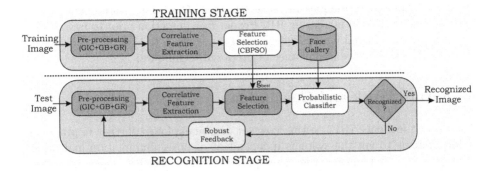

Fig. 4. Block diagram of proposed FR system

$$P_{standard} = PAL1 + PAL2 - P(AL1 \cap AL2) \tag{6}$$

$$P_{RF} = \frac{2}{Total\ Number\ Of\ Classes} \tag{7}$$

5   Experiments

5.1   Color FERET Database

FERET database[12] which stands for Facial Recognition Technology database mainly accounts for pose, scale and costume variations. Table 1 depicts the performance of the proposed FR system. Fig. 4 shows the process of generating the feature set where DWT+SWT is used as extractors (correlative extraction).

Important features are extracted out using Correlative Binary Particle Swarm Optimization (CBPSO) [13],[14]. BPSO with cross correlation as the fitness function is termed as CBPSO. RR without PC+RF is 81.43%, and with them is 98.32%. This shows the importance of proposed PC and RF. We used Matlab 2012b[11] to simulate the results in a system (2nd gen i7, 2.2Ghz).

Table 1. FERET comparison for 8:12 training to testing ratio for 35 classes

Method	RR(%)
DFT+DCT[15]	80.23
Selective Illumination Enhancement[16]	85.41
Laplacian Gradient Masking[17]	85.71
K-Means Clustering[18]	86.14
Proposed Method	**98.32**

5.2   LFW Dataset

LFW dataset [19] is chosen to show that PC also increases RR even for wild datasets and is invariant of discrepancies of the images. RR is less compared to other datasets as face has to be detected or background must be removed. Without the use of PC we will get very less RR (for 8:12, RR is

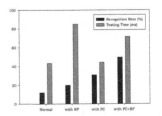

Fig. 5. Impact of Proposed techniques on LFW

19.92%). Proposed techniques increase RR to 49.52%. This shows that PC is playing a major role in increasing RR. Proper face detection prior to face recognition is not considered here as the importance is given to show the variance of PC on RR. Fig. 5 shows importance of proposed techniques.

6 Future Work and Conclusion

The Robust feedback (RF) testing concept can be exploited further in practical vision systems to get the systems independent of pose, illumination and other discrepancies as explained earlier. Here pose is tackled using RF, which generated a considerable increase in the RR.

Proposed Probabilistic Classifier (PC) opens a door in the classification domain using the simplest probability models to decrease the size of the databases. Also, PC increases the RR to substantially high value. Intricacy of machine when working on large dataset can be completely eliminated using this approach. The design of post classifier or an efficient class separator is the future work required. Proposed classifier can be extended for real-time video applications.

References

1. Tou, J.T., Gonzalez, R.C.: Pattern recognition principles (1974)
2. Brent, R.P.: Algorithms for minimization without derivatives. Courier Dover Publications (1973)
3. Kulkarni, S.R., Lugosi, G., Venkatesh, S.S.: Learning pattern classification-a survey. IEEE Transactions on Information Theory 44(6) (1998)
4. Jain, A.K., Duin, R.P.W., Mao, J.: Statistical pattern recognition: A review. IEEE TPAMI 22(1), 4–37 (2000)
5. Duda, R.O., Hart, P.E., Stork, D.G.: Pattern classification. John Wiley & Sons (2012)
6. Chellappa, R., Wilson, C.L., Sirohey, S.: Human and machine recognition of faces: A survey. Proceedings of the IEEE 83(5), 705–741 (1995)
7. Zhang, X., Gao, Y.: Face recognition across pose: A review. Pattern Recognition 42(11), 2876–2896 (2009)
8. Belhumeur, P.N., Hespanha, J.P., Kriegman, D.J.: Eigenfaces vs. fisherfaces: Recognition using class specific linear projection. IEEE TPAMI 19 (1997)
9. Ortiz, E.G., Becker, B.C.: Face recognition for web-scale datasets. Comput. Vis. Image Understand (2013), http://dx.doi.org/10.1016/j.cviu.2013.09.004
10. Kollar, D., Friedman, N.: Probabilistic graphical models: principles and techniques. The MIT Press (2009)
11. MATLAB, http://www.mathworks.com
12. Color FERET, http://www.nist.gov/itl/iad/ig/colorferet.cfm.
13. Kennedy, J., Eberhart, R.C.: A discrete binary version of the particle swarm algorithm. In: IEEE International Conference on Systems, Man, and Cybernetics, vol. 5 (1997)
14. Ramadan, R.M., Abdel-Kader, R.F.: Face recognition using particle swarm optimization-based selected features. International Journal of Signal Processing, Image Processing and Pattern Recognition 2(2), 51–65 (2009)

15. Deepa, G., Keerthi, R., Meghana, N., Manikantan, K.: Face recognition using spectrum-based feature extraction. Applied Soft Computing 12(9), 2913–2923 (2012)
16. Vidya, V., Farheen, N., Manikantan, K., Ramachandran, S.: Face Recognition using Threshold Based DWT Feature Extraction and Selective Illumination Enhancement Technique. Procedia Technology 6, 334–343 (2012)
17. Murthy, N.N., Raghunandana, R., Manikantan, K., Ramachandran, S.: Face recognition using DWT thresholding based feature extraction with laplacian-gradient masking as a pre-processing technique. In: Proceedings of the CUBE International Information Technology Conference, pp. 82–89. ACM (2012)
18. Surabhi, A., Parekh, S.T., Manikantan, K., Ramachandran, S.: Background removal using k-means clustering as a preprocessing technique for DWT based Face Recognition. In: 2012 International Conference on Communication, Information & Computing Technology (ICCICT), pp. 1–6. IEEE (2012)
19. Gary, B., Huang, M., Ramesh, T.: Labeled Faces in the Wild: A Database for studying Face Recognition in Unconstrained Environments

Fuzzy Membership Estimation Using ANN: A Case Study in CTG Analysis

Sahana Das[1], Kaushik Roy[2], and Chanchal Kumar Saha[3]

[1] Narula Institute of Technology, Kolkata-700109, India
sahana.das73@yahoo.com
[2] West Bengal State University, Department of Computer Science, Kolkata-700126, India
kaushik.mrg@gmail.com
[3] Biraj Mohini Matrisadan and Hospital, Barrackpore, Kolkata-700122, India
chanchal1069@yahoo.com

Abstract. The Cardiotocograph (CTG) is being used by the obstetricians since 1960s as a means for recording (graphy) the heart beat (cardio) and the uterine contraction pressure (toco) of the mother, to evaluate the well being of the fetus. One of the major features of fetal heart rate (FHR) is its baseline,the accurate classification which is of utmost importance as all the other parameters of CTG rely on it. Inherent vagueness in the assessment given by the physicians can probably be modeled using fuzzy logic. It is one of the most trusted tools to handle uncertainty intrinsically present in the linguistic expression of human. The main challenge in designing a fuzzy logic based system is to design its membership function. In this paper we have presented a ANN based technique for the design of Fuzzy Membership Function (FMF) of FHR and used it in Fuzzy Unordered Rule Induction Algorithm (FURIA) in order to classify the CTG. The results obtained show significant improvement in classification over non FMF based technique.

Keywords: Cardiotocography, Fetal Heart Rate (FHR), Artificial Neural Network (ANN), FURIA, Fuzzy Membership Function.

1 Introduction

This Medical decision making in today's world depend on a large extent on the automated analysis of medical data. This helps improve medical diagnosis and treatment in large extent. Some of these systems render well to intelligent classification of data [1]. Cardiotocography is one such medical decision making system that had been in use since 1960s, but only recently researches had been carried out to automate its decision making capability using various artificial intelligent techniques. Thus, though the traditional mode of interpretation of CTG by visual analysis is not time consuming, the accuracy of it is largely dependent on the knowledge and experience of the clinician and many a time varies from clinician to clinician. An automated system for decision support may be able to eliminate some of the shortcomings of visual interpretation.

© Springer International Publishing Switzerland 2015
S.C. Satapathy et al. (eds.), *Proc. of the 3rd Int. Conf. on Front. of Intell. Comput. (FICTA) 2014*
– *Vol. 1*, Advances in Intelligent Systems and Computing 327, DOI: 10.1007/978-3-319-11933-5_25

The features of FHR are Baseline, Variability, Acceleration and Deceleration. Each of these can be classified as Reassuring, Non-Reassuring and Abnormal. In order for proper interpretation of CTG, each of these features needs to be separately classified. Baseline is the most fundamental feature of FHR as all the other features are directly dependent on it. It is a common belief that robustness of the algorithm for the estimation and classification of FHR baseline is largely responsible for the efficiency of the automated fetal monitoring system.

Last few decades had seen several attempts at automation of the CTG analysis and especially the estimation of baseline. These range from classical approaches that employ non-linear filtering of FHR signals to estimate baseline to intelligent approaches based on Artificial Neural Network. System 8000 was developed in 1991 for antenatal analysis based on the work of Dawes. This system was later upgraded to SonicAid FetalCare, which is used commercially for FHR analysis [2]. 2CTG2 was developed based on the work of Magenes [3]. NST-EXPERT, which was developed in 1995 by Alonso-Betanzos is an expert system that is capable of making a diagnosis and proposing a treatment [4]. CAFE is an extension of NST-EXPERT and was developed in 2002 by Gujjaro-Berdinas and Alonso-Betanzos. It uses neuro-fuzzy approach to automate all the tasks associated with the analysis of CTG [5]. SisPorto was developed over a span of 14 years at the University of Porto, Portugal. It uses an expert system to estimate the individual aspects of CTG. The current version of SisPorto that is in use is Omniview SisPorto 3.5 [6]. So far only Sonicaid, which is based on traditional crisp method of CTG analysis, is used commercially worldwide, but the others failed to get any wide acceptance due to lack of specificity and poor practicality in every day use. We have exploited the neuro-fuzzy based tools capability to tolerate ambiguity and uncertainty that are inherent in any medical analysis.

1.1   Cardiotocography

Heart rate of fetus is controlled predominantly by autonomic nervous system whose activity basically depends on the oxygen content of the fetal blood. Thus, analysis of FHR signals can predict asphyxia [7].

CTG is non-invasive, cost-effective tool for checking the fetal status. Its induction to the clinical practice limited the occurrence of fetal problems leading to a decline in child morbidity and mortality [8]. Accurate interpretation of CTG are impeded by several factors: Crisp rules used in FIGO guidelines are not sufficient to capture the inherent vagueness present in the interpretation of CTG.

To analyze a CTG, each of its features is identified and then classified using rules derived from the guideline of FIGO. But since these guidelines are solely based on experimental observations, they lack precision and hence sometimes give rise to difference in opinion of the medical practitioners [9]. Also, great care need to be taken to interpret the data at the boundaries. These factors may lead to misdiagnosis which can cause lifelong affliction such as cerebral palsy that occurs due to hypoxia or even death. Many of these deaths can be prevented if the abnormal FHR pattern can be correctly recognized [10].

Baseline is the resting level of fetal heart rate. Acceleration of FHR is a temporary increase of fetal heart rate 15bpm above the baseline and lasting for 15 secs or longer. This occurs in response to fetal movements and signifies fetal central nervous system alertness and fetal well-being. The temporary decreases of FHR at least 15 bpm below the baseline and lasting for 15secs or longer is called Deceleration. It usually suggests such hazardous events as compression of the umbilical cord.

The baseline of FHR is established by approximating the average FHR rounded to increments of 5 beats per minute (bpm) during a 10-minute observation. Periods of acceleration, deceleration, and marked fetal heart rate variability (> 25 bpm) are to be excluded to define the baseline. There must be at least 2 minutes of recognizable baseline segments which need not be contiguous in any 10-minute window, or the baseline for that period is undetermined [11].

The 3-tier classification of CTG trace is as follows: A CTG trace is said to be *Normal* (N) if all the four features of FHR fall into reassuring category. *Suspicious* (S) if one of the features of FHR fall into non-reassuring category, while the others fall into reassuring category. *Pathological* (P) if two or more features fall into non-reassuring category or more than one feature fall into abnormal category [11].

1.2   Fuzzy Membership Estimation Using Artificial Neural Network

ANN is capable of handling quantitative information, while fuzzy systems can handle qualitative data. Unlike crisp system partitions in fuzzy classifications are soft and continuous. Thus, the combination of ANN and fuzzy should have a more robust decision making system.

Input data points are divided into classes C_1, C_2,.....,C_m. We assign a full membership of 1 if a data point completely belongs to a class. A neural network is created that uses the data points and the corresponding membership values for different classes to train itself. As the input values are passed through the ANN the errors if any are estimated. The process is iterated until error is within some user defined limit. This way the membership values of the entire set of data points are calculated. During the training phase ANN learns relationships between given set of inputs and outputs by changing the weights. The best known training algorithm is the Back Propagation Algorithm. It minimizes the difference between resulting and actual output [12]. Trained ANN is capable of generalization i.e. it can produce a set of outputs from inputs that are not their in the training set.

1.3   FURIA

FURIA is an extension of a modern rule learner RIPPER (Repeated Incremental Pruning to Produce Error Reduction). It is capable of dealing with datasets with large number of features. A list of rules favors a default class, which may induce a bias. Its learning is based on fuzzy rules instead of conventional rules and it generates rule set instead of rule lists [13]. A set of rules is generated for each class using one-vs-rest strategy. Thus, each class is separated from all the other classes. As a result, there is no default rule and the order of the rule is not relevant.

The rest of the paper is organized as follows: Sec. 2 deals with challenges in CTG interpretation followed by methodology which is explained in Sec 3 and discussion are given in Sec 4. Sec 5 contains the conclusion.

2   Challenges in CTG Interpretation

Sometimes the conclusions given by the experts are expressed as a fuzzy estimation. This mainly arises when expert's knowledge is subject to the uncertainty of the outcome and experts put across their opinion in imprecise terms. This also may lead to disagreement between experts. Such dilemma is often encountered in bio-medical signal analysis and pattern recognition [14]. Resolution uncertainty and ambiguity are inherent in the interpretation of CTG signal, giving rise to possible disagreement between physicians. Fuzziness is associated in the language used by the physicians to convey the conclusion, e.g.

```
if baseline exhibits bradycardia with low variability
then the CTG signal is abnormal
```

To estimate the fuzzy membership function from numerical data using multi-layer network, the fuzzy estimation provided by the experts should first be converted to numerical values.

3   Methodology

For our work we have used CTG data set from UCI Irvine Machine learning Data Repository[15]. The baseline values of the given CTG data are fuzzified using Fuzzy Membership Function (FMF). From the given 2126 data points, 105 unique data points are identified. Linguistic values of the fuzzy linguistic variable Baseline can be expressed as:

Reassuring (R) : $110 \leq$ baseline≤ 160
Non-Reassuring (NR) : $100 \leq$ baseline≤ 109
$\qquad\qquad\qquad 161 \leq$ baseline≤ 180
Abnormal (Ab) : baseline≤ 100
$\qquad\qquad$ baseline ≥ 180

Let x be a N-dimensional vector, with N being the number of data points and y be a M-dimensional vector with M being the number of classes. Thus,

$$x = [x_1, x_2, \ldots, x_n]^T \text{ and }$$

$$y = [y_1, y_2, \ldots, y_m]^T$$

Let $\{(x_n, y_n)\}$ denote a set of labeled data, where x_n is the instance associated with the classification y_n.

The proposed neuro-fuzzy classification technique estimates the membership value of each data point and assigns them to a class accordingly. The data points are assigned a membership value of unity, in accordance with the classification provided by the obstetricians. The one bit assignment of membership value is done as follows:

$$\forall\ x_i\ \in\ \mathbf{x}\ni\ y_i\in\ \mathbf{y}\ \text{such that}\ y_i = 1\ \text{and}\ y_j = 0 \forall\ i \neq j \tag{1}$$

The proposed NN consists of one input layer with of number of neurons equal to the number of data points, one hidden layer with five neurons and a output layer with number of neurons equal to three. Input x_i from the training data set are passed through the ANN. Input to the hidden layer is calculated as:

$$f_j(\overline{x_j}) = \sum_{i=0}^{N} x_i w_{ij} + w_j \tag{2}$$

where, $\overline{x}_j$ is the sum of the inputs to node j, w_j is the bias term of node j and w_{ij} is the weight associated with the link connecting node i to node j. N=105 in our case.

Output of the jth node in the hidden layer is calculated using Sigmoid function according to the equation below:

$$O_j = \frac{1}{1 + \exp[-(\sum f(\overline{x}_j)]} \tag{3}$$

The error function of the jth input-output pair is given by

$$E_j = \sum_{i=1}^{N} \| o_j - a_j \|^2 \tag{4}$$

where, o_j is the ideal output and a_j is the actual output. For an element with error E the weight may be updated as:

$$w_i(new) = w_i(old) + \eta E x_i$$

where, η is the learning rate. Classical pattern recognition presumes that the classes are mutually exclusive. Part of the result obtained is shown in Table 1. The membership function thus obtained is given in Figure1.

4   Results and Discussion

Using the five-fold cross validation with FURIA we have obtained 92.14% classification of FHR with only 4.61% error compared to 94.59% classification with 5.40% error obtained without applying the membership value of baseline in Reassuring(R), Non-Reassuring(NR) and Abnormal(Ab) categories. The comparisons of the same are given in Table 2 and confusion matrix for overall FHR classification is given in Table 3.

Though the automated classification shows a higher accuracy the error is much less using the FMF based classification. This is even more evident from the confusion matrix. Evaluation is done by WEKA on Intel Core 2 Duo CPU E7500, 2.93GHz with Ubuntu 12.04 OS with 5 fold cross validation.

FMF based approach has higher recognition accuracy, especially for Pathological data. For FMF based approach FURIA has recognized approximately 4% Pathological data as Normal whereas in non-FMF based interpretation approach about 7.4% Pathological data are identified as Normal. Obtaining precise identification for baseline and hence the accurate classification of FHR is of utmost importance in analysis of CTG data.

Table 1. Estimated Membership Values of FHR Baseline

Baseline	NR	R	Ab
110	0.144174	0.843447	0.000000
113	0.000008	0.999986	0.000000
114	0.000006	0.999989	0.000000
115	0.000006	0.999989	0.000000
116	0.000006	0.999989	0.000000
120	0.000006	0.999989	0.000000
121	0.000006	0.999989	0.000000
122	0.000006	0.999989	0.000000
176	0.999999	0.000000	0.000001
177	0.999987	0.000000	0.000001
178	0.999537	0.000000	0.000388
179	0.942712	0.000000	0.054488
180	0.056736	0.000000	0.947901
181	0.000287	0.000000	0.999770
182	0.000004	0.000000	0.999997

Table 2. Comparison between FMF Based and Non-FMF Based Classification

	Classified using FMF	Without FMF [16]
Accuracy	92.14%	94.59%
Error	**4.61%**	**5.40%**
Time	19.15	19.02
No. of Rules	33	39

Table 3. Confusion Matrix for FMF Based Classification

Classification	N	S	P
N	1629	21	4
S	58	198	2
P	7	6	159

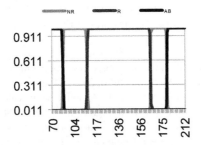

Fig. 1. Trapezoidal membership function

5   Conclusion and Future Scope

We have successfully employed the ANN based technique to design the FMF and used it in FHR classification. Overall accuracy of 92.14% with a recognition accuracy of 92.44% for abnormal cases was obtained. Encouraged by this result we plan to fuzzify all the parameters of FHR to increase its performance. We have compared our result with [16] and found that by applying FMF the error rates for pathological cases are decreased by almost 0.79% whereas for detection of abnormal cases improved by about 4.9% which is of paramount importance as it can help save some child from lifelong disability or even death. This is because if a normal or suspicious case is recognized as suspicious or abnormal it may lead to unnecessary c-section but the reverse could endanger the life of would be mother or fetus or both. We are working on improving the accuracy of classification, especially for pathological cases since it very important from clinical point of view to recognize the fetal distress accurately.

Acknowledgments. The authors would like to thank Biraj Mohini Matrisadan and Hospital authority for allowing us to use their instruments and infrastructure to collect the data. The authors would also like to extend their thanks to the patients for their kind cooperation.

References

1. Yilmaz, E., Kilikçier, C.: Determination of Fetal State from Cardiotocogram using LS-SVM with Particle Swarm Optimization and Binary Decision Tree. J. Comp. Math. Methods in Med. 2013, 1–8 (2013)
2. Dawes, G.S., Visser, G.H., Goodman, J.D., Redman, C.W.: Numerical analysis of the human fetal heart rate: the quality of ultrasound records. Am. J. Obstet. Gynecol. 141(1), 43–52 (1981)
3. Magenes, G., et al.: Improving the fetal cardiotocographic monitoring by advanced signal processing. In: 25th IEEE Annual International Conference of Engineering in Medicine and Biology Society, vol. 3, pp. 2295–2298. IEEE Press, Italy (2003)
4. Alonso-Betanzos, A., Guijarro-Berdiñas, B., Moret-Bonillo, V., López-Gonzalez, S.: The NST-EXPERT project: the need to evolve. J. Artif. Intell. Med. 7(4), 297–313 (1995)

5. Guijarro-Berdiñas, B., Alonso-Betanzos, A.: Empirical Evaluation of a Hybrid Intelligent Monitoring System using Different Measures of Effectiveness. J. Artif. Intell. Med. 24(1), 1–96 (2002)
6. de Campos, A., Sousa, P., Costa, A., Bernardes, J.: Omniview-SisPorto® 3.5 - A central Fetal Monitoring Station With Online Alerts Based on Computerized Cardiotocogram+ST Event Analysis. J. Perinatal Medicine 36(3), 260–264 (2008)
7. Helgason, H., Abrey, P., Gharib, C., et al.: Adaptive Multiscale Complexity of Fetal Heat Rate. IEEE Transactions on Biomedical Engineering 58(8), 2186–2193 (2011)
8. Das, S., Roy, K., Saha, C.K.: A Novel Approach for Extraction and Analysis of Variability of Baseline. In: IEEE International Conference on Recent Trends in Information Systems, pp. 336–339. IEEE Press, Kolkata (2011)
9. Skinner, J.F., Garibaldi, J.M., Ifeachor, E.C.: A Fuzzy System for Fetal Heart Rate Assessment. In: Reusch, B. (ed.) Fuzzy Days 1999. LNCS, vol. 1625, pp. 20–29. Springer, Heidelberg (1999)
10. Chinnasamy, S., Muthasamy, C., et al.: An Outlier Based Bi-Level Neural Network Classification Systemfor Improved Classification of Cardiotocogram Data. J. Life Science 10(1), 244–251 (2013)
11. Macones, G.A., et al.: The 2008 National Institute of Child Health and Human Development Workshop Report on Electronic Fetal Monitoring: Update on Definitions, Interpretation, and Research Guidelines. J. Am. College of Obstet. & Gynecol. 112, 661–666 (2008)
12. Altiparmak, F., Dengiz, B., Smith, A.E.: A General Neural Network Model for Estimating Telecommunications Network for Reliability. IEEE Transactions on Reliability 58(1), 2–9 (2009)
13. Chen, C.Y., Chen, J.C., Yu, C., et al.: A Comparative Study of a New Cardiotocography Analysis Program. In: IEEE Annual Conference of Med. Biol. Soc., pp. 2567–2570. IEEE Eng., Taiwan (2009)
14. Langellė, R., Devoux, T.: Training Multilayer Perceptrons Layer Using an Objective Functions for Internal Representations. J. Neural Networks 58(8), 83–98 (1996)
15. UCI Irvine Data Repository, http://archive.ics.uci.edu/ml/datasets/Cardiotocography
16. Das, S., Roy, K., Saha, C.K.: Application of FURIA in the Classification of Cardiotocograph. In: IEEE- International Conference on Research and Development Prospects on Engineering and Technology, pp. 120–124. IEEE Press, Chennai (2013)

A Comparative Study of Fractal Dimension Based Age Group Classification of Facial Images with Different Testing Strategies

Anuradha Yarlagadda[1,*], J.V.R. Murthy[2], and M.H.M. Krishna Prasad[2]

[1] Department of CSE, JNTUH, Hyderabad
[2] Department of CSE,UCEK, JNTUK, Kakinada

Abstract. The demand of estimation of age from facial images has tremendous applications in real world scenario like law enforcement, security control, and human computer interaction etc. However despite advances in automatic age estimation, the computer based age classification has become prevalent. The present paper evaluates the method of age group classification based on the Correlation Fractal Dimension (FD) of facial image using different validation techniques. To reduce variability, multiple rounds of cross validation are performed using different partitions to the data. The expected level of fit of the model classifying facial images into four categories based on FD value of a facial edge is estimated using multiple cross-validation techniques. The simulation is carried out and results are analyzed on different images from FG-NET database, Google database and from the scanned photographs as these are random in nature and help to indicate the efficiency and reliability of the proposed method. It is also a successful demonstration that Correlation Fractal Dimension of a facial edge is sufficient for a classification task with high percentage of classification accuracy.

Keywords: Age Group classification, Correlation Fractal Dimension, facial image, canny edge, facial edge image, cross validation.

1 Introduction

Automatic age estimation and predicting future faces have rarely been explored. With human age progression face features changes. Humans can identify very informative facts from facial images, which include identifying, age, gender etc. The identification of different features of face images has been well explored in real-world applications [1], including passports and driving licenses. Despite the broad exploration of person identification from face images, there is only a limited amount of research [2] on how to accurately estimate and use the demographic information contained in face images such as age, gender, and ethnicity. This laid foundation for interesting research topics on gender classification [3], facial image recognition [4], predicting future faces [5], and reconstructing faces from specified features [6] and so on. As human age

---

* Corresponding author.

© Springer International Publishing Switzerland 2015
S.C. Satapathy et al. (eds.), *Proc. of the 3rd Int. Conf. on Front. of Intell. Comput. (FICTA) 2014*
– *Vol. 1*, Advances in Intelligent Systems and Computing 327, DOI: 10.1007/978-3-319-11933-5_26

estimation supports many potential application areas, identification of age by computers has become prevalent. Fu and Huang [7] estimated the age on the holistic appearance of the image. Chao et al. [8] made classification based on Label-sensitive relevant component analysis and Chang et al. [9] considered Ordinal hyper plane ranking. A hierarchical age estimator [10] is proposed for automatic age estimation. Age groups for classification including babies, young adults, middle-aged adults, and old adults is given by Wen-Bing et al.,[11]. Age group classification on facial images based on crania-facial development theory and skin wrinkle analysis [12], considered only three age-groups babies, young adults, and senior adults. The calculations are done based on crania-facial development theory and skin wrinkle analysis. While studying physical changes obtained by ageing of human being, many researchers tried to classify facial images into various groups [13]. Sirovich and Kirby [14] classified images into two categories, babies and adults. Neural networks are used for discriminating facial age generation [15][16]. Classification of Facial image of human into four categories based on Fractal Dimension value of the facial skin [17] is a new concept developed and this paper focuses on finding the accuracy of the model based on different test strategies. In this paper the focus is on studying the efficiency of the proposed method [17], Correlation Fractal Dimension based age group classification using multiple cross validation techniques. Different testing and validation techniques such as hold out method and tenfold cross validation method are investigated and the comparison analysis is presented. From the simulation it is observed that our proposed method [17] passes with good classification efficiency through rigorous methods of testing successfully and hence robust in nature.

The rest of the paper is organized as follows. Section 2 deals the proposed age group classification method. Different cross validation techniques and their results are discussed in section 3 and section 4 presents conclusions.

2   Proposed Method

Age group classification of the proposed method [17] is done by using facial edge of an image. The rapid wrinkle changes in the skin are exploited by edges of facial image. The following steps are proposed to estimates the Correlation Fractal Dimension (FD) value derived from the facial edges.

Step 1: Consider the original color image
Step 2 : Original image is cropped based on the location of the eyes
Step 3: Cropped image is converted to a gray scale image
Step4: Facial edges of the gray scale image are extracted as given in Fig 1.
Step 5: Estimate the fractal dimension value of the facial edge of an image
Step 6 : Classify the age group of the facial image based on the correlation fractal dimension

The original facial image is cropped based on the two eyes location in the second step. In the step 3, if the images are color images then those are converted into a gray scale facial image by using HSV color model. In the fourth step, extract the edges of facial image by using canny edge operator. In the fifth step, calculate the Correlation

Fractal dimension value. In the last step a new algorithm is derived for an efficient age group classification system based on the Correlation Fractal Dimension.

Recent literature revel various color models in color image processing. In order to extract facial image features from color image information, the proposed method utilized the HSV color space. In the RGB model, images are represented by three components, one for each primary color – red, green and blue. Hue is a color attribute and represents a dominant color. Saturation is an expression of the relative purity or the degree to which a pure color is diluted by white light. HSV color space describes more accurately the perceptual color relationship than RGB color space because it is adopted with a non-linear transform. The proposed method uses HSV color space model conversion, because the present study is aimed to classify the human age in to four groups with a gap of 15 years.

This paper [17] found that edges are relatively a good choice for obtaining facial image attributes or contents. The facial image edge detection is the process of locating sharp discontinuities in a facial image. The discontinuities are unexpected changes in pixel intensity which differentiate boundaries of objects in a scene. The paper [17] utilizes the canny edge detection algorithm to detect the edges of the facial image [18][19][20] The Canny edge detection algorithm is the optimal edge detector.

2.1   Calculate the Fractal Dimension Value

Fractal is self-similar objects. Inherently, fractals also have a degree of self-similarity. This means that a small part of a fractal object may resemble the entire fractal object. Fractals are Geometric primitive such as self-similar and irregular in nature. Fractal Geometry was introduced by Mandelbrot [21]. The correlation fractal dimension (FD) is the defining characteristic of a fractal which has been used as a measure of similarity. The fractal- based methods have been applied to many areas of digital image processing, such as, image synthesis, image compression and image analysis [22][23]. The present paper analyses the results more extensively on the method for classifying the facial edge image into four categories such as child (0-15), young adults (15-30), middle-aged adults (31-50), and senior adults (> 50)[17].

Fig. 1. Facial edge images of original facial images

Many application areas of digital image processing used fractal dimension [24] and demonstrated that finding the fractal dimension on colored images is not giving better results for classification and hence in the proposed method [17] correlation fractal dimension value, is calculated on edges of the facial image which proved to give better results. Correlation Fractal Dimension value is estimated using algorithm given by [25]. Given a dataset that has the self-similarity property in the range of scales [r1, r2], its Correlation Fractal dimension D2 for this range is measured as given in following equation 1. The algorithm for computation of Correlation Fractal Dimension is shown in Algorithm 1.

$$D_2 = \frac{\partial \log \sum_i S_{r,i}}{\partial \log r}, r \in [r1, r2] \tag{1}$$

Where $S_{r,i}$ is the squared sum of occupancy with which the pixel fall in the ith cell when the original space is divided into grid cells with sides of length r.

Algorithm 1 : Computation of Fractal Dimension [17]

Step 1: Read a 2-Dimensional facial edge image (FI)
Step 2: Find the size of the Image i.e. number of Rows (R) and Columns (C)
Step 3: if R is greater than C, r is assigned to R otherwise r is assigned to C
Step 4: Compute the Correlation Fractal Dimension value using the equation 1.
End

The algorithm for age group classification using correlation fractal dimension is shown in algorithm 2

Algorithm 2: Age Group Classification Using Correlation Fractal Dimension

Let fractal dimension value is treated as FDV

if (FDV < 1.46)
        print (facial image age is Child (0-15))
else if (FDV < 1.49)
        print (facial image age is Young age(16-30))
else if (FDV < 1.54)
        print (facial image age is middle-age(31-50))
else
        print (facial image age is Senior age(> 50))
end.

3   Results and Discussions

The objective of our classification is to categorize the images of the dataset into four different categories based on age where the child class is between 0 and 15 years, Young Age is between 15 and 30 years, Middle Age from 31 to 50 years and Senior Age is above 50 years. This section projects the detailed presentation of the results obtained using different cross validation techniques. The images for age group classification are collected from multiple data sources like 1002 facial images from FG-NET database, 500 images from Google database and 600 images from the scanned photographs leading to a total of 2102 sample facial images. A few of them are shown in Fig 2. FG-NET consists of 1,002 images of 82 individuals. The average number of images per individual is 12. Although the age of subjects in FG-NET ranges from 0-69 years, over 50% of the subjects in FG-NET are between the ages 0 and 13.The Google database consists of thousands of randomly chosen facial images.

Fig. 2. Sample images of FG-Net aging database

To prove the efficiency of the proposed method the results are tested with hold out and tenfold cross validation testing strategies

3.1   Hold Out Result Analysis Method (HORM)

In Hold Out Result Method (HORM) the total 2102 images are divided into two sets where two third for training the algorithm and one third for testing. Training set consists of 700 FG-Net aging data base, 300 images from Google and 400 images collected from scanned images leading to a total of 1400 images for which FD values are calculated using the algorithm 1. Second set consists of 302 images from FG-Net database, 200 images from Google and 200 images from scanned photographs, totally 702 images. Second set is treated as a test database where the correlation fractal dimension values is found to classify the images based on the proposed algorithm [17]. The classification results of the FG-Net ageing, Google and scanned images in the test database are listed out in the tables 1, 2 and 3 respectively. The classification graph of the test database in Hold out method is shown in figure 3. From HORM method it is observed that middle aged human faces got 100 percentage classifications irrespective of image. The FG-Net ageing database got 100% classification results in all categories. In HORM, Scanned images got low percentage of classification due to poor quality of the scanned images.

Table 1. Classification results of the FG-NET database

| Category | FG-NET database | | | |
	Total	Correctly classified	not correctly classified	% of Classification
Child	89	89	0	100
Young adults	67	67	0	100
Middle Adults	85	85	0	100
Senior Adults	61	61	0	100

Table 2. Classification results of the Google database

| Category | Google database | | | |
	Total	Correctly classified	not correctly classified	% of Classification
Child	43	42	1	97.67
Young adults	59	59	0	100
Middle Adults	45	45	0	100
Senior Adults	53	52	1	98.11

Table 3. Classification results of the Scanned image database

| Category | Scanned images | | | |
	Total	Correctly classified	not correctly classified	% of Classification
Child	56	56	0	100
Young adults	51	50	1	98.03
Middle Adults	47	47	0	100
Senior Adults	46	45	1	97.8

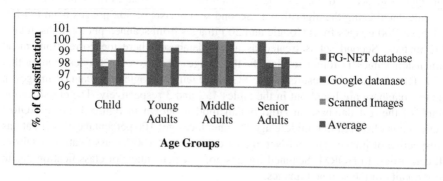

Fig. 3. Classification graph of the Hold out method

3.2   Ten Fold Cross Validation Method (TFVM)

In TFVM results analysis method, the entire 2102 images are divided into ten data sets, every time nine subsets are taken for calculating FD during our training phase and one subset is used for testing. In this process each and every subset is considered while training. The classification accuracies are separately computed in each round and finally the algebraic average of ten rounds are presented as percentage of correct classification.   Each set consists of 210 images of four categories facial images i.e. Child, Young adults, Middle Adults, senior Adults from FG-Net, Google, and scanned images. Form this analysis we strengthen the proposed algorithm. The classification results of the proposed method in ten rounds are evaluated and are listed out in tables 4 to 13 respectively. The overall classification results of the TFVM are listed out in table 14 and corresponding round wise classification graph in shown in figure 4.

Table 4. Classification results of proposed method in round 1 of TFVM

| Category | Total | Round 1 | | |
		Correctly classified	not correctly classified	% of Classification
Child	46	45	1	97.82
Young adults	43	43	0	100
Middle Adults	64	64	0	100
Senior Adults	57	56	1	98.24

Table 5. Classification results of proposed method in round 2 of TFVM

| Category | Total | Round 2 | | |
		Correctly classified	not correctly classified	% of Classification
Child	46	45	0	100
Young adults	43	42	1	97.67
Middle Adults	64	64	0	100
Senior Adults	57	56	1	98.24

Table 6. Classification results of proposed method in round 3 of TFVM

| Category | Total | Round 3 | | |
		Correctly classified	not correctly classified	% of Classification
Child	66	66	0	100
Young adults	48	47	1	97.91
Middle Adults	51	50	1	98.03
Senior Adults	45	45	0	100

Table 7. Classification results of proposed method in round 4 of TFVM

Category	Total	Correctly classified	Round 4 not correctly classified	% of Classification
Child	66	65	1	98.48
Young adults	37	36	1	97.29
Middle Adults	47	46	1	97.87
Senior Adults	60	59	1	98.33

Table 8. Classification results of proposed method in round 5 of TFVM

Category	Total	Correctly classified	Round 5 not correctly classified	% of Classification
Child	39	39	0	100
Young adults	43	42	1	97.67
Middle Adults	58	58	0	98.28
Senior Adults	74	73	1	98.65

Table 9. Classification results of proposed method in round 6 of TFVM

Category	Total	Correctly classified	Round 6 not correctly classified	% of Classification
Child	38	37	1	97.36
Young adults	47	46	1	97.87
Middle Adults	74	73	1	98.67
Senior Adults	51	51	0	100

Table 10. Classification results of proposed method in round 7 of TFVM

Category	Total	Correctly classified	Round 7 not correctly classified	% of Classification
Child	62	62	0	100
Young adults	53	53	0	100
Middle Adults	45	45	0	100
Senior Adults	50	49	1	98.00

Table 11. Classification results of proposed method in round 8 of TFVM

| Category | Total | Round 8 | | % of Classification |
		Correctly classified	not correctly classified	
Child	48	47	1	97.96
Young adults	56	55	1	98.25
Middle Adults	61	60	1	98.36
Senior Adults	45	44	1	97.77

Table 12. Classification results of proposed method in round 9 of TFVM

| Category | Total | Round 9 | | % of Classification |
		Correctly classified	not correctly classified	
Child	52	51	1	98.11
Young adults	47	46	1	97.87
Middle Adults	46	46	0	100
Senior Adults	65	64	1	98.46

Table 13. Classification results of proposed method in round 10 of TFVM

| Category | Total | Round 10 | | % of Classification |
		Correctly classified	not correctly classified	
Child	60	59	1	98.36
Young adults	51	51	0	100
Middle Adults	42	41	1	97.67
Senior Adults	57	56	1	98.28

Table 14. Overall % Classification of TFVM method

Category	Total	Correctly classified	not correctly classified	% of Classification
Child	522	516	6	98.85
Young adults	468	461	7	98.50
Middle Adults	552	547	5	99.09
Senior Adults	560	553	7	98.75
Total	2102	2077	25	98.81

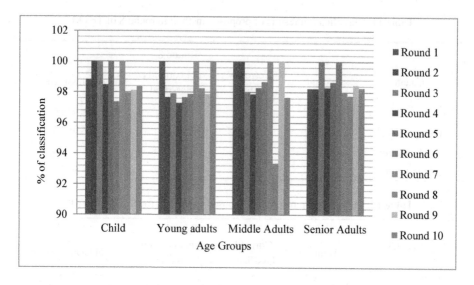

Fig. 4. Round wise classification graph of TFVM method

Comparison of the Proposed Method with Other Existing Methods

The proposed method of age classification is compared with the existing methods [26][11][27]. The method proposed in [27] identified facial image using RBF Neural Network Classifier. The method proposed in [11] is based on two geometric features and three wrinkle features of facial image. The method proposed in [26] classifies the facial image into either child or adult based on Primitive Patterns with Grain Components on Local Diagonal Pattern (LDP). The graphical representation of the percentage mean classification rate for the proposed method and other existing methods are shown in Fig.5.

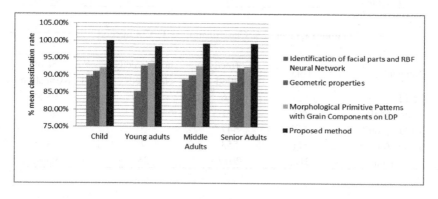

Fig. 5. Comparison graph of proposed method with other existing methods

4   Conclusions

In this paper the results obtained using random testing [31] strategy is found to match well with other two strategies studied here. In all the three strategies the percentage of correct classification is very close to each other in all the considered data sets. For example the mean correct classification in random testing [31] is 99.19 and hold out method is 98.89 and Ten fold testing is 98.85 .The mean standard deviation of all the methods is 0.185. From the above discussion it can be clearly observed that our proposed algorithm based on correlation fractal dimension to detect age from facial images is a robust algorithm. This claim is proved from the analysis of percentage of correct classification using different testing strategies. The performance of the present system is more effective for the FG-NET aging database when compare with Google Images and scanned images. Our method can be further extended for images with noise.

References

1. Li, S.Z., Jain, A.K. (eds.): Handbook of face recognition, 2nd edn. Springer, London (2011)
2. Kumar, N., Berg, A., Belhumeur, P., Nayar, S.: Describable visual attributes for face verification and image search. IEEE Transaction on PAMI 33(10), 1962–1977 (2011)
3. Atkinson, P.M., Lewis, P.: Geostatistical classification for remote sensing: An introduction. Computers and Geosciences 26, 361–371 (2000)
4. Ahonen, T., Hadid, A., Pietikäinen, M.: Face recognition with local binary patterns. In: Pajdla, T., Matas, J(G.) (eds.) ECCV 2004. LNCS, vol. 3021, pp. 469–481. Springer, Heidelberg (2004)
5. Ahonen, T., Hadid, A., Pietikainen, M.: Face description with local binary patterns: application to face recognition. IEEE Transactions on Pattern Analysis and Machine Intelligence 28(12), 2037–2041 (2006)
6. Chandra, M., VijayaKumar, V., Damodaram, A.: Adulthood classification based on geometrical facial features. ICGST (2009)
7. Fu, Y., Huang, T.: Human age estimation with regression on discriminative aging manifold. IEEE Trans. Multimedia 10(4), 578–584 (2008)
8. Chao, W.-L., Liu, J.-Z., Ding, J.-J.: Facial age estimation based on label-sensitive learning and age-oriented regression. Pattern Recognition 46(3), 628–641 (2013)
9. Chang, K.-Y., Chen, C.-S., Hung, Y.-P.: Ordinal hyperplanes ranker with cost sensitivies for age estimation. In: IEEE CVPR, pp. 585–592 (2011)
10. Choi, S.E., Lee, Y.J., Lee, S.J., Park, K.R., Kim, J.: Age estimation using a hierarchical classifier based on global and local facial features. Pattern Recognition 44(6), 1262–1281 (2011)
11. Wen-Bing, H., Cheng-Ping, L., Chun-Wen, C.: Classification of Age Groups Based on Facial Features. Tamkang Journal of Science and Engineering 4(3), 183–192 (2001)
12. Young, H.K., Niels-da-Vitoria, L.: Age Classification from Facial Images. Computer Vision and Image Understanding 74(1), 1–21 (1999)
13. Todd, J.T., Mark, L.S., Shaw, R.E., Pittenger, J.B.: The perception of human growth. Scientific American 242(2), 132–144 (1980)

14. Sirovich, L., Kirby, M.: Low-dimensional procedure for the characterization of human face. J. Opt. Am. A 7(3), 519–524 (1987)
15. Hasegawa, H., Simizu, E.: Discrimination of facial age generation using neural networks. T. IEE Japan 117-C(12), 1897–1898 (1997)
16. Kosugi, M.: Human-face recognition using mosaic pattern and neural networks. IEICE Trans. J76-D-II(6), 1132–1139 (1993)
17. Anuradha, Y., Murthy, J.V.R., Krishnaprasad, M.H.M.: A Novel Method for Human Age Group Classification based on Correlation Fractal Dimension of Facial Edges. Communicated to International Journal of Saud Arab (Elsevier publication) and is Under Review (2014)
18. Canny, J.F.: Finding edges and lines in image, Master's thesis, MIT (1983)
19. Canny, J.F.: A computational approach to edge detection. IEEE Trans. Pattern Anal. Machine Intell. PAMI-8(6), 679–697 (1986)
20. Raman, M., Himanshu, A.: Study and Comparison of Various Image Edge Detection Techniques. International Journal of Image Processing 3(1), 1–11 (2009)
21. Mandelbrot, B.B.: The Fractal Geometry of Nature. W.H.Freeman, New York (1982)
22. Matthews, J.: An introduction to edge detection: The sobel edge detector (2002)
23. Pentland, A.P.: Fractal-Based Description of Natural Scenes. IEEE PAM I -6(6) (1984)
24. Paul, S.A.: Fractal and chaos. IOP publishing (2005)
25. Anuradha, Y., Murthy, J.V.R., Krishnaprasad, M.H.M.: Estimating Correlation Dimension using Multi Layered Grid and Damped Window Model over Data Streams. Elsevier Procedia Technology 10, 797–804 (2013)
26. Sujatha, B., Vijayakumar, V., Rama, B.M.: Morphological Primitive Patterns with Grain Components on LDP for Child and Adult Age Classification. International Journal of Computer Applications 21(3), 0975-8887 (2011)
27. Yazdi, M., Mardani-Samani, S., Bordbar, M., Mobaraki, R.: Age Classification based on RBF Neural Network. Canadian Journal on Image Processing and Computer Vision 3(2), 38–42 (2012)

A Case Study of Evaluation Factors for Biomedical Images Using Neural Networks

Abdul Khader Jilani Saudagar

Department of Information Systems,
College of Computer and Information Sciences,
Al Imam Mohammad Ibn Saud Islamic University (IMSIU),
P.O. Box 5701, Riyadh 11432, Saudi Arabia
saudagar_jilani@ccis.imamu.edu.sa

Abstract. The overall aim of the research is to compare the retrieved image with the original image with respect to the evaluation factors such as MAE, MSE, PSNR and RMSE which reflects the quality of biomedical image for telemedicine with minimum percentage of error at the recipient side. This paper presents spectral coding technique for biomedical images using neural networks in-order to accomplish the above objectives. This work is in continuity of ongoing research project aimed at developing a system for efficient image compression approach for telemedicine in Saudi Arabia. This work compares the efficiency of proposed technique against existing image compression techniques viz JPEG2000 and improved BPNN. To my knowledge, the research is the primary in providing a comparative study of evaluation factors with other techniques used in compression of biomedical images. This work is explored and tested on biomedical images such as X-rays, CT, MRI, PET etc.

Keywords: Mean Absolute Error, Mean Squared Error, Peak Signal to Noise Ratio, Root Mean Squared Error, Artificial Intelligence, Quality.

1   Introduction

Medical Image Compression (MIC) is basic but important factor in the Telemedicine where medical image samples are transferred over a channel from one location to other for remote analysis. In such cases the observer needs accurate information as near to as the original sample to have correct decision. However in current scenario to transmit this large volume of image data, a higher resource such as high bandwidth is needed. Improving allocated bandwidth is not an economical solution; hence advanced compression approaches are to be developed so as to compress the medical samples at the source side and high quality image is retrieved at the destination side.

For the compression of medical images various image compression approaches were proposed in past. The conventional image compression approaches such as JPEG [1], JPEG-2000 [2], SPHIT [3], EBCOT [4], Lifting scheme [5] etc. were proposed earlier. These approaches are majorly categorized under lossy or losseless compression schemes. In lossy compression [6] the information are not accurately

S.C. Satapathy et al. (eds.), *Proc. of the 3rd Int. Conf. on Front. of Intell. Comput. (FICTA) 2014*
– *Vol. 1*, Advances in Intelligent Systems and Computing 327, DOI: 10.1007/978-3-319-11933-5_27

retrieved at the receiver side resulting in low PSNR. These methods are basically suitable for faster transmission approach. In various scenarios where degradation of image is not tolerable, lossless compression schemes were proposed. Lossless compression scheme [7] is a method that allows the exact accurate original data to be reconstructed from the compressed data. A scheme such as wavelet-based compression with adaptive prediction [8] is a lossless approach of image compression. This scheme is mainly used to achieve higher compression ratio. For obtaining a lossless compression in [9] a lifting scheme is suggested based on adaptive threshold.

Lossy and lossless compression schemes were found limited while applying over medical image processing. In order to retrieve the sample with highest accuracy and faster transmission, artificial intelligence based approaches were proposed. Artificial Neural Networks (ANN) has been applied to medical image compression problems, due to their superiority over traditional methods when dealing with noisy or incomplete data. Artificial Neural Networks (ANN) approaches are accurate in making decision but are computationally effective. Lanzarini. L et.al [10] have presented a technique in medical application of images compression using neural networks, which allows to carry out both compression and decompression of the images with a fixed ratio of 8:1 and a loss of 2%. Here back propagation network is created for correspondence functional calculations of input and output patterns.

Similar approach in [11] with back propagation algorithm using Feed Forward Neural (FFN) network is suggested. In this method medical image compression is carried out by calculating coupling weights and activation values of each neuron in the hidden layer. This method found to be better in terms of PSNR compared to conventional JPEG approach.   S. Anna Durai, & E. Anna Saro [12] have suggested another compression technique using back propagation method with Cumulative Distributed function (CDF). This approach is based on mapping the pixels by estimating the CDF values. However, the decompressed image is fuzzy which is not suggested in Medical applications.

To improve the retrieval accuracy, Chee Wan [13] had proposed a neural network approach based on preservation of edges. In this network, quantization levels are used to represent the compressed patterns. The average Mean square value is calculated to achieve the compression ratio. In [14], a lossless medical compression technique based on neural network with improved back propagation method is proposed. From the analysis, it is found that the system exhibits significant performance in compression with low PSNR. Khashman [15] has proposed a medical compression using a neural network in which a Haar wavelet compression with nine compression ratios and a supervised neural network that learns to associate the image intensity (pixel values) with a single optimum compression ratio.   The limitation of this method is image quality is not good which is not tolerable in medical Processing applications. To improve the image quality, in [16] Neural network with multi-resolution method is suggested. This method uses a filter bank that can synthesize the signal accurately from only the reference coefficients will be well suited for low-bit rate coding where the detail coefficients are coarsely quantized. This approach shows advantages over the conventional approaches for compression at low bitrates, although its performance suffers at high bitrates. For achieving higher bit rates Jianxun & Huang [17] presents neural network concept with principal component analysis. Convergence speed is high for this technique but the image quality is poor. A similar technique is proposed in [18]. The technique includes steps to break down large images into smaller windows and to eliminate redundant information. From the analysis this technique results in achieving higher compression ratio with the cost of high complexity.

Cottrell, Munro and Zipser [19] developed a multilayered perceptron neural network with back propagation as the error learning function. This technique results in optimal compression ratio. Khashman and Dimililer [20] have presented neural network for image compression by DCT transform. Here compression is achieved by DCT coefficients and a supervised neural network that learns to associate the grey image intensity (pixel values) with a single optimum compression ratio. More recently, different image compression techniques were combined with neural network classifier for various applications [21,22,23]. However, none of these works has achieved optimum compression ratio. To get higher compression ratio, neural network with bipolar coding [24] was proposed. The Bipolar Coding technique using feed forward back propagation neural network converts decimal values into its equivalent binary code and reconvert in decompression phase. Besides higher compression ratio it also preserves the quality of the image. In [25] similar image compression technique for neural network with GA was suggested. This method mainly focuses on GA algorithm which uses XOR classification and mapping of small data for compression.

Gaidhane [26] has suggested a neural network based image compression technique with MLP algorithm for better faster transmission. In this technique some of the information below the threshold value is removed or replace by zero and therefore more information removed from the feature vector matrix and hence from image data which results in poor image quality. A similar concept was suggested in [27] which is called as vector quantization in which a set of code vectors is generated using the self-organizing feature map algorithm. Then, the set of blocks associated with each code vector is modeled by a cubic surface for better perceptual fidelity of the reconstructed images. Allaf [28] also suggested similar method neural network image compression technique. The performance of the suggested method in terms of PSNR, convergence speed and compression ratio are satisfactory.

For achieving better results [29] suggests a novel technique i.e. neural network with bipolar interpolation to balance the tradeoff of speed and quality. In this technique, compression is achieved by selecting primitive and non-primitive regions to interpolate them. This method found superior to conventional methods in some aspects, such as the clarity and the smoothness in the edge regions as well as the visual quality of the interpolated images. Hui and Yongxue [30] were presented similar neural network concept with haar wavelet and reconstruct the medical image by wavelet packet. It is based on the fact that wavelet packet domain of the same orientation are often similar, and thus coded by similar code words with a vector quantization algorithm. A neural network approach with arithmetic coding using perceptron neural network to compress pixel into single value is explained in [31].

A counter propagation neural network has been used to successfully compress and decompress image data. The network also shows robustness for various classes of images.

Mishra and Zaheeruddin [32] have suggested a new fuzzy neural network for medical image compression. This process is based on approximation problem in which it involves determining or learning the input- output relations using numeric input-output data for image compression application. A similar concept was proposed in [33] in which neural network is designed with modified preprocessing algorithm. The method was divided into two phases. In the first part presents the BS-CROI method of image selection and back propagation image compression in which it is different from traditional ROI. It is found from analysis that, the reconstructed image by this method was promising in terms of PSNR and MSE.

This concept is extended in [34] for better retrieval image. In this work, neural networks are designed to a combination of cascaded networks with one node in the hidden layer. A redistribution of the gray levels in the training phase is implemented in a random fashion to make the minimization of the mean square error applicable to a broad range of images. From the analysis it is found that the performance superiority of cascaded neural networks compared to that of fixed architecture training paradigms especially at high compression ratios. With the existing approaches for compression, the application for image compression based on advanced intelligence approaches using neural network is observed to be an effective approach for compression. The approach for medical image compression using neural network is developed in this work. In this research the effectiveness of neural network approach to biomedical image compression is focused.

2   Methodology

For medical image compression, in this work an ANN based image compression architecture is developed. In ANN based compression system the image is coded with respect to its pixel values and pixel coordinate. In [35] an approach for medical image compression based on BPNN is proposed. The approach is developed as improved BPNN and is compared with conventional JPEG based coding system. In such an approach an image is first read into a matrix of dimension $m \times n$ and the co similar pixel coefficients are searched forming a pair of pixel value of its counts. This approach is similar to the approach of run length coding for the obtained co-similar pairs, a NN process is carried out, wherein these pairs are given as input to the NN system. The process of NN processing for image compression is briefed as:

2.1   Image Compression Process

Step 1: Input image is converted to matrix format (I) containing $X_{m,n}$,where m is row and n is column.

Step 2: Using (I), pixel values and the number of occurrences of the neighbouring pixel values are counted and represented by pair values (P) as follows.

$$P = (U_1, V_1)(U_2, V_2)(U_3, V_3) \dots (U_i, V_j). \tag{1}$$

Where,

U = Pixel values.

V = Number of occurrences of the neighboring pixel values.

Step 3: The pair values (P) obtained from the above step can be represented in sequence order (S).

$$S = U_1, V_1, U_2, V_2, U_3, V_3 \dots U_i, V_j. \tag{2}$$

Step 4: The sequence order (S) can be provided as an input (X_i) to the Multi-Layer Feed-Forward Back propagation Neural Network.

$$X_i = X_1, X_2, X_3 \dots X_n. \tag{3}$$

Step 5: Calculate weight (W_{ji}) using the formula.

$$W_{ji} = \sum_{i=1}^n X_i X_i^T \quad \text{Where, } 1 \le j \le k. \tag{4}$$
$$X_i \text{ is the input Layer.}$$

Step 6: The Hidden Layer of the Multi-Layer Feed-Forward Back propagation Neural Network is created by using the formula (H_j).

$$H_i = \sum_{i=1}^n W_{ij} X_i \quad \text{Where, } 1 \le j \le k. \tag{5}$$
$$X_i \text{ is the input Layer.}$$
$$H_i = H_1, H_2, H_3, \dots H_k$$

The result of the H_j obtained refers the compressed file.

2.2   Image Decompression Process

Step 1: Get (H_j) of the Multi-Layer Feed-Forward Back propagation Neural Network

$$H_j = H_1, H_2, H_3, \dots H_k \tag{6}$$

Step 2: Calculate the weight (W_{ij}) using the formula.

$$W_{ij} = \sum_{j=1}^k H_j H_j^T \quad \text{Where, } 1 \le i \le k \tag{7}$$
$$H_i \text{ is the Hidden Layer.}$$

Step 3: The Output Layer of the Multi - Layer Feed - Forward Back propagation Neural Network is created by using the formula (Y_i).

$$Y_i = \sum_{j=1}^k W_{ij} H_j \quad \text{Where, } 1 \le i \le n. \tag{8}$$
$$H_i \text{ is the Hidden Layer.}$$
$$Y_i = Y_1, Y_2, Y_3, \dots Y_n$$

Step 4: The Output Layer (Y_i) is represented in Sequence Order (S).

$$S = U_1, V_1, U_2, V_2, U_3, V_3 \dots U_i, V_j. \tag{9}$$

Step 5: The Sequence Order (S) Value can be represented in Pair Values (P). Each Pair represents the Pixel Value and the number of occurrences of the neighbouring pixel values.

$$P = (U_1, V_1)(U_2, V_2)(U_3, V_3) \dots (U_i, V_j). \tag{10}$$

Where,

U = Pixel values.

V = Number of occurrences of the neighbouring pixel values.

Step 6: All the Pair Values (P) represented in Pixel Values are converted into Matrix Format (I).

Step 7: Now the matrix format (I) is converted into the image file format.

Due to this conversion the retrieval accuracy is lower. To improve such estimation accuracy the image must be processed in spectral domain rather than processing on direct pixel values. Where-in in this conventional approach image is directly processed, so the finer details of biomedical image samples are more or less astray. The multi resolution information is not observed in previous approach of JPEG2000. So there needs a coding technique which presents a high resolution coding resulting in higher estimation accuracy than the JPEG system. The approach of such spectral coding is adopted for medical image compression into NN based coding. In this approach the medical image are first processed to extract the spectral coefficient over which NN is applied and this novel technique of coding results in higher efficiency when compared with existing approaches.

The proposed approach is as outlined below:

In image compression process, the input image is not processed directly, instead the input image after converted to matrix format and is decomposed to four multi resolution components C_1(horizontal coefficients), C_2 (vertical coefficients), C_3 (diagonal coefficients) and C_4 (approximate coefficients).

Step 1: Input image is converted to matrix format (I), where $I = f(X_{m,n})$, where m is row and n is column.

Step 2: Using (I), decompose the image (I), in multi resolution components C_1, C_2, C_3, C_4

Where

C_1 is horizontal coefficients

C_2 is vertical coefficients

C_3 is diagonal coefficients

C_4 is approximate coefficients

Further the steps from 2 to 6 of image compression process and steps from 2 to 7 of decompression process mentioned in the conventional approach are repeated over the coefficient C_i, i =1,2,3,4.

The following equations are used in-order to calculate the MAE, MSE, PSNR and RMSE.

$$MAE = \frac{1}{mn}\sum_0^{m-1}\sum_0^{n-1}\|f(i,j) - g(i,j)\| \tag{11}$$

$$MSE = \frac{1}{mn}\sum_0^{m-1}\sum_0^{n-1}\|f(i,j) - g(i,j)\|^2 \tag{12}$$

$$PSNR = 20\log_{10}\left(\frac{MAX_f}{\sqrt{MSE}}\right) \tag{13}$$

$$RMSE = \sqrt{\frac{1}{mn}\sum_0^{m-1}\sum_0^{n-1}\|f(i,j) - g(i,j)\|^2} \tag{14}$$

f represents the matrix data of our first image.

g represents the matrix data of our second image.

m represents the numbers of rows of pixels of the images and i represents the index of that row.

n represents the number of columns of pixels of the image and j represents the index of that column.

MAX_f is the maximum signal value.

2.3   Proposed System Architecture

The functional description of the proposed block diagram (Fig. 1) is as follows

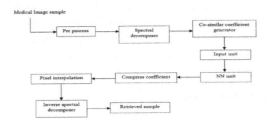

Fig. 1. Proposed block diagram

Pre process unit: This unit reads the medical sample and extracts the gray pixel intensity for processing. The read samples are passed as pixel array as output of this block and passed for decomposition in spectral decompose unit.

Spectral decomposer unit: This unit reads the gray coefficients and performs a pyramidical decomposition to extract the spectral resolutions for given input sample. The decomposition structured is a 2 dimensional recursive filter bank units, performing DWT operation. The recursive operation is carried out by the recursive filtration using pairs of successive high and low pass filter.

Co-similar coefficient generator unit: for the obtained coefficient after spectral decomposition, the coefficients which reflect similar spectral coefficients are segregated, these coefficients are called redundant pixel in the image. The suppression of co-similar coefficient results in first level compression based on redundant information. For the obtained co-similar coefficients a neural network modelling is developed.

Input unit: This unit reads the selected coefficient and normalizes the coefficients to pass to neural network. The unit extracts the coefficient in a column wise manner and is normalized to maximum pixel value.

NN unit: This unit realizes a feed forward neural network using the command 'newff' in matlab tool. The NN unit extract the min-max value of given input and creates a feed forward neural network taking least mean learning algorithm. A tangential sigmoid driving function is used as a kernel function for creating this network. The network is created for converging to the error with a goal of 0.1 and with number of epochs=50. The created network is trained with these coefficient values based on the given input and the created feed forward network.

Compress coefficient: The coded coefficient after the neural network process is stored into a buffer called compressed coefficient. This formulates an array logic wherein the coded output of the NN is stored for future usage.

Pixel interpolation: The compressed data is processed back in this unit. Wherein the simulated result of the created neural network is normalized back to its original scale based on the obtained simulated output of the neural network. The retrieved pixel coefficients are rearranged depending on the sequence order as obtained from the encoding side.

Inverse spectral decomposer: The coefficient obtained from the above units are processed back, where the coefficients are passed back as resolutional information to successive high and low pass filter. The recursive output of each level of filtration is added to the other level filtration result and is recursively filtered to obtain final retrieved level. An inverse DWT approach is followed in this unit.

3   Results and Discussion

For the evaluation of the suggested approach a simulation model is developed using MATLAB and is tested on 12 original gray-scale sample of medical images (Fig. 2) such as human nerve cells, human body organs etc., of different dimensions collected from hospitals in King Fahd Medical City, Riyadh, Saudi Arabia with 500 dpi resolution. The training error plot for neural system developed is as shown in (Fig. 3). The observations of first five samples Q1, Q2, Q3, Q4 and Q5 are as follows with original processing sample read with various specifications, followed by output image using conventional JPEG2000 approach, then output image after applying improved BPNN approach and at last the retrieved image after applying proposed spectral-BPNN approach. In [36,37] the authors compare the proposed approach with the existing approaches with respect to compression ratio and other factors were left for future work. The observations are as shown in (Fig. 4), (Fig. 5), (Fig. 6), (Fig. 7) and (Fig. 8).

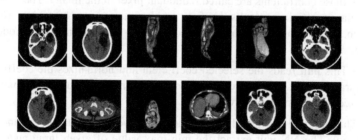

Fig. 2. Original image samples Q1, Q2, Q3, Q4, Q5, Q6, Q7, Q8, Q9, Q10, Q11 and Q12

Simulation results

Image Type   : Medical Image
File Type   : TIFF
Test sample   : Q1, Q2, Q3, Q4 and Q5
Resolution   : 512 X 512

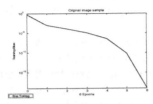

Fig. 3. Training error plot for neural system developed

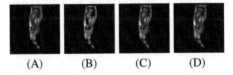

(A)   (B)   (C)   (D)

Fig. 4. (A) Original processing sample Q1; (B) retrieved image using JPEG2000; (C) retrieved image using improved BPNN; (D) retrieved image using proposed spectral-BPNN

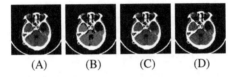

(A)   (B)   (C)   (D)

Fig. 5. (A) Original processing sample Q2; (B) retrieved image using JPEG2000; (C) retrieved image using improved BPNN; (D) retrieved image using proposed spectral-BPNN

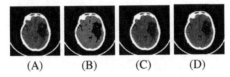

(A)   (B)   (C)   (D)

Fig. 6. (A) Original processing sample Q3; (B) retrieved image using JPEG2000; (C) retrieved image using improved BPNN; (D) retrieved image using proposed spectral-BPNN

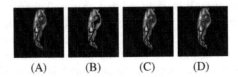

(A)   (B)   (C)   (D)

Fig. 7. (A) Original processing sample Q4; (B) retrieved image using JPEG2000; (C) retrieved image using improved BPNN; (D) retrieved image using proposed spectral-BPNN

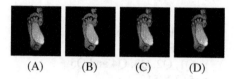

Fig. 8. (A) Original processing sample Q5; (B) retrieved image using JPEG2000; (C) retrieved image using improved BPNN; (D) retrieved image using proposed spectral-BPNN

Comparison plots of five biomedical image samples Q1, Q2, Q3, Q4 and Q5 on X-axis with respect to their observed values of MAE (Fig. 9 A), MSE (Fig. 9 B), PSNR (Fig. 9 C) and RMSE (Fig. 9 D) on Y-axis for JPEG2000, improved BPNN and proposed spectral-BPNN approach are illustrated below. It is found that the proposed spectral-BPNN is more efficient than JPEG2000 and improved BPNN in all evaluation factors.

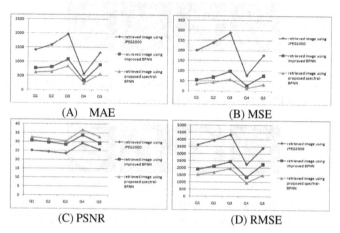

Fig. 9. Comparison values of original image sample with JPEG2000, improved BPNN and spectral-BPNN

4   Conclusion

This research work implements an enhanced image coding system and a comparative study of evaluation factors for biomedical images compared to the existing JPEG2000 and other coding techniques. It is observed that proposed approach is able to achieve good quality performance with a relatively simple algorithm. Since ANN also has the desirable properties resulting from its successive approximation quantization, different topologies were applied to solve the problem. The results obtained from proposed novel neural networks approach found much better results when compared to conventional approaches.

Since this work mainly focuses on gray-scale images, in the future, it can be extended to color medical images by considering regional information such as texture, boundary information etc., and the observed results can be compared with other standard compression schemes which are used for compression in biomedical imaging.

Acknowledgments. I am very much thankful to Al Imam Mohammad Ibn Saud Islamic University for providing an environment to carry out this research successfully. This research is financially supported by Scientific Research Deanship, Al Imam Mohammad Ibn Saud Islamic University, Riyadh, Saudi Arabia with fund number 330909.

References

1. Wallace, G.K.: The JPEG still picture compression standard. IEEE Trans. on Cons. Elec. 38(1) (1992)
2. Usevitch, B.E.: A tutorial on modern lossy wavelet image compression: Foundations of JPEG 2000. IEEE Sig. Proc. Mag. 18(5), 22–35 (2001)
3. Jyotheswar, J., Mahapatra, S.: Efficient FPGA implementation of DWT and modified SPIHT for lossless image compression. J. of Sys. Arch. 53(7), 369–378 (2007)
4. Taubman, D.: High performance scalable image compression with EBCOT. IEEE Trans. on Img. Proc. 9(7), 1158–1170 (2000)
5. Srikala, P., Umar, S.: Neural network based image compression with lifting scheme and RLC. Int. J. of Res. Engg. and Tech. 1(1), 13–19 (2012)
6. Matsuoka, R., Sone, M., Fukue, K., Cho, K., Shimoda, H.: Quantitative Analysis of Image Quality Of Lossy Compression Images. International Society for Photogrammetry and Remote Sensing (2013),
http://www.isprs.org/proceedings/XXXV/congress/comm3/papers/348.pdf
7. Chang, S.G., Yovanof, G.S.: A simple block-based lossless image compression scheme. Proc. of the 13th Asilomar Conf. on Sig., Sys. and Comp. 1, 591–595 (1996)
8. Chen, Y.T., Tseng, D.C., Chang, P.C.: Wavelet-based medical image compression with adaptive prediction. In: Proc. of the Int. Symp. on Intelli. Sig. Proc. and Comm. Sys. (ISPACS 2005), pp. 825–828 (2005)
9. Spires, W.: Lossless Image Compression Via the Lifting Scheme (2005),
http://www.cs.ucf.edu/~wspires/lossless_img_lifting.pdf
10. Laura, L., Camacho, M.T.V., Badran, A., Armando, D.G.: Images Compression for Medical Diagnosis Using Neural Networks (1990),
http://journal.info.unlp.edu.ar/journal/journal2/papers/image.pdf
11. Yeo, W.K., Yap, D.F.W., Oh, T.H., Andito, D.P., Kok, S.L., Ho, Y.H., Suaidi, M.K.: Grayscale medical image compression using feedforward neural networks. In: Proceedings of the IEEE Conf. on Comp. Appl. and Ind. Elec (ICCAIE 2011), pp. 633–638 (2011)
12. Durai, S.A., Saro, E.A.: Image compression with backpropagation neural network using cumulative distribution function. Int. J. of Engg. and Appl. Sci. 3(4), 185–189 (2007)
13. Wan, T.C., Kabuka, M.: Edge preserving image compression for magnetic resonance Images using DANN-based neural networks. In: Med. Img. Proc. of SPIE, vol. 2164 (1994)

14. Liang, J.-Y., Chen, C.-S., Huang, C.-H., Liu, L.: Lossless compression of medical images using Hilbert space-filling curves. Comp. Med. Img. and Grap. 32(3), 174–182 (2008)
15. Khashman, A., Dimililer, K.: Medical radiographs compression using neural networks and haar wavelet. In: Proc. of the IEEE EUROCON 2009, pp. 1448–1453 (2009)
16. Northan, B., Dony, R.D.: Image compression with a multiresolution neural network. Canadian J. of Electrical and Comp. Engg. 31(1), 49–58 (2006)
17. Mi, J., Huang, D.: Image compression using principal component neural network. In: Proc. of the 8th Int. Conf. on Cont., Auto., Robo. and Vis., pp. 698–701 (2008)
18. Kulkarni, S., Verma, B., Blumenstein, M.: Image compression using a direct solution method based neural network. In: Proc. of the 10th Australian Joint Conf. on Arti. Intelli., pp. 114–119 (1997)
19. Cottrell, G., Munro, P., Zipser, D.: Image Compression by Back Propagation: An Example of Extensional Programming. Adv. in Cogn. Sci. (1989)
20. Khashman, A., Dimililer, K.: Neural networks arbitration for optimum DCT image compression. In: Proc. of the Int. Conf. on Comp. as a Tool (EUROCON 2007), pp. 151–156 (2007)
21. Ma, L., Khorasani, K.: Adaptive constructive neural networks using hermite polynomials for image compression. In: Wang, J., Liao, X.-F., Yi, Z. (eds.) ISNN 2005. LNCS, vol. 3497, pp. 713–722. Springer, Heidelberg (2005)
22. Karlik, B.: Medical image compression by using vector quantization neural network (VQNN). Neu. Net.World 16(4), 341–348 (2006)
23. Zhou, Y., Zhang, C., Zhang, Z.: Improved variance-based fractal image compression using neural networks. In: Wang, J., Yi, Z., Żurada, J.M., Lu, B.-L., Yin, H. (eds.) ISNN 2006. LNCS, vol. 3972, pp. 575–580. Springer, Heidelberg (2006)
24. Tripathi, P.: Image compression enhancement using bipolar coding with LM algorithm in artificial neural network. Int. J. of Sci. and Res. Publications 2(8) (2012)
25. Rajput, G.G., Singh, M.K.: Modeling of neural image compression using GA and BP: a comparative approach. Int. J. of Adv. Comp. Sci. and Appl., 26–34 (2011)
26. Gaidhane, V., Singh, V., Kumar, M.: Image compression using PCA and improved technique with MLP neural network. In: Proc. of the 2nd Int. Conf. on Adv. in Recent Technologies in Comm. and Computing (ARTCom 2010), pp. 106–110 (2010)
27. Laha, A., Pal, N.R., Chanda, B.: Design of vector quantizer for image compression using self-organizing feature map and surface fitting. IEEE Trans. on Img. Proc. 13(10), 1291–1303 (2004)
28. AL-Allaf, O.N.A.: Improving the performance of backpropagation neural network algorithm for image compression/ decompression system. J. of Comp. Sci. 6(11), 1347–1354 (2010)
29. Lin, C.-T., Fan, K.-W., Pu, H.-C., Lu, S.-M., Liang, S.-F.: An HVS-directed neural-network-based image resolution enhancement scheme for image resizing. IEEE Trans. on Fuzzy Sys. 15(4), 605–615 (2007)
30. Hui, G., Yongxue, W.: Wavelet packet and neural network basismedical image compression. In: Proc. of the Int. Conf. on E-Product E-Service and E-Entertainment, pp. 1–3 (2010)
31. Rekha, S.V.: A segmented wavelet inspired neural network approach to compress images. IOSR J. of Comp. Engg. 2(6), 36–42 (2012)
32. Mishra, A., Zaheeruddin, Z.: Hybrid fuzzy neural network based still image compression. In: Proc. of the Int. Conf. on Computational Intelli. and Comm. Net. (CICN 2010), pp. 116–121 (2010)

33. Vijideva, R.: Neural network-wavelet based dicom image compression and progressive transmission. Int. J. of Engg. Sci. & Adv. Tech. 2(4), 702–710 (2012)
34. Obiegbu, C.: Image compression using cascaded neural networks (M.S. thesis) (2003), http://scholarworks.uno.edu
35. Senthilkumaran, N., Suguna, J.: Neural network technique for lossless image compression using X-ray images. Int. J. of Comp. and Elec. Engg. 3(1) (2011)
36. Saudagar, A.K.J., Shathry, O.A.: Neural network based image compression approach to improve the quality of biomedical image for telemedicine. British J. of Applied Sci. and Tech. 4(3), 510–524 (2014)
37. Saudagar, A.K.J.: Minimize the Percentage of Noise in Biomedical Images Using Neural Networks. The Scient. World J. 2014, Article ID 757146

Classification of Web Logs Using Hybrid Functional Link Artificial Neural Networks

Ajit Kumar Behera[1], Ch. Sanjeev Kumar Dash[2], and Satchidananda Dehuri[3]

[1] Department of Computer Application
[2] Department of Computer Science and Engineering,
Silicon Institute of Technology,
Silicon Hills, Patia, Bhubaneswar, 751024
ajit_behera@hotmail.com, sanjeev_dash@yahoo.com
[3] Department of Information and Communication Technology,
Fakir Mohan University, Vyasa Vihar,
Balasore-756019, Odisha, India
satchi.lapa@gmail.com

Abstract. Over the decades, researchers are striving to understand the web usage pattern of a user and are also extremely important for the owners of a website. In this paper, a hybrid analyzer is proposed to find out the browsing patterns of a user. Moreover, the pattern which is revealed from this surge of web access logs must be useful, motivating, and logical. A smooth functional link artificial neural network has been used to classify the web pages based on access time and region. The accuracy and smoothness of the network is taken birth by suitably tuning the parameters of functional link neural network using differential evolution. In specific, the differential evolution is used to fine tune the weight vector of this hybrid network and some trigonometric functions are used in functional expansion unit. The simulation result shows that the proposed learning mechanism is evidently producing better classification accuracy.

Keywords: Differential evolution, Functional link artificial neural network, Classification, Web log.

1   Introduction

We live in a time of unprecedented access to cheap and vast amounts of web resources, which is producing a big leap forward in the field of web mining. This overabundance and diversity of resources has ignited the need for developing an automatic mining technique for the WWW, thereby giving rise to the term web mining [1].

Every website contains multiple web pages. Every web page has: 1) contents which can be in any form e.g. text, graphics, multimedia, etc; 2) links from one page to another; and 3) users accessing the web pages. Based on this, the area of web mining can be categorized content, structure, and usage mining as illustrated in Figure 1.

© Springer International Publishing Switzerland 2015   255
S.C. Satapathy et al. (eds.), *Proc. of the 3rd Int. Conf. on Front. of Intell. Comput. (FICTA) 2014*
– *Vol. 1*, Advances in Intelligent Systems and Computing 327, DOI: 10.1007/978-3-319-11933-5_28

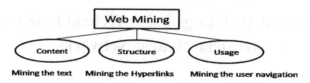

Fig. 1. Categories of Web Mining

Mining the contents such as text, graphs, pictures, etc. relevance to the search query from web pages is called "Content Mining". Mining the relationships between web pages is called "Structure Mining". Mining the web access logs is called "Web Usage Mining".

Web server record and mount up data about user interactions, upon request, it retrieves. Analyzing the web access logs of different web sites can help to understand the user behavior and the web structure, thereby improving the design of this colossal collection of resources. To proceed towards a semi-automatic hybrid web analyzer, obviating the need for human intervention, we need to incorporate and embed computational or artificial intelligence into web analyzing tools. However, the problem of developing semi-automated tools in order to find, extract, filter, and evaluate the users desired information from unlabeled, scattered, and diverse web access logs data is far from being solved. Machine learning algorithm offers an approach to solve above mentioned problem. FLANN a member of computational intelligence family could be used as a classifier [2]. The proposed method is combining the idea of simple FLANN classifier [3] and Differential Evolution (DE) an emerging evolutionary computation technique. In this method the weights of FLANN are trained using differential evolution.

2   A Framework of FLANN in Web Access Logs

The overall framework of our work has been described in Figure 2. Subsection 2.1 describes the web usage mining and different steps required for preprocessing the dataset highlighted in the framework. Subsection 2.4 briefly describes FLANN and 2.5 describe about the DE for classification of web access logs.

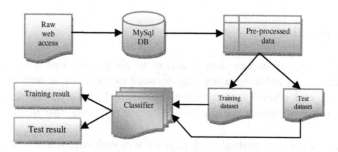

Fig. 2. Block diagram of web log classification

2.1   Web Usage Mining and Preprocessing

User navigates through a range of web pages while visiting a website. For each access to a web page, a record is created in log files which are managed by web servers. The raw data from access log file cannot be used for mining directly. These records must be pre-processed before using. Pre-processing as mention in [4] involves, i) Data cleaning, ii) User identification and iii) Session identification. After pre-processing, these records can be divided into training and test data set for classification. We have followed the database approach for pre-processing work, where all the raw records are inserted into a table and the pre-processing tasks such as data cleaning, user identification and session identification are done using SQL queries. This approach gives better flexibility and faster execution.

Initially the raw web access log records are scanned from log files and inserted into a table which is designed in MySQL database. As we know query execution is easier and the time taken for query execution is very less in comparison to file processing. Once all records are inserted successfully, then the preprocessing steps can be executed.

In this paper we have used the web access log data from www.silicon.ac.in. Silicon Institute of Technology is one of the premiere technical institutes in Odisha, India. We have collected the records between 22-Oct-2010 04:15:00 to 24-Oct-2010 04:05:48 and the size of the file is 12842 records.

2.2   Data Transformation

We have used java program which reads each line from the log file and insert into the table in MySQL database. The different fields in log record and their value are described as in Table 1.

Table 1. Field name & value of a HTTP request

Value	Field Name
122.163.111.210	IP Address
-	Client ID
-	User ID
[22/Oct/2010:04:15:00 -0400]	Request Date & Time
"GET /sitsbp/index.html HTTP/1.1"	Request Method, URL, Protocol
200	Response Status
4520	Bytes Sent
"http://www.google.co.in/search?hl=en&rlz=1R2ADFA_e nIN388&q=silicon+institute+of+technology&btnG=Searc h&meta=&aq=f&aqi=&aql=&oq=&gs_rfai="	Referrer
"Mozilla/4.0 (compatible; MSIE 6.0; Windows NT 5.1; GTB0.0; SV1; .NET CLR 2.0.50727; .NET CLR 3.0.04506.30; .NET CLR 3.0.4506.2152; .NET CLR 3.5.30729; RediffIE8)"	User Agent

Initially we split the line into words separated by blank space by which we get the clientIP, clientID, userID. As the request date and time is written in pair of square brackets [] so we find the index of '[' and ']' then by using getChars() method we get the characters between these index. The rest part is stored in another variable and using another loop and searching character by character we retrieve the requested URL, request status, method, user agent etc.

2.3   Data Cleaning

For data cleaning we need to remove all the records which contain the request for files like jpg, gif, css, js etc. For this we execute the delete query. Before the data cleaning the total number of records were 12842 which is reduced to 1749 records by data cleaning, which are the actual page access records. Every unique page were assigned a PageID. The total number of unique web pages was found to be 97.

2.4   FLANN

FLANNs are a new generation NNs without hidden units introduced by Klassen and Pao in 1988. Despite their simple structure, FLANNs can capture non-linear input–output relationships, provided that they are fed with an adequate set of polynomial inputs, or the functions might be a subset of a complete set of orthonormal basis functions spanning in an n-dimensional representation space [5]. FLANNs can be used for non-linear prediction and classification. In this construction, Subsections 2.4.1 and 2.4.2 are briefing some related works on FLANNs for classification and non-linear prediction.

FLANN architecture [6], [7] uses a single layer feed forward neural network by removing the concept of hidden layers. The learning of a FLANN may be considered as approximating or interpolating a continuous, multivariate function f(X) by an approximating function. A set of basis functions Φ and a fixed number of weight parameters W are used to represent a FLANN. With a specific choice of a set of basis functions, the problem is then to find the weight parameters W that provides the best possible approximation of $f(.)$ on the set of input-output examples. Hence the most important thing is that how to choose the basis functions to obtain better approximation.

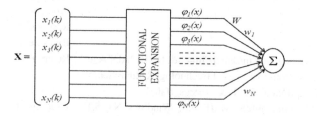

Fig. 3. FLANN architecture

2.4.1 FLANNs for Classification

For selecting appropriate number of polynomials as a functional input to the network genetic algorithm could be used as proposed by Sierra et al.[8], which was applied to the classification problem. However their main concern was selection of optimal set of functional links to construct the classifier. In contrast, the proposed method gives much emphasis on how to develop the learning skill of the classifier. Misra and Dehuri [9] have used a FLANN for classification problem in data mining with a hope to get a compact classifier with less computational complexity and faster learning. With a motivation to restrict certain limitations, Dehuri, et al. [10] have coupled genetic algorithm based feature selection with FLANN (GFLANN).

2.5   Differential Evolution

Differential evolution (DE) [11], [12], [13] is a population based stochastic algorithm that typically operates on real encoding of chromosomes. Like EAs, DE maintains a set of potential solutions which are then perturbed in an effort to uncover yet better solutions to a problem in hand. In DE, individuals are perturbed based on the scaled difference of two randomly chosen individuals of the current population. One of the advantages of this approach is that the resulting 'step' size and orientation during the perturbation process automatically adapts to the fitness function landscape.

    With the years of journey in the development process of DE, many variants have been developed [11]. The variants of the DE algorithm are described using the shorthand DE/x/y/z, where x specifies choice of base vector, y is the number of difference vectors used, and z denotes the crossover scheme. We primarily describe a version "classic DE" or DE/rand/1/bin scheme.

    Classic DE keeps a population of n, d-dimensional vectors $x_i = \left(x_{i1}, x_{i2}, ..., x_{id} \right)$, $i = 1...n$ each of which is evaluated using a fitness function $f(.)$. During the search process, each individual (i) is iteratively refined with the following steps:

Mutation: Select a donor vector which encodes a solution, using randomly selected members of the population.

Crossover: Develop a trial vector by combining the donor vector with target vector i.

Selection: Determine whether the newly-created trial vector replaces i in the population or not.

    For each vector $x_i(t)$, a donor vector $v_i(t+1)$ is generated as equation (1).

$$v_i \left(t+1 \right) = x_k \left(t \right) + f_m * \left(x_l \left(t \right) - x_m \left(t \right) \right), \tag{1}$$

where these vectors are disjoint, which we can achieve by having i $\neq k \neq$ l$\neq$ m $\in$ $1,...,n$ are mutually distinct, randomly selected indices such that all the indices $\neq i$. Selecting three indices randomly implies that all members of the current population have the same chance of being selected, and therefore controlling the creation of the different vectors. The difference between vectors x_l and x_m is multiplied by a scaling parameter f_m called mutation factor such as $f_m \in [0,2]$. The mutation factor controls the amplification of the difference between x_l and x_m which is used to avoid stagnation of the search process.

A remarkable feature of the mutation step in DE is that of its self-scaling nature. The size of mutation factor along each dimension stems explicitly from the location of the individuals in the current population.

Following the creation of the donor vector, a trial vector $u_i(t+1)$ is obtained by equation (2).

$$u_i(t+1) = \begin{cases} v_{ip}(t+1), if\,((rand \leq c_r)\ or\ (i = rand(ind)) \\ x_{ip}(t),\quad if\,((rand > c_r)\,and\ i \neq rand(ind)) \end{cases}, \qquad (2)$$

where $p = 1,2,.....,d$, $rand$ is a random number generated in the range (0, 1), c_r is the crossover constant from the range [0, 1), and $rand(ind)$ is an index which is randomly chosen from the range (1, 2, ... ,d). The resulting trial vector replaces its parent if it has higher fitness, otherwise the parent survives unaffected into the next iteration. Finally, if the fitness of the trial vector goes beyond that of the fitness of the parent then the parent is replaced as described in equation (3).

$$x_i(t+1) = \begin{cases} u_i(t+1), if\,(f\,(u_i(t+1)) > f(x_i(t)) \\ x_i(t),\qquad Otherwise \end{cases}. \qquad (3)$$

3   Proposed Method

With an objective to design a smooth and accurate hybrid classifier, the proposed approach is harnessing the best attribute of DE for parameter optimization in FLANN classifier [14]. Figure 4 illustrates the overall architecture of the approach.

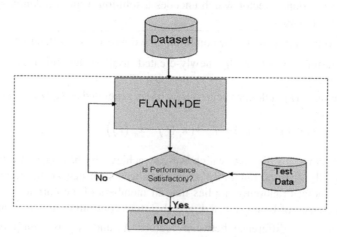

Fig. 4. Architecture of Proposed Method

In our work, we have used trigonometric function for mapping the d features though there are other functions available. Here we use five functions out of which one is linear and four are trigonometric. In general, let $f_1, f_2, ..., f_k$ is the number of functions used to expand each feature value of the patterns. The network has the ability to learn through DE. During the training, the network is repeatedly presented with the training data with FLANN where weights adjusted by DE till the desired input output mapping occurs. This process is repeated for each chromosomes of the DE and subsequently each chromosome will be assigned a fitness value based on its performance. Using this fitness value the usual process of DE is executed until some good topology with an acceptable predictive accuracy is achieved

4 Experimental Study

4.1 Description of Dataset and Parameters

For classification purpose we have categorized the access time into 2 categories, i.e. from 00:00 hours to 12:00 hours 1 and 12:00 to 24:00 as 2. Similarly we have grouped all the region of access into 2, i.e. IP addresses accessing from within India as 1 and IP addresses accessing from outside India as 2. To find the region from IP address we have used the website www.ip2location.com [15].

Our class attribute is frequent. For this we have considered two values i.e. 1 for not frequent and 2 for frequent. This dataset includes 237 instances and each having 4 attributes e.g. ID, request count, timeid, locationid. The instances are classified into 2 classes (i.e., every instance either belongs to class 1 or 2). Class 1 has 129 instances, and class 2 contains 108. None of the attributes contain any missing values.

4.2 Results and Analysis

In this experimental set up, we have considered for a comparative study of DE-RBFN [16] with the proposed method DE-FLANN. Tables 2 and 3 illustrate the results obtained from DE-FLANN and DE-RBFN, respectively. The classification accuracy of the proposed method with varying mutation factors is given as part of parametric analysis.

Table 2. Classification accuracy obtained for DE-FLANN with C_r=0.8 and varying mutation factor

Mutation Factor	Accuracy
0.1	72.8682
0.2	88.3721
0.3	68.9922
0.4	91.4729
0.5	86.8217
0.6	86.0465
0.7	68.9922
0.8	55.0388
0.9	68.9922

Further, this illustrative experimental results in Tables 2 and 3 is presented for a fixed crossover rate (i.e., $C_r = 0.8$). The reason is that, our multiple independent runs with several crossover rates recommend its suitability.

Table 3. Classification accuracy obtained for DE-RBFN with $C_r=0.8$ and varying mutation factor

Mutation Factor	Accuracy
0.1	72.3181
0.2	84.7690
0.3	67.8792
0.4	84.2472
0.5	84.3482
0.6	82.2046
0.7	68.0220
0.8	61.3034
0.9	67.6599

5   Conclusion and Future Work

In this paper, we have crafted the integrated framework of differential evolution and FLANN for weblog data analysis and compared its performance with DE-RBFN. The experimental results shows that DE-FLANN gives better results as compared the DE-RBFN with varying mutation factor. It is also observed that we got better accuracy at mutation factor 0.4. As we have applied the classification on the attributes like region and time in this paper, similarly other attributes like user agent and referrer can also be consider in future for classification and driving the analysis in a fruitful way.

References

1. Chen, H., Zong, X., Lee, C.-W., Yeh, J.-H.: World Wide Web Usage Mining Systems and Technologies. Journal of Systemics, Cybernetics and Informatics 1(4), 53–59 (2003)
2. Pao, Y.H., Philips, S.M.: The Functional Link Net Learning Optimal Control. Neurocomputing 9(2), 149–164 (1995)
3. Misra, B.B., Dehuri, S.: Functional Link Neural Network for Classification Task in Data mining. J. Comput. Sci. 3(12), 948–955 (2007)
4. Srivastava, J., Cooley, R., Deshpande, M., Tan, P.N.: Web Usage Mining: Discovery and Applications of Usage Patterns from Web Data. SIGKDD Explorations 1(2), 1–12 (2000)
5. Pao, Y.H., Takefuji, Y.: Functional Link Net Computing: Theory, system, Architecture and Functionalities. IEEE Comput., 76–79 (1992)
6. Patra, J.C., Pal, N.R.: A Functional Link Neural Network for Adaptive Channel Equalization. Signal Process. 43(2), 181–195 (1995)
7. Patra, J.C., et al.: Modelling of an Intelligent Pressure Sensor using Functional Link Artificial Neural Networks. ISA Trans. 39, 15–27 (2000)

8. Sierra, A., Macias, J.A., Corbacho, F.: Evolution of Functional Link Networks. IEEE Transactions on Evolutionary Computation 5(1), 54–65 (2001)
9. Misra, B.B., Dehuri, S.: Functional Link Neural Network for Classification Task in Data mining. J. Comput. Sci. 3(12), 948–955 (2007)
10. Dehuri, S., Cho, S.B.: Evolutionary Optimized Features in Functional Link Neural Networks for Classification. Expert Systems with Applications 37(6), 4379–4391 (2010)
11. Price, K., Storn, R., Lampinen, J.: Differential Evolution: A Practical Approach to Global Optimization. Springer (2005)
12. Stron, R., Price, K.: Differential Evolution-A Simple and Efficient Adaptive Scheme for Global Optimization over Continuous Spaces. Technical Report TR-05-012: Int. Comp. Sci. Institute, Berkely (1995)
13. Stron, R., Price, K.: Differential Evolution- Simple and Efficient Heuristic for Global Optimization over Continuous Spaces. Int. J. Global Optimization 11, 341–359 (1997)
14. Dehuri, S., Roy, R., Cho, S.B., Ghosh, A.: An Improved Swarm Optimized Functional Link Artificial Neural Network (ISO-FLANN) for Classification. J. Systems and Software 85, 1333–1345 (2012)
15. http://www.ip2location.com
16. Dash, C. S.K., Behera, A.K., Pandia, M.K., Dehuri, S.: Neural Networks Training Based on Differential Evolution in Radial Basis Function Networks for Classification of Web Logs. In: Hota, C., Srimani, P.K. (eds.) ICDCIT 2013. LNCS, vol. 7753, pp. 183–194. Springer, Heidelberg (2013)

Hybrid Technique for Frequent Pattern Extraction from Sequential Database

Rajalakshmi Selvaraj[1], Venu Madhav Kuthadi[2], and Tshilidzi Marwala[3]

[1] Faculty of Engineering and the Built Environment, University of Johannesburg,
South Africa & Department of Computer Science, BIUST, Botswana
[2] Department of AIS, University of Johannesburg, South Africa
[3] Faculty of Engineering and the Built Environment, University of Johannesburg, South Africa

Abstract. Data mining has became a familiar tool for mining stored value from the large scale databases that are known as Sequential Database. These databases has large number of itemsets that can arrive frequently and sequentially, it can also predict the users behaviors. The evaluation of user behavior is done by using Sequential pattern mining where the frequent patterns extracted with several limitations. Even the previous sequential pattern techniques used some limitations to extract those frequent patterns but these techniques does not generated the more reliable patterns .Thus, it is very complex to the decision makers for evaluation of user behavior. In this paper, to solve this problem a technique called hybrid pattern is used which has both time based limitation and space limitation and it is used to extract more feasible pattern from sequential database. Initially, the space limitation is applied to break the sequential database using the maximum and minimum threshold values. To this end, the time based limitation is applied to extract more feasible patterns where a bury-time arrival rate is computed to extract the reliable patterns.

Keywords: Sequential pattern, Data mining, Hybrid pattern mining.

1 Introduction

In the Sequence database, the sequential pattern mining establishes the frequent subsequences as patterns and also it is an essential data mining issue in the broad applications which includes the learning of client purchase behavior, DNA sequences, Disease treatments, Web access patterns, Scientific experiments, Natural disasters etc,.[1]. More efforts are put up to develop an effective algorithm for the purpose of searching frequent sequential patterns. Existing sequential pattern mining algorithm is divided into two approaches and they are of Apriori which is of candidate-generation [2] [3] [4] [5] and pattern-growth approach [6] [7] [8] [9].

Initially, the issues of sequential pattern mining were proposed by Srikant and by Agrawal in [3]. "A set of sequences is mentioned in which each sequences has a list of elements and those elements has a set of items and also an user least support threshold is given. The sequential pattern mining is to identify the entire frequent subsequences i.e. "Those Subsequences which has occurrence frequency but not less than the minimum support in the set of sequences".

Many sequential pattern mining algorithms won't deal with the intervals among consecutive item. The mine conventional sequential patterns deals particularly with the items order [10]. The sequential pattern which includes possibilities of time among two consecutive items gives essential information rather than the convectional sequential pattern. So, it is very important in more real time applications.

Considering an example i.e. a product of items A and B are sold in a supermarket. Also, the supermarket transaction is identified the possibility of the product that are been sold within 1, 2, 3 days and rest of the days once a customer has product A. As a result, existing research work are taken with various variations, sequence patterns with their probabilities of time is not shown.

Prefix Span Algorithm [9]: It is a recognizable pattern growth approach and it separates the database in to small probable databases and resolves them as it doesn't require to create any candidates sequence and doesn't require to scan the database at many times. Therefore it is faster than the Apriori-like Algorithm. Moreover, it is very clear that the Traditional Prefix Span algorithm functions slowly when there are huge numbers of frequent subsequences [9].

P-Prefix Span algorithm [6]: It is also a recognizable pattern growth approach. In this the frequent patterns are identified as per the possibility of inter-arrival time. Moreover, the frequent occasion of sequence pattern affect the sequence database, so this problem can increase the difficultly to obtain the end results.

This paper introduces a latest technique known as Hybrid Pattern mining algorithm in order to solve the mining problems. This hybrid pattern mining algorithm will extract the feasible patterns from sequential database by different specified limitation of the user. At first, the proposed algorithm will divide the sequential database according to the user specified max-min space limitations.

The sequence database is classified by hybrid pattern mining technique which is been identified as a continuous sequential patterns using probability of buying time of frequent items. The proposed algorithm is executed as to reestimate and support to search for frequent patterns using the estimated probability of end time. A new metric namely called probability-based pattern evaluation is introduced in this paper. The main aim of sequential pattern mining is to obtain whole patterns and their probability from the given set of transactions. The probabilities are obtained from a predefined time-period length that must be greater than least probability threshold. Also, to extract those type of sequential patterns, the probabilities should be verified when a item is mutated to sequential pattern such as if frequent item χ is to be combined to a frequent sequential pattern (ω), T represents the duration of time among the two repeated items as well as the least-probability threshold and is described as P (T $\leq$ 20) = 50%. Let $T_{\chi| (\omega) \text{ is}}$ to explain the time period among the occurrence of frequent item χ and the last occurrence item in (ω). If P ($T_{\chi|<\omega>} \leq$ 20) > 50%, then frequent item χ will be combined to (ω) creates a novel sequential pattern (ω'), or else, frequent item χ which cannot be create a novel pattern. The estimation of time probability is depended on the space constraints and Least-Prop-Threshold, however proposed algorithm requires less calculation time than existing techniques and searched patterns are more reliable.

The remaining paper is framed as: In section 2, the related work is discussed. In section 3, the proposed algorithm is discussed. In section 3, the result of proposed algorithm is discussed. At last in section 4, conclusion is discussed.

2   Related Work

The important problem of sequential pattern mining is the huge processing cost because of extracting the patterns from huge amount of data. Numerous of algorithms are introduced to expedite mining process. In that, SPAM [8], P-Prefix span [6], GSP [3], Prefix span [9], AprioriAll [2] and SPADE [11] are the representatives.

AprioriAll [2] is well-known three phase algorithm where it initially identifies the entire item sets with min support (frequent itemsets) and alters the database so that every transaction is altered with the set of entire frequent itemsets and then it identifies the sequential database. Actually, there are two issues in this approach. The first issue is that it is very expensive to perform the data transformation on the fly at every pass while finding the sequential database and also to modify the database and to store the modified database is unfeasible or else impractical in lots of applications as it doubles the disk space needs which are too expensive for huge database. The second issue is that it doesn't show any feasible to integrate sliding windows if it is probable to enlarge this algorithm in order to manage the time limitation and taxonomies.

In GSP algorithm, Agrawal and Srikant [2] implemented a bottom up method where this method creates initially frequent 1-sequences after that it creates frequent 2-sequences until the pattern has discovered. Also, this method creates the candidate n-sequences which are too from the frequent (n-1) –sequences in the every iteration. Also, it is according to the anti-monotone property where entire subsets of a frequent item sequence must be in frequent at all times. The candidate n-sequences are determined by their frequent which is measured by their supportive counts where in the situation of all iteration. For the purpose of supportive counts, this algorithm is expensive in order to decompose the item sequences.

Huge number of candidate sequences is the other issue of this GSP algorithm. In order to solve this issue [11] proposed the lattice idea using with the algorithm namely called SPADE where it differentiate the candidate sequence into several set of items, also is stored the every sets successfully in the primary memory itself. Moreover, the algorithm namely called GSP that utilizes ID List method in order to minimize the cost of measuring supportive counts. The sequence of ID pairs is kept in the ID list where it signifies the exact position in the database. Moreover, it doesn't create reliable patterns.

The FP-tree [7] is a well-known short come of AprioriAll to generate the huge sequential patterns. This method is utilized a FP-tree based data structure. The tree structure which is utilizing the prefix method for reorder the all transaction or frequent items. In subsequences, if the sequence of various items orders is larger than its re-arrangement method of the sequences doesn't execute properly. For the purpose of dealing the FP-Tree issues, Free Span [8] is utilized. The main concept of this Free

Span is to create the projected or prefix based sequential databases of particular frequent item patterns. Then, these frequent patterns have the capability to grow by just linking the frequent subset items from the small projected or prefix databases. This Free Span can get the entire frequent patterns by this way. The sequence of created candidate patterns are very less than the received candidate patterns from the combining technique as well as the database scanning cost is very cheap.

Many researchers attempted to solve many realistic requirements in the recent years. In [13] [14] [15] for an example, they are analyzed the entire issues of sequential patterns mining along with their quantities. Most of the real-time applications contain the entire size information which is obtained in data. Moreover, this information may be neglected by many more existing algorithms.

3   Proposed System

For the purpose of sequential pattern mining, an effective algorithm known as Hybrid Pattern Technique is proposed in this paper. In the frequent sequential pattern among two consecutive events, not exactly the time-period even its possibilities are revealed in mining process. It considers the space limitation and the time-period among two consecutive items or events because of every infinite sequence itemsets occurs in the sequence database.

Notations

In this research, a sequence of items is described as an ordered itemsets in list manner also it is indicated as (I_1, I_2, in). The parentheses '(' and ')' are used to enclose the itemset which are from the same items. Consider an example, $\{(2,3),1,4,(1,5,6),3\}$ is the sequence of items then it also have the following itemsets: $\{2,3\}, \{1\}, \{4\}, \{1,5,6\},$ and $\{3\}$. As a result, if one item is present in the item set then the parentheses can be neglected. Also, in this research, an itemset is indicated as flowing like the every item is combined with a transaction time and the frequent items also occurred in the similar transaction time. A item sequence is indicated as $(s_1,t_1), (s_2,t_2), \ldots, (s_n, t_n)$, where s_j is a item and t_j indicates the transaction time of s_j occurs, $1 \leq j \leq n$, and $t_1 \leq t_2 \leq \cdots \leq t_n$. Also, a sequential pattern can be represented as $(\omega) = (\omega_1, \omega_2, \omega_n)$. The following definitions are described with help of the above mentioned representation,

Definition-1: A sequence is S= $\{[s_1, t_{s1}], [s_2, t_{s2}] \ldots [s_n, t_{sn}]\}$, and a pattern P = $\{p1, p2, pm\}$, p is time-rest prefix of S and m$\leq$n

Definition-2: A sequence database is DS= $\{S_1, S_2 \ldots S_\infty\}$ i.e. a sequential database having

Definition-3: A time-rest subsequence P′ of sequence P is indicates a prefix or projection of P w.r.t the time-rest prefix R if (a) P′ has time-rest prefix R and (b) there presents not in proper time-rest super sequence P″ of P′ such that P′ is a time-rest subsequence of P and has a time-rest prefix R. If one eliminates direct manner, the time-rest prefix R from the projection P′, the new sequence is called the postfix of P with respect to time-rest prefix R.

Definition 4: Let T^{Last} indicates the final transaction occurrence of a sequence. Database DS. $S = \{[s_1, t_{s1}], [s_2, t_{s2}] \ldots [s_n, t_{sn}]\}$ is a sequence in DS. /S $(t_1, t2)$/ indicates the entire items that represent as sequence S which are occur among transaction time t_1 and t_2. Assume that, a time-rest subsequence of S is represented as $P = \{[p1, tp1], [p_2, tp_2, [p_m, t_{pm}]\}$ and χ is a one of the item that do not depended on $\{S\ (t_{pm},t_{qn})\}$. The correct transaction time of χ does not know to us based on the present sequence S. The probable censoring time of item χ w.r.t pattern p which is described as T^{Last} – tpm to realize the time interval probability of two frequent items.

Table 1. Notations utilized in Hybrid algorithm

Parameter	Expansion
$<\omega>$	A time-rest prefix
$|(\omega)|$	Length $<\omega>$
DS_{exis}	Continuous sequence database
DS_{new}	Divided Database of DS_{exist}
$DS|_{<\omega>}$	Projected Database of DS_{new} w.r.t $<\omega>$
μ	Least-Prop-Threshold
T_p	Time Duration
Þ	Least probability threshold
Min_s	Minimum space threshold
Max_s	Maximum space threshold

3.1   Hybrid Mining Algorithm

Space Constraints

The space constraint is applied on the item sequence id which is from sequential consecutive pattern. It remove out the sequence id of frequent pattern then it attempt to discover the given minima and maxima sequence id based frequent pattern. Consider an example, 0 to 10 sequential id which is sequential database and then specify lower space 1 and maximum space is 5. Then they obtained pattern can be be 012345 sequence id.

Time Constraints

A time constraint is work on the sequential frequent pattern i.e. time constraints remove the unnecessary subsequence

Pattern from the extracted subsequence patterns based on Least-Prop-Threshold and time probability threshold. The following steps are used to extract the feasible patterns from the sequential database.

3.1.1   Steps for Hybrid Mining Process

Step 1: Divide the sequential database space using minima and maxima threshold values

Step 2: Scan the sequential database for generating frequent itemsets

Step 3: Evaluate the arrival rate between each items

Step 4: Evaluate the Bury-Arrival time probability for frequent items at each time t.

Step 5: The Bury-arrival probability is greater than the time probability threshold, append the frequent item into pattern.

Step 6: Repeat the step-2 to 5, until obtain the specified reliable patterns

3.1.2   Hybrid Mining Algorithm

Consider a sequential database DS and pattern pattern (ω), we use projected or prefix appended database DS | (ω)| to indicate the list of postfixes in DS w.r.t pattern (ω). The quantity of items represents the size of pattern (ω) which is indicated | (ω)|. The following pseudo code has represents our proposed hybrid pattern mining algorithm.

Algorithm: Hybrid Algorithm
Input: DS_{exist}, DS_{new}, μ, T_p, Þ, Min_s, Max_s;
Output: A set of feasible frequent patterns
Iteration: Hybrid ($<\omega>$, |(ω)|, $DS|_{<\omega>}$)
1. Sequence count= Min_s;
2. For (sequence_count; sequence_count<Max_s, sequence_count++)
3. {
4. Select sequence pattern from DS_{exist}
5. Insert sequence pattern into DS_{new}
6. }
//end for
7. Scan $DS|_{<\omega>}$ one time, Extract all frequent items
8. Total_item=All_items_present in DS
9. if (|(ω)|=0)
10. {
11. Every frequent item χ, combine χ to $<\omega>$ as $<\omega'>$
12. }
//end if
13. if(|(ω)|>0)
14. {
15. for (χ;$\chi\leq$ Total_item; χ++)
16. δ=**arrival_rate**($<\omega>$,$DS|_{<\omega>}$,χ)
17. BATP=1-$e^{\delta td}$
BATP- Burry-Arrival Time Probability
18. if(BATP> Þ)
19. {
20. Combine χ to $<\omega>$ as $<\omega'>$
21. }
//end if
22. }

end for

23. for (Until all $<\omega'>$)

24. {

25. Construct Projected Database $DS|_{<\omega'>}$ w.r.t $<\omega'>$

26. Recall Hybrid ($<\omega'>$, $|(\omega')|$, $DS|_{<\omega'>}$)

27. }

Iteration: Arrival rate($<\omega>$, $DS|_{<\omega>}$,χ)

Input: $<\omega>$, $DS|_{<\omega>}$,χ

Output: δ-arrival rate of item χ occurrence after final item in $<\omega>$ χ-frequent item

1. tf= the transaction time of final item in $<\omega>$

2. r=0;

r-total number of sub sequence

3. α_1=0;

α_1- Difference between transaction time of final time in (ω) and initial item χ.

4. α_2=0;

α_2. the censoring time of censored item χ

5. for (every postfix in $DS|_{<\omega>}$)

6. if ($\chi \notin$ items in present postfix)

7. {

8. $\alpha_{2=}\alpha_2$+ (Transaction time of Final occurrence item-Transaction time of initial occurrence item)

10. }

11. else

12. {

13. r=r+1;

14. t_χ =transaction time of item χ

$\alpha_1=\alpha_1$+(t_χ. Transaction time of initial occurrence next item)

15. }   //end if

16. }   //end for

17. δ= r/($\alpha_1+\alpha_2$)

18. return δ;

4   Performance Evaluation and Experimental Results

Hybrid Algorithm is implemented in the Java language and also it is tested with AMD Sempron (tm) 1.60 GHz CPU, MS Windows XP operating system and 1GB memory on a computer system in order to calculate the performance of the proposed work. To store the datasets My SQL server 2005 is utilized.

The following are the threshold values that are applied in the Hybrid Technique.

1.   Minimum space =1

2.   Maximum space =4

3.   Least-Prop-Threshold =2

4.   Least Time Probability=0.3

5.   Expected   Time Period =7 days

Table 2 represents the infinite datasets which are occurred in the Sequential Database. Table 3 represents the total number of patterns, arrival rate, Bury-arrival rate and selected frequent patterns which are established by the Proposed Hybrid algorithm. This proposed Hybrid algorithm provides better performance than the other various pattern mining algorithms like a-priori, FP-tree, Prefix-Span and P-Prefix-Span. Table 3 represents the performance of Proposed and Existing approach.

Table 2. ∞ - Sequential Pattern Database

Sequence Id	Datasets
0	{(4,12),(3,4),(5,6),(2,9)}
1	{(2,2),(3,2),(1,4),(4,7),(1,8),(3,15)}
2	{(1,1),(4,5),(6,5),(2,12)}
3	{(1,1),(2,1),(3,1),(6,4),(6,6),(2,8),(3,9)}
4	{(3,5),(1,7),(2,15),(4,18),(6,18)}
5	{(3,5,(4,6),(6,7)}
∞	....

Table 3. Extracted Patterns and their Burry Arrival Time Probability

Pattern	Arrival rate	Bury-Arrival Time Probability	Patterns Extraction
{1,1}	0.033	0.208	-
{1,2}	0.087	0.456	Extract
{1,3}	0.072	0.399	Extract
{1,4}	0.045	0.273	-
{1,6}	0.142	0.632	Extract

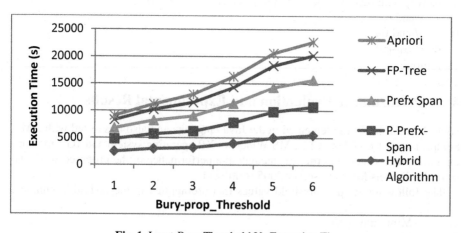

Fig. 1. Least-Prop-Threshold Vs Execution Time

Figure 1 represents the entire relationship among the Least-Prop Threshold of pattern generation and also their execution time. When the threshold value is increased, then the execution time is also increased. It means that the number of items mutation is increased. When compared with other existing algorithm, the proposed approach takes very less execution time.

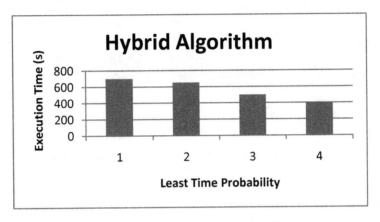

Fig. 2. Least-Time-Threshold Vs Execution Time

Figure 2 represents the relationship of proposed approach among the least time probability threshold of pattern extraction and also their execution time. The high probability can be reduced the execution time of pattern generation because of the generated Bury-prop-arrival rate is greater than the least time probability.

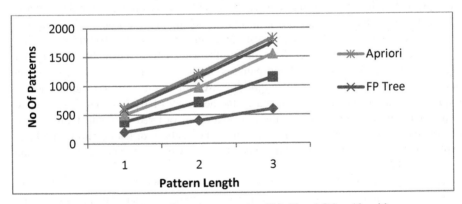

Fig. 3. Comparison of pattern generation Hybrid and Other Algorithm

Figure 3 represents the number of pattern generation among the proposed Hybrid Algorithm and other existing Algorithm. The proposed Hybrid Algorithm creates very limited amount of frequent patterns than other existing algorithms. If the pattern length increases, then the amount of patterns is also increased Figure 4 represents the execution time of Proposed Hybrid Algorithm and other existing algorithms.

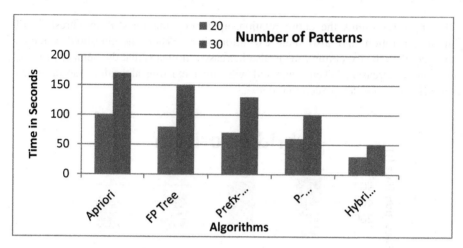

Fig. 4. Comparison of Time execution of Hybrid and Other Algorithm

The proposed Hybrid Algorithm takes very less time when compared to other algorithm like A-priori, FP-tree, Prefix Span and P-Prefix-Span. If the number of pattern increase then the time also increases as the number of patterns is propositional to time. The performance of proposed Hybrid Algorithm is very proficient in computation speed, storage space and time.

5   Conclusion

In sequential pattern mining, many researches focuses only on the Symbolic Patterns where the numerical investigation is primarily belongs to the range of forecasting investigation and trend analysis. A latest algorithm namely called Hybrid Algorithm is proposed in this paper which is mainly utilized for extracting realistic sequential patterns where it focus on symbolic patterns at the same time it also focus on the numerical analysis. As well as this algorithm uses minimum and maximum space of sequence database where it can increases the accuracy of final output. The reliable sequential pattern not only yields the information of ordered frequent items, it also obtains the detail probability of arrival items time. This information shows the brief explanation of the derived patterns characteristics which are very critical for the decision makers.

The users can identify the Least-Prop-Threshold and Least time probability threshold to establish the feasible sequential patterns as per the algorithm. The proposed algorithm minimizes candidate patterns amount   with help of the threshold of least time probability which makes the proposed technique is greater than the other mining algorithms.

The experimental results represents that the Proposed Hybrid algorithm is very effective and very suitable technique for mining the sequential pattern. Thereafter, the proposed approach can also be utilized with more constraints for extracting the feasible patterns.

References

1. Agrawal, R., Srikant, R.: Fast algorithms for mining association rules. In: 20th VLDB Conference, Chile, pp. 487–499 (1994)
2. Agrawal, R., Srikant, R.: Mining sequential patterns. In: 11th International Conference on Data Engineering, Taiwan, pp. 3–14 (1995)
3. Ayres, J., Flannick, J., Gehrke, J., And Yiu, T.: Sequential pattern mining using a bitmap representation. In: 8th ACM SIGKDD International Conference on Knowledge Discovery and Data Mining, pp. 429–435. ACM, New York (2002)
4. Chen, Y.L., Chiang, M.C., Ko, M.T.: Discovering time-interval sequential patterns in sequence databases. Expert Systems with Applications 25, 343–354 (2003)
5. Chen, Q., Dayal, U., Han, J., Hsu, M.C., Mortazavi-Asl, B., Pei, J.: FreeSpan: frequent pattern-projected sequential pattern mining. In: International Conference on Knowledge Discovery and Data Mining, USA, pp. 355–359 (2000)
6. Chen, Q., Dayal, U., Han, J., Hsu, M.C., Mortazavi-Asl, B., Pei, J., Pinto, H., Wang, J.: Mining sequential patterns by pattern-growth: the prefixspan approach. IEEE Transactions on Knowledge and Data Engineering 16, 1424–1440 (2004)
7. Jou, C., Shyur, H.-J., Chang, K.: A data mining approach to discovering reliable sequential patterns. The Journal of Systems and Software 86, 2196–2203 (2013)
8. Han, J., Pei, J., Yin, Y.: Mining frequent patterns without candidate generation. In: Proceedings of ACM SIGMOD International Conference of Data, Dallas, TX, pp. 1–12 (2000)
9. Pei, J., Han, J., Pinto, H.: PrefixSpan: Mining Sequential Patterns Efficiently by Prefix-Projected Pattern Growth. In: 11th IEEE International conference on Data Engineering, Germany, pp. 215–224 (2001)
10. Kim, C., Lim, J.H., Ng, R.T., Shim, K.: SQUIRE: sequential pattern mining with quantities. Journal of Systems and Software 80, 1726–1745 (2007)
11. Orlando, S., Perego, R., Silvestri, C.: A new algorithm for gap constrained sequence mining. In: ACM Symposium on Applied Computing, New York, pp. 540–547 (2004)
12. Toroslu, I.H.: Repetition support and mining cyclic patterns. Expert Systems with Applications 25, 303–311 (2003)
13. Zaki, M.: SPADE: An efficient algorithm for mining frequent sequences. Machine Learning 42(2), 31–60 (2001)
14. Kuthadi, V.M.: A new data stream mining algorithm for interestingness-rich association rules. Journal of Computer Information Systems 53(3), 14–27 (2013)
15. Selvaraj, R., Kuthadi, V.M.: A modified hiding high utility item first algorithm with item selector for hiding sensitive item sets. International Journal of Innovative Computing, Information and Control 9(12), 4851–4862 (2013)

Application of Fuzzy C-Means Clustering Method to Classify Wheat Leaf Images Based on the Presence of Rust Disease

Diptesh Majumdar[1], Arya Ghosh[2], Dipak Kumar Kole[3],
Aruna Chakraborty[3], and Dwijesh Dutta Majumder[4]

[1] Department of Computer Science and Engineering, Indian Institute of Technology,
Guwahati, Assam, India
[2] Department of Information Technology,
Indian Institute of Engineering Science and Technology, Shibpur, West Bengal, India
[3] Department of Computer Science and Engineering,
St. Thomas' College of Engineering & Technology, Kidderpore, Kolkata, West Bengal, India
[4] Professor Emeritus in Electronics and Communication Sciences Unit,
Indian Statistical Institute,
Director in Institute of Cybernetics Systems and Information Technology,
Kolkata, West Bengal, India

Abstract. This paper presents a novel and efficient way to detect the presence and identification of disease in wheat leaf from its image. The system applies FCM on data-points consisting of selected features of a set of Wheat Leaf images. In the first step, number of clusters is fixed to 2, in order to divide the input into sets of diseased and undiseased leaf images. The diseased leaf set is further classified into 4 sets corresponding to possibility of occurrence of known 4 types of disease, by applying FCM on this set with number of clusters fixed to 4. We have proposed an efficient method for selection of feature set based on inter and intra-class variance. Although testing has been done only on wheat leaf images, this method can also be applied on other leaf images through careful selection of the feature set.

Keywords: Image Processing, Feature Extraction, Fuzzy C-Means Clustering.

1 Introduction

Wheat Leaf rust, Puccinia recondite, usually does not cause spectacular damage, but on a world-wide basis it probably causes more damage than the other wheat rusts. In India, average losses of 3% have been estimated, although higher losses occur in certain areas if the cultivars are susceptible to leaf rust.

Researches in Image Processing and Analysis for Detection of Plant Diseases have grown immensely over the past decade. Various methods have been devised that are used to study plant diseases/traits using Image Processing. The methods studied are aimed at increasing throughput & reducing subjectiveness arising from human experts in detecting the plant diseases [1].

© Springer International Publishing Switzerland 2015   277
S.C. Satapathy et al. (eds.), *Proc. of the 3rd Int. Conf. on Front. of Intell. Comput. (FICTA) 2014*
– *Vol. 1*, Advances in Intelligent Systems and Computing 327, DOI: 10.1007/978-3-319-11933-5_30

As shown in Fig. 1, the general system for detection and recognition of disease in plant leaf consists of three main components: Image analyzer, Feature extraction and classifier.

The processing that is done by using these components is divided into two phases. The first processing phase is the offline phase or Training Phase. In this phase, a set of input images of leaves (diseased and normal) were processed by image analyzer and certain features were extracted. Then these features were given as input to the classifier, and along with it, the information whether the image is that of a diseased or a normal leaf. The classifier then learns the relation among the features extracted and the possible conclusion about the presence of the disease. Thus the system is trained.

The second processing phase is the online phase, in which the features of a specified image is extracted by image analyzer and then tested by the classifier whether the leaf is diseased or not, according to the information provided to it in the learning phase (offline phase).

Now, assume there is a large set of images of plant leaves, and we need to determine the type of the disease, if there is any, the leaves are infected with. If we go by the existing system, we have to take the images one by one, feed it to the input of the system, get the output and continue the steps for the next image until there is none. This cycle may be time consuming if the test is conducted for a large number of leaves, which is the case in most practical situations. The algorithm that we propose overcomes this shortcoming- it is very easy to comprehend and is designed to work for a large set of images.

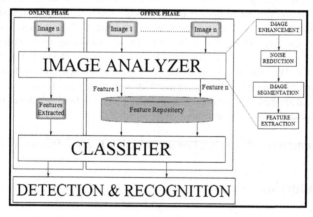

Fig. 1. Architecture of a General System for Detection and Recognition of Disease in Plant Leaf

The Paper has been organized as follows. In Section 2, we go through the type of diseases a wheat leaf may be infected with. Section 3 describes Architecture of the proposed system. Selection of appropriate features is detailed in Section 4. Section 5 explains the basics and application of Fuzzy C-Means Algorithm in our system. Finally we show the results in Section 6, and reach our conclusion in Section 7.

For this work, more than 300 wheat leaf images were collected from a wheat farm at Field Crop Research Station, Department of Agriculture, Govt of West Bengal, Burdwan. Then the leaves where classifed according to the severity of infection in the leaves by two experienced doctors, Dr. Amitava Ghosh (Ex. Economic Botanist IX, West Bengal) and Dr. P. K. Maity (Chief Agronomist & Ex-Officio Joint Director of Agriculture, West Bengal).

2   Recognition of Wheat Leaf Diseases and Gradations

A wheat leaf can be infected with the following four diseases:

   I. Powdery Mildew: Elliptical patches of white fungal growth appear on both leaf surfaces.
   II. Septoria Leaf Spot: Yellow flecks first appear on the lower leaves. Later, yellow to red-brown or gray-brown spots or blotches may develop on all above-ground parts of wheat.
   III. Tan spot or Yellow Leaf: Oval-shaped tan spots up to 12 mm in length appear on the leaves.
   IV. Snow Molds: Leaves may be partly or entirely dried and appear brown or bleached. When the crowns are attacked, the plants are usually killed. When the crowns are unharmed, new leaves emerge among the damaged leaves and the wheat plants often recover.

All four diseases mentioned above can be considered to be wheat leaf rusts with various degrees of infections.

A study by Pathologists Marsalis and Goldberg [2] reveals that Wheat Leaf rust disease symptoms begin as small, circular to oval yellow spots on infected tissues of the upper leaf surface. As the disease progresses, the spots develop into orange colored pustules which may be surrounded by a yellow halo (Fig. 2).

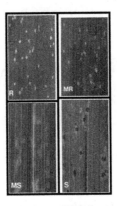

Fig. 2. Relative resistances of wheat to leaf rust: R=resistant, MR= moderately resistant, MS= moderately susceptible, and S= susceptible. Source: Rust Scoring Guide, Research Institute for Plant Protection, Wageningen, Netherlands.

Accordingly, our proposed system is able to identify a wheat leaf as diseased or not, and if found to be diseased, it can determine the severity of the infection by labeling it as R, MR, MS or S.

3   Architecture of the System

The first step in making the proposed system to be working properly is selection of the appropriate set of features. In order to do that, we have considered the most common features used in such applications, namely, Entropy, Median, Mode, Variance [3], Standard Deviation, Number of Zeros and connected components in the Binary Image, Number of Peaks in the Histogram, Color Moments [4, 5] and Texture based features [6, 7] like Inertia, Correlation, Energy and Homogeneity. Then we used a simple algorithm based on inter-class and intra-class variance (refer to Section 4) to identify the most effective among these features.

Next, we take the set of images of wheat leaf, which is required to be divided into separate clusters. We extract the selected features from the input set of images. The feature vectors are then fed to the Fuzzy C-Means Clustering Algorithm (refer to Section 5) fixing number of clusters to 2. This divides the set of wheat leaf images into sets of diseased and undiseased leaf images. Next, we consider only the feature vectors of the set of diseased leaves. We again run the Fuzzy C-Means Clustering Algorithm on this set, fixing the number of clusters to 4. Thus, we obtain 4 sets partitioned according to the degree of infection of the system.

The architecture of the system is schematically represented in Fig. 3a. The "Feature Extraction and Fuzzy C-Means" block in Fig. 3a has been expanded in Fig. 3b.

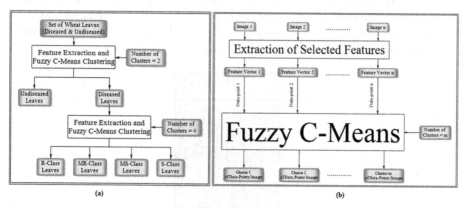

(a)   (b)

Fig. 3. (a) Architecture of the System (b) Feature Extraction followed by Fuzzy C-Means Clustering

4   Feature Selection Algorithm

The algorithm selects the most suitable features from the set of all features mentioned in the previous section. This activity decreases the size of each feature vector to an optimum level, and thus enhances performance of the FCM. The steps of the algorithm are as follows.

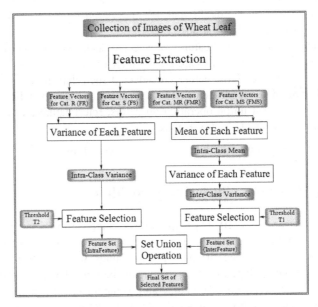

Fig. 4. Feature Selection Process

1. Collect 300 or more images of wheat leaf such that 50% of wheat leaf images belong to R-Class, 25% to S-Class, 13% to MR-Class and 12% to MS-Class.
2. Extract the 25 Features of each of the images collected and stored in 4 separate files for 4 categories of image. The file containing features of all the images belong to R Class is named FR, and the others are named FS, FMR and FMS respectively.
3. Calculate the variance of all the feature values belonging to one feature in a particular class-file. This operation is done for each of the 4 files and stored in 1 file named Intra_Class_Var. This file will contain 4 rows corresponding to 4 classes of image, and 25 columns corresponding to each of the features. Hence, the value at position i,j specifies the variance of feature j of images belonging to class i. These values are called Intra-class variance.
4. Calculate the mean of all the feature values belonging to one feature in a particular class-file. This operation is done for each of the 4 files and stored in 1 file named Intra_Class_Mean. This file will contain 4 rows corresponding to 4 categories of image, and 25 columns corresponding to each of the features. Hence, the value at position i,j specifies the mean of feature j of images belonging to class i. These values are called Intra-class Mean.

5. Calculate the variance of mean values (all rows) corresponding to each feature (one column) in the file Intra_Class_Mean. Thus we get 25 values of variance and store these values in a file named Inter_Class_Var.
6. Choose two thresholds T1 and T2, one based on inter-class-variance, and another on intra-class-variance.
7. A set of features InterFeature are selected such that the inter-class variance value for all those features are greater than T1.
8. A set of features IntraFeature are selected such that the intra-class variance values of all 4 categories for all those features are less than T2.
9. We perform set union operation on 2 sets, InterFeature and IntraFeature to get the final selected set of features.

The schematic representation of the algorithm has been shown in Fig. 4.

5   Fuzzy C-Means Clustering Algorithm

The concept of Fuzzy sets [10, 11] is as follows. Suppose there are n elements in a set, and we need to distribute these n elements into c clusters. Each of these n elements will have c degree of membership values which will correspond to c clusters. Among these c values, the element will be considered to be a part of that cluster which will have the highest membership value for that element.

In our case, the set consists of all the feature vectors of the images. We have applied Fuzzy c-means Clustering Algorithm [12] which iteratively minimizes the objective function:

$$J_m = \sum_{i=1}^{N} \sum_{j=1}^{C} u_{ij}{}^m \parallel x_i - c_j \parallel^2 \tag{1}$$

Where m is any real number greater than 1, u_{ij} is the degree of membership of feature vector xi in the cluster j, xi is the ith of d-dimensional measured data, cj is the d-dimension center of the cluster, and $\parallel * \parallel$ is any norm expressing the similarity between any measured data and the center.

In our application, we have considered 4 clusters corresponding to 4 categories of disease- R, MR, MS and S. This operation categorizes the feature vectors into 4 clusters. From the result we can conclude that an image I belonging to cluster C implies the wheat leaf in image I is infected with disease belonging to category C.

6   Results

6.1   Feature Selection

As shown in the description of architecture of the system, the first step required for proper functioning of the proposed algorithm is the proper selection of the appropriate features. We first collected 310 wheat leaf images and extracted all 25 features, as described in Section 4. Then we apply the algorithm formulated in Section 5, to get a set of 7 features which are designated as the best among all the features in this particular application. For this experiment, the Intra-Class Threshold was set to 0.6 and the Inter-Class Threshold was set to 0.03. The 7 features that were selected from the system are shown in Table 1.

6.2 Fuzzy C-Means Results

In the first step, we run FCM on feature vectors of 310 wheat leaf images fixing C=2. We get a set of diseased wheat leaf images in one cluster. Next, we run FCM only on the set of diseased leaves, fixing C=4. Classification into diseased and undiseased leaves was found to be accurate in 88% of the cases. Recognition of type of disease was accurate in 56% of the cases.

Table 1. Selected Features and their Sample Values

Wheat Leaf Images				
Category	**R**	**MR**	**MS**	**S**
Mode	0.11695	0.0035	0	0.13443
Variance	341.77	0.0587	301.94	73.155
Standard Deviation	5846.1	7.6602	5495	2704.7
No. of 0s in Binary Image	89.084	0.5634	18.972	44.876
No. of Connected Components	0.0055036	0.0032	0.0080	0.2457
S Plane Moment3	-0.07605	0	0.1442	-0.034339
V Plane Moment 1	2.2459	0	0.3411	0.018235

7   Conclusion

This paper demonstrates the power of Fuzzy C-Means Clustering Algorithm as a classifier in applications of identification of disease in plant leaves. It is simpler and faster than other contemporary integrated image processing approaches used for disease identification. The only drawback is that the output of the system (partitioned set of images) can be ambiguous in some cases. For example, suppose if the input set of diseased leaves does not cover all possible type of infections, the output can still be separated into 4 clusters. This drawback can be avoided through better selection of the features. An improved AdaBoost algorithm [8] or KPCA based feature selection [9] can also be adopted on the large feature set to select the most significant features.

References

1. Patil, J.K., Kumar, R.: Advances in Image Processing for Detection of Plant Diseases. Journal of Advanced Bio. Informatics Applications and Research 2(2), 135–141 (2011) ISSN 0976-2604
2. Halkidi, M., Batistakis, Y., Vazirgiannis, M.: Cluster validity methods: Part I. SIGMOD Record (September 2002)
3. Gonzalez, R.C., Woods, R.E.: Digital Image Processing, 2nd edn. Prentice-Hall, Inc. (2002)
4. Powbunthorn, K., Abudullakasim, W., Unartngam, J.: Assessment of the severity of Brown Leaf Spot Disease in Cassava using Image Analysis. In: The International Conference of the Thai Society of Agricultural Engineering, Chiangmai, Thailand, August 4-5 (2012)
5. Patil, J.K., Kumar, R.: Color Feature Extraction of Tomato Leaf Diseases. International Journal of Engineering Trends and Technology 2(2) (2011)
6. Majumder, D.D., Chanda, B.: Digital Image Processing and Analysis. Prentince Hall of India Private Limited (September 2007)
7. Patil, J.K., Kumar, R.: Feature Extraction of Diseased Leaf Images. Journal of Signal and Image Processing 3(1), 60–63 (2012) ISSN: 0976-8882 & E-ISSN: 0976-8890
8. Zhang, M., Meng, Q.: Citrus canker detection based on leaf images Analysis. IEEE (2010)
9. Tian, J., Hu, Q., Ma, X., Han, M.: An Improved KPCA/GA-SVM Classification Model for Plant Leaf Disease Recognition. Journal of Computational Information Systems 8(18), 7737–7745 (2012)
10. Majumder, D.D., Pal, S.K.: Concepts of Fuzzy Sets and its Application in Pattern Recognition Problems. In: Proc. Nat. Conf. of CSI, Hyderabad, India (January 1976)
11. Majumder, D.D., Majumder, S.: Fuzzy Set in Pattern Recognition and Image Analysis. In: Dutta Majumder, D. (ed.) Advances in Inform. Sc. and Tech., pp. 50–69. Statistical Publishing Society, ISI (1982)
12. Bezdek, J.C., Ehrlich, R., Full, W.: FCM: the Fuzzy c-mean Clustering Algorithm. Computer and Geoscience 10, 191–203 (1984)

Hepatitis Disease Diagnosis Using Multiple Imputation and Neural Network with Rough Set Feature Reduction

Malay Mitra and R.K. Samanta

Expert Systems Laboratory, Department of Computer Science and Application,
University of North Bengal, Raja Rammuhunpur, 734013, W.B., India
{malay.mitra68,rksamantark}@gmail.com

Abstract. Intelligent automated decision support systems are found to be useful for early detection of hepatitis for augmenting survivability. We present here an intelligent system for hepatitis disease diagnosis using UCI data set for experiment. We use multiple imputation technique for managing missing values in the UCI data set. One of the potential tools in this context is neural network for classification. For better diagnostic classification accuracy, various feature selection techniques are deployed as prerequisite. These features are considered to be more informative to the doctors for taking final decision. This work attempts rough set-based feature selection (RS) technique. For classification, we use incremental back propagation learning network (IBPLN), and Levenberg-Marquardt (LM) classification tested on UCI data base. We compare classification results in terms of classification accuracy, specificity, sensitivity and receiver-operating characteristics curve area(AUC).

Keywords: Hepatitis, *UCI database, M*ultiple Imputations, RS, Neural networks, Incremental back propagation, Levenberg-Marquardt Classification.

1 Introduction

Hepatitis is an inflammation of the liver without pinpointing any specific reason [1]. Mainly, there are three viruses responsible for causing hepatitis: A –virus (HAV), B-virus (HBV), and C- virus (HCV). Recent years, it has become one of the major diseases worldwide. More causalities are reported worldwide due to this cause. At present, there are about 2 million people infected with hepatitis B virus, and 200,000 to 300,000 people infected with hepatitis C [2]. One of the major problems for the diagnosis of hepatitis for a physician [1] as most people suffering from hepatitis B and hepatitis C do not show any major symptom. Generally, a physician takes decision by evaluating the current test results of a patient and/or by comparing the present patient with previous similar patients using heuristics; leaving a scope of error in diagnosis. To mitigate the problem, an automated intelligent decision support system can be helpful. We, propose an intelligent system for hepatitis diagnosis using a method based on rough-set based feature selection (RS) for feature reduction, and IBPLN, and LM classification. We use multiple imputation technique for missing values in the data set. This is a novel approach compared to the previous studies on hepatitis disease diagnosis [1-4].

© Springer International Publishing Switzerland 2015
S.C. Satapathy et al. (eds.), *Proc. of the 3rd Int. Conf. on Front. of Intell. Comput. (FICTA) 2014*
– *Vol. 1*, Advances in Intelligent Systems and Computing 327, DOI: 10.1007/978-3-319-11933-5_31

Section 2 presents UCI hepatitis database. Section 3 explains multiple imputation. Section 4 presents RS algorithm. Section 5 presents the preliminaries of ANN along with the two algorithms we use in this study. Section 6 presents the application. Section 7 presents the results. Lastly, our conclusions are summarized.

2   UCI Hepatitis Database

In this study, we use hepatitis data set from UCI Machine Learning Repository [5]. The data set contains 19 attributes plus one attribute for class (binary). It contains 155 samples to two different classes ('die'- 32 cases; 'live'- 123 cases). The details of the dataset are presented in Table 1.

Table 1. The attributes of hepatitis disease database of UCI

Attribute number	Attribute name	Attribute values	Attribute number	Attribute name	Attribute values
1.	Class	DIE, LIVE	11.	SPLEEN PALPABLE	no, yes
2.	AGE	10,20,30,40,50,60,70,80	12.	SPIDERS	no, yes
3.	SEX	Male, female	13.	ASCITES	no, yes
4.	STEROID	no, yes	14.	VARICES	no, yes
5.	ANTIVIRALS	no, yes	15.	BILIRUBIN	0.39,0.80,1.20,2.00,3.00,4.0
6.	FATIGUE	no, yes	16.	ALK PHOSPHATE	33, 88, 120, 160, 200, 250
7.	MALAISE	no, yes	17.	SGOT	13, 100, 200, 300, 400, 500
8.	ANOREXIA	no, yes	18.	ALBUMIN	2.1, 3.0, 3.8, 4.5, 5.0, 6.0
9.	LIVER BIG	no, yes	19.	PROTIME	10,20,30,40,50,60,70,80,90
10.	LIVER FIRM	no, yes	20.	HISTOLOGY	no, yes

In this data set, there are a number of missing values. For the missing value management, we use multiple imputation technique as described in the following section 3.

3   Multiple Imputations

Missing value problem may adversely affect the derived results if not addressed properly. There are different methods which are being used; e.g., Delete the records containing missing data; Use attribute mean; Use attribute median; Use a global constant to fill in for missing which seem not relevant to the decision attribute; Use a data mining method. We use a bootstrap-based EMB algorithm [6] performing multiple imputation. A schematic view of the approach to multiple imputation with the EMB algorithm is shown in Fig. 1. In multiple imputation technique, EMB algorithm combines EM algorithm with bootstrap approach to take draws from the posterior which is computed by considering assumed MAR (missing at random), likelihood, law of expectations and a flat prior. For each draw, the data is bootstrapped to simulate estimation uncertainty and then run EM [7] algorithm to find the mode of the posterior for the bootstrapped data and fundamental uncertainty [8]. After having draws imputations are done and m sets of imputed data are generated, analysis are performed which produce m separate results. These separate results are combined by arithmetic mean [6], [9] to have final results.

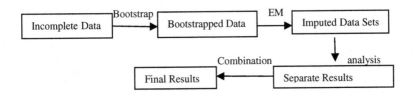

Fig. 1. A schematic view of multiple imputation

4   Rough SET-Based Feature Subset Selection (RS)

Rough sets theory was first presented by Pawlak in the year 1980's [10]. Rough set is a formal approximation of a crisp set in terms of a pair of sets which give lower approximation with positive region and upper approximation with negative region. In between there a boundary. Let there be an information system $I = (U,A)$ (attribute – value system), where U be the universe of discourse and is a non-empty set of finite objects; A is a non-empty finite set of attributes. With any $P \in A$, there is an associated equivalence relation IND(P), called P – indiscernibility relation. Let $X \in U$ be a target set. The target set X can be approximated using only the information contained within P by constructing P-lower ($\underline{P}X$) and P-upper (PX) approximation of X. The tuple ($\underline{P}X$, PX) is called rough set. The accuracy of the rough-set representation of the set X can be given [11] by the following:

$$\kappa^P = |\underline{P}X| / |PX| \tag{1}$$

5   Artificial Neural Networks

5.1   Preliminaries

Artificial neural networks (ANN) mimic the workings of the neurons of human brain. The neurons are connected to one another by connection links. Each link has a weight. A simple McCulloch-Pitts model of a neuron [12], the basis of ANN was presented in the year 1943. In the literatures, different forms of ANNs are there for modeling different tasks.

Modeling with ANN involves two important tasks, namely, *design* and *training* the network. The design of a networks involves (1) fixing the number of layers, (2) the number of neurons for each layer, (3) the node function for each neuron, (4) whether feedback or feedforward, and (5) the connectivity pattern between the layers and the neurons. All these adjustments are to be taken care of for improved performance of the system. The training phase or the learning phase involves adjustments of weights as well as threshold values from a set of training examples. The kind of learning law was first proposed by Donald Hebb [13]. Currently, there are hundreds of such leaning algorithms in the literature [14], but the most well-known among them are backpropagation [15], [16], ART [17], and RBF networks [18].

5.2   Incremental Backpropagation Learning Networks

The normal backpropagation network is not an incremental by its nature [19]. The network learns by the backpropagation rule of Rumelhart et al.[20] under the constraint that the change to each weight for each instance is bounded, which is achieved by introducing a scaling factor s which scales down all weight adjustments so that all of them are within bounds. The learning rule is now.

$$\Delta W_{ij}(k) = s(k) \, \eta \delta_j(k) O_i(k) \tag{2}$$

where W_{ij} is the weight from unit i to unit j, η ($0 < \eta < 1$) is a trial-independent learning rate, δ_j is the error gradient at unit j, O_i is the activation level at unit i, and the parameter k denotes the k-th iteration. In the incremental learning scheme, initial weights prior to learning any new instance represent knowledge accumulated so far. The IBPLN proceeds as proposed in [19].

5.3   Levenberg-Marquardt (LM) Algorithm

LM algorithm is basically an iterative method that locates the minimum of a multivariate function that is expressed as the sum of squares of non-linear real-valued functions [21], [22]. LM algoritjm finds a solution even if it starts far off the final minimum. During the iterations, the new configuration of weights in step k+1 is calculated as follows

$$w(k+1) = w(k) - (J^T J + \lambda I)^{-1} J^T \varepsilon(k) \tag{3}$$

where J – the Jacobian matrix, λ - adjustable parameter, ε - error vector.

6   Applications

Basically, this study consists of three stages: Generation of five sets of data using multiple imputation; and then the feature extraction and reduction phase by rough-set-based feature selection(RS); and then classification phase by incremental back propagation neural networks (IBPLN), and Levenberg-Marquardt (LM) algorithm. The schematic view of our system is shown in Fig. 2.

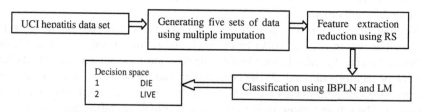

Fig. 2. Block diagram of a system for hepatitis disease diagnosis

6.1 Data Preprocessing

Data preprocessing is the primary step for any model development. We completely randomize the data sets with missing records. Then, we apply multiple imputation technique using Amelia [6] which generates five sets of filled data. To avoid outliers during multiple imputations, we apply certain range specifications after consulting a specialist [23]. Next, we apply RS using ROSETTA [24]. We use Johnson's algorithm to extract reducts, which extracts a single reduct not a set of reducts like other algorithms. We extracted all reducts from the five imputed files under the option 'Full' and 'Full' with 'Modulo decision' and got ten sets of reduct. After this extraction process we chose all the attributes appeared in these reducts as the reduced features. The seven reduced features obtained in this way is shown in Table 2. The data sets are partitioned into three: Training set (68%), Validation set (16%), and Test set (16%).

Table 2. Reduced hepatitis attributes

#	Name of the attributes	#	Name of the attributes
1.	AGE	5.	SGOT
2.	STEROID	6.	ALBUMIN
3.	BILIRUBIN	7.	PROTIME
4.	ALK PHOSPHATE		

6.2 Network Architecture

Balancing the trade-off between accuracy and generalizability is the prime characteristic of selecting a model. The ANN model selection includes choice of network architecture and feature selection. The hold-out data set called the *validation set* would be useful helping all these decisions successful [25]. In our networks, we use logistic function of the form $F(x) = 1/(1+e^{-x})$ in the hidden and output nodes. Theoretically, a network with one hidden layer and logistic function as the activation function at the hidden and output nodes is capable of approximating any function arbitrarily closely, provided that the number of hidden nodes are large enough [26]. So, we use one input layer, one hidden layer, and one output layer. For fixing the number of neurons in the hidden layer, we use the formula of Goa[27] modified by Huang et al. [28].

$$s= \sqrt{(0.43mn + 0.12n^2 + 2.54m + 0.77n + 0.35)} + 0.51 \tag{4}$$

where s is the number of neuron, m is the number of input, n is the number of output. In the present study, m = 7, n =2; and hence s= 6 after round off.

7 Modelling Results

The classification algorithms using two combinations were implemented in Alyuda NeuroIntelligence [29]. Table 3 shows the network parameters. As a overtraining control measure, we retain the copy of the network with the lowest validation error.

Table 3. Network parameters applying to WBCD

Classifiers	Network structure			Epochs(retrain)	Numbers patterns		
	I	HL	O		Training	Validation	Testing
RS + IBPLN	7	6	1	2000(10)	109	23	23
RS + LM	7	6	1	2000(10)	109	23	23

7.1   Performance Evaluation Methods

As the performance measure, we compute classification accuracy, sensitivity, specificity and AUC. The area under ROC curve (AUC) is an important measure of classification performance that is being used in biomedical research to assess the performance of diagnostic tests [30]. AUC close to one indicates more reliable diagnostic result [30] and it is considered one of the best methods for comparing classifiers in two-class problems.

7.2   Experimental Results

The compiled results from 80 simulations of our studies are shown in Table 4A and Table 4 B.

Table 4A. Results from 80 simulations

Imputation Number(Method)	Test set (CCR%)				Specificity			
	Highest (freq)	Lowest (freq)	Avg	Grand Avg	Highest (freq)	Lowest (freq)	Avg	Grand Avg
IMP-1(RS+IBPLN)	100(4)	78.26(2)	88.97		100(38)	33.33(5)	80.18	
IMP-2(RS+IBPLN)	100(7)	78.26(2)	91.47		100(38)	25(1)	79.97	
IMP-3(RS+IBPLN)	100(11)	73.91(1)	91.19	90.48	100(39)	25(1)	80.15	79.34
IMP-4(RS+IBPLN)	100(8)	78.26(2)	91.25		100(38)	20(1)	81.74	
IMP-5(RS+IBPLN)	100(3)	78.26(6)	89.51		100(31)	25(4)	74.66	
IMP-1(RS+LM)	100(9)	82.61(2)	92.39		100(45)	50(6)	87.45	
IMP-2(RS+LM)	100(13)	86.96(11)	93.97		100(42)	33.33(1)	85.05	
IMP-3(RS+LM)	100(13)	86.96(6)	94.35	**93.34**	100(53)	33.33(1)	89.15	**85.81**
IMP-4(RS+LM)	100(7)	86.96(12)	93.2		100(35)	33.33(3)	82.86	
IMP-5(RS+LM)	100(8)	86.96(10)	92.77		100(42)	33.33(2)	84.42	

Table 4B. Results from 80 simulations

Imputation	Sensitivity				AUC			
Number(Method)	Highest (freq)	Lowest (freq)	Avg	Grand avg	Highest (freq)	Lowest (freq)	Avg	Grand avg
IMP-1(RS+IBPLN)	100(11)	80(1)	90.8		100(11)	51(1)	89.43	
IMP-2(RS+IBPLN)	100(18)	83.33(2)	93.54		100(11)	53(1)	90.4	
IMP-3(RS+IBPLN)	100(25)	80.95(1)	93.36	92.78	100(18)	54(2)	91.61	
IMP-4(RS+IBPLN)	100(18)	77.27(1)	93.6		100(12)	56(1)	91.83	90.72
IMP-5(RS+IBPLN)	100(14)	78.95(1)	92.62		100(6)	70(1)	90.34	
IMP-1(RS+LM)	100(17)	84.21(1)	93.73		100(16)	74(4)	90.46	
IMP-2(RS+LM)	100(32)	83.33(2)	95.92		100(18)	70(1)	92.24	
IMP-3(RS+LM)	100(30)	85(1)	95.43	**94.98**	100(26)	52(1)	92.25	**92.52**
IMP-4(RS+LM)	100(26)	85(2)	95.36		100(21)	71(1)	93.69	
IMP-5(RS+LM)	100(19)	83.33(1)	94.44		100(12)	79(1)	93.94	

The following observations are noted below:

- Out of the two methods, RS+LM shows better performance in terms of CCR, Specificity, Sensitivity, and AUC.
- All of the methods provide 100% classification accuracy as the highest level and the lowest classification performance is 82.61% for RS + LM simulation results using 68% training, 16% validation, and 16% testing data sets.

We now compare our results with other studies [4] as shown in Table 5. It is evident from the table that our method offers comparable results with 80 simulations at present.

Table 5. Comparison of accuracies of different methods applied in hepatitis disease diagnosis

Hepatitis disease	KNN	Naïve Bays	SVM	FDT	PSO	CBR-PSO	RS- LM
Best	89.86%	86.35%	90.31%	78.15%	89.46%	94.58%	**100%**
Lowest	70.29%	66.94%	65.22%	61.49%	75.35%	77.16%	**82.61%**
Average	83.45%	82.05%	86.92%	75.39%	82.66%	92.83%	**93.34%**

8   Conclusion

This work has explored two classification techniques, namely, IBPLN and LM, of intelligent diagnostic systems for hepatitis diagnosis. For feature selection and reduction, we use RS technique. A few of the previous works pointed out the lowest performance of their systems. We argue that the lowest performance should also be a judging parameter for the performance of a system. Moreover, much of the authors of their studies on UCI hepatitis data do not indicate whether their results represent the best simulation product or an average of several simulations. So, we present here the highest, lowest, and the average behavior of the methods used.

This work provides a novel approach compared to other results including Neshat et al. [4]. In a significant number of cases, specificity, sensitivity, and AUC have reached 100%. While this work does not claim the highest performance achiever, but, at the same time reports that a combination of seven features (i.e., 'AGE', 'STEROID', 'BILLIRUBIN', 'ALK PHOSPHATE', 'SGOT', 'ALBUMIN', 'PROTIEN') derived using RS might be worthwhile to scrutiny when the final decision is made by the doctors. The results of this research demonstrated that the techniques used here could be applied to other diseases.

References

1. Polat, K., Gunes, S.: Hepatitis disease diagnosis using a new hybrid system based on feature selection (FS) and artificial immune recognition system with fuzzy resource allocation. Digital Signal Processing 16(6), 889–901 (2006)
2. Rezaee, K., et al.: An Intelligent Diagnostic System for Detection of Hepatitis using Multi-Layer Perceptron and Colonial Competitive Algorithm. The J. of Mathematics and Computer Science 4(2), 237–245 (2012)
3. Calisir, D., Dogantekin, E.: A new intelligent hepatitis diagnosis system: PCA-LSSVM. Expert Systems with Applications 38, 10705–10708 (2011)
4. Neshat, M., Sargolzaei, M., Nadjaran, A.N., Masoumi, A.: Hepatitis Disease Diagnosis using Hybrid Case Based Reasoning and Particle Swarm Optimization. ISRN Artificial Intelligence (2012), doi:10.5402/2012/609718
5. http://archieve.ics.uci.edu/ml/datasets/Hepatitis
6. Honaker, J., King, G., Blackwell, M.: 2011, AMELIA II: A Program for Missing Data (accessed: 3rd, November 2011), http://gking.harvard.edu/amelia/
7. Dempster, A.P., Laird, N.M., Rubin, D.B.: Maximum Likelihood from Incomplete Data via the EM Algorithm. Journal of the Royal Statistical Society. Series B 39(1), 1–38 (1977)
8. Honaker, J., King, G.: What to do About Missing Values in Time Series Cross-Section Data. American J. of Political Science 54(2), 561–581 (2010)
9. King, G., Tomaz, M., Wittenberg, J.: Making the Most of Statistical Analyses: Improving and Presentation. American Journal of Political Science 44(2), 341–355 (2000)
10. Pawlak, Z.: Rough sets. Int. J. of Parallel Programming 11(5), 341–356 (1982)
11. Pawlak, Z.: Rough Sets: Theoretical Aspects of Reasoning About Data. Klwer Academic Publishing, Dordrecht (1991)
12. McCulloch, W., Pitts, W.: A logical calculus of the ideas immanent in nervous activity. Bulletin of Mathematical Biophysics 7, 115–133 (1943)
13. Hebb, D.O.: The Organization of Behavior, a Neuropsychological Theory. John Wiley, New York (1949)
14. Roy, A.: Artificial Neural Networks- A Science in Trouble. SIGKDD Explorations 1(2), 33–38 (2000)
15. Rumelhart, D.E., McClelland, J.L. (eds.): Parallel Distributed Processing: Explorations in Microstructures of Cognition, vol. 1, pp. 318–362. Foundations, MIT Press, Cambridge (1986)
16. Rumelhart, D.E.: The Architecture of Mind: A Connectionist Approach. In: Haugeland, J. (ed.) Mind_design II, ch. 8, pp. 205–232. MIT Press (1986)

17. Grossberg, S.: Nonlinear Neural Networks: Principles, Mechanisms, and Architectures. Neural Networks 1, 17–61 (1988)
18. Moody, J., Darken, C.: Fast Learning in Networks of Locally-Tuned Processing Units. Neural Computation 1, 281–294 (1989)
19. Fu, L., Hsu, H., Principe, J.C.: Incremental Backpropagation Learning Networks. IEEE Trans. on Neural Networks 7(3), 757–761 (1996)
20. Rumelhart, D.E., Hinton, G.E., Williams, R.J.: Learning internal representation by error propagation. In: Parallel Distributed Processing: Explorations in the Microstructures of Cognition, vol. 1, MIT Press, MA (1986)
21. Levenberg, K.: A method for the solution of certain non-linear problems in least squares. Quarterly in Applied Mathematics 2(2), 164–168 (1944)
22. Marquardt, D.W.: An algorithm for the least-squares estimation of nonlinear parameters. SIAM Journal of Applied Mathematics 11(2), 431–441 (1963)
23. Amitava Basu, M.D.: Pathologist, India (personal communication)
24. Hall, E., Frank, G., Holmes, B., Pfahringer, P., Reutemann, I., Witten, H.: The WEKA Data Mining Software: An Update. SIGKDD Explorations 11(1) (2009)
25. Hung, M.S., Shankar, M., Hu, M.Y.: Estimating Breast Cancer Risks Using Neural Networks. J. Operational Research Society 52, 1–10 (2001)
26. Hornik, K., Stinchcombe, M., White, H.: Multilayer feedforward networks are universal approximator. Neural Network 2, 359–366 (1991)
27. Goa, D.: On structures of supervised linear basis function feedforward three-layered neural networks. Chin. J. Comput. 21(1), 80–86 (1998)
28. Huang, M.L., Hung, Y.H., Chen, W.Y.: Neural network classifier with entropy based feature selection on breast cancer diagnosis. J. Med. Syst. 34(5), 865–873 (2010)
29. Alyuda NeuroIntelligence 2.2, http://www.alyuda.com
30. Bradley, A.P.: The use of the area under the ROC curve in the evaluation of machine learning algorithms. Pattern Recognition 30(7), 1145–1159 (1997)

Design and Analysis of Fuzzy Logic and Neural Network Based Transmission Power Control Techniques for Energy Efficient Wireless Sensor Networks

Ramakrishnan Sabitha, Krishna T. Bhuma, and Thangavelu Thyagarajan

Dept. of Instrumentation Engineering, MIT Campus, Anna University, Chennai, India
Sabitha.ramakrishnan@gmail.com, bhumatk2000@yahoo.com,
thyagu_vel@yahoo.co.in

Abstract. In this paper, we present transmission power control algorithms, based on soft computing techniques, for reducing the energy consumption in wireless sensor network, without affecting its throughput. Two algorithms are designed, one using Artificial Neural Network (ANN) and the other using Fuzzy Logic Control (FLC). The algorithms show marked improvement in performance when compared to the conventional Medium Access Control protocol standard IEEE 802.15.4. We also show the effects of optimizing the proposed methods further using Genetic Algorithm

Keywords: Wireless sensor network, Transmission Power Control, Fuzzy Logic Control, Artificial Neural Network, Genetic Algorithm.

1   Introduction

Energy conservation in a Wireless Sensor Network (WSN) can be achieved considerably through Transmission Power Control (TPC) techniques while processing power control techniques can offer only a marginal saving of energy [1]. Another interesting fact is that energy efficient protocols using TPC are more effective in Medium Access Control (MAC) layer than in network layer [2]. The double benefit of TPC in MAC layer is that, on one hand, reduced transmission power can extend the battery life, while on the other hand, increased transmission power can improve link reliability and offer higher bit/baud ratio. However, due to dynamic channel characteristics, the transmission channel cannot be modelled precisely and its characteristics can be determined accurately. Hence, in this paper, we propose two transmission power control technique using Artificial Neural Network (ANN) and Fuzzy Logic Control (FLC) which do not depend on the channel model or characteristics for deciding the required transmission power. The paper is organized as follows. Section 2 gives a brief account of previous research on TPC techniques. The WSN scenario chosen for study is detailed in Section 3. FLC based TPC algorithm using FLC and ANN are discussed in Section 4 and Section 5 respectively. In Section 6, we analyze the performance of the proposed methods and finally in Section 7, we present the conclusions.

© Springer International Publishing Switzerland 2015

S.C. Satapathy et al. (eds.), *Proc. of the 3rd Int. Conf. on Front. of Intell. Comput. (FICTA) 2014*
– *Vol. 1*, Advances in Intelligent Systems and Computing 327, DOI: 10.1007/978-3-319-11933-5_32

2   Related Work

Fuzzy logic control has been utilized in WSN for sampling period decision [4], Cluster head election [5] Congestion detection and control [6]. FLC has also been used for TPC by using the Received Signal Strength Index (RSSI) and Link Quality Index (LQI) as inputs [7] or RSSI and source transmission power as inputs [8]. However, fuzzy logic requires an extensive knowledge base for efficient performance. Several other TPC techniques have also been proposed in literature. Control-theoretic based TPC technique [9] using online parameter estimation and dynamic TPC algorithm using the RSSI and LQI of neighbor nodes [10] are two examples. There is a strong relation between ANN and WSN [11] which promises a major impact on the research in hybrid technology, specifically in WSN. To quote a few, ANN based energy efficient path discovery, nodes clustering, cluster-head selection, data aggregation / fusion, data association, context / data classification and data prediction [12], topological clustering using a self organizing map neural network [13] and clustering / categorizing using unsupervised learning methods [14] leading to robust sensor data. Estimation techniques such as Kalman Filter can be used to predict the RSSI which in turn can be used to decide the required transmission power [15]. However, it requires a system identification tool for formulating the estimation algorithm. ANN has the advantage that it neither requires a knowledge base as in the case of FLC nor a system model as in the case of Kalman filter for its implementation.

3   WSN Specifications Chosen for Study

The WSN nodes are assumed to be mobile with a random way point mobility of 0 to 1m/s. The maximum Transmission power (P_t) is controlled in the range [0.1mW, 1 mW] conforming to IEEE 802.15.4 standards [16]. The transmission channel chosen is a fading channel with log-normal shadowing [17] as defined in Equation 1.

$$\log(P_r) = \log\left[P_t G_t G_r \left(\frac{\lambda}{4\pi}\right)^2 \left(\frac{d_o}{Dist}\right)^3 \right] + X_\sigma \cdot \tag{1}$$

where P_r is the RSSI, λ the wavelength of signal, '*Dist*' the separation between ED and COORD, G_t and G_r the Gain of transmitting and receiving antennas and X_σ the Additive White Gaussian Noise (AWGN) with zero mean and standard deviation σ. Constant Bit Rate (CBR) traffic with 250 Kbps data rate and packet length of 64 bytes are assumed. While the actual receiver threshold of a practical WSN is around -92dBm, we assume a threshold value of -85dBm which provides a 7dBm tolerance in noise floor. The communication between ED and COORD is shown in Fig. 1.

Fig. 1. Communication between End Device and Coordinator in WSN with TPC

The TPC algorithm is implemented in the COORD and it calculates the required transmission power (Ptreq) based on the transmission power used by ED (Ptsrc) to transmit its data packet and the RSSI with which it is received by the COORD. The COORD sends the Ptreq to ED in its acknowledgement so that the ED sends its next data packet at that power level. The next section presents the FLC based TPC algorithm (FLC-TPC) and the GA optimized FLC based TPC algorithm (FLC-GA-TPC) used by the COORD for calculating Ptreq.

4   Fuzzy Logic Based Transmission Power Control in WSN

FLC is a heuristic algorithm applied when mathematical models do not exist or traditional system models become too complex. Since a wireless channel cannot be modeled accurately, FLC is a natural choice for transmission power control. The design parameters chosen for FLC-TPC are shown in Table 1.

Table 1. Design parameters for FLC-TPC

Parameter	Input			Output
	RSSI	**Ptsrc**	**Error**	**Ptreq**
Range (dBm)	[-110,0]	[-10,0]	[-40,30]	[-10,0]
Membership functions	5	10	5	10
linguistic rules		80		
Decision making		Min-Max method		
Defuzzification		Centre of gravity method		

The flow chart for the FLC based TPC algorithm (FLC-TPC) implemented in COORD is shown in Fig. 2.

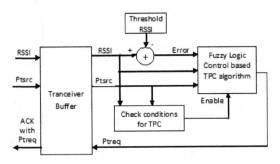

Fig. 2. Flow chart for FLC based TPC algorithm

The transceiver buffer of COORD considers the received RSSI and the Ptsrc of the ED along with an additional 'Error' input calculated as given in Equation 2.

$$Error = RSSI_{Threshold} - RSSI \qquad (2)$$

where $RSSI_{Threshold}$ = -85dBm. Packets received with RSSI below $RSSI_{Threshold}$ will be dropped. There is no scope for TPC under the following conditions:

Condition 1: RSSI < -85dBm (Below threshold) & Ptsrc = 0dBm ($Ptsrc_{max}$)
Condition 2: RSSI >> -85dBm (very high) & Ptsrc = -10dBm ($Ptsrc_{min}$).

Under all other conditions, the TPC algorithm is executed to decide the Ptreq. Table 2 shows the Rule Base matrix for FLC-TPC with 80 linguistic rules. We have considered Mamdani model, according to which the output generated by the inference engine is converted into crisp value of Ptreq using the centre of gravity method.

Table 2. Rule Base Matrix for Ptreq

RSSI/Error Ptsrc	HM / LM	H / L	M / L	M / M	L / M	L / H	LM / H	LM / HM
LM	LM	LM	LM	LM	LM	LM	LM	LM
LM-L	LM-L	LM L	LM-L	LM-L	LM-L	LM-L	LM-L	LM-L
L	L	L	L	L	L	L	L	L
L-M	L-M	L-M	L-M	L-M	L-M	L-M	L-M	L-M
M	M	M	M	M	M	M	M	M
M-H	M-H	M-H	M-H	M-H	M-H	M-H	M-H	M-H
H	H	H	H	H	H	H	H	H
H-HM	H-HM	H-HM	H-HM	H-HM	H-HM	H-HM	H-HM	H-HM
HM	HM	HM	HM	HM	HM	HM	HM	HM
HM1	HM1	HM1	HM1	HM1	HM1	HM1	HM1	HM1

The membership functions of Ptsrc are further optimized using Genetic Algorithm (GA) in order to maximize the energy efficiency. The various Parameters used for the GA optimized FLC based TPC (FLC-GA-TPC) are given in Table 3.

Table 3. Parameters used for FLC-GA-TPC algorithm

Parameter	Value
Population size	10
String length	12
Fitness function	1/(error+1)
Selection	Ranking selection
Crossover	Simple crossover
Probability of crossover	0.99
Mutation	Single value mutation
Probability of mutation	0.01
Coding method	Real coded

The next section presents the design of ANN based TPC (ANN-TPC) which does not require the knowledge base like FLC-TPC.

5   Artificial Neural Network Based TPC in WSN

ANN is an arithmetic algorithm capable of learning complicated mappings between the inputs and outputs according to supervised training or it can classify input data in an unsupervised manner. The block diagram of the ANN-TPC is shown in Fig. 3 (a).

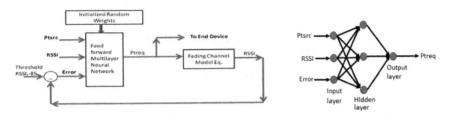

Fig. 3. Block diagram of ANN-TPC (*left*) and ANN Structure (*right*)

Back Propagation learning Algorithm (BPA) is used for training the neural network. BPA is a gradient descent algorithm which works with the difference in error, based on which it modifies the weights of the network. It is simple, allows quick convergence to a local minima and is easy to implement in real time scenarios. The learning rate assigned for the network is 0.5. Sigmoid bipolar activation function is chosen for the output, as given in Equation 3.

$$\text{Output} = \frac{(1-\exp{(-net)})}{(1+\exp{(-net)})} \ . \tag{3}$$

where net=$\sum W_i\, X_i$, X_i being the ith input to the neurons and W_i the weight associated with the connection between neurons. We use a Feedforward Multilayer Neural Network as shown in Fig. 3 (b). The weights for the network are initialized with random values. The number of neurons in the hidden layer is decided by trial and error to achieve minimum Integral of Time Weighted Absolute Error (ITAE). It is observed that convergence is achieved at around 5500 iterations. The value of Ptreq will be iteratively corrected by ANN-TPC algorithm until an RSSI just greater than RSSI_{Threshold} is achieved. At this point, the Ptreq value is communicated to ED and the ANN controller goes to the idle mode. The flowchart of the ANN-TPC algorithm is shown in Fig. 4. The back propagation learning algorithm may get trapped at a local minimum. Since, GA is used to arrive at the global minimum. This optimized ANN-TPC algorithm named as ANN-GA-TPC reduces the number of iterations in the ANN for convergence. It is observed that the number of iterations are reduced to 1420 thus, reducing the initial latency in establishing energy efficiency. The design parameters of ANN-GA-TPC are the same as given in Table 3 except that the population size chosen is 32 and the string length is 16.

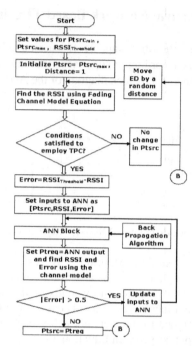

Fig. 4. Flow chart of the ANN-TPC algorithm

6   Performance Analysis of the Proposed TPC Algorithms

The simulation results are studied for both the soft computing methods under three different channel noise conditions, namely: Low (60dBm SNR/sample), Medium (40dBm SNR/sample) and High (20dBm SNR/Sample). The simulation results for the proposed TPC algorithms are given in Fig. 5 for these three channel noise conditions.

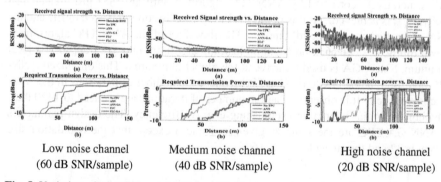

Low noise channel	Medium noise channel	High noise channel
(60 dB SNR/sample)	(40 dB SNR/sample)	(20 dB SNR/sample)

Fig. 5. Variation of (a) RSSI and (b) Ptreq with distance under various channel conditions for the proposed TPC algorithms

When compared to the conventional IEEE 802.15.4 Medium Access Control (MAC) protocol, all the TPC algorithms offer considerable saving in the transmission power while maintaining the RSSI above threshold. It can also be observed that ANN-TPC offers better power efficiency and smoother transition in Ptreq values, than FLC-TPC algorithm. The GA optimization contributes only a meagre improvement in terms of energy efficiency in both ANN as well as FLC. As the noise level in the channel increases, more fluctuations occur in both Ptreq and RSSI values.

Three parameters, namely, throughput, average power used and average energy consumed per successful transmission are considered for comparing the performance of the proposed algorithms. Throughput is calculated by using Equation 4.

$$\text{Throughput} = \frac{\text{Packets delivered}}{\text{Packets transmitted}} \times 100 \ . \tag{4}$$

The average transmission power is calculated using the Equation 5.

$$P_{avg} = \frac{1}{N} \sum_{i=1}^{N} P_i \ . \tag{5}$$

where P_i is the power used for individual transmissions and N is the total number of transmissions. The average energy consumed per successful transmission is given by Equation 6.

$$E_{avg} = \frac{1}{K} \sum_{i=1}^{N} P_i \, x t_d \ . \tag{6}$$

where N is the total number of transmissions, K is the number of successful transmissions, t_d is the duration of transmission. Fig. 6 shows the performance of all the proposed TPC algorithms.

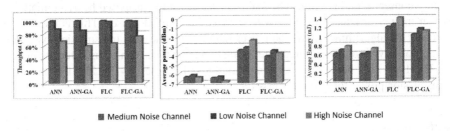

■ Medium Noise Channel   ■ Low Noise Channel   ■ High Noise Channel

Fig. 6. Performance of the proposed TPC algorithms

An important observation is that under high noise channel conditions, FLC-TPC and FLC-GA-TPC offer better throughputs than ANN-TPC and GA-ANN-TPC. However, the average power utilized is much higher. The quantitative results are tabulated in Table 4. It can be noted that ANN-GA-TPC uses the least value of average power. However the difference from ANN-GA-TPC is only a very small percentage compared to ANN-TPC. Both FLC-TPC and FLC-GA-TPC require much larger average transmission power.

Table 4. Throughput obtained by the TPC algorithms

Channel noise	Throughput (%)				Average power (dBm)				Average energy (mJ)			
	ANN	ANN GA	FLC	FLC GA	ANN	ANN GA	FLC	FLC GA	ANN	ANN GA	FLC	FLC GA
Low	100	100	100	100	-6.39	-6.48	-3.5	-4.2	0.60	0.59	1.19	1.03
Medium	86.8	84.4	98.5	99.0	-6.16	-6.35	-3.2	-3.6	0.69	0.63	1.25	1.15
High	67.4	59.4	64.0	75.0	-6.40	-6.80	-2.4	-2.4	0.77	0.72	1.39	1.10

7   Conclusion

In this paper, energy efficiency using soft computing based transmission power control techniques have been discussed, namely Neural Network based transmission power control and Fuzzy Logic based transmission power control.   Both the algorithms are optimized using genetic algorithm for improving the performance. Thus, four different TPC algorithms, namely, FLC-TPC, FLC-GA-TPC, ANN-TPC and ANN-GA-TPC have been designed. For a chosen WSN scenario, the performance of the proposed algorithms are analyzed with respect to the network parameters, namely, throughput, average power and average energy per successful transmission. It is observed that ANN-TPC outperforms FLC-TPC excellently in terms of energy saving.   Optimization with GA helps the ANN-TPC in bringing down the number of iterations for convergence.   It does not contribute much to improve the energy efficiency.   In a highly noisy channel, the performance of FLC-TPC is found to offer a marginally better throughput than ANN-TPC, however, at the cost of high transmission power. Thus, ANN-TPC provides an efficient and optimal solution for energy conservation in WSN. The major limitation of the proposed heuristic methods is that the performance degrades with increase in channel noise. This is due to the fact that the algorithms decide the power required for the next transmission based on current observations. Therefore, state estimation techniques can be used for prediction of RSSI of the next transmission accurately in order to decide the required transmission power for that transmission.

References

1. Karl, H., Willig, A.: Protocols and Architectures for Wireless Sensor Networks, 1st edn. Wiley, Europe (2007)
2. Pottie, G., Kaiser, W.: Wireless Integrated Network Sensors. Communications of the ACM 43(5), 51–58 (2000)
3. Correia, H., Macedo, F., Santos, L., Loureiro, A., Nogueira, S.: Transmission Power Control Techniques for Wireless Sensor Networks. Computer Networks Journal 51, 4765–4779 (2007)
4. Xia, F., Zhao, W., Sun, Y., Tian, Y.: Fuzzy Logic Control Based QoS Management in Wireless Sensor/Actuator Networks. Sensors 2007 7(12), 3179–3191 (2007)

5. Gupta, I., Riordan, D., Sampalli, S.: Cluster-head election using fuzzy logic for wireless sensor networks. In: Proceedings of the 3rd Annual IEEE conference on Communication Networks and Services Research, pp. 255–260 (2005)
6. Wei, J., Fan, B., Sun, Y.: A congestion control scheme based on fuzzy logic for wireless sensor networks. In: Proceedings of the 9th International Conference on Fuzzy Systems and Knowledge Discovery (FSKD), pp. 501–504 (2012)
7. Jin, S., Fu, J., Xu, L.: The transmission power control method for wireless sensor networks based on LQI and RSSI. In: Xiao, T., Zhang, L., Fei, M. (eds.) AsiaSim 2012, Part II. CCIS, vol. 324, pp. 37–44. Springer, Heidelberg (2012)
8. Sabitha, R., Thyagarajan, T.: Fuzzy logic-based transmission power control algorithm for energy efficient MAC protocol in wireless sensor networks. International Journal of Communication Networks and Distributed Systems (IJCNDS) 9(3/4), 247–265 (2012)
9. Fu, Y., Sha, M., Hackmann, G., Lu, C.: Practical Control of Transmission Power for Wireless Sensor Networks. In: IEEE International Conference on Network Protocols – ICNP 2012 (2012)
10. Ping, J., Kun, S., Hsieh, Y., Cheng, Y.: Distributed Trasmission Power Control Algorithm for Wireless Sensor Networks. Journal of Information Science and Engineering 25, 1447–1463 (2009)
11. Oldewurtel, F., Mahonen, P.: Neural Wireless Sensor Networks. In: Procs. of the Intl Conference on Systems and Networks Communication, ICSNC 2006 (2006)
12. Enami, N., Moghadam, R., Dadashtabar, K., Hoseini, M.: Neural Network Based Energy Efficiency In Wireless Sensor Networks A Survey. International Journal of Computer Science & Engineering Survey (IJCSES) 1(1), 39–55 (2010)
13. Azimi, M., Ramezanpor, M.: A Robust Algorithm For Management of Energy Consumption In Wireless Sensor Networks Using Som Neural Networks. Journal of Academic and Applied Studies 2(3), 1–14 (2012)
14. Kulakov, A., Davcev, D., Trajkovski, G.: Implementing Artificial Neural-Networks in Wireless Sensor Networks. In: Proceedings of the IEEE Symposium on Advances in Wired and Wireless Communication, pp. 94–97 (2005)
15. Masood, M., Khan, A.: A Kalman filter based adaptive on demand transmission power control (AODTPC) algorithm for wireless sensor networks. In: Proceedings of the International Conference on Emerging Technologies (ICET), pp. 1–6 (2012)
16. IEEE 802.15.4. IEEE Standard for Local and Metropolitan Area Network - Low Rate Wireless Personal Area Networks (LRWPAN). IEEE Standards Association (2011)
17. Rappaport, T.: Wireless Communication Principles And Practices, 2nd edn. Pearson Education Inc., India (2002)

Conceptual Multidimensional Modeling
for Data Warehouses: A Survey

Anjana Gosain and Jaspreeti Singh

University School of Information and Communication Technology,
Guru Gobind Singh Indraprastha University, New Delhi - 110078, India
{anjana_gosain,jaspreeti_singh}@yahoo.com

Abstract. Conceptual multidimensional modeling aims at providing high level of abstraction to describe the data warehouse process and architecture, independent of implementation issues. It is widely accepted as one of the major parts of overall data warehouse development process. In the last several years, there has been a lot of work devoted to conceptual multidimensional modeling for data warehouses. This paper presents a survey of various proposed conceptual multidimensional models for core as well as advanced features supported. Hereafter, a comparison of the models is done based on the criteria broadly categorized as: general aspects, relationship aspects and implementation aspects, showing the evolution in this area. General aspects involve basic features of the multidimensional model. The relationships among various constructs used in the multidimensional model are referred under relationship criteria. The implementation criteria relate aspects such as mathematical formalism and guidelines to define complex multidimensional structures.

Keywords: Conceptual Modeling, Data Warehouse, Design, Multidimensional.

1 Introduction

Multidimensionality in data warehousing is a design technique which represents data as if placed in multidimensional space. It specifies two essential notions: the fact and the dimension [1], [2], [3]. The benefits of using multidimensional modeling are: (i) It is close to the way data analysts think and, therefore, helps them understand data better and (ii) It supports performance improvements as it allows system designers to predict end users' intentions [4]. To reflect the real-world situations, special attention should be given to define relationships between facts and dimensions as well as between various levels of aggregation in a dimension hierarchy [5]. Also, consideration to summarizability issues needs to be ensured [6], to avoid incorrect results while computing aggregate values in data analysis tools such as OLAP.

The importance of conceptual modeling in data warehousing is widely recognized [4], [7], [8]. It aims at providing a high level of abstraction to describe the data warehouse process and architecture, independent of implementation issues. For this reason, a variety of approaches for conceptual modeling of multidimensional systems

© Springer International Publishing Switzerland 2015

305

S.C. Satapathy et al. (eds.), *Proc. of the 3rd Int. Conf. on Front. of Intell. Comput. (FICTA) 2014*
– *Vol. 1*, Advances in Intelligent Systems and Computing 327, DOI: 10.1007/978-3-319-11933-5_33

have been proposed (*e.g.* [9], [10], [11], [12], [13], [14]) to represent main structural and dynamic properties. However, it should be noted that no consensus on formalism for conceptual multidimensional modeling is yet established [4], [15], [16].

In this paper we focus on conceptual models for OLAP applications. A large number of models exist in the field of conceptual multidimensional modeling in data warehousing. We have categorized the features of these models into the following three aspects: general, relationship and implementation. A number of surveys about multidimensional modeling have been carried out in the literature. For instance, [17] provides a summary about using web data in DW, [16] consists of a survey about methodologies used in multidimensional modeling, [6] provides a survey about summarizability issues in multidimensional modeling, [18] is a survey which cope only with hierarchies in DW, and [19] provides classification framework for various multidimensional models proposed till early 2000's. However, none of the surveys has incorporated the entire core as well as advanced features related to multidimensional modeling. We try to attempt a survey which spans most of the related features of multidimensional modeling, with special consideration to the aspects related to dimensions, their associated hierarchies and the relationships among the constructs provided in the conceptual model. Apart from most of the models already taken under consideration, we take into account those models that were not previously considered by any of these surveys.

The remainder of this paper is structured as follows: section 2 briefly describes each conceptual multidimensional model and its constructs. Section 3 presents the criteria used to compare the models surveyed. A tabular comparison is presented on the basis of above mentioned criteria, showing the evolution in this area. Finally, section 4 conclude the paper.

2   State-of-the-Art

We discuss below various approaches to conceptual multidimensional modeling proposed in the last several years. To present these approaches, the chronological order by year is followed to show the evolution of this area along the way. We have included an alias for each approach for an easier identification.

Notably, providing graphical support is a desirable characteristic of conceptual models so that it can be easily understood by system designers as well as end-users [20]. So, the approaches to multidimensional modeling that have no graphical support and are aimed at only formalism of cubes, hierarchies, etc. ([9], [10], [13], [21], [22], [23], [24]) are not included in this survey.

2.1   Go98: Golfarelli M., Maio D. and Rizzi S. (1998)

Golfarelli *et.al* in [11], [25], [26], [27] proposed *Dimensional Fact Model* (*DFM*) which emphasize on the distinction between dimensions and measures at the conceptual level. A *dimensional scheme* comprises a set of *fact schemes* which further include facts, measures, dimensions and hierarchies. A directed tree, linked by -to-one relationship between *dimension attributes*, represents a *hierarchy*. The terms *compatible* and *strict compatible* are introduced to define relation between the overlapping *fact schemes* sharing *dimension attributes*.

2.2   Tr98: Trujillo J. and Palomar M. (1998)

Ref [28] introduces Object Oriented multidimensional data model named *OOMD* which was further renamed as *GOLD* [12]. This approach [12], [29], [30] introduces some constraints and extends graphical elements of UML. A dimensional schema consists of dimension classes (DC), fact classes (FC), cube classes (CC) and views. A *classification hierarchy* is defined to distinguish between *Attribute Roll-up Relation Paths* (ARRP) and *Attribute Classification Paths* (ACP). The authors present the behavioral patterns [31] and a CASE tool [32] with reference to the model.

2.3   Sa98: Sapia C., Blaschka M., Hofling G., and Dinter B. (1998)

In this paper, the authors [33] present *Multidimensional Entity Relationship (ME/R)* model, a specialization of the E/R one. The key considerations are specialization, minimal extension and representation of multidimensional semantics. The authors introduced an entity set *i.e. dimension level*, and two relationship sets connecting it *i.e. fact* and *rolls-up to*.

2.4   Tr99: Tryfona N., Busborg F., and Christiansen J.G.B. (1999)

Tryfona *et.al* [34] addresses the set of requirements from users' point of view and proposes model named *starER* combining E/R model constructs with the star schema. The requirements that need to be handled are: to represent *facts* and their *properties*; connect the temporal dimensions to facts; represent *objects* and capture their *properties* and *associations*; record *associations* between *objects* and *facts*; and distinguish *dimensions* and categorize them into *hierarchies*. Based on these requirements, various constructs of the model are: *fact set, entity set, relationship set* and *attribute*.

2.5   Fr99: Franconi E. and Sattler U. (1999)

The authors [35] introduce an extended E/R formalism to define *Data Warehouse Conceptual Data Model (CDWDM)* that allows representing the structure of *multidimensional aggregations*. The semantics of the multidimensional data model introduced in [10] are extended to allow description of the components of *aggregations*, besides the relationships between the properties of the *components* and that of the *aggregation* itself.

2.6   Hu0: Husemann B., Lechtenborger J., and Vossen G. (2000)

The authors in [14] proposes a phase-oriented design process to derive DW schema from the operational one. In the context of conceptual design, a phase model is provided to derive a graphical multidimensional diagram comprising *fact schemata* with their related *measures* and *dimension lattices*. The extension of this work [36] is a logical design phase presenting transformation process to produce view definitions from fact schemata.

2.7   Ts1: Tsois A., Karayannidis N., Sellis T. (2001)

The authors [37] propose a user-centric model named *Multidimensional Aggregation Cube (MAC)*. It describes multidimensional constructs as *dimension levels, drilling relationships, dimension paths, dimensions, cubes* and *attributes*. The classes of *dimension members* are represented as *dimension levels*. A *dimension path* is represented as a set of *drilling relationships* and a *dimension* is formed by defining meaningful groups of *dimension paths* which may share common levels. Finally, a *multidimensional aggregation cube* is an n-way relationship among dimension domains with its associated attributes.

2.8   Lu2: Luján-Mora S., Trujillo J. (2002)

The paper [38] presents an O-O approach which is based on the previously proposed model [12] and extends UML. [39] specify how to apply packages provided by UML, not restricting only to flat diagrams. The *fact classes* and *dimension classes* are defined to specify structural properties. The *fact classes* are composite classes in shared aggregation relationships with its associated *dimension classes*. The multidimensional modeling guidelines for using UML packages are provided. This approach is capable of providing support for major relevant multidimensional features such as *degenerate dimensions*, non-strict hierarchies etc.

2.9   Ab6: Abello A., Samos J. and Saltor F. (2006)

Ref [40] also presents a model, namely, YAM^2, following UML extension mechanism. A preliminary version of this work is presented in [41]. One of the remarkable features of YAM^2 is the way it represents several semantically related star schemas. The structure in the model consists of *nodes* and its associated *arcs*. Six kinds of nodes (*i.e., Fact, Dimension, Cell, Level, Measure*, and *Descriptor*) are defined. A *Fact* and its associated *Dimensions* compose a *Star*. In turn, the *arcs* represent different kinds of relationships between various elements at different *levels*. In line with the necessary conditions for summarizability [42], it supports various kinds of hierarchies. Ref [43] provides the implementation of the proposed operations (*Drill-across, Projection, ChangeBase, Roll-up* and *Dice*) on a relational DBMS. Furthermore, based on the list of requirements presented in [21], [44] comparison of this model with other contributions in this area is provided.

2.10   Lu6: Lujan-Mora S., Trujillo J. and Song Y. (2006)

Ref [45] includes extension to the previously presented approach [38], enriching it with new properties to obtain an improved proposal. The concept of dimension is re-defined to make it more understandable and readable by end users. The new properties considered in this paper, apart from those mentioned in [38], include specifying roles on associations in classification hierarchies, degenerate facts and same dimension in different roles related to one fact. Moreover, it explicitly defines the navigability of an association in a classification hierarchy to define roll-up or drill-down path.

2.11 Pr6: Prat Nicolas., Akoka J., and Wattiau I.C. (2006)

In [46], the authors present a UML-based methodology to derive the conceptual, logical and physical schema of the DWs. At each phase, a metamodel and a set of transformations to perform the mapping between metamodels is provided. Referring the conceptual design, *dimension levels* are grouped into *hierarchies* to define *dimensions*. Different *dimension levels* are linked by parent-child relationship. Besides, an *isTime attribute* distinguishes temporal dimension level from non-temporal one. Furthermore, *facts* are composed of *measures* and *dimensioning* refers to the relationship between a *fact* and each of its dimension *levels*.

2.12 Ma6: Malonowski E. and Zimanyi E. (2006)

Besides the classification of different kinds of hierarchies in [47], the authors [48] also present a *MultiDimER* model which is an extension of E/R model. A *MultiDimER* schema comprises a finite set of *dimensions* and *fact relationships*. Each *dimension* is either a *level* or one or more *hierarchies*. A *hierarchy* consists of several related *levels*, the last *level* being a *leaf*. Further, the relationships among these *levels* in a *hierarchy* are characterized by *cardinality* and the *analysis criterion*, where the latter expresses different structures used in analysis process. Finally, a *fact relationship* describes an n-ary relationship among *leafs*. Furthermore, a whole section is devoted to refinement of general mapping rules by incorporating relational mapping for different hierarchies.

2.13 Ri7: Rizzi, S. (2007)

In [20], Rizzi extends *DFM* [11], that was first proposed by Golfarelli *et. al*, to provide support for some of the advanced modeling features such as descriptive and cross-dimension attributes; shared and incomplete hierarchies; multiple and optional arcs; and additivity. Besides, the author proposes a comprehensive set of solutions for conceptual modeling based on *DFM*.

2.14 Ma7: Mansmann S. (2007)

Ref [49] aim to integrate the requirements proposed by several authors for multidimensional modeling and based on that extend *DFM* [11] to propose a semantically rich graphical notation *X-DFM (Extended Dimensional Fact Model)* capable of handling complex multidimensional data. It is based on their previous work [50], which provides a framework for classifying and modeling complex data. With respect to the graphical notations, the author highlights a number of deficiencies in DFM, and accordingly provides graphical constructs to resolve the issues such as modeling of non-dimensional attributes, different types of relationships, etc.

2.15 Ka8: Kamble A. (2008)

The paper [51] presents formal aspects of the conceptual data model first introduced in [52]. The model, namely *CGMD*, combines the semantics of E/R model with logic-based formalism proposed in [53], and extends the idea of conceptual data model first proposed in [35]. One of the most interesting features of this model is computation of aggregations from pre-computed aggregations. The two main extensions to E/R schema are *simple aggregated entities* and *multidimensional aggregated entities*. An n-dimensional aggregation, represented by an aggregated weak entity, consists of n dimensions and their associated levels along with a fact involved in aggregation.

2.16   Sa9: Sarkar A., Choudhury S., Chaki N. and Bhattacharya S. (2009)

Ref [54] presents a graph semantic based O-O multidimensional model called *GOOMD* which extends the idea of Graph Data Model [55]. A preliminary version of this work is presented in [56]. The multidimensional schema is viewed as a graph in a layered manner. The basic components of the model are: a set of distinct *attributes;* a set of *measures*; *elementary semantic group;* and *contextual semantic group (CSG).* The *dimensional semantic group and fact semantic group* are specialized forms of CSG.

3   Comparison Criteria

This section describes the criteria to compare multidimensional models reviewed in the previous section, for core as well as advanced features supported. We group these criteria as follows: general aspects, relationship aspects, and implementation aspects. General aspects involve basic features such as approach used in modeling (E/R, UML, etc.), derivability of facts and dimensions, sharing of dimensions, sharing of levels in a hierarchy etc. The relationship among various constructs used in multidimensional modeling is referred under relationship criteria. The last category *i.e.* implementation relates aspects such as mathematical formalism, mapping from conceptual to logical level, guidelines to define complex multidimensional structures etc. These criteria have been devised in an iterative fashion by analyzing the modeling approaches surveyed and also integrating the requirements/properties proposed in [21], [40], [44]. Thereafter, the features captured in each modeling approach are mapped onto above mentioned criteria.

3.1   General Aspects

- Technique: This column shows the association of the model to some method for conceptual design and classifies the models based on E/R, O-O (particularly, UML), or an ad hoc model.
- Derivability: The schema should include ways to define concepts based on other concepts. This includes derived measures [44] as well as derived dimension attributes [45].
- Additivity: The model should show how to obtain a concept at coarser granularity and which aggregation functions can be applied to a measure [21], [44].

- Non-hierarchical dimensions: Sometimes it is useful to provide other characteristics of the analysis dimensions that do not define hierarchy. Such dimensions can be used to delimit the query results [44].
- Cross-dimensional attributes: Either dimension or descriptive attribute can be represented as a cross-dimension attribute that is determined by the combination of two or more attributes, possibly belonging to different hierarchies [20].
- Degenerate dimensions: The data model should allow defining a dimension that has no content except for its primary key. Such dimensions are useful when seeking association between facts in DW and the original data sources [3], [57].
- Sharing dimensions: The model should support sharing of dimensions as it allows analyzing measures present in different facts [40], [45], [48].
- Sharing few hierarchy levels: To ensure correct aggregation, the distinction between specific and shared hierarchy levels should be clearly represented in the data model [58].
- Multiple alternative hierarchies: To model many real-world situations, these hierarchies are needed as dimensions can contain more than one natural hierarchy [3].
- Parallel hierarchies: When considering different analysis criteria for a dimension's associated hierarchies, parallel hierarchies should be supported [48], [49].
- Different roles of a dimension with the same fact: The data model should allow the same dimension in different roles with the same fact, thereby facilitating the design process such as automatic generation of the implementation or providing users with different views of cubes [45].

3.2   Relationship Aspects

- Fact-dimension relationship: The grain of fact is determined by the relationship between fact and dimension, which may not be the "classical" one and include incomplete association and non-strictness [5], [59].
- Intra-dimensional relationship: The data model should allow featuring real-world situations using relationships between levels of aggregation in a dimension hierarchy [21], [60]. Often, these hierarchies may be non-strict (*i.e.*, can have many-to-many relationships between the different levels in a dimension), non-onto (not balanced) or non-covering (*i.e.* "skip" one or more levels). The generalization/specialization relationship between the levels can be modeled via generalized hierarchies, which include a special case commonly referred to as non-covering hierarchies. The existence of such hierarchies may affect summarizability, which needs to be ensured in the data model [6].
- Inter-dimensional relationship: The data model should exhibit semantic relationship between dimensions, possibly as associations and generalizations [40].
- Fact constellation: It should be possible to explicitly represent more than one fact in the same multidimensional data model [40], [45].

Table 1. Comparison of conceptual multidimensional models[1]

Comparison Criteria		Go9	Tr98	Sa98	Tr99	Fr99	Hu0	Ts1	Lu2	Ab6	Lu6	Pr6	Ma6	Ri7	Ma7	Ka8	Sa9
General Aspects	Technique	Ad.	UML	E/R	E/R	E/R	Ad.	Ad.	UML	UML	UML	UML	E/R	Ad.	Ad.	E/R	Ad.
	Atomic meas.	✓	✓	✓	✓	✓	✓	✓	✓	✓	✓	✓	✓	✓	✓	✓	✓
	Deri. meas.	-	✓	-	-	-	-	-	✓	✓	✓	-	-	-	✓	✓	-
	Deri. dim. attributes	-	-	-	✓	-	-	-	-	-	-	-	-	-	-	-	-
	Additivity	✓	✓	-	-	-	-	-	✓	✓	✓	✓	✓	✓	✓	✓	✓
	Cross-dim. attributes	-	-	-	-	-	-	-	-	-	-	-	-	✓	✓	-	-
	Degenerate facts	-	-	-	-	-	-	-	-	-	✓	-	-	-	✓	-	-
	Degenerate dim.	-	✓	✓	-	-	-	-	-	-	-	-	-	-	✓	-	✓
	Sharing dim.	-	✓	-	-	-	-	✓	✓	✓	✓	✓	✓	✓	✓	✓	-
	Sharing levels	-	✓	✓	-	-	-	-	✓	✓	✓	-	✓	-	✓	✓	✓
	Parallel hier.	-	-	-	-	-	-	-	-	-	-	-	-	-	✓	-	-
	Diff. roles of dim.	-	-	-	-	-	-	-	-	-	-	-	-	-	✓	-	-
Relationship Aspects — F dr	Incompleteness	-	-	-	-	-	✓	✓	✓	✓	✓	-	✓	✓	✓	-	✓
Intra dim.	Non-strictness	-	✓	-	✓	-	✓	✓	✓	✓	✓	-	✓	-	✓	✓	-
	Non-onto	-	-	-	-	-	-	-	-	-	-	-	-	-	-	✓	-
	Non-covering	-	✓	-	✓	-	-	-	✓	✓	✓	-	✓	-	✓	-	-
	Non-strictness	-	✓	-	✓	-	✓	-	✓	✓	-	-	✓	-	-	✓	✓
Inter dim.	Generalization	-	-	-	-	-	-	-	✓	✓	-	✓	-	-	✓	✓	-
	Generalization	-	-	-	-	-	-	-	✓	✓	-	-	-	-	-	✓	-
	Association	-	-	-	-	-	-	-	-	✓	-	-	-	-	✓	-	-
	Fact constellation	-	✓	-	-	✓	-	-	✓	✓	✓	✓	✓	-	✓	✓	✓
	Math. constructs	-	✓	✓	-	-	-	✓	✓	-	✓	-	✓	-	✓	✓	-
	Trans. of hier.	-	-	-	-	-	-	-	-	-	✓	-	-	-	-	✓	-
Impl.	Guidelines	✓	-	-	✓	-	✓	-	✓	-	✓	✓	✓	✓	✓	-	-

[1] Tick: supported in the model, Hyphen: not supported or not explained how to support it, meas.: measure, deri.: derived, dim.: dimension, hier.: hierarchy, trans.: transformation, Fdr.: fact-dimension relationship, Impl.: Implementation

3.3   Implementation Aspects

• Mathematical construct used for the operations: This column shows whether there is some mathematical formalism [44] (like, query language, an algebra or calculus) associated to the model to define the operations over data.

• Transformation of hierarchies from conceptual to logical level: Providing the mapping from conceptual to logical level on the basis of well-known rules facilitates in the design process and further in the implementation of the data model.

• Guidelines/ Methodology: Apart from specifying graphical notation, providing guidance or a structured methodology on how to define complex multidimensional structures plays an important role in the successful implementation of the designed data model [6].

Based on the criteria mentioned above, we present a detailed comparison of each of the conceptual multidimensional models reviewed in this paper (see Table 1). There, rows correspond to criteria introduced above and columns correspond to each model studied. Cells contain information about the model as to whether it supports given criterion or not.

4   Conclusion

This paper discuss various conceptual multidimensional data models proposed in the last several years. We have introduced a set of criteria to facilitate the comparison of these models for core as well as advanced features supported. The criteria have been devised in an iterative fashion by analyzing the modeling approaches surveyed and also integrating the requirements/properties proposed by various authors. These criteria are grouped as: general aspects, relationship aspects, and implementation aspects. A comparison of features captured in each modeling approach is carried out by mapping them onto different criteria. Looking at the comparison one can realize that more recent the models are, they capture semantics in a better way. This can be inferred as a trend to semantically enrich conceptual multidimensional models.

References

1. Inmon, W.: Building the Data Warehouse, 4th edn. John Wiley & Sons, Inc., New York (2005)
2. Chaudhuri, S., Dayal, U.: An overview of data warehousing and OLAP technology. ACM Sigmod Record 26(1), 65–74 (1997)
3. Kimball, R.: The data warehouse toolkit. John Wiley & Sons (2006)
4. Rizzi, S., Abelló, A., Lechtenbörger, J., Trujillo, J.: Research in data warehouse modeling and design: dead or alive? In: Proceedings of the 9th ACM International Workshop on Data Warehousing and OLAP, pp. 3–10. ACM (2006)
5. Mazón, J.N., Lechtenbörger, J., Trujillo, J.: Solving summarizability problems in fact-dimension relationships for multidimensional models. In: Proceedings of the ACM 11th International Workshop on Data Warehousing and OLAP, pp. 57–64. ACM (2008)

6. Mazón, J.N., Lechtenbörger, J., Trujillo, J.: A survey on summarizability issues in multidimensional modeling. Data & Knowledge Engineering 68(12), 1452–1469 (2009)
7. Jarke, M., Lenzerini, M., Vassiliou, Y., Vassiliadis, P.: Fundamentals of Data Warehouses, 2nd edn. Springer (2003)
8. Herreman, D., Schau, D., Bell, R., Kim, E., Valencic, A.: Data modeling techniques for data warehousing. IBM (1998)
9. Lehner, W.: Modeling large scale OLAP scenarios. In: Schek, H.-J., Saltor, F., Ramos, I., Alonso, G. (eds.) EDBT 1998. LNCS, vol. 1377, pp. 153–167. Springer, Heidelberg (1998)
10. Cabibbo, L., Torlone, R.: A logical approach to multidimensional databases. In: Schek, H.-J., Saltor, F., Ramos, I., Alonso, G. (eds.) EDBT 1998. LNCS, vol. 1377, pp. 183–197. Springer, Heidelberg (1998)
11. Golfarelli, M., Maio, D., Rizzi, S.: The dimensional fact model: a conceptual model for data warehouses. International Journal of Cooperative Information Systems 7(02-03), 215–247 (1998)
12. Trujillo, J.: The GOLD model: An Object Oriented multidimensional data model for multidimensional databases. In: Proc. of the ECOOP Workshop for Ph.D. Students in Object Oriented Systems, Lisbon, Portugal (1999)
13. Nguyen, T.B., Tjoa, A.M., Wagner, R.R.: An object oriented multidimensional data model for OLAP. In: Lu, H., Zhou, A. (eds.) WAIM 2000. LNCS, vol. 1846, pp. 69–82. Springer, Heidelberg (2000)
14. Hüsemann, B., Lechtenbörger, J., Vossen, G.: Conceptual data warehouse design. In: Proc. of 2nd Int. Workshop on Design and Management of Data Warehouses (DMDW), Stockholm, Sweden, pp. 6–1 (2000)
15. Sen, A., Sinha, A.P.: A comparison of data warehousing methodologies. Communications of the ACM 48(3), 79–84 (2005)
16. Romero, O., Abelló, A.: A survey of multidimensional modeling methodologies. International Journal of Data Warehousing and Mining (IJDWM) 5(2), 1–23 (2009)
17. Pérez, J.M., Berlanga, R., Aramburu, M.J., Pedersen, T.B.: Integrating data warehouses with web data: A survey. IEEE Transactions Knowledge and Data Engineering 20(7), 940–955 (2008)
18. Talwar, K., Gosain, A.: Hierarchy Classification for Data Warehouse: A Survey. Procedia Technology 6, 460–468 (2012)
19. Abelló, A., Samos, J., Saltor, F.: A data warehouse multidimensional data models classification. In: Proceedings of the International Conference on Database and Expert Systems Applications, Munich, pp. 668–677 (2000)
20. Rizzi, S.: Conceptual modeling solutions for the data warehouse. Data Warehouses and OLAP: Concepts. Architectures and Solutions, 1–26 (2007)
21. Pedersen, T.B., Jensen, C.S., Dyreson, C.E.: A foundation for capturing and querying complex multidimensional data. Information Systems 26(5), 383–423 (2001)
22. Jensen, C.S., Kligys, A., Pedersen, T.B., Timko, I.: Multidimensional data modeling for location-based services. The VLDB Journal-The International Journal on Very Large Data Bases 13(1), 1–21 (2004)
23. Sánchez, A.: Multidimensional conceptual model IDEA. Ph. D. Thesis, in spanish (2001)
24. Konovalov, A.: Object-oriented data model for data warehouse. In: Manolopoulos, Y., Návrat, P. (eds.) ADBIS 2002. LNCS, vol. 2435, pp. 319–325. Springer, Heidelberg (2002)
25. Golfarelli, M., Maio, D., Rizzi, S.: Conceptual design of data warehouses from E/R schemes. In: Proceedings of the Thirty-First Hawaii International Conference on System Sciences, vol. 7, pp. 334–343. IEEE (1998)

26. Golfarelli, M., Rizzi, S.: A methodological framework for data warehouse design. In: Proceedings of the 1st ACM International Workshop on Data Warehousing and OLAP, pp. 3–9. ACM (1998)

27. Golfarelli, M., Rizzi, S.: Designing the data warehouse: Key steps and crucial issues. Journal of Computer Science and Information Management 2(3), 88–100 (1999)

28. Trujillo, J., Palomar, M.: An object oriented approach to multidimensional database conceptual modeling (OOMD). In: Proceedings of the 1st ACM international workshop on Data warehousing and OLAP, pp. 16–21. ACM (1998)

29. Trujillo, J., Gómez, J., Palomar, M.S.: Modeling the behavior of OLAP applications using an UML compliant approach. In: Yakhno, T. (ed.) ADVIS 2000. LNCS, vol. 1909, pp. 14–23. Springer, Heidelberg (2000)

30. Trujillo, J., Palomar, M.S., Gómez, J.: Applying object-oriented conceptual modeling techniques to the design of multidimensional databases and OLAP applications. In: Lu, H., Zhou, A. (eds.) WAIM 2000. LNCS, vol. 1846, pp. 83–94. Springer, Heidelberg (2000)

31. Trujillo, J., Palomar, M., Gómez, J.: Detecting patterns and OLAP operations in the GOLD model. In: Proceedings of the 2nd ACM International Workshop on Data Warehousing and OLAP, pp. 48–53. ACM (1999)

32. Trujillo, J., Palomar, M., Gomez, J., Song, I.Y.: Designing data warehouses with OO conceptual models. IEEE Computer 34(12), 66–75 (2001)

33. Sapia, C., Blaschka, M., Höfling, G., Dinter, B.: Extending the E/R model for the multidimensional paradigm. In: Kambayashi, Y., Lee, D.-L., Lim, E.-p., Mohania, M., Masunaga, Y. (eds.) ER Workshops 1998. LNCS, vol. 1552, pp. 105–116. Springer, Heidelberg (1999)

34. Tryfona, N., Busborg, F., Borch Christiansen, J.G.: starER: a conceptual model for data warehouse design. In: Proceedings of the 2nd ACM International Workshop on Data Warehousing and OLAP, pp. 3–8. ACM (1999)

35. Franconi, E., Sattler, U.: A data warehouse conceptual data model for multidimensional aggregation: a preliminary report. In: Proc. of the Workshop on Design and Management of Data Warehouses (1999)

36. Lechtenbörger, J.: Data warehouse schema design, vol. 79. IOS Press (2001)

37. Tsois, A., Karayannidis, N., Sellis, T.K.: MAC: Conceptual data modeling for OLAP. In: Proceedings of the International Workshop on Design and Management of Data Warehouses, vol. 1, p. 5 (2001)

38. Luján-Mora, S., Trujillo, J., Song, I.Y.: Extending the UML for multidimensional modeling. In: Jézéquel, J.-M., Hussmann, H., Cook, S. (eds.) UML 2002. LNCS, vol. 2460, pp. 290–304. Springer, Heidelberg (2002)

39. Luján-Mora, S., Trujillo, J., Song, I.Y.: Multidimensional modeling with UML package diagrams. In: Spaccapietra, S., March, S.T., Kambayashi, Y. (eds.) ER 2002. LNCS, vol. 2503, pp. 199–213. Springer, Heidelberg (2002)

40. Abelló, A., Samos, J., Saltor, F.: YAM2: a multidimensional conceptual model extending UML. Information Systems 31(6), 541–567 (2006)

41. Abelló, A., Samos, J., Saltor, F.: YAM2 (Yet Another Multidimensional Model): An Extension of UML. In: Database Engineering and Applications Symposium, International, pp. 172–172. IEEE Computer Society (2002)

42. Lenz, H.J., Shoshani, A.: Summarizability in OLAP and statistical data bases. In: Proceedings of Ninth International Conference on Scientific and Statistical Database Management, pp. 132–143. IEEE (1997)

43. Abelló, A., Samos, J., Saltor, F.: Implementing operations to navigate semantic star schemas. In: Proceedings of the 6th ACM International Workshop on Data Warehousing and OLAP, pp. 56–62. ACM (2003)
44. Blaschka, M., Sapia, C., Hofling, G., Dinter, B.: Finding your way through multidimensional data models. In: Proceedings of Ninth International Workshop on Database and Expert Systems Applications, pp. 198–203. IEEE Computer Society (1998)
45. Luján-Mora, S., Trujillo, J., Song, I.Y.: A UML profile for multidimensional modeling in data warehouses. Data & Knowledge Engineering 59(3), 725–769 (2006)
46. Prat, N., Akoka, J., Comyn-Wattiau, I.: A UML-based data warehouse design method. Decision Support Systems 42(3), 1449–1473 (2006)
47. Malinowski, E., Zimányi, E.: OLAP hierarchies: A conceptual perspective. In: Persson, A., Stirna, J. (eds.) CAiSE 2004. LNCS, vol. 3084, pp. 477–491. Springer, Heidelberg (2004)
48. Malinowski, E., Zimányi, E.: Hierarchies in a multidimensional model: From conceptual modeling to logical representation. Data & Knowledge Engineering 59(2), 348–377 (2006)
49. Mansmann, S., Scholl, M.H.: Extending the Multidimensional Data Model to Handle Complex Data. JCSE 1(2), 125–160 (2007)
50. Mansmann, S., Scholl, M.H.: Empowering the OLAP technology to support complex dimension hierarchies. International Journal of Data Warehousing and Mining (IJDWM) 3(4), 31–50 (2007)
51. Kamble, A.S.: A conceptual model for multidimensional data. In: Proceedings of the Fifth Asia-Pacific Conference on Conceptual Modelling, vol. 79, pp. 29–38. Australian Computer Society, Inc. (2008)
52. Franconi, E., Kamble, A.: A Data Warehouse Conceptual Data Model. In: Proceedings of the International Conference on SSDBM, vol. 4, pp. 435–436 (2004)
53. Franconi, E., Kamble, A.: The GMD Data Model and Algebra for Multidimensional Information. In: Persson, A., Stirna, J. (eds.) CAiSE 2004. LNCS, vol. 3084, pp. 446–462. Springer, Heidelberg (2004)
54. Sarkar, A., Choudhury, S., Chaki, N., Bhattacharya, S.: Conceptual Level Design of Object Oriented Data Warehouse: Graph Semantic Based Model. International Journal of Computer Science (INFOCOMP) 8(4), 60–70 (2009)
55. Choudhury, S., Chaki, N., Bhattacharya, S.: Gdm: a new graph based data model using functional abstraction. Journal of Computer Science and Technology 21(3), 430–438 (2006)
56. Sarkar, A., Bhattacharya, S.: The graph object oriented multidimensional data model: A conceptual perspective. In: Proceedings of the 16th International Conference on Software Engineering and Data Engineering, pp. 165–170 (2007)
57. Giovinazzo, W.A.: Object-oriented data warehouse design: building a star schema. Prentice Hall PTR (2000)
58. Hurtado, C.A., Mendelzon, A.O.: Reasoning about summarizability in heterogeneous multidimensional schemas. In: Proceedings of the 21st ACM Int. Conf. on Management of Data and Symposium on Principle of Databases Systems, pp. 169–179 (2001)
59. Rowen, W., Song, I.Y., Medsker, C., Ewen, E.: An analysis of many-to-many relationships between fact and dimension tables in dimensional modeling. In: Proceedings of the International Workshop on Design and Management of Data Warehouses (DMDW 2001), Interlaken, Switzerland (2001)
60. Jagadish, H.V., Lakshmanan, L.V.S., Srivastava, D.: What can hierarchies do for data warehouses? In: Proceedings of the 25th International Conference on Very Large Databases, pp. 530–541 (1999)

Optimized Clustering Method
for CT Brain Image Segmentation

Amlan Jyoti[1], Mihir N. Mohanty[1,*], Sidhant Kumar Kar[2], and B.N. Biswal[3]

[1] ITER, Sikshya O Anusandhan University, Bhubaneswar, Odisha, India
[2] Capegemini, Chennai
[3] Bhubaneswar Enggineering College, Bhubaneswar, Odisha, India
{amlanjyoti1,sidhu343}@gmail.com,
mihirmohanty@soauniversity.ac.in,
bhabendra_biswal@ yahoo.co.in

Abstract. Though image segmentation is a fundamental task in image analysis; it plays a vital role in the area of image processing. Its value increases in case of medical diagnostics through medical images like X-ray, PET, CT and MRI. In this paper, an attempt is taken to analyse a CT brain image. It has been segmented for a particular patch in the brain CT image that may be one of the tumours in the brain. The purpose of segmentation is to partition an image into meaningful regions with respect to a particular application. Image segmentation is a method of separating the image from the background, read the contents and isolating it. In this paper both the concept of clustering and thresholding technique with edge based segmentation methods like *sobel, prewitt* edge detectors is applied. Then the result is optimized using GA for efficient minimization of the objective function and for improved classification of clusters. Further the segmented result is passed through a Gaussian filter to obtain a smoothed image.

Keywords: Image Segmentation, Clustering, Thresholding, Genetic Algorithm, Smoothing.

1 Introduction

Over last two decades bio-image analysis and processing occupied an important position. Image segmentation is the process of distinguishing the objects and background in an image. It is an essential preprocessing task for many applications that depend on computer vision such as medical imaging, locating objects in satellite images, machine vision, finger print and face recognition, agricultural imaging and other many applications. The accuracy of image segmentation stage would have a great impact on the effectiveness of subsequent stages of the image processing. Image segmentation problem has been studied by many researchers for several years; however, due to the characteristics of the images such as their different modal

---

* Corresponding author.

© Springer International Publishing Switzerland 2015
S.C. Satapathy et al. (eds.), *Proc. of the 3rd Int. Conf. on Front. of Intell. Comput. (FICTA) 2014*
– *Vol. 1*, Advances in Intelligent Systems and Computing 327, DOI: 10.1007/978-3-319-11933-5_34

histograms, the problem of image segmentation is still an open research issue and so further investigation is needed.

Image Segmentation is the process of partitioning a digital image into multiple regions or sets of pixels. Partitions are different objects in image which have the same texture or colour. All of the pixels in a region are similar with respect to some characteristic or computed property, such as colour, intensity, or texture. Adjacent regions are significantly different with respect to the same characteristics. The critical step in image interpretation is separation of the image into object and background. Segmentation subdivides an image into its constituent regions or objects.

Segmentation separates an image into its component parts or objects. The level to which the separation is carried depends on the problem being solved. Segmentation algorithms for images generally based on the discontinuity and similarity of image intensity values. Discontinuity approach is to partition an image based on abrupt changes in intensity and similarity is based on partitioning an image into regions that are similar according to a set of predefined criteria. Thus the choice of image segmentation technique is depends on the problem being considered.

Identifying specific organs or other features in medical images requires a considerable amount of expertise concerning the shapes and locations of anatomical features. Such segmentation is typically performed manually by expert physicians as part of treatment planning and diagnosis. Due to the increasing amount of available data and the complexity of features of interest, it is becoming essential to develop automated segmentation methods to assist and speed-up image-understanding tasks. Medical imaging is performed in various modalities, such as magnetic resonance imaging (MRI), computed tomography (CT), ultrasound, etc. Several automated methods have been developed to process the acquired images and identify features of interest, including intensity-based methods, region-growing methods and deformable contour models. Intensity-based methods identify local features such as edges and texture in order to extract regions of interest. Region-growing methods start from a seed-point (usually placed manually) on the image and perform the segmentation task by clustering neighborhood pixels using a similarity criterion. Deformable contour models are shape-based feature search procedures in which a closed contour deforms until a balance is reached between its internal energy (smoothness of the curve) and external energy (local region statistics such as first and second order moments of pixel intensity). The genetic algorithm framework brings considerable flexibility into the segmentation procedure by incorporating both shape and texture information. In the following sections we describe our algorithm in depth and relate our methodology to previous work in this area.

Various methods have been proposed by the authors in [1-7], where out of many clustering methods; the segmentation method is carried out mainly focusing on the fuzzy clustering algorithm for effective and accurate feature extraction even in the presence of noise. In [12], comparative analysis using K-means and C-means algorithm is considered where author has concluded that C-means is better in terms of speed and accuracy over K-means. Basing on these evaluations and our domain knowledge, we choose to evaluate using C-means soft clustering algorithm. Many authors have hybridized the fuzzy C-means algorithm using different methods like watershed method, ant bee colony optimization, region growing method, etc. for the segmentation purpose [8-11, 13-15]. In [16], the generalized segmentation approach using morphology is providing an effective segmentation based on speed and accuracy.

In this paper an attempt is made to segment an image using fuzzy c-means clustering algorithm with its modified objective function by considering the approach of genetic algorithm. This proposed method minimizes the objective function better than the conventional FCM and gives superior quality of segmented result.

This paper is organized as follows. Section 2 depicts the methods for image segmentation. The result is discussed in section 3 and finally in section 4 concludes the work.

2   Proposed Method for Segmentation

Segmentation is a popular pre-processing operation for any processing method of images. Therefore, it is critical among all other procedures. Its purpose is to identify and delineate the specified entities in the anatomical/pathological organ image. The technique is based on pattern recognition. Among various methods for data analysis, clustering method is most efficient one, as it can find hidden structure data. A partition of a set of N patterns in an N-dimensional feature space must be found in such a manner that those patterns in a given cluster are more similar to each other. As the image considered here is monochromatic, the algorithm is based on the following image gray level properties [17-18].

a. **Discontinuity**: the objective is to find hard changes on gray level, using this information as the method to edge detection; and

b. **Similarity**: closest pixels are very similar. Some of the main challenges to the scientific community are related to the development of techniques that realize the automatic or unsupervised image segmentation. In controlled environment the image segmentation process is easily achieved than in a non-controlled environment, where light and other circumstances affect physical process of image acquisition.

Image segmentation applications contemplate many areas of Computer Graphics. In the case of Computer Vision, one of the objectives is make robots move in a semi or non-controlled environment, and realize tasks like find and interact with specific objects. Another area of interest is the automatic vehicle guiding. On Image Understanding and Analysis there is Content Based Image Retrieval, that aims to develop efficient search engines that can find items on an image database by using a reference image, detecting similarities.

2.1   Clustering

In a very simple level of abstraction, the image segmentation process is very close to the clustering problem. To find clusters in a data set is to find relations amongst unlabelled data. The "relation" means that some data are in some way next to another that they can be grouped. As in [19-20], the algorithm is described as follows.

The C-means clustering method is optimized using the evolutionary computing approach as GA approach and the procedure is given as follows:

Algorithm for Image Segmentation

1. The CT brain image has been considered for the work as input image.
2. The point of interest is a particular area within the brain image, as it is of the patient case. For this reason, the ROI was extracted for analysis of the image. The ROI is evaluated by statistical method from the pixel by considering the boundary [21].
3. Then the image was de-noised using a Gaussian filter and the model is given as:

$$H(u,v) = e^{-D^2(u,v)/2D_0^2}$$

4. The image matrix was then converted to intensity matrix with values ranging from 0 to 1 to prepare for the vital clustering stage.

5. It follows the fuzzy c-means algorithm for analysis. In this case the objective function, j_m considered as:

$$j_m(U,v) = \sum_{k=1}^{N} \sum_{i=1}^{c} (u_{ik})^m \|y_k - V_i\|_A^2,$$

Where,

$y = \{y_1, y_2 \ldots \ldots y_N\} \subset R^n$ = the data;
c = number of clusters in Y; $2 \leq c < n$,
m = weighting exponent; $1 \leq m < \infty$,
U = fuzzy c-partition of Y; $U \in M_{fc}$
$v = (v_1, v_2 \ldots \ldots v_c)$ = vectors of centers,
$v_i = (v_{i1}, v_{i2}, .., v_{in})$ = center of cluster i,
$\| \| _A$ = induced A-norm on R^n
A = positive-definite (n × n) weight matrix.

and can be optimized. The optimization is carried out as evolutionary computing method as genetic algorithm and is described in the following section.

6. The optimized value is used as an initial value for the initialization of FCM, and the process was carried out as described in step 5.
7. The image was binarized using threshold value [16].
8. The unwanted boundary is removed.
9. Then the edge detection was performed using well-known "Sobel" operator to get the segmented image [22].
10. The final segmented image was obtained by smoothing using a Gaussian filter as described in step 3.

2.2   Optimization Using Genetic Algorithm

The optimization technique of GA optimizes the fuzzy C-Means data and can be termed of GFCM algorithm. The major points for optimization are as follows:

1. *Coding.* It refers to how to encode the solution (the chromosome); one way of doing this is the *string-of-group-numbers encoding* where for Z coded solutions (partitions), represented by strings of length N, each element of each string (an allele) contains a cluster number.

2. *Initialization.* The initial population $P0$ is defined randomly: each allele is initialized to a cluster number. The next population $Pi+1$ is defined in terms of the selection, mutation and the C-means operator.

3. *Selection.* Chromosomes from a previous population are chosen randomly according to a distribution.

4. *Crossover.* It is a probabilistic process that exchanges information between two parent chromosomes for generating two new (descendant) chromosomes.

5. *Mutation.* The mutation operator changes an allele value depending on the distances of the cluster centroids from the corresponding pattern.

6. *C-Means Operator (CMO).* This operator is used to speed up the convergence process and is related to one step of the classical C-means algorithm. Given a chromosome, each allele is replaced in order to be closer to its centroid.

A Specified number of iterations with a required fitness value to bring this generation cycle to an end.

The fitness computation of each individual in that population was calculated, considering the penalty for covered and uncovered points. A point x is called a covered point if $x \in S_j$, S_j is a region that contains connected points around the center C_j, and x is called is uncovered if $x \in S_j$, S_j [10].

$$\text{Fitness} = \alpha \sum_{i=1}^{n} \sum_{j=1}^{k} \left\| C_j - R_j(X_i, y_i) \right\|^2 + \text{NCR}$$

Where NCR is a penalty for uncover points.

The Euclidian distance term represents the shortest distances between the centroid c_j, $j = 1, 2...k$, and all pixels p_i, $i = 1, 2, ...N$, of a region R_j.

If $d_{ij} < d_{euclid}$, then it finalize the point x_j, as a covered value, else it is uncovered.

The minimum of the first term is obtained when all pixels fall in regions with the Center c_j. Some pixels which are uncovered by regions are represented by NCR and should be minimized for optimal value of fitness, which can be calculated as:

$$\text{NCR} = \alpha \sum_{i=1}^{m} \left\| \text{Median} - R(x_i, y_i) \right\|^2$$

Where m is the number of uncovered points.

3   Result and Discussion

Fig1 (a) shows the imageacquired for input.As a patient case is considered here, our interest is that particular area where the tumor is located. So that portion is to be extracted from the entire image by statistical approach, which is said to be our region of interest, which is shown in Fig 1(b). Hence forth all the methods will be applied on this part of the image. So, we initially de-noised it by considering a Gaussian filter. The output of the de-noising process is shown in Fig 1 (c).

After application of GFCM algorithm, the threshold value is obtained and is shown in Fig 1 (d). The resultant binary image is processed through the popular "sobel" operator to get its edge, obtaining our segmented image, as shown in Fig 1 (e). This segmented image can be further processed by smoothing it due to discontinuity, noise and outliers as in [23], through a Gaussian based low pass filter, which is shown in Fig 1 (f).

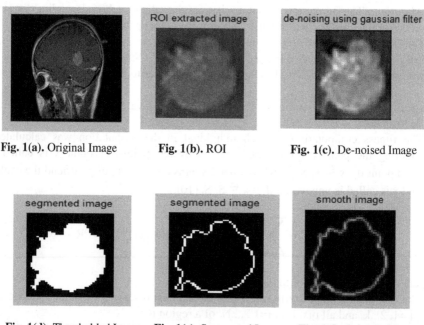

Fig. 1(a). Original Image   **Fig. 1(b).** ROI   **Fig. 1(c).** De-noised Image

Fig. 1(d). Thresholded Image   **Fig. 1(e).** Segmented Image   **Fig. 1(f).** Smoothed Image

The data points when clustered using the fuzzy C-means can be plotted as in Fig. 2 (a), and Fig.2 (b) is depicting the example for classifying covered and uncovered data points.After the genetic algorithm is applied and the fuzzy clustering method is evaluated again, its cluster-point distribution is shown in Fig. 2 (c).

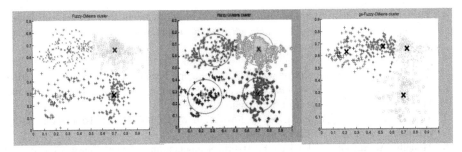

Fig. 2(a). FCM   **Fig. 2(b).** Classification of   **Fig. 2(c).** After GFCM
Covered & Uncovered points

Different results for non-optimized and of optimized methods are as follows:

    a) Using conventional FCM,

    Iteration count = 28, obj. fcn = 8.647208

    b) Using modified GFCM,

    Iteration count = 17, obj. fcn = 6.268363

4   Conclusion

In this paper, an optimized method for image segmentation using GFCM is described. We have analysed the fuzzy based clustering method. Again the same has been optimized using evolutionary computing as genetic algorithm. The proposed segmentation scheme focuses on the clustering and optimized clustering approach. As shown in our experimental results, the algorithm generates visually meaningful segmentation results. It demonstrated that proposed method is efficient for medical image analysis.Finally, it is to mention that automated segmentation methods can never replace doctors but they will likely become vital elements of medical image interpretation. Thus, there are numerous challenges to improve clinical decision making based on automated processing for engineers, mathematicians, physicists and physicians working to advance the field of image segmentation and its analysis.

References

1. Bezdek, J.C., Ehrlich, R., Full, W.: FCM: The Fuzzy C-Means Clustering Algorithm. Computers& Geosciences 10(2-3), 191–203 (1984)
2. Belhassen, S., Zaidia, H.: A novel fuzzy C-means algorithm for unsupervised heterogeneous tumorquantification in PET. Med. Phys. 37(3) (March 2010)
3. Chuang, K.-S., Tzeng, H.-L., Wu, S., Chen, T.-J.: Fuzzy c-means clustering with spatial information for image segmentation. Computerized Medical Imaging and Graphics 30, 9–15 (2006)

324   A. Jyoti et al.

4. Yucheng, L., Yubin, L.: An Algorithm of Image Segmentation Based on Fuzzy Mathematical Morphology. In: IEEE Conference International Forum on Information Technology and Applications (2009)
5. Wang, Z., Yang, M.: A Fast Clustering Algorithm in Image Segmentation. In: 2nd International Conference on Computer Engineering and Technology, vol. 6, pp. 1–5 (2010)
6. Li, C., Li, Y., Wu, X.: Novel Fuzzy C-means Segmentation Algorithm for Image with the Spatial Neighborhoods. In: IEEE Conference (2012)
7. Thyagarajan, R., Murugavalli, S.: Segmentation of Digital Breast Tomograms using Clustering Techniques. IEEE Conference (2012)
8. Mouelhi, A., Sayadi, M., Fnaiech, F.: Hybrid Segmentation of Breast Cancer Cell Images Using a New Fuzzy Active Contour Model and an Enhanced Watershed Method. In: IEEE Conference,CoDIT-2013 (2013)
9. Chabrier, S., Rosenberger, C., Emile, B., Laurent, H., Biswas, U., Naskar, M.K.: Optimization-Based Image Segmentation by Genetic Algorithms. EURASIP Journal on Image and Video Processing 10 pages (2008)
10. Zanaty, E.A., Ghiduk, A.S.: A Novel Approach Based on Genetic Algorithms and Region Growing for Magnetic Resonance Image (MRI) Segmentation. ComSIS 10(3) (June 2013)
11. Griffin, J.D.: Methods for Reducing Search and Evaluating Fitness Functions in Genetic Algorithms for the Snake-In-The-Box Problem., thesis, Athens, Georgia (2009)
12. Sinha, K., Sinha, G.R.: Comparative Analysis of Optimized k-Means and c-Means Clustering Methods for Segmentation of Brain MRI Images for Tumor Extraction. In: Emerging Research in Compting,Information,Communicatipon and Applications(ERCICA) Conference (2013)
13. Ganesan, R., Radhakrishnan, S.: Segmentation of Computed Tomography Brain Images Using Genetic Algorithm. International Journal of Soft Computing, 157–161 (2009)
14. Menon, N.R., Karnan, M., Sivakumar, R.: Brain Tumor Segmentation In MRI Image UsingUnsupervised Artificial Bee Colony And FCMClustering. International Journal of Computer Science and Management Research 2(5) (May 2013)
15. Halder, A., Pramanik, S., Kar, A.: Dynamic Image Segmentation using Fuzzy C-Means based Genetic Algorithm. International Journal of Computer Applications 28(6), 975–8887 (2011)
16. Jyoti, A., Mohanty, M.N., Mallick, P.K.: Morphological Based Segmentation of Brain Image forTumor Detection. In: IEEE Conference International Conference on Electronics and Communication Systems, ICECS 2014 (2014)
17. Pratt, W.: Digital Image Processing, 3rd ed., Wiley (2001)
18. Gonzalez, R.C., Woods, R.E.: Digital Image Processing, 3rd edn. PHI (2008)
19. Jain, A.K.: Digital Image Processing. PHI (1999)
20. Tripathy, A., Mohanty, A.K., Mohanty, M.N.: Electronic Nose for Black Tea Quality Evaluation Using Kernel Based Clustering Approach. Int. Journ. of Image Processing (IJIP) 6(2) (April 2012)
21. Kar, S.K., Mohanty, M.N.: Statistical Approach for Color Image Detection. In: IEEE Int. Conf. on Computer Communication and Informatics (ICCCI), January 4-6, pp. 1–4 (2013)
22. Behera, S., Mohanty, M.N., Patnaik, S.: A comparative analysis on edge detection of colloid cyst: A medical imaging approach. In: Patnaik, S., Yang, Y.-M. (eds.) Soft Computing Techniques in Vision Sci. SCI, vol. 395, pp. 63–86. Springer, Heidelberg (2012)
23. Dash, S., Mohanty, M.N.: A Comparative Analysis for WLMS Algorithm in System Identification. In: IEEE Conf., ICECT, Kanyakumari, April 6-8 (2012)

Rotational-Invariant Texture Analysis
Using Radon and Polar Complex Exponential Transform

Satya P. Singh[1], Shabana Urooj[2], and Aime Lay Ekuakille[3]

[1] Electronics Engineering Department
Galgotias College of Engineering & Technology
Gr. Noida, UP, India
satya002u@gmail.com
[2] Electrical Engineering Department, School of Engineering
Gautam Buddha University
Gr. Noida, UP, India
shabanaurooj@ieee.org
[3] Department of Innovation Engineering
University of Salento Lecce, Italy
aime.lay.ekuakille@unisalento.it

Abstract. Rotational invariant texture analysis technique using Radon and PCET is proposed in this stab. Rotation invariance is achieved within the Radon space. Translation invariance is achieved by normalization of moment of PCET. A k- nearest neighbor classifier is employed for classifying the texture. To test and evaluate the proposed method several sets of texture was evaluated. The evaluation is achieved with different scaling, translation and rotation under different noisy conditions. Correct classification percentage is calculated for various noise conditions. Experimental result shows pre-eminence of the employed method as compared to the recent invariant texture analysis methods.

Keywords: Radon transforms Polar complex exponential transform, Geometric invariance, classifier.

1 Introduction

Texture analysis a critical and challenging issue in the field of image processing with applications in remote sensing, medical image sensing, and content based image retrieval (CBIR) and Forest imaginary. Ideally, the image obtained by image sensor should be invariant under geometric deformation (e.g. scaling, rotation, translation and illumination). However, most of the proposed techniques assume that the image object of image sensor is acquired by the same viewpoint, which is not always the case [1]. In many practical applications, it is almost impossible to adjust the image sensor for capturing the image with same scaling, rotation and translation to each other. A good survey of invariant analysis of texture may be found in [2]. An important class of geometric degradation we often encounter in reality is image scaling, translation and rotational image which is caused by object motion,

© Springer International Publishing Switzerland 2015
S.C. Satapathy et al. (eds.), *Proc. of the 3rd Int. Conf. on Front. of Intell. Comput. (FICTA) 2014*
– *Vol. 1*, Advances in Intelligent Systems and Computing 327, DOI: 10.1007/978-3-319-11933-5_35

misalignment of image sensor or vibration. Khotanzad et al. used Zernike moment to scale, translation a rotational invariant features [3]. A comparison for moment features and, moment invariant was experimented.   Recently, Dai et al. [4] constructed a set of invariant PZM to convolution with circularly symmetric point spread function. Xiao [5] proposed a radial moment (combination of Jacobi-Fourier moment) for invariant features. These radial orthogonal moments were invariant to scale and rotation. A systematic theory of invariant moment for pattern recognition can be found in [6]. A Non-Linear Approach to ECG Signal Processing using Morphological Filters [16] and improvement of Masses Detection in Digital Mammograms [17] [20], [21] is discussed by V Bhateja et al. Robust Polynomial Filtering Framework approach for Mammographic Image Enhancement is reported recently [17] [19] . Effective evaluation of tumor region in brain MR images using hybrid segmentation is discussed in [18].

Motivation: Polar Harmonic Transform (PHT) is recently introduced by Yup et al [7]. Due to their simple kernel, less complexity and significantly reduced computational time compare to other moment such as Zernike Moment (ZM) [10] and Pseudo Zernike Moment (PZM) [11], they might be used efficiently for geometric invariant Radon Transformed images. Since recent introduction of PHT in literature, they have not received much attention of the researchers in the field of invariant moment analysis. Recently Li et al. [8], uses PHT to get rotational invariance in watermarking. Above all, there are two major drawback of PHT in comparison with ZM and PZM, i.e. higher reconstruction error using PCT and lower classification percentage [7]. To overcome these drawbacks, the rotation invariance is achieved within the Radon space and the PHT is applied.

2   Basic Materials

2.1   Radon Transform

The Radon Transform (RT) [9] of 2-D function $f(x, y)$ is defined as line integral of f for all lines ℓ defined by parameter θ . The 2-D RT defined as:

$$\Re(r, \theta)[f(x, y)] = \int_{-\infty}^{+\infty} \int_{-\infty}^{+\infty} f(x, y)\delta(r - x\cos\theta - y\sin\theta\, dxdy) \tag{1}$$

According to Fourier slice theorem this transform is invertible. This theorem states that the FT of the projection in the spatial domain is identical to the profile 2-D Fourier domain.

2.2   Polar Complex Exponential Transform (PCET)

PCET [7] is a set of complex polynomial which represents a complex orthogonal set over the interior of unit circle defined by $x^2+y^2=1$. Let these polynomial is denoted by

$H_{nl}(x,y)$, the PCET of order n with repetition l and $|\ell| = |n| = 0,1,....\infty$ may be define as:

$$M(n,l) = \frac{1}{\pi} \int_0^{2\pi} \int_0^1 [H_{n,l}(r,\theta)]^* f(r,\theta) r dr d\theta \qquad (2)$$

Where, r is the Length of vector from origin to a particular pixel (x, y) within unit circle defined by $x^2 + y^2 = 1$. θ is the angle between 'x' axis and vector length r in clockwise direction. n,l Positive and negative integers and $[.]^*$ Denotes the complex conjugate of the basis H_{nl}, (r,θ).

3   Proposed Method

3.1   Rotation Invariant Features for Radon Image

Let us consider that the computed tomography image is rotated through an angle ϕ and represented by f'. According to the rotational property of *Radon transform* the original image f and the rotated image f' can be related as:

$$f'(r,\theta) = f(r, \theta - \phi) \qquad (3)$$

The polar harmonic moment can be map into polar coordinate polar coordinate from its Cartesian plane x-y by changing the variables in double integral from such as:

$$\iint_C F(x, y) dx dy = \iint_P F[p(r,\theta), q(r,\theta)] \frac{\partial(x, y)}{\partial(r,\theta)} dr d\theta \qquad (4)$$

Here the term $\partial(x, y) / \partial(r,\theta)$ denotes the Jacobean transformation. Substituting x $= r \cos \theta$ and y= r sin θ, the Jacobean simplifies in r, therefore,

$$M(n,l) = \frac{1}{\pi} \int_0^{2\pi} \int_0^1 [H_{n,l}(r,\theta)]^* f(r,\theta) r dr d\theta \qquad (5)$$

The above equation could be expressed in radial and exponential term, such as:

$$M(n,l) = \frac{1}{\pi} \int_0^{2\pi} \int_0^1 R_n(r) e^{-i\ell\theta} f(r,\theta) r dr d\theta \qquad (6)$$

Now let us consider that the image is rotated in same coordinate. Its PHT could be written as:

$$M'(n,l) = \frac{1}{\pi} \int_0^{2\pi} \int_0^1 R_n(r) e^{-i\ell\theta} f(r, \theta - \phi) r dr d\theta \tag{7}$$

By change in variable $\theta' = \theta - \phi$

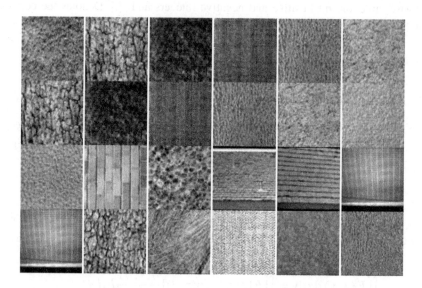

Fig. 1. Some example of Texture images used for data set

	Angle of Rotation (in degree)				
Data Set	0	30	60	90	120
D12					
D94					

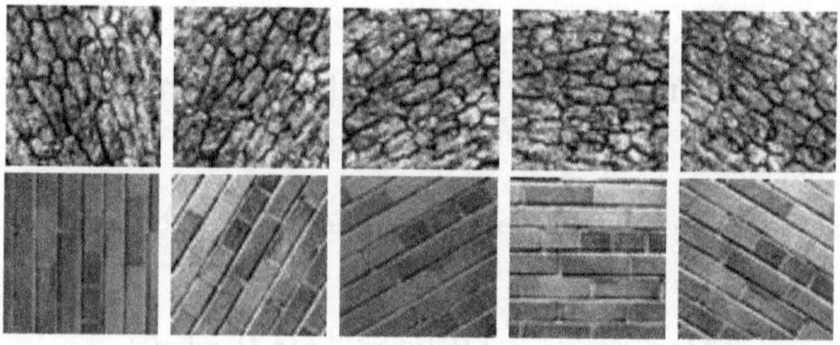

Fig. 2. Rotated Texture images sample

$$M'(n,\ell) = \frac{1}{\pi} \int_0^{2\pi} \int_0^1 R_n(r) e^{-i\ell(\theta'+\phi)} f(r,\theta') r\, dr\, d\theta' \qquad (8)$$

$$M'(n,\ell) = M(n,\ell) e^{-i\ell\phi} \qquad (9)$$

Equation (9) shows that PCET have a simple rotational property. If, somehow, we suppress the exponential term, we get the rotational invariance. The simplest method to get absolute invariant is taking magnitude of both sides, i.e.

$$\left| M'(n,\ell) \right| = \left| M(n,\ell) \right| \qquad (10)$$

$$M'(n,\ell) \left| M'(n,\ell) \right| * = (n,\ell) \left| M(n,\ell) \right| * \qquad (11)$$

This simple process leads to conclusion that magnitude of PCET remains identical to those of rotated image. One can concern $\left| M(n,\ell) \right|$ with $n, \ell \geq 0$, but this results in loss of information which contained in rotated phase. From above equation it could be observed that $M_{n,\ell}$ is the PCET of $f(r,\theta)$. Now let's define a new moment:

$$
\begin{aligned}
M''_{n,\ell} &= e^{i\ell \arg[M_{0,1}]} M(n,\ell) \\
&= e^{i\ell \arg[e^{-i\phi}M_{0,1}]} M(n,\ell) e^{-i\ell\phi} M(n,\ell) \\
&= e^{i\ell \arg[M_{0,1}]} M(n,\ell) \\
&= M''_{n,\ell}
\end{aligned}
\qquad (12)
$$

It proves that the proposed moments for radon image are invariant to rotation without loss of information in phase. The same procedure can be applied to PCT and PST.

4   Results and Discussion

This section describes the experiments conducted to test the performance of the proposed method in terms of correct classification percentage, mean square error, noise robustness against Gaussian noise and computational complexity. All the experiments were carried out in the MATLAB R2009b environment on core i3 processor with 4GB RAM. To test the proposed method, 60 textured images are selected from the Brodatz texture database [15] (Some samples of these images are shown in Fig.1). Each image is 512X512 pixels in dimension. These images are cropped and resized to 128X128 pixels. Each sub-image is individually normalized to zero mean and unit variance. In the second dataset (Fig. 2), each image is rotated by an angle 20^0, 40^0, 60^0, 80^0, 100^0, 120^0, 140^0, 160^0, 180^0. The total number of test image is thus 2160: 60 (number of original images) X9 (number of rotation angles) X4 (number of distortion level).

Two different dataset are used to test correct classification percentage (C_{cp}). All the experiment was conducted on image size 512X512. Each image is converted to gray scale to get the standard image size 128X128. Therefore, the resultant radius is N=64 pixels. The sampling rate limited to $N\pi \leq 200$ [13]. The performance of different

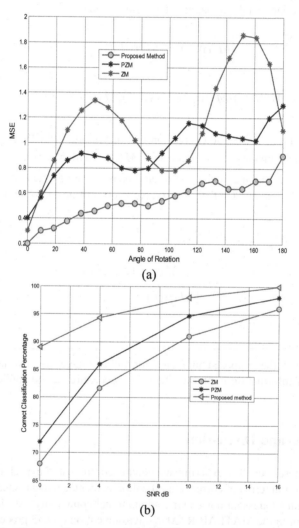

(a)

(b)

Fig. 3. (a) Relation between Rotation angle and MSE and (b) Average Percentage of correct classification for different SNR and k=1, 3,5,7,9 for the k- NN Classifier

number of projection is computed and the best one is chosen. The optimal number of projections is estimated to calculate the C_{cp}. Features are being extracted for different order of moment and different number of projections for angles between 0 to 180 degrees. The test image was corrupted with a Gaussian noise with variance 0.05, 0.10, 0.15, 0.20. Then the noise contained images are classified according to their respective moments. To achieve best performance, it is required to optimize the projection. The k-NN classifier with Euclidean distance is used to classify the appropriate class [14]. Results are shown in Fig. 3 (a). The performance of the proposed method for rotational invariant was experimented. For scale invariance, the image was scaled with a scaling factor 0 to 1.2 with an interval 0.1. To achieve the best result, the image is extracted into 60 different images and average of 60 images was taken.

As compared to ZM and PZM [12], the proposed method shows high consistency in classification. It should be noted that the proposed method is more effective for directional texture. For small scale variation, Radon based PCET out performs for scale factor greater than 0.6 compare to 0.8 in PZM. PZM and ZM shows some degradation in performance when the scale factor exceeds 0.8 and 1.2 respectively. Mean Square Error (MSE) is calculated for different rotational angle. To calculate MSE, method reported in [7] is being used. The MSE is calculated with different rotation angle (20^0, 40^0, 60^0, 80^0, 100^0, 120^0, 140^0, 160^0, 180^0) as shown in Fig. 3 (b). The proposed method shows almost uniform MSE for each rotation. The Radon base PHT shows higher complexity and reflects that it takes bit more computational time for calculating each moment as compared to PZM and ZM.

5   Conclusions

In this paper, a new technique of geometric invariant (Rotation, translation and scaling) using Radon transform and Polar Harmonic Transform for image sensors have been proposed. The properties of Radon Transform are studied and Radon Transform is calculated for unit disk area inside the image. The scaling and translation is achieved by normalizing the moment. Rotation invariant is experimented using the property of Radon transform and Polar Harmonic Transform. The method is examined in different noise and its merits are illustrated. It is also compared with leading invariant moment methods. The proposed method outperformed the method of Zernike moment and Pseudo Zernike moment in terms of correct classification percentage, sensitivity, complexity and noise robustness. The correct classification percentages is calculated and compared with existing research. Some other possible applications of the proposed method are image retrieval, remote sensing, data mining, MRI and PET scan, face detection and texture classification. It is considered that the proposed method may increase the efficiency and accuracy of image sensors and sensor interface standards.

References

1. Al-Shaykh, O.K., Doherty, J.F.: Invariant image analysis based on Radon transform and SVD. IEEE Transactions on Circuits and Systems II: Analog and Digital Signal Processing 43(2), 123–133 (1996)
2. Jianguo, Z., Tan, T.: Brief review of invariant texture analysis methods. Pattern Recognition 35(3), 735–747 (2002)
3. Alireza, K., Hong, Y.H.: Invariant image recognition by Zernike moments. IEEE Transactions on Pattern Analysis and Machine Intelligence 12(5), 489–497 (1990)
4. Dai, X., Liu, T., Shu, H., Luo, L.: Pseudo-zernike moment invariants to blur degradation and their use in image recognition. In: Yang, J., Fang, F., Sun, C. (eds.) IScIDE 2012. LNCS, vol. 7751, pp. 90–97. Springer, Heidelberg (2013)
5. Bin, X., Wang, G.-Y.: Generic radial orthogonal moment invariants for invariant image recognition. Journal of Visual Communication and Image Representation 24(7), 1002–1008 (2013)
6. Yajun, L.: Reforming the theory of invariant moments for pattern recognition. Pattern Recognition 25(7), 723–730 (1992)
7. Pew-Thian, Y., Jiang, X., Kot, A.C.: Two-dimensional polar harmonic transforms for invariant image representation. IEEE Transactions on Pattern Analysis and Machine Intelligence 32(7), 1259–1270 (2010)
8. Li, L., et al.: Geometrically invariant image watermarking using Polar Harmonic Transforms. Information Sciences 199, 1–19 (2012)
9. Easton, R.L.: The Radon Transform. Fourier Methods in Imaging, 371–420 (2010)
10. Noll, R.J.: Zernike polynomials and atmospheric turbulence. JOsA 66(3), 207–211 (1976)
11. Chong, C.-W., Raveendran, P., Mukundan, R.: The scale invariants of pseudo-Zernike moments. Pattern Analysis & Applications 6(3), 176–184 (2003)
12. Hongqing, Z., et al.: Combined invariants to blur and rotation using Zernike moment descriptors. Pattern Analysis and Applications 13(3), 309–319 (2010)
13. Kourosh, J.-K., Soltanian-Zadeh, H.: Rotation-invariant multiresolution texture analysis using Radon and wavelet transforms. IEEE Transactions on Image Processing 14(6), 783–795 (2005)
14. Henley, W.E., Hand, D.J.: A k-nearest-neighbour classifier for assessing consumer credit risk. The Statistician, 77–95 (1996)
15. Brodatz, P.T.: A Photographic Album for Artists and Designers. Dover, New York (1966)
16. Vikrant, B., Rishendra, V., Rini, M.: A Non-Linear Approach to ECG Signal Processing using Morphological Filters. International Journal of Measurement Technologies & Instrumentation Engineering 3(3), 46–59 (2013), doi:10.4018/ijmtie.2013070104
17. Vikrant, B., Mukul, M., Urooj, S.: A Robust Polynomial Filtering Framework for Mammographic Image Enhancement from Biomedical Sensors. IEEE Sensors Journal 13(11) (November 2013), doi:10.1109/JSEN.2013.2279003
18. Verma, K., Urooj, S., Rituvijay: Effective evaluation of tumour region in brain MR images using hybrid segmentation. In: IEEE International Conference on Computing for Sustainable Global Development INDIACom 2014, March 5-7, IEEE, BVICAM New Delhi (2014), doi:10.1109/IndiaCom.2014.6828024
19. Vikrant, B., Urooj, S., Mishra, M., Pandey, A., Ekuakille, A.L.: A Polynomial Filtering Model for Enhancement of Mammogram Lesions. In: IEEE International Symposium on Medical Measurements and Its Applications MeMeA 2013, pp. 97–100 (2013), doi:10.1109/MeMeA.2013.6549714

20. Vikrant, B., Urooj, S., Ekuakille, A.L.: Improvement of Masses Detection in Digital Mammograms employing Non-Linear Filtering. In: IEEE International Conference iMAC4s 2013, pp. 406–408 (2013), DOI: 978-1-4673-5090-7/13
21. Bhateja, V., Devi, S., Urooj, S.: An Evaluation of Edge Detection Algorithms for Mammographic Calcifications. LNEE, vol. 222, pp. 487–498. Springer, India (2012) ISBN 978-81-322-0999-7 ISBN 978-81-322-1000-9 ISSN 1876-1100

Moving Object Detection by Fuzzy Aggregation Using Low Rank Weightage Representation

A. Gayathri and A. Srinivasan

Dept. of IT, MNM Jain Engg. College,
Chennai, India
{gaybalahari,asrini30}@gmail.com

Abstract. We envisage a new algorithm, to detect moving objects having dynamic and challenging background conditions, by applying low rank weightage and fuzzy aggregated multi-feature similarity method. Model level fuzzy aggregation measure driven background model maintenance is used to ensure more robustness. The model and current feature vectors are evaluated between corresponding elements to find out the similarity functions. To compute fuzzy similarities from the ordered similarity function values for each model concepts of Sugeno and Choquet integrals are incorporated in our algorithm. A fuzzy integral set is using model updating and foreground/background classification decision methods. Sugeno Integral calculates only minimum and maximum weightage. We use choquet concept because it has the same functionality as Sugeno but it also uses additional operations like arithmetic mean and Ordered Weighted Averaging (OWA). Here we explain to segment the object by fuzzy aggregation with low rank weightage concept for extracting moving objects with accurate shape in dynamic background. PSNR, MSE and SSIM values are calculated to do performance evaluation.

Keywords: Image Segmentation, Low rank Weightage, Sugeno and Choquet Integrals, Moving Object Detection.

1 Introduction

The method to recognize instances of real-time objects namely human faces, cycles, etc. is through object detection in the form of images or videos. To detect the moving object one of the popular approach is background subtraction [1]. Background models are maintained for every pixel. Also, the foreground/background taxonomy decision depends on the similarity between the model based on the current features and stored models.

All possible dynamic situation of background model should preferably be associated with a model that evolves from the data [2]. Several background models for every pixel can capably manage dynamic situations like rippling water, swaying vegetation, etc., else which would be wrongly classified as moving foreground. Every

© Springer International Publishing Switzerland 2015
S.C. Satapathy et al. (eds.), *Proc. of the 3rd Int. Conf. on Front. of Intell. Comput. (FICTA) 2014*
– *Vol. 1*, Advances in Intelligent Systems and Computing 327, DOI: 10.1007/978-3-319-11933-5_36

model could be a single feature that is pixel-based [3], [4], [5] or region-based [6], [7], [8].

Under different situation all the individual feature has some limitations and advantages as explained in [9]. When comparing, the multiple features [10], [11], [12] are more complex than single feature based techniques, which exploits the advantages of different types of features. The background subtraction approach is used to maintain multi-feature based multi-background models for each pixel and implementing the same may leads to many practical difficulties. The first vital point is to deal with proper choice of features. Pixel-based features are not suitable for dynamic background because it exhibits some inherent weaknesses. Region-based features have vice versa characteristics.

Statistical Texture (ST) i.e. energy, texture mean and local homogeneity are consider as region features and intensity (I) as the pixel feature. Consecutive major issue to be addressed in multi-feature model is to achieve an aggregated measure of multiple features of assorted nature for every model to make an accurate classification assessment out of multiple models.

2   Background Work

To begin with, the related topics for this paper are identified as Preprocessing, background modeling, background/foreground detection, motion modelling, motion segmentation, data validation and background subtraction.   Most relevant papers relating to the identified topics are analyzed for its approach, methodology, merits and demerits.

Based on survey steered by D. Gorilla on visual analysis of human movement, proposes a Bayesian framework, for characterizing background appearance incorporating spectral, spatial and temporal features. Based on Bayes decision rule, and statistical principal features, background and foreground classification are derived.   This approach helps to extract foreground regions by filling holes and removing small regions. It is difficult to maintain accuracy of difference between image referenced and background.

C. Rougier et.al in their study on human shape deformation using fall detection for robust video surveillance, proposed extracting information relating to 3-D structure, orientation, spatial property of a generic object. This leads to detect moving objects during various instances of time for a given video sequence. This approach is an optimum solution for video sequences having low illumination. This method is not suitable for frames having distinctive peaks in the histograms.

Pujala Chiranjeevi and Somnath Sengupta proposed Advanced Fuzzy Aggregation based Background Subtraction (AFABS) to extract moving object from dynamic background with accurate shape and minimal error. This method is to be further tunned up for heavily dynamic background situation to enhance its performance.

Xiaowei Zhou, Yang and Weichuan proposed a novel framework Detecting Contiguous Outliers in the Low-rank Representation (DECOLOR) for segmenting moving objects from image sequence. For dealing with complex background image sequences, this method uses, low rank outlier detection modeling to avoid complication in motion computation.

Problem Definition. With reference to above literature survey, this paper poised to provide a suitable solution for video sequence having heavily dynamic background situation.   Detecting a moving object based on its dynamic background and to segment the object by applying an efficient algorithm and tracking frame by frame are the prime objective of this paper. Subsequently following are its other objectives :

(i) With multi-feature fuzzy aggregation concept for dynamic behavior of the background a novel process is envisaged to detect outlier using low-rank weightage representation.
(ii) The dynamic behavior of background represented by choquet & sugeno integrals and a robust labeling foreground/background obtained in terms of f-measures and
(iii) The effectiveness of Low Rank Weightage based multi-feature fuzzy aggregation method is experimented on a vast variety of video image sequence with challenging background situation.

3   Proposed System

The core objective of this paper is to propose tacking of moving object frame-by-frame, analyze its behavior and segmentation. Existing systems uses pixel based method by low rank representation in static backgrounds and rank is given based on the object changes between the initial and next frame. This helps to detect moving object and subsequently segmented. Proposed system uses pixel and region (fuzzy logic) based method by low rank weightage representation in dynamic backgrounds. The weightage with rank is given based on the challenging backgrounds in certain region for comparing the frames and segmentation of object.

Fuzzy Logic: Boolean logic is the basic approach used in this method. Aggregation which is collection of several things grouped together or considering as whole, and applied to fuzzy sets, it means several fuzzy sets are grouped together to form a single fuzzy set. By reducing set of number into a distinctive representative number, having function of mathematical object in aggregation operator.

3.1   System Architecture

Block diagrammatic representation of the proposed method is show in Fig – 1. This method uses region and pixel based fuzzy technique for detecting moving object from a given video sequence. Initially video sequence are segmented into a number of homogeneous regions using the concept of fuzzy edge modelling to handle spatial ambiguities the object boundaries.

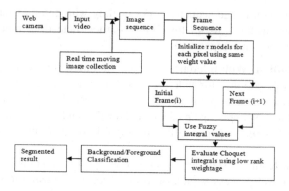

Fig. 1. Proposed method schematic diagram

3.2   Background Model

In video sequence, periodical motion of dynamic texture or illumination change causes background intensity changes else remains unchanged. Low rank Matrix B is formed, by correlating linearly background images with each other. No additional assumptions are considered except low-rank property of the background scene. Imposing the constraint in Equation 2 :

B Rank (B) <= K ; Where, K is predefined constraint implying complexity of background model.

3.3   Using Soft Impute Algorithm Estimating Low Rank Matrix

Comparing pixels of current and next frame with neighboring pixels and rank is given with priority to changing pixels. Soft impute algorithm is used to estimate low rank matrix and its combined with Singular Value Decomposition (SVD) to compute real or complex matrix by factorization.

3.4   Estimation of Outliers Support

Using Markov Random Field (MRF) and the Active contour the outliers are computed to detect the boundaries of the moving object. Imposing smoothness of spatial and temporal is possible by inter-relating of all pairs of nodes in G. This can be isolated into sub-graphs of individual frame and separately for each image the graphs cut can be operated.

3.5   Parameter Tuning

Alpha parameter is used to control the complexity of image background. Smaller value B is given by large alpha and rough estimate i.e. K value is set in this algorithm for background model. This algorithm starts from alpha value and if Rank (B) < K, after each run of SOFT-IMPUTE, reducing by a factor until rank (B) > K. Outliers support sparsely controlled by beta value, process the algorithm by the same and decrement by given factor after every iteration.

3.6   Proposed Method Consists Following Steps

Step 1: Model initialization
Step 2: Advanced Fuzzy aggregation background Subtraction (AFABS)
Step 3: Foreground detection

Model Initialization. Model initialization is the primary step for background subtraction. Initialize the model using first frame of the video model. Pixels are initialized with feature vector from neighborhood pixels. Neighboring pixels share its feature vectors for their model initialization.

Advanced Fuzzy Aggregation Background Subtraction (AFABS). Efficient fusions of information are contributed by the individual features and used by multi-feature based background subtraction techniques. Information fusion is the process of combining these features into a single datum and the fusion is achieved through aggregation operators, which are mathematical functions. Parameterization is required for aggregation operator to define additional information and features for this process i.e weighted mean and weighted maximum. Sugeno and Choquet integrals have interactions among the features that explained below.

Algorithm AFABS

Step1: For each pixel at t=0, initialize r models and initialize models
weights to the same values for t>0 to the end of sequence do for
each pixel
Step2: Form a feature vector X
Step3: Normalize the models weights
Step4: Select top b high weighted models
Step5: Evaluate the similarity function for each model with X
Step6: Evaluate the Fuzzy Integral set(f_1,......f_r)
Step7: [Val , S]=max(f_1,....f_r)
Step8: if(Val < T_p) do
Step8.1: Pixel label=foreground
Step8.2: Update the model using case-1
else do   case-2
Step8.3: if(S>b)pixel label=foreground
Step8.4: else pixel label=background
Step8.5: Update the matched model
Step8.6: Update models weights
end
end

For the changes to be incorporated in the image background the model of pixel should be updated.

case 1. if the max(f1,...Fr)<Tp, the vector of model feature with lowest weight is substituted by vector of current feature.

case 2. if the max(f1,....,Fr)>=Tp, the best matching model that is the model feature vector having the maximum integral value is updated with the current feature vector.

Sugeno Integrals. The sugeno integrals are monotonically non decreasing and continuous operators this suggest that the fuzzy integral is forever comprised between "Maximum Weight" and "Minimum Weight". In sugeno integrals it is used to satisfy a similar property with MIN and MAX. This method is flexible than probability and controlled by its additives property.

Choquet Integrals. The choquet integrals are monotonically non decreasing idempotent and continuous operators. If the fuzzy measure in an additive then the choquet integral is converted into Arithmetic mean whose weights g ({xi}). For Positive linear transformation Choquet integral method is an appropriate method which includes properties of Weight Arithmetic Sum and OWA operators. Fulfillment of multiple criteria relating with aggregating rates are provided by OWA operator, unified as one operator combining conjunctive and disjunctive behavior.

Foreground Detection: An object which moves differently from the background is called as foreground object and of contiguous small size pieces. Intensity changes are not fixed into low-rank background model that is provided by foreground motion and can be identified as low-rank outliers representation.

By applying Markov Random Field Model, entries of binary states of foreground support can be denoted as "S". Considering the same as graph Equation 3, wherein

$$S = (V, E); \text{Where,}$$

V - denotes set of vertices in the sequence as m*n pixels.
E – denotes temporally neighboring pixels.

4   Evaluation of Existing System and Proposed System

Existing and proposed methodologies are evaluated for its performance against using four videos namely People1, People2, Cars6 and Cars7. For evaluating these videos, its properties are given in Table 1.

Table 1. Video Properties

Video Name	File Size (KBPS)	No. of Frames per Sec.
People1	2614	40
People2	2076	34
Cars6	1843	25
Cars7	1637	34

PSNR, MSE, SSIM, F-Measure, Precision, and Recall values are calculated based on existing method (DECOLOR) and Proposed method (Low Rank Based Advanced Fuzzy Aggregation Background Subtraction LR AFABS) for videos as mentioned in Table 1 and are tabulated in Table 2

Table 2. Matrices Value for Existing (DECOLOR) and Proposed (LR AFABS) Methods

Video Name	Method	PSNR	MSE	SSIM	F-Measure	Precision	Recall
People 1	DECOLOR	27.08	0.20	0.80	0.90	81.80	90.80
	LR AFABS	28.29	0.18	0.90	0.95	83.30	92.30
People 2	DECOLOR	26.38	0.19	0.70	0.82	80.80	89.30
	LR AFABS	27.89	0.17	0.80	0.85	82.30	89.90
Car 6	DECOLOR	28.63	0.16	0.80	0.92	84.10	91.50
	LR AFABS	29.56	0.15	0.90	0.94	85.80	94.20
Car 7	DECOLOR	29.58	0.18	0.70	0.82	80.80	87.80
	LR AFABS	30.29	0.17	0.80	0.86	85.30	89.30

From the above table the proposed system has certain improvements in all the metrics. The PSNR Value is increased to 5% in all cases. MSE value decreases as expected and in the proposed system same has decreased by **2% of existing method value**. For video 1 and 4 the compression ratio is fixed as 5% the MSE value differ by 2% against existing value. For video 2 and 3 the compression ratio is fixed as 10% MSE value decreases by 5%. F-measure is calculated by Precision and Recall Value and it increases as expected in the proposed method.

Considering the above outcome, it may be concluded that, it is prominently identifiable as MSE value decreases while increasing the compression value for the proposed and existing work.

5   Conclusion

A model level low rank weightage based fuzzy aggregation multi-feature similarity background subtraction algorithm using intensity and ST features is presented and its superiority over other features pixel level fusion is exposed visually and numerically for both Sugeno and Choquet integrals. Qualitative and quantitative experiments are carried out to show the effectiveness of handling various challenging situations by comparing with the PSNR, MSE, SSIM values. Advantages of Pixel and Region based features are inherited in Low rank weightage AFABS, which helps to extract moving objects in accurate and effective shape even in dynamic background with least error.

References

1. Yilmaz, A., Javed, O., Shah, M.: Object Tracking: A Survey. ACM Computing Surveys 38(4), 1–45 (2006)
2. Moeslund, T., Hilton, A., Kruger, V.: A Survey of Advances in Vision-Based Human Motion Capture and Analysis, Computer Vision and Image Understanding. Computer Vision and Image Understanding 104(2/3), 90–126 (2006)
3. Chiu, C., Ku, M., Liang, L.: A robust object segmentation system using a probability-based background extraction algorithm. IEEE Trans. Circle Syst. Video 20(4), 518–528 (2010)
4. Stauffer, C., Grimson, W.E.L.: Adaptive background mixture models for real-time tracking. Proc. IEEE Conf. CVPR 2, 246–252 (1999)
5. Tang, Z., Miao, Z.: Fast background subtraction and shadow elimination using improved Gaussian mixture model. In: Proc. IEEE Workshop Haptic, Audio, Visual Environ. Games, pp. 541–544 (2007)
6. Zhang, S., Yao, H., Liu, S.: Dynamic background modeling and subtraction using spatio-temporal local binary patterns. In: Proc. 15th IEEE ICIP, pp. 1556–1559 (2008)
7. Heikkila, H., Pietikainen, M.: A texture-based method for modeling the background and detecting moving objects. IEEE Trans. Pattern Anal. Mach. Intell. 28(4), 657–662 (2006)
8. Wang, L., Pan, C.: Fast and effective background subtraction based on ELBP. In: Proc. ICASSP, pp. 1394–1397 (2010)
9. Chiranjeevi, P., Sengupta, S.: Moving object detection in the presence of dynamic backgrounds using intensity and textural features. Electron. Imag. 20(4), 043009-1–043009-11 (2011)
10. Zhang, S., Yao, H., Liu, S., Chen, X., Gao, W.: A covariance-based method for dynamic background subtraction. In: Proc. ICPR, pp. 1–4 (2008)
11. Chiranjeevi, P., Sengupta, S.: Spatially correlated background subtraction, based on adaptive background maintenance. J. Vis. Commun. Image Represent. 23(6), 948–957 (2012)
12. Jabri, S., Duric, Z., Wechsler, H., Rosenfeld, A.: Detection and location of people using adaptive fusion of color and edge information. In: Proc. 15th ICPR, vol. 4, pp. 627–630 (2000)
13. Chen, Y., Chen, C., Huang, C., Hung, Y.: Efficient hierarchical method for background subtraction. Pattern Recognit 40(10), 2706–2715 (2007)
14. Wan, Q., Wang, Y.: Background subtraction based on adaptive nonparametric model. In: Proc. 7th WCICA, pp. 5960–5965 (2008)

Enhanced BitApriori Algorithm: An Intelligent Approach for Mining Frequent Itemset

Zubair Khan[1], Faisal Haseen[1],
Syed Tarif Abbas Rizvi[1], and Mohammad ShabbirAlam[2]

[1] Department of Computer Science and Engineering, Invertis University,
Bareilly, Uttar Pradesh, India
[2] Department of Computer Science, Jazan University, K.S.A
{zubair.762001,faisal.haseen31,
tarifabbas786110,amushabbir}@gmail.com

Abstract. Data mining is the field in which the most of the new researches and discoveries are being made and in it frequent mining of itemsets is the very critical and preliminary task. Apriori is the algorithm which is mostly used for this very purpose. Apriori also suffers from some problems such as finding the support count, which is a very time consuming procedure. To overcome the above stated problem BitApriori algorithm was devised. Though this problem was eradicated, but this algorithm suffers from a memory scarcity problem and to overcome this problem in the paper here a new Enhanced BitApriori algorithm is devised which performs better than its predecessors through the experimental results.

Keywords: BitAproiri, Enhanced BitAproiri, Eclat, FP Growth, Trie.

1 Introduction

Data mining [11] is the process of retrieving the data from very vast databases. In data mining the most frequently used technology is association rules based discovery of data. The first algorithm used for the purpose of finding frequent itemsets was given by Agarwal *et al.* [2] in 1993. After this algorithm was being proposed many other modifications were made to this algorithm and the most prominent were only considered here for the comparison purposes.

The special form of data mining which relies on association rules is cut up into two distinct forms: firstly, exploring entire possible frequent itemsets and thereafter producing association rules from the entire databases. A very crucial role is being played by frequent itemsets in association rules mining. This process consumes a lot of time. Due to the fact that these association rules are very simple and understandable and hence they have been used in various fields of businesses. The three most remarkable and noticeable algorithms are Apriori [3], Eclat, FP growth [5].As all of these three algorithms are very effective most of the algorithms developed are variations of these three algorithms.. Due to their effectiveness many of the developed algorithms are modifications of these three algorithms.

© Springer International Publishing Switzerland 2015
S.C. Satapathy et al. (eds.), *Proc. of the 3rd Int. Conf. on Front. of Intell. Comput. (FICTA) 2014*
– *Vol. 1*, Advances in Intelligent Systems and Computing 327, DOI: 10.1007/978-3-319-11933-5_37

The first algorithm was given by Agarwal *et al.* [2].After one year Apriori [3] was devised by the same authors which became the most famous of the algorithms in this field. After the emergence of Apriori [3] many modifications of it were proposed and most noticeable out of them were given by Pork *et al.* [6] which involves DHP (Direct Hashing and Pruning).Then Brin *et al.* [7] Proposed a new algorithm which involves dynamic itemsets count. During the year 1996 Zaki *et al.* [4] proposed Eclat algorithm. Eclat differs from other algorithms as in this algorithm the database is vertically represented Dong *et al.* [9] proposed an algorithm BitTable F1 with the introduction of data structure BitTable. Apriori [4] algorithm suffers from long execution time for the process of support count and also it consumes a large quantity of memory. Many other modifications were also made, but they were not very effective.

2   Problem Statement

Let $\{C_1, C_2... C_m\}$ be a collection of items, and D is a database transaction set where every transaction R is a collection of items or set such that $R \subseteq C$. Every transaction C is associated with an identifier; frequent item-sets mining can be defined as stated below. Let X be a collection of items or set, if transaction R contain X if $X \subseteq R$. An association rule is an implication of the form $X \Rightarrow Y$, where $X \subset I$, $Y \subset I$, $X \neq \phi$, $Y \neq \phi$, and $X \cap Y = \phi$. $X \Rightarrow Y$ is an association rule then it can be found in the database transaction set D with the support s, where s is the transaction's percentage in D that contain $X \cup Y$. This is taken to be the probability, $P (X \cup Y)$. If $X \Rightarrow Y$ is an association rule then in the transaction set D it has the confidence c, where c is used to show the percentage of transactions in D that contains A that also having B. The conditional probability can be defined by the rule $X \Rightarrow Y$ is $P (Y/X)$.

$$\text{Support} (X \Rightarrow Y) = P (X \cup Y) \tag{1}$$

$$\text{Confidence} (X \Rightarrow Y) = P (X/Y) \tag{2}$$

Only those rules having both a minimum support threshold (min_ sup) and a minimum confidence threshold (min_ conf) are known as strong. We used to show support and confidence value between 0% and 100%, instead of 0 to 1.0. A set or collection of items is known as an item-set. An item-set that contains k number of items is a k-item-set [12].This is the problem statement of the algorithm.

3   Two Important Properties

The most crucial properties used in the Enhanced BitAproiri algorithm for equal support pruning are described below:

3.1   Property 1

Let P, Q, R $\subseteq$ I and P$\subset$Q. If
Proof . For P$\subset$Q, so cover S (Q) $\subseteq$ cover S(P).

If supp (P) = supp (Q), then according to
definition of supp, | coverS (P) |=| coverS (Q) |.
So cover S(P) = cover S(Q),

coverS (P) $\cap$ coverS (R) = coverS (B) $\cap$ coverS (R) can
Be deduced for coverS (P$\cup$R)= coverS (P) $\cap$ coverS (R) ,
coverS (Q$\cup$R)= coverS (Q) $\cap$ coverS (R) , that means coverS (P$\cup$R)= coverS(Q$\cup$R)
,so
Supp (Q$\cup$R) = supp (P$\cup$R).

If supp (P$\cup$ {i}) = P, so the supp (P $\cup$ {i} $\cup$R) = supp (P$\cup$R). If these are equal
then generation of supersets are not required. The requirement is only that we want to
save the information, so the superset of P$\cup$ {i} is represented by a corresponding
superset of P.

3.2   Property 2

Let Q be the prefix of item-set Q$\cup$r, Where |r| =1. If Q has a subset P such that

|P|+1=|Q| and supp (P $\cup$r) = supp (P),
then Supp (Q$\cup$r) = supp (P).
Proof derived from Property 1.
   Based on property 2, equal-support pruning could be formulated. The equal-support pruning has two important steps. First, before counting the support of a candidate item-set Q$\cup$r is being counted , support of the item-set P$\cup$r is equal to the support of the frequent item-set Q, if one of its subset P has the same support with P$\cup$r. The frequent item-set Q$\cup$r is pruned, and put in the equal-support set of its parent, frequent item-set Q. Secondly, before inserting the new frequent item-set is in the Trie, Trie is firstly pruned and then it is inserted into the equal-support set of parent, if its parent support is equal to it.

4   The Enhanced BitApriori Algorithm

The Enhanced BitApriori algorithm is like a BitAproiri [1] algorithm, but the main differences is that in BitAproiri [1] used the Bitwise "And" operation on binary strings and in Enhanced BitAproiri Bitwise "Xor" operation is used on binary strings. The code for Enhanced BitAproiri is given in program code 1. By utilizing of this code first frequent item-sets will be found out by scanning the database once, after those pair i.e. 2-frequent item-sets will be finding out by scanning the database again with the help of program code 2. After finding the pair i.e. 2- frequent item-sets the data structure Trie is generated, The Trie is generated till the n-layer which is the last layer. Every node p, in the (n-1)-layer of the Trie, is merged with one of its right side sibling nodes q. Entire (n-1)-subsets are evaluated. If any one of the (n-1)-subsets falls in the equal-support set, then the item residing in node q is inserted into the equal-support set of p. If any of the (n-1)-subsets are found not to be frequent; all of its supersets are also deemed not frequent. If none of the (n-1)-subsets is infrequent, then the operation "Xor" will be performed between the string of node p and q is performed. The support of the candidate item-sets is shown by the number of "1" in the evaluated string. If the support is found to be in equivalence to p, the item residing in node q is incremented in the equal-support set of p. If the support is found not equal to p, or is found to be bigger than the minimum support, a son node of p, recording the support and the item in node q, is being inserted in the Trie.

```
public void order_frequent_ITEMS()
{
try
  {

Statement stmt=conn.createStatement();

String query="SELECT * FROM transactions";

ResultSet
rs=stmt.executeQuery(query);
while(rs.next())
{

count1++;
}
}
catch(Exception ex)
{

System.out.println("SORRY  NOT  ESTABLISHED  :::::::::::::
jhgjhg ::::: "+ex.getMessage());
  }
System.out.println(count1);

for(int i=0;i<count1;i++)
{
vv[i]=new Vector();
}
try
{
Statement st=conn.createStatement();
String      query="SELECT     *      FROM
sort_frequent_items";
```

```
ResultSet rs2=st.executeQuery(query);  while(rs2.next()
}
{
String str=rs2.getString(1);

compare_frequent_items(str);
}
}

catch(Exception ex){
System.out.println("SORRY        NOT            ESTABLISHED
...........".+ex.getMessage());}
}
```

The example of how this Enhanced BitAproiri works is shown with the help of table 1, 2 and figure 1and 2.The various items and their ordered pairs are shown in the table below. Here the min_support is found out by counting the number of items from the items list given below

Table 1. Data itemsets

Tid	No. of Items	Ordered items
1	ABCDI	ABCD
2	ABCDF	ABCF
3	ABCDL	ABDG
4	ACDKGH	ACDG
5	ABCEDJ	ABCD

Now suppose that the minimum support is 2 then items having minimum support more than or equal to 2 are selected and other items having lower support are deleted. The items are shown in the table 2 below.

Table 2. Frequent Items

ITEMS Count	A	B	C	D
SUPPORT	5	4	5	5

In table 1 first frequent item-sets is find out and in table 3 second frequent item-sets is find out by scanning the database, after finding the second frequent item-sets data structure Trie will be generated with the binary string established on each leaf of the Trie, see in figure 1.Trie after generation (1) is established in figure 1 and Trie after generation (2) is established in figure 2.

Table 3. 2-Frequent itemsets

Tid	Ordered items	AB	AC	AD	BC	BD	CD
1	ABCD	1	1	1	1	1	1
2	ABCF	1	1	0	1	0	0
3	ABDG	1	0	1	0	1	0
4	ACDG	0	1	1	0	0	1
5	ABCD	1	1	1	1	1	1
support		4	4	4	3	3	3

Now according to this data first layer of the Trie is generated. It is shown below.

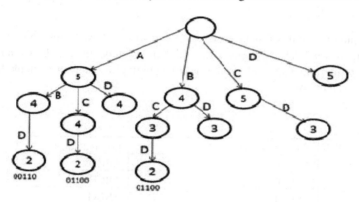

Fig. 1. Trie after generation (1)

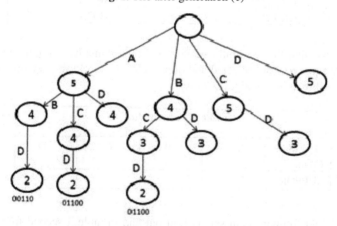

Fig. 2. Trie after generation (2)

Now to generate the third layer of the Trie the item-sets (A, B, C) is checked and since all of its subsets are frequent and there is no equal support pruning and hence "Xor" operation is employed and the result is written at the end of the node. Similarly then the item-set (A, C, D) and all of its subsets and "Xor" operation is employed similarly and the result is written at the end of the node. Similar operation is performed on item-set (B, C, D) and the result is written in the similar way as well.

For the generation of the fourth layer item-set (A, B, C, D) is checked and since the result i.e. "Xor" operation yields 01010 and since the support is equal to the (A, B, D) so there is no need to add the new son node and D is put into the equal pruning set and then in fifth layer generation of Trie there is only one node from the parent node so the height if the Trie does not increases and the frequent items are written out according to the Trie.

5   Experimental Results

The performances of the various implementations of Apriori [4] algorithm are shown in the table 6. From the table it is clearly visible that every Enhanced implementations is out running the previous one when it comes to the execution time of theirs. The table contains five datasets and their values. The values consists of five datasets namely Mushroom, Chess, Connect, Pumsb* and Pumsb.. The table shows various implementations at the different values of minimum support.

Table 4. Execution time in sec of each dataset with min_supp

Min_supp	Operation	mushroom	Chess	Connect	Pusmsb*	Pusmsb
40%	Enhanced-BitApriori	0.6198	0.3907	0.6719	10.875	11.006
	BitApriori	0.6498	0.4207	0.7219	11.875	12.006
	Apriori	13.511	9.515	14.646	237.07	239.91
50%	Enhanced-BitApriori	0.6198	0.3907	0.6719	10.875	11.006
	BitApriori	0.6942	0.4375	0.7525	12.18	12.326
	Apriori	10.908	6.875	11.825	191.4	193.69
60%	Enhanced-BitApriori	0.5702	0.3594	0.6182	10.005	10.125
	BitApriori	0.595	0.375	0.645	10.44	10.565
	Apriori	8.206	5.172	8.89	143.98	145.708
70%	Enhanced-BitApriori	0.4959	0.3125	0.5375	8.70	8.805
	BitApriori	0.5207	0.3282	0.5644	9.137	9.245
	Apriori	5.478	3.46	5.94	96.135	97.28
80%	Enhanced-BitApriori	0.4215	0.2657	0.4569	7.395	7.484
	BitApriori	0.4463	0.2813	0.4838	7.83	7.924
	Apriori	2.503	1.57	2.690	43.5425	44.064

6   Conclusion

This paper defines a new technique for the BitAproiri [1] and it is shown by both practical as well as the theoretical explanation that this algorithm out performs the previous versions in terms of the execution time. This higher efficiency is achieved through the use of "Xor" gate in the binary strings.

This paper can be very beneficial to the scholars who wants further enhancement in the algorithm. For the future aspects of the paper the efficiency can yet be improved by applying the new gates and by improving the technique of removing the infrequent data items.

References

1. Zheng, J.: An efficient algorithm for frequent itemsets in data mining: Department of Management Sciences. City University of Hong Kong, Hong Kong (2010)
2. Agrawal, R., Imielinski, T., Swami, A.: Mining association rules between sets of items in large databases. In: Proceedings of the ACM SIGMOD Conference on Management of Data, pp. 207–216 (1993)
3. Agrawal, R., Srikant, R.: Fast algorithms for mining association rules. In: The International Conference on Very Large Databases, pp. 487–499 (1994)
4. Zaki, M.J., Parthasarathy, S., Ogihara, M., Li, W.: New algorithms for fast discovery of association rules. In: Proceedings of the 3rd International Conference on Knowledge Discovery and Data Mining, pp. 283–296 (1997)
5. Han, J., Pei, J., Yin, Y.: Mining frequent patterns without candidate generation. In: Proceedings of the 2000 ACM SIGMOD International Conference on Management of Data, pp. 1–12. ACM Press (2000)
6. Pork, J.S., Chen, M.S., Yu, P.S.: An effective hash based algorithm for mining association rules. ACM SIGMOD, 175–186 (1995)
7. Brin, S., Motwani, R., Ullman, J.D., Tsur, S.: Dynamic itemset counting and implication rules for market basket data. In: Proceedings of the ACM SIGMOD International Conference on Management of Data, pp. 255–264 (1997)
8. Brin, S., Motwani, R., Silverstein, C.: Beyond market baskets: generalizing association rules to correlations. In: Proceedings of the ACM SIGMOD International Conference on Management of Data, Tuscon, Arizona, pp. 265–276 (1997)
9. Dong, J., Han, M.: BitTableFI an efficient mining frequent itemsets algorithm. Knowledge Based Systems 20(4), 329–335 (2007)
10. Frequent Itemset Mining Implementations Repository,
    http://fimi.cs.helsinki.fi
11. Kantardzic, M.: Data mining Concepts, Models, Methods, and Algorithms. Wiley Inter-science, NJ (2003)
12. Han, J., Kamber, M.: Data Mining Concepts and Techniques. Morgan Kaufmann, San Francisco (2000)

Haralick Features Based Automated Glaucoma Classification Using Back Propagation Neural Network

Sourav Samanta[1], Sk. Saddam Ahmed[2], Mohammed Abdel-Megeed M. Salem[3],
Siddhartha Sankar Nath[2], Nilanjan Dey[4], and Sheli Sinha Chowdhury[4]

[1] Dept. of CSE, University Institute of Technology, BU, Burdwan, India
sourav.uit@gmail.com
[2] Dept. of CSE, JIS College of Engineering, Kalyani, West Bengal, India
{sksaddamahmed,siddharthanath008}@gmail.com
[3] Computer and Information Sciences, Ain Shams University, Abbassia, Cairo, Egypt
salem@cis.asu.edu.eg
[4] Dept. of ETCE, Jadavpur University, Kolkata, West Bengal, India
neelanjandey@gmail.com, shelism@rediffmail.com

Abstract. According to recent researches, glaucoma, an optic nerve disease, is considered as one of the major causes which can lead to blindness. It has affected a huge number of people worldwide. Rise in intraocular pressure of the eye leads to the disease resulting in progressive and permanent visual loss. Texture of normal retinal image and glaucoma image is different. Here texture property of the total image has been extracted from both with and without glaucoma image. In this work, Haralick features have been used to distinguish between normal and glaucoma affected retina. Extracted features have been utilized to train the back propagation neural network. Classification of glaucoma affected eye is successfully achieved with an accuracy of 96%.

Keywords: Glaucoma, Classifier, Neural Network, Texture feature extraction.

1   Introduction

Glaucoma is an optic nerve disease caused due to the rise in intraocular pressure of the eye. The disease results in a progressive and permanent visual loss. Due to malfunction or malformation of the drainage system intraocular pressure increases. A small space in the front portion of the eye is called the anterior chamber. Aqueous humor is a clear fluid flowing into or out from the anterior chamber. The optic nerve that sends signal to the brain gets damaged by the increasing intraocular pressure within the eye [1]. Thus the ability to recognize images and making a correct vision gets severely affected. The techniques that one can use to detect glaucoma are described as follows: Assessment of damaged optic nerve head is an effective procedure where IOP measurement is superior and promising. A trained professional is required for optic head assessment. As manual assessment is time consuming, costly and subjective, so it would be beneficial if the optic nerve head assessment process can be automated. One strategy for automatic assessment is the binary classification between glaucomatous and healthy subjects [2]. Similarly the assessment of abnormal visual field where retinal images feature extraction process can be categorized as either texture feature extraction or structural feature extraction. The aim of this study is to extract the texture property of the total image from both with or without glaucoma and then select a set of features which are sufficient to distinguish between normal and glaucoma affected image. Neural network is often used to classify and analyze automatically the normal and the abnormal eye images. Therefore the classified properties have been used to train the back propagation neural network along with the training of classifier. On performing an analysis on the rate of classification, it has been shown that our proposed system detects glaucoma with 96% accuracy.

2   Literature Survey

A survey was performed to get an estimate of the number of people who are suffering or will suffer with glaucoma worldwide in 2010 and 2020 [3]. By using prevalence models, a review was published. In order to setup the prevalence model for OAG and ACG on the basis of ethnicity, sex and age the standard definitions must be satisfied, and the data should be proportional to sample size of each study. In order to obtain the approximate number of patients suffering from glaucoma, UN world population projections for 2010 and 2020 were combined in the model. In 2010, there will be 60.5 million people having OAG and ACG and increasing to 79.6 million by 2020 and among these 74% people will have OAG. Comparatively, in 2010, women will comprise 70% of ACG, 55% of OAG and 59% of all glaucoma patients. In 2010, bilateral blindness will be present in 3.9 million people with ACG and 4.5 million people with OAG and rising to 5.3 and 5.9 million in 2020 respectively. The second leading reason of worldwide blindness is glaucoma affecting women and Asians. A total of 3280 persons were examined in the Singapore Malay eye study [4], belonging to the age group between 40 to 80 years. This was a cross sectional survey based on population, done to examine the prevalence and types of glaucoma found in Asian Malay people. All participants underwent clinical standardized examination including dilated optic disc assessment, Goldman application tonometry, and slit-lamp bio microscopy. Visual field examination (24-2 SITA standard, Humphrey Visual Field Analyser II), repeat applications tonometry and gonioscopy, was conducted on persons who were suspected to have glaucoma. The criteria for Glaucoma were defined as per the International Society for Geographical and Epidemiologic Ophthalmology criteria. 4.6% people of 3280 participants had been diagnosed with glaucoma. An age and sex-standardized prevalence of 3.4% was observed. The age and sex-standardized prevalence of primary open-angle glaucoma was found to be 2.5%, primary angle-closure glaucoma 0.12% and that of secondary glaucoma 0.61%. The prevalence of glaucoma among old persons and Malay individuals in Singapore, aged above 40 years, is 3.4% in comparison with ethnic Chinese people in Singapore and other Asian racial/ethnic groups. Primary open angle glaucoma is the main form of glaucoma which is affecting the Chinese, Caucasians, and African population and previously more than 90% of glaucoma cases detected belonged to this category.

3   Methodology

Classification of retinal image can be performed by the following methods. To predict glaucoma or non-glaucoma/normal retinal image, several different techniques are used.

I.   Image Acquisition

Image capturing is the first phase in the fundus digital image analysis. It is generally acquired by back-mounted fundal camera (mydriatic or non-mydriatic). Digital cameras use an image sensor like Direct digital sensors which are either a Complementary Metal Oxide Semiconductor Active Pixel Sensor (CMOS-APS) or Charge-Coupled Device (CCD) (Gonzalez and Woods, 1992) [5].

In 2004, S. Y. Lee proposed that fundus camera system integrated with digital camera (D60,CanonInc) were used to acquire RNFL photographs. During image acquisition, we used green filters for enhancing the RNFL on the fundus image [6][7].

II.   Feature Extraction

In image processing or in pattern recognition, feature extraction is the process of reducing the image along its dimensionality. Haralick (1973) [8] constructed a co-occurrence matrix based fourteen statistical features defined over an image. Co-occurrence matrix elements are calculated at four directions 0^0, 45^0, 90^0, 135^0 and the mean of these values will result statistical features. The first thirteen Haralick texture features (1) to (13) are optimized, symbolized as p_i, where i represent the feature number from 1 to 13. Feature [9] is the basis of SGLD (space gray level dependence) matrices [9]. Feature number 14 (Maximum Correlation Coefficient) is not calculated. Let, H is a spatial dependence matrix. $H(i,j)$ is an element at (i,j) location in the spatial matrix(i,j) be the $(i,j)^{th}$ entry in a normalized gray tone spatial matrix dependence matrix $=H(i,j)/R$, where, R is a normalizing constant. Further, let $h_x(i)$ be the ith entry in the marginal probability matrix obtained by summation of rows of $h(i,j)$.).It is represented mathematically $as:$ $\sum\limits_{j=1}^{N_g} H(i,j)$

Similarly, $h_y(j)$ represents the jth entry in the marginal probability matrix obtained by summing the columns of $h(i,j)$. It is represented mathematically as: $\sum\limits_{i=1}^{N_g} H(i,j)$

$$h_{x+y}(k) = \sum_{\substack{i=1 \\ i+j=k}}^{N_g} \sum_{j=1}^{N_g} h(i,j), \quad \text{Where, } k=2, 3,\ldots\ldots, 2N_g$$

$$h_{x-y}(k) = \sum_{\substack{i=1 \\ |i-j|=k}}^{N_g} \sum_{j=1}^{N_g} h(i,j), \quad \text{Where, } k=0,1\ldots N_g\text{-}1$$

where, Ng is the number of distinct gray levels in the quantized image. $h_{x+y}(k)$ is the probability of occurrence matrix coordinates summing to $x+y$ and k is the index. $h_{x-y}(k)$ is the probability of occurrence matrix coordinates summing to x-y and k is the index.

Let h_x, h_y are the partial probability density (PDF) function. μ_x, μ_y are the mean of h_x, h_y and σ_x, σ_y are the standard deviations of h_x and h_y.

Using this notation, we can determine the following Haralick features:

Table 1. Haralick Features

Feature	Equation
Angular Second Moment	$P_1 = \sum_i \sum_j h(i,j)^2$
Contrast	$P_2 = \sum_{n=0}^{N_2-1} n^2 \{\sum_{i=1}^{N_2}\sum_{j=1}^{N_2} h(i,j)\}, \|i-j\|=n$ Where n is an integer
Correlation	$P_3 = \dfrac{\sum_i \sum_j (ij)h(i,j)^2 - \mu_x \mu_y}{\sigma_x \sigma_y}$
Sum of Squares: Variance	$P_4 = \sum_i \sum_j (i-\mu)^2 h(i,j)$
Inverse Difference Moment	$P_5 = \sum_i \sum_j \dfrac{1}{1+(i-j)^2} h(i,j)$
Sum Average	$P_6 = \sum_{i=2}^{2N_2} ih_{x+y}(i)$ Where x and y are the coordinates (row and column) of an entry in the co occurrence matrix
Sum Variance	$P_7 = \sum_{i=2}^{2N_2} (i-f_8)^2 h_{x+y}(i)$
Sum Entropy	$P_8 = -\sum_{i=2}^{2N_2} p_{x+y}(i) \log\{h_{x+y}(i)\} = P_g$
Entropy	$P_9 = -\sum_i \sum_j h(i,j) \log(h(i,j))$
Difference Variance	$P_{10} = \mathrm{variance\ of\ } h_{x-y}$
Difference Entropy	$P_{11} = -\sum_{i=0}^{N_2-1} h_{x-y}(i) \log\{h_{x-y}(i)\}$
Information on Measure of Correlation 1	$P_{12} = \dfrac{GXY - GXY_1}{\max\{GX, GY\}}$
Information on Measure of Correlation 2	$P_{13} = (1-\exp[-2(GXY2 - GXY1)])^{\frac{1}{2}}$ Where $GXY = -\sum_i \sum_j h(i,j) \log(h(i,j))$ GX and GY are the entropies of h_x and h_y, the partial probability density function. $GXY_1 = -\sum_i \sum_j h(i,j) \log\{h_x(i)h_y(j)\}$ $GXY_2 = -\sum_i \sum_j h_x(i)h_y(j) \log\{h_x(i)h_y(j)\}$

III.   Image Classification

Image classification is the ability to separate glaucoma and non glaucoma region by applying feature based image extraction method. Classifiers achieve great results when the underlying model applied for separation fits well with the sample distribution of data. As the underlying distribution of data is unknown so different classifiers can be tested. Mostly back propagation network (BPN) can be considered to be quintessential neural net. Back propagation [10] is the learning or training algorithm rather than a separate network itself. A BPN generally learns by example i.e. an algorithmic example mentioning what the network should do can be given and the changes made in the network weight so that, it will give desired output for a particular input when the training is complete. Back propagation networks are ideal for simple pattern recognition and mapping tasks [11].

BPN Algorithm

Initially all the inputs are applied and corresponding outputs are worked out. Here the initial outputs could be anything as initial weights are random values. The error of neuron B is found. Error is (What we want – What we get), in other words: ErrorB = OutputB (1-OutputB)(TargetB – OutputB)

Next the weight is changed. Let WAB+ be the new (trained)weight and W_{AB} $W_{AB}^{+}= W_{AB} + (Error_B x$ $Output_A)$.

The errors for the hidden layer neurons are calculated. But, these errors can't be calculated directly; therefore it is required to back propagate them from the output layer. So we need to take errors from the output neuron and run backward through its weights to obtain the hidden layer errors. After obtaining the errors for the hidden layer, neuron go to step 3 to change hidden layer weights. By repeating this method a network can be trained with any number of layers.

4   Proposed Method

Experimental dataset contains both normal and glaucoma images. Total 28 (fourteen Haralick features and its mean values) features are used to train a back propagation neural network (BPN). Fig.1 shows the proposed system diagram using BPN classifier. In this study, 90% of the overall dataset is used for training, 5% for validation and rest 5% for testing.

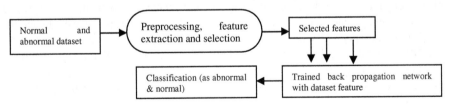

Fig. 1. Proposed System for Haralick Feature Based Glaucoma Detection using BPN

5   Result and Discussion

In this study, Haralick feature based glaucoma classification using back propagation network has been proposed. The experimental image database (RIM-ONE) contains 455 images [12] [13]. Out of these 200 images are for glaucoma affected and 255 images are normal. In our current research study, 321 images are chosen randomly out of the 455 images. Amongst them, 115 are glaucoma affected and 206 are normal. First all images are converted into grayscale images. Then from each image, Haralick features have been extracted. The extracted features are used to train the BPN. In training stage, 90% of the total dataset has been used for training the network, 5 % for validation and 5% for testing purpose. Effective classification accuracy (96%) is reported in Table 2. The major advantages of the proposed system are that the features of the entire image have been used without selecting the region of interest (ROI). This method eliminates the complexity of ROI selection. This system works well even if ROI is not very much prominent. In this current area of research, high risk in proposed system denotes glaucoma and low risk in proposed system denotes non glaucoma. While defining the standard performance metrics for sensitivity analysis were adapted by defining True Positive (TP), False Positive (FP), True Negative (TN), and False Negative (FN) as introduced in [16]. TN: is the total number of non glaucoma cases that are classified by the proposed system as non glaucoma cases. TN rate = Negatives correctly classified (TN) / Total negatives (N). TP: is the total number of glaucoma cases that are classified as glaucoma cases. TP rate = Positives correctly classified (TP) / Total positives (P). FN: is defined as the total number of glaucoma cases which are classified as non glaucoma cases. FN rate = Negatives incorrectly classified (FN) / Total positives (P). FP: is total number of non glaucoma cases that are classified as glaucoma cases. FP rate = Negatives incorrectly classified (FP) / Total negatives (N). Using these quantities, more metric measures are computed. Sensitivity is known as the probability that a classifier will give a glaucoma label from the RIM-ONE database, and is computed as TP/(TP+FN) (called as precision). Specificity is a measure of the probability that the classifier will result in a non glaucoma label when used on low risk glaucoma patients and is calculated as: TN/(TN+FP). Positive predictive value (PPV), which is defined as the probability of patient's label as glaucoma correctly diagnosed, is denoted as TP/ (TP+FP). Negative predictive value (NPV), defined as the probability of patient's label as glaucoma incorrectly diagnosed and is calculated as TN/ (TN+FN). The ratio of the number of correctly classified glaucoma patients to the glaucoma patients is termed as Accuracy, and is calculated as (TP+TN)/ (P+N) [14]. From Table 2, we can see that sensitivity and specificity obtained for the test are 99% and 90% respectively. PPV and NPV obtained for the test is 94% and 99%. Since, we know that a high specificity and high sensitivity indicates an ideal test scenario. A positive outcome in this case means condition is likely and a negative outcome in this case means condition is unlikely.

Table 2. Summary of the resulted analysis measures (Features(28): Haralick)

Sensitivity (%)	99.5146	Specificity (%)	90.4348	PPV (%)	94.9074
NPV (%)	99.0476	Accuracy %	96.262		

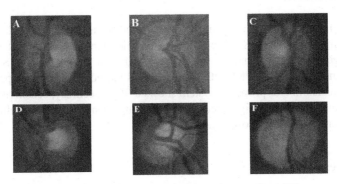

Fig. 2. Normal: (A, B, and C); Glaucoma: (D, E, and F)

A confusion matrix (Table 3) illustrates data regarding output and target classifications which are conducted by a classification system. Obtained results report the confusion matrix of our current study.

Table 3. Confusion Matrix

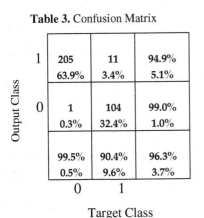

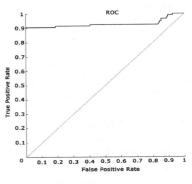

Fig. 3. ROC Curve

The ROC curve shown in Fig. 3 helps in measuring the performance of classifiers. The false positive rate is denoted on the X axis and the true positive rate on the Y axis by the plot on the ROC graph. The (0,1) point denotes the perfect classifier. It correctly performs accurate classification for all the positive and negative cases. It is denoted by the (0,1) point that the true positive rate is 1 (all) and the false positive rate is 0 (none). All cases are predicted to be negative by a classifier denoted by the

(0,0) point, whereas a classifier which predicts each and every case to be positive is denoted by the point (1,1). The classifier which represents that it is inaccurate for all the classifications is the point (1, 0).

6   Conclusion

In this current study, back propagation neural network is used to classify glaucoma and non glaucoma retinal images. Here, different Haralick features are extracted from the whole image without any manual selection of region of interest (ROI). Selection of features from ROI is always a lengthy task which is biased by person's experience and degree of precision. Our proposed system shows a 96 % accuracy, sensitivity is 99.5% and specificity is 90%. A high accuracy along with sensitivity and specificity shows the effectiveness of our proposed system.

References

[1] Varma, R., et al.: Disease progression and the need for neuroprotection in glaucoma management. Am. J. Manage Care 14, S15–S19 (2008)
[2] Bock, R., Meier, J., Michelson, G., Nyúl, L.G., Hornegger, J.: Classifying glaucoma with image-based features from fundus photographs. In: Hamprecht, F.A., Schnörr, C., Jähne, B. (eds.) DAGM 2007. LNCS, vol. 4713, pp. 355–364. Springer, Heidelberg (2007)
[3] Quigley, H.A., Broman, A.T.: The number of people with glaucoma worldwide in 2010 and 2020. Br. J. Ophthal. 90(3), 262–267 (2006)
[4] The prevalence and types of glaucoma in Malay people: the Singapore Malay eye study. Invest Ophth. Vis. Sci. 49(9), 3846–3851 (2008)
[5] Gonzalez, R.C., Woods, R.E.: Digital Image Processing, 3rd edn., pp. 849–861. Pearson, USA (2010)
[6] Lee, S.Y., Kim, K.K.: Automated Quantification of Retinal Nerve Fiber Layer Atrophy in Fundus Photograph. In: Proceedings of the 26th Annual International Conference of the IEEE EMBS, San Francisco, CA (2000)
[7] Haralick, R.M., Shanmugam, K.: Textural features for Image classification. IEEE Tran. SMC. 3(6), 610–621 (1973)
[8] Tahir, M.A., Bouridane, A., Kurugollu, F.: An FPGA Based Coprocessor for GLCM and Haralick Texture Features and Their Application in Prostate Cancer Classification, vol. 43(2), pp. 205–215. Springer Science (2005)
[9] Chandrika, S., Nirmala, K.: Comparative Analysis of CDR Detection for Glaucoma Diagnosis. Int. J. of Com. Sc. & App. 2(4) (2013)
[10] Haykin, S.: Neural Networks-A comprehensive foundation. Mac. Press, NY (1994)
[11] Alayon, S., Gonzalez de la Rosa, M., Fumero, F.J., Sigut Saavedra, J.F., Sanchez, J.L.: Variability between experts in defining the edge and area of the optic nerve head. Archivos de la Sociedad Española de Oftalmología (English Edition) 88(5), 168–173 (2013)
[12] Fumero, F., Alayon, S., Sanchez, J.L., Sigut, J., Gonzalez-Hernandez, M.: RIM-ONE: An open retinal image database for optic nerve evaluation. In: 2011 24th Int. Sym. on CBMS, pp. 1–6 (2011)
[13] Fawcett, T.: An introduction to ROC analysis. Patt. Rec. Lett. 27(8), 861–874 (2006)

ANN Based Adaptive Detection of ECG Features from Respiratory, Pleythsmographic and ABP Signals

Vinay Kumar[1], Md. Azharuddin Laskar[2], Yumnam Shantikumar Singh[3], Swanirbhar Majumdar[4], and Subir Kumar Sarkar[5]

[1] Department of ECE, NIT Meghalaya, Shillong
vinay.kumar@nitm.ac.in
[2] Department of ECE, Tezpur University, Assam
azharuddin.laskar.12@gmail.com
[3] Department of ECE, NIT Manipur, Imphal
yshantikumar99@gmail.com
[4] Department of ECE, NERIST, Itanagar, Arunachal Pradesh
swanirbhar@ieee.org
[5] Department of ETCE, Jadavpur University, West Bengal
subirsarkar@ieee.org

Abstract. This paper presents the prediction of ECG features using artificial neural networks from respiratory, plethysmographic and arterial blood pressure(ABP) signals. One cardiac cycle of ECG signal consists of P-QRS-T wave. This process of feature prediction determines the amplitudes and intervals in the ECG signal for subsequent analysis. The amplitude and interval values of ECG signal determine the functioning of heart for every human. This process is based on artificial neural network (ANN) and other signal analysis technique. In this process a feed forward multilayer perceptron network has been designed using back propagation algorithm. ECG signal is predicted from this network from the application of the respiratory, plethysmographic and ABP data to its input layer. For analyzing the data, a five point differentiation is done on the signal, so as to note the slope change of the resulting graph. Points with zero slopes were considered as the end of respective waves. The algorithm is tested with physionet database. The training and simulation results of the network have been obtained from Matlab7® software.

Keywords: Artificial Neural Network (ANN), Multilayer Perceptrons, Back Propagation Algorithm, Five Point Differentiation, ECG, Respiratory, Plethysmographic and Arterial Blood Pressure(ABP).

1   Introduction

The investigation of the ECG has been extensively used for diagnosing many cardiac diseases. The ECG works by detecting and amplifying the tiny electrical changes on the skin that are caused when the heart muscle "depolarizes" during each heart beat. One cardiac cycle in an ECG signal consists of the P-QRS-T wave. Fig. 1 shows a sample ECG signal. At rest, each heart muscle cell has a charge across its outer wall,

© Springer International Publishing Switzerland 2015
S.C. Satapathy et al. (eds.), *Proc. of the 3rd Int. Conf. on Front. of Intell. Comput. (FICTA) 2014*
– *Vol. 1*, Advances in Intelligent Systems and Computing 327, DOI: 10.1007/978-3-319-11933-5_39

or cell membrane. Reducing this charge towards zero is called de- polarization, which activates the mechanisms in the cell that causes it to contract. During each heartbeat a healthy heart will have an orderly progression of a wave of depolarization that is triggered by the cells in the sinoatrial node, spreads out through the atrium, passes through "intrinsic conduction pathways" and then spreads all over the ventricles. This is detected as tiny rises and falls in the voltage between two electrodes placed either side of the heart which is displayed as a wavy line either on a screen or on paper. This display indicates the overall rhythm of the heart and weaknesses in different parts of the heart muscle [2].ECG signal are essentially responsible for monitoring and diagnosis of patients. The predicted feature from the ECG signal plays a vital in diagnosing cardiac disease. The purpose of feature prediction is to find some properties within ECG signal that would allow successful abnormality detection and efficient diagnosis [3].

1.1   Cardiac Cycle

The mechanical pumping action of the heart is produced by cardiac muscle cells that contain contractile proteins. The timing and synchronization of contraction of these myocardial cells are controlled by noncontractile cells of the pacemaking and conduction system. Impulses generated within these specialized cells create a rhythmic repetition of events called cardiac cycles. Each cycle includes electrical and mechanical activation (systole) and recovery (diastole).The electrical recording from inside a single myocardial cell progresses through a cardiac cycle as illustrated in Fig. 1. During electrical diastole, the cell has a baseline negative electrical potential and is also in mechanical diastole with separation of the contractile proteins. An electrical impulse arriving at the cell allows positively charged ions to cross the cell membrane causing its depolarization. This movement of ions initiates electrical systole which is characterized by an action potential. This electrical event then initiates mechanical systole in which the contractile proteins within the myocardial cell slide over each other thereby shortening the cell. Electrical systole continues until the positively charged ions are pumped out of the cell causing its repolarization. The electrical potential then returns to its negative resting level. This return of electrical diastole causes the contractile proteins within the cell to separate. The cell is then capable of being reactivated if another electrical impulse arrives at its membrane.

1.2   Multilayer Perceptrons Used in Neural Networks

A neuron is an information-processing unit that is fundamental to the operation of a neural network. An artificial neural network (ANN) is actually a network made of such neurons. The input signal propagates through the network in a forward direction, on a layer-by-layer basis. These neural networks are commonly referred to as multilayer perceptrons (MLPs). MLPs have been applied successfully to solve some difficult and diverse problems by training them in a supervised manner with a highly popular algorithm known as the error back-propagation algorithm. This algorithm is based on the error-correction learning rule. As such, it may be viewed as a generalization of an equally popular adaptive filtering algorithm: the ubiquitous least-mean-square (LMS) algorithm. The error back-propagation algorithm is also referred to in the literature as the back-propagation algorithm, or simply back-prop. Henceforth we will refer to it as the back-propagation algorithm. The learning process performed with the algorithm is called back-propagation learning. The back-propagation algorithm is considered to have converged when the absolute rate of change in the average squared error per epoch is sufficiently small.

1.3   Numerical Differentiation

Numerical differentiation is a technique of numerical analysis to produce an estimate of the derivative of a mathematical function or function subroutine using values from the function and perhaps other knowledge about the function. Higher order methods for approximating the derivative, as well as methods for higher derivatives exist. Below is the five point method for the first derivative which has been used in this work:

$$f'(x) \approx \frac{-f(x+2h)+8f(x+h)-8f(x-h)+f(x\_2h)}{12h} \tag{1}$$

2   Design Algorithm Used in Analysis of ECG Signals

The network has been designed using Matlab7® software. The number of hidden layers has been set to two, with 64 and 32 neurons respectively. The number of epochs and the goal has been set to 103 and 10^{-7} respectively.

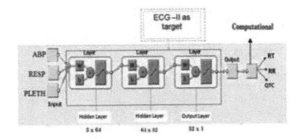

Fig. 1. The training neural network with inputs of ABP, respiration and plethysmograph

3   Results and Discussions

The Matlab program for analysis is then run using the designed network on eleven persons. It first simulates for the segmented input data consisting of respiratory, plethysmographic and ABP signals to yield the corresponding segment of ECG. The process of five point differentiation is applied on these signals and then heart rate and QT interval are calculated from each of it using the program .The results are listed below in Table-1.

Table 1. Result of analysis before removal of noise

Patients' No.	Parameters of ECG signal simulated from network		Parameters of original ECG		Parameters of Plethysmographic
	Heart rate	QT interval	Heart rate	QT interval	Heart rate
1.	75.95	0.56	75.48	0.54	75.95
2.	72.70	0.39	77.98	0.17	77.45
3.	83.85	0.73	73.19	0.44	81.95
4.	74.55	0.40	72.76	0.53	73.63
5.	73.63	0.56	68.46	0.44	73.19
6.	74.53	0.50	73.29	0.45	73.19
7.	72.33	0.41	72.33	0.48	73.63
8.	73.19	0.55	75.95	0.49	71.91
9.	74.53	0.59	73.63	0.45	78.95
10.	72.55	0.62	85.19	0.45	85.19
11.	84.51	0.46	82.87	0.50	81.95

To detail a case, we have applied the respiratory, plethysmographicic and arterial blood pressure signals of patient no.3 to the designed network, at the input layer. These signals are shown below in figure 2.

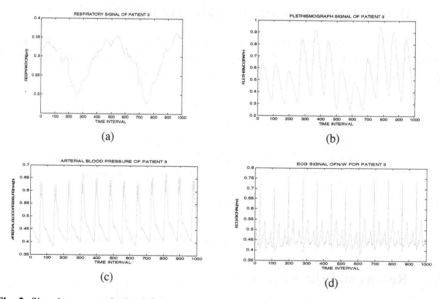

(a) (b)

(c) (d)

Fig. 2. Signal patterns obtained from one particular patient (patient no 3) (a) Respiratory; (b) Plethysmographic; (c) ABP; and (d) ECG signal obtained

The five point differentiation on the signal for the extraction of its features, as shown in figure 3. The heart rate has been compared for the 3 different types of signal i.e. the Plethesmograph, ECG and the neural network based predicted ECG signal (obtained from ABP, respiratory signal and plethesmograph). Whereas the QT interval obtained from the neural network based predicted ECG signal is compared with that of the originally recorded ECG signal. These were again compared after the removal of the respective noise via filtering. The graphs are show in figure 4 with applied curve fitting. The goodness of fits are tabulated for all the four figures of figure 4 in Tables 2 and 3.

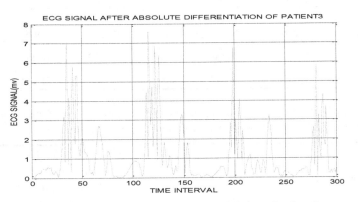

Fig. 3. ECG signal after absolute differentiation of patient 3

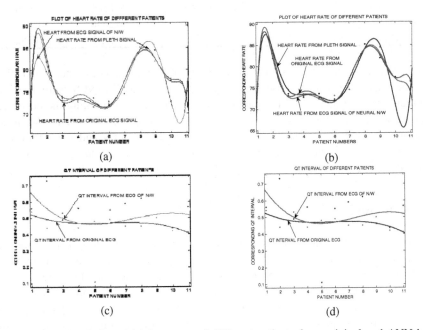

Fig. 4. Curve fitted plot of (a) heart rate of different patients from original and ANN based ECG and plethesmograph ; (b) heart rate of different patients from noise removed original and ANN based ECG and noise removed plethesmograph; (c) comparison of QT interval obtained from ECG original and neural network; and (d) comparison of QT interval obtained from noise removed ECG original and neural network

Table 2. Goodness of fit for the curve predicting heart rates via curve fitting. The data has been compared for both the noisy and noise removed version of the same.

Goodness of fit	Best fit for raw data, before noise removal for the heart rate			Best fit for the data after removal of noise for the heart rate		
	ANN predicted ECG	ECG	Pleth	ANN predicted ECG	ECG	Pleth
SSE	7.565	4.566	6.96	7.565	4.566	6.96
R- square	0.9684	0.9793	0.9664	0.9684	0.9793	0.9664
Adjusted R- square	0.8422	0.8966	0.9318	0.8422	0.8966	0.9318
RMSE	1.945	1.511	1.865	1.945	1.511	1.865

Table 3. Goodness of fit for the curve predicting QT intervals via curve fitting. The data has been compared for both the noisy and noise removed version of the same.

Goodness of fit	Best fit for raw data, before noise removal for QT		Best fit for the data after removal of noise for QT	
	ANN predicted ECG	ECG	ANN predicted ECG	ECG
SSE	0.2151	0.01025	0.2151	0.01025
R- square	0.1218	0.4446	0.1218	0.4446
Adjusted R- square	-0.2546	0.2065	-0.2546	0.2065
RMSE	0.1753	0.03826	0.1753	0.03826

The presence of noise in the data provided by physionet cannot be neglected. But since the characteristics of noise present is not important SPLINES were used to smoothen the input data of all the patients before training and adapting the designed neural network. Here the system generates the same set of SSE, R- square, Adjusted R- square and RMSE for the goodness of fit for both the noisy and noise removed version of data. This encourages the curve fitting based equation prediction. The curve-fitting done is via usage of polynomial based prediction of the curve.

4   Conclusion

This paper used the trial and error optimization technique to design a multilayer perceptron network. This process will do till the error was reduced to a small value. ECG features were used to train two networks, one with the available data and the other with smoothened data. After simulation, analysis was done on the five point differentiated signals, to obtain heart rate and the QT interval for several patients. The ECG II signals were predicted to an appreciable extent for several patients by each of the networks. The heart rates and the QT intervals as calculated by performing the analysis on these signals were found to be matching with that obtained from the analysis of the original ECG II and the plethysmographic signals of the respective patients. For a comparative study on the data obtained from the analysis of simulated signals and original signals curve fitting was done. The results obtained for heart rates and the QT intervals depicted good consistency of the networks in giving the right ECG signal.

This feature prediction is a humble step towards the ECG analysis which is a very important field in medical science. This feature prediction however could have been extended further to automate it using GUI. Other optimization techniques like genetic algorithm could have been used to better the network.

References

1. Neural Networks by Simon Haykin
2. Mahmoodabadi, S.Z., Ahmadian, A., Abolhasani, M.D.: ECG feature extraction using Daubechies wavelets. In: Proceedings of Fifth Iasted Interational Conference Visualization, Imaging and Image Processing, Spain
3. Majumder, S., Pal, S., Dutta, P.K., Ray, A.K.: Wavelet and Empirical Mode Decomposition Based QT Interval Analysis of ECG Signal. In: Proceedings of 2nd IEEE International Conference on Intelligent Human Computer interaction (IHCI 2010), pp. 247–254 (2010) ISBN No: 978-81-8489-540-7
4. Mitra, M., Pal, S., Majumdar, S.: ECG Signal Processing for Analysis of Abnormalities based on QT Interval- A Novel Approach. In: Proceeding of International Conference MS 2007, India, December 3-5 (2007)
5. Ahnve, S.: Correlation of the QT interval for heart rate: review of different formulas and the use of Bazett's formula in myocardial infarction. Am Heart J 109, 568–574 (1985)
6. Alexakis, C., Nyongesa, H.O., Saatchi, R., Harris, N.D., Davis, C., Emery, C., Heller, S.R.: Detection of Hypoglycaemia-Induced Cardiac Arrhythmias Using Neural and Static Classifiers. Neural Networks and Expert Systems in Medicine (2003)
7. McLaughlin, N.B., Campbell, R.W.F., Murray, A.: Comparison of automatic QT measurement techniques in the normal 12 lead electrocardiogram. Br.Heart J. 74, 84–89 (1995)
8. MathWorks Inc. Wavelet Toolbox for Use with MATLAB (2007)
9. Physionet ECG database
10. Mateo, J., Sanchez, C., Torres, A., Cervigon, R., Rieta, J.J.: Neural Network Based Canceller for Powerline Interference in ECG Signal. IEEE Transactions on Signal Processing 42(2)
11. Majumder, S., Pal, S., Mitra, M.: Time Plane, Feature Extraction of ECG wave and Abnormality Detection ISBN-13: 978-3-8473-3977-9, ISBN-10: 384733977X, EAN: 9783847339779

A Survey on Breast Cancer Scenario and Prediction Strategy

Ashutosh Kumar Dubey, Umesh Gupta, and Sonal Jain

JK Lakshmipat University, Jaipur, India
ashutoshdubey123@gmail.com,
{umeshgupta,sonaljain}@jklu.edu.in

Abstract. The breast cancer is one of the most critical cancer types that are found in women. Significant number of researches has been going on this disease. Post diagnosis treatments results into side effect on the patient's health. The only way to survive from this disease is its early detection. This paper is aimed to survey the frameworks proposed by researchers to detect the breast cancer at initial stage leading to its treatment with minimum side effects. The paper also suggests improvements in the direction of development of such frameworks.

Keywords: Breast Cancer, Prediction Strategy, Optimization, Data Mining.

1   Introduction

Breast cancer is by approximately the most suitable minister to murrain surrounded by women with an estimated 1.38 and 1.67 million new cancer cases diagnosed in 2008(23% of all cancers) and 2012 respectively (25% of all cancers) and ranks second overall (10.9% of all cancers)[1][4]. It is suitable the surpass regular complaint both in well-ripened and growth acumen upon surrounding 6, 90,000 new cases estimated in each region (population ratio 1:4) [3]. Incidence rates vary from 19.3 per 100,000 women in Eastern Africa to 89.7 per 100,000 women in Western Europe, and are high (greater than 80 per 100,000) in developed regions of the world (except Japan) and low (less than 40 per 100,000) in most of the developing regions [2] [3].The range of mortality rates is much less (approximately 6-19 per 100,000) because of the most favorable survival of breast cancer in (high-incidence) developed regions [3]. From the overall ratio breast cancer ranks as the fifth cause of death (458,000-deaths), but it is still the most frequent cause of cancer death in women in both developing (269,000-deaths, 12.7% of total) and developed regions, where the estimated 189,000-deaths is almost equal to the estimated number of deaths from lung cancer (188,000-deaths) [3]. The worldwide incidence and mortality is shown in table 1(incidence and mortality 2008 and 2012). Most of the cancer types can be curable if the cancer can be detected in the early stages [5]; early and precise detection is a critical factor in

© Springer International Publishing Switzerland 2015
S.C. Satapathy et al. (eds.), *Proc. of the 3rd Int. Conf. on Front. of Intell. Comput. (FICTA) 2014*
– *Vol. 1*, Advances in Intelligent Systems and Computing 327, DOI: 10.1007/978-3-319-11933-5_40

selecting the proper and effective treatment for the disease [6]. In this paper, a survey has been done to find the optimal solution in the direction of breast cancer [7]. So in this regard survey of data mining and optimization techniques are also done which are notably in cancer identification and prediction [8].

Table 1. Breast Cancer Incidence and Mortality Worldwide [3] [4]

Estimated age-standardized rates (World) (per 100,000) Estimated numbers (thousands)	Cases 2008	Deaths 2008	Cases 2012	Deaths 2012
World	1384	458	1677	522
More developed regions	692	189	794	198
Less developed regions	691	269	883	324
WHO Africa region (AFRO)	68	37	100	49
WHO Americas region (PAHO)	320	82	408	92
WHO East Mediterranean region (EMRO)	61	31	99	42
WHO Europe region (EURO)	450	139	500	143
WHO South-East Asia region (SEARO)	203	93	240	110
WHO Western Pacific region (WPRO)	279	73	330	86
IARC membership (24 countries)	740	214	940	257
United States of America	182	40	233	44
China	169	44	187	48
India	115	53	145	70
European Union (EU-28)	332	89	367	91

The rest of the paper is arranged as follows: Section 2 introduces related work; Section 3 shows the problem domains and analysis; Section 4 shows the conclusions and future work.

2   Related Works

In 2007, Tewolde et al., [9] ingenuousness, efficaciousness and flexibility of the Particle Swarm Optimization (PSO) method to propose single and multi-surface based data separation methods for classification of Breast Cancer Data were suggested. In 2010, Pang et al., [10] how the age, year and sex affect the probability of getting breast cancer by applying a new impact analysis approach was suggested. The impact analysis is an approach that is used to interpret the association of the impact factor with the acceptance inconstant, and to represent the result in terms of estimated probability [10]. The prudence has demonstrated the female incidence rate of breast cancer has the sharp increase starting from the age 15. In 2011, Malpani et al., [11] a method of data pre-processing and two different associations rule mining approaches for discovering breast cancer regulatory mechanisms of gene module were proposed. It is involved with two independent data sources: (a) a single breast cancer patient profile data file, (b) a candidate enhance information data file. It is useful also in the case of other data mining applications. In 2011, Modiri et al.,[12] particle swarm optimization (PSO) algorithm was used to estimate the permittivity's of the tissue layers at microwave frequency band. In addition, breast cancer is an appropriate candidate of microwave radiometry (MWR) due to the breast's exclusive physiology. They evaluated several algorithms for analyzing the measurement data and solving the inverse scattering problem in microwave radiometry (MWR), and different levels of accuracy. They develop two distinct algorithms for the two considered scenarios. In the first scenario, they assume no prior knowledge of the tissue under the test, whereas, in the second scenario, prior knowledge is assumed. Yet, in the favor of these researches underestimate the loss encountered by the test data samples; the methods are not valid for body tissue case. In 2011, Sbeity et al., [13], optimal uses of different medicines are necessary for controlling the number of clinically observed tumor cells was suggested. For the optimal solution, finding in the direction of cancer therapy mathematical model was suggested and provide a comparative study. The accomplishing of the optimization methods depends on the choice of the objective function and its functional relationship to the control parameters [13]. In 2012, Wang et al.,[14] a novel cancer selection method; Association Rule-based SVM-classifier (ARSVM) was proposed. It is a classifier that combines support vector machine and association rule. It can be used as feature extraction approach to catch the relation(non-linear) among different genes, and support vector machine is used to classify the transformed gene expression data [14]. It achieves both high classification accuracy and good biological interpretability. In 2011, Liu et al.,[15] a classifier using discrete particle swarm optimization (DPSO) with an additional new rule pruning procedure for detecting lung cancer and breast cancer was proposed. It improves the classification accuracy, and it is effective in making cancer prediction. In 2012, Dubey et al., [16] an efficient method for knowledge discovery which is based on subset and superset approach was proposed. A frequent superset means it contains more transactions than the minimum support. It utilizes the concept that if the itemset is not frequent, but the superset may be frequent which is considering for the further data mining task. By this approach, improved association can be achieved. In 2013, Zibakhsh et al., [17] a new memetic algorithm was proposed. It is capable of extracting interpretable and accurate fuzzy if–then rules from cancer data. In comparison to classic memetic algorithms, it enhances the rule discovery process significantly. In 2013, Nahar et al., [18] sick and healthy factors which contribute to heart disease for males and females were investigated. Association rule mining is used to identify these factors and the UCI Cleveland dataset, a biological database, is considered along with the three-rule generation algorithms Apriori, Predictive Apriori and Tertius. The resting ECG being either normal or hyper and slope being flat are potentially high-risk factors for women only. For men, on the in rotation deal out, desolate an abstemious rule expressing resting ECG being hyper was shown to be a significant factor. So for women, resting the ECG status is a key distinct factor for heart disease prediction. In 2013, li et al., [19] a bionic optimization algorithm based dimension reduction method named Ant Colony Optimization-Selection (ACO-S) for high-dimensional datasets was presented. The match close-fisted role of rove ACO-S has a notable ability to generate a gene subset with the smallest size and salient features while yielding high-classification accuracy. In 2009, Wang et al., [20] structured Support vector machine (SVM) model for determining mammographic region for normal or cancerous by considering the cluster structures in the training set was suggested. It achieves better detection performance compared with a tested SVM classifier in terms of the area under the ROC curve. In 2013, Yu et al., [21] ACO Sampling method based on the idea of Ant Colony Optimization (ACO) to address the problem of class imbalance problem frequently occurs was proposed. The above study shows that there is the need of hybrid methodology as a single methodology is not sufficient or convincing for early stage detection.

3   Problem Domains and Analysis

In 2011, National Breast Cancer Coalition (NBCC) [22] issued a progress report on breast cancer named "Breast Cancer Deadline 2020". According to the report, breast cancer will be totally uprooted by 2020 and in 90 % cases the death of breast cancer patient died due to spreading of breast cancer in other parts of body. All the breast cancer is not of the same type and their division can be done through biology of the tumor [22]. So the types can be diagnoses differently for better results. It means the symptoms of breast cancer are not the same in all patients, in view of this fact it is very necessary to characterize them and give separate treatment[22][23]. So there is the need of clustering and classification techniques. According to the estimation by National Cancer Institute (NCI) in the United States more than 288,000 women and 2,140 men developed invasive in 2011, and 39,520 women and 450 men died from the disease [24]. It can be increases in 2012 with over 290,000 women and 2,190 men predicted to receive a diagnosis [25]. In 2012, NBCC investigates to use computational and bioinformatics approaches to carry out a systematic analysis of existing and developing genomic, proteomic, glycaemic or immune system profiling data within the context of human breast cancer [22]. The detail incidence and mortality rate of US can be better analyzed by figure 1 and figure 2. The death rates shown in figure 2 shows the alarming condition. Age is also a major factor for the breast cancer. Breast cancer diagnosed in women in UK was reported in [26]. According to this report 50 % cases are diagnosed in the women in the age of 50-69. 13 % of the patients were diagnosed in the age of less than 50, 40 % diagnosed in the age between 50-69 and 47 % are diagnosed over age 70[23]. Means increasing age is also one of the strongest risk factor for breast cancer. It is shown in Figure 3. According to [27] if breast cancer is detected in the primary stage then the chances of survival is more. The chances of survival in the case small tumour diagnosis are more. The survival in the case of small tumours at diagnosis (<=10mm)[27] is around 98 % compared with 73 % for women with large tumours(>=30 mm)[28][23].

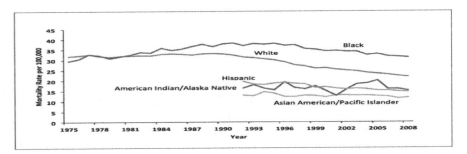

Fig. 1. Female Breast Cancer Incidence Rates by Race and Ethnicity, US, 1975-2008[22][23][25]

[Rates are age-adjusted to the 2000 US standard population]
Data source: Surveillance, Epidemiology, and End Results (SEER) Program, 1975-2008, Division of Cancer Control and Population Science, National Cancer Institute, 2012. Data for whites and blacks are from the SEER 9 registries. Data for other race/ethnicities are from the SEER13 registries. Hispanics and Non-Hispanics are not mutually exclusive from whites, blacks, Asian/Pacific Islanders, and American Indians/Alaska Natives.

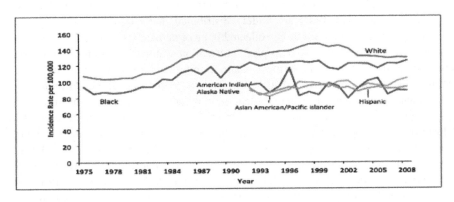

Fig. 2. Female Breast Cancer Mortality Rates by Race and Ethnicity, US, 1975-2008[22][23][25]

[Rates are age-adjusted to the 2000 US standard population]
Data source: US Mortality Files, National Center for Health Statistics, CDC. Rates for American Indian/Alaska Native are based on the CHSDA (Contract Health Service Delivery Area) counties.

Other risk factor is gender. Breast cancer chances in woman are high in comparison to the men [23]. It may affluence more frequently in western population. It had been analyzed that breast cancer cases can be high if it is a hereditary disease. Regular habits of taken alcohol intake can also be the strongest cause. Women diagnosed with invasive breast cancer have an increased risk of developing another breast cancer. Dense breast tissue on mammography is also emerging as a strong risk factor [23]. Overweight or obese increase the chances of breast cancer. Area is also a

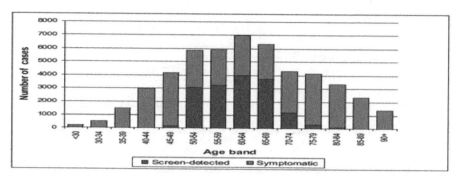

Fig. 3. Age Profile and route of presentation for women diagnosed with breast cancer in 2007[26]

major cause[23].A survey on 30 % Australian women diagnosed with breast cancer which are lived in major cities and find the chances are high in urban area in comparison to the rural area[23].

Data mining have great potential for exploring the hidden patterns in the Text sets of the medical domain. These patterns can be utilized for clinical diagnosis. Even so, the open raw medical Texts are widely distributed, heterogeneous in nature, and voluminous. These data need to be collected in an organized form. This collected data can be then integrated to form a hospital information system. Data mining technology provides a user-oriented betterment. Data mining and Observations both crack opinion discovering patterns and structures in data. Statistics deals round mongrel in large quantity unequalled, whereas data mining deals with heterogeneous fields. Association rules mining which is appropriate for finding factors can be used [18]. Clustering is also helpful because of different categorization is needed and treated the category differently. Classification is also needed as there are several factors like age, sex, gender, hereditary, alcohol intake and overweight can affect breast cancer. So there is need of strong classifier. Rule pruning sermon the business of over fitting the training data by removing the irrelevant terms from the rule, and improves the predictive power of the rule, and in the meantime simplifies it[15][29][30]. According to[15][8][30] conventional pruning procedure taken out at a time to examine the rule quality[8], for rule which there are multiple limitation conditions in one attribute, the influence of individual parameter inside each attribute is overlooked, and thus it is worth examine each parameter separately. Optimization can be used to generate the rules. In this direction different researchers use different optimization techniques. In [15] and [8] particle swarm optimization (DPSO) was used. Genetic algorithm was suggested in [31]. It can reduce both the cost and the time of rule discovery. In [19] a bionic optimization algorithm based dimension reduction method named Ant Colony Optimization-Selection (ACO-S) was suggested. It can not only select a feature subset of smallest size, but also achieve the best classification accuracy. So as per the analysis and discussion there is the need of a hybrid algorithm which can be a combination of Association Rule Mining, Data Clustering and Optimization for early stage breast cancer detection.

4   Conclusions and Future Work

In this research paper survey of related research papers has been presented and discussed the methodology used in breast cancer detection. It is concluded that age, sex, regular habits of taken alcohol, urban/rural region and weight are the major factors of Breast Cancer. The chances of breast cancer cases can be high if it is a hereditary disease. For analysis, there is a need of huge amount of data so data mining may be used. The symptoms of breast cancer are not the same in all patients, in view of this fact it is very necessary to characterize them and give separate treatment. For grouping of alike symptoms clustering can be used. The best results in different conditions may be discovered by optimizing the results.

As per our observation and discussion there are lots of factors which affect breast cancer in different ways. So if the methodologies are clubbed in a single framework the chances of finding better solution will be increased. So a hybrid framework consists of classification, clustering, association and optimization will prove to be better in the above situation.

References

1. http://www.breastcancerindia.net/
2. http://www.asianscientist.com/health-medicine/late-breast-cancer-diagnosis-linked-survival-rates-rural-india-2013/
3. Jemal, A., Bray, F., Center, M.M., Ferlay, J., Ward, E., Forman, D.: Global cancer statistics. A Cancer Journal for Clinicians 61(2), 69–90 (2011)
4. GLOBOCAN: Estimated Cancer Incidence, Mortality and Prevalence Worldwide in (2012)
5. Ott, J.J., Ullrich, A., Miller, A.B.: The importance of early symptom recognition in the context of early detection and cancer survival. European Journal of Cancer 45(16), 2743–2748 (2009)
6. Shital, S., Kusiak, A.: Cancer gene search with data-mining and genetic algorithms. Computers in Biology and Medicine 37(2), 251–261 (2007)
7. Wei-Chang, Y., Chang, W.-W., Chung, Y.Y.: A new hybrid approach for mining breast cancer pattern using discrete particle swarm optimization and statistical method. Expert Systems with Applications 36(4), 8204–8211 (2009)
8. Tiago, S., Silva, A., Neves, A.: Particle swarm based data mining algorithms for classification tasks. Parallel Computing 30(5), 767–783 (2004)
9. Tewolde, G.S., Hanna, D.M.: Particle swarm optimization for classification of breast cancer data using single and multisurface methods of data separation. In: IEEE International Conference on Electro-Information Technology, pp. 443–446. IEEE (2007)
10. Kwok-Pan, P., Ali, A.S.: Finding association of impact factor for breast cancer patient-A novel statistical approach. In: Second International Conference on Environmental and Computer Science (ICECS), pp. 1–5. IEEE (2010)
11. Rakhi, M., Lu, M., Zhang, D., Sung, W.K.: Mining transcriptional association rules from breast cancer profile data. In: IEEE International Conference on Information Reuse and Integration (IRI), pp. 154–159. IEEE (2011)

12. Arezoo, M., Kiasaleh, K.: Permittivity estimation for breast cancer detection using particle swarm optimization algorithm. In: Annual International Conference of the IEEE EMBS, pp. 1359–1362. IEEE (2011)
13. Hode, S., Younes, R., Topsu, S., Mougharbel, I.: Comparative study of the optimization theory for cancer treatment. In: International Conference on Biomedical Engineering and Informatics (BMEI), vol. 2, pp. 927–933. IEEE (2011)
14. MeiHua, W., Su, X., Liu, F., Cai, R.: A cancer classification method based on association rules. In: International Conference on Fuzzy Systems and Knowledge Discovery (FSKD), pp. 1094–1098. IEEE (2012)
15. Yao, L., Chung, Y.Y.: Mining cancer data with discrete particle swarm optimization and rule pruning. In: International Symposium on IT in Medicine and Education (ITME), vol. 2, pp. 31–34. IEEE (2011)
16. Kumar, D.A., Agarwal, V., Khandagre, Y.: Knowledge discovery with a subset-superset approach for Mining Heterogeneous Data with dynamic support. In: CSI Sixth International Conference on Software Engineering (CONSEG), pp. 1–6. IEEE (2012)
17. Zibakhsh, A., Saniee Abadeh, M.: Gene selection for cancer tumor detection using a novel memetic algorithm with a multi-view fitness function. Engineering Applications of Artificial Intelligence 26(4), 1274–1281 (2013)
18. Jesmin, N., Imam, T., Tickle, K.S., Chen, Y.-P.P.: Association rule mining to detect factors which contribute to heart disease in males and females. Expert Systems with Applications 40(4), 1086–1093 (2013)
19. Ying, L., Wang, G., Chen, H., Shi, L., Qin, L.: An ant colony optimization based dimension reduction method for high-dimensional datasets. Journal of Bionic Engineering 10(2), 231–241 (2013)
20. Defeng, W., Shi, L., Heng, P.A.: Automatic detection of breast cancers in mammograms using structured support vector machines. Neurocomputing 72(13), 3296–3302 (2009)
21. Hualong, Y., Ni, J., Zhao, J.: ACOSampling: An ant colony optimization-based undersampling method for classifying imbalanced DNA microarray data. Neurocomputing 101, 309–318 (2013)
22. http://www.breastcancerdeadline2020.org/about-the-deadline/progress-reports/2012-progress-report.html
23. Jacques, F., Shin, H.-R., Bray, F., Forman, D., Mathers, C., Parkin, D.M.: Estimates of worldwide burden of cancer in 2008: GLOBOCAN 2008. International Journal of Cancer 127(12), 2893–2917 (2010)
24. Cancer Australia. Report to the nation - breast cancer 2012, Cancer Australia, Surry Hills, NSW (2012)
25. Howlader, N., Noone, A.M., Krapcho, M., Neyman, N., Aminou, R., Altekruse, S.F., Kosary, C.L., et al.: SEER cancer statistics review(vintage 2009 populations), pp. 1975–2009. National Cancer Institute, Bethesda (2012)
26. Lawrence, G., Kearins, O., Lagord, C., Cheung, S., Sidhu, J., Sagar, C.: The Second All Breast Cancer Report. London: West Midlands Cancer Intelligence Unit. NCIN (2011)
27. Silverstein, M.J.S.M., Wyatt, J., Weber, G., Moore, R., Halpern, E., Kopans, D.B., Hughes, K.: Predicting the survival of patients with breast carcinoma using tumor size. Cancer 95(4), 713–723 (2002)
28. Australian Institute of Health and Welfare, Australian Association of Cancer Registries & National Breast and Ovarian Cancer Centre. Breast cancer survival by size and nodal status in Australia. Cancer series no. 39. Cat. no. CAN 34. Canberra: AIHW (2007)

29. Witten, I.H., Frank, E.: Data Mining: Practical machine learning tools and techniques. Morgan Kaufmann (2005)
30. Wang, Z., Sun, X., Zhang, D.: A PSO-based classification rule mining algorithm. In: Huang, D.-S., Heutte, L., Loog, M. (eds.) ICIC 2007. LNCS (LNAI), vol. 4682, pp. 377–384. Springer, Heidelberg (2007)
31. Behrouz, M.-B., Barmaki, R., Nasiri, M.: Mining numerical association rules via multi-objective genetic algorithms. Information Sciences 233, 15–24 (2013)

Closed-Set Text-Independent Speaker Identification System Using Multiple ANN Classifiers

Munmi Dutta, Chayashree Patgiri,
Mousmita Sarma, and Kandarpa Kumar Sarma

Department of Electronics & Communication Engineering, Gauhati University,
Guwahati-14, Assam, India
{munmid1,chayashreepatgiri21,go4mou,kandarpaks}@gmail.com

Abstract. This paper presents an Artificial Neural Network (ANN) based algorithm design to identify speakers of specific dialect using features obtained from various speaker dependent parameters of voiced speech. It is evident that speakers can be identified from their voiced sounds which have higher energy. Voice sounds are extracted from continuous speech signal from a set of trained male and female speakers. Here, feature vectors are generated from the speaker specific characteristics like pitch, linear prediction (LP) residual and empirical mode decomposition (EMD) residual of the speech. Using these feature vectors, three different ANN classifiers are designed using Multilayer Perceptron (MLP) and Recurrent Neural Network (RNN) to identify the speakers along with the dialect of the speaker. From the experiment, it is found that a hybrid classifier designed by combining all three classifiers correctly identifies more than 90% of the enrolled speakers.

Keywords: Artificial neural network (ANN), Multi-layer perceptron (MLP), Recurrent neural network (RNN), Pitch, Linear prediction (LP), Empirical mode decomposition (EMD).

1 Introduction

Speaker identification is divided into two categories: closed-set and open-set. A closed-set speaker identification system identifies the speaker as one of those enrolled, even if he or she is not actually enrolled in the system. On the other hand, an open-set speaker identification system should be able to determine whether a speaker is enrolled or not (impostor) and, if enrolled, determine his or her identity. The task can also be divided into text-dependent and text-independent identification. The difference is that in the first case the system knows the text spoken by the person while in the second case the system must be able to recognize the speaker from any text [1] [2] [3]. The purpose of a speaker identification system described in this paper is to determine the identity of an unknown speaker along with his or her dialectal characteristics. This paper focuses on the closed-set text-independent dialectal speaker identification based

© Springer International Publishing Switzerland 2015   377
S.C. Satapathy et al. (eds.), *Proc. of the 3rd Int. Conf. on Front. of Intell. Comput. (FICTA) 2014*
– Vol. 1, Advances in Intelligent Systems and Computing 327, DOI: 10.1007/978-3-319-11933-5_41

on voiced frames extracted from the continuous speech signal. Voiced part in a continuous speech signal has higher energy. Hence, voiced frames are used to extract different amounts of speaker discriminative information in situations where acoustic information is noise corrupted. In this work, we have trained three different ANN classifiers which categorize the set of speakers. To train the classifier feature vectors are created using speakers source information obtained from pitch, Linear Prediction (LP) residual and Empirical Mode Decomposition (EMD) residual obtained from the voiced frame.

Speech contains enormous information that allows a listener to determine speaker identity. In speaker recognition system, this information contained by the speech is used to generate the feature vector that identifies the speaker. Many methods have been used in speaker recognition system based on linear prediction coefficient [1], Mel frequency cepstral coefficient (MFCC) [2] and discrete cosine transform (DCT) [4] etc. Few studies attempt to derive speaker-specific model using predominantly the source characteristics of the speech production, and use this model for speaker recognition [5] [6]. In this paper, three speaker specific characteristics, LP residual and EMD residual along with pitch are considered for identification purpose. Artificial Neural Networks (ANN) plays an important role in pattern recognition. Various ANN structures and training algorithms are like Multi Layer Perceptron (MLP), Time Delay Neural Networks (TDNN), Back Propagation (BP), Radial Basis Function Networks (RBFN) are observed to be applied in speaker recognition purpose in the last two decades A few recently reported works are [5] [7] [8] [9] [10] [11] [12]. ANN models are composed of many non-linear computational elements operating in parallel and arranged in a form similar to that of biological neural network. The brains impressive superiority while dealing with a wide range of cognitive skills like speech recognition has motivated researchers to explore the possibilities of ANN models in the field of speech recognition [13]. These studies are motivated by the fact that human neural network like models may ultimately lead to identical performances on such complex tasks.

Speech is a time varying signal. To identify speaker from speech signal, identification system should have the ability to capture temporal information. Among the supervised learning ANNs, the RNNs has the dynamic structure with a capability learning temporal information and hence are suitable for speech based applications like speaker identification. In this work we have designed three classifiers one using MLP and other two using RNN. The MLP based classifier is trained with a feature vector created using the pitch value obtained in normal, loud and angry mode for each speaker. On the other hand the first RNN classifier is trained using LP residual and second RNN classifier is trained using EMD residual. These features are captured from the voiced part of the speech signal uttered by the speaker. Short time energy and short time zero-crossing rate is used to identify and segment the voiced part from continuous speech signal of three different modes of speaking.

Sentence1: অসম আমাৰ ৰূপহী গুণবোৰ নাই শেষ ।

ɔxɒm amar rupɒɦi gunɒrɔ nai xex

Sentence2: ঐৰাৱতে ঔটেঙা খায় ।

ɔirawɒte ɔuteɲa khaj

Sentence3: বসন্ত ঋতুত প্ৰকৃতি সুন্দৰ হয় ।

bɒxɒntɒ ritut prɒkriti xundɒr ɦɒj

Fig. 1. Sentences and their respective transcriptions

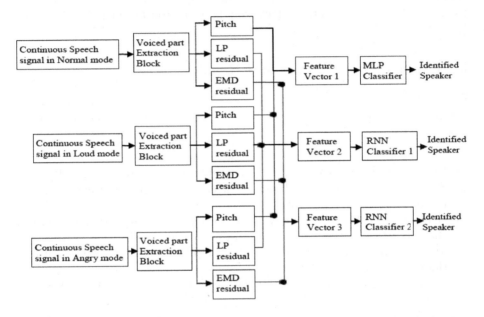

Fig. 2. Block diagram of speaker identification system based on ANN Classifier

The description included here is organized as below. A brief account about the relevant details of the speaker database collected for the work is described in Section 2. The experimental details and results are included in Section 3. Section 4 concludes the description.

2 Speaker Database

The Assamese language is the easternmost member of the Indo-European family tree. The speakers enrolled into the system are native speakers from Assamese language including male and female. Assamese, which is the most widely spoken language in the north eastern part of India, has four dialect groups of namely Eastern, Central, Kamrupi and Goalpariya groups [14] [15]. Assamese phonemic inventory consists of eight vowels and twenty-one consonants depending upon the analysis. Among the consonants presence of fifteen voiced phonemes are observed [14]. A few phonetically balanced sentences are prepared containing all the voiced sounds of Assamese language and are recorded by the trained speakers in a noise free environment. Three such sentences and their respective transcriptions are shown in Fig. 1. Each speaker repeats the sentences 10 times in three different moods. The mood variations covered normal mood, loud mood and angry mood, thus yielding a total of 90 continuous speech samples per speaker. For recording, the speech analysis software Wavesurfer and a PC headset is used. The sampling frequency is taken to be 16000 Hz. From the recorded samples, voiced part of continuous speech samples are extracted and after that pitch, LP residuals and EMD residuals are calculated, which were later used in training the MLP and RNN.

3   Experimental Details and Results

Fig. 2 shows the block diagram of the speaker identification system using ANN classifiers. The steps involved on the proposed work can be summarized below-

1. Recording of continuous Assamese speech sentences from a few speakers covering all the four dialects of Assamese language.
2. Extraction of voiced part from the recorded speech sample using STE and STZ and analysis of dialectal speaker specific characteristics using pitch, LP residual and EMD residual.
3. Generation of three different types of feature vectors from each of the said characteristics for the training and testing of said classifiers.
4. Training each of the classifier with 60 samples per dialectal speaker using those feature vectors.
5. Testing of identification algorithm with 60 samples per dialectal speaker.

3.1   Voiced Part Extraction from Continuous Speech Signal and Feature Vector Generation

Voiced part is extracted from the recorded speech samples using zero crossing rate and short term energy calculated on frame by frame basis. The short time energy and short time zero crossing rate of the speech signal can be effectively used to distinguish between voiced and unvoiced region of a speech signal. Since, the voiced speech has higher energy and lowers zero crossing rate, whereas the unvoiced speech has lower energy and higher zero crossing rate [16]. Voiced speech is segmented considering each speech segment of length 20 msec and the computation is done 10 msec overlap. After getting the voiced part, different speaker dependent parameters are determined for identification purpose of the dialectal speakers. Speaker dependent parameters used here are pitch, LP residual and EMD residual of speech signal. For each of these three parameters, three speaker identification models are designed. Feature vectors for three models are generated using separately and each feature vector contains speaker specific parameter in normal, loud and angry mood of speaking style respectively. For each speaker

60 feature vectors are used, so a total of 960 feature vectors for 16 speakers which form the pattern vector for each of the network.

3.2 Result of MLP Classifier

MLP is a feedforward ANN with one or more layers between input and output layer. Pitch frequency is determined from the extracted voiced sample using Simplified Inverse Filtering Technique (SIFT)[17]. Feature vectors are generated using pitch values of the dialectal speakers in three different moods that are normal, loud and angry. Such a feature vector is shown in Table 1. These feature vectors are applied to the MLP classifier. The network is trained with 60 feature vectors per speaker and for testing purpose 60 feature vectors were used per speakers.

Table 1. Examples of Feature Vector generated using pitch values

Gender	Normal (Hz)	Loud (Hz)	Angry (Hz)
Male	150.9434	275.8621	222.2222
Female	242.4242	347.8261	123.0769

Table 2. Recognition rate for MLP Classifier, RNN Classifier 1 and RNN Classifier 2

Individual Speakers	MLP Classifier		RNN Classifier 1		RNN Classifier 2	
	Correct Recognition (%)	Faulty Recognition (%)	Correct Recognition (%)	Faulty Recognition (%)	Correct Recognition (%)	Faulty Recognition (%)
Speaker 1	92	8	88	12	90	10
Speaker 2	95	5	92	8	92	8
Speaker 3	97	3	95	5	97	3
Speaker 4	93	7	90	10	93	7
Speaker 5	97	3	92	8	95	5
Speaker 6	95	5	90	10	95	5
Speaker 7	97	3	87	13	93	7
Speaker 8	95	5	85	15	97	3
Speaker 9	95	5	90	10	90	10
Speaker 10	97	3	92	8	93	7
Speaker 11	98	2	93	7	98	2
Speaker 12	92	8	90	10	95	5
Speaker 13	95	5	87	13	93	7
Speaker 14	97	3	90	10	90	10
Speaker 15	93	7	92	8	93	7
Speaker 16	95	5	88	12	95	5

Table 3. Recognition rate for MLP Classifier, RNN Classifier 1 and RNN Classifier 2 over dialect

Dialect	MLP Classifier		RNN Classifier 1		RNN Classifier 2	
	Correct Recog-nition (%)	Faulty Recog-nition (%)	Correct Recog-nition (%)	Faulty Recog-nition (%)	Correct Recog-nition (%)	Faulty Recog-nition (%)
Eastern	68	32	60	40	62	38
Central	65	35	63	37	60	40
Kamrupi	67	33	63	37	67	33
Goalpariya	68	32	65	35	68	32

Table 2 shows the recognition rate for MLP classifier where pitch values obtained from voiced speech in three different moods are used to form the feature vectors. Overall success rate obtained in MLP for speaker identification is approximately 95%. Table 3 shows overall success rate obtained in MLP for dialect classification of Assamese language of the speakers which is approximately 67%.

3.3   Result of RNN Classifier 1

In the second step of speaker identification model RNN classifier is used. Here, feature vectors generated from the LP residuals of the voiced speech are presented to the pattern layer of the RNN classifier. Initially the gradient descent with adaptive learning rate backpropagation algorithm is used to train the RNN with 3 hidden layer and 40 feature vectors. But recognition rate observed to be somewhat lower and requires more time. Then the Resilient Backpropagation (RBP) and Levenberg-Marquardt (LM) training algorithms are adopted for training with 20 more feature vectors, which increased the success rate to 80%. Finally Scaled Conjugate Gradient (SCG) and Bayesian Regularization (BA) algorithms is also used, which further increases the success rate to an acceptable mark and is found to be the best among all the four algorithms in terms of success rate. A comparison between all the training algorithms and their respective success rate is shown in Table 4 . Table 2 shows the identification percentage of each speaker using RNN classifier 1. It is observed that the overall recognition rate is 90% considering LP residual and RNN with three hidden layer and 60 feature vector trained with SCG learning algorithm . Table 3 shows the dialectal classification of speech sample using LP residual which is approximately 63%.

3.4   Result of RNN Classifier 2

Here, feature vectors generated from the EMD residuals of the voiced speech are applied to the RNN classifier 2. The empirical mode decomposition (EMD), a signal processing technique particularly suitable for non-linear and non-stationary series, has been proposed as a new tool for data analysis. The EMD, first introduced by N. E. Huang et al. in 1998 [18], adaptively decomposes a signal into

Table 4. Comparison of different training algorithms

Training Algorithm	Recognition Rate (%)	Training Time
Levenberg-Marquardt	64	50.20 seconds
Resilient Backpropagation	80	15.76 seconds
Scaled Conjugate Gradient	90	372.70 seconds
Bayesian Regularization	92	2510.60 seconds

Table 5. Recognition rate for Hybrid Classifier

Individual Speakers	Correct Recognition (%)	Faulty Recognition (%)	Individual Speakers	Correct Recognition (%)	Faulty Recognition (%)
Speaker 1	95	5	Speaker 9	97	3
Speaker 2	97	3	Speaker 10	98	2
Speaker 3	100	0	Speaker 11	97	3
Speaker 4	98	2	Speaker 12	98	2
Speaker 5	97	3	Speaker 13	100	0
Speaker 6	100	0	Speaker 14	97	3
Speaker 7	95	5	Speaker 15	98	2
Speaker 8	98	2	Speaker 16	98	2

Table 6. Recognition rate for hybrid Classifier over dialect

Dialect	Correct Recognition (%)	Faulty Recognition (%)
Eastern	70	30
Central	70	30
Kamrupi	75	25
Goalpariya	68	32

oscillating components called intrinsic mode functions (IMF) and a residue in accordance with different frequency bands. The design specification of the classifier 2 is same with classifier 1 i.e. RNN with three hidden layer and 60 feature vector trained with SCG learning algorithm. Table 2 shows the performance of the system using RNN classifier 2 where EMD residual is used. Overall success rate obtained for EMD residual is approximately 94% where as Table 3 shows overall performance of classifier 2 over four different dialects of Assamese which is approximately 64%.

3.5   Result of Hybrid Classifier

Here, a hybrid model is designed where a new decision block is added to take global decision depending on the local decisions taken by the three classifiers. Same feature vector is applied to all the three classifiers and decision is taken in favor of a speaker when decision of any two classifiers are same. The success rate obtained from the hybrid classifier is shown in the Table 5. The advantage of this hybrid scheme is that the correct speaker discrimination capability of the three classifiers designed using three different feature vectors are used in simultaneously. Thus if one of the classifier fails to identify one speaker then the other two can compensate so that correct identification result can be obtained. Thus, the proposed system shows effective performance in categorizing the set of speakers using source information obtained from pitch, LP and EMD residual parameters obtained from the voiced frame. The system tested for Assamese language with different dialect groups namely Eastern, Central, Kamrupi and Goalpariya establishes the effectiveness to a certain extend of a hybrid ANN classifier framework for closed set, text-independent speaker identification application.

4   Conclusion

In this paper, we have reported some experimental work on dialectal speaker identification of Assamese language using speaker specific feature obtained from voiced sounds and three ANN based classifiers. The speaker specific information is obtained from pitch, LP residual and EMD residual. It has been observed that the hybrid classifier obtained by combining the three classifiers shows around 98% success rates in case of the enrolled speakers whereas dialectal classification of Assamese speaker shows 71%. The work shows certain new directions towards design of voiced sound based bio-inspired speaker identification system. To improve the performance of the dialectal classification by adopting some mathematical and signal processing domain approach is the future challenge of the work.

References

1. Badran, E.F.M.F., Selim, H.: Speaker Recognition Using Artificial Neural Networks Based on Vowel phonemes. In: In Proceedings of 5th International Conference on Signal Processing Proceedings (ICSP 2000), pp. 796–802 (2000)
2. Hossaina, M., Ahmedb, B., Asrafic, M.: A Real Time Speaker Identification using Artificial Neural Network. In: Proceedings of 10th International Conference on Computer and Information Technology (ICCIT 2007), Dhaka, pp. 1–5 (2007)
3. Campbell Jr., J.P.: Speaker Recognition: A Tutorial. Proceedings of the IEEE 85(9) (1997)
4. Vyas, G., Kumari, B.: Speaker Recognition System Based On MFCC and DCT. International Journal of Engineering and Advanced Technology (IJEAT) 2(5) (2013)
5. Yegnanarayana, B., Sharat Reddy, K., Kishore, S.P.: Source and System Features for Speaker Recognition using AANN models. In: Proceedings of IEEE International Conference of Acoustics, Speech, Signal Processing, Salt Lake City, pp. 409–412 (2001)
6. Pati, D., Prasanna, S.R.M.: Speaker Recognition from Excitation Source Perspective. IETE Technical Review 27(2) (2010)

7. Sarma, M., Sarma, K.K.: Speaker Identification Model for Assamese Language using a Neural Framework. In: Proceedings of International Joint Conferrence on Neural Network (IJCNN), Dallas, Tx, United States (2013)
8. Sarma, M., Sarma, K.K.: Vowel Phoneme Segmentation for Speaker Identification Using an ANN-Based Framework. Journal of Intelligent Systems 22(2), 111–130 (2013)
9. Lajish, V.L., Kumar, R.K.S., Vivek, P.: Speaker Identification using a Nonlinear Speech Model and ANN. International Journal of Advanced Information Technology 2(5), 15–24 (2012)
10. Ranjan, R., Singh, S.K., Shukla, A., Tiwari, R.: Text-Dependent Multilingual Speaker Identification for Indian Languages Using Artificial Neural Network. In: Proceedings of 3rd International Conference on Emerging Trends in Engineering and Technology, pp. 632–635. ABV-IIITM, Gwalior (2010)
11. Bennani, Y., Gallnari, P.: Neural networks for discrimination and modelization of speakers. Speech Communication 17, 159–175 (1995)
12. Mak, M.W., Allen, W.G., Sexton, G.G.: Speaker identification using multilayer perceptron and radial basis function networks. Neurocomputing 6(1), 99–117 (1994)
13. Hu, Y.H., Hwang, J.N.: Handbook of Neural Network Signal Processing. The Electrical Engineering and Applied Signal Processing Series. CRC Press, USA (2002)
14. Goswami, G.C.: Structure of Assamese, 1st edn. Department of Publication, Gauhati University, Guwahati, Assam, India (1982)
15. Assamese, Resource Centre for Indian Language Technology Solutions, Indian Institute of Technology, Guwahati
16. Rabiner, L.R., Schafer, R.W.: Digital Processing of Speech Signals. Pearson Education, New Delhi (2009)
17. Markel, J.D.: The SIFT algorithm for fundamental frequency estimation. IEEE Transaction on Audio and Electroacoustics AU-20, 367–377 (1972)
18. Huang, N.E., Shen, Z., Long, S.R., Wu, M.C., Shin, H.H., Zheng, Q.: The empirical mode decomposition method and the Hilbert spectrum for nonlinear and non-stationary time series analysis. Proceedings of the Royal Society of London, Series A, 903–995 (1998)

Application of Data Mining Techniques for Defect Detection and Classification

B.V. Ajay Prakash[1], D.V. Ashoka[2], and V.N. Manjunath Aradya[3]

[1] Dept. of Information Science and Engineering, SJBIT, Bangalore
[2] Dept. of Computer Science and Engineering, JSSATE, Bangalore
[3] Dept. of Master of Computer Application, SJCE, Mysore
ajayprakas@gmail.com

Abstract. In order to overcome the software development challenges like delivering a project on time`, developing quality software products and reducing development cost, software industries commonly uses defect detection software tools to manage quality in software products. Defects are detected and classified based on their severity, this can be automated in order to reduce the development time and cost. Nowadays to extract useful knowledge from large software repositories engineers and researchers are using data mining techniques. In this paper, software defect detection and classification method is proposed and data mining techniques are integrated to identify, classify the defects from large software repository. Based on defects severity proposed method discussed in this paper focuses on three layers: core, abstraction and application layer. The designed method is evaluated using the parameters precision and recall.

Keywords: Software Bugs Tracking, Data Mining Techniques, Software Quality Assurance, Bug Tracking, Bug Classification.

1 Introduction

A Software defects/bugs can be defined as "state in a software product that fail to meet the software requirement specification or customer expectations". In other words defect is a failure in coding or logic error that makes a program to produce unexpected results. In software development life cycle (SDLC), defect management is the integral part of development and testing phases. Defect management process consists of defect prevention, defect identification, defect reporting, defect classification and prediction. Defect prevention activities are to find errors in software requirements and design documents, to review the algorithm implementations, defects logging and documentation, root cause analysis. Defect identification is to identity the code rule violations as the code is developed, changed and modified. Defect reporting is to clearly describe the problem associated with particular module in software product so that developer can fix it easily. Defect classification is the process of classifying the defects based on severity to show the scale of negative impact on quality of software. Defect prediction objective is to predict number of defects or

© Springer International Publishing Switzerland 2015
S.C. Satapathy et al. (eds.), *Proc. of the 3rd Int. Conf. on Front. of Intell. Comput. (FICTA) 2014*
– *Vol. 1*, Advances in Intelligent Systems and Computing 327, DOI: 10.1007/978-3-319-11933-5_42

bugs in software product, before the deployment of software product, also to estimate the likely delivered quality and maintenance effort. Defects can be in different phases of SDLC, table 1 shows the percentage of defect introduced in each phase.

Table 1. Division of percentage of Defects Introduced into Software in each Phase of SDLC, (source: Computer Finance Magazine)

Software Development Phases	Percent of Defects Introduced
Requirements	20 %
Design	25 %
Coding	35 %
User Manuals	12 %
Bad Fixes	8 %

Nowadays different data mining techniques are used in defect management to extract useful data from software historical data. Data mining is the process of extracting patterns from data. Data mining, or knowledge discovery, is the computer-assisted process of digging through and analyzing enormous sets of data and then extracting the meaning of the data. A Typical knowledge discovery process (KDP) is shown in Figure 2. KDP may consist of the following steps: data selection, data cleaning, data transformation, pattern searching (data mining), and finding presentation, finding interpretation and evaluation. Data collection phase is to extract the data relevant to data mining analysis. The data should be stored in a database where data analysis will be applied.

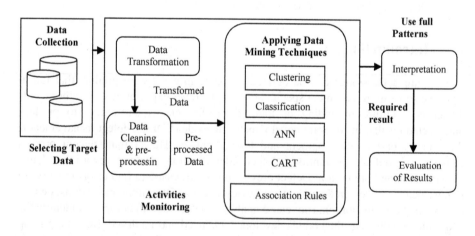

Fig. 1. Knowledge Discovery Process

Data cleaning and preprocessing and data transformation phase of KDP involves data cleansing and preparation of data, converting the data suitable for processing and obtaining valid results to achieve the desired results. Activity monitoring module deals with real time information. The purpose of data mining (DM) phase is to analyze the data using appropriate algorithms to discover meaningful patterns and rules to produce predictive models. Data mining is the most important phase of KDP cycle. After building data warehouse data mining algorithms such as clustering, classification, artificial neural networks, rule association, decision tree and classification and regression trees (CART) or chi-square automatic induction (CHAID) are applied. Interpretation and evaluation of results is the final phase involves making useful decisions by interpreting and evaluating results obtained from applying knowledge discovery techniques.

Developing a software product without defect is difficult, project managers and software engineers put more effort to minimize the defects. If the defects are detected after delivering the software product the project budget and resources cost will be more compared to cost of early detection. Software defects can be tracked and managed using software tools such as Bugzilla [15] Perforce [16], Trac [17], Fossil [18]. Software repositories contains historical data such as code bases, execution traces, historical code changes, contains a wealth of information about a software project's status, progress, and evolution [1]. Software defects data will be in formats like CSV (comma separated value), HTML (Hyper Text Markup Language) or XML (Extensible Markup Language) as software repositories. New technologies and techniques are required to reuse the existing knowledge from software repositories.

This paper is organized as follows: Section 2 gives detailed survey of different knowledge discovery techniques applied to various aspects of the defect management activities while section 3 proposed method to identify and classify the defects. Section 4 presents the results we have obtained so far. Finally, in conclusion and future work, we conclude the paper and present our goals for future research.

2   Related Work

Several research works has been carried out on defect management process by many authors. Honar and Jahromi [13] introduced a new framework for call graph construction to be used for program analyses , they choose (ASM and Soot) as a byte code reader for their environment to store information about the structure of the codes such as classes, methods, files and statements. Kim et. al [2] proposed new techniques for classifying the change requests as buggy or clean. Antoniol et al. [3] highlighted that not all bug reports are associated to software problems, but also change request bug reports. Fluri et. al. [4], proposed an semi-automated approach using agglomerative hierarchical clustering to discover patterns from source code changes types. Fenton et. al [16], presented an empirical study modeled by pareto principle on two versions of large-scale industrial software and shown the results of distribution failure and faults in software product.   Jalbert and Weimer [5] have proposed method to identify he duplicate bug reports automatically from software bug repositories. Cotroneo et. al. [6] performed failure analysis of Java Virtual Machine (JVM). Guo et al [9] explored factors which are affecting in fixing the bugs for Windows Vista, and Windows 7 systems. Davor et. al [8], have proposed an approach for automatic bug tracking using text categorization. They proposed a prototype for bug assignment to developer using supervised Bayesian learning algorithm. The evaluation shows that their prototype can correctly predict 30% of the report assignments to developers. Lutz et. al [12], analyzed the causes and impact of defects in requirements during testing phase, resulting from non documented changes and guidelines are given to respond to the situations. Kalinowski et al [7], aware that defect rates can be reduced by 50%, rework can be reduced, quality and performance is improved using Defect Causal Analysis (DCA). Then DCA is enhanced and named it as Defect Prevention Based Process Improvement (DPPI) to conduct, measure.

3   Proposed Method

The proposed method basically classified into four major steps: defects identification phase, applying data mining techniques and defects classification based on severity measures. The block diagram of proposed method is as shown in Fig 2.

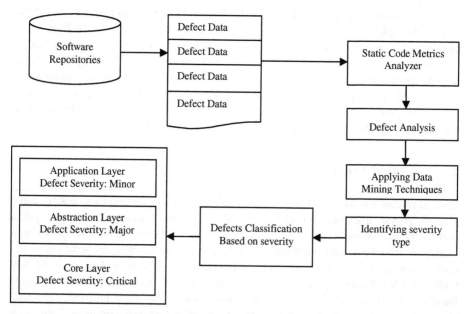

Fig. 2. Steps in defects identification and classification

We have collected online available Promise repository NASA MDP software defect data sets [20]. Defect data sets are retrieved and stored in the file system. For pre-processing the defect data attributes are parsed and some attributes are selected for measurements using various software metrics. Based on bugs severity proposed method is divided into three layers: core, abstraction and application layer. Core layer contains the core functionality of the system, the defects in this layer affects critical functionality of the system and leads to failure of the project. An abstraction layer defect affects major functionality or major data of the system. Application layer contains user oriented functionality responsible for managing user interaction with the system defects in this layer affects minor functionality of the system.

Steps to identify defects and classification

Step1: Collect the data sets from different defect data sources (Eg.: Promise repository, NASA MDP repository)

Step 2: static code metrics are used to measure the source code, various metrics used are:
McCabe Metrics; Cyclomatic Complexity and Design Complexity, Halstead Metrics; Halstead Content, Halstead Difficulty, Halstead Effort, Halstead Error Estimate, Halstead Length, Halstead Level, Halstead Programming Time and Halstead Volume, LOC Metrics; Lines of Total Code, LOC Blank, Branch Count, LOC Comments, Number of Operands, Number of Unique Operands and Number of Unique Operators, and lastly Defect Metrics; Error Count, Error Density, Number of Defects (with severity and priority information).

Step 3: Defects data are extracted from software repositories and stored in to the database for pre-processing
    1.#defect Analyser
    2.Start
    3.prev = NULL
    4.while (iteration < 25)
    5.{
    6. if (iteration!=1)
    7.{
    8.   cur = getcurrentchanges( );
    9.sold = comparechanges(prev, cur)
    10.if(sold >= accptablelimit)
    11.{
    12. Buglist = findbugs(changed modules)
    13.Updatewith severity(buglist)
    14.}
    15.}
    16.prev = cur;
Step 4: Applying data mining techniques to identify the defects severity and classify
    17.#defect severity identified ( )
    18.Builds static module mapping

19. Hash < module name, module type>
20. Enum { module type} Core Module, Abstraction Layer, Application
    Layer
21. Get module type (module name)
Step 5: Defect Classification
22. get moduletype(module name)
23. {
24. return hashtable get(name)
25. }

We have used weka 3.6 software tool to apply the data mining techniques Using Naïve Bayes, J48 and Multilayer Perceptron (MLP) classification techniques, we have applied the classification for input attributes and the setting the test mode to 10-fold validation.

4   Performance Evaluation

4.1   Accuracy

Accuracy for defect classification can be calculated as ratio of number of defects correctly classified to the total number of defects. The equations 1 and 2 are used to calculate the accuracy:

$$Accuracy = \frac{TP + TN}{TP + TN + FP + FN} \qquad (1)$$

$$Accuracy(\%) = \frac{CorrectlyClassfiedSoftwareDefects}{TotalSoftwareDefects} * 100 \qquad (2)$$

Where, true positive (TP), indicates a positive value that the source code has correctly classified as positives. TN is true negatives the proposed method identified defect as negative. FP is false positive, negative values the method identified as positives and FN is false negatives, positives values that the method identified as negative.

4.2   Precision

Precision for a class is the ratio of the number of correctly classified software defects and the actual number of software defects which was assigned to the type. Precision is the measurement of correctness and is also defined as the ratio of the true positives (TP) to total positives (TP + FP) and is calculated using Equation (3).

$$Pr\,ecision = \frac{TP}{TP + FP} \qquad (3)$$

4.3   F-Measure

F-Measure is a measure which combines the harmonic mean of recall and precision it is calculated using equation 4:

$$F = 2 * \frac{Precision.recall}{Precision + recall} \qquad (4)$$

From the NASA MDP, we have applied Naïve Bayes, J48 and Multilayer Perceptron (MLP) classification techniques on various data sets like CM1, JM1, KC1, KC3, MC1, MC2, MW1, PC1, PC2 and PC3. The performance results are shown in table 2. MLP achieves significantly better results compare to other two classifies i.e. J48 and Naïve Bayes. Our proposed method divides the source code defects into three layers namely core, abstraction and application. Core layer corresponds to condition which can cause failure of the software products or harm the life of the end user these defects are termed as critical defects. Abstraction layer corresponds to condition which can cause major functionality of the software, these defects are termed as major defects. Application layer corresponds to condition which contains user oriented functionality responsible for managing user interaction with the system defects in this layer affects minor functionality of the system.

4.4   Results

Table 2. Performance of Various Classifiers

Data Sets	Classifiers	TP Rate	FP Rate	Precision	Recall	F-Measure
	J48	0.817	0.717	0.801	0.817	0.808
CM1	Naive Bayes	0.792	0.66	0.804	0.792	0.798
	MLP	0.847	0.794	0.797	0.847	0.816
	J48	0.764	0.628	0.725	0.764	0.736
JM1	Naive Bayes	0.782	0.651	0.739	0.782	0.742
	MLP	0.79	0.711	0.75	0.79	0.727
	J48	0.748	0.504	0.725	0.748	0.729
KC1	Naive Bayes	0.727	0.538	0.699	0.727	0.706
	MLP	0.758	0.582	0.733	0.758	0.714
	J48	0.794	0.562	0.776	0.794	0.783
KC3	Naive Bayes	0.789	0.52	0.782	0.789	0.785
	MLP	0.768	0.611	0.75	0.768	0.758
	J48	0.974	0.892	0.963	0.974	0.967
MC1	Naive Bayes	0.889	0.661	0.962	0.889	0.922
	MLP	0.974	0.913	0.962	0.974	0.967
	J48	0.712	0.374	0.705	0.712	0.707
MC2	Naive Bayes	0.72	0.432	0.712	0.72	0.696
	MLP	0.68	0.423	0.669	0.68	0.672
	J48	0.87	0.76	0.838	0.87	0.85
MW1	Naive Bayes	0.814	0.414	0.874	0.814	0.837
	MLP	0.866	0.668	0.852	0.866	0.858

Table 2. (*continued*)

	J48	0.901	0.722	0.88	0.90	0.88
PC1	Naive Bayes	0.881	0.59	0.886	0.881	0.883
	MLP	0.921	0.616	0.907	0.921	0.911
	J48	0.972	0.979	0.957	0.972	0.965
PC2	Naive Bayes	0.907	0.858	0.959	0.907	0.932
	MLP	0.976	0.979	0.957	0.976	0.967
	J48	0.85	0.655	0.831	0.85	0.839
PC3	Naive Bayes	0.357	0.162	0.859	0.357	0.409
	MLP	0.863	0.762	0.821	0.863	0.834

5   Conclusion and Future Work

As the requirement becomes complex IT industries needs new techniques and approaches to reduce the development time and cost, also to satisfy the customer's needs. As a result, Data mining techniques are widely used to extract useful knowledge from large software repositories to adopt it in bug detection to improve software quality. This paper, adapted software defects detection and classification methods by integrating data mining techniques to identify, classify the defects from large software repository. Work discussed here uses Core, abstraction and application layer based on defects severity in the software product for classification. The designed method discussed here is evaluated using the parameters precision, recall and F-measure. MLP achieves significantly better results compare to other two classifies i.e. J48 and Naïve Bayes. As a future work, different machine learning algorithms may be included in the experiments on different data sets.

References

1. Ajay Prakash, B.V., Ashoka, D.V., Aradhya, V.N.: Application of data mining techniques for software reuse process. Procedia Technology 4, 384–389 (2012)
2. Kim, S., Whitehead Jr., E.J., Zhang, Y.: Classifying Software Changes: Clean or Buggy? IEEE Transactions on Software Engineering 34(2), 181–196 (2008)
3. Antoniol, G., Ayari, K., Penta, M.D., Khomh, F., Guéhéneuc, Y.G.: Is It a Bug or an Enhancement? A Text-Based Approach to Classify Change Requests. In: Proceedings of the 2008 Conference of the Center for Advanced Studies on Collaborative Research, New York, pp. 304–318 (2008)
4. Fluri, B., Giger, E., Gall, H.C.: Discovering Patterns of Change Types. In: Proceedings of the 23rd International Conference on Automated Software Engineering (ASE), L'Aquila, September 15-19, pp. 463–466 (2008)
5. Jalbert, N., Weimer, W.: Automated Duplicate Detection for Bug Tracking Systems. In: IEEE International Conference on Dependable Systems & Networks, Anchorage, June 24-27, pp. 52–61 (2008)
6. Cotroneo, D., Orlando, S., Russo, S.: Failure Classification and Analysis of the Java Virtual Machine. In: Proceedings of the 26th IEEE International Conference on Distributed Computing Systems, Lisboa, July 4-7, pp. 1–10 (2006)

7. Kalinowski, M., Mendes, E., Card, D.N., Travassos, G.H.: Applying DPPI: A Defect Causal Analysis Approach Using Bayesian Networks. In: Ali Babar, M., Vierimaa, M., Oivo, M. (eds.) PROFES 2010. LNCS, vol. 6156, pp. 92–106. Springer, Heidelberg (2010)
8. Čubranić, D.: Automatic bug triage using text categorization. In: SEKE 2004: Proceedings of the Sixteenth International Conference on Software Engineering & Knowledge Engineering (2004)
9. Guo, P.J., Zimmermann, T., Nagappan, N., Murphy, B.: Characterizing and Predicting which Bugs Get Fixed: An Empirical Study of Microsoft Windows. In: ACM International Conference on Software Engineering, Cape Town, May 1-8, pp. 495–504 (2010)
10. Fenton, N.E., Neil, M.: A critique of software defect prediction models. IEEE Transactions on Software Engineering 25, 675–689 (1999)
11. Yorozu, Y., Hirano, M., Oka, K., Tagawa, Y.: Electron spectroscopy studies on magneto-optical media and plastic substrate interfaces (Translation Journals style). IEEE Transl. J. Magn. Jpn. 2, 740–741 (1982)
12. Lutz, R.R., Mikulski, C.: Requirements discovery during the testing of safety-critical software. In: Presented at the Proceedings of the 25th International Conference on Software Engineering, Portland, Oregon, pp. 578–583 (2003)
13. Honar, E., Jahromi, M.: A Framework for Call Graph Construction, Student thesis at School of Computer Science. Physics and Mathematics (2010)
14. Boetticher, G., Menzies, T., Ostrand, T.: PROMISE repository of empirical software engineering data, West Virginia University, Department of Computer Science (2007), http://promisedata.org/repository
15. Bugzilla, http://www.bugzilla.org
16. Perforce, http://www.perforce.com
17. Trac Integrated SCM & Project Management, http://trac.edgewall.org
18. Fossil, http://www.fossil-scm.org

Sentence Completion Using Text Prediction Systems

Kavita Asnani, Douglas Vaz, Tanay PrabhuDesai, Surabhi Borgikar,
Megha Bisht, Sharvari Bhosale, and Nikhil Balaji

Department of Computer Engineering
Padre Conceição College of Engineering
Goa, India
{kavitapcce,Doug.A.Vaz}@gmail.com

Abstract. Text Prediction for sentence completion is a widely used method to enhance the speed of communication as well as reducing the total time taken to compose text. This paper briefly describes the approaches, design and implementation issues involved, as well as the factors and parameters that determine effectiveness of a system. The information is then used to build a software system, capable of modeling text data, in order to generate predictions in real-time. By using a pure statistical approach, we generate N-gram models that are adaptive to users by applying instance based learning. Details of the software development method, used to prototype and iteratively build a highly effective system, are provided.

Keywords: text prediction, statistical NLP, N-gram model.

1 Introduction

Text prediction for sentence completion refers to the capability of adding or filling in words, given a context or prefix. It is one of the most popular methods to enhance the rate of communication. Such techniques are especially helpful to people with cognitive or motor impairments that limit their use of conventional modes of text composition[2]. Prior to the emergence and widespread use of modern computers, printed boards containing characters and words were used as an aid. This necessitated knowledge of the language morphology and syntax, as well as the context of the conversation. The same technique may be reproduced programmatically. Prior studies have shown that statistical models demonstrate a high precision for sentence completeion[1]. A key point is that text composition happens in real-time, and any effective algorithm must be able to calculate predictions faster than performing manual input.

The underlying software must be flexible enough to work with user interfaces that are customized for different applications and tastes. A single global coordination among all applications is preferred as fragmentation may be stressful for a user that needs to use several applications if each predicts differently. Therefore,

© Springer International Publishing Switzerland 2015   397
S.C. Satapathy et al. (eds.), *Proc. of the 3rd Int. Conf. on Front. of Intell. Comput. (FICTA) 2014*
– Vol. 1, Advances in Intelligent Systems and Computing 327, DOI: 10.1007/978-3-319-11933-5_43

the challenge is to develop a system that can model natural languages, personalize these models based on a user's style, provide a flexible interface so that several applications can interact with it and performs adequately for real-time use. An operational prototyping approach is used to construct software based on theoretical computational algorithms. The implementation is then tested against speed and memory constraints, iteratively building a high-performance system.

2   Related Work

There exist several vastly different methods for text prediction, making it difficult to select one that best fits a given application[3]. The approaches used may be broadly categorized as follows.

2.1   Syntactic

This method involves using the syntactic properties of a natural language to make predictions. There are two common strategies used for implementing such a system:

The first method is to use probability tables. Two probabilities are maintained, viz. the probability of appearance of each word and the relative probability of appearance of every syntactic category to a preceding category. Words in the language vocabulary are marked by a category and frequency or probability of occurrence. Adaptation of these systems is performed by updating the probabilities of the table and the frequencies in the lexicon.

Another method involves the use of grammars along with NLP techniques.

The syntactic approach has several issues with regard to adaptations. Proposals tend to appear in different places of order as the frequency changes. For new words to be added, the semantic tag as well as other morpho-syntactic information may need to be provided.

2.2   Semantic

The semantic approach to text prediction is the hardest with a high computational complexity as compared to the other approaches.

Each word has an associated semantic category (or a set of semantic categories), similar to the syntactic categories of the previous approach. The remaining characteristics (working method, complexity, dictionary structure, adaptations, etc.) are very similar to the syntactic approach using grammars.

The major difference here is the use of semantic categories. Semantics is the study of meaning. It focuses on the relation between signifiers, like words and phrases, signs, and what they stand for. Specifying semantic categories for symbols is a non-trivial task. Automatic allocation methods, such as the use of semantic clustering and classification[5], are infeasible for a real-time system.

2.3 Statistical

As demonstrated in the past, sentence completion can be achieved using statistical methods[1]. This approach uses *Statistical Language Modeling* (SLM) as the basis for generating predictions. SLM is the science of building models that estimate the prior probabilities of word strings. Large amounts of text are used to automatically determine the Language Models (LM) parameters[6,7]. Algorithms base the output on the probability or frequency distributions of words and symbols over a text corpus.

Several reasons make this approach favorable:

- Text modeling does not require any additional information other than the frequencies or probabilities of words, which is easily obtained from a plain text corpus.
- The model can be personalized and adapted for a user by considering text used as input to the system. Weights may be added to give user text a higher priority.
- Modeling algorithms used may be language independent since neither the syntactic or semantic meanings of symbols are taken into account.

One way to look at the statistical or empirical approach is by comparing it with the way humans learn natural languages. An assumption is made that a baby does not begin with detailed sets of principles and procedures specific to the various components of language and other cognitive domains. The brain then learns a language by recognizing patterns and generalizing from the rich sensory inputs. This is known as the mechanism of cognition[8]. A direct implementation of this for text prediction is the *Cogent Confabulation* method[9].

Confabulation uses unsupervised machine learning algorithms for constructing a large knowledge base (KB) that gathers language sense by extracting semantic connection between words from a large vocabulary. The drawback of this approach is the high costs of time and memory requirements, making it unsuitable for real-time adaptations to the prediction model.

An alternate for implementing statistical models is the use of word tries. Language Models based on this make use of *N-grams*, which is a contiguous sequence of n items from a given sequence of text. Hence the models are commonly referred to as N-Gram models. The N-Gram probabilities are computed by enumerating relative frequency on a large dataset. It is based on Markov's assumption that "The probability of a word depends only on the probability of a limited history". This is called the *Maximum Likelihood Estimation* (MLE).

N-Gram Models have the added advantage of instance based learning, and show a high level of accuracy within the context of the training corpus.[1]

$$\text{Unigrams: } P(W_1) = \frac{F(W_1)}{N} \tag{1}$$

$$\text{Bigrams: } P(W_2 \mid W_1) = \frac{F(W_1, W_2)}{F(W_1)} \tag{2}$$

$$\text{N-grams: } P\left(W_n \mid W_{n-N+1}^{n-1}\right) = \frac{F\left(W_{n-N+1}^{n-1}, W_n\right)}{F\left(W_{n-N+1}^{n-1}\right)} \tag{3}$$

Where:

$P=$ Probability of a word, $F=$ Frequency of a word, $N=$ Total number of words

However, such models tend to be large, consisting of billions of n-grams introducing storage challenges[10].

The rest of the paper focuses on implementing a highly efficient system that uses a pure statistical approach to construct practical N-gram Language Models and use these for predicting words.

3   Design Issues

Several design issues with regard to the software and corpora need to be resolved before going into implementation details:

3.1   Handling Sparse Symbols

In statistical models, a sparse number of words occur frequently whereas an enormous number of words are seen only once. To estimate the likelihood of unseen N-grams, *smoothing* is performed. Smoothing is the process of decreasing the probability of seen words, in order to account for unseen words. Laplacian Smoothing was used while constructing the language model.[13]

3.2   Data Structures for Model Representation

The choice of data structures has a huge impact on the performance and capabilities of the system. Non-linear data structures such as graphs make it easy to represent models but incur the additional costs involved in marshaling and unmarshaling that is necessary for storage.

Hash tables are an attractive option as they provide constant time lookups by mapping a prefix to it's possible completions. However the low time complexity does not justify the high space complexity. While developing the prediction system it was noticed that hash tables for a large number of N-grams exhausts the available memory on typical consumer devices.

Therefore we present a representation based on Implicit Tries[15] that was found to be best suited for use in generating real-time predictions with adaptations.

3.3 Language and Corpus Considerations

Statistical models offer a representation of the vocabulary of a language. A likely question would be "how many words does a language have?". Regardless of whether this is deterministic[1], we are more interested in the parts of the language that are commonly used.

Consider the vocabulary provided by the Oxford English Corpus (OEC)[16]:

Table 1. Use of English words in the Oxford English Corpus

Vocabulary Size	% of Content	Example Lemmas
10	25%	*the, of, and, to, that, have*
100	50%	*from, because, go, me, our, well, way*
1000	75%	*girl, win, decide, huge, difficult, series*
7000	90%	*tackle, peak, crude, purely, dude, modest*
50,000	95%	*saboteur, autocracy, calyx, conformist*
>1,000,000	99%	*laggardly, endobenthic, pomological*

The distribution of word usage suggests that associated information entropy is lower than expected. A large amount of space may be saved by encoding lemmas (words) using variable length code words[17].

Indirectly, this also reduces the cost of performing sequential lookups in the model when the sequence is sorted by probability of occurrence.

4   Implementation Details

The target software system was implemented as a system service with a shared library to access the service. This design choice enables the writing of separate interfaces.

In order to support a large number of locales, inputs were handled as Unicode strings. These were then tokenized using boundary analysis to identify words and characters. Internally, all string comparisons were performed on canonical forms to ensure accuracy.

We now look at the detailed designs involved in implementation:

4.1 Obtaining N-Gram Sequences

The first step in model construction is obtaining all N-gram sequences from text data. The general procedure is as follows:

---

[1] The main argument here is what should be regarded as a "word" since so many dialects, technical terms, slang words etc. exist.

```
tokens := list of words obtained from the input;
grams[1..N];

for i := 1 to N:
    j := 1;
    while j+i <= length(tokens):
        grams[i].append(tokens[j..j+i])
        j = j + 1
```

For a constant depth N, the time complexity is linear bound on input size n i.e. $O(n)$

Our implementation considers a fixed N of 5 as per the configuration used by Google Books Ngram Corpus (Michel et al.,. 2011). The British Academic Written English (BAWE) Corpus[12] was used to train the model.

Once N-gram sequences have been obtained, the challenge lies in picking a suitable representation.

4.2   Representation Using Hash Tables

Since the aim is to build a real-time prediction system, we start with a structure that offers the least search time (constant time complexity). A two-level hash scheme is used. *token*

1. Let *WordCount* be a hash tables which map a string (key) to a number (value). Let *Model* be a hash table which maps a prefix string to key-value pairs of suffix to N-Gram probability.
2. For each element *token* in the N-gram sequence of text being modeled:
   (a) If *token* is not in *WordCount*, add a new element with value 1.
   (b) If *token* exists in *WordCount*, increment it's value.
3. For every key in *WordCount*, calculate it's probabilty.
4. Add N-Gram sequences to *Model* with their coresponding frequencies.

The table is densely loaded. Space utilization could be up to 31% more than space of entries in the table.

4.3   Representation Using Tries

Tries offer constant time lookups that depends only on the depth. Besides, they are more space efficient than hash tables and may be recursively constructed.

Modeling:

1. Each node consists of *word* and *frequency* and a list *ChildNodes*.
2. Set em CurrentNode to the root.
3. For each word in the text string:

(a) If the word is in *ChildNodes*, increment it's frequency.
(b) Else, insert a new node into *ChildNodes* with frequency 1.
4. Repeat for every element in *ChildNodes* while limiting the recursion depth to N=5.

The string trie obtained may be reduced to a compressed prefix trie while still maintaining fast insertion and lookups.

4.4   Representation Using Sorted Arrays

1. Tries may be maintained implicitly using sorted arrays[13].
2. With this approach, search occurs with a time complexity of $O(n)$, where n is the number of nodes in the array. As stated in Section 3.3, the search space is drastically reduced due to a frequency sorted representation.
3. Over time, frequently used words are promoted towards the beginning of the array. User words not previously in the model were weighed to be promoted faster than words added during initial training.
4. Worst case insertions occur when a new word is to be added to the model.

5   Observations

For the Hash Table implementation in Section 4.2, although search and insertions happen in constant time, overall physical memory utilized by the model was over 3 GB, rendering it infeasible for practical use.

In Section 4.3, space utilization for Tries was 37% of the Hash Table implementation, which is still impractical for use reads into physical memory. Lookups are slower than in hash tables, requiring at most 5 comparisons. Also, serializing the model for persistence required the traversal of every node in the tree.

Implicit Array Tries described in Section 4.4 may be maintained with 32-bits per node (inclusive of encoding strings to integers), using approximately 50 MB of memory. Lookups were reasonably fast regardless of the linear search being performed. Since arrays occupy contiguous blocks of memory, writing to and reading from the secondary storage requires just one operation per array.

6   Conclusion

An effective word prediction system was developed using N-Gram Language Models. Based on observations, we conclude that (a) a statistical approach with instance based learning performs well, improving accuracy over time, and (b) that Sorted Array Tries, a space efficient linear data structures may be utilized to provide a real-time prediction system. It's also noted that content of the text corpus has a strong effect on the results. A favorable corpus may be selected based on the entropy of it's text.

Future Enhancements: Loading the entire model into memory greatly simplifies the task of searching and updating by paying a large price for space. The use of an index so that only requested branches be read from secondary memory will further reduce the space used.

References

1. Bickel, S., Haider, P., Scheffer, T.: Learning to complete sentences. In: Gama, J., Camacho, R., Brazdil, P.B., Jorge, A.M., Torgo, L. (eds.) ECML 2005. LNCS (LNAI), vol. 3720, pp. 497–504. Springer, Heidelberg (2005)
2. Zagler, W., Beck, C.: FASTY - faster typing for disabled persons. In: Proceedings of the European Conference on Medical and Biological Engineering (2002)
3. Garay-Vitoria, N., Abascal, J.: Text prediction systems: a survey. Universal Access in the Information Society 4(3), 188–203 (2006)
4. Garay-Vitoria, N., Abascal, J.: A Comparison of Prediction Techniques to Enhance the Communication Rate. In: Stary, C., Stephanidis, C. (eds.) UI4ALL 2004. LNCS, vol. 3196, pp. 400–417. Springer, Heidelberg (2004)
5. Baker, L.D., McCallum, A.K.: Distributional clustering of words for text classification. In: Proceedings of the 21st ACM-SIGIR International Conference on Research and Development in Information Retrieval (SIGIR) (1998)
6. Rosenfeld, R.: Two decades of statistical language modeling: Where do we go from here? Proceedings of the IEEE 88 (2000)
7. Rosenfeld, R.: Adaptive statistical language modeling: a maximum entropy approach. Diss. IBM (2005)
8. Hecht-Nielsen, R.: Confabulation Theory: The Mechanism of Thought. Springer (August 2007)
9. Qiu, Q., et al.: Confabulation based sentence completion for machine reading. In: 2011 IEEE Symposium on Computational Intelligence, Cognitive Algorithms, Mind, and Brain (CCMB). IEEE (2011)
10. Brants, T., Popat, A.C., Xu, P., Och, F.J., Dean, J.: Large language models in machine translation. In: Proceedings of the Conference on Empirical Methods in Natural Language Processing (2007)
11. Manning, C.D., Schütze, H.: Foundations of Statistical Natural Language Processing. MIT Press (1999) ISBN 0-262-13360-1
12. British Academic Written English (BAWE) corpus, Hilary Nesi, Sheena Gardner (Centre for Applied Linguistics, Warwick), Paul Thompson (Department of Applied Linguistics, Reading) and Paul Wickens (Westminster Institute of Education, Oxford Brookes)
13. Chen, S.F., Goodman, J.: An Empirical Studies of Smoothing Techniques for Language Modelling. TR-10-98 August (1998)
14. Magnuson, T., Hunnicutt, S.: Measuring the effectiveness of word prediction: The advantage of long-term use. Speech, Music and Hearing 43, 57–67 (2002)
15. Pauls, A., Klein, D.: Faster and smaller n-gram language models. In: Proceedings of the 49th Annual Meeting of the Association for Computational Linguistics: Human Language Technologies, vol. 1. Association for Computational Linguistics (2011)
16. Oxford English Corpus: Facts about the language. Oxford University Press
17. Shannon, C.E.: A mathematical theory of communication. The Bell System Technical Journal 27(3), 379–423 (1948)

Real Time Intrusion Detection and Prevention System

Poonam Sinai Kenkre, Anusha Pai, and Louella Colaco

Depatment of Information Technology,
Padre Conceicao College of Engineering,
Goa University
{pookamat,anusha.pai}@gmail.com,
lmesquita@rediffmail.com

Abstract. Major challenge for organizations in today's era is to meet the security needs. Techniques for logging data, detecting intrusions, preventing intrusions have been evolving for years. This paper presents a solution to combine logging, and network based intrusion detection and prevention system. The system has been developed considering the Software Engineering framework of requirements analysis, design, implementation, and testing. For IPS open source tool snort is configured in inline mode, so that sensors captures packets and drops them in case of any suspicious activity is observed. Signatures for Detection and prevention help IPS in alerting the administrator and dropping the packet. Logging of dropped packets is done using Splunk. In the process it provides for cost effective, customizable and scalable solution alternative to Organizations.

Keywords: Software Engineering in IPS, Intrusion Detection, Inline mode, sensors, Intrusion Prevention.

1   Introduction

Requirements of customers have to be translated into a software product; the process used to achieve the above target is be called as software Engineering. This process helps to achieve the target in efficient manner with proper use of resources and budgeted cost criteria. The software engineering terminology used in this paper is for the application of real time intrusion detection and prevention system development.

Society increasingly relies on computing environments ranging from simple home networks, commonly attached to high speed Internet connections, to the largest enterprise networks spanning the entire globe. This increased reliance and convenience; coupled with the fact that attacks are concurrently becoming more prevalent has consequently elevated the need to have security controls in place to minimize risk as much as possible. An intrusion detection system (IDS) is software that automates the intrusion detection process. Network-Based IDS (NIDS) monitors network traffic for particular network segments or devices and analyzes the network activity to identify suspicious activity. Intrusion Prevention System (IPS) is used to actively drop packets of data or disconnect connections that contain unauthorized data. Intrusion detection systems are classified in accordance with the objects of the

© Springer International Publishing Switzerland 2015   405
S.C. Satapathy et al. (eds.), *Proc. of the 3rd Int. Conf. on Front. of Intell. Comput. (FICTA) 2014*
– *Vol. 1*, Advances in Intelligent Systems and Computing 327, DOI: 10.1007/978-3-319-11933-5_44

system detection, they are divided into the host-based, the network based, and the hybrid IDS. In accordance with system architecture, they are divided into centralized and distributed IDS; and finally in accordance with the detection type, they can be divided into anomaly-based model and misuse-based model IDS.

2   Current Contributions from Research Community

Software Engineering is an area of research for long period. Many researchers have done studies on applications on software engineering for real time networks intrusions. Lots of research on application of software engineering practices is done by various authors. Authors in [1] have compared various software life cycle models in terms of their common, distinct and unique features. Comparison of different software development models is discussed by authors in [2]. Authors here also discussed a practical approach to implement these models. Using of architecture designs in the software development life cycle is shown by the authors in [3]. As there is rapid change in the field of real time technology, researchers have focused on integrating software engineering practices with real time system development.

There has been many advancements in the field of network intrusion detection and prevention. Authors in [4] include building the compiling environment on windows and analyzing the workflow and rule tree, they also further explain how VC++ can be used to analyze and modify snort. They made use of Cygwin simulation environment of Linux in windows. Authors of [5] developed a NIDS using Java programming language. They simulated land attack, the flooding attack and the death's ping attack. The sniffer mode is done using Jpcap library. Novel string matching technique in [6] is an optimization of other matching algorithms. Authors have shown that, algorithm breaks the string into small sets of state machines. Each state machine recognizes the subset of string. If any suspicious behavior occurs then the system broadcasts the information about intruder to every module (state machine) which holds the database in order to define rules and compares the signatures of intruder with predefined detected signatures. Authors in [7],provides a mechanism where IDS is acting like a centralized system. IDS is installed on a single system and all traffic is diverted through this server machine. So less overhead in installing the system on different hosts. Authors in [8] have presented a paper on Security Patterns for Intrusion Detection Systems. They have presented a pattern for abstract IDS that define their general features and patterns for Signature-Based IDS and Behavior-Based IDS. Authors in [9] have presented four different algorithms namely, Multilayer Perception, Radial Base Function, Logistic Regression and Voted Perception. These data mining classifier algorithms help in classifying attacks. Authors conclude that Multilayer Perceptron feed forward neural network, has got highest classification accuracy and lowest error rate as compared to other neural classifier algorithm that are discussed. Feature reduction technique is also discussed by the authors in [9] to enhance the results. In this paper pure signature based intrusion detection and prevention system is discussed. Snort along with its rules is configured to prevent intrusions from further succeeding in network.

3   Case Study on Intrusion Detection and Prevention System

Intrusion detection and prevention system that is developed, follows phases of waterfall model. Phases are explained below.

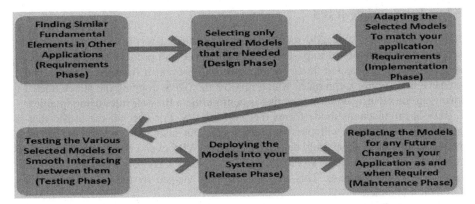

Fig. 1. Intrusion Detection System via Software Engineering Framework

Requirement Phase: The different software tools required to make IDPS were analyzed. Also attacks that are possible and different ways of handling those attacks are seen and analyzed. Snort is selected as intrusion detection and prevention tool.

Design Phase: Design model for IDPS is developed, which has both intrusion detection and intrusion prevention. Snort components are selected to get the IDPS working.

Implementation Phase: All the components are combined together to make the application ready. Snort is configured to work in promiscuous mode to provide basis for intrusion detection. Also, Splunk is configured, such that packets captured by snort have to be logged into Splunk. Modify the Snort configuration file, so that snort recognizes the path for rules (signatures), also, Snort should know HOME_NET IP address and EXTERNAL_NET IP address. Signatures are written for the attacks to avoid intrusions.   (Eg: a intruder is trying to do telnet on snort IDS machine, then IDS should trigger an alert to administrator).This alert gets logged by default to var/log/snort and also in splunk. To make snort act as IPS, it is configured in inline mode, here snort alerts the admin about the intrusion and also drops the packet from entering the network. For this, Iptables are configured, rules are written in Iptables such that all the packets snort is capturing is queued, and given to the snort IPS for prevention.

Testing Phase: All the components of Snort, Splunk are tested for their functioning. Snort is tested for packet capture. Signatures are tested with the captured packets. In case of intrusion, alerts are generated from snort, that are then logged into output module-Splunk. All these modules are tested and results are achieved.

Release Phase: Set up of network is done with four machines connected in network. One machine is completely blocked, that is, any request from that machine is dropped. Another machine is treated as external machine, and rules are applied for packets coming from external IP machines. All alerts are logged into Splunk.

3.1   The Principle of IPS

IDPS designed will typically record information related to observed events, notify security administrators of important observed events, and produce reports. IDPS can also respond to a detected threat by attempting to prevent it from succeeding. There are several response techniques, which involve the IDPS stopping the attack itself by changing the security environment that is configuring a firewall rules using Iptables.

IDS in promiscuous mode forms the base for IPS. For IDS, all the traffic that is generated on network will pass through sensor but also pass through the internal network. This mechanism only detects the attacks. Whenever a packet that is analyzed matches with the signature stored on sensor machine, alert gets triggered and event gets logged into database. This alert can be only viewed by the administrator as a intimation of attack.

Fig No 2, Describes working of Intrusion Prevention system. Here all packets compulsory pass through sensor machine that acts as IPS machine. Therefore in case of any attacks taking place, its signature will be triggered, alert will be sent to administrator and also response action will be taken on the packet, depending upon the rules written in the firewall of the IPS machine.

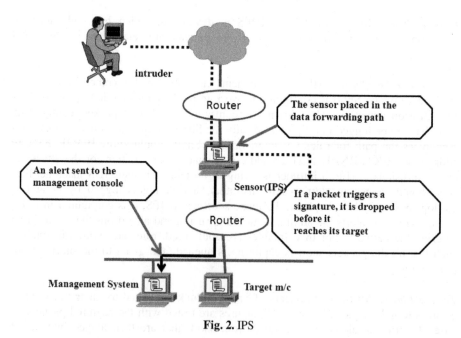

Fig. 2. IPS

3.2   The Functions of the System

3.2.1   Intrusion Detection and Prevention

The first function of IPS is to assist network and system administrators to analyze their installed IT components in the network and to detect potential vulnerabilities. For this purpose IPS uses open source software tool Snort which is utilizing a rule-driven language. Snort is used primarily to monitor network traffic and generate alerts when threats are detected. Snort is configured in inline mode and deployed on a server that forwards/routes network traffic as opposed to only sniffing network traffic and, Snort "alert" rules are changed into "drop/reject" rules. Iptables firewall application is included and so, Snort in Inline mode interacts with Iptables to receive and process network traffic. Appropriate Iptables rules are used to direct network traffic to Snort Inline for inspection according to Snort rules. Given this interaction between Snort Inline and Iptables, successful configuration of Snort Inline depends on successful configuration of Iptables.

3.2.3   Scenario of Network

A network of four machines is established. One machine (SM1) on which snort is installed acts as server machine. Remaining three machines (M2, M3, and M4) are connected to SM1 machine via modem. All machines are assigned static IP's. Snort.conf file is updated to store the IP's of machines. SM1 is treated as HOME_NET and remaining are EXTERNAL_NET. Also, rules files of different protocols are saved.

4   Results

Testing of IDS: For testing purpose, signatures for following protocol are discussed in this paper. FTP, TCP, ICMP, Telnet, and HTTP. Whenever EXTERNAL_NET machine wants to perform (for eg: FTP) then snort would send an alert regarding same, but the packet would penetrate in the network. These alerts by default gets logged into var/log/snort, and also, in Splunk. Splunk is data analysis tool that is available. Splunk shows alerts that are logged (fig 3), in detail. It also gives wide coverage of different alerts that are captured by snort. Analysis on top Destination IP's (fig 4) from where the packet originates can be given, also the top Signatures (fig 5) that are hit most of the time can be analyzed. Count for the events can be also found.

Testing of IPS: In IDS, since snort only detects, intruder gets the privilege of entering the network and cause damage to it. Therefore in IPS, the main aim is to restrict the packet to enter the network. For this purpose the packets that snort captures had to be queued and sent to Iptables, where the rules were cross checked with captured packets. If any rule is matched then, that packet is dropped and also reported to Splunk.

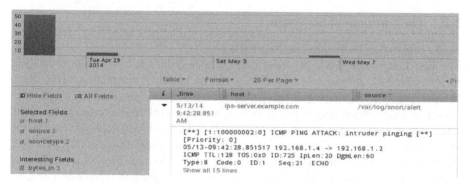

Fig. 3. Logged Alert with Overall Alert Logged Graph

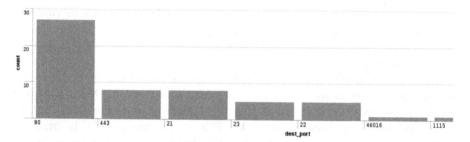

Fig. 4. Top Destination Ports

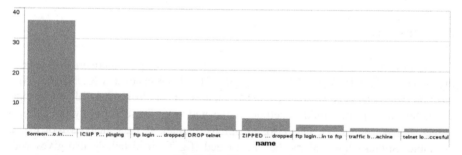

Fig. 5. Top Signatures

5   Conclusion

Integration of software engineering practices in development of real time intrusion detection and prevention system has been discussed in this paper. IDPS helps in prohibiting the suspicious packets that may cause harm to network. This approach provides a means to a cost effective and customizable solution to small organizations. Limitation of this work is, introduction of new attacks coming into existence. Signatures based IDPS will only recognize the attacks for which signatures have already been written, In case the Signature is not present or there is little change in packet data and contents of signature, the intruder packet will be sent to inside network, hence causing harm. Future prospects of this paper would be, to write signatures for other protocols and viruses, Also, a separate tool may be designed to generate graphs, and reports of snort alerts.

References

1. Rodríguez, L.C., Mora, M., Alvarez, F.J.: A Descriptive Comparative Study of the Evolution of Process Models of Software Development Life Cycles (PM-SDLCs). In: Proc. of Mexican International Conference on Computer Science 2009, pp. 298–303 (2009)
2. Unphon, H.: Making use of architecture throughout the software life cycle - How the build hierarchy can facilitate product line development. In: Proc. of ICSE Workshop on Sharing and Reusing Architectural Knowledge 2009, pp. 41–48 (2009)
3. Wild, C., Maly, K., Zhang, C., Roberts, C.C., Rosca, D., Taylor, T.: Software engineering life cycle support-decision based systems development. In: Proceedings of IEEE Region Ninth Annual International Conference TENCON 1994, vol. 2, pp. 781–784 (1994)
4. Zhou, Z., Zhongwen, C., Tiecheng, Z., Xiaohui, G.: The study on network intrusion detection system of Snort. In: 2nd International Conference on Networking and Digital Society (ICNDS), vol. 2, pp. 194–196 (2010)
5. Ezin, E.C., Djihountry, H.A.: Java-Based Intrusion Detection System in a Wired Network (IJCSIS) International Journal of Computer Science and Information Security 9(11), 33–40 (2011)
6. Tan, L., Sherwood, T.: A High Throughput String Matching Architecture for Intrusion Detection and Prevention. In: Proceedings of the 32nd Annual International Symposium on Computer Architecture, pp. 112–122 (2005)
7. Ahmed, M., Pal, R., Hossain, M., Bikas, A.N., Hasan, K.: NIDS:A network based approach to intrusion detection and prevention. In: 2009 International Association of Computer Science and Information Technology - Spring Conference, IACSIT-SC, pp. 141–144 (2009)
8. Kumar, A., Fernandez, E.B.: Security Patterns for Intrusion Detection Systems. In: 1st LACCEI International Symposium on Software Architecture and Patterns (LACCEI-ISAP-MiniPLoP 2012), Panama City, Panama, July 23-27 (2012)
9. Singh, S., Bansal, M.: Improvement of Intrusion Detection System in Data Mining using Neural Network. International Journal of Advanced Research in Computer Science and Software Engineering 3(9) (September 2013)

Early Detection and Classification of Breast Cancer

V. Vishrutha and M. Ravishankar

Department of Information Science and Engineering,
Dayananda Sagar College of Engineering, Bangalore, India
{v.vishrutha18,ravishankarmcn}@gmail.com

Abstract. Breast cancer is one of the most common cancers among women. About two out of three invasive breast cancers are found in women with age 55 or older. A Mammogram (low energy X ray of breast) done to detect breast cancer in the early stage when it is not possible feel a lump in the breast. In this paper we have proposed a method to detect microcalcifications and circumscribed masses and also classify them as Benign or malignant. The proposed method consists of three steps: The first step is to find region of interest (ROI). The second step is wavelet and texture feature extraction of ROI. The third step is classification of detected abnormality as benign or malignant using Support vector machine (SVM) classifier. The proposed method was evaluated using Mini Mammographic Image Analysis Society (Mini-MIAS) dataset. The proposed method has achieved 92% accuracy.

Keywords: Breast cancer, circumscribed mass, early detection, microclacification, texture analysis, wavelet feature extraction.

1   Introduction

Breast cancer is a type of cancer originating from breast tissue, most commonly from the inner lining of milk ducts or the lobules of breast and then metastasizes to other areas of the body. In India, over 100,000 women are newly diagnosed with breast cancer every year; and has overtaken cervical cancer to become the leading cause for death among women in metropolitan cities [1]. Breast Cancer is second most common cancer all over world [2]. More than 60% of breast cancers are detected in the advanced stage and hence death rate from breast cancer are also high [3]. Hence early detection of breast cancer is essential in effective treatment and in reducing the number of deaths caused by breast cancer.

Early breast cancer is subdivided into two major categories, microcalcifications and circumscribed masses. Microcalcifications ($Ca_5(PO_4)OH$) are tiny deposits of calcium[4]. They appear as localized high-intensity regions (bright spots) on mammograms. The diameter of microcalcifications range from 0.1mm to 1mm and have an average diameter of about 0.3 mm [5].Circumscribed masses are tumors which are formed When an healthy cell DNA is damaged resulting in unchecked growth of mutated cells. A tumor can be benign or malignant. Benign tumors are not considered cancerous. Their cells are close to normal in appearance, they grow

© Springer International Publishing Switzerland 2015
S.C. Satapathy et al. (eds.), *Proc. of the 3rd Int. Conf. on Front. of Intell. Comput. (FICTA) 2014*
– *Vol. 1*, Advances in Intelligent Systems and Computing 327, DOI: 10.1007/978-3-319-11933-5_45

slowly, and they do not invade nearby tissues or spread to other parts of the body. Malignant tumors are cancerous. Left unchecked, malignant cells eventually can spread beyond the original tumor to other parts of the body.

There are several imaging techniques for examination of the breast, including magnetic resonance imaging, ultrasound imaging, and X-ray imaging. Mammogram is considered most reliable method in early detection of breast cancer. Mammography uses low dose X-ray (around 30kVp) rays to produce an image that allows visualization of the internal structure of the breast. On average, mammography will detect about 80–90% of the breast cancers in women without symptoms. Mammography offers high-quality images at a low radiation dose, and is currently the only widely accepted imaging method used for routine breast cancer screening. There are two types of mammography: one is film mammography and the other is digital mammography. In film mammography, the image is created directly on film, whereas digital mammography takes an electronic image of the breast and stores it directly on a computer.

Hence in this paper we propose an automated computer aided diagnosis system that helps radiologists in breast cancer detection and classification in digital mammograms

2   Mammogram Database

The Mammography Image Analysis Society (MIAS) is an organization of UK research groups which have produced a digital mammography database. The images are in gray scale file format (PGM – Portable Gray Map). The original MIAS Database (digitized at 50 micron pixel edge) has been reduced to 200 micron pixel edge and clipped/padded so that every image is 1024 pixels x 1024 pixels known as the mini-MIAS database. We have used mini-MIAS database as it contains complete information about abnormalities of each mammographic image [6].

3   Methodology

The proposed system involves the following four stages

Image preprocessing: In preprocessing the noises such as salt and pepper noise, label of mammogram image and black background in a mammogram are removed [7].

ROI selection: During the region of interest (ROI) selection the tumor region is identified based on highest intensity point and this region will be allocated for feature extraction.

Feature extraction and selection: Features such as energy values of wavelets and mean and standard deviation are calculated using gray level co-occurrence Matrix (GLCM) based texture analysis

Classification: Based on the selected features, the suspicious regions are classified using support vector Machine (SVM) classifier.

These four stages are illustrated in Fig.1

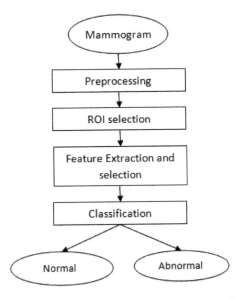

Fig. 1. Block Diagram of CAD system for breast cancer detection

3.1   Preprocessing

In preprocessing, the mammogram image is processed to remove noise and other abnormalities present. It consists of following steps:

Removal of Noise: The mammogram is preprocessed using median filter. The median filter is a nonlinear digital filtering technique that is used to remove noises such as speckle noise [8][9].

Removal of Label: The connected component technique is used to remove the label present in a mammogram image. The largest region is the breast region and all the other smaller regions are eliminated.

Removal of Black Background: For a mammogram image, the sum of intensities for each column is calculated. If sum of intensity of a specific column is less than threshold value then the column will be deleted.

$$\text{For each column} \sum_{i=0}^{j=n} I(i,j) \leq \text{Threshold} \qquad (1)$$

Where I (i, j) is the intensity of pixel i, j and n, m are the dimensions of mammogram image [10].

3.2   ROI Selection

In this stage, the output from pre-processing stage is used as input to find the ROI which is done using region growing technique. Seed point S is chosen as highest intensity value in the image which happens to be the tumor and appropriate threshold value T is chosen in order to find the ROI [11][12].

3.3   Feature Extraction and Selection

In this stage features are extracted using Wavelet analysis and texture analysis

Wavelet Analysis Based Feature Analysis
A Wavelet is time Amplitude representation of a signal and it is a waveform of limited duration with an average value of zero [13]. In our study we use Daubechies wavelets. The energy is calculated using following steps:

Decomposition of wavelets: The wavelet is decomposed into decomposition vector, C and the corresponding bookkeeping matrix, S.

Detailed Co-efficient Extraction: The decomposition vector C will consist of the three detailed coefficients - horizontal coefficient, H, vertical coefficient, V and diagonal coefficient, D. These coefficients are extracted from decomposition vector for scale 2 to scale 5.

Normalization: The three detailed coefficients vectors (H, V and D) are normalized by dividing each vector by it maximum value. Normalized results in value ranging from 0 to 1 [14].

Energy computation: The energy is computed for each vector by squaring every element in the vector. These energy values are considered as features for the classification process.

Texture Based Feature Analysis
Texture analysis of mammograms helps to identify texture feature information about the spatial distribution of tonal variations and describes the pattern of variation in gray level values in a neighborhood. Gray Level co-occurrence Matrix (GLCM) is used to extract texture information from images. The GLCM characterizes the spatial distribution of gray levels in an image. The features that are used for classification are: mean which is a measure of average intensity and standard deviation which is measure of average contrast submission.

$$\text{Mean} \quad \mu = \sum_{i=0}^{n} Z_i \, p(Z_i) \tag{2}$$

$$\text{Standard deviation} \quad \sigma = \sqrt{\mu \, (Z)} \tag{3}$$

Where Z is intensity of pixel and P(Z) is probability of occurrence of pixel Z

3.4   Classification

In this stage the features obtained from previous stage are converted to feature vectors. These feature vectors are used for differentiating between a microcalcification and a circumscribed mass and they are also further classified into benign or malignant case as shown in fig.2. Classification is done using an SVM classifier.

A support vector machine constructs a hyper-plane or set of hyper-planes in a high or infinite-dimensional space, which is used for classification. During testing an image the feature vectors are marked in feature plane and they belong to one of the two categories. A good separation is achieved by the hyper-plane that has the largest distance to the nearest functional margin

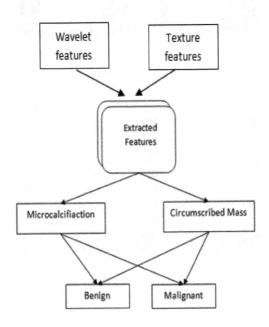

Fig. 2. Block diagram of Classification stages

4   Results

Presented below are the results of the proposed method carried out on mini-MIAS dataset. Fig.3(a) shows the original image and Fig.3(b) shows the preprocessing step which involves label removal. This is followed by black background removal which is shown in Fig.4(a). Fig.4(b) shows the detected abnormality which will be the ROI for feature extraction.Fig.4(c) shows the segmented ROI. The wavelet and texture features are extracted for the ROI. The accuracy of proposed method is 92% which is found using the following formula

$$\text{Accuracy} = \frac{TP+TN}{TP+FP+TN+FN} \tag{4}$$

Where TP is the number of True Positives, TN is the number of True Negative, FP is the number of False Positives, and FN is the number of False Negatives

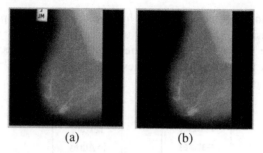

Fig. 3. Normal Mammogram, (a)input image with label; (b) image after label removal

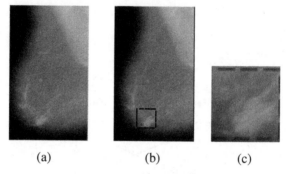

Fig. 4. (a)preprocessed image; (b)ROI indetified; (c) segmented image

5   Conclusion

The proposed method using combination of wavelet features and texture features improves the accuracy of CAD system for detection of early stage breast cancer. We successfully proposed and implemented a method to detect microcalcifications and circumscribed masses which indicate early stage breast cancer and classify them as benign and malignant using SVM classifier with an accuracy of 92%. SVM classifier provides more accurate and unique results. This research can be further developed to work with different wavelet based features and with more texture features which can result in higher efficiencies.

References

[1] Survey by Indian cancer society, Indian Cancer Society (2013)
[2] Padmanabhan, S., Sundarajan, R.: Enhanced Accuracy of Breast Cancer Detection in Digital Mammograms using wavelet analysis. IEEE Trans. Imag. Proc. (2012)
[3] Spandana, P., Rao, K.M.M., Jwalasrikala, J.: Novel Image Processing Techniques for Early Detection of Breast Cancer. In: Matlab and Lab View Implementation. IEEE Point-of-Care Healthcare Technologies (PHT), Bangalore, India, pp. 16–18 (2013)

[4] Malar, E., Kandaswamy, A., Chakravarthy, D., Giri Dharan, A.: A Novel Approach for Detection and Classification of Mammographic Microcalcifications using Wavelet Analysis and Extremelearning Machine. Computers in Biology and Medicine 42, 898–905 (2012)

[5] Dash, J.K., Sahoo, L.: Wavelet Based Feaures of Circular Scan Lines for Mammographic Mass Classification. In: 1st International Conference on Recent Advances in Information Technology, RAIT (2012)

[6] ftp://peipa.essex.ac.uk

[7] Kidsumran, V., Chiracharit, W.: Contrast Enhancement Mammograms using Denoising in Wavelet Coefficients. In: 10th International Joint Conference on Computer Science and Software Engineering (2013)

[8] Mustra, M., Grgic, M.: Enhancement of Micro-calcifications in Digital Mammograms. In: IWSSIP, pp. 11–12 (2012)

[9] Mencattini, A., Salmeri, M., Lojacono, R., Frigerio, M., Caselli, F.: Mammographic Images Enhancement and Denoising for Breast Cancer Detection using Dyadic Wavelet Processing. IEEE Transactions on Instrumentation and Measurement 57(7) (2008)

[10] Maitra, I.K., Nag, S., Bandyopadhyay, S.K.: Technique for Preprocessing of Digital Mammogram. Computer Methods and Programs in Biomedicine 7, 175–188 (2012)

[11] Dakovic, M., Mijovic, S.: Basic Feature Extractions from Mammograms. In: Mediterranean Conference on Embedded Computing, MECO (2012)

[12] Peter, C.: Tay, Yiming Ma.: A Novel Microcalcification Shape Metric To Classify Regions of Interests. Ieee (2010)

[13] Tsai, N.-C., Chen, H.-W., Hsu, S.-L.: Computer-aided diagnosis for early-stage breast cancer by using Wavelet Transform. Computerized Medical Imaging and Graphics 35, 1–8 (2011)

[14] Hamad, N.B., Ellauze, M., Salim, M.: Wavelets Investigation for Computer Aided Detection of Microcalcification in Breast Cancer. In: IEEE SSIAI (2010)

Trusted Service Discovery against Identity Spoofing in Ubiquitous Computing Environment

Swarnali Hazra and S.K. Setua

Computer Science and Engineering, University of Calcutta, India
{swarnali.hazra,sksetua}@gmail.com

Abstract. Ubiquitous Computing environment mingles mobile computing and smart spaces, invisibility, localized scalability and uneven conditioning. In Ubiquitous network, service stations that are registered under respective service providers, invisibly available in real space to the user for providing user's desired services. In ubiquitous network, service discovery by user for getting their desired service can be disrupted by identity spoofing attacker. Identity spoofing attacker can spoof the identity (IP address, ID) of a service station under a service provider and can falsify that service provider with its spoofed identity. At this point, ubiquitous network requires some level of trust to be established between service provider and service stations. In this paper, we have provided a Trusted Service Discovery against Identity Spoofing (TSD-IS) with its computational and architectural model in service discovery of ubiquitous network.

Keywords: Service discovery, identity spoofing attack, direct trust, indirect trust, trustor, trustee.

1 Introduction

In ubiquitous environment, service stations are available invisibly at smart space from everywhere at any time to serve the user requirements through service providers. Service providers may be considered as directory agents or registries or discovery servers. Ubiquitous network is integrated with service providers with its registered service stations, users and other networking components. A user initiates service discovery [3] by sending service request to service providers for getting its desired service from available service station. A Service provider sends service reply with identity information (IP address, ID) of its registered service station which is available to give user requested service. With received identity information user communicates that service station. An identity spoofing attacker can take advantage of this process. Attacker can block a legitimate service station from network and spoof the identity of that service station. When service provider provides identity information of that blocked service station in reply of user request, user is redirected to identity spoofing attacker.

To prevent this type of identity spoofing in service discovery, we have proposed TSD-IS (Trusted Service Discovery against Identity Spoofing attack) in ubiquitous computing environment. The primary objective of our proposal is to isolate attackers and service stations those are compromised by attackers from the environment. Here, trust [2] is a measure of belief or disbelief level of one entity to another in network based on context. With context awareness, direct and indirect interactions are used to compute final trust of trustee by trustor. Here, a service station is considered as trustee (TE) which claims that unavailable service is currently available to give. On other hand, service provider is considered as trustor (TR) under which TE is registered. Depending on final trust of TE, TR believes or disbelieves TE to consider or to isolate in service discovery process. In our proposal, trust based security approach is supported by its architectural model which manages service discovery with provisioning of desired services along with associate trust computation and respective decision.

Related works are discussed in section 2. Section 3 describes identity spoofing attack. In section 4, Trust model is explained. In section 5, proposed TSD-IS is discussed based on our trust model with trust computation. Simulation results of our experiments are presented in section 6. Section 7 concludes the presented work.

2   Related Work

In [5], authors proposed a trust approach based on Bayesian inference model to evaluate trustworthiness with respect to forwarding packets without modifying source IP address. Here judge router is introduced to samples traffics and compute trust. In [8], using Deterministic Packet Marking, authors proposed an IP address traceback method which tolerates the arbitrary server failure for distributed system based on online voting. In [7], authors introduced attribute based encryption to secure unknown services communication channels with anonymous services in a decentralized manner in contrast of PKI based solutions. In this paper scalability of such solution is discusses for decentralized discovery protocols. In [1], authors proposed Context Aware Middleware for Ubiquitous robotic companion Systems which offers context sentient active service applications and responds in timely manner to contextual information. To evaluate the proposed system, authors have applied the proposed active services to the TV domain. In [4], the human notion of trust based an access control model is introduced to secure pervasive computing environment depending on adaptive trust and recommendation. The design outlined in [6] for ubiquitous computing environment, provides an infrastructure and communication protocol for presenting services to heterogeneous mobile clients.

3   Identity Spoofing Attack

Identity spoofing attackers block the target service station by jamming and it spoofs that service station's identity (IP address and ID) to pretend as its own identity. This pretension will disrupt the service discovery process for user. In Fig. 1(a) and Fig. 1(b), SP is service provider under which service stations SS_1 SS_2 SS_3 are registered. User, U requests SP for its desired service. SP checks its service availability list and finds that SS_1 is available to give user requested service. SP sends back service reply with SS_1's identity to U. In Fig. 1(a), a network part is considered without attacker. If there is no attacker to block SS_1 and to spoof its identity, U will be directed to legitimate SS_1 considering SP forwarded SS_1's identity. In Fig. 1(b), a network part is considered with attacker. In this case, legitimate SS$_1$ is already blocked from network due to jamming by attacker A as well as SS$_1$'s identity is spoofed by A. When U initiates the communication with SP forwarded SS$_1$'s identity, U is directed to A with spoofed identity in place of legitimate SS$_1$. Such a way the service discovery initiated by user U is affected with service spoofing attacker A and U can not get its desired service.

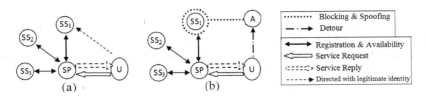

Fig. 1. Example network part including identity spoofing attack

4   Trust Model

To defend against identity spoofing in service discovery, we have introduced trust based security approach at every service provider. Specific modules of our trust model compute different levels of trust and defend identity spoofing. Our Trust Model is comprised of Service Manager, Computation Manager and Decision Manager.

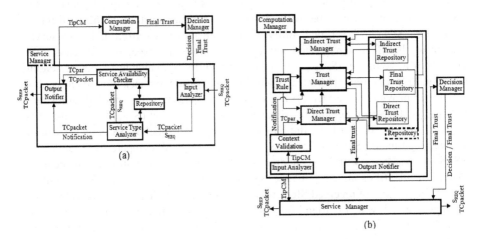

Fig. 2. Detailed architecture of trust model

Service Manager (Fig. 2(a)) deals with service discovery and trust computation based packets. Trust computation related parametric values are sent by service manager to Computation Manager as TipCM (Trust Input for Computation Manager) for trust computation. Service availability checker of Service Manager checks availability list from Repository to find available service station for necessary service of user. Input analyzer analyzes inputs and output Analyzer analyzes outputs for sending respective input output to dedicated sub module.

In Computation Manager module (Fig. 2(b)), final trust value is computed through several levels of context aware trust computations. Computation Manager stores different levels of computed trust values and notifications in respective repositories for current and further computations. Input Analyzer analyzes inputs of trust computation and management where as Output Notifier notifies the outputs to Decision Manager. Direct Trust Manager, Indirect Trust Manager and Final Trust Manager follow the trust rules for computing direct, indirect and final trust respectively. Different levels of trust values vary from 0.0 to 0.1.

Computed final trust value is received by Decision Manager from Computation Manager. Decision Manager takes belief-disbelief decision on basis of received final trust value. Decision Manager sends its taken decision to Service Manager for isolation or inclusion of the trutsee. Decision Manager sends the received final trust to Service Manager for notification.

5   TSD-IS: Trusted Service Discovery against Identity Spoofing

In ubiquitous environment, service stations are registered under respective service providers. When a service station (SS) intends to be a member of ubiquitous network to provide its services, it needs to register itself under a service provider (SP). The requisite information for registration are service station ID, interface, service ID, service type, service description and MAC address of SS. SP always maintains available services of its registered SSs in service availability list .

When user required a service, user requests SPs with service request packet SReq. SReq consists of service type, service description, time constrain along with related parameters. SPs check for availability of the requisite service from their own service availability list. If user requested service is available, the respective SP sends a service check alive packet ($S_{CHK\text{-}ALV}$) directed to identity of available SS before sending back SR_{EP} (service reply packet) to user. $S_{CHK\ ALV}$ is used to check the aliveness of SS which will provide the service. If the identity of respective SS is spoofed, in response to $S_{CHK\text{-}ALV}$, the spoofing attacker sends back an alive packet (S_{ALV}) to SP with MAC address. Attacker do not have the physical control over blocked SS, attacker can not spoofed legitimate SS's MAC address. On getting S_{ALV}, SP compare MAC address of SS in received S_{ALV} with MAC address of SS in service availability list. If there is a match, SP identifies aliveness of the SS and sends back SR_{EP} with SS's identity to user. With identity of received SR_{EP}, user communicates with SS to get service.

Registered SS which is found as available to give user requested service in service availablity list, considered as trustee (TE). SP under which TE is registered, considered as trustor (TR).

Attacker's misbehavior is defined as follows:

- Misbehavior (M): Attacker blocks and spoofs the identity of TE (SS) but can not spoof original MAC address of TE. As a consequence, attacker sends a different MAC address in comparison to legitimate SS to TR (SP) with S_{ALV} packet. This is considered as misbehavior M of TE.

Depending on considered misbehaviors M, Contexts C is defined as follows:

- Context (C): It is the context of receiving a different MAC address instead legitimate TE's MAC address.

We have considered the symbol ${}^{C}[_{A}(T)_{B}]^{t}$, where T denoted the trust evaluated by A for B depending on the context C at time instant t. TR computes final trust with consideration of direct trust and indirect trust, described as follows.

5.1 Direct Trust

If TR's Service Manager identifies any mismatch of MAC addresses between S_{ALV} and service availability list, TR identifies the misbehavior M on context C. On mismatch, TR's Service Manager sends a negative acknowledgement N_{ACK} to TR's direct trust manager in Computation Manager module. But on match, TR's Service Manager sends a positive acknowledgement P_{ACK}. On getting N_{ACK} or P_{ACK}, TR's direct trust manager in Computation Manager module evaluate the ${}^{C}[_{TR}(MDT_{cur})_{TE}]^{t}$ (current TR monitored direct trust) as per equation (1).

$$
{}^{C}[_{TR}(MDT_{cur})_{TE}]^{t} =
$$

$$
\begin{cases}
For\ MAC\ ad\ dress\ mismatch\ of\ service\ availabili\ ty\ list\ an\ d\ S_{ALV}\ ,\ 0.1 & (1)\\
For\ MAC\ ad\ dress\ match\ of\ s\ ervice\ ava\ ilability\ list\ and\ S_{ALV}\ , & 0.9
\end{cases}
$$

TR's direct trust manager stores ${}^{C}[_{TR}(MDT_{cur})_{TE}]^{t}$ in direct trust repository for further computation.

In TR's direct trust repository, previously evaluated TR monitored direct trust values are classified as continuous positive, continuous negative, discrete positive and discrete negative. Continuous positive denotes successive increase in computed direct trust values and continuous negative denotes successive decrease. On the other hand, discrete positive denotes distinct increase in computed direct trust values and discrete negative denotes distinct decrease.

TR's direct trust manager aggregates the previously evaluated continuous positive direct trust values. Aggregative effect is denoted as ${}^{C}[_{TR}(DTcp_{old})_{TE}]^{t}$, computed according to equation (2).

$$
{}^{C}[_{TR}(DTcp_{old})_{TE}]^{t} = \frac{1}{tcp}\sum_{i=1}^{tcp}\{{}^{C}[_{TR}(MDTcpi_{old})_{TE}]^{t} \times e^{-(t-t_{o})}\} \qquad (2)
$$

${}^{C}[_{TR}(MDTcpi_{old})_{TE}]^{t}$ = stored i^{th} continuous positive TR monitored direct trust for TE. $1 \leq i \leq tcp$. Here tcp is total numbers of stored continuous positive trusts for TE. $e^{-(t-t0)}$ is time decaying function, where t_{0} and t are initial and computing time instants.

TR's direct trust manager aggregates the stored continuous negative direct trust values. Aggregative effect is denoted as ${}^{C}[_{TR}(DTcn_{old})_{TE}]^{t}$ which is computed according to equation (3).

$$
{}^{C}[_{TR}(DTcn_{old})_{TE}]^{t} = \frac{1}{tcn}\sum_{i=1}^{tcn}\{{}^{C}[_{TR}(MDTcni_{old})_{TE}]^{t} \times e^{-(t-t_{o})}\} \qquad (3)
$$

$^C[_{TR}(MDTcni_{old})_{TE}]^t$ = Stored i^{th} continuous negative TR monitored direct trust for TE. $1 \le i \le tcn$. Here tcn is total number of stored continuous negative direct trusts for TE.

Direct trust manager computes the combined effect of stored all continuous direct trust (positive, negative). Combined trust is denoted as $^C[_{TR}(DTc)_{TE}]^t$ which is computed according to equation (4).

$$^C[_{TR}(DTc)_{TE}]^t = \frac{tcp}{tc} \times {}^C[_{TR}(DTcp_{cur})_{TE}]^t + \frac{tcn}{tc} \times {}^C[_{TR}(DTcn_{old})_{TE}]^t \quad (4)$$

Here tc is total number of continuous trusts (positive, negative) for TE, stored at TR.

TR's direct trust manager aggregates the stored discrete positive direct trust. Aggregative effect is denoted as $^C[_{TR}(DTdp_{old})_{TE}]^t$, computed according to equation (5).

$$^C[_{TR}(DTdp_{old})_{TE}]^t = \frac{1}{tdp} \sum_{i=1}^{tdp} \{ {}^C[_{TR}(MDTdpi_{old})_{TE}]^t \times e^{-(t-t_o)} \} \quad (5)$$

$^C[_{TR}(MDTdpi_{old})_{TE}]^t$ = Stored i^{th} discrete positive TR monitored direct trust for TE. $1 \le i \le tcp$. Here tdp is total numbers of stored discrete positive direct trust for TE.

TR's direct trust manager aggregates the stored discrete negative direct trust. Combine effect is denoted as $^C[_{TR}(DTdn_{old})_{TE}]^t$, computed according to equation (6).

$$^C[_{TR}(DTdn_{old})_{TE}]^t = \frac{1}{tdn} \sum_{i=1}^{tdn} \{ {}^C[_{TR}(MDTdni_{old})_{TE}]^t \times e^{-(t-t_o)} \} \quad (6)$$

$^C[_{TR}(MDTdni_{old})_{TE}]^t$ = Stored i^{th} discrete negative TR monitored direct trust for TE. $1 \le i \le tcn$. Here tdn is total number of stored discrete negative direct trusts for TE.

Direct trust manager computes the combined effect of stored all discrete direct trust (positive, negative). Combined trust is denoted as $^C[_{TR}(DTd)_{TE}]^t$ which is computed according to equation (7).

$$^C[_{TR}(DTd)_{TE}]^t = \frac{tdp}{td} \times {}^C[_{TR}(DTdp_{old})_{TE}]^t + \frac{tdn}{td} \times {}^C[_{TR}(DTdn_{old})_{TE}]^t \quad (7)$$

Here td is the total numbers of stored discrete trusts (positive, negative) of TR on TE.

TR's direct trust manager computes aggregative effect ($^C[_{TR}(ADT_{old})_{TE}]^t$) of old direct trusts with $^C[_{TR}(DTc)_{TE}]^t$ and $^C[_{TR}(DTd)_{TE}]^t$ according to equation (8).

$$^C[_{TR}(ADT_{old})_{TE}]^t = 0.5 \times {}^C[_{TR}(DTc)_{TE}]^t + 0.5 \times {}^C[_{TR}(DTd)_{TE}]^t \} \quad (8)$$

Direct trust manager computes final direct trust ($^C[_{TR}(FDT)_{TE}]^t$) as per equation (9).

$$^C[_{TR}(FDT)_{TE}]^t = 0.5 \times {}^C[_{TR}(MDT_{cur})_{TE}]^t + 0.5 \times {}^C[_{TR}(ADT_{old})_{TE}]^t \} \quad (9)$$

TR's Direct Trust Manager stores $^C[_{TR}(FDT)_{TE}]^t$ in Direct Trust Repository.

5.2   Indirect Trust

In indirect trust repository, notifications from recommenders (who notifies its computed final trust about TE) about TE are stored. ($^C[_{Ri}(NT)_{TE}]^t$) is denoted as i^{th} recommender notified final trust value. Here $1 \leq i \leq n$, n is total number of recommenders. TR's indirect trust manager computes final indirect trust considering all received notifications. final indirect trust $^C[_{TR}(FIT)_{TE}]^t$ is computed as per equation (10).

$$^C[_{TR}(FIT)_{TE}]^t = \frac{\sum_{i=1}^{n}\{^C[_{Ri}(NT)_{TE}]^t \times e^{-(t-t_o)}\} \times \{^C[_{TR}(FT_{last})_{Ri}]^t \times e^{-(t-t_o)}\}}{\sum_{i=1}^{n}\{^C[_{TR}(FT_{last})_{Ri}]^t \times e^{-(t-t_o)}\}} \qquad (10)$$

$^C[_{TR}(FT_{last})_{Ri}]^t$ is lastly computed final trust value by TR about Ri. Indirect trust manager stores this $^C[_{TR}(FIT)_{TE}]^t$ in indirect trust repository.

5.3   Final Trust

TR's final trust manager computes final trust ($^C[_{TR}(FT)_{TE}]^t$) using $^C[_{TR}(FDT)_{TE}]^t$ and $^C[_{TR}(FIT)_{TE}]^t$ from their respective repositories as per equation (11).

$$^C[_{TR}(FT)_{TE}]^t = 0.5 \times {}^C[_{TR}(FDT)_{TE}]^t + 0.5 \times {}^C[_{TR}(FIT)_{TE}]^t\} \qquad .(11)$$

5.4   Decision and Reaction

TR's final trust manager sends $^C[_{TR}(FT)_{TE}]^t$ to Decision Manager. Decision Manager compares $^C[_{TR}(FT)_{TE}]^t$ with uncertainty point. Uncertainty point depends on application. For this work, uncertainty point is considered as 0.5. If $^C[_{TR}(FT)_{TE}]^t > 0.5$, Decision Manager takes belief decision otherwise disbelieves it. Decision Manager sends decision to Service Manager. With disbelief decision, TR's Service Manager does not forward TE's identity in S_{REP} and isolates it.

6   Simulation

Simulation experiments evaluate the efficiency of proposed TSD-IS against normal service discovery (NSD). The considered network provides 200 different types of services. Out of these service types, each of 120 types are provided by three service providers , each of another 60 types of services are provided by two service providers and each of remaining 20 types of services are provided by one service provider. Fig.3 (a) shows attack success rate which is function of numbers of attacked users among users who requested for service. Fig. 3 (b) shows service acquiring rate is function of percentage to get service properly with respect to requested service. Network is simulated with 100 simulation runs and maximum requesting user 20, 30, 40.

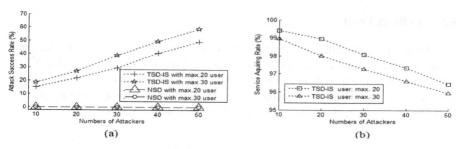

Fig. 3. Simulation results

7   Conclusion

Our proposed TSD-IS efficiently defends ubiquitous network against identity spoofing attack. The suggested approach isolates attackers during service discovery process using our new trust model. Our Simulation results have shown the efficient performance of proposed TSD-IS against normal service discovery. Future scope of this work includes the defense mechanism for other attacks in ubiquitous network and extension of the same with a probabilistic approach.

References

1. Moon, A., Kim, H., Kim, H., Lee, S.: Context-Aware Active Services in Ubiquitous Computing Environments. ETRI Journal 29(2) (2007)
2. Xiu, D., Liu, Z.: A Formal Definition for Trust in Distributed Systems. In: Zhou, J., López, J., Deng, R.H., Bao, F. (eds.) ISC 2005. LNCS, vol. 3650, pp. 482–489. Springer, Heidelberg (2005)
3. Zhu, F., Mutka, M.W., Ni, L.M.: Service Discovery in Pervasive Computing Environment. IEEE Pervasive Computing, 81–90 (October-December 2005)
4. Kim, J., Baek, J., Kim, K., Zhou, J.: A Privacy-Preserving Secure Service Discovery Protocol for Ubiquitous Computing Environments. In: Camenisch, J., Lambrinoudakis, C. (eds.) EuroPKI 2010. LNCS, vol. 6711, pp. 45–60. Springer, Heidelberg (2011)
5. Gonzalez, J.M., Anwar, M., Joshi, J.B.D.: A Trust-based Approach against IP-spoofing Attacks. In: Ninth Annual International Conference on Privacy, Security and Trust, pp. 978–971. IEEE (2011) print ISBN : 978-1-4577-0582-3, 978-1-4577-0584-7
6. Kagal, L., Korolev, V., Avancha, S., Joshi, A., Finin, T., Yesha, Y.: A Highly Adaptable Infrastructure for Service Discovery and Management in Ubiquitous Computing. Techinal Report: TR-CS-01-06, Department of Computer Science and Electrical Engineering, University of Maryland Baltimore County (2001)
7. Trabelsi, S., Pazzaglia, J.-C., Roudier, Y.: Secure Web Service Discovery: Overcoming Challenges of Ubiquitous Computing, pp. 35–43 (December 2006) print ISBN: 0-7695-2737-X
8. Soundar Rajam, V.K., Sundaresa Pandian, B., Mercy Shalinie, S., Pandey, T.: Application of IP Traceback Mechanism to OnlineVoting System. In: Thilagam, P.S., Pais, A.R., Chandrasekaran, K., Balakrishnan, N. (eds.) ADCONS 2011. LNCS, vol. 7135, pp. 621–630. Springer, Heidelberg (2012)

Architectural Recommendations
in Building a Network Based Secure, Scalable
and Interoperable Internet of Things Middleware

Shiju Sathyadevan, Krishnashree Achuthan, and Jayaraj Poroor

Amrita Center for Cybersecurity Systems and Networks,
Amrita Vishwa Vidyapeetham,
Kollam, India
shiju.s@am.amrita.edu,
krishna@amrita.edu,
jayaraj.poroor@gmail.com

Abstract. With every device capable of emitting data, the need to amass and process them in real time or near real time, along with ever growing user requirement of, information-on-demand, have paved the way for the sudden surge in the development of the theme Internet of Things (IoT). Even though this whole new technology looks fascinating in theory, its practical implementation and ongoing sustenance is something that will need a lot of thought, effort and careful planning. Several cloud based and network based cloud platforms/middleware solutions are available but a lot of them are either extremely complex to set up, or needs standardized solutions to be applied across all participating devices or would leave behind vivid security loopholes that can't be curbed with ease considering the overall capability of the devices involved. Based on a previous research effort conducted by our team, detailed scrutiny of prominent existing IoT platforms/middleware solutions were performed. This effort has resulted in defining the core problem matrix which if addressed adequately could result in the development of a well balanced IoT middleware solution. This paper uses the problem matrix so identified as the roadmap in defining a Secure, Scalable, Interoperable Internet of Things Middleware.

Keywords: IoT, Internet of Things, IoT Middleware, IoT Architecture, Interoperability, Scalability, MATCHES, Event Channels, Concurrent processing.

1   Introduction

The Internet has evolved over the years from a network of computer clusters to one that also encompasses heterogeneous devices capable of consuming and generating data at random pace. This expanded network is referred to as an 'Internet of Things' (IoT). There is no clear definition that would mark the boundaries of its operational spectrum. From its initial inception through RFID devices, IoT has broadened its scope to include road traffic monitoring systems, building surveillance cameras, home utility-metering devices to personal medical devices. In short, IoT spans nearly every

© Springer International Publishing Switzerland 2015
S.C. Satapathy et al. (eds.), *Proc. of the 3rd Int. Conf. on Front. of Intell. Comput. (FICTA) 2014*
– *Vol. 1*, Advances in Intelligent Systems and Computing 327, DOI: 10.1007/978-3-319-11933-5_47

walk of life and is clearly the way for the future Internet. It is estimated that there are over 1.5 million IoT-enabled devices in the planet as of now and this number is growing at a very rapid pace. These devices are from a diverse list of manufacturers with their own protocols and data formats making the standardization task tedious and error-prone in a short span of time. What is clear is that current Internet and supporting technologies would need to reshape themselves so that they can accommodate these devices.

A fundamental problem in the Internet of Things lies in the difference (in power, bandwidth, and processing capabilities) between traditional hosts and small devices on the network. Although they are interconnected with one another, these devices generally are not able to understand one another due to differences in data formats and communication protocols employed. If the interoperability of these devices were possible on a semantic level, we would be in a better position to render intelligent services with minimal or no human intervention. Currently, the wealth of information emanating from these devices has not been fully assimilated and we have not yet realized their full benefit.

This paper presents an architectural framework for developing a semantically interoperable Internet of Things middleware. In developing this model, we paid special attention to three important criteria's: security scalability and interoperability; but all these achieved through minimal changes to the current mannerism in which "things" operate and communicate. The intended applications require that sensitive user information be communicated in a secure manner. Since devices have limited resources, it is important that we have light-weight mechanisms for authentication and authorization. While this is relatively easier to accomplish on a small scale, the challenge is to extend this to the massive scale of IoT which typically involving millions of devices. Our research therefore focuses in addressing the unique naming/addressing of devices for identification, device authentication, provides language support for programming heterogeneous devices and to process the data generated in a concurrent fashion, and also develop analytics for instantaneous decision making.

Lastly, we also address the important problem of how to tap, tame, tackle and thereby derive useful information and intelligence from the enormous amount of data that is generated by these constantly interacting devices. In some instances these decision aiding intelligence should be instantaneous whereas in other cases latencies due to processing are tolerable. Considering the diverse domains and the sheer size of the generated data that need to be processed on the fly or mined incrementally when stored in large databases, the need for a high responsive 'big data' analytical platform is inevitable.

Paper recommends the need for the following components that will provide a complete end to end IoT solution:

a) Design and implementation of a language based on asynchronous pi-calculus. This would provide good programming abstractions to achieve true concurrency offered by multi-cores for concurrent data processing.

b) An efficient addressing scheme for IoT objects in order to uniquely identify and address them.

c) Middleware communication layer for interfacing incompatible devices, while ensuring that the identity/discovery of heterogeneous object, communication between objects, and handling of multi-object level triggers initiated by any event raised by a participating object in a secure and controlled fashion.

d) A multichannel Real-time Big Data Analytical Engine capable of handling streaming data from multiple channels/domains at the same time process them on the fly to deliver instantaneous decisions.

The remainder of the paper is organized as follows. Section 2 looks into research efforts in related topics. Section 3 advocates the need for an unified IoT middleware solution. Section 4 details the key features and capabilities of that the newly proposed middleware should encompass

2   Related Work

Various IoT middleware solutions are analyzed and focuses on various technical challenges[3]. Middleware systems are then classified into different categories based on the domain in which it is being used. Core technical challenges in designing an IoT middleware is discussed in detail. The key functional components of different IoT middleware solutions has been proposed in [4]. This helps researchers to understand the way current IoT middleware solutions works and also exposes their existing issues and gaps. The paper then goes ahead comparing and categorizing various IoT middleware solutions based on their features and associates each of those middleware solutions to appropriate application domain. IoT middleware solutions proposed are extremely weak in handling security and privacy issues. Both these challenges are well addressed in [5] where IoT is categorized into different areas of interest like communication, sensors, actuators, storage, devices, processing, localization and identification. Technological details and core security issues associated with each of these areas are discussed. This paper also highlights those areas that require further research. Detailed study of M2M (Machine to Machine) platforms has been done in [6]. Different M2M/Platforms based on object, people, environment and enterprise has been analyzed in detail.

3   Need for an Unified IoT Middleware

The term "Internet of Things (IoT)" initially was coined in a paper about Electronic Product Code (EPC) way back in 2001 by David Brook but first came into limelight when Auto-ID center envisaged the use of EPC network for tracking goods in supply chains. Ever since then IoT has been taken up by researchers and practitioners more aggressively trying to find ways to build an intelligent, highly scalable, interoperable information super highway that is robust enough to handle billions of diverse objects and the huge volume of data that these objects will generate on an ongoing basis. As the technology started to evolve over the years IoT has now been envisaged as not just a tool that can help better manage and control business processes but will alleviate the comfort level of one's life as such when put in use at its full potential.

There are several challenges that IoT has brought with it. Most critical of the lot is that being an ubiquitous environment, enforcing a common standard across the board is practically impossible. Heterogeneous objects should be able to continue to remain connected supporting failovers or should be able plug on demand (PoD) to the network when it needs to dispense off or receive data. Building a network infrastructure capable of handling:

- Massive volume of objects without degrading the performance of the network.
- Constant object movements resulting in erratic event and data generation.
- Identification and selection of right services/objects to communicate from the mist of these several participating devices so as to obtain the most precise and up to date information.

This paper aims to develop a complete deployable IoT suite providing an end to end architectural framework that will effectively and efficiently marry all participating heterogeneous objects encapsulating the complexity of the same within the framework, at the same time stretch itself to scale up to the demands posed in the future by this every growing network.

4   Proposed IoT Middleware and Its Components

The interest generated by IoT is creating an outburst of activities in various research arenas to shape this conceptual model to a fully working physical model. The path to IoT realization is not an easy task as the mere surface skim through its ideology itself exposes the challenges that it will pose to those researchers who would want to see this operational in actuality. Today IoT as such exist as disjoint sets of custom build networks. But in order to extract the full potential of IoT and to put it to real human benefit there need to be an organized effort in defining a standard that will enable these disjoint networks to talk to each other in a secure fashion so that appropriate intelligence can be derived out of the huge data mass these objects generates supporting instantaneous decision making. Following are the four key deployable IoT sub-components that this paper is proposing to develop in an effort to build a holistic IoT enabled Information Superhighway.

4.1   Matches: Design and Implementation of a Language for IoT Systems Based on Asynchronous π-calculus

The internet of things (IoT) promises to saturate our world with physical objects embedded with sensors and tiny computing devices. Such systems are continually generating event data from embedded sensors, including producing real-time data streams.

In order to take advantage of this scenario, these events must be concurrently processed by applications running in computing systems ranging from embedded to server systems. There is a lack of fundamental research and development in proper programming abstractions for such systems. Good programming abstractions would allow us to easily take advantage of true concurrency offered by multi-cores for concurrent data processing. We may look up to the rich developments in the concurrency theory for gaining insights into suitable abstractions. Process calculi are concurrent formal languages for specifying and reasoning about concurrent, communicating systems. The π-calculus is a process calculus for mobile systems.

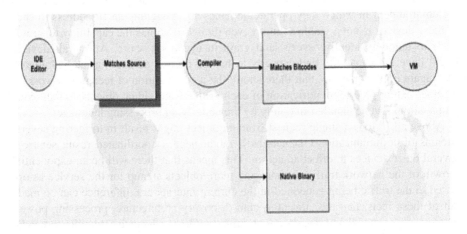

Fig. 1. Tool Chain Implementation for "Matches"

In the context of IoT, we propose to investigate programming abstractions for concurrent event-processing systems based on the asynchronous π-calculus. The proposed language is tentatively named *MATCHES* (Mobile, Asynchronous Typed Channels for Event Streams). The process-based model offered by the π-calculus would allow us to quickly take advantage of multi-cores wherever available for rapid event processing. The asynchronous channels have underlying bounded FIFO queues. When queue is full either the channel must block (e.g., when specifying in-memory queues) or the operation simply returns with no effect (lossy, distributed channel) or it simply pushes the oldest entry out (real-time data streams). A receive operation on an empty channel either returns a default value (polling mode) or blocks. The channels may be configured for any of these behaviors. The use of asynchronous FIFO channels is the key to providing a viable channel abstraction for supporting real-time data streams.

There is already some work in designing practical programming languages based on π-calculus – namely, occam-π [9], which is a programming language based on the synchronous π-calculus. Though our work would draw insights from the experience in developing occam-π, our language is fundamentally different owing to the use of asynchronous channels. The proposed tool chain implementation for the language is shown in Fig 1.

4.2   IoT Object Identification Scheme

Global human population is likely to hit 7.6 billion by 2020 with 50 billion connected devices, computing to an average of 6.58 connected devices per person [Cisco IBSG, April 2011). These numbers show that there are more communicating devices than human being. Defining a naming/addressing scheme for the IoT is indeed a challenge considering the operational restrictions and the sheer nature of the objects that will be participating within this model. All of these devices operating in this framework will be internet enabled and should be able to communicate in all possible dimensions, if the environment in which they operate demands so. Assigning an IP address to each of these devices is not a viable solution even though IPv6 has the capability to stretch itself to uniquely identify every sand grain in every sea shore. Above all devices might have the capability and capacity to handle normal addressing methodologies. This again implies the fact that there should be a combination of techniques working together aiding precise identification of each of these individual objects so that event propagations can be handled smoothly by the underlying processing languages.

Above all in IoT, a single request from an object might result in triggering several streams of communication channels as it might need a coordinated result set from several other devices to effect an action. This means that there will be an exponential growth of the network traffic as more and more objects signup for the service as opposed to the traffic being generated in the current internet era. Inference can be made out of these facts, that the current system discovery architecture, processing power, memory usage model and response time might not have adequate capability to handle this traffic explosion.

Any addressing/naming schemes, security mechanisms and communication protocols proposed for IoT should be feather weigh, as several of these objects have constrained capacity and capabilities in term of resources available to process them. In order to accomplish this, new addressing schemes or a combination of existing proven addressing schemes that is flexible, easily scalable but still complying with existing standard need to be defined and tested out so that an optimal solution can be drafted.

Present day implementation of IoT uses Electronic Product Code (EPC), Object Naming Service (ONS), XML based markup language to define the data format for various communication performed by the participating objects and a light weighted IPv6 protocol stack. For certain participating objects dependency on a central authority might hinter uninterrupted connectivity where in which object clustering to form IoT local area networks similar to that implemented in wireless sensor networks (WSN) and Mobile Ad-hoc Network (MANET) have to be embraced. Another alternative is to extend the first option with more specific object level unique ID using objectID@URI notation where objectID will uniquely identify an object at a specific URI. Considering the volume of objects that will start flogging into the IoT pipe following its full blown outburst, it is not a feasible solution to assign IPv6 address to each one of them. Object clustering into a master slave model is one solution that we would want to research on as part of this IoT initiative which will then act as the addressing standard easily deployable onto existing and new built objects capable of communicating over a medium. Not all devices have the capability to reach out to the nearest access points and might not need to constantly radiate data for analysis. Such devices will have a local address (EPC) which is tied to a device capable of reaching out to the world by assigning public IPv6 address. Possibility of marrying IPv6 and EPC will a solution worth trying. We could extend the EPCglobal framework by defining a customized identification scheme for each object in an IoT platform as detailed in our work in [10].

4.3   Semantic Object Communication Broker (SOCoB)

SOCoB will act as the middle layer for the proposed IoT architecture. Of the many IoT deployment challenges listed during the initial part of this document, handling the diverse nature of various participating objects within the IoT frame seems to be one of the most critical defy that need to be resolved for the success of any Interoperable, scalable IoT standard/platform. As more and more devices make its way to this Information superhighway, proliferation of the diverse technologies with different specifications will pose more compatibility issues. Any solid architectural solution requires secure granting of permission to pair with permitted objects in order to facilitate multi-object interactions and sharing of information there by facilitating highly critical decision making activities. "IoT Object Identification Scheme" detailed in the previous section very much entwine with this module and play equal importance towards the success of the whole model from an architectural perspective. That's where the newly proposed Semantic Object Communication broker will chip in to act as the middleware for IoT making it flexible enough to accommodate this ever growing diverse network.

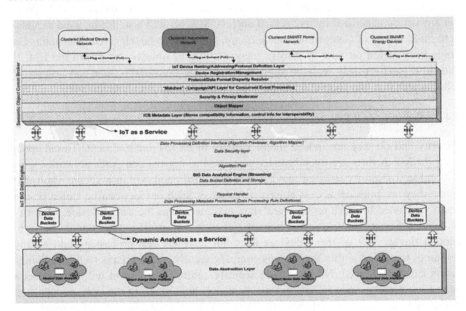

Fig. 2. Proposed High Level IoT Architecture

Consumers and participants of IoT information superhighway should be able to avail this capability as a service (IoT as a Service) which can be utilized when needed similar to what is being offered as PaaS and SaaS in a cloud Environment. Any application or object that wishes to hook up to this information superhighway should be able to take advantage of the service of this layer in the course of fulfilling its objective irrespective of its vendor, embedded software, protocol used and data format followed. Complexity of this network is escalated further as an event triggered from one or multitude of objects can result in an outburst of event series creating an Event Mesh (EM) involving several objects that are within the native network or spanning across several external networks. Object Mapper component will initiate the process flow as per the EM definition.

Figure 2 details the various components that will constitute the overall architectural layout of the proposed IoT model. Semantic Object Communication Broker (SOCoB) will ensure to offer compatibility to any device that would want to avail the same as a service by encapsulating the complexities of protocol and data format translations. It will act as the broker to ensure that any machine-to-machine or machine-to-human interactions are serviced to the need. SOCoB can be offered to its consumers as a cloud service or can be installed locally on a server which will then hold the metadata information necessary to effectively resolve disparity in protocol and data formats. SOCoB can also be implemented in an intelligent gateway device which is hosted outside the middleware layer. Such an implementation will ensure that a large percentage of the task such as protocol resolution, first level of authentication etc could be performed at the gateway itself, thus avoiding the stress on the middleware. Following any action generated by its internal objects, SOCoB will also use its local metadata to build up list of concurrent events, if any, that need to be triggered. Metadata also holds information on how to resolve unavailability of objects that are in the EM. If data from such critical decision support objects are unavailable then the last set of reliable data that was recorded will be used consulting the BiG Data Engine.

4.4   Security and Privacy Moderator

Security and privacy are indeed major worries in an IoT environment. Threat spectrum for IoT devices are even broader compared to normal computer networks. This is because a majority of these devices are not positioned within a secure vault and hence they are physical access is easy to achieve. Such devices can be stolen or can easily be cloned to make replicas and then masquerade as the actual device. Majority of devices does not have a secure way of authenticating itself to the middleware or to another device. Lot of these devices does not have enough processing and energy resources to embed within themselves complex security mechanisms so that authentication can be achieved. No one single security control mechanism will not protect a device adequately. Security control mechanisms should kick in the very moment the device is turned on to establishing the initial trust to initiate any sort of communication over a trusted channel through to sustaining the same through techniques that cannot be tampered with. Implanting device level secure booting, applying appropriate access control mechanisms, ensure fool proof device authentication, secure patch and software updates etc are some of the critical areas where solid solutions are much needed. Hence security can never be considered as an add-onto any participating devices, rather treated as its integral part for it to function reliably. Security need to be implemented even to the lowest level like internal messaging queues which acts as the bridge between the IoT middleware and BigData processing frameworks.

4.5   Big Data and Real-Time Analytics

As the data generated by the IoT systems is going to be massive and continuous, it demands a solid big data platform to effectively store, process and analyze these data streams coming from various devices or sensors methodically for real-time analytics and reporting. This module outlines the big data platform hosted on cloud plus real-time analytic engine to support this specific need of the overall IoT system. Key areas which needs to be addressed by this module are:

1. All integrated, highly scalable Big Data framework that can process large volume of static and streaming data, process them in-memory or in-database but encapsulating the deployment and maintenance complexities from the user.
2. Multi source data adaptors that can hook up to different input sources or systems, Queues, data feeds, data formats like XML/JSON feeds etc.
3. Support of several secure messaging queues that have user level authentication and topic based security as well.
4. Built-in Data mining engines which performs incremental and rule based data mining to bring out intelligence from the data streams.
5. Customizable mining rule maps which will grow as per the intelligence it gathers using machine learning techniques or algorithms.
6. Real-time in memory analysis and reporting.
7. Predictive data analysis and auto-alert management and monitoring.
8. Real-time data score cards and visually interactive dashboards available on PC and mobile platforms.
9. User definable dynamic dashboards which can be built and managed by the data owners.

4.6   Data Interface Module

The application expects to input data streaming from any vertical or application domain for doing real-time analytics and reporting. The layer is designed to do the function of hooking up the data input streams coming from these various sources. As the need here is to do analysis of in-motion data before it is persisted anywhere, makes it important to have adaptable data connector instances to be created to link with each data source as separate stream. The SIL system layer will allow creating multiple data connector instances as per the requirement of the end-user.

4.7   Analytics Module

This forms the core part of the system and focus on the analytic part of the proposed architecture. As the speed at which the streaming data has to be sliced, diced and incrementally mined to produce real-time insights, mandates scalable architecture, provisioning independent instances of the core analytic components to be able to take the data input stream from each of the source interface layer connections. The analytic engine will support both real-time and stored data analysis post-transaction. One of the vital features of this component will be to deliver on-the-fly real-time analysis of

the data making use of the power of in-memory computing where the data gets analyzed even before it is stored on the disk. Fig.4. shows low level architectural layout
for handling bigdata.

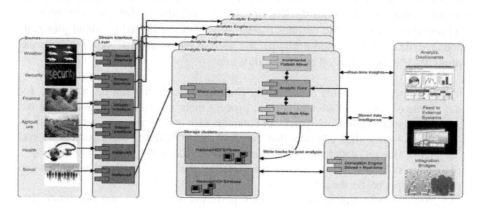

Fig. 3. Low Level Big Data Analytical Engine Architecture

4.8   Data Visualization Module

This is the reporting layer where various data visualization tools and Dashboards are
provided. The real-time analytical intelligence derived will be presented in meaningful alerts, tickers, score cards, charts, etc here. There can be data feeds coming from
the Analytical Engine to external consumer systems as required. For enabling seamless integration with external systems, intelligent connectors or bridges will be provided which allows continuous flow of data request and data pumping as required by
the client requirements.

5   Conclusion

Architecture detailed in this paper is based on actual research implementation. Current IoT solutions have several weak spots that does not fully qualify them to be the
IoT solution for the future. Either their deployment complexity, or lack of reliability
due to vivid, easy to capitalize security loop holes makes them less attractive to its
consumers. Proposed architecture will expect no changes to the way devices interact
today. Devices could continue to use the same protocols to communicate without the
need to embed any additional software. Such complexities can be either handled by
intermediate intelligent devices or by the middleware itself. Security mechanisms are
proposed at various levels. Device level security was achieved through actual implementation of techniques like Physically Unclonable functions (PUF). Queue level
security (topic based authentication) was achieved in queues like MQTT, ActiveMQ
etc. Intelligent gateways developed could ensure security of communication channels
and also in offloading a lot of tasks from the middleware reducing the overall traffic.
Event mesh creation and management could be easily handled through the development of the new programming language MATCHES without the developer worrying

too much about the overall performance and in achieving concurrency. It could concurrently process data streams ensuring the full utilization of all available cores in a machine. All in one Big Data processing framework could integrate various data processing engines under one roof, managed by the framework itself, thereby encapsulating the complexities of the same from the user. Proposed model in this paper not only could nullify a lot of serious shortfalls that the current IoT solutions carry with them, but also ensure that the solution can be easily deployed.

References

1. IERC, About IoT,
   http://www.internet-of-things-research.eu/about_iot.htm
2. Atzori, L., Iera, A., Morabito, G.: The Internet of Things: A survey. Computer Networks 54, 2787–2805 (2010)
3. Chaqfeh, M.A., Mohamed, N.: Challenges in Middleware Solutions for the Internet of Things. In: 2012 International Conference on Collaboration Technologies and Systems (CTS), May 21-25 (2012)
4. Bandyopadhyay, S., Sengupta, M., Maiti, S., Dutta, S.: Role of Middleware for Internet of things: A Study. International Journal of Computer Science & Engineering Survey (IJCSES) 2(3) (August 2011), doi:10.5121/ijcses.2011.2307 94
5. Mayer, C.P.: Security and Privacy Challenges in the Internet of Things. In: Workshops der Wissenschaftlichen Konferenz Kommunikation in Verteilten Systemen (WowKiVS 2009) (2009)
6. Castro, M., Jara, A.J., Skarmeta, A.F.: An analysis of M2M platforms: challenges and opportunities for the Internet of Things. In: 2012 Sixth International Conference on Innovative Mobile and Internet Services in Ubiquitous Computing (IMIS). IEEE (2012)
7. Kominers, P.: Interoperability Case Study: Internet of Things (IoT). Berkman Center Research Publication No. 2012-10 (April 1, 2012), available at SSRN:
   http://ssrn.com/abstract=2046984 or http://dx.doi.org/10.2139/ssrn.2046984
8. Sha, L., Gopalakrishnan, S., Liu, X., Wang, Q.: Cyber-physical Systems: A New Frontier. In: 2008 IEEE International Conference on Sensor Networks, Ubiquitous, and Trustworthy Computing (2008)
9. PLAS, An Occam-pi Quick Reference, https://www.cs.kent.ac.uk/research/groups/plas/wiki/OccamPiReference/
10. Sathyadevan, S., Akhila, C.A., Jinesh, M.K.: Customizing EPCglobal to fit Local ONS Requirements. In: 2014 International Conference on Intelligent Computing, Communication & Devices (ICCD 2014) (2014)
11. Weber, R.H.: Internet of Things–New security and privacy challenges. Computer Law & Security Review 26(1), 23–30 (2010)
12. Kowatsch, T., Maass, W.: Critical Privacy Factors of Internet of Things Services: An Empirical Investigation with Domain Experts. In: Rahman, H., Mesquita, A., Ramos, I., Pernici, B. (eds.) MCIS 2012. Lecture Notes in Business Information Processing, vol. 129, pp. 200–211. Springer, Heidelberg (2012)

Modeling and Simulation of Quarter Car Semi Active Suspension System Using LQR Controller

K. Dhananjay Rao[1] and Shambhu Kumar[2]

[1] School of Engineering and Technology, Centurion University, Odisha, India
kdhananjayrao@gmail.com
[2] Jadavpur University, Kolkata, India
kumarshambhu544@gmail.com

Abstract. In this paper design of the Linear Quadratic Regulator (LQR) for Quarter car semi active suspension system has been done. Current automobile suspension systems use passive components only by utilizing spring and damping coefficient with fixed rates. The vehicle suspension systems are typically rated by its ability to provide good road handling and improve passenger comfort. In order to improve comfort and ride quality of a vehicle, four parameters are needed to be acknowledged. Those four parameters are sprung mass acceleration, sprung mass displacement, unsprung displacement and suspension deflection. This paper uses a new approach in designing the suspension system which is semi-active suspension. Here, the hydraulic damper is replaced by a magneto-rheological damper and a controller is developed for controlling the damping force of the suspension system. The semi-active suspension with controller reduces the sprung mass acceleration and displacement hence improving the passengers comfort.

Keywords: Linear Quadratic Regulator (LQR), Bryson's Rule of Tuning, Quarter car semi active suspension system.

1 Introduction

A vehicle suspension system performs two major tasks. It should isolate the vehicle body from external road disturbances for the sake of passenger comfort and control the vehicle body attitude and maintain a firm contact between the road and the tyre to provide guidance along the track. A Basic automobile suspension that is known as a passive suspension system consists of an energy storing element normally a spring and an energy dissipating element normally a shock absorber [10].

The main weakness of the passive suspension is that it is unable to improve both ride comfort and safety factor simultaneously. There is always a trade-off between vehicle ride comfort and safety factor [2, 5, 9]. To improve the ride comfort, the safety factor must be sacrificed, and vice versa. One way to overcome such a problem, the car suspension system must be controlled.

© Springer International Publishing Switzerland 2015   441
S.C. Satapathy et al. (eds.), *Proc. of the 3rd Int. Conf. on Front. of Intell. Comput. (FICTA) 2014*
− *Vol. 1*, Advances in Intelligent Systems and Computing 327, DOI: 10.1007/978-3-319-11933-5_48

Thus to design and analyze the car suspension system controller, high fidelity mathematical model for capturing the realistic dynamic of a car suspension system is necessary [7, 8].

In this paper, a semi-active suspension system is proposed [1, 7]. The semi-active suspension system is developed based on the passive suspension system. A variable MR Damper is installed parallel with the passive suspension. This MR Damper is controlled by LQR controller.

2   Quarter Car Semi Active Suspension System Modelling

The mathematical modelling of a two degree of freedom quarter car body for a semi-active suspension system is being carried out by using basic laws of mechanics.

Modelling of suspension system has been taking into account the following observations.

- The suspension system modelled here is considered two degree of freedom system and assumed to be a linear or approximately linear system for a quarter cars.
- Some minor forces (including backlash in vehicle body and movement, flex in the various linkages, joints and gear system,) are neglected for reducing the complexity of the system because effect of these forces is minimal due to low intensity. Hence these left out for the system model.
- Tyre material has damping property as well as stiffness.

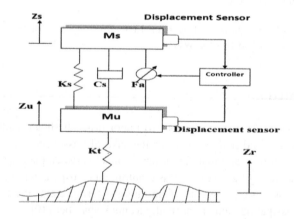

Fig. 1. Quarter car semi active suspension model

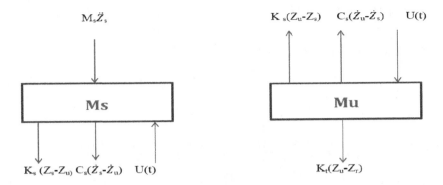

Fig. 2. Free body Diagram

From Figure 2, we have the following equations,

$$M_s \ddot{Z}_s + K_s \left(Z_s - Z_u \right) + C_s \left(\dot{Z}_{s}\dot{Z}_u \right) + U(t) = 0$$

$$\ddot{Z}_s + K_s (Z_s - Z_u) + C_s \left(\dot{Z}_{s}\dot{Z}_u \right) = -U(t) \tag{1}$$

$$M_u \ddot{Z}_u + K_s \left(Z_u - Z_s \right) + C_s \left(\dot{Z}_{u}\dot{Z}_s \right) + K_t (Z_u - Z_r) = U(t) \tag{2}$$

Where,

M_s = mass of the wheel /unsprung mass (kg)
M_u = mass of the car body/sprung mass (kg)
r = road disturbance/road profile
Z_r = wheel displacement (m)
Z_s = car body displacement (m)
K_s = stiffness of car body spring (N/m)
K_t = stiffness of tire (N/m)
C_s = damper (Ns/m)

After choosing State variables as,

$$x_1(t) = Z_s(t) - Z_u(t)$$
$$x_2(t) = Z_u(t) - Z_r(t)$$
$$x_3(t) = \dot{Z}_s(t)$$
$$x_4(t) = \dot{Z}_u(t)$$

Where,

$\left(Z_s - Z_u \right)$ = Suspension Deflection
$\left(Z_u - Z_s \right)$ = Tyre Deflection
$\dot{Z}_s$ = Car body Velocity
$\dot{Z}_u$ = Wheel Velocity

From equation (1), we have

$$M_s \dot{x}_3(t) + C_s [x_3(t) - x_4(t)] + K_s [x_1(t)] = -U(t)$$

From equation (2), we have

$$M_u \dot{x}_4(t) + C_s[x_4(t) - x_3(t)] - K_s[x_1(t)] + K_t[x_2(t)] = U(t)$$

Disturbance caused by road roughness,

$$W(t) = \dot{Z}_r(t)$$

Therefore,

$$\dot{x}_1(t) = x_3(t) - x_4(t)$$
$$\dot{x}_2(t) = x_4(t) - W(t)$$
$$\dot{x}_3(t) = -\frac{K_s * x_1(t)}{M_s} - C_s * \frac{x_3(t)}{M_s} + C_s * \frac{x_4(t)}{M_s} - \frac{U(t)}{M_s}$$
$$\dot{x}_4(t) = -\frac{K_s * x_1(t)}{M_u} - K_t * \frac{x_2(t)}{M_u} + C_s * \frac{x_3(t)}{M_u} - C_s * \frac{x_4(t)}{M_u} + \frac{U(t)}{M_u}$$

State space equation can be written as form,

$$\dot{x}(t) = Ax(t) + BU(t)$$

$$\begin{bmatrix} \dot{x}_1 \\ \dot{x}_2 \\ \dot{x}_3 \\ \dot{x}_4 \end{bmatrix} = \begin{bmatrix} 0 & 0 & 1 & -1 \\ 0 & 0 & 0 & 1 \\ -K_s/M_s & 0 & -C_s/M_s & C_s/M_s \\ K_s/M_u & -K_t/M_s & C_s/M_s & -C_s/M_s \end{bmatrix} \begin{bmatrix} x_1 \\ x_2 \\ x_3 \\ x_4 \end{bmatrix} + \begin{bmatrix} 0 \\ 0 \\ -1/M_s \\ 1/M_u \end{bmatrix} U + \begin{bmatrix} 0 \\ -1 \\ 0 \\ 0 \end{bmatrix} W \quad (3)$$

Where,

$$A = \begin{bmatrix} 0 & 0 & 1 & -1 \\ 0 & 0 & 0 & 1 \\ -K_s/M_s & 0 & -C_s/M_s & C_s/M_s \\ K_s/M_u & -K_t/M_s & C_s/M_s & -C_s/M_s \end{bmatrix}$$

$$B = \begin{bmatrix} 0 \\ 0 \\ -1/M_s \\ 1/M_u \end{bmatrix} \quad Bw = \begin{bmatrix} 0 \\ -1 \\ 0 \\ 0 \end{bmatrix}$$

$$C = \begin{bmatrix} 1 & 0 & 0 & 0 \\ 0 & 1 & 0 & 0 \\ 0 & 0 & 1 & 0 \\ 0 & 0 & 0 & 1 \end{bmatrix}, D = \begin{bmatrix} 0 & 0 & 0 & 0 \end{bmatrix}$$

Table 1. Parameters used in system simulation

S.N.	Parameter	Symbol	Quatities
1	Mass of vehicle body	Ms	504.5kg
2	Mass of the tyre and suspention	Mu	62kg
3	Coefficient of suspension spring	Ks	13100N/m
4	Coefficient of tyre material	Kt	252000 N/m
5	Damping coefficient of the dampers	Cs	400 N-s/m

The parameter values are taken from [7] and are listed in Table 1.

3   LQR Controller Design

Consider a state variable feedback regulator for the system given as

$$u(t) = Kx(t)$$

K is the state feedback gain matrix.

The optimization procedure consists of determining the control input U, which minimizes the performance index J. J represents the controller input limitation as well as the performance characteristic requirement. The optimal controller of given system is defined as controller design which minimizes the following performance index.

$$J = \frac{1}{2} \int_0^t (x^t Q x + u^t R u) \, dt$$

The matrix gain K is represented by:

$$K = R^{-1} B^T P$$

The matrix P must satisfy the reduced-matrix equation given as

$$A^T P + PA - PBR^{-1}B^T P + Q = 0$$

Then the feedback regulator U

$$u(t) = -(R^{-1}B^T P)x(t)$$

$$u(t) = -Kx(t)$$

Fig.3 shows the block diagram using LQR controller,

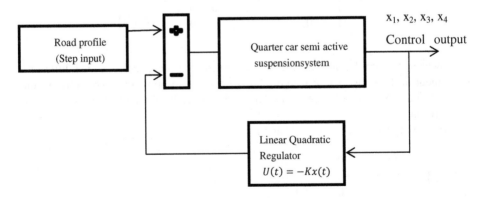

Fig. 3. A schematic Diagram for LQR controller Design

The LQR controller has a function to adjust the damping coefficient of the variable shock absorber in order to keep the car body always stable. Adjustable process is based on the characteristic of the road surface.

3.1   Bryson's Rule for Tuning

The selection of Q and R determines the optimality in the optimal control law [3]. The choice of these matrices depends only on the designer. Generally, preferred method for determining the values for these matrices is the method of trial and error in simulation. As a rule of thumb, Q and R matrices are chosen to be diagonal. In general, for a small input, a large R matrix is needed. For a state to be small in magnitude, the corresponding diagonal element should be large. Another correlation between the matrices and output is that, for a fixed Q matrix, a decrease in R matrix's values will decrease the transition time and the overshoot but this action will increase the rise time and the steady state error. In the other condition, where R is kept fixed but Q decreases, the transition time and overshoot will increase, in contrast to this effect the rise time and steady state error will decrease.

Here LQR control strategy is used for controller. Then the weighing matrices Q and R have to be determined. When not knowing Q and R values, a rule of thumb, Bryson's rule, may be give them values according to following equations [3,4].

$$Q_{ii} = \frac{1}{Max(x_{ii}^2)}$$

$$R = \frac{1}{Max(u^2)}$$

The maximum value of state is found by simulating with no input. R can initially set to 1 and then tuned by finding maximum input when a controller is included in the simulation.

Using this method matrices Q and R are obtained as follows:

$$Q = \begin{bmatrix} 0.000865 & 0 & 0 & 0 \\ 0 & 1.8114 & 0 & 0 \\ 0 & 0 & 0.011\ 31 & 0 \\ 0 & 0 & 0 & 65.03 \end{bmatrix},$$

$$R = [1]$$

However, by simulating with the gain obtained from this, results shows little improvement in damping. These weigh matrices are not so optimal; to get better result we tune Q and R manually, and found that a dramatically different Q and R gave far better result.

After tuning finally we choose Q and R values are as following:

$$Q = \begin{bmatrix} 0.000865 & 0 & 0 & 0 \\ 0 & 1.8114 & 0 & 0 \\ 0 & 0 & 0.01131 & 0 \\ 0 & 0 & 0 & 65.03 \end{bmatrix}$$

$$R = [0.000009]$$

4   Simulation Results

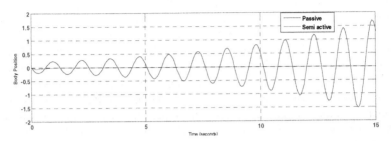

Fig. 4. Time response of vehicle body position

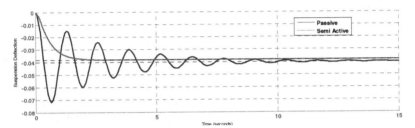

Fig. 5. Time response of vehicle suspension Deflection

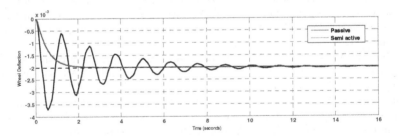

Fig. 6. Time response of vehicle wheel deflection

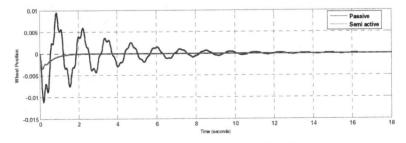

Fig. 7. Time response of vehicle wheel position

5   Conclusion

Implementation of Linear Quadratic Regulator control strategy in linear system of semi active suspension for a half car model is studied successfully. The designed matrix for feedback gain is also presented. The crucial step is to vary the value of matrix Q and matrix R. It is because there is effect at the transients output if matrix Q too large and there also effect at the usage of control action if matrix R is too large.

Finally comparison between semi-active and passive suspension system is presented and their dynamic characteristics are also compared. It has been observed that performances is improved in reference with the performance criteria like settling time and Peak overshoot for body acceleration, wheel deflection, wheel position, suspension deflection and body position. This performance improvement in turn will increase the passenger comfort level and ensure the stability of vehicle.

References

1. Kurczyk, S., Pawełczyk, M.: Fuzzy Control for Semi-Active Vehicle Suspension. Journal of Low Frequency Noise, Vibration and Active Control 32, 3217–3226 (2013)
2. Zuo, L., Zhang, P.-S.: Energy harvesting, ride comfort, and road handling of regenerative vehicle suspensions. Journal of Vibration and Acoustics 135(1), 011002 (2013)
3. Dharan, A., Storhaug, S.H.O., Karimi, H.R.: LQG Control of a Semi-active Suspension System equipped with MR rotary brake. In: Proceedings of the 11th WSEAS International Conference on Instrumentation, Measurement, Circuits and Systems, and Proceedings of the 12th WSEAS International Conference on Robotics, Control and Manufacturing Technology, and Proceedings of the 12th WSEAS International Conference on Multimedia Systems & Signal Processing. World Scientific and Engineering Academy and Society (WSEAS) (2012)
4. Al-Younes, Y.M., Al-Jarrah, M.A., Jhemi, A.A.: Linear vs. nonlinear control techniques for a quadrotor vehicle. 2010 7th International Symposium on IEEE Mechatronics and its Applications (ISMA) (2010)
5. Biglarbegian, M., Melek, W., Golnaraghi, F.: Intelligent Control of Vehicle Semi-Active Suspension Systems for improved Ride Comfort and Road handling. In: Proc. Fuzzy Information on the North American Annual Meeting, p. 1924 (June 2006)
6. Paulides, J.J.H., Encica, L., Lomonova, E.A., Vandenput, A.J.A.: Design Consid-erations for a Semi-Active Electromagnetic Suspension System. IEEE Transactions on Magnetics 42(10) (2006)
7. Haiping, D., Kam, Y.S., James, L.: Semi-active H infinity control of Vehicle suspension with Magnetorheological dampers. Journals of Sound and Vibration 283, 981–996 (2005)
8. Tan, H.-S., Bradshaw, T.: Model Identification of an Automotive Hydraulic Active Suspension System. In: Proc. of American Control Conference, New Mexico, vol. 5, p. 29202924 (1997)
9. HueiPeng, S.R., Ulsoy, A.G.: A Novel Active Suspension Design Technique Simulation and Experimental Results. In: Proc. of AACC (1997)
10. Smith, M.C.: Achievable Dynamic Response for Automotive Active Suspension. Vehicle System Dynamics 1, 134 (1995)

Algorithm for Autonomous Cruise Control System – Real Time Scheduling and Control Aspects

Annam Swetha[1], N.R. Anisha Asmy[1], V. Radhamani Pillay[1], V. Kumaresh[1],
R. Saravana Prabu[1], Sasikumar Punnekkat[2], and Santanu Dasgupta[3]

[1] Amrita Vishwa Vidyapeetham (University), India
[2] Malardalen University, Sweden
[3] Mohandas College of Engineering & Technology, India
{swethaannam4,anishaasmy,kumrnkv,saravanaprabu17,
sdasgupta100}@gmail.com, vr_pillay@cb.amrita.edu,
sasikumar.punnekkat@mdh.se
http://rtime.felk.cvut.cz/scheduling-toolbox/

Abstract. Advances in embedded system design, has led to development of autonomous systems in cruise control where real-time scheduling and control aspects have to be integrated. A synchronization of scheduling and control aspects requires consideration of task temporal attributes and control dynamics. In this paper an algorithm has been developed and simulated in Matlab with TORSCHE[1] toolbox that enables a system designer to explore control strategies, timing aspects during real-time scheduling, various interactions between control, scheduling and real-time constraints. P, PI control strategies have been used and simulation runs monitor the real-time performance. A GUI with real-time scheduling and control dynamics display, facilitates a fast prototyping design. To validate the algorithm for autonomous cruise system, different real-time scenarios are considered for speed and safe distance.

Keywords: Autonomous cruise control prototype, Safety critical, Real-time scheduling, Cruise speed, Safe distance.

1 Introduction

Modern automotives integrate a wide variety of sensors and actuators with embedded systems and run in highly dynamic environments, adapting their behavior at runtime (in response to frequent changes in their environment). There exists a shift in automotives from human controlled to self directed and represent safety critical real-time systems. In such systems an integrated scheduling and control aspects have to be studied to ensure safe functionality. Autonomous Cruise Control (ACC) system is one such system deployed in modern automotives for providing assistance to the driver.

ACC maintains constant velocity, safe distance with respect to the other vehicles and road conditions and hence becomes a Driver Assistance System (DAS) [7]. Based on the information from the sensor, controller sends the command signals to the throttle to either regulate the vehicle speed to a given set value or maintain a safe

© Springer International Publishing Switzerland 2015
S.C. Satapathy et al. (eds.), *Proc. of the 3rd Int. Conf. on Front. of Intell. Comput. (FICTA) 2014*
– *Vol. 1*, Advances in Intelligent Systems and Computing 327, DOI: 10.1007/978-3-319-11933-5_49

distance with leading vehicle along with sending the status information to the driver (Fig. 1) [8].

A synchronization of scheduling and control aspects (P/PI) requires consideration of complex timing requirements of both task temporal attributes and control dynamics [9]. To highlight various aspects of cruise system over a cruise distance range of 50m, four strategies have been formulated, taking cognizance of the presence of lead vehicle and its variations in speed. A real-time scheduling algorithm for such a system has been designed and implemented in Matlab. A GUI has been developed for monitoring the cruise system and its real-time scheduling over one complete hyperperiod. It displays the physical parameters of ACC like cruise speed, current speed of host vehicle and lead vehicle and the distance between them.

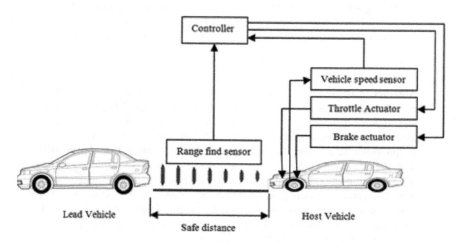

Fig. 1. Autonomous Cruise Control system

Rest of the paper is organised as follows, literature review and backround study of cruise system, its control, scheduling and their integration aspects are discussed in Section 2. The description of system model, its implementation with simulation environment is explained in Section 3. Results and discussions are presented in Section 4. Section 5 deals with conclusion and future scope of the work.

2   Some Aspects of ACC–Literature Review and Background Study

The first cruise prototype was developed in early nineties [1] with the purpose of keeping the velocity of the vehicle at a constant level and providing driver assistance. Further technology created an additional capability of maintaining safe distance between vehicles in the same lane [2]. It controls the accelerator and vehicle brakes to maintain desired time gap from the lead vehicle and is widely designated as Autonomous/Adaptive Cruise Control (ACC) system. Real-time scheduling ensures that deadlines are met so that cruise speed and safe distance are guaranteed to prevent collisions in such a safety critical framework.

The control aspects of this system consist of input data collection, processing and actuation. Physical input data is received by different sensors and the control algorithm can include P/PI/PD/PID control [3]. Different scheduling strategies of kernel for periodic scheduling of safety critical real-time systems include static table driven scheduling, cyclic scheduling and priority driven preemptive approach. Server based approach with a constant bandwidth is proposed for serving aperiodic tasks [4]. Dynamic planning based approach given in [5] explains the feasibility of scheduling aperiodic task with previously guaranteed tasks. The response latency between reading sensor values and actuation, jitter are issues which have lead to research in the co-design of scheduling and control system [6]. The time attributes of the cruise control [10] with precedence constraints forms the basis for the implementation of the algorithm for ACC. The critical functioning of the cruise system is based on real-time scheduling of control and actuating tasks which have been deemed to be critical tasks [11], [12].

3   Design and Development of Control and Scheduling Algorithm (*Co_SA*)

The schematic representation of the system is presented in Fig.2, where the Controller implements the *Co_SA* to maintain cruise speed and safe distance in host vehicle with and without the lead vehicle. The real-time constraints of system are activation periods, response time, precedence constraints, input output delays and jitter. To avoid the deterioration of control performance due to jitter, and to meet safety compliance, flexibility is introduced by the kernel in the activation time of the control task. A control task is instantiated every time after receiving input from the sensor tasks. The scheduling algorithm implemented in the controller ensures all critical tasks for control and actuation are executed without missed deadlines and P/PI strategies are employed. A cruise control system model [12] has been considered for the control aspect of the system.

On activating the Cruise On switch, with the current speed above 40m/s, controller sets the current speed as the desired speed. Vehicle speed sensor continuously senses the current speed of the vehicle and feeds it as input to the controller for every 3time units. Range find sensor is employed to sense the speed of front vehicle and to maintain a safe distance between the vehicles. In the processor, control algorithm compares the current speed with the desired speed set by the driver and produces corresponding actuating signal. A synchronization of scheduling of the cruise tasks in the real-time kernel with the control thread is implemented with the global clock. On pressing the Acc or Dcc switch, controller changes from cruise mode to normal mode and accelerates or decelerates at a rate of 2 m/s^2. Assumption is made such that the response latency between input and output signals at every mode is considered.

Simulation has been carried out under different operating modes

i)   Normal Mode: Host vehicle acceleration and deceleration
ii) Cruise Mode: Maintaining desired set speed of driver in the absence of lead vehicle.
iii) Safe distance mode: Maintaining longitudinal safe distance with respect to lead vehicle.
iv) Lead vehicle mode: Varying speed with respect to the speed of lead vehicle

The *Co_SA,* integrating control and scheduling aspects has been formulated, simulated to evaluate the performance under different modes for a cruise range of 50m. For the initial distance of 10m, system is made to run under normal mode with acceleration and deceleration switches being enabled. After reaching a minimum speed of 40m/s, cruise mode can be initialized to maintain a constant speed, and the controller executes the cruise tasks maintaining precedence constraints with P/PI control.

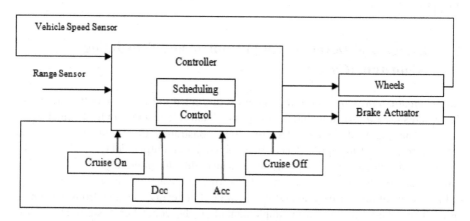

Fig. 2. Schematic representation – ACC

A flow chart of the formulated approach is shown in Fig. 3. The variable D represents the instantaneous cruise distance and is initialized to 0, similarly current speed is initialized to 0 m/s. Global clock provides synchronization between control dynamics and real-time scheduling.

In onboard ACC of this system *Co_SA* has been implemented in Matlab using Time Optimisation Resource and SCHEduling (TORSCHE) toolbox. Table 1 gives the task set of ACC which includes sensor tasks, control tasks and actuating tasks. The control task and actuating tasks are critical tasks and are precedence constrained with sensing tasks.

Matlab GUI indicates the working of ACC and is shown in Fig. 4, where the Driver Assistance System (DAS) represents the cruise system to maintain constant speed and safe distance. Control switches initiate the cruise mode and display includes current speed and desired speed of vehicle along with safe distance maintained with the lead vehicle. The GUI provides visualization of real-time scheduling and the steady state response of the control algorithms. An offline analysis and simulation for the cruise control model has been done in Matlab. This operation simulates the controller and system behavior in the execution window of controller task.

The velocity of the lead vehicle is assumed to start at an initial value of 60 m/s. The initial global longitudinal positions of lead vehicle and host vehicle are assumed to be 130m and 0 m, respectively. This means that the host vehicle is initially out of range sensor.

In this section, four scenarios based on the speed of the lead vehicle and host vehicle are described.

Mode 0: *Normal mode.* System made to run under normal mode for an initial distance of 10 m with acceleration and deceleration switches being enabled. On pressing the corresponding Acc and Dcc switches, system accelerates or decelerates at a constant rate of 2 m/s^2.

Mode 1: *Cruise mode.* With no lead vehicle in front of the host vehicle within the sensor range, host vehicle is under the velocity control mode or the conventional cruise control mode. On reaching a desired speed if cruise switch is pressed the vehicle maintains a same speed.

Mode 2: *Safe distance mode.* With the detection of lead vehicle in the safe distance range sensor notifies the host vehicle and reduces its speed until the distance is maintained.

Mode 3: *Lead vehicle mode.* With a slow moving lead vehicle, host vehicle starts to decrease its velocity in order to maintain desired spacing between the vehicles.

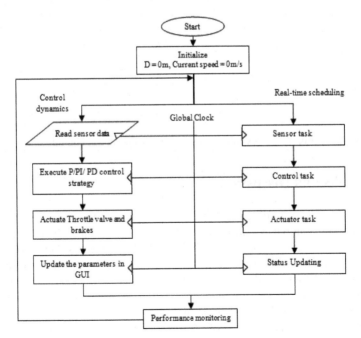

Fig. 3. Flow cart of integrated Control and Scheduling Algorithm (*Co_SA*)

Table 1. Task set – ACC

SI.No	Tasks	C_i (time units)	T_i (time units)
1.	Monitoring the Speed τ_{NC1}	3	15
2.	Monitoring acceleration τ_{NC2}	2	10
3.	Monitoring the CCS clutch τ_{NC3}	2	10
4.	Monitoring the brakes τ_{NC4}	3	15
5.	Monitoring proximity sensor τ_{NC5}	2	15
6.	Computing the control values τ_{C6}	10	55
7.	Actuating the throttle valves τ_{C7}	5	30
8.	Updating the parameters in τ_{NC8}	10	15

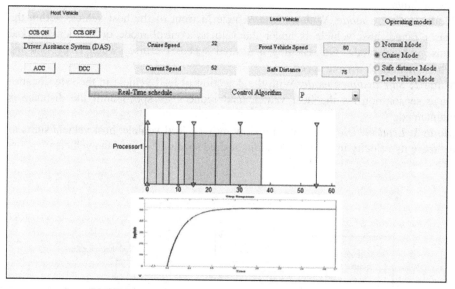

Fig. 4. GUI - Cruise system with scheduling and control

4   Results and Discussion

The speed with respect to distance graph is given in Fig.5, where the *normal mode* is represented for an initial distance of 10m with acceleration switch being enabled where the speed increases and reaches 52m/s at 0.19 time units. With cruise switch being activated, system transfers to *cruise mode* from 10m for a distance of 15m maintaining a constant speed of 52m/s. At 0.5 time units where the safe distance is assumed to be violated, system changes to *Safe distance mode* to retain 50m of distance between the lead vehicle by reducing its speed to 27m/s and continues till 1.48 time units. From the distance of 40 m host vehicle speed, varies based on the speed of lead vehicle till 1.66 time units.

Simulation results of different operational modes with scheduling and control have been given in Fig. 6. P/PI control has been implemented for the control task during the window of operation for specified speed and time attributes in cruise and safe distance mode. Simulation runs of the control task (12-27 time units), in cruise speed and safe distance have been implemented with P/PI control and the zoomed out time interval of the control task is shown in Fig.6. In mode1 speed control (52m/s) is achieved by P/PI control with rise time as 0.15 and 0.05 time units respectively. In mode 2 with host vehicle speed (52m/s) and lead vehicle speed (40 m/s), safe distance is maintained with the response latency of 0.13 time units. P/PI control in mode 3 increases the speed of host vehicle to 40 m/s with rise time of 0.15,0.05 due to increase in the speed of lead vehicle by 20m/s without violating the safe distance condition.

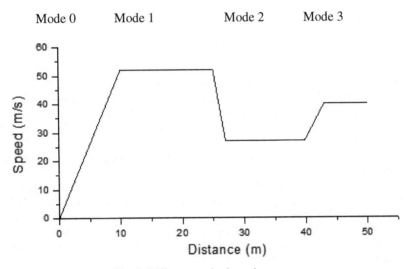

Fig. 5. Different modes in cruise range

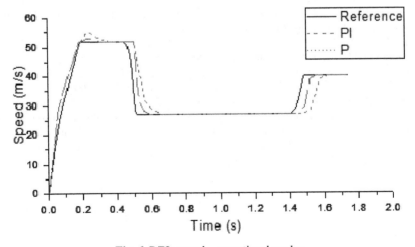

Fig. 6. P/PI control – operational modes

5   Conclusion

An algorithm has been developed and simulated in Matlab for an autonomous cruise control system. The complex timing requirements of temporal attributes and control performance which involve synchronization of scheduling and control aspects with P/PI strategies have been implemented. A Matlab GUI provides visualization of the system with physical parameters and the speed profile of the ACC under different modes. Other aspects of autonomous vehicle systems can be studied and integrated using this system. Simulation of more realistic implementation of online scheduling and controlling can be done using advanced Matlab Real-Time toolbox. Practical problems associated with timing variations due to real-time scheduling and control can be studied.

References

1. Björnander, S., Grunske, L.: Adaptive Cruise Controllers – A Literature Review. Technical Report, Faculty of Information and Communications Technologies (August 8, 2008)
2. Vahidi, A., Eskandarian, A.: Research advances in intelligent collision avoidance and adaptive cruise control. IEEE Transactions on Intelligent Transportation Systems 4(3), 143–153 (2003) ISSN 1524-9050
3. Cervin, A.: Integrated Control and Real-Time Scheduling, Thesis report, Lund Institute of Technology, Lund (2003)
4. Audsley, N., Burns, A., Davis, R., Tindell, K., Welling, A.: Real-time system scheduling. In: Predictably Dependable Computer Systems, pp. 41–52. Springer, Heidelberg (1995)
5. Ramamritham, K., Stankovic, J.A.: Scheduling Algorithms and Operating Systems Support for Real-Time Systems. Proceedings of IEEE 82(1) (January 1994)
6. Kakade, R.S.: Automotive Cruise Control System. Thesis report, Indian Institute of Technology, Bombay (July 2007)
7. KPIT, Autonomous Vehicles, TechTalk@KPIT. Journal of KPIT Technologies Limited 6(4) (October-December 2013)
8. Naus, G.J.L.: Model-based control for automotive application, Ph.D. thesis, Eindhoven University of Technology, Eindhoven, Netherlands (2010)
9. Henriksson, D.: Resource-Constrained Embedded Control and Computing Systems, Thesis report, Lund Institute of Technology, Lund (2006)
10. Staines, A.S.: Modelling and Analysis of Real Time Control Systems: A Cruise Control System Case Study. Recent Advances in Technology. InTech Open (2009)
11. Swetha, A., Radhamani, P.V., Punnekkat, S.: Design, Analysis and Implementation of Improved Adaptive Fault Tolerant Model for Cruise Control Multiprocessor System. International Journal of Computer Applications (0975 – 8887) 86(15) (January 2014)
12. Control Tutorial for Matlab & Simulink, http://ctms.engin.umich.edu

Enhancing Flexibility for ABE through the Use of Cipher Policy Scheme with Multiple Mediators

Rasal Shraddha[*] and Tidke Bharat

Department of Computer Science,
Pune University, Maharashtra, India
rasalshraddha@hotmail.com

Abstract. The existing work in cryptography had just way of sharing keys between senders and receivers which was made through signature storage provided for the user's public key. But increasing number of users was giving a great challenge for using such certificate storage and key distribution is on the other hand a difficult job. Identity Based Encryption (IBE) has been proposed to overcome traditional risk which had again created the time consuming environment due to its one-to-one communication phenomenon as personal information only a way to use the keys. This problem had been solved by Attribute Based Encryption (ABE) through single mediator by providing multicast communication. This concept was based on Key-Policy ABE (KP-ABE) as well as could not provide the revocation phenomenon for keys. So this paper aims to increase the level of encryption using MAMM (Multiple Authority Multiple Mediators) with the use of CP-ABE (Cipher Policy ABE) instead of KP-ABE that will give two level decryption by hierarchical mediators' tree as well as user's specific keys with forward access level and will provide flexibility in data transfer service.

Keywords: Attributes set, Cipher-text policy, Encryption, Multi-Authority, Mediator.

1 Introduction

Data privacy is the most important feature in today's computer world. In traditional system, data security had done through encryption and decryption through just certificates that binds user's keys. This had minimized by Shamir [1] who proposed new concept of Identity Based Encryption (IBE) that had used the user's own information (example: email id). It has limitedly used for one-to-one communication which has overcame by Fuzzy IBE[3] so called as an Attribute based encryption (ABE), that could encrypts the document for all the users having specific set of attributes. When the numbers of users were huge, use of multi-authority had a challenging job for doing such work. This had made possible by using Cipher-policy based encryption technique of cryptography for Multi-Authority environment.

---

[*] Corresponding author.

© Springer International Publishing Switzerland 2015
S.C. Satapathy et al. (eds.), *Proc. of the 3rd Int. Conf. on Front. of Intell. Comput. (FICTA) 2014*
– *Vol. 1*, Advances in Intelligent Systems and Computing 327, DOI: 10.1007/978-3-319-11933-5_50

This has an additional structure called as "Revocation" which basically integrates the user's abolition for using the services with level of access that maintains the higher level of data privacy in encryption. Example: As the access level is in ascending order from student to admin. The level of access is more for admin and reduces to student. If the user is student using the services and leaves the service and again added as staff say then revocation makes possibility to change the assess policy. This means that an access policy is dynamic every time depending on type of user. IBE scheme was proposed first to eliminate the certificate storage and then Attribute Base Encryption proposed new security technology that motivates to survey on best encryption techniques for leveled data privacy using cipher text policy (CP-ABE). Such cipher-texts can be decrypted by anyone with a set of attributes that fits the policy. This work gives details about using services through strong encryption level using multiple mediators with CP-ABE. Further Section 2 consists literature survey followed by research methodology in Section 3 and Section 4 consists implementation details then conclusion.

2   Literature Survey

2.1   Fuzzy Identity Based Encryption

Shamir [1] proposed a new concept initially called as an Identity Based Encryption (IBE) an exciting alternative to public-key encryption as it eliminates the need for a Public Key Infrastructure (PKI) for which the practical implementation was done in 2001 but it have some limitations which didn't make fully satisfactory work in Encryption of message as it is only limited for one to one communication to provide error tolerance property, new approach is proposed by Sahani and Waters [3] in 2005 called Fuzzy Identity Based Encryption (FIBE) to provide the multicast communication. In fuzzy IBE [3][4] identity of a user is given as a set of attributes. The name "Attributes" given such that number of attributes for each user are located so as to maintain the access level. It had given an idea that how an IBE system encrypts data in multiple hierarchical-identities in a in forward secure manner, e.g. {company, branches, departments etc.}. It has shown that how their techniques are useful to prevent collision attacks in attribute-based encryption. Fuzzy IBE was based on the set of attributes based on group of bilinear group for the pairing of elements called as a bilinear map [10]. For such scheme a fantastic option was proposed by V. Miller [5] in 1985 through the use of elliptic curves which had an arithmetic structure as well as it is better to fit for solving the problems of traditional system such as Diffie-Hellman or ElGamal. Use of pairings used was good idea that solved the discrete algorithm difficulties and rules out the simpler bilinear map for cryptosystem. Bilinear map has been created using two groups say G1 and G2.The group G1 is selected that contains points on elliptic curves over function Fq. The order of G1 was taken as prime l. When l=q, there exist an adaptive pairing that sends G1 to the additive group G2 = (Fq, +). This pairing have key parameter called as security parameter. When this r (security parameter) was small, the pairing was computed efficiently [6]. Problem occurred when the value became large [3][7]. Two pairings types are defined on elliptic curves that are Weil pairing and Tate pairing. Since at some value of r Weil pairing was unable to reach the optimum value, Tate pairing was used that provided an optimum solution as well has had less cost. Frey, Muller and Ruck[8][9][10] proposed to use it as a replacement for the Weil pairing. Hence the construction of pairing is done as bi-linear map on elliptic curve to select a set of attributes. The preliminaries regarding to the security parameter used on elliptic curve for proper pairing as bi-linear map, the multicast communication is possibly carried out through the set of attributes for users hence as variant to the Fuzzy IBE further leads to more advanced Attribute based IBE. For the set of attributes for user, an authority was required in any ways to make the management between the attributes for single authority attributes or multi authority attributes depending on the data. Single authority attributes can be easily managed between the users and providers [14][15] but the single authority had some problems such as inefficiency, non scalability and non applicability which were the open problems proposed by M. Pirretti [16], by Shucheng Yu [17], by V.Goyal[18] respectively. The most challenging and interesting job was to manage the security for multi-authority attributes as there were many applications which requires multiple authorities. Designing a multi-Authority ABE scheme was an interesting open problem first proposed by Sahani and Waters in [4] and later solved by Melissa Chase [13] in 2007.

2.2   Multi-Authority ABE

The focus had made on multi-authority which was having its own procedure to follow the tasks and then combining the results. It allowed any number of attribute authorities to be corrupted, and guarantee the security of encryption as long as the required attributes could not be obtained exclusively from those authorities and the trusted authority remains honest. But this work again had some problems regarding the collision. The Sahani and Water [4] had prevented the collusion within authorities, so different keys obtained from any one authority could not combine. For example suppose cipher-text is given which requires attributes from authority 1 and authority 2. If Alice has all the appropriate attributes from authority 1 and Bob has all the appropriate attributes from authority 2, they still should not be able to combine their keys and decrypt. But in multi authority based concept, secrets were necessarily divided between multiple authorities and this had carried out between authorities independently. Here for these purpose two main techniques were used: The first was, every user have a kind of a global identifier (GID) such that: (1) no user can argue on another user's identifier, and (2) all authorities have rights to verify a user's identifier that creates their own public as well as private key. Thus, the GID could be SSN which is randomly generated. Example given in [13] shows why it is required to have such a credentials: In the first, Bob requests keys for authority 1 and Alice requests keys for authority 2. In the second Bob requests attribute set A1 from authority 1 and attribute set A2 from authority 2. The global identifier (GID) [13] allowed authorities to distinguish between such cases in order to prevent collision.

Each authority acts as pseudorandom function (PRF) which it was used to randomize the secret keys it has given out. A PRF was proving the guaranty that, on the one hand, the secret keys for each user were derived deterministically, but, at the same time they appeared completely random. When a user requested a secret key, the authority computed the PRF on the user's GID and then used the result as the secret in Sahani and Water's key generation. This idea was worthwhile that broken up the secret throughout multiple authorities based on the user's GID, such that each authority have its own such that each authority was doing its own work independently as only the GID is provided. The use of PRFs mean that each user's secret keys are independent of any other user's keys and collusion is impossible [13]. This was again done through more complex access structure in order to decrypt the cipher-text. Multi-Authority Attribute Based Encryption was worked out along with the access structure using PRF with independent GID.

2.3   Key Revocation

Boneh and Franklin [3] was first suggested simple "Revocation" concept for IBE in random oracle [21] told about expiration date for keys for maintaining the access control. Another approach towards revocation in CP-ABE was proxy re-encryption technique that made use of the proxy servers [17].The proxy servers could be dishonest or could be compromised and hence the scheme was not very secure. In 2011, with the aim of providing encryption based access control in social networks the authors in [22] proposed the concept of proxy re-keying for minimizing the trust on proxy servers to enable efficient revocation. These approaches were limited to revoking a predefined number of attributes also providing the limitation of existing approaches for revocation in ABE includes inefficiency [3][16], non scalability [17], unreliability and non-applicability to CP-ABE. Furthermore in 2012 Riddhi Mankad et al [25] were combined the multi-authority scheme with revocation concept where she implemented the combinations of authority and mediator with final combination proposed was MASM [25] (Multi-authority Single Mediator) which eliminated the problem about revocation of previous scheme.

3   Our Approach

Proposed approach has been implemented using figure 1 and generated results as shown in figure 2 using run-time data from server database. The existing system has been providing facility for static data environment but Multi-Mediator scheme has used dynamic data environment and using 256 encryption techniques for key generation. The proposed system is giving better results than previous system in speed as well as time. Similarly, Like encryption use of multiple mediators is nice idea useful at the time of decryption as well such that user required to perform dual decryption to access its final data. On the other hand, use of access policy and different mediators based on type of users is strong idea for increasing level of privacy for user that requires more encryption modes for creating highly secure data.

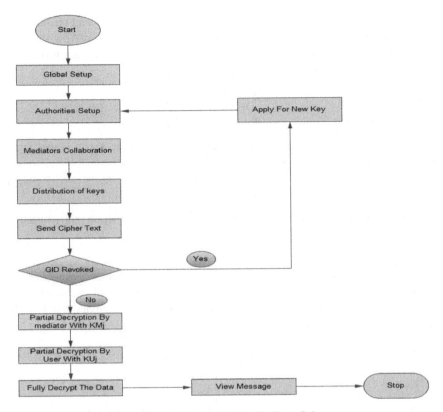

Fig. 1. Flowchart for Multiple Mediator Scheme

- Global Setup {GPk,GSk}: Trusted authority (TA) carried outs this step using users credential to generate global public key GPk and global secrete key GSk where GSk is only known to TA.

  setup={g,α, β , r, H } , where parameter 0<α<10, 0<β<10 , 0<r20

  GSk=g α, where g parameter is generated using Global identifier thts is unique id and α, β , r be randomly generated.

- Authority Setup {PKj, SK$_j$}: This setup is made to generate attribute level keys. Authority i take GPk and attribute set j according to type of user and generates public key and secret key say PKj and SKj respectively.

- User-key setup {K$_0$}: User requests a unique global identifier GID from TA that was calculated by MD5 hashing algorithm using user's credentials. TA uses this GID GSk to generate K$_0$ component.

$$r=H \ (GID)$$
$$Ski=\{r, K0, T\} \tag{1}$$

- Attribute Key Generation: An authority checks for j attributes, user type along with GID, GPk and SKj and generates attribute level keys using SKj. To provide different level of decryption this key is divided into user key Ukj and mediator key Mkj. These keys generated using Advanced encryption Standard(AES)algorithm and are user specific depending on type of user so the data encrypted using these keys is decrypted only if it satisfies these key policies.

$$DH \; spec = getL()$$
$$DH \; spec = getP()$$
$$Key = (p, l, g, H(K0))$$

K_0-Encryption $\{C_1\}$: Message M is initially encrypted using K_0 on requesting to authority. It also used policy say T based on CP-ABE scheme.

$$C1 = \{M, K0\} \qquad (2)$$
$$Lock_{in} = C1$$

- PKj-Encryption $\{C_2\}$: Cipher-text C_1 again encrypted using PKj on requesting from authority and gives cipher-text say C2.

$$C2 = \{C1, PKj\} \qquad (3)$$

- Partial Decryption$\{C'_1\}$: Decryption is made initially by requesting user level key Ukj from authority where user send it to mediator. Mediator will use its own Mkj and appends it to users key $\{Ukj + Mkj\}$ and functions for cipher-text C_2. This operation performs partial decryption and gives C'_1.

$$if(Policy \; T)$$
$$\{Key = \{Ukj + Mkj\} \qquad (4)$$
$$If(valid \; Key)$$
$$\{C'1 = \{C2, key\}\} \qquad (5)$$

- K_0-Decryption: Finally user requests global parameter K_0 to TA to generate final message M or original data.

$$M = \{C'1, K0 = g^{(+)/r}\} \qquad (6)$$

3.1   Comparison Results

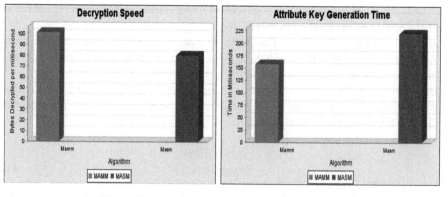

Fig. 2. Decryption speed and Key Generation speed

Table 1. Camparison Table for Performance Measure

Measuring terms	Current Scheme	Multiple mediator Scheme
Average Time for attribute key generation for byte of data	260.0ms	221.7143ms
Number of Bytes Decrypted/ms	80.1359 bytes/ms	86.7314 bytes
Average encryption Time(T) based on graphs	(Te)ms	[(Te)-10]ms
Policy Level(P)	[P]attributes	[P+2]attributes
Bytes Decryption(D)	(D)bytes/ms	[(D)+30]bytes/ms
Key generation Speed	moderate speed	high speed service

4   Conclusion

Existing techniques and their implementation about ABE scheme gives an idea that there are still limitations in the area. This paper shows results for Multiple Authority with multiple mediators using Cipher policy scheme in which even one media-tor gets compromised the service will still in working using higher level mediator access. So these approaches showing a more scope in encryption techniques under the distributed environment which provides a brief knowledge about future work in the same to enhance and improve the performance specifically in the area of cloud computing.

References

1. Shamir, A.: Identity Based Cryptosystems and Signature schemes. Department of Applied Mathematics (1998)
2. Chatterji, S., Sarkar, P.: Identity based Encryption, pp. 17–50. Springer NewYork library (2010) ISBN 978-1-4419-9382-3 e-ISBN 978-1-4419-9383-0, doi:10.1007/978-1-4419-9383-0
3. Boldyreva, A., Goyal, V.: Identity-based Encryption with Efficient Revocation (2008)
4. Boneh, D., Franklin, M.: Identity-based encryption from the Weil pairing. In: Kilian, J. (ed.) CRYPTO 2001. LNCS, vol. 2139, pp. 213–229. Springer, Heidelberg (2001)
5. Sahai, A., Waters, B.: Fuzzy identity based encryption. In: Cramer, R. (ed.) EUROCRYPT 2005. LNCS, vol. 3494, pp. 457–473. Springer, Heidelberg (2005)
6. Rivest, R.: The MD5 message digest algorithm. MIT Laboratory for computer Science and RSA Data security Inc. (April 1992)
7. Miller, V.S.: Use of elliptic curves in Cryptography. In: Williams, H.C. (ed.) CRYPTO 1985. LNCS, vol. 218, pp. 417–426. Springer, Heidelberg (1986)
8. Joux, A.: The Weil and Tate Pairings as Building Blocks for Public Key Cryptosystems. In: Fieker, C., Kohel, D.R. (eds.) ANTS 2002. LNCS, vol. 2369, pp. 20–32. Springer, Heidelberg (2002)

9. Boneh, D., Lynn, B., Shacham, H.: Short signatures from the Weil pairing. In: Boyd, C. (ed.) ASIACRYPT 2001. LNCS, vol. 2248, pp. 514–532. Springer, Heidelberg (2001)

10. Barreto, P.S.L.M., Kim, H.Y., Lynn, B., Scott, M.: Efficient algorithms for pairing-based cryptosystems. In: Yung, M. (ed.) CRYPTO 2002. LNCS, vol. 2442, pp. 354–369. Springer, Heidelberg (2002)

11. Frey, G., Muller, M., Ruck, H.-G.: The Tate pairing and the discrete logarithm applied to elliptic curve cryptosystems. IEEE Transactions on Information Theory 45(5), 1717–1718 (1999)

12. Galbraith, S.D., Harrison, K., Soldera, D.: Implementing the Tate pairing, Volume (2002)

13. Yang, P., Cao, Z., Dong, X.: Fuzzy identity based signature (2008)

14. Lewko, A., Waters, B.: Decentralizing attribute-based encryption. In: Paterson, K.G. (ed.) EUROCRYPT 2011. LNCS, vol. 6632, pp. 568–588. Springer, Heidelberg (2011)

15. Chase, M.: Multi-authority attribute-based encryption. In: Vadhan, S.P. (ed.) TCC 2007. LNCS, vol. 4392, pp. 515–534. Springer, Heidelberg (2007)

16. Waters, B.: Ciphertext policy attribute based encryption an expressive, Efficient and provably secure realization. In: Catalano, D., Fazio, N., Gennaro, R., Nicolosi, A. (eds.) PKC 2011. LNCS, vol. 6571, pp. 53–70. Springer, Heidelberg (2011)

17. Ibraimi, L., Petkovic, M., Nikova, S., Hartel, P., Jonker, W.: Mediated Ciphertext Policy attribute-based encryption and its application. In: Youm, H.Y., Yung, M. (eds.) WISA 2009. LNCS, vol. 5932, pp. 309–323. Springer, Heidelberg (2009)

18. Pirretti, M., Traynor, P., McDaniel, P.: Waters, Secure attribute-based Systems. In: ACM CCS 2006, vol. 6377 (2006)

19. Yu, S., Wang, C., Ren, K., Lou, W.: Attribute based data Sharing with attribute revocation. In: ASIACCS 2010 (2010)

20. Chase, M.: Multi-authority Attribute Based Encryption. Computer Science Department Brown University Providence, RI 02912 (2007)

21. Goyal, V., Pandey, O., Sahai, A., Waters, B.: Attribute based encryption for Fine Grained access control of encrypted data. In: ACM Conference on Computer and Communications Security, pp. 88–98 (2006)

22. John, B., Amit, S., Brent, W.: Ciphertext-policy attribute-Based encryption. In: IEEE Symposium on Security and Privacy, pp. 321–334 (2007)

23. Jahid, S., Mittal, P., Borisov, N.: Easier: Encryption-based access Control in social networks with efficient revocation. In: ASIACCS 2011 (March 2011)

24. Mankad, R., Jinwala, D.: Investigating multi authority attribute based Encryption with revocation. NIT Surat (2012)

25. Bellare, M., Rogaway, P.: Random oracles are practical: A paradigm for Designing efficient protocols. In: ACM conference on Computer and Communications Security (ACM CCS), pp. 62–73 (1993)

26. Goyal, V.: Reducing Trust in the PKG in Identity Based Cryptosystems. In: Menezes, A. (ed.) CRYPTO 2007. LNCS, vol. 4622, pp. 430–447. Springer, Heidelberg (2007)

27. Sami, M., Macchetti, M., Regazzoni, F.: Speeding Security on the Intel Strong ARM. In: Embedded Intel Solutions, pp. 31–33 (Summer 2005)

Mean Interleaved Round Robin Algorithm: A Novel CPU Scheduling Algorithm for Time Sharing Systems

R.N.D.S.S. Kiran[1], Ch. Srinivasa Rao[2], and Manchem Srinivasa Rao[3]

[1] Swarnandhra College of Engineering and Technology, Narsapur, India 534280
[2] SVKP & Dr. KS Raju Arts & Science College, Penugonda, India 534320
[3] Swarnandhra College of Engineering and Technology, Narsapur, India 534280
{scetseta,chiraparapu,cnumaster}@gmail.com

Abstract. In round robin scheduling algorithm, the scheduler preempts a process when time slice (δ) expires and picks up the next processes in the ready queue for execution. A potential problem that we observed with the traditional round robin scheduling algorithm is: when the time slice increases, both the average waiting time and the turnaround time are also increasing instead of decreasing. This paper proposes a remedy for the observed problem which works as follows calculate the mean burst time (m) of all the processes in the ready queue. Then insert m at its appropriate position in the sorted ready queue. With this the ready queue is divided into two parts: the first part contains the burst times, b1, b2,...b_{j-1}, which are smaller than m and the second part contains burst times $b_{i+1}, b_{i+2}....b_{n+1}$, which are greater than or equal to m. now pickup the process p_1 with burst time b_1 from the sorted ready queue, assign it to CPU and execute it for one time slice (δ). Next pickup the process p_{i+1} with burst time b_{i+1} from the sorted ready queue and execute it for another δ. This process is repeated until all the jobs in the ready queue complete their execution. Then average waiting time and average turnaround time of all the processes in the ready queue are computed. When the proposed method is applied, it observed that even if the time slice increases, both the average waiting time and the turnaround time are found to be also decreasing.

Keywords: Time sharing systems, Round Robin algorithm, Mean Interleaved Round Robin Algorithm, MDRR, Preemptive Scheduling.

1 Introduction

The purpose of a CPU scheduling algorithm is to execute more than one process at a time and transmit multiple flows of control simultaneously. CPU scheduling is the process of determining which program should be allocated to CPU and how long. CPU scheduling forms the basis for multi programmed operating systems. The multi programmed operating systems maximize CPU utilization by way of having some processes executing at all times. When one process needs to wait the CPU scheduler selects another process from the ready queue for execution on the CPU. Thus CPU will not be allowed to sit idle. That is how CPU utilization is maximized with the multi programming technique. In this way, every time a process has to wait another process takes over use of the CPU. This pattern repeats until all the processes are finish their execution.

1.1   Type of Scheduling Algorithms

Scheduling algorithms of two major types: (1) preemptive scheduling algorithms, (2) non preemptive scheduling algorithms, a preemptive priority scheduling algorithm preempts a low priority job and allocates CPU to a higher priority job when the higher priority job enters while the lower priority job is executing. A non preemptive scheduling algorithm continues the execution of currently running process to its completion without preemting it.

1.1.1   Preemptive Scheduling Algorithms
A scheduling decision intervenes while a process is executing. Thus, preemptive scheduling may force a process in execution to release the CPU, so that the execution of another process can be undertaken. The following are the algorithms which use preemptive scheduling:

   a)   Round Robin (RR) scheduling algorithm
   b)   Priority scheduling algorithm

a) **Round Robin (RR) Scheduling:** The Round Robin algorithm is designed for time sharing systems. The primary objectives of Round Robin algorithm are interactive use, good response time and sharing of the resource equitable among processes. It is similar to FCFS but pre-emption is added to switch between processes. The processes are allocated a small unit of time, known as time quantum in rotation until the completion of all processes.

b) **Priority Scheduling:** In priority scheduling each process in the system is assign a priority level and the scheduler always chooses the highest priority process.

1.1.2   Non Preemptive Scheduling Algorithms
In this type of scheduling, a scheduled process always completes before another scheduling decision is made. Therefore, finishing order of the processes is also same as their scheduling order. The scheduling algorithms which use non preemptive scheduling are:

   a)   First Come First Served (FCFS) Scheduling algorithm
   b)   Shortest Job Next (SJN) Scheduling algorithm

a) **First Come First Severed (FCFS):** Is the most popular non preemptive scheduling algorithm which schedules the processes in such a way   that the processes which request right of CPU are allocated first.

b) **Shortest Job Next (SJN):** In SJN scheduling whenever a new job is to be admitted, the shortest of the arrived jobs is selected and given the CPU time.

1.2   Scheduling Criteria

Scheduling Criteria [1] are used to determine which algorithm to use in a particular situation.   Many criteria have been suggested for comparing CPU Scheduling algorithms.   An algorithm is judged to be the best based on the characteristics that are used for comparison.   The criteria include the following:

CPU Utilization: Refers to a computer's usage of processing resources, or the amount of work handled by a CPU.   In single tasking environment the processors utilization is low, in multi tasking and time sharing systems the CPU utilization must be high.

Throughput: Throughput was conceived to evaluate the productivity of computer processors. This was generally calculated in terms of jobs or tasks per second and millions of instructions per second.

Turnaround Time: The interval time from the submission of a process to the time of completion is the turnaround time. Turnaround time is the sum of the periods spent waiting to get into memory, waiting in the ready queue, executing on the CPU, and doing I/O.

Waiting Time: CPU scheduling algorithm affects only the amount of time that a process spends waiting in the ready queue. So, waiting time is the sum of the periods spent waiting in the ready queue.

Response Time: Response time is measure of the time from the submission of a request until the first response is produced. Thus, response time is the time it takes to start responding, not the time it takes to output the response.

1.3   Real Time Systems

Real time systems are intended to respond to events that occur in real time. Real time system is defined as a system in which the time slice defined to compute and respond to user inputs is so small that it controls the environment. Real time systems are used when there are strict time requirements on the operation of a processor or the flow of events. The real time system can be used as a control device in a dedicated application. Real time operating system has pre defined, pre determined time constraints otherwise the system will fail.

1.4   Time Sharing Systems

Time sharing systems provide a uniform way of execution for process in the system by means of defining a small time slice there by allow many users to share the CPU simultaneously. The time sharing system provides the direct access to a large number of users where CPU time is divided among all the users on scheduled basis. The operating system allocates a slice of time to each user. When this time is expired, it passes control to the next user on the system. The time allowed is extremely small and the users are given the impression that they each have their own CPU and they are the sole owner of the CPU.

2   Literature Survey

The rest of the paper is divided into the following sections: Literature Survey (Section2), Proposed Method (Section3), Results and Discussion (Section4) and Conclusions.

2.1   Round Robin Scheduling Algorithm [1]

Round Robin architecture is a preemptive version of first come first served scheduling algorithm. The processes are arranged in the ready queue in first come first served manner and the processor executes the process from the ready queue based on time slice. If the time slice ends and the process are still executing on the processor the scheduler will forcibly pre-empt the executing process and keeps it at the end of the ready queue then the scheduler will allocate the processor to the next process in the ready queue. The preempted process will make its way to the beginning of the ready queue and will be executed by the processor from the point of interruption.

2.2   Optimizing CPU Scheduling for Real Time Applications Using Mean Difference Round Robin Scheduling [4]

This section discusses the methodology adapted to develop MDRR Algorithm. The proposed algorithm calculates the mean burst time of all the processes in the ready queue. Next, it finds out the difference between a process burst times and the calculated mean burst time. This step is repeated for all the processes in the ready queue. Then, the proposed algorithm find out the process having the largest difference value and assigns it to CPU, and execute it for one time slice. Once the time slice of the process expires, the next process with the largest difference value is picked up from the ready queue and executed for one time slice. The process is repeated for all the processes in the ready queue.

2.3   An Optimized Round Robin Scheduling Algorithm
for CPU Scheduling [3]

The proposed algorithm will be executed in three phases which are given as follows:

Phase 1: Allocate every process to CPU, a single time by applying RR scheduling with a initial time quantum (say k units).

Phase 2: After completing first cycle perform the following steps:

  a) Double the initial time quantum (2k units).
  b) Select the shortest process from the waiting queue and assign to CPU.
  c) After that we have to select the next shortest process for execution by excluding the already executed one in this phase.

Phase3: For the complete execution of all the processes we have to repeat phase 1 and 2 cycle.

3   Proposed Method

In this paper we propose a novel CPU scheduling method called "Mean Interleaved Round Robin (MIRR) algorithm", which is a remedy for a potential problem that we observed in the conventional Round Robin scheduling algorithm. The working of the proposed method is described in section 3.1.

3.1   Pseudo Code of MIRR Algorithm

Input: Queue of Processes, Burst Times, Time Slice, Context Switch Time.

Output: Average waiting time for all the Processes, Average Turnaround time for all the processes.

Method:

Step 1: calculate the mean burst time (m) of all the processes in the ready queue.

Step 2: sort the burst times of all the processes in the ready queue in ascending order.

Step 3: Then insert m at its appropriate position in the sorted ready queue. With this the ready queue is divided into two parts: the first part contains the burst times, b1,b2,...bi-1, which are smaller than m and the second part contains burst times bi+1,bi+2....bn+1, which are greater than or equal to m.

Step 4: Now pickup the process p1 with burst time b1 form the sorted ready queue, assign it to CPU and execute it for one time slice (δ).

Step 5: Next pickup the process pi+1 with burst time bi+1 from the sorted ready queue and execute it for another δ.

Step 6: Repeat step 4 for process p2 with burst time b2 in the sorted ready queue.

Step 7: Perform step 5 for pi+2 process with burst time bi+2 in the ready queue.

Step 8: Repeat step 4 to step 7 until all the processes in the ready queue are exhausted.

Step 9: Compute average waiting time of all the processes in the ready queue using the formula

Average Waiting Time (WT) = $\sum W_i / N$, where W_i is the waiting time of ith process in the ready queue and N is the total number of processes in the queue.

Step 10: Compute average turnaround time of all the processes in the ready queue using the formula.

Average Turnaround Time (TAT) = $\sum T_i / N$, where T_i is the waiting time of ith process in the ready queue and N is the total number of processes in the queue.

Step 11: Stop.

4 Results and Discussion

In this section we discuss the results produced by our proposed algorithm, the standard Round Robin algorithm with the help of the following example:

Example: Consider a system with four processes with names p1,p2,p3,p4, assumed to have arrived at time 0,1,2 and p4 enters after 12 msec into ready queue with the length of CPU bust time given in milliseconds as 24,26,14,18 respectively, time quantum(δ) is taken as 8,10,12,14 milliseconds, and context switch time is 0.2 milliseconds.

Table 1. Processes with their burst times and arrival times given in milliseconds

Name of the process	Burst Time (msec)	Arrival Time (msec)
P1	24	0
P2	26	1
P3	14	2
P4	18	12

Average waiting time (AWT) can be computed using the formula given below:

$$AWT = \sum Wi / N . \tag{1}$$

Where Wi is the sum of waiting time of all the processes and N is the total number of processes.

Average Turnaround Time (ATT) can be computed using the formula given below:

$$ATT = \sum Ti / N . \tag{2}$$

Where Ti is the Turnaround Time of all the processes and N is the total number of processes.

The following table shows the comparative study of the performance of the standard Round Robin and the proposed Scheduling Algorithm.

Table 2. Comparison of the performance of the standard Round Robin and the proposed Mean Interleaved Round Robin Scheduling Algorithm

COMPARISON	Time Quantum 8 msec	Time Quantum 10 msec	Time Quantum 12 msec	Time Quantum 14 msec
RR AVG WT	47.25	49.25	49.75	42.75
MIRR AVG WT	41.25	41.75	37.25	32.75
RR AVG TAT	69.25	69.75	70.25	63.25
MIRR AVG TAT	61.75	62.25	58.25	53.25

The below graphs shown the comparative study of the performance of the standard Round Robin and the proposed scheduling Algorithm. In the following graph x-axis represents time slice and y-axis represents average waiting time and average turnaround time.

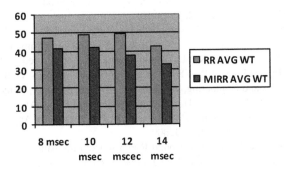

Fig. 1. Comparison of average waiting time

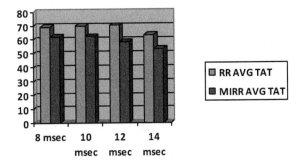

Fig. 2. Comparison of average turnaround time

The computed best case time complexity of both RR, MIRR algorithms is o (n) and the worst case time complexity of both RR, MIRR algorithms is o (n^2).

5   Conclusions

In this paper, a novel CPU scheduling algorithm namely Mean Interleaved Round Robin (MIRR) has been discussed. MIRR algorithm resolves the observed problem associated with conventional round robin algorithm. The proposed algorithm is found to have produced optimized results when compared with other scheduling algorithms like RR. In our future work we will further improve the MIRR algorithm so that it will be effectively applied to embedded systems for improving their performance.

References

1. Silberschatz, A., Galvin, P.B., Gagne, G.: Operating system principles, 7th edn.
2. Dhamdhere, D.M.: Operating Systems A concept-based approach. Tata McGraw Hill
3. Sinth, A., Goyal, P., Batra, S.: An optimized Round Robin Scheduling Algorithm for CPU Scheduling. IJCSE 02(07), 2383–2385 (2010)
4. Kiran, R.N.D.S.S., Babu, P.V., Krishna, B.B.M.: Optimizing CPU scheduling for real time applications using mean-difference round robin (MDRR) algorithm. In: Satapathy, S.C., Avadahani, P.S., Udgata, S.K., Lakshminarayana, S., et al. (eds.) ICT and Critical Infrastructure: Proceedings of the 48th Annual Convention of CSI - Volume I. Advances in Intelligent Systems and Computing, vol. 248, pp. 713–721. Springer, Heidelberg (2014)
5. Matarneh, R.J.: Self-Adjustment Time Quantum in Round Robin Algorithm Depending on Burst Time of the Now Running Processes. American Journal of Applied Sciences 6(10), 1831–1837 (2009) ISSN 1546-9239©2009
6. Hiranwal, S., Roy, K.C.: Adaptive Round Robin scheduling using shortest burst approach, based on smart time slice. International Journal of computer Science and Communication 2(2), 219–326 (2011)

ART Network Based Text Independent Speaker Recognition System for Dynamically Growing Speech Database

A. Jose Albin[*], N.M. Nandhitha, and S. Emalda Roslin

Sathyabama University, Jeppiaar Nagar, Chennai 600119, Tamil Nadu, India
{josealbin2001,nandhi_n_m,roemi_mich}@yahoo.co.in

Abstract. Automated recognizing a speaker from the speech signals is the foremost application in forensics. Speaker recognition system involves two phases namely feature extraction and a classifier system. Features extracted from the speech signals are fed to an already trained classifier system that identifies the speaker. Major challenge occurs when the database is periodically updated which necessitates retaining the classifier with new set of exemplars includes the old and new datasets. As training the neural network is computationally intensive, Back Propagation system is not ideal for speaker recognition system (updation). Hence it necessitates an efficient speaker recognition system that doesn't forget the old database but adjusts to the new set of data. In this paper an Adaptive Resonance Theory (ART) based speaker recognition system is proposed that is capable of functioning well even in the case of periodic updation.

Keywords: Feature extraction, Classification, DWT, S transform, ART.

1 Introduction

Speaker recognition system is used to identify a speaker whose voice in the database best matches with the input speech signal. In recent years computer aided speaker recognition systems are most commonly used in forensics department to determine the identity of the antisocial elements, access control, transaction authentication and speech data management. Feature extraction and classification are the two main phases of a speaker recognition system. Feature extraction technique basically involves transforming the original speech signal into uncorrelated co-efficient using an appropriate non-stationary signal analysis tool and statistical parameters to characterize the features. The proposed classifier accepts the features and determines the identity of the speaker. Conventionally Back Propagation Network (BPN) and Probabilistic Neural Networks (PNN) are used as classifiers. These classifiers are best suited for static database in which the speaker is effectively identified from the key

---

[*] Corresponding author.

© Springer International Publishing Switzerland 2015
S.C. Satapathy et al. (eds.), *Proc. of the 3rd Int. Conf. on Front. of Intell. Comput. (FICTA) 2014*
– Vol. 1, Advances in Intelligent Systems and Computing 327, DOI: 10.1007/978-3-319-11933-5_52

signal features. On the other hand in the case of dynamically growing database, these networks forget the previously trained dataset or rigorous retraining is required before classifying the speaker. Hence dynamically growing database requires a classifier that retains the previous dataset and updates the growing database. In this paper, Adaptive Resonance Theory (ART) is used for classifier design as it has plasticity-elasticity stabilization.

This paper is organized as follows; Section 2 gives the overview of the related works. The proposed methodology is described in section 3. In section 4, the results and discussions of the proposed work is dealt. Section 5 deals with the conclusion and future work.

2   Overview of Related Works

Considerable research is extensively carried out in the area of computer aided speaker recognition. Preethi (2012) developed a speech recognition system using DWT and BPN for isolated spoken words. The following are the steps involved in her approach; Normalization of the input signal, decomposition of the input using Daubechies 8 wavelet, obtain cross correlation between the input and the template in the database, output the best match. It has also been proved that 90 % efficiency can be obtained using this hybrid approach [2]. Li et al (2012) proposed a novel automatic speaker age and gender identification approach. They have proposed seven different methods for acoustic and prosodic level information fusion for achieving good performance in classification process. They also proved that pitch, spectral harmonic energy and formant are the most effective features for identification of the emotional state of the speaker [3]. Saundade and Kurle (2013) proposed a text independent speaker recognition system in which Mel frequency cepstrum coefficients (MFCC) is used to process the input signal and for speaker identification, vector quantization is used. MFCC is computed from the continuous speech signals upon undergoing the following steps frame blocking, windowing, Fast Fourier transform, Mel frequency wrapping [4]. Srinivasan (2011) proposed a new speech recognition technique using the classifier Hidden Markov model (HMM). For analyzing the input speech signal a powerful interface named as wave surfer is used and the parameters namely Spectrogram, Pitch and Power panes are considered. The claims that the proposed technique works efficiently even in noisy environment [5]. Moisa (2010) proposed a speaker recognition method for authentication purpose of an Industrial robot. Mel ceptral analysis is made to identify the authenticated speaker. Also the authors considered the speech parameters namely zero crossing rate (ZCR), energy, autocorrelation, average, magnitude difference function. Speech or silence detection algorithm is also implemented based on the energy and ZCR values. Since the proposed work is for isolated words, the resources used for recognition is very less [6]. Anand et al (2012) describes a text independent voice recognition system for security systems. A Graphical User Interface (GUI) is also developed. The authors used MFCC for feature extraction and vector quantization for feature matching. The proposed algorithm provides a minimum error rate by achieving 96% best match. Khoury et al (2013) developed an open source speaker recognition toolbox called SPEAR. For classification and modeling, the authors used Gaussian Mixture Models, inter-session variability, joint factor analysis and total variability. MFCC and LFCC features were included in the Spear tool box. Singh (2014) evaluated the performance of MFCC by applying K means clustering in two different experiments. Direct matching with the speech features was done in the first experiment. In the second method, a vector quantization codebook was created to match the speech features. Minimum Euclidean distance was the measure used in both the experiments for speaker recognition. Vaishnavi et al (2014) identified an efficient speaker recognition system by comparing the results obtained in two different techniques. In the first technique, features namely entropy and first three formant frequencies were extracted from wavelet decomposition and cascaded feed forward back propagation neural network was used for classification. The second technique used MFCC feature and support vector machine (SVM) as classifier. They also proved that the wavelet transform based techniques gave better result than the other techniques.

However in all these cases, the classifiers required computationally intensive retraining which increases the computational time. Hence it necessitates an efficient speaker recognition system that automatically updates its database with every search.

3   Proposed Methodology

Artificial Neural Networks (ANNs) are trained either using supervised learning or unsupervised learning. Adaptive Resonance Theory (ART) networks use supervised learning where the exemplars with inputs and output are used for training the network. They are capable of self organizing in real time environment and they can able to perform a stable recognition even by getting different input patterns beyond those stored already. As ART 1 is suited for Binary classification/prediction, ART 2 is used in this paper for speaker identification [9][10]. The automated speaker recognition system involves the following steps: Speech signal acquisition, Feature extraction and Classification. A research database has been created with 10 speech signals each of 10 different speakers. Features were extracted using the statistical moments on the approximation and detailed coefficients of the decomposed speech signals using Discrete Wavelet Transform with Discrete Meyer wavelet. In addition to these features statistical moments on the stock well coefficients are also determined. Adaptive Resonance Theory (ART) is chosen as the classifier. Choice of ART is justified due to its plasticity Elasticity feature. It means that ART not only adjusts itself to the new data set but also remembers the old data set. The proposed classifier is trained with three different data sets (DWT features, Stockwell features, DWT and Stockwell features). Among the exemplars different datasets are used for training and testing the network. Performance of the proposed technique is measured in terms of sensitivity.

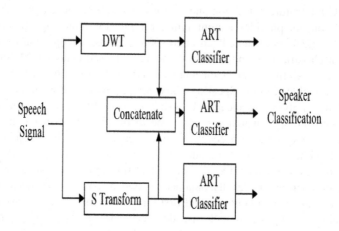

Fig. 1. Block Diagram of the Proposed Work

Case 1: Feature extraction using S transform features

Stockwell transform provides the localized time frequency components in contrast to frequency averaged or time averaged amplitude. These features are aggregated using statistical features namely Mean, Standard Deviation, Kurtosis and Skewness for real and imaginary parts were extracted. These statistical features describe the overall inform of the co-efficients and the distribution of these co-efficients. These features are given to an ART network for classification. From the classifier output the sensitivity is calculated. A sample of one speech signal from all 10 speakers with its corresponding extracted features using S transform is listed in Table 1.

Case 2: Feature extraction using DWT features

Using DWT, the speech signal gets decomposed to approximation coefficients and detailed coefficients. Discrete Meyer (dmey) wavelet is used for decomposition as it has been already found that dmey transform provides features with higher inter-class variance and lesser intra-class variance [11]. In addition to the statistical features (namely mean, skewness, kurtosis), energy, entropy, ZCR, second order and third order moment were extracted from both approximation and detailed coefficients. Hence a total of 16 features were given to an ART network for speaker classification. Table 2 shows a sample of one speech signal corresponding to each speaker and its features.

Case 3: Feature extraction using DWT and S transform features

Eight features extracted from S transform and 16 features extracted from DWT were combined together. As a total, 24 features were given to an ART network for classification i.e. the values from table 1 and 2 were concatenated to form the input data set of an ART network. In all the above cases, performance of the proposed work is evaluated using Sensitivity [12].

$$\text{Sensitivity } (\%) = TP / (TP + FN) \tag{1}$$

Where TP is true positive i.e. identifying speaker 1as speaker 1 and so on. FN is false negative i.e. identifying speaker 1 as some other speaker and so on. Sensitivity based on DWT, S transform and DWT&S transform is listed in table 3.

4   Results and Discussions

Feature extraction and classification are performed on all the three different cases, over the signals stored in the database. The actual and the desired output of ART for the selected set of features in all the first two cases are shown in Tables 1-2. In Table 1, the actual and the desired output for the speakers 1, 4, 5, 6 and 9 are same. This shows that the proposed technique using DWT features identifies 5 speakers correctly out of 10 speakers. Using S transform features, speakers 1, 2, 5 and 8 were identified correctly as shown in Table 2. By combining the DWT and S transform features, the speakers 1, 4, 5 and 6 were correctly identified. Sensitivity of the proposed techniques for all the cases is listed in Table 3. From the Table 3, it is clear that for case 2 i.e. using DWT features, the sensitivity is high for maximum number of speakers, whereas for the other two cases, the sensitivity is comparatively less. However sensitivity of the DWT based features should also be increased further by choosing highly appropriate features. The graph showing comparison on the sensitivity of all the three cases is shown in Figure 2.

Table 1. Sample database of the features Extracted using DWT

Approximation Coefficient								Detail Coefficient								Output	
Mean	Standard Deviation	Kurtosis	Skewness	Energy	ZCR	Second Order Moment	Third Order Moment	Mean	Standard Deviation	Kurtosis	Skewness	Energy	ZCR	Second Order Moment	Third Order Moment	Desired output	Actual output
-0.00239	0.006442	0.037107	-0.03385	0.012123	0.110909	4.14E-05	-2.85E-08	2.37E-07	1.02E-05	1.42E-02	-0.00266	0.000238	0.525455	1.03E-10	-2.31E-17	1	1
-0.00062	0.002365	0.196848	0	0.004313	0.241818	5.38E-06	-2.72E-08	-3.01E-07	9.97E-06	6.47E-02	0.023559	0.000234	0.381818	9.93E-11	1.93E-16	2	3
3.70E-05	0.001403	0.079881	0.00183	0.002476	0.118182	1.96E-06	1.59E-11	-1.38E-07	1.08E-05	1.21E-02	0.013207	0.000254	0.576364	1.17E-10	1.39E-16	3	7
2.38E-05	0.069055	0.20029	0.036662	0.121815	0.12	0.00476	3.80E-05	-9.54E-08	9.18E-06	1.98E-02	-0.01816	0.000215	0.565455	8.40E-11	-1.16E-16	4	4
7.46E-07	0.000207	0.397919	0.196412	0.000364	0.194545	4.26E-06	5.43E-12	-1.96E-07	4.81E-06	3.17E-02	-0.02711	0.000113	0.549091	2.31E-11	-2.49E-17	5	5
1.10E-04	0.012832	0.094253	-0.15472	0.022636	0.152727	0.000164	-1.03E-06	-4.78E-07	1.65E-04	1.12E+0	-0.9157	0.003828	0.518182	2.66E-08	-3.30E-11	6	6
2.25E-04	0.003888	0.099469	-0.1215	0.006871	0.110909	1.51E-05	-2.25E-08	8.11E-08	1.05E-05	5.78E-02	0.090693	0.000245	0.521818	1.08E-10	8.59E-16	7	10
-0.02178	0.441093	0.07686	-0.00781	0.77905	0.109091	0.194209	-0.00211	5.67E-06	0.001271	9.64E-01	0.645468	0.02977	0.630909	1.61E-06	1.09E-08	8	7
7.32E-05	0.00226	0.08462	0.052192	0.003989	0.129091	5.10E-06	1.90E-09	1.81E-07	9.43E-06	2.67E-02	-0.03062	0.000221	0.478182	8.88E-11	-2.12E-16	9	9
2.82E-05	0.002495	0.105616	-0.07628	0.004401	0.178182	6.21E-06	-3.73E-09	-1.34E-07	1.40E-05	1.91E-01	0.064948	0.000329	0.516364	1.96E-10	1.48E-15	10	8

Table 2. Sample database of the features Extracted using S Transform

Mean (real)	Variance	Skewness (real)	Kurtosis (real)	Mean (Imag.)	Skewness (Imag.)	Kurtosis (Imag.)	Desired output	Actual output
-0.00215	0.001027	9.76E-09	9.76E-09	-1.09E-01	-0.10933	0.000136	1	1
-0.00056	0.000276	1.23E-08	1.23E-08	-1.93E-02	-0.0193	0.000884	2	2
-0.00128	0.001037	1.87E-10	1.87E-10	5.45E-02	0.054501	0.000388	3	2
0.000818	-0.01332	0.000421	0.000421	-2.10E-02	-0.02102	-3.50E-05	4	10
-1.30E-06	-1.10E-05	1.29E-12	1.29E-12	-5.67E-01	-0.56672	-2.80E-05	5	5
-5.70E-05	0.002273	2.68E-06	2.68E-06	-1.09E-02	-0.0109	0.000283	6	2
0.001717	0.000732	4.89E-09	4.89E-09	-3.64E-02	-0.03638	0.000154	7	9
-7.60E-08	-3.30E-05	7.57E-12	7.57E-12	-3.84E-04	-0.00038	9.12E-05	8	8
0.000215	-0.00184	2.16E-09	2.16E-09	-2.81E-02	-0.02815	0.000744	9	3
-0.00096	0.00256	8.19E-09	8.19E-09	-6.03E-02	-0.06031	0.000197	10	2

Table 3. Sensitivity of the proposed technique

Speaker	Stockwell Features	DWT Features	Stockwell & DWT Features
1	0	60	80
2	40	20	0
3	20	20	20
4	20	80	80
5	0	100	60
6	0	20	20
7	0	60	20
8	40	20	20
9	0	0	0
10	20	40	0

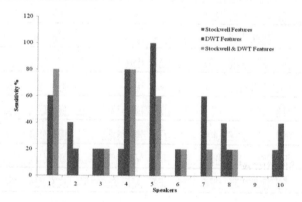

Fig. 2. Comparative Study of the proposed technique using ART

5   Conclusion and Future Direction

An efficient classifier has to be identified for an accurate speaker identification system that involves dynamic updation. The various feature extraction techniques namely, S transform, DWT and S transform / DWT were considered. The extracted features from all the above mentioned techniques are given as inputs to an ART network for classification. The performance of the proposed work is evaluated using sensitivity. A comparative study is also made on the sensitivity obtained using all the techniques.

References

1. Yang, Q.: Automatic Speaker Recognition. In: csl.anthropomatik.kit.edu/./MMMK-PP12-SpeakerRecognition-SS (2013)
2. Sharma, P.: Intelligent Voice Recognition System Based on Acoustic and Speaking Fundamental Frequency Characteristics. Journal of Engineering, Computers & Applied Sciences (JEC&AS) 1(2), 1–8 (2012), ISSN No: 2319-5606
3. Li, M., Han, K.J., Narayanan, S.: Automatic speaker age and gender recognition using acoustic and prosodic level information fusion. In: Computer Speech and Language, vol. 27, pp. 151–167. Elsevier Ltd (2012), doi:10.1016/j.csl.2012.01.008
4. Saundade, M., Kurle, P.: Speech Recognition using Digital Signal Processing. International Journal of Electronics, Communication & Soft Computing Science and Engineering 2(6), 31–34 (2013), ISSN: 2277-9477
5. Srinivasan, A.: Speech Recognition Using Hidden Markov Model. Applied Mathematical Sciences 5(79), 3943–3948 (2011)
6. Moisa, C., Silaghi, H., Silaghi, A.: Speech and Speaker Recognition for the Command of an Industrial Robot. In: Mathematical Methods And Computational Techniques In Electrical Engineering, pp. 31–36 (2010) ISSN: 1792-5967
7. Anand, R., Singh, J., Tiwari, M., Jains, V., Rathore, S.: Biometrics Security Technology with Speaker Recognition. International Journal of Advanced Research in Computer Engineering & Technology (IJARCET) 1(10), 232–236 (2012)
8. Bansal, N.A.: Speaker Recognition Using MFCC Front End Analysis and VQ Modeling Technique for Hindi Words using MATLAB. International Journal of Computer Applications 45(24), 48–52 (2012)
9. Heins, L.G., Tauritz, D.R.: Adaptive Resonance Theory (ART): An Introduction. In: Internal report 95-35, Department of Computer Science, Leiden University (1995)
10. Tanaka, T., Weitzenfeld, A.: Adaptive Resonance Theory, http://cannes.itam.mx/Alfredo/English/publications/nslbook/mitpress/157.17.ch08.pdf
11. Albin, A.J., Nandhitha, N.M., Roslin, S.E.: Text Independent Speaker Recognition System using Back Propagation Network with Wavelet Features. In: IEEE International Conference on Communication and Signal Processing, pp. 942–946 (2014)
12. http://en.wikipedia.org/wiki/Sensitivity_and_specificity
13. Khoury, E.: SPEAR: An Open Source Toolbox For Speaker Recognition Based On BOB. idiap.ch/downloads/papers/./Khoury_ICASSP_2014.pdf (2014)

14. Singh, A.K., Singh, R., Dwivedi, A.: Mel frequency cepstral coefficients based text independent Automatic Speaker Recognition using matlab. In: IEEE International Conference on Optimization, Reliabilty, and Information Technology, pp. 524–527 (2014)
15. Vaishnavi, A., Raju, B.C., Prathiksha, G., Reddy, L.H., Kumar, C.S.: Comparison of Two Speaker Recognition Systems. International Journal of Engineering and Advanced Technology (IJEAT) 3(4) (2014)

Cryptanalysis of Image Encryption Based on Permutation-Substitution Using Chaotic Map and Latin Square Image Cipher

Musheer Ahmad and Faiyaz Ahmad

Department of Computer Engineering, Faculty of Engineering and Technology,
Jamia Millia Islamia, New Delhi 110025, India

Abstract. Recently Panduranga *et al.* suggested an image encryption algorithm based on permutation-substitution architecture using chaotic map and Latin square. According to the proposal, the pixels of plain-image are firstly scrambled according to permutation vector, extracted from chaotic sequence, in permutation phase. In substitution phase, the permuted image is substituted by XOR operation with key image generated from a keyed Latin square. The algorithm has the ability to adapt and encrypt any plain-image with unequal width and height. Moreover, it also exhibits the features of high entropy, low pixels correlation, large key space, high key sensitivity, etc. However, a careful analysis of Panduranga *et al.* algorithm unveils few security flaws which make it susceptible to cryptographic attack. In this paper, we analyze its security and proposed a chosen plaintext-attack to break the algorithm completely. It is shown that the plain-image can be successfully recovered without knowing the secret key. The simulation of proposed attack demonstrates that Panduranga *et al.* algorithm is not at all secure for practical encryption of sensitive digital images.

Keywords: Image encryption, Latin square, chaotic map, permutation, substitution, cryptanalysis.

1 Introduction

Cryptanalysis is the science of breaching cryptographic security systems with an aim to recover plaintext without an access to the secret key. Successful cryptanalysis may recover the plaintext or the secret key. It is needed to find weaknesses in the security system that eventually may leads to the previous results [1, 2]. Cryptanalysis is co-evolved together with cryptography, and the contest can be traced through the history of cryptography. The new security systems being designed to replace old broken designs and new cryptanalytic techniques are invented to crack the improved security systems. In practice, both the cryptanalysis and cryptography are two equally significant sides of same coin. In order to create secure and cryptographically strong systems, it is aimed and recommended to design against possible cryptanalysis [2-4]. An attempted cryptanalysis is called an attack. An encryption algorithm is said to be

© Springer International Publishing Switzerland 2015   481
S.C. Satapathy et al. (eds.), *Proc. of the 3rd Int. Conf. on Front. of Intell. Comput. (FICTA) 2014*
– *Vol. 1*, Advances in Intelligent Systems and Computing 327, DOI: 10.1007/978-3-319-11933-5_53

insecure if it is vulnerable to any of the classical and other type of cryptographic attacks. The ultimate objective of an attacker is to find a way to recover secret key or plaintext in lesser time or storage than the brute-force attack [5]. The four classical attacks in cryptanalysis [1], in context to image encryption, are the following:

(i)   *Ciphertext-only attack*: the attacker only has access to some ciphertext images that can be utilized to recover the plaintext image.

(ii)   *Known-plaintext attack*: the attacker can obtain some plaintext images and the corresponding ciphertext images.

(iii)   *Chosen-plaintext attack*: the attacker can have temporary access to encryption machine and choose some specially designed plaintext images to produce corresponding ciphertext images, and

(iv)   *Chosen-ciphertext attack*:   the attacker can have temporary access to the decryption machine and choose some specially designed ciphertext images to obtain the corresponding plaintext images.

In past two decades, an exponential number of image encryption algorithms have been suggested by the researchers with an aim to provide sufficient security to images. However, the careful security examinations of some, seemingly complex and secure, image encryption algorithms discover that they are insecure and cryptographically weak. Their inherent inevitable flaws make them prone to even classical attacks and they can be easily breakable with simple statistical methods. As a result, many image encryption algorithms have been successfully broken by cryptanalysts under various design-specific attacks [6-14]. Recently, Panduranga *et al.* [15] proposed an image encryption algorithm based on permutation-substitution architecture. The authors utilized chaotic sequence to derive permutation vector to shuffle pixels in permutation phase and keyed Latin square to mask the pixels in substitution phase. The algorithm has great encryption strength with flat histograms, high information entropy, low pixels correlations, high key sensitivity, etc. However, the security investigation exposes the inherent defect of plain-image independency on the generation of permutation vector during first phase and key image through Latin square on second phase of algorithm. As a result, the algorithm fails to resist the proposed chosen-plaintext attack which completely breaks the algorithm.

The organization of rest of this paper is as follows: Section 2 provides Panduranga *et al.* image encryption algorithm. Section 3 analyzes and discusses the complete break of encryption algorithm under proposed chosen plaintext-attack. The simulation of cryptanalysis is also demonstrated in the same Section. The conclusions of the work are made in Section 4.

2   Review of Panduranga *et al.* Algorithm

The Panduranga *et al.* image encryption algorithm is based on the famous permutation-substitution architecture which is the base of most of the chaos-based image encryption proposals suggested in literature. The working block diagram of their algorithm is depicted in Figure 1. The first phase is permutation phase meant for the plaintext confusion. In this phase, plain image is permuted using chaotic sequences generated from 1D chaotic Logistic map.

$$x(n+1) = \lambda \times x(n) \times (1 - x(n)) \tag{1}$$

Where, x is map's variable, $x(0)$ is initial condition, λ is map's parameter. The Logistic map exhibits chaotic dynamics for $\lambda \in (3.5699456, 4]$ and $x(n) \in (0, 1)$ for all $n \geq 0$. The chaotic Logistic map is iterated with the provided initial values of $x(0)$ and λ to get a chaotic sequence of desired length. The sequence is sorted in ascending order in a 1D array. Then the indexes of values of sorted array in previous chaotic sequence are evaluated and stored to get a permutation vector T. Index value represents the position of chaotic value in chaotic sequence. According to this index values pixels of (reshaped 1D) plain-image are shuffled to receive a permuted image. The second phase is substitution phase, which is meant for plaintext diffusion. In this phase, a Latin square is generated with a 256-bit external secret key K. A Latin square provides a key image of size 256×256, in which each row and column holds value ranging from 0 to 255 is repeated exactly once in the respective row and column. The final encrypted image is obtained by masking permuted image with key image through bitwise XOR operation. The Panduranga *et al.* employed a 256-bit external secret key to generate initial value $x(0)$ of chaotic map and the key image of size 256×256. To generate a Latin square with a 256-bit key, the authors utilized the algorithm presented by Wu *et al.* in [16]. To extract $x(0)$ from external key, the following transformation is reported by authors.

Let $K = b_0 b_1 b_2 b_3 \ldots \ldots \ldots \ldots \ldots b_{254} b_{255}$ be the external 256-bit secret key.

$$x(0) = x(0) + \sum_{i=0}^{255} \frac{b_i}{2^i} x(0) = mod(x(0), 1)$$

The successful decryption of encrypted image is performed by applying above steps in reverse order with correct secret key. The readers are advised to refer to Ref. [14] for a detailed explanation of algorithm under study.

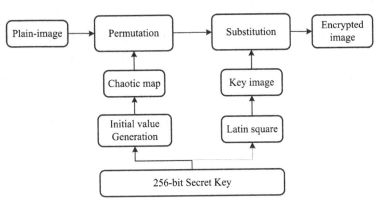

Fig. 1. Synoptic of Panduranga *et al.* image encryption algorithm

3   Cryptanalysis

In modern cryptography, there is a fundamental principle enounced by Auguste Kerckhoff in 19-th century. Kerckhoff's principle says that "*A cryptographic system should be secure even if everything about the system, except the key, is public knowledge*" [1, 17]. The principle entails that the attacker knows complete design and working of encryption algorithm except the secret key, i.e. everything about encryption algorithm including its implementation is public except the secret key which is private. In other words, the attacker has temporary access to encryption and decryption machinery. The sole objective of attacker is to recover the plaintext without knowing secret key. Recovering the plain-image is as good as knowing the secret key. The following inherent flaws are found in the image encryption algorithm under scrutiny.

- In permutation phase, the permutation vector T is completely dependent to initial condition of chaotic map, which in turn depend on the external secret key, and has nothing to do with the pending plain-image. As a result, same permutation vector is generated every time when encrypting distinct plain-images.
- In substitution phase, the permuted image is substituted with the key image generated from keyed Latin square. The key image solely depends on 256-bit external secret key used to generate it. Neither the substitution of permuted image nor the generation of key image depends on pending plain-image. Consequently, the same key image is generated by algorithm when encrypting distinct plain-images.

We, as cryptanalysts, exploit the above analytical information to execute the proposed chosen plaintext attack. The complete cryptanalysis of Pandurange *et al.* encryption algorithm with proposed attack is described as follows:

Let P be the plain-image which is to be recovered from its received encrypted image C. The successful execution of proposed attack needs specially designed plain-images Z and P_1 of same size as C.

$$
Z = \begin{pmatrix} 0 & 0 & 0 & \cdots & 0 \\ 0 & 0 & 0 & \cdots & 0 \\ 0 & 0 & 0 & \cdots & 0 \\ \vdots & \vdots & \vdots & \cdots & \vdots \\ 0 & 0 & 0 & \cdots & 0 \end{pmatrix}
\qquad
P_1 = \begin{pmatrix} 1 & 2 & 3 & \cdots & 255 & 0 & \cdots & 0 \\ 0 & 0 & 0 & \cdots & 0 & 0 & \cdots & 0 \\ 0 & 0 & 0 & \cdots & 0 & 0 & \cdots & 0 \\ \vdots & \vdots & \vdots & \vdots & \vdots & \vdots & \vdots & \vdots \\ 0 & 0 & 0 & \cdots & 0 & 0 & \cdots & 0 \end{pmatrix}
$$

Algorithm #1: *CPA-attack( )*

Input   :   Images Z, P_1 and received encrypted image C
Output   :   Recovered image P

begin

    $K = GetSubKey(Z)$

    $D = Exclusive\text{-}OR(C, K)$

    $T = GetPermSeq(Z, P_1)$

    $D = Reshape(D, 1, MN)$

    for $i = 1$ to MN

        $P(i) = D(t_i)$

    end

    $P = Reshape(P, M, N)$

end

Algorithm #2: *GetSubKey( )*

Input   :   Chosen zero image Z
Output   :   Key image K used for substitution

begin

    $K = Encrypt(Z)$

end

Algorithm #3: *GetPermSeq( )*

Input   :   Chosen image Z and P_1
Output   :   Permutation sequence $T = \{t_1,\ t_2,\ t_3,\ ...,\ t_{MN}\}$

begin

    $K = GetSubKey(Z)$

    for $n = 1$ to $\left\lceil \dfrac{M \times N}{255} \right\rceil$

        $C_1 = Encrypt(P_1)$

        $D_1 = Exclusive\text{-}OR(C_1, K)$

        $D_1 = Reshape(D_1, 1, MN)$

        for $j = 1$ to 255

                $i = 255 \times (n - 1) + j$

                $t_i = Find\text{-}index(D_1, j)$

        end

        $P_1 = Rotate\text{-}right(P_1, 255)$

    end

end

Where method *Encrypt(x)* encrypts the input plain-image x according to (i.e. implementation of) Panduranga *et al.* algorithm, *Exclusive-OR(x, y)* performs bitwise exclusive-OR operations on inputs x and y, *Reshape(x, y, z)* transforms the current dimension of input vector x to $y \times z$ dimension, *Find-index(x, y)* returns the index of element y in vector x and *Rotate-right(x, y)* rotates 1D vector x in right direction by y number of positions.

A detailed description of proposed cryptanalysis on an encrypted image of size 4×4 is provided in Table 1. The simulation results of complete cryptanalysis of Panduranga *et al.* algorithm, on a benchmark *Lena* image of size 256×256, are shown in Figure 2. Hence, the theoretical and simulation analyses highlight that the image encryption algorithm under study is insecure and can be cryptanalyze successfully with proposed chosen-plaintext attack.

Table 1. Detailed description of attack on an image of size 4×4

$$
P = \begin{pmatrix} 124 & 38 & 231 & 89 \\ 17 & 173 & 186 & 92 \\ 56 & 48 & 118 & 215 \\ 101 & 29 & 198 & 65 \end{pmatrix}
\qquad
C = \begin{pmatrix} 66 & 114 & 190 & 99 \\ 186 & 7 & 188 & 33 \\ 118 & 207 & 105 & 162 \\ 135 & 113 & 76 & 79 \end{pmatrix}
$$

$$
Z = \begin{pmatrix} 0 & 0 & 0 & 0 \\ 0 & 0 & 0 & 0 \\ 0 & 0 & 0 & 0 \\ 0 & 0 & 0 & 0 \end{pmatrix}
\qquad
P_1 = \begin{pmatrix} 1 & 2 & 3 & 4 \\ 5 & 6 & 7 & 8 \\ 9 & 10 & 11 & 12 \\ 13 & 14 & 15 & 16 \end{pmatrix}
$$

$$
K_{partial} = \begin{pmatrix} 132 & 51 & 19 & 83 \\ 204 & 123 & 91 & 155 \\ 47 & 222 & 190 & 254 \\ 154 & 73 & 41 & 105 \end{pmatrix}
\qquad
D = \begin{pmatrix} 198 & 65 & 173 & 48 \\ 118 & 124 & 231 & 186 \\ 89 & 17 & 215 & 92 \\ 29 & 56 & 101 & 38 \end{pmatrix}
$$

$$T = \{6, 16, 7, 9, 10, 3, 8, 12, 14, 4, 5, 11, 15, 13, 1, 2\}$$

$$
P = \begin{pmatrix} 124 & 38 & 231 & 89 \\ 17 & 173 & 186 & 92 \\ 56 & 48 & 118 & 215 \\ 101 & 29 & 198 & 65 \end{pmatrix}
$$

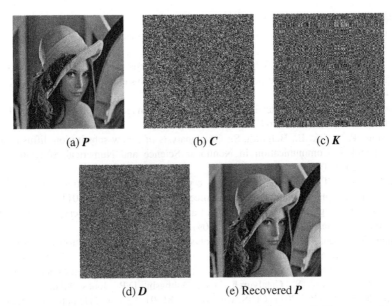

(a) *P*   (b) *C*   (c) *K*

(d) *D*   (e) Recovered *P*

Fig. 2. Simulation of proposed chosen-plaintext attack (a) plain-image *P* (b) encrypted image *C* of *P* (c) key image *K* (d) permuted image *D* and (e) recovered image *P* from *D* through recovered sequence *T*

4   Conclusion

This paper proposes to break a recent image encryption algorithm based on permutation-substitution architecture and chaotic map proposed by Panduranga *et al.* The chaos is employed to yield permutation vector for pixels permutation. The algorithm uses the feature of keyed Latin square to generate key image for pixels substitution. The inevitable flaw of algorithm lies in the fact that the generation of permutation vector and key image are both solely depends on secret key and independent to pending plain-image. As a result, it is found that the algorithm is insecure and susceptible to proposed attack. We have shown that plain-image can be recovered with chosen-plaintext attack without having any knowledge of secret key. This work demonstrates cryptanalysis and finds that the Panduranga *et al.* encryption algorithm is completely insecure and doesn't suitable for the protection of sensitive digital images.

References

1. Schneier, B.: Applied Cryptography: Protocols Algorithms and Source Code in C. Wiley, New York (1996)
2. Bard, G.V.: Algebraic Cryptanalysis. Springer, Berlin (2009)
3. Cryptanalysis (last access on June 02, 2014),
   http://en.wikipedia.org/wiki/Cryptanalysis

4. Military Cryptanalysis Part I- National Security Agency (last access on June 02, 2014), http://www.nsa.gov/public_info/_files/military_cryptanalysis/mil_crypt_I.pdf

5. Hermassi, H., Rhouma, R., Belghith, S.: Security analysis of image cryptosystems only or partially based on a chaotic permutation. Journal of Systems and Software 85(9), 2133–2144 (2012)

6. Çokal, C., Solak, E.: Cryptanalysis of a chaos-based image encryption algorithm. Physics Letters A 373(15), 1357–1360 (2009)

7. Rhouma, R., Solak, E., Belghith, S.: Cryptanalysis of a new substitution-diffusion based image cipher. Communication in Nonlinear Science and Numerical Simulation 15(7), 1887–1892 (2010)

8. Li, C., Lo, K.T.: Optimal quantitative cryptanalysis of permutation-only multimedia ciphers against plaintext attacks. Signal Processing 91(4), 949–954 (2011)

9. Rhouma, R., Belghith, S.: Cryptanalysis of a spatiotemporal chaotic cryptosystem. Chaos, Solitons & Fractals 41(4), 1718–1722 (2009)

10. Wang, X., Luan, D., Bao, X.: Cryptanalysis of an image encryption algorithm using Chebyshev generator. Digital Signal Processing 25, 244–247 (2014)

11. Ahmad, M.: Cryptanalysis of chaos based secure satellite imagery cryptosystem. In: Aluru, S., Bandyopadhyay, S., Catalyurek, U.V., Dubhashi, D.P., Jones, P.H., Parashar, M., Schmidt, B. (eds.) IC3 2011. CCIS, vol. 168, pp. 81–91. Springer, Heidelberg (2011)

12. Rhouma, R., Belghith, S.: Cryptanalysis of a new image encryption algorithm based on hyper-chaos. Physics Letters A 372(38), 5973–5978 (2008)

13. Özkaynak, F., Özer, A.B., Yavuz, S.: Cryptanalysis of a novel image encryption scheme based on improved hyperchaotic sequences. Optics Communications 285(2), 4946–4948 (2012)

14. Solak, E., Rhouma, R., Belghith, S.: Cryptanalysis of a multi-chaotic systems based image cryptosystem. Optics Communications 283(2), 232–236 (2010)

15. Panduranga, H.T., Kumar, N., Kiran, S.K.: Image encryption based on permutation-substitution using chaotic map and Latin square image cipher. The European Physical Journal-Special Topics 223(8), 1663–1677 (2014)

16. Wu, Y., Zhou, Y., Noonan, J.P., Agaian, S.: Design of image cipher using latin squares. Information Sciences 264, 317–339 (2014)

17. Kerckhoffs's principle (last access on May 12, 2014), http://crypto-it.net/eng/theory/kerckhoffs.html

Windowed Huffman Coding with Limited Distinct Symbols by Least Recently Used Symbol Removable

Utpal Nandi[1] and Jyotsna Kumar Mandal[2]

[1] Dept. of Computer Science, Bangabasi College,
Kolkata-700009, West Bengal, India
[2] Dept. of Computer Sc. & Engg., University of Kalyani,
Nadia-741235, West Bengal, India
{nandi.3utpal,jkm.cse}@gmail.com

Abstract. In this paper, a variant of windowed Huffman coding with limited distinct symbols is proposed. Symbols most recently processed are stored in a window where the number of distinct symbols cannot exceed a specified threshold. But, if the number of distinct symbols exceeds a specified threshold, least recently used symbol is removed from the window instead of oldest symbol of the window. Then, another variant of the method is proposed where instead of single window two windows are used. The most recently processed symbols are kept in a small primary window buffer. A comparatively large secondary window buffer is used to store more past processed symbols. The first proposed technique significantly improves the compression rates of most of the file type. The compression rates of the second proposed technique are not so poor than its counter parts.

Keywords: Data compression, Huffman tree, Windowed Huffman coding, compression ratio.

1 Introduction

In first pass of two pass Huffman coding [4] the probabilities of symbols are calculated, Huffman tree is constructed and symbol codes are obtained. In the second pass, symbol codes are assigned to symbols. But, the limitations of this coding and its variants [3,4,6,7,8] are the scanning the symbols twice and frequency table transmission with compressed data. One pass dynamic Huffman coding [5,10,11] eliminates these problem by constructing Huffman tree using probabilities of symbols already encoded and no transmission of frequency table is required. The algorithm calculates the frequency of symbols of the entire encoded data. But, the probabilities of symbols may change in different segment of the source data during encoding. To solve these problem Residual Huffman algorithm [10] and Windowed Huffman coding [2] are introduced. Windowed Huffman coding removes the past encoded symbols by using a fixed-size window buffer that keeps only recently encoded symbols. It restricts the number of symbols to be used to construct Huffman tree. But, the height of the Huffman tree depends on the number of distinct symbols , not

© Springer International Publishing Switzerland 2015
S.C. Satapathy et al. (eds.), *Proc. of the 3rd Int. Conf. on Front. of Intell. Comput. (FICTA) 2014*
– *Vol. 1*, Advances in Intelligent Systems and Computing 327, DOI: 10.1007/978-3-319-11933-5_54

on total number of symbols. Again, the algorithm maintains fixed size window buffer. The efficiency of the algorithm depends on the window buffer size. But, it is difficults to find a proper window size for all files. Because, the proper the window size is file content dependent. Adaptive version of the algorithm sets window size by using different policies. But, the available policies are unable to find the optimal window size during encoding. To overcome the above limitations of Windowed Huffman algorithm, two new variants are introduced known as Windowed Huffman algorithm with limited distinct symbols (WHDS) [1] and Windowed Huffman algorithm with more than one window (WHMW) [1]. This WHDS algorithm also keeps a window. It restricts the number of distinct symbols within the window buffer. But, total number of symbols may change. The algorithm generates comparatively shorter code of symbols. The WHMW algorithm keeps two window buffers. A primary window contains most recently compressed symbols. The secondary window contains comparatively more past symbols. In WHDS and WHMW techniques, if the number of distinct symbols exceeds a specified threshold, oldest symbol is removed repeatedly until the number of distinct symbols within the specified threshold. That is the technique used First In First Out (FIFO) rule to remove a symbol if the window is full. Now the problem is that the symbol selected by FIFO to be removed from window may be a highly probable symbol of occurrence. If this symbol is removed from the window, there is a high chance to bring it back again into the window during the encoding process. To overcome such limitations, two new compression techniques are proposed in section 2. The results have been given in section 3 and the conclusions are drawn in section 4. The references are noted at end.

2   The Proposed Compression Techniques

To overcome the above limitations of Windowed Huffman algorithm with limited distinct symbols (WHDS), two new variants are proposed known as Windowed Huffman algorithm with limited distinct symbols and least recently used symbol removing (WHDSLRU) and Windowed Huffman algorithm with more than one window and least recently used symbol removing (WHMWLRU) discussed in section 2.1 and 2.2 respectively.

2.1   The Proposed WHDSLRU Technique

The proposed technique maintains a window buffer to keep recently processed symbols. The window size is not restricted by the number of symbols but number of distinct symbols similar with WHDS technique. But, if the number of distinct symbols exceeds a specified threshold, least recently used symbol is removed from the window instead of oldest symbol of the window. That is, the proposed technique uses least recently uses (LRU) scheme rather than first in first out (FIFO). The procedure of the proposed technique is given in Fig. 1. Initially, a Huffman tree is constructed with a 0- node and a empty window buffer is taken. A symbol from input file/stream is read as new_symbol and encoded using Huffman tree. The Huffman tree is updated by inserting new_symbol. The new_symbol is inserted into the window buffer. Then, a checking is done in the window. If the number of distinct symbols exceeds the specified threshold (T), then the least recently used symbol (LRU_symbol) is removed from window buffer and the Huffman tree is updated by removing the LRU_symbol. This process is continued until end of file/stream.

Initialize the Huffman tree with a 0- node;
Take a empty window buffer;
  While (not end of file/stream) do
    Read a next symbol as new_symbol from input file/stream;
    Encode {or decode} the read symbol using Huffman tree;
    Update the Huffman tree by inserting new_symbol;
    Insert the new_symbol into the window buffer;
    If (Number of distinct symbols > T), then
      While (Number of distinct symbols > T) do
        Select the least recently used symbol as LRU_symbol using LRU scheme from window buffer;
        Remove the LRU_symbol from window buffer ;
        Update the Huffman tree by removing the LRU_symbol;
      End_while;
    End_if ;
  End_while;
End:

Fig. 1. The proposed WHDSLRU technique

2.2   The Proposed WHMWLRU Technique

In the proposed WHDSLRU technique, symbols within the window are encoded by Huffman code. But, symbols not in the window are encoded by ESCAPE code followed by 8 bit ASCII. To reduce this problem associated with WHDSLRU technique, a variant is proposed known as WHMWLRU where instead of a single window buffer, two window buffers are used. First one i.e. the primary window buffer keeps most recently encoded symbols similar to WHDSLRU technique. Another window buffer i.e. the secondary window buffer keeps comparatively more encoded symbols. The technique reduces the use of ESCAPE codes using this secondary window buffer. Symbols to be encoded found in secondary window but not in primary window are encoded by Huffman codes where the Huffman tree that is constructed by the symbols of the secondary window. Symbols not in the both window buffers are encoded by ESCAPE code followed by 8 bit ASCII. But, extra overhead associated with this technique is to maintain two window buffers. The procedure of the proposed algorithm is given in Fig. 2 where specified threshold of primary and secondary window buffers are T1 and T2 respectively.

Take two empty window buffers as the primary and secondary.
Initialize the Huffman tree for primary (H_tree1) and secondary
(H_tree2) window with a 0- node.
While (not end of file/stream) do
   Read a next symbol as new_symbol from input file/stream and
   encode {or decode} the same;
   Update the first Huffman tree (H_tree1) and second Huffman tree
   (H_tree2) by inserting new_symbol;
   Insert new_symbol into the both primary and secondary window;
   If (No. of distinct symbols of primary window> T1), then
     While (No. of distinct symbols of primary window> T1) do
       Select the least recently used symbol as LRU_symbol using
       LRU scheme and remove it from primary window buffer;
       Update the H_tree1 by removing the LRU_symbol;
     End_while;
   End_if ;
   If (No.of distinct symbols of secondary window> T2), then
     While (No. of distinct symbols of secondary window> T2) do
       Select the least recently used symbol as LRU_symbol using
       LRU scheme and remove it from secondary window buffer;
       Update the H_tree2 by removing the LRU_symbol;
     End_while;
   End_if;
End while:

Fig. 2. The proposed WHDSLRU technique

3   Results

For experimental purpose, seven different types of file have been taken as input. The compression ratios of Huffman, Adaptive Huffman, Windowed Huffman, WHDS, WHMW and proposed WHDSLRU, WHMWLRU techniques are compared in Table 3. The graphical representation of this comparison is shown in Fig. 4. The performances in term of compression ratio of the proposed WHDSLRU are significantly better for EXE, CORE, TIFF files and also well for other file types than other Huffman based techniques. The proposed WHMWLRU technique offers better performance for DOC file particularly.

Table 1. Comparison of compression ratios in different techniques

File name	Huffman	Adaptive Huffman	Windowed Huffman	WHDS	WHMW	Proposed WHDSLRU	Proposed WHMWLRU
Doc	41.3	41.5	40.4	40.9	40.5	41.4	41.6
Exe	18.2	18.9	20.1	20.4	20.2	20.5	20.2
Hybrid	29.9	30.1	42.5	43.2	41.7	43.6	41.9
Core	10.9	12.8	12.5	13.6	11.8	13.8	12.1
NTSC	20.4	20.5	29.4	31.1	30.2	31.1	29.5
Tiff	17.1	18.2	23.3	23.7	21.1	23.9	21.3
C++	41.1	41.6	41.6	42.5	41.8	42.8	41.7

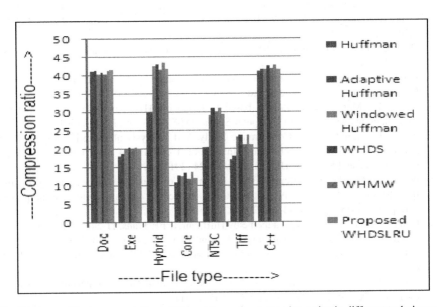

Fig. 3. The graphical representation of Comparison of compression ratios in different techniques

4   Conclusions

The proposed WHDSLRU and WHMWLRU techniques not only restrict the number of distinct symbols in the window buffer but also remove least recently used symbol from the window buffer if necessary and reduce the problem associated with WHDS and WHMW techniques. The WHDSLRU offers comparatively better compression ratio for most of the file types than its counter parts. The proposed WHMWLRU reduces the use of ESCAPE codes of WHDSLRU technique by maintaining another window buffer. But, this extra window buffer increases overhead. The performances of WHMWLRU are not so well for most of the file types. Further investigation is required to enhance more the performance and to use these for compression of images also.

Acknowledgements. The authors extend sincere thanks to the department of Computer Science and Engineering, University of Kalyani and PURSE Scheme of DST, Govt. of India and Bangabasi College, Kolkata, West Bengal, India for using the infrastructure facilities for developing the technique.

References

1. Nandi, U., Mandal, J.K.: Windowed Huffman coding with limited distinct symbol. In: 2nd International Conference on Computer, Communication, Control and Information Technology(C3IT), Hooghly, West Bengal, India, vol. 4, pp. 589–594 (2012)
2. Huang, H.-C., Wu, J.-L.: Windowed Huffman coding algorithm with size adaptation. IEE Proceedings-I 140(2), 109–113 (1993)
3. Nandi, U., Mandal, J.K.: Region based Huffman (RB H) Compression Technique with Code Interchange. Malayasian Journal of Computer Science (MJCS) 23(2), 111–120 (2010)
4. Huffman, D.A.: A method for the construction of minimum redundancy codes. Proc. IRE, 1098–1101 (1952)
5. Faller, N.: An adaptive system for data compression. In: Record of the 7th Asilomar Conference on Circuits, System, and Computers, pp. 593–597 (1973)
6. Nandi, U., Mandal, J.K.: Region based Huffman (RBH) Compression Technique with Code Interchange. Journal of Computer Science(MJCS) 23(2), 111–120 (2010)
7. Nandi, U., Mandal, J.K.: Comparative Study And Analysis of Adaptive Region Based Huffman Compression Techniques. International Journal of Information Technology Convergence and Services (IJITCS) 2(4), 17–24 (2012)
8. Nandi, U., Mandal, J.K.: Region Based Huffman Compression with region wise multiple interchanging of codes. Advancement of modelling & simulation techniques in enterprises(AMSE) 17(2), 44–58 (2012)
9. Gallager, R.G.: Variations on a theme by Huffman. IEEE Trans. IT 24(6), 668–674 (1978)
10. Knuth, D.E. Dynamic Huffman coding. J. Algorithms, (6), p 163-170 (1985).
11. Vitter, J.S.: Dynamic Huffman coding. ACM Trans. Math. SoJt-ware 15(2), 158–167 (1989)
12. Huang, H.-C., Wu, J.-L.: Design and analysts of residual Huffman algorith. In: National Computer Symposium, Taiwan, pp. 200–205 (1991)
13. Nandi, U., Mandal, J.K.: Adaptive region based huffman compression technique with selective code interchanging. In: Meghanathan, N., Nagamalai, D., Chaki, N. (eds.) Advances in Computing & Inform. Technology. Advances in Intelligent Systems and Computing, vol. 176, pp. 739–748. Springer, Heidelberg (2012)
14. Nandi, U., Mandal, J.K.: A Compression Technique Based on Optimality of LZW code(OLZW). In: The Third IEEE International Conference on Computer & Communication Technology (ICCCT-2012), Allahabad, India, pp. 166–170 (2012)

Theme Interception Sequence Learning: Deflecting Rubber-Hose Attacks Using Implicit Learning

Shyam Diwakar[1], Priya Chellaiah[2], Bipin Nair[1], and Krishnashree Achuthan[2]

[1] Amrita School of Biotechnology, Amrita Vishwa Vidyapeetham (Amrita University),
Amritapuri, Clappana P.O., Kollam, 690 525, Kerala, India
[2] Amrita Center for Cybersecurity, Amrita Vishwa Vidyapeetham (Amrita University),
Amritapuri, Clappana P.O., Kollam, 690 525, Kerala, India
shyam@amrita.edu

Abstract. Existing cryptographic systems use strong passwords but several techniques are vulnerable to rubber-hose attacks, wherein the user is forced to reveal the secret key. This paper specifies a defense technique against rubber-hose attacks by taking advantage of image sequence-based theme selection, dependent on a user's personal construct and active implicit learning. In this paper, an attempt to allow the human brain to generate the password via a computer task of arranging themed images through which the user learns a password without any conscious knowledge of the learned pattern. Although used in authentication, users cannot be coerced into revealing the secret key since the user has no direct knowledge on the choice of the learned secret. We also show that theme interception sequence learning tool works significantly well with mixed user age groups and can be used as a secondary layer of security where human user authentication remains a priority.

Keywords: Rubber-Hose Attack, Implicit learning, TISL, Authentication.

1   Introduction

Secret key that involves a user to securely work with a system plays a major role in most cryptographic models. Even though these passwords can be stored securely, they remain vulnerable to eavesdropping, dictionary attacks, social engineering and shoulder surfing. Studies have shown that selecting the password is also related to human cognition and psychology [1]. Knowledge-based passwords may be vulnerable to coercion attacks. Rubber-hose cryptanalysis [2] is an easy way to defeat cryptography in some scenarios.

Securing passwords is one of the most significant problems today. Long and difficult-to-guess passwords usually add to user-related memory and storage issues. Although biometric authentication could be used [3,4,5] to overcome this, post-attacks, re-validation is not easy in biometric techniques as data is physiology-dependent. Images allow the human ability to understand and remember pictorial representations better than the alpha-numeric complexes of meaningless strings. Although image-based authentication schemes allow users to learn strong secret passwords that are easily memorable [6,7,8,9], these systems are not resistant to

© Springer International Publishing Switzerland 2015
S.C. Satapathy et al. (eds.), *Proc. of the 3rd Int. Conf. on Front. of Intell. Comput. (FICTA) 2014*
– *Vol. 1*, Advances in Intelligent Systems and Computing 327, DOI: 10.1007/978-3-319-11933-5_55

rubber-hose attacks due to recall awareness. Previous studies have discussed about incoercible multiparty computation and schemes of deniable encryption [10,11].

1.1   Rubber-Hose Attacks and Existing Techniques

Rubber-Hose attacks are a type of coercion attacks in which a user is physically forced by the attacker until the secret password is retrieved. Reports on the Syrian war shows that people were systematically tortured and killed using rubber belts [12] while extracting critical information.

This paper showcases the design of an anti-rubber hose attack scheme and the implementation of the tool using implicit learning, centered on cognitive psychology [13,14,15]. The technique was based on sequential arrangement of image strips [6,7] and the characteristics of associative memory recall. Representing an information using images, specifically with comic strips, increases the attention, comprehension, recall and adherence ability in human subjects. This correlation between the images and the memory construct was used in the implementation of our technique [16,17,18]. Knowledge learned implicitly are not consciously accessible to or may not be described explicitly by the person being trained [13,14,15,19,20]. Everyday examples of this phenomenon include swimming, riding bicycle, playing golf, playing musical instruments etc. These tasks are implicitly understood although explicit declaration may not be straight forward.

To overcome the problem with recall-based authentication, some recognition-based authentication schemes used images instead of textual passwords. Human ubiquity to graphical information has been better with graphical sequences than with texts and numbers [6, 7, 8,9]. Pictures or comic strips are also used in several fields such as health organizations [21], educational institutions [22], and in others [23] to enhance communication or to motivate learning. We used image sequences in order to take advantage of human ability to elicit planned emotions as a way of authenticating human user and user's emotional individuality. Implicit learning plays a major role in designing coercion-resistant security systems [2]. In techniques like ours, a user learns the secret key through implicit learning that may be detected during authentication but may not be described explicitly by an observer or the user. As part of this anti-coercion tool, users were initially trained to do a special task called Theme Interception Sequence Learning (TISL).

1.2   TISL Threat Model and Design

The tool was designed to be used as an alternate or secondary authentication system which identifies a human user, based on a pre-validated sequence of image patterns. The technique also allows physical presence of user to be validated at the time of authentication. Our tool was implemented in two phases: Training and Authentication. Training was done using a computer task that results in implicit learning of specific sequence of images which became the secret key for authentication. The server stores the trained pattern and used it to authenticate the user successfully. The attacker may not identify the same pattern unlike the user since the attacker is unaware of the training sequence.

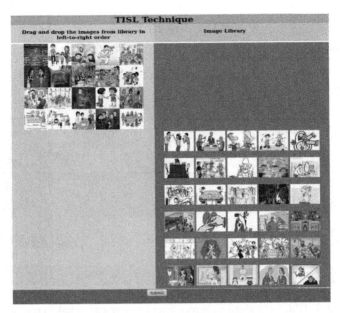

Fig. 1. TISL training. Screenshot of training task in progress

2   Theme Interception Based Task

The technique proposed in this paper is referred to as Theme Interception Sequence Learning (TISL) task. TISL task was based on the idea of sequence learning [24]. In this implementation, we have used various sets of comic strip series with different themed sequences. Theme interception sequence learning was based on Serial Interception Sequence Learning (SISL) [2]. In SISL task, the users developed sensitivity to patterned symbol information without any conscious knowledge of the learned information. In TISL unlike SISL, symbol sequences were not used. In a TISL task, users were sensitized implicitly to the structured information by intercepting image sequences. An image library containing different themes was presented on the right side and an empty grid was presented on the left side of the computer screen (see Fig. 1).The task-level objective for the user was to intercept the object. Interception was performed by selecting images from same theme and re-arranging onto the grid provided on the left side as per user choice. During the 20 to 25 minute training process, a user performed several TISL tasks and the sets of the image library followed covertly embedded repeated themes in 75% of the trials. Each image in the image library represented an individual event within a set in a specific theme. The image library for the authentication task (see Fig. 2) contained different sets including both trained as well as untrained sets of comic strip images. The sequences for authentication from within the trained sets were chosen randomly so that the user would get different sets during each authentication task. Participants implicitly identified the themes by repeatedly performing the TISL task.

Fig. 2. TISL authentication. Screenshot of authentication task in progress

3 Implementation of an Anti-Rubber-Hose Attack Tool

The concept behind TISL task was to store a secret key 'sequence' in the human brain that could be retrieved during authentication, but may not be described verbally. It was possible to train users to remember a sequence that could be used as an alternative to password-based authentication. The image sequence themes included real life scenes like, going on a trip, daily routine of an office staff, preparing food, having dinner, playing with friends, caring pets, celebrating birthdays, festivals, marriage function etc. Each image inside a set represents an event that occurred in that theme. These events could be arranged in any order. A fixed order for these events was avoided. User choice was defined by the perceived personal or impersonal constructs as identified by the observer. Such sequence orders may be different for different users for the same theme and thereby allowing uniqueness in the pass sequence.

The user identification system was designed in two phases: training and authentication. For training task, each user was presented with an image library consisting of several themes (see Table 1). User learned five themes from the set $T = \{T_1, T_2, T_3, T_4, T_{5...} T_n\}$. Each theme contained several sets such as $T_i = \{S_{jk}\}$ where i = j = 1 to n and k = 1 to m. There were five images within each set representing five events for every theme as $S_{jm} = \{x_{ef}\}$ where e = 1 to n and f = 1 to 5 . Two sets from each theme were selected to form the image library. The image library for each user contained 10 sets. These sets were selected randomly in order to avoid multiple serial sequences from same theme while a user performs a training task. This helped to avoid repeated sequence similarities and unique passwords per user. User's task was to arrange the five events within a set as apt order based on user's thought construct and perception. This order, $O(S_{jk}) = \{x_{ef}\}$ was saved and was different for various user in our tests.

Table 1. TISL sample themes and some events within each theme

Theme	Example events
Going for a trip	Planning, packing the luggage, getting into the bus, enjoying the trip, bbq camp-fire
Having Dinner	A boy playing with friends, return home, finish homework, have dinner, go to sleep
Preparing food	Planning the food item, getting the required ingredients, cleaning the ingredients, cook food, serve food
Caring pets	Giving water, clean the shed, giving food , playing with the pet, going for a walk
Playing with friends	Forming teams, game planning, getting the required things to play, playing, return home

Training. User learns 10 secret keys by performing the TISL task repeatedly in a trusted environment. The following procedure was used to train the user:

1. $\forall \ U_l$ where U denotes the user and $l = 1$ to n; Choose five different themes from the set T.
2. Randomly select two sets from the selected themes T_i and form the image library.
3. Display an empty grid on the left side of the screen to arrange the images.
4. Choose any set S_{jk} from the library and arrange the images x_{ef} with their own choice within the set in the grid provided on left side of the screen.
5. The arrangement of these events could be performed using any permutations. For ex. one of the possible orders of these events could be $O(S_{jk}) = \{ \ x_{e3}, \ x_{e5}, \ x_{e1}, \ x_{e4}, \ x_{e2} \}$. The participant should match with this order while performing authentication at a later time.
6. Repeat step 4 for the remaining sets in the library then submit the training. Record the sequence and store as trained order.

During training, a total of 10x5=50 images were presented to the user which took 20-25 minutes to arrange. The system recorded the final order of the images.

Authentication. For authentication, the trained user was presented with the TISL task where the image library contains sets from trained as well as untrained themes for comparison. Untrained sets were chosen from the set $M = \{M_1, M_2, M_3....M_n\}$. The untrained sets comprised of the themes similar as those selected for training. This reduces possibility of guess-based predictions for a hacker although a trained user would identify the sets because of having seen the images earlier during training. The user validates identity by also exhibiting better performance on the trained sets than untrained sets. The procedure used for authentication was as follows:

1. $\forall$ U_l where U denotes the user and $l = 1$ to n; Choose two sets from $\{S_{jk}\}$ where $j = 1$ to n and $k = 1$ to m for the trained themes T_i.
2. Randomly choose three untrained sets from the set M and form the image library by combining the sets selected in step1.
3. Choose five random numbers and randomize the sets in the image library and present it to the user.
4. User is required to identify the two trained sets and perform the TISL task for the two identified sets.
5. The TISL task is performed by arranging the images from the selected sets in the order $O(S_{op})$.
6. If the user identifies the two trained sets correctly and also if $O(S_{op}) = = O(S_{jk})$, then the system declares that the authentication succeeded otherwise it shows authentication failure.

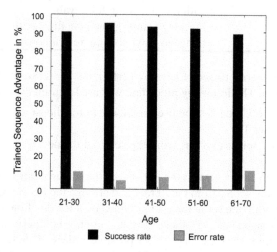

Fig. 3. Authentication success rates among varied age-groups. Percentage of successful authentication was higher than that of failure percentage. X-axis represents age of the participant's and y-axis represents the averaged percentage of authentication success.

Age-Based User Analysis. An assessment test was conducted immediately after authentication task to evaluate the role of implicit learning based success among users and to test that authentication's dependence on user age. The technique was performed by a subject group of 25 participants of age between 21 and 70 (see Fig. 3). The training procedure described in the previous section was used and included a total of 50 trials that took 20-25 minutes to complete. Participants completed the TISL authentication test immediately after training phase while estimating the amount of learning-based recall in each participant. In this immediate test, most of the participants successfully identified two trained sets and performed TISL task. Our tests suggest training and authentication was not dependent on age group of users since that percentage of successful authentications was higher than that of percentage of failed cases.

4   Conclusion and Future Work

Implicit learning and user behavior were crucial to train and authenticate in this image theme based technique. Although extensive tests need to be done, we have shown that success rate for training and authentication remained similar for age ranges 21-70. Since the tool uses a set of cartoon images as different themes to generate the secret key, it may take a significant amount of time for training as well as authentication task. The objective of such schemes was to deter common rubber hose attacks that relies on explicit authentication and to serve as a secondary layer of protection in secure systems where human user authentication remains a priority. This scheme requires the physical presence of the user so remote authentication is not possible. Adversary may guess the sequence order since the sequence length used for authentication is of short strips (images per sequence =5). This may be prevented by increasing the sequence length and by selecting the themes that are more related to the user's social cyber life. This scheme may be used by participants of any age group as it was implemented based on human implicit learning of events and themes in order to generate the secret key. The problem of forgetting the password may be overcome by performing the training task when required.

This implemented technique used a special 'task' to learn the sequence and hence was titled, Theme Interception Sequence Learning (TISL). The tool needs to be improved via other cognitive tests such as recall assessment to study the dependency of implicit learning on age of the participants, the rate at which the learned secrets are forgotten, the frequency required to refresh the training session etc. The training task will be modeled by adding different levels such as high, intermediate and sparse to study the effectiveness of sequence length in implicit learning.

Acknowledgments. This project derives direction and ideas from the Chancellor of Amrita University, Sri Mata Amritanandamayi Devi. Authors would like to acknowledge all participants who have contributed to this use-case study.

References

1. Anderson, R.: Usability and Psychology. Security Engineering, 17–62 (2008)
2. Bojinov, H., Sanchez, D., Reber, P., Boneh, D., Lincoln, P.: Neuroscience Meets Cryptography: Designing Crypto Primitives Secure Against Rubber Hose Attacks. USENIX Security (2012)
3. Kale, A., Rajagopalan, A.N., Cuntoor, N., Krueger, V., Chellappa, R.: Identification of humans using gait. IEEE Transactions on Image Processing 13, 1163–1173 (2002)
4. Bhattacharyya, D., Ranjan, R., Alisherov, F.A., Choi, M.: Biometric Authentication: A Review. International Journal of u- and e- Service, Science and Technology 2(3) (September 2009)
5. Monrose, F., Reiter, M., Wetzel, S.: Password hardening based on keystroke dynamics. Int. J. of Inf. Sec. 1(2), 69–83 (2002)

6. Sonkar, S.K., Paikrao, R.L., Kumar, A.: Graphical Password Authentication Scheme Based On Color Image Gallery. International Journal of Engineering and Innovative Technology (IJEIT) 2(4) (October 2012)

7. Wiedenbeck, S., Waters, J., Birget, J.C., Brodskiy, A., Memon, N.: Authentication Using Graphical Passwords: Basic Results. In: HumanCompute Interaction International, Las Vegas, NV (2005)

8. Almuairfi, S., Veeraraghavan, P., Chilamkurti, N.: A novel image-based implicit password authentication system (IPAS) for mobile and non-mobile devices. Mathematical and Computer Modelling 58, 8–116 (2013)

9. Dunphy, P., Yan, J.: Do Background Images Improve Draw a Secret" Graphical Passwords? In: Proceedings of the 14th ACM Conference on Computer and Communications Security, pp. 36–47 (2007)

10. Canetti, R., Gennaro, R.: Incoercible multiparty computation. In: Proceedings of the 37th Annual Symposium on Foundations of Computer Science, pp. 504–513 (1996)

11. Canetti, R., Dwork, C., Naor, M., Ostrovsky, R.: Deniable encryption. In: Kaliski Jr., B.S. (ed.) CRYPTO 1997. LNCS, vol. 1294, pp. 90–104. Springer, Heidelberg (1997)

12. Krever, M., Elwazer, S.: Gruesome Syria photos may prove torture by Assad regime. CNN.com. Cable News Network (January 22, 2014)

13. Destrebecqz, A., Cleeremans, A.: Can sequence learning be implicit? new evidence with the process dissociation procedure. Psychonomic Bulletin & Review 8, 343–350 (2001)

14. Cleeremans, A., Jiménez, L.: Implicit learning and consciousness: A graded, dynamic perspective. Implicit learning and consciousness, 1–40 (2002)

15. Sanchez, D., Gobel, E., Reber, P.: Performing the unexplainable: Implicit task performance reveals individually reliable sequence learning without explicit knowledge. Psychonomic Bulletin & Review 17, 790–796 (2010)

16. McVicker, C.: Comic Strips as a Text Structure for Learning to Read. International Reading Association The Reading Teacher 61(1), 85–88 (2007)

17. Megawati, F., Anugerahwati, M.: Comic strips:A Study on the Teaching of Writing Narrative Texts to Idonesian EFL Students. TEFLIN Journal: A Publication on the Teaching and Learning of English 23(2) (2012)

18. Pierson, M.R., Glaeser, B.C.: Using Comic Strip Conversations to Increase Social Satisfaction and Decrease Loneliness in Students with Autism Spectrum Disorder. Education and Training in Developmental Disabilities 42(4), 460–466 (2007)

19. Denning, T., Bowers, K., van Dijk, M., Juels, A.: Exploring implicit memory for painless password recovery. In: Proceedings of the SIGCHI Conference on Human Factors in Computing Systems, pp. 2615–2618. ACM (2011)

20. Weinshall, D., Kirkpatrick, S.: Passwords you'll never forget, but can't recall. CHI Extended Abstracts, 1399–1402 (2004)

21. Houts, P., Doak, C., Doak, L., Loscalzo, M.: The role of pictures in improving health communication: a review of research on attention, comprehension, recall, and adherence. Patient education and counseling, 61(2), 173 -190 (2006)

22. Bolton-Gary, C.: Connecting through Comics: Expanding Opportunities for Teaching and Learning. Online Submission (2012)

23. Ginman, M., von Ungern-Sternberg, S.: Cartoons as information. Journal of Information Science, 29(1) 69-77 (2003)

24. Sun, R., Giles, L.: Sequence Learning: From Recognition and Prediction to Sequential Decision Making. Intelligent Systems, IEEE 16(4), 67–70 (2001)

A Novel Approach for Stereo-Matching Based on Feature Correspondence

Sonu Thomas[1], Suresh Yerva[2], and T.R. Swapna[1]

[1] Dept. of Computer Science and Engineering,
Amrita Vishwa Vidyapeetham, Coimbatore, India
sonupt4u@gmail.com, tr_swapna@cb.amrita.edu
[2] Centre for Research in Engineering Sciences and Technology(CREST),
KPIT Technologies Ltd., Pune, India
suresh.yerva@kpit.com

Abstract. A dense two-frame stereo matching technique generally uses an image pair as input in addition to the knowledge of disparity range. For real-time computer vision systems, however, there is lots of information that can enhance stereo-correspondence, e.g. feature points. Feature correspondence is essential for computer vision applications that require structure and motion recovery. For these applications, disparity of reliable feature points can be used in stereo-matching to produce better disparity images. Our proposed approach deals with adding the feature correspondences effectively to dense two-frame stereo correspondence framework. The experimental results show that the proposed approach produces better result compared to that of the original algorithm RecursiveBF [1].

Keywords: Local stereo matching, cost aggregation, feature correspondences.

1 Introduction

Dense two-frame stereo-matching has been a challenging research area for the past few decades. It aims at producing disparity map from an image pair which is obtained by rectifying different views of the same scene. The disparity image represents the horizontal displacement of corresponding points in the rectified input images. According to taxonomy of stereo matching algorithm by Scharstein and Szeliski [2], the stereo algorithms consist of following four steps: Matching cost computation, Cost (support) aggregation, Disparity computation/optimization and Disparity refinement.

There are generally two classes of stereo-matching algorithms: local and global approaches. Local approaches compute disparity of a given point based on intensity values of its neighbors. Global approaches represent the stereo-matching problem as energy minimization function and perform optimization explicitly making smoothness assumption. Generally, global approaches yield more accurate result, however these are computationally intensive and require high number of iterations to produce better result. Local methods are fast and more compatible for real-time applications.

S.C. Satapathy et al. (eds.), *Proc. of the 3rd Int. Conf. on Front. of Intell. Comput. (FICTA) 2014*
– *Vol. 1*, Advances in Intelligent Systems and Computing 327, DOI: 10.1007/978-3-319-11933-5_56

The matching cost computation creates a cost volume where the cost values of each pixel are placed along disparity axis. Conceptually choosing the index along disparity axis where the cost value is lowest should give the disparity of a pixel (Winner Take All strategy). However, the matching cost functions are not appropriate enough to produce accurate result with only these two steps, because there can be multiple low cost values for ambiguous regions such as uniform regions or repetitive regions. Aggregating the cost over a predefined window centered at a given pixel can reduce the appearance of multiple minima and hence make the disparity computation much accurate, provided the support window must contain pixels belonging to same disparity.

Among local approaches, Adaptive Weight Support, initially proposed by Yoon and Kweon [3], has been a better method to overcome edge flattening problem, where the object shape is not preserved. The main assumptions are that the neighbor pixels similar to center pixel belongs to same disparity of the center pixel (frontal-parallel surfaces) and spatial distance contributes to disparity difference (slanted surfaces). The adaptive approaches ensure that pixels that are closer in color and space will influence the center pixel's disparity more. Yoon and Keon adopted bilateral filter weights and Hosni et al. [4] used geodesic weights. Though these approaches produce accurate result comparable to that of global methods, computational complexity is quadratically proportional to support size. Because of its linear time complexity, the recursive implementation of bilateral filter by Yang [1] has attracted attention in the domain of edge preserving filters and stereo matching.

The proposed method exploits the knowledge of some known disparity for local stereo matching such as feature correspondence points to make the disparity map more accurate. The feature correspondences are generally computed for structure and motion related applications. Our algorithm adds the feature correspondences to matching cost volume and applies edge preserving filter for cost aggregation. This makes sure that the feature points contribute to the disparity estimation and thereby provide better solution for ambiguous regions near-by.

The rest of this paper is organized as follows: algorithm steps are explained in detail in section 2. In Section 3, the results of the proposed algorithm are shown. Section 4 concludes the paper with discussion and summary.

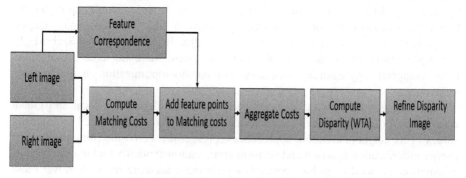

Fig. 1. Block diagram for algorithm outline

2   Algorithm

Fig. 1 shows the algorithm outline. The proposed algorithm involves the steps similar to the dense two-frame stereo-matching architecture [2] with an additional step of adding feature points. In the following section, we explain each step of the algorithm in detail.

2.1   Matching Cost Computation

The Matching cost computation uses both intensity and gradient information from the images. Truncated absolute difference is found to be a simple and effective measure for pixel comparison in stereo-correspondence, as it works robust to unknown radiometric distortions and light sources.

The cost value $C(p,d)$ is obtained by comparing pixel p of reference image I_R with pixel $p'(=p+(s*d))$ of the target image I_T, where $s=-1$ if left image is the reference image and $s=1$,otherwise.

$$C(p,d) = \alpha * \min(\|I_R(p) - I_T(p')\|, T_c)$$
$$+ (1 - \alpha) * \min(\|\nabla_x I_R(p) - \nabla_x I_T(p')\|, T_g) \qquad (1)$$

Here ∇_x is gradient along horizontal direction, α is color weight coefficient with value in the interval (0, 1), T_c and T_g are truncation limit for color and gradient respectively.

2.2   Feature Points Inclusion

This step is the essence of the algorithm. The cost volume C and n feature points are the inputs to this step.

$$f_i = (p_i \quad d_i), 1 \le i \le n \qquad (2)$$

In equation (2), f_i vector contains pixel position p and associated disparity d respectively. These feature points are included to cost volume C to generate feature added cost volume C_f.

As explained earlier, the general motto of cost aggregation is to provide clear estimate of minimum for the cost vector of each pixel. It must be taken care, if the inclusion of the feature points has to contribute to the disparity estimation. Applying edge preserving smoothening filters to each pixel slice of the cost volume as cost aggregation strategy can serve the purpose. Smoothening filters generally perform weighted aggregation of pixel values over a window, where the weights can be based on color similarity, spatial similarity, etc.

For each feature point f with disparity d, a cost volume vector C_f is generated based on equation. (3),

$$C_f(p_f,d) = \begin{cases} 0 & \text{, if } d = d_f \\ \max(C_d) & \text{, else} \end{cases} \quad (3)$$

Where a feature point f with disparity will have cost value 0 at index d and maximum value (of that x-y slice) at other indices.

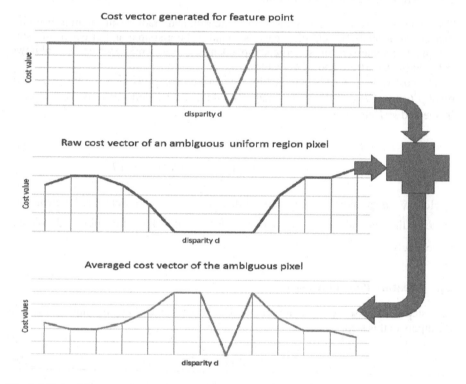

Fig. 2. Effect of feature points in aggregation- The figure shows the effect of aggregating a cost values (blue color) of feature point with that of a pixel in an ambiguous region such as uniform region(red color). The cost vector of both pixels is subjected to cost aggregation by an averaging filter and the result cost vector is shown (green color).

In order to understand the effect of feature points in cost aggregation, let us consider fixed window aggregation where the cost value of the center pixel will be replaced by the average cost value of pixels in an n x m window. Fig. 2 shows the cost volume vectors of a feature point, a pixel from an ambiguous region (such as a uniform region) and the cost vector of the ambiguous pixel after aggregation. If both pixels are at same disparity, disparity estimation should assign same disparity to ambiguous pixel which is equal to that of the feature point. However, the presence of consecutive multiple minima cause bad result with Winner Takes All (WTA) strategy. After averaging, the cost volume vector of the feature point has clearly made the disparity estimation much easier for the ambiguous pixel. Evidently, this effect will be highly dependent on density of feature points in the support window. Further improvement of result depends upon the smoothening filter used. If the smoothening filter considers the spatial and color distance, the weights on costs vector of feature point will make sure that these costs affect only pixels closer in color and space. Hence the generated cost vector of a feature point is made independent of the aggregation method being used, and it improves the results by using better aggregation methods.

2.3   Cost Aggregation

The cost aggregation strategy that we have adopted uses Yang's recursive bilateral filter [1] because of its less memory complexity and efficient computational complexity. Bilateral filter proposed by Tomasi and Manduchi [5] has many applications in fields of computer vision and computer graphics due to its edge-preserving properties. It contains two filter kernels: a range kernel and a spatial kernel. These two kernels can be Gaussian functions. There are many fast implementations [6, 7, 8, 9] for this filter. Yang proposed a recursive version concentrating on Gaussian range filtering kernel.

In equation (4), a new recursive Gaussian range filtering kernel $R_{k,i}$ is proposed which accumulates range distance between every two consecutive pixels from k to i.

$$R_{k,i} = \prod_{j=k}^{i-1} R_{j,j+1} = \prod_{j=k}^{i-1} \exp(-\frac{\left|x_j - x_{j+1}\right|^2}{2\sigma_R^2}) = \exp(-\frac{\sum_{j=k}^{i-1}\left|x_j - x_{j+1}\right|^2}{2\sigma_R^2}) \tag{4}$$

The recursive system is modified as in equation (5),

$$y_i = \sum_{k=0}^{i} R_{i,k}\left(\sum_{m=0}^{n-1} \lambda_{i-m-ka_m}\right)x_k , \tag{5}$$

where
$$\lambda_i = \begin{cases} 1 & i = 0, \\ \sum_{k=1}^{\min(i,n)} - b_k\lambda_{i-k} & i > 0, \\ 0 & i < 0, \end{cases} \tag{6}$$

The initial condition is set to $y_0 = a_0 x_0$ and $x_i = 0$, when $i < 0$. Here $R_{k,i}$ is the range kernel and $S_{i,k} = \sum_{m=0}^{n-1} \lambda_{i-m-ka_m}$ is the spatial kernel.

2.4   Disparity Computation

The disparity computation uses the Winner Takes All (WTA) strategy. The result disparity value d_p for pixel p is obtained as following:

$$d_p = \underset{d \in D}{\operatorname{argmin}} \, C'(p,d) \tag{7}$$

Where D refers to set of disparities allowed.

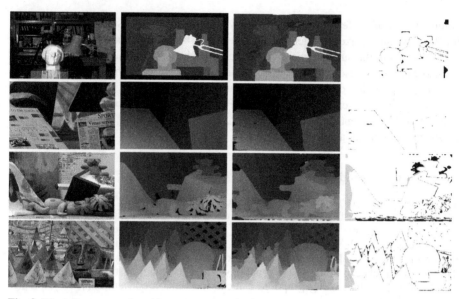

Fig. 3. Disparity map produced by proposed method- The first two columns contain the input and the ground truth images from Middlebury stereo test-bed [11], second column contains the ground truth, third column contains the disparity map results produced by proposed algorithm and fourth column contains error map for the result.

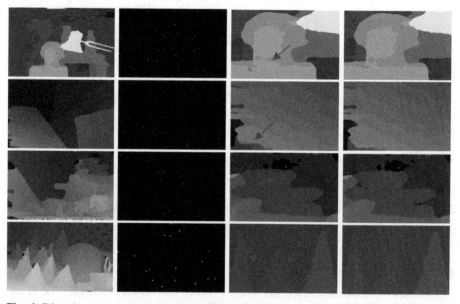

Fig. 4. Disparity map result comparison- The first column contains the disparity map results produced by the previous method Recursive bilateral filtering [1]. Second, third and fourth columns contain the zoomed region of the feature map where the feature points are marked with white values, of result images of previous and the proposed method respectively. The red arrows in the third column show the improvements.

2.5   Disparity Refinement

The raw disparity obtained may contain errors in occluded regions. Left-right consistency check is performed to handle occlusions. There are two disparity maps produced- left disparity image d_L where reference image is left image and right disparity image d_R where right image is the reference image.

$$|d_L(p) - d_R(p - (d_L(p),0)| < T = 1 \qquad (8)$$

The pixels which fail the equation (8) are marked as "occluded pixels" and occlusion mask is created marking their pixel positions as white. Based on the occlusion mask, the "non-occluded" pixels are subjected to cost aggregation to produce better disparity map.

3   Results and Discussions

The stereo-matching algorithm was implemented and tested using an Intel Core i5-3470, 3.20 GHz PC. The recursive bilateral filter implementation is available at author's page [10]. The Fig. 3 shows the results for the proposed approach. From Fig. 4, we can clearly see that the regions pointed out by red arrows have accurate disparity. It is evident that the feature points closer to them have contributed to better estimation of disparity for these pixels. The quantitative analysis also confirms the statement. The table (1) shows the quantitative evaluation result of our approach using Middlebury Stereo evaluation test-bed [11]. Since feature points usually are close to edges and depth discontinuities, these improve disparity result at these regions. We can see that our approach has improved the result of the original version of the filter implementation [1]. Since the feature adding module is designed such that it emphasizes the goal of the cost (support) aggregation, any state-of-the-art cost aggregation can be adopted to produce the better and reliable disparity. In order to prove our method performs better, we adopted bilateral filter.

Table 1. Middlebury stereo ranking

Alg.	Avg. Rank	Tsukuba			Venus			Teddy			Cones			Avg. percent of bad pixels
		Non	All	Disc	Non	All	Disc	Non	All	Disc	Non	All	Disc	
Proposed	59.8	1.89	2.55	7.51	0.28	0.71	2.31	6.38	12.2	14.3	2.82	8.93	7.94	5.65
Previous	62.3	1.96	2.61	7.68	0.30	0.76	2.44	6.49	12.3	14.6	2.82	8.93	7.97	6.04

4   Conclusion

This work presents a novel approach for making effective use of feature correspondences in the local stereo-matching. The feature correspondences are essential whenever an application requires motion and structure recovery. The feature correspondences are added to two-frame dense stereo-matching framework in simple and efficient way to improve the disparity result, especially at regions near depth discontinuity. We experimented on cost aggregation using bilateral filter and produced better results than the original method. The design of the approach is in such a way that it can be coupled with other cost aggregation methods. As future scope, we will focus on designing a novel cost aggregation method which makes use of the known reliable disparity.

Acknowledgments. We would like to thank Dr. Vinay Vaidya, Chief Technical Officer, Mr. Krishnan Kutty, Technical Fellow, CREST, KPIT Technologies Ltd., Pune and other colleagues for their support and encouragement during the project.

References

1. Yang, Q.: Recursive bilateral filtering. In: Fitzgibbon, A., Lazebnik, S., Perona, P., Sato, Y., Schmid, C. (eds.) ECCV 2012, Part I. LNCS, vol. 7572, pp. 399–413. Springer, Heidelberg (2012)
2. Scharstein, D., Szeliski, R.: A taxonomy and evaluation of dense two-frame stereo correspondence algorithms. International Journal of Computer Vision 47(1-3), 7–42 (2002)
3. Yoon, K.J., Kweon, I.S.: Adaptive support-weight approach for correspondence search. IEEE Transactions on Pattern Analysis and Machine Intelligence 28(4), 650–656 (2006)
4. Hosni, A., Bleyer, M., Gelautz, M., Rhemann, C.: Local stereo matching using geodesic support weights. In: 2009 16th IEEE International Conference on Image Processing (ICIP), pp. 2093–2096. IEEE (November 2009)
5. Tomasi, C., Manduchi, R.: Bilateral filtering for gray and color images. In: Sixth International Conference on Computer Vision, pp. 839–846. IEEE (January 1998)
6. Paris, S., Durand, F.: A fast approximation of the bilateral filter using a signal processing approach. In: Leonardis, A., Bischof, H., Pinz, A. (eds.) ECCV 2006. LNCS, vol. 3954, pp. 568–580. Springer, Heidelberg (2006)
7. Porikli, F.: Constant time O (1) bilateral filtering. In: IEEE Conference on Computer Vision and Pattern Recognition, CVPR 2008, pp. 1–8. IEEE (June 2008)
8. Yang, Q., Tan, K.H., Ahuja, N.: Real-time O (1) bilateral filtering. In: IEEE Conference on Computer Vision and Pattern Recognition, CVPR 2009, pp. 557–564. IEEE (June 2009)
9. Adams, A., Gelfand, N., Dolson, J., Levoy, M.: Gaussian KD-trees for fast high-dimensional filtering. ACM Transactions on Graphics (TOG) 28(3), 21 (2009)
10. Yang, Q.-X.: Recursive bilateral filtering,
   http://www.cs.cityu.edu.hk/~qiyang/publications/eccv-12/
   (retrieved August 7, 2014)
11. Scharstein, D., Szeliski, R.: Middlebury stereo evaluation-version 2 (March 22, 2011),
   http://vision.middlebury.edu/stereo/eval
12. Pham, C.C., Jeon, J.W.: Domain transformation-based efficient cost aggregation for local stereo matching. IEEE Transactions on Circuits and Systems for Video Technology 23(7), 1119–1130 (2013)
13. Rhemann, C., Hosni, A., Bleyer, M., Rother, C., Gelautz, M.: Fast cost-volume filtering for visual correspondence and beyond. In: 2011 IEEE Conference on Computer Vision and Pattern Recognition (CVPR), pp. 3017–3024. IEEE (June 2011)

Multi Resonant Structures on Microstrip Patch for Broadband Characteristics

P. Satish Rama Chowdary[1], A. Mallikarjuna Prasad[2], and P. Mallikarjuna Rao[3]

[1] Dept. of ECE, Raghu Institute of Technology, Visakhapatnam, India
[2] Dept of ECE, UCoE, JNTUK, Kakinada, India
[3] Dept. of ECE, AUCE, Andhra University, Visakhapatnam, India
psr_satish@yahoo.com

Abstract. Multi resonant structure is one of the solutions for broadband characteristics in patch antennas. Thick substrate materials with low dielectric constant are also favourable for such characteristics. In this paper, the combination of E and U shaped slots on the patch antenna along with above mentioned substrate material properties is designed, fabricated and analysed. The characteristics of the designed antenna are studied with respect to the reports like reflection coefficient, VSWR plot, and radiation pattern. The proposed antenna exhibited a wide bandwidth of 4GHz. The effect of the material properties and the physical parameters on the broadbanding are also observed.

Keywords: Broadbanding, Patch antenna, Multiresonant, EU slot.

1 Introduction

Patch antennas have revolutionized the field of communications by making it possible to accommodate antenna in a hand held device. This is possible because of the conformal and low profile characteristics of it. Also patch antennas have such advantages as low cost, light weight, easy to fabricate, compatibility with integrated circuits. Besides these advantages the main drawback of the patch antenna is, having very narrow bandwidth which confines its use to very few commercial applications.

Consider a communication system in which has an amplifier tuned to a particular operating frequency is not exactly coinciding with the resonating frequency of the patch antenna. Because of the inherent narrow bandwidth of the patch antenna it will not be able to provide enough signal amplitude required for the amplifier at its operating frequency, there arises the need to increase the bandwidth of the patch. Extensive research has been carried out so far to enhance the bandwidth of the patch antenna and suggested various methods such as increasing patch height, reducing substrate permittivity, using multiple resonators etc. to widen the bandwidth of the patch. One of the best methods to increase the bandwidth, which has lower increment in antenna volume behind, is the use of a slot in various shapes in the single layer single patch antenna as reported by number of researchers [1-3].

Here we have considered a square patch in which E and U slots are carved, thus forming multiple resonators in the patch. Also the substrate with lowest dielectric

S.C. Satapathy et al. (eds.), *Proc. of the 3rd Int. Conf. on Front. of Intell. Comput. (FICTA) 2014*
– Vol. 1, Advances in Intelligent Systems and Computing 327, DOI: 10.1007/978-3-319-11933-5_57

constant (air) and maximum possible height are taken. The variation of bandwidth with dielectric constant and height of the substrate is observed from the plots of reflection co-efficient curve. Also the radiation patterns at different frequencies within the broadened bandwidth are also provided.

2   Multiple Resonant Structures

Broadband characteristics in patch antennas can be achieved by coupling additional microstrip patches directly to its radiating or non-radiating edges[4] - [5]. In this section the formation of U-slot and E-slot in the patch is explained with air substrate. Embedding a suitable U-shaped slot [11] in the antenna's radiating patch is a very effective method for achieving a wide bandwidth for a probe-fed microstrip antenna with a thick air substrate. By using an E-shaped patch instead of a U-slotted patch, similar broadband operation can be obtained. The E-shaped patch is formed by inserting a pair of wide slits at the boundary of a microstrip patch. Figure 1(a) through Fig 1(b) shows the geometry of a broadband probe-fed rectangular microstrip antenna with a U-shaped and E-shaped slot. Fig.1(c) describes the probe feed technique for the above mentioned patch antennas.

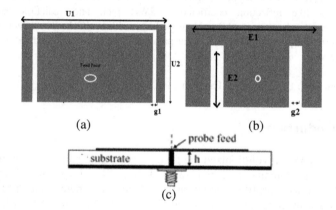

Fig. 1. Geometry of a broadband probe-fed rectangular microstrip antenna with a) U-shaped slot, b) E-shaped slot and c) probe feed system

In the U-shaped antenna [1], a narrow slit of $g1$ width is carved along the three sides of the patch. This results in an inner rectangular patch and outer U-strip patch. The three control parameters $U1$, $U2$ and $g1$ are used to extract the broadband characteristics from this multiresonant structure by properly tuning them over a range of values. In the E-shaped patch the two wide slits have the same length and width and are inserted at the bottom edge of the patch [1] [3]. The separation of the two wide slits is $g2$ and the two slits are placed symmetrically with respect to the patch's centerline (y axis). Thus the three parameters E1,E2 and g2 are used to tune the antenna for improved broadband characteristics. In addition to these geometries in which the U-slot or the E-slot is said to be oriented symmetrically in the rectangular patch and we have asymmetrically placed U and E slots. Though they do not have much impact on the broadband characteristics of the antenna but can bring significant changes to the resonant characteristics as well as its directive properties [6]- [11]. These asymmetrically placed slots bring changes in the current distribution on the surface of the patch.

3   EU Slot Antenna

The geometry of the EU slot patch antenna is as shown in the Fig.2(a). The corresponding simulated geometry in HFSS is presented in Fig.2(b). The proposed geometry is a combination of both the U-shaped slot and E-shape slot presented in the Fig.1(a)&(b). Both the slots are designed on the patch layer of the rectangular microstrip patch antenna of dimensions *U1xU2*. The E-shaped slot is carved inside the rectangular patch with a gap of *g2* with dimensions *E1* x *E2*. The feed point is selected along the y-axis on the centre arm of the inner E-shaped slot.

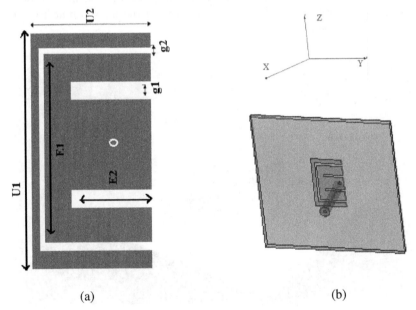

(a)   (b)

Fig. 2. Representation of (a) EU slot geometry (b) Simulated EU slot with probe feed

The antenna is designed using efficient CAD tool integrated in HFSS. The overall experimentation is divided into 3 cases. The Dielectric constant (ε_r) and the substrate height (subH) are varied with every case to analyse the impact of these parameters and compare the results. For all the cases the outer rectangular patch is chosen to have dimensions of 21mm x 13mm and inner E-slot has E1=16mm and E2=11mm with g1=g2=1mm. A considerable ground plane is created as a lower layer of the patch. For both the patch and the ground plane copper material is chosen. The choice of the feed point plays a vital role in determining the depth of the dip in reflection coefficient plot as well as the efficiency and gain of the antenna. As this is not a regular geometry we have to perform parametric sweep with changing feed point and determine the feed point where we have deeper dip in S11 plot and terminal impedance is close to 50 ohms. This is where the efficiency of the simulation tools comes into the picture. HFSS has an efficient optimetrics tool integrated in it. It has tuning and parametric sweep options. For this work tuning option is considered. A discrete solution type is considered for taking the results in HFSS. Though the solution type is time consuming, it is considered as the accurate and free from solution frequency based analysis. In the following three cases the same strategy is adopted to find the best feed point.

3.1   Case-1: Low ε_r and Thickest Substrate

For a very low ε_r air substrate is chosen. Keeping in view of the fabrication ease a subatrate height (subH) of 3.2mm is considered. The feed point is obtained following the steps mentioned above. The fabricated prototype of the air substrate EU slot antenna is as shown in the Fig.3.

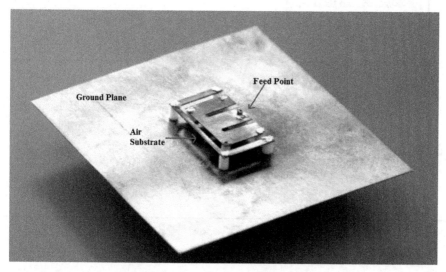

Fig. 3. Fabricated EU slot antenna Prototype

The reflection coefficient (RC) and the VSWR curves of the simulated EU slot are presented in the Fig 4 and Fig 5. It can be inferred from the RC plot that there exists a wide and continuous bandwidth between 8.7 GHz and 12.8GHz. The colored area in the plot represents the same. The corresponding VSWR plot is used to support the same. The measured RC plot for a wide sweep of 0.5 GHz to 20 GHz is presented in the Fig 6. The measured results are in good agreement with the simulated results.

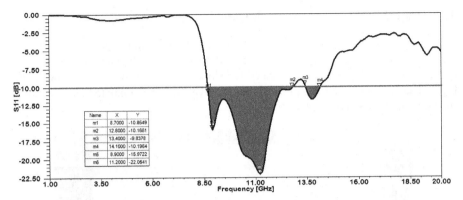

Fig. 4. Reflection Coefficient plot of simulated EU slot antenna with air substrate

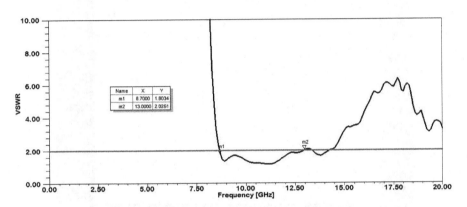

Fig. 5. VSWR plot of simulated EU slot antenna with air substrate

Frequency (GHz) vs S11(dB)

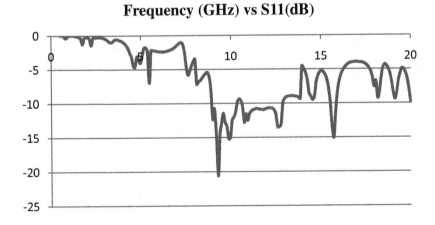

Fig. 6. Reflection Coefficient plot of fabricated EU slot antenna with air substrate

3.2   Case-2: High ε_r and Thin Substrate

Arlon 320C material is chosen basing on the availability with the fabrication lab. The substrate is sandwiched between two copper layers and has thickness of 1.6mm with ε_r of 3.2. The prototype of the fabricated EU slot on Arlon substrate is as shown in the Fig.7. The corresponding RC plots of the prototype and the simulated antenna are as shown in the Fig. 8 and Fig.9.

Fig. 7. Fabricated prototype of the EU slot antenna on Arlon substrate

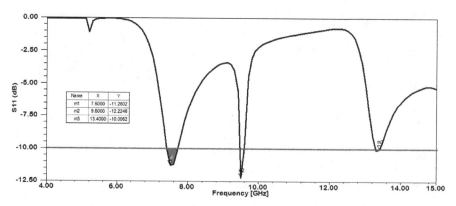

Fig. 8. RC of simulated EU slot antenna on Arlon substrate

Frequency (GHz) vs S11(DB)

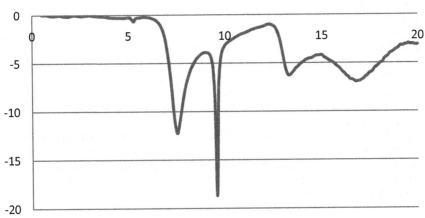

Fig. 9. RC plot of fabricated EU slot antenna on Arlon

From the RC plots of the fabricated and simulated antennas it can be inferred that the geometry shows a multi resonant characteristics which is due to the resonant structures namely the U-slot and E-slot shapes. Both the measured results and the simulated results are in good agreement with each other.

4   Conclusion

The EU shaped antenna is simulated and its characteristics are validated using the practical measurements taken on the fabricated prototype. Both the simulated and the measured parameters are in good agreement. The multi resonant characteristics of the EU slot antenna are observed with Arlon 320C substrate where the dielectric constant and the substrate height are moderate. The broadbanding characteristics are obtained by increasing the substrate height and decreasing the ε_r. The same is verified with the case-1 in section 3. A wider and continuous bandwidth is observed when high subH and low ε_r are considered along with E and U slots on the patch surface.

References

1. Garg, R., Bhartia, P., Bahl, I., Ittipiboon, A.: Microstrip Antenna Design Handbook, pp. 534–538. Artech House, Boston London (2001)
2. Krauss, J.D.: Antennas, 3rd edn. McGraw-Hill, New York (2002)
3. Pozar, D.M.: Microwave Engineering, 3rd edn. Wiley, New York (2005)
4. Lin, W., Xie, X.-Q., Bo, Y., Feng, L.: Design of a broadband E-shaped microstrip antenna. In: Cross Strait Quad-Regional Radio Science and Wireless Technology Conference (CSQRWC), July 26-30, vol. 1, pp. 386–388 (2011)

5. Pauria, I.B., Kumar, S., Sharma, S.: Design and Simulation of E-Shape Microstrip Patch Antenna for Wideband Applications. International Journal of Soft Computing and Engineering (IJSCE) 2(3), 2231–2307 (2012)

6. Lin, W., Xie, X.-Q., Bo, Y., Feng, L.: Design of a broadband E-shaped microstrip antenna. In: Cross Strait Quad-Regional Radio Science and Wireless Technology Conference (CSQRWC), July 26-30, vol. 1, pp. 386–388 (2011)

7. Patel, S.K., Kosta, Y.P.: E-shape Microstrip Patch Antenna Design for GPS Application. In: IEEE International Conference on Current Trends In Technology, Nuicone 2011 (2011)

8. Razzaqi, A.A., Mustaqim, M., Khawaja, B.A.: Wideband E-shaped antenna design for WLAN applications. In: 2013 IEEE 9th International Conference on Emerging Technologies (ICET), December 9-10, pp. 1–6 (2013)

9. Yang, F., Zhang, X.X., Ye, X., Rahmat-Samii, Y.: Wide-band E-shaped patch antennas for wireless communications. IEEE Trans. Antennas Propagat. 49(7), 1094–1100 (2001)

10. Ali, M., Khawaja, B.A., Tarar, M.A., Mustaqim, M.: A Dual Band U-Slot Printed Antenna Array for LTE and WiMAX Applications. Wiley Microwave and Optical Technology Letters 55(12), 2879–2883 (2013)

11. Neyestanak, A.A.L., Kashani, F.H., Barkeshli, K.: W-shaped enhanced-bandwidth patch antenna for wireless communication. Wireless Personal Communications 43(4), 1257–1265 (2007)

SVM and HMM Modeling Techniques for Speech Recognition Using LPCC and MFCC Features

S. Ananthi and P. Dhanalakshmi

Dept. of Computer Science and Engineering,
Annamalai University, Chidambaram,Tamil Nadu, India
{ananti.anglaise,abidhana01}@gmail.com

Abstract. Speech Recognition approach intends to recognize the text from the speech utterance which can be more helpful to the people with hearing disabled. Support Vector Machine (SVM) and Hidden Markov Model (HMM) are widely used techniques for speech recognition system. Acoustic features namely Linear Predictive Coding (LPC), Linear Prediction Cepstral Coefficient (LPCC) and Mel Frequency Cepstral Coefficients (MFCC) are extracted. Modeling techniques such as SVM and HMM were used to model each individual word thus owing to 620 models which are trained to the system. Each isolated word segment from the test sentence is matched against these models for finding the semantic representation of the test input speech. The performance of the system is evaluated for the words related to computer domain and the system shows an accuracy of 91.46% for SVM 98.92% for HMM. From the exhaustive analysis, it is evident that HMM performs better than other modeling techniques such as SVM.

Keywords: Acoustic Feature Extraction, Hidden Markov Model (HMM), Isolated word recognition, LPCC, MFCC, Support Vector Machine (SVM), Voice Activity Detection (VAD).

1 Introduction

During recent years, plenty of researchers have investigated in the field of speech processing because of its increasing attention in many real world applications. Many solutions have emerged in the past few decades to ease the day to day life of differently abled people [1]. The proposed work focuses on Speech Recognition (SR) system. It is exceedingly applicable for disabled persons. Speech disorders or hearing disorders are a type of communication disorder where 'normal' speech or hearing is disrupted.

Proposed work aims to develop a system which has to convert spoken word into text using SVM and HMM modeling techniques using acoustic features namely LPCC and MFCC. In this work the temporal envelop through RMS energy of the signal is derived for segregating individual words out of the continuous speeches using voice activity detection method.

© Springer International Publishing Switzerland 2015
S.C. Satapathy et al. (eds.), *Proc. of the 3rd Int. Conf. on Front. of Intell. Comput. (FICTA) 2014*
– *Vol. 1*, Advances in Intelligent Systems and Computing 327, DOI: 10.1007/978-3-319-11933-5_58

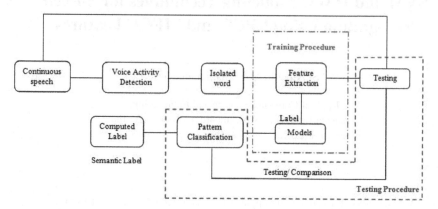

Fig. 1. Framework of the proposed system

Features for each isolated word are extracted and those models were trained. During training process each isolated word is separated into 20ms overlapping windows for extracting 19 LPCCs and 39 MFCCs features respectively. SVM and HMM modeling techniques were used to model each individual utterance. It leads to 620 models and these 620 models were trained to the system. Thus each isolated word segment from the test sentence is matched against these models for finding the semantic representation of the test input dialogue. The frame work of the proposed system is shown in Fig.1.

2   Voice Activity Detection

Voice Activity Detection Algorithms are language independent and specifically afford the concern information regarding where the speech signal is present ie., it identifies where the speech is voiced, unvoiced or sustained. These details help to deactivate the process during non-speech segment in a speech [2]. It makes the smooth progress of speech processing. Isolated words in an audio speech were exploited using the long pauses in a dialog which is shown in Fig. 2. The spectral and temporal envelop of a signal provide maximum information about the signal content. In this work the temporal envelop through RMS energy of the signal is derived for separating/segregating individual words out of the long speeches. Based on the threshold value the temporal environment is analyzed to find the region of words.

Isolated words in an audio speech were exploited using the long pauses in a dialog. The spectral and temporal envelop of a signal provide maximum information about the signal content. In this work the temporal envelop through RMS energy of the signal is derived for separating/segregating individual words out of the long speeches. Based on the threshold value the temporal environment is analyzed to find the region of words.

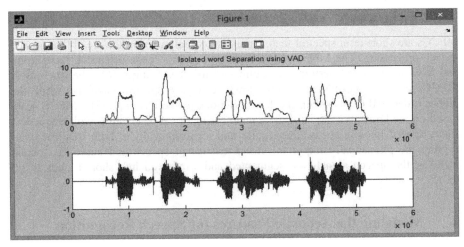

Fig. 2. Isolated Separation

RMS over the window of size which is shown in Eqn. 1. wlen is the length of the window. The optimum threshold is chosen through trial and error. In this work 0.5 is chosen as the threshold over the RMS energy window of 20ms. When the energy envelope exceeds the pre-defined threshold value then that sample is marked as the beginning of the segment. Likewise the adjacent sample which falls below the threshold is termed as the end of the segment.

$$RMS = \sqrt{x^2 \otimes wlen} \qquad (1)$$

Similarly other segment in the energy environment is found by differentiating the threshold signal. Finally, that value is used to extract the isolated words samples from that original speech.

3   Acoustic Feature Extraction

The purpose of feature extraction is to compress the speech signal into a vector that is representative of the class of acoustic event it is trying to characterize.

3.1   Linear Prediction Cepstral Coefficients

The Cepstral coefficients are the coefficients of Fourier transform representation of the algorithm magnitude spectrum. In speech recognition, pitch, LPCC, MFCC are the important features used to extract Speech processing applications [3]. The popular feature extraction technique LPCC is used for extracting specific features from the speech signal. LPCC is a common transform used to gain information from an audio signal [4]. They are calculated so as to minimize the prediction error. In this work the most frequently used LPCC parameters are considered to determine the best feature set for the speech database [5]. A 19 dimensional weighted linear LPCC for each frame is used as a feature vector.

3.2   Mel Frequency Cepstral Coefficients

MFCCs are short-term spectral features and are widely used in the area of audio and speech processing. The MFCCs have been applied in a range of speech recognition and have shown good performance compared to other features [6]. The low orders MFCCs contain information of the slowly changing spectral envelope while the higher orders MFCCs explain the fast variations of the envelope [7]. MFCC's are based on the known variation of the human ear's critical bandwidths with frequency, filters spaced linearly at low frequencies and logarithmically at high frequencies to capture the phonetically important characteristics of speech and audio. To obtain MFCCs, the speech signals are segmented and windowed into short frames of 20 msec.

4   Modeling the Acoustic Features for Speech Recognition

4.1   Support Vector Machine

Support vector machine (SVM) is a kernel-based technique which is successfully applied in the pattern recognition area and, is based on the principle of structural risk minimization (SRM) [8]. SVM constructs a linear model to estimate the decision function using non-linear class boundaries based on support vectors [9]. If the data are linearly separated, SVM trains linear machines for an optimal hyperplane that separates the data without error and into the maximum distance between the hyperplane and the closest training points. The training points that are closest to the optimal separating hyperplane are called support vectors. Through some nonlinear mapping SVM maps the input patterns into a higher dimensional feature space. SVM generally applies to linear boundaries. Three types of SVM kernel functions were used in this work are Polynomial, Gaussian and Sigmoidal respectively.

4.2   Hidden Markov Model

HMM is powerful statistical tool which is widely used in pattern recognition [10] [11]. Especially, the HMM has been developed extensively in speech recognition system over the last three decades [12]. There are two main reasons for choosing HMM in speech processing. First, the transition and duration parameters in HMM may properly reveal the evolution of features over time, which is very important in modeling speech/audio perception. Second, there are many kinds of variations of the HMM as well as experiences of using them which are developed in speech recognition researches. This makes HMM a mature technique to be applied in this research. Steps involved in implementing training and testing procedure is shown below.

Training Procedure:
1. Transition probability and emission probability is randomly initialized based on the observation (o=19)
2. Based on the randomized emission and transition probability the model is trained for the given word feature vector.

3. The new transition and emission probability are estimated based on the word vector.
4. The HMM for the individual word is represented by the parameters namely mean, co-variance, mixmat, emission, transition and LL.

Testing Procedure:

1. Loglikely hood ratio for each and every frame in the isolated word is computed against every model. The most likely model is considered as a winner.
2. Based on the frequency of winning HMM models the most occurring model is considered as the markov model representing that word's feat vector.

5   Evaluation

5.1   Database

Under controlled environment domain restricted speech sentences were recorded at sampling freq 16kHZ. A database of 240 different sentences containing words relevant to computer related terms were recorded from different speakers (6 Males and 6 Females). Each speech clip consists of 1 to 2 minutes. Thus utterances of individual words ranges around 1-2 seconds. Database description is shown in Table 1. Recorded word samples relevant to various domains are shown in Table 2.

Table 1. Database description

| Total Number of Speakers : 12 |
| Number of Sentences Recorded : 240 |
| No. of Isolated word in Recorded Sentences: 620 |

Table 2. Recorded word samples relevant to various domains

Terms	Speech Processing	Pattern Recognition	Image Processing	Medical Processing	Prepositions
No. of words	195	134	126	125	40

5.2   Feature Extraction and Modeling

In this work the pre-emphasized signal containing the continuous speech is taken for testing. Through VAD the isolated words are extracted from the sentences. Thus frames which are unvoiced excitations are removed by thresholding the segment size (100ms). Features such as LPCC and MFCC are extracted from each frame of size 320 window with an overlap of 120 samples. Thus it leads to 19 LPCCs and 39 MFCCs respectively which are used individually to represent the isolated word segment. During training process each isolated word is separated into 20ms overlapping windows for extracting 19 LPCCs and 39 MFCCs features respectively.

For specific usage of domains, the storage of entire words or sentences allows the system for high quality outcomes. In this work, words or sentences related to Speech processing, Pattern Recognition, Image Processing, Medical Processing and finally prepositions were stored in the database. By providing various initialization techniques for each modeling technique, the overall recognition accuracy is increased. Kernels in SVM and Gaussians in HMM are analyzed.

6   Experimental Results

Using VAD isolated words in a speech is separated which is discussed in section 2. N SVMs are created for each isolated word. For training, 620 isolated words from were considered. Hence, this results in 620 feature vectors each of 19 dimensional LPCC and 39 dimensional MFCC for 620 isolated words respectively.

The training process analyzes speech training data to find an optimal way to classify speech frames into their respective classes. The derived support vectors are used to classify speech data. For testing, 19 dimensional LPCC and 39 dimensional MFCC feature vectors (1 sec of speech file) were given as input to SVM model and the distance between each of the feature vectors and the SVM hyperplane is obtained. The average distance is calculated for each model.

The text corresponding to the query speech is decided based on the maximum distance. The same process is repeated for different query speech, and the performance is studied. The performances of SR for different kernels are compared for LPC, LPCC and MFCC acoustic features.

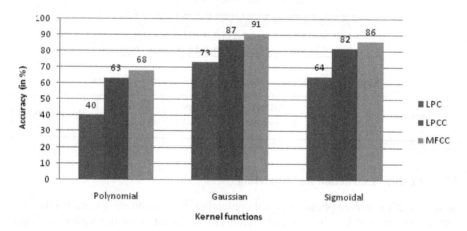

Fig. 3. Performance of SVM

In Hidden Markov Model each unit speech utterance is mapped on to states of HMM. Based on the given speech input the individual models in the HMM is matched to find the path. The path which follows maximum probability is selected as the appropriate model. By providing 4 Gaussians the results are premium as shown in graph. Gaussian in HMM is varied and its performance is analyzed. 4 gaussian in HMM provides better results when compared to other 1 and 2 gaussian.

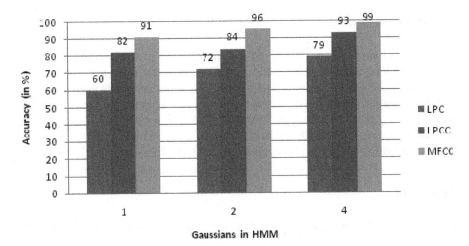

Fig. 4. Performance of HMM

Table 3. Confusion Matrix for the spoken word using SVM with MFCC Feature

Terms	Speech Processing	Pattern Recognition	Image Processing	Medical Processing	Prepositions
Speech Processing	183	7	5	0	1
Pattern Recognition	9	121	3	1	0
Image Processing	7	5	110	3	1
Medical Processing	3	3	1	117	0
Prepositions	1	0	2	0	37

Table 4. Confusion Matrix for the spoken word using HMM with MFCC Feature

Terms	Speech Processing	Pattern Recognition	Image Processing	Medical Processing	Prepositions
Speech Processing	190	3	2	0	1
Pattern Recognition	3	130	1	0	0
Image Processing	1	2	122	1	0
Medical Processing	0	1	2	122	0
Prepositions	0	0	0	0	40

Table 3 and 4 discusses the results of SVM and HMM using MFCC. For the chosen domain and database, SVM shows an accuracy of 93.7% for SP, PR results in 90.3%, IP results in 87.3%, MP as 93.6% and Preposition results in 92.5% and HMM shows an accuracy of 96.94% for SP, PR results in 97.01%, IP results in 96.83%, MP as 97.6% and Preposition results in 100%.

7   Conclusion

A system has been developed to convert spoken word into text using SVM and HMM modeling techniques using LPCC and MFCC features are extracted to model the words. Voice Activity Detection (VAD) is used for segregating individual words out of the continuous speeches. Features for each isolated word are extracted and those models were trained successfully. SVM and HMM were used to model each individual utterance. From the exhaustive analysis, SVM shows an accuracy of 91.46% for MFCC and HMM shows an accuracy of 98.92% using MFCC.

References

1. Anusuya, M.A., Katti, S.K.: Front end analysis of speech recognition: A review. Int. J. Speech Technology 14, 99–145 (2011)
2. Ramírez, J., Górriz, J.M., Segura, J.C.: Voice Activity Detection- Fundamentals and Speech Recognition System Robustness. Robust Speech Recognition and Understanding, 1–22 (2007) ISBN 978-3-902613-08-0
3. Das, B.P., Parekh, R.: Recognition of Isolated words using features based on LPC, MFCC, ZCR and STE with Neural Network Classifiers. International Journal of Modern Engineering Research (IJMER) 2(3), 854–858 (2012)
4. Herscher, M.B., Cox, R.B.: An adaptive isolated word recognition system. In: Proc. Conf. on Speech Communication and Processing, Newton, MA, pp. 89–92 (1972)
5. Thiang, W.S.: Speech recognition using linear predictive coding and artificial neural network for controlling movement of mobile root. In: International Conference on Information and Electronics Engineering (IPCSIT), vol. 6, pp. 179–183 (2011)
6. Bhattacharjee, U.: A comparative Study of LPCC and MFCC Features for the Recognition of Assamese Phonemes. International Journal of Engineering Research and Technology (IJERT) 2(1), 1–6 (2013)
7. Zolnay, A., Schulueter, R., Ney, H.: Acoustic feature combination for robust speech recognition. IEEE Transactions on Acoustics, Speech and Signal Processing, 457–460 (2005)
8. Chapelle, O., Vapnik, V., Bousquet, O., Mukherjee, S.: Choosing multiple parameters for support vector machines. Machine Learning 46, 131–159 (2002)
9. Duan, K., Keerthi, S., Poo, A.: Evaluation of simple performance measures for tuning SVM hyperparameters. Neuro Computing 51, 41–59 (2003)
10. Bourouba, H., Bedda, M., Djemil, R.: Isolated words recognition system based on Hybrid approach DTW/GHMM. Informatica 30, 373–384 (2006)
11. Baum, L.E.: An inequality and associated maximization technique in statistical estimation for probabilistic functions of Markov processes. Equalities 3, 1–8 (1972)
12. Reynals, S., Morgan, N., Bourland, H., Franco, R.: Connectionist probability estimators in HMM Speech Recognition. IEEE Trans. on Speech and Audio Processing 2(1), 161–174 (1994)

Energy Efficient Distributed Topology Control Technique with Edge Pruning

Mohasin Tamboli[1], Suresh Limkar[2], and Maroti Kalbande[3]

[1] Department of Information Technology, GHRCEM, Pune
[2] Department of Computer Engineering, AISSMS IOIT, Pune
[3] Department of Computer Engineering, JSPM, Hadapsar, Pune
mohasin.tamboli@raisoni.net,
{sureshlimkar,maruti.patil}@gmail.com

Abstract. Topology control mechanism plays a vital role while designing wireless sensor network and mobile ad-hoc control. Topology control mechanism should create topology with lower transmission power, sparser connectivity, with smaller node degree. Proposed paper presents analysis of different topology control mechanisms at present. Energy conservation is one of the main aim behind the topology control .This paper presents distributed topology control mechanism which is energy efficient and provides increases network lifetime. Proposed topology control algorithm gives better results in energy conservation as compared to existing algorithm.

Keywords: SBYaoGG, WSN, Topology Control, Energy Efficient, MANET.

1   Introduction

In the last decade growth of latest technology in wireless communications and electronics is responsible for the development of low-power, low cost, multifunctional sensor nodes that are small in size and converse untethered in short distances. These tiny sensor nodes consist of communicating components data processing unit and sensing module. The sensor node position in the sensor network need not be engineered or predestined. This implies arbitrary deployment of nodes in battle fields, inaccessible terrains and disaster relief operations. WSNs are powerful in that they are open to support a lot of live and real world applications that vary considerably in terms of their requirements and characteristics. In many industrial application network of sensors are exist which provide the ability to monitor and control the environment in real time.

2   Literature Survey

I.F. Akyildiz, W. Su,Y. Sankarasubramaniam et al., [1] discuss as latest technological growth in electronics and wireless communication which enables the development of low cost sensor network. In many areas like health, military, home these sensor

© Springer International Publishing Switzerland 2015   527
S.C. Satapathy et al. (eds.), *Proc. of the 3rd Int. Conf. on Front. of Intell. Comput. (FICTA) 2014*
– *Vol. 1*, Advances in Intelligent Systems and Computing 327, DOI: 10.1007/978-3-319-11933-5_59

network are used. Most of the application domain, there are poles apart technical problem exists that researchers are currently trying to resolve. Many state of art of the sensor network which are currently used is captured and protocol stack layer section related solutions are discussed Also points out the open research issues and intends to spark new interests and developments in this field.

A. Willig [2] discusses importance of selecting interesting research areas and promising areas in the design of protocols and the systems for wireless industrial communications which gives the increasing age of many industrial systems. E. Yoo et al [3] explains the collaborative nature of industrial wireless sensor networks brings several advantages over conventional wired industrial monitoring and control mechanism, including processing capability, self-organization, fast deployment and flexibility in organizing. In particular, energy harvesting methods, radio transmission technologies, and cross-layer design issues for WSN has been discussed.

V. C. Gungor and F. C. Lambert [4] explains the opportunities and challenges of hybrid network architecture. More specifically, Internet based Virtual Private Networks, power line communications, satellite communications and wireless communications are de-scribed in detail. The motivation is to provide a better understanding of the hybrid network architecture that can provide heterogeneous electric sys-tem automation application requirements. V. C. Gungor, B. Lu et al [5] gives an overview of the application of WSNs for electric power systems along with their opportunities and challenges and opens up future work in many unexploited research areas in diverse smart-grid applications. Then, it presents a comprehensive experimental study on the statistical characterization of the wireless channel in different electric power-system environments.

I.F.Akyildiz, T.Melodia et al [6] discussed the state of the art survey of algorithms, communication protocols, and hardware development for wireless multimedia sensor networks. Architectures for wireless multimedia sensor networks are explored, along with their advantages and shortcoming. A. Kouba, R. Severino[7] proposes H-NAMe, a very simple yet extremely efficient hidden-node avoidance mechanism for WSNs. Algorithm relies on splitting each cluster of a network into disjoint groups of nodes that scales to multiple clusters that guarantees no interference between overlapping clusters. S.Adee[8] explains H-NAMe, which is tested with IEEE 802.15.4/ZigBee, which is one of the most widespread communication technologies for WSNs.

L. Lobello and E. Toscano [9] presents a topology management protocol which changed dynamically i.e. dynamic by nature that overcomes the fixed approach introducing support for node joining at runtime , event driven data transmissions and providing a novel adaptive technique which tends to energy balancing among nodes thus increase network lifetime can be achieved . P. Santi [10] states topology control problems related to wireless ad hoc and sensor net-works, and explained survey state-of-the-art solutions which have been proposed to deal with controlling the topology of nodes. N. Li, J. C. Hou [11] presents topology control algorithm in wireless multi hop networks named a minimum spanning tree (MST)-based algorithm, called local mini-mum spanning tree (LMST). In the proposed algorithm, Independently each node builds its LMST and keeps only on-tree nodes which are one-hop away as its neighbors in the finally generated topology. Many important properties of LMST are proved analytically. LMST can increase the capacity of network as well as reduce the energy consumption proved by simulation.

L. Li, J. Y. Halpern [12] proposed CBTC i.e. cone-based distributed topology-control algorithm which works on directional information. The basic idea of the proposed algorithm is with minimum power each node transmits required to ensure that in every cone of aroung, there exist some nodes which can reach with degree power. Also propose few optimizations which guarantees of reduction in power consumption and confirm that they seize on to network connectivity. D. M. Blough, M. Leoncini et al[13] propose algorithm where each wireless node can select communication neighbor locally and accordingly adjust its transmission power so that all nodes together self form energy efficient topology for both uni-cast and broad-cast mechanism. Further shows that a small constant value say k is the expected average degree of all sensor nodes.

P. Santi and D. M. Blough [14] propose for energy efficient networking a dynamic, stable, distributed clustering. Effect of transmission power and mobility on network stability is evaluated from simulation. T. J. Kwon and M. Gerla [15] propose an approach where improvement of the network lifetime by determining the optimal assignment of nodes to cluster-heads. Authors presented different tactics to cluster-heads, which based on the minimum transmission power criterion.

V. Kawadia and P. R. Kumar [16] analyzed GAF i.e. Geographical adaptive fidelity (GAF) algorithm which minimizes energy consumption in ad hoc wireless networks . In the proposed method by identifying equivalent nodes and turning off nodes which are unnecessary energy conservation takes place. From simulations it is proved that network lifetime improves proportionally to node density. To extend overall system lifetime the nodes can also coordinate to exploit the redundancy provided by high density . A. Cerpa and D. Estrin [17] proposed ASCENT the large number of nodes deployed in these systems will preclude design-time pre configuration. Therefore, self-configuration of nodes happens to establish a topology that provides communication under rigorous energy constraints. ASCENT works on the basic idea that, only a subset of the nodes are necessary to generate a routing forwarding backbone as density increases . In ASCENT, each node assesses its connectivity and adapts its participation in the multi hop network topology based on the measured area. Results show that while providing adequate connectivity the system achieves linear improvement in energy conserving as a function of the density and the convergence time required in case of failure of nodes.

Dimitrios Katsaros, Yannis Manolopoulos et al [18] Presents a Proximity Graph of a set V on n distinct points in the plane is a straight-line drawing with V as vertex set, so 2 vertices on the region could be connected only if that region should be empty. Region is called as proximity region. i.e. it does not contain any other element of V. Proximity graphs have the property of being connected if the original graph is connected, which maintains worst case connectivity. Yao Graph is denoted by $YG_k(G)$ where k is an integer parameter greater than or equal to 6.At each node u k cones are defined by separating them equally. shortest edge from u suppose u v should be selected amongst all the existing edges. Ties are broken randomly.

Generated graph is called Yao Graph. Reverse Yao graph can be drawn by changing the order of edge vu. The relative network graph know as (RNG) connects a straight line between two points if and only if there is no other point in the set i.e closer to both the points which made straight line than they are two each other.

Table 1. Survey of topology control technique

Sr. No.	Technique	Advantage	Disadvantage
1	MST[11]	Increase network capacity,reduce energy consumption	Network may disconnected
2	CBTC[12]	Deals efficiently with recon- figuration and asynchrony	Less chance of minimum power consumption
3	GAF[16]	Efficient communication	Depend on location information system to calculate grid and allocation of nodes in that.
4	ASCENT[17]	Adaptive to react on dynamic event , self-reconfigurable	Due to unbalanced load distribution fast energy loss
5	Gabriel Graph[18]	Gabriel Graph Produces low Interference	Firstly, they might have a high degree as high as n-1. Second is its the large stretch factor, as large as $O(\sqrt{n})$

3   Proposed Model

Notice that the nodes do not connect to each other directly but through the wireless channel module(s). The arrows signify message passing from one module to another. When a node has a packet to send this goes to the wireless channel which then decides which nodes should receive the packet. The nodes are also linked through the physical processes that they monitor. There can be multiple physical processes, representing the multiple sensing devices (multiple sensing modalities) that a node has, as well as multiple wireless channels to represent the multiple radios (operating in an orthogonal manner, e.g., different frequencies or different codes) that a node might have. The node module is actually a composite one. Figure 1   shows the internal structure of the node composite module. The solid arrows signify message passing and the dashed arrows signify an interface between the modules with simple function calling. For instance, most of the modules call a function of the resource manager to signal that energy has been consumed.

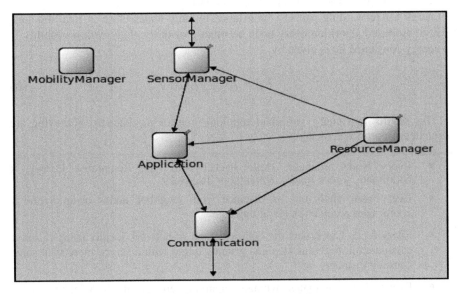

Fig. 1. The node composite module

In proposed method network topology is modelled as Graph. Original Graph G(S,E) where set S of Vertex(nodes) , E is set of edges(communication links), Gabriel graph GG(s', e') is a graph such that s' ⊆ S; e' ⊆ E, and Edge PQ ⊆ e' only when unit disc of which PQ is a diameter contains no other elements of S.

$$ARE = i/|GG| * \Sigma \, RE \, (i) \; \forall \; i \subseteq GG \qquad (1)$$

Where |GG| is number of nodes in GG, are is average remaining energy, RE(i) is remaining energy of node i Final Topology graph FTG (N,C) is graph such that

$$\{\exists N | N \subseteq s', RE \, of \, N >= ARE\} \qquad (2)$$

$$\{\exists C \subseteq e' \, | \, \forall \, P \, Q \subseteq C, P, Q \subseteq N\} \qquad (3)$$

Let m, is number of nodes in S i.e. m = |S| and n, number of nodes in final topology n = |N|, then P, probability of node being in final topology is given by

$$P = n/m \qquad (4)$$

Energy consumed by node Ec is given by

$$Ec = Es + Ef \qquad (5)$$

Where Es is energy required for sending and receiving packet for itself, Ef is energy required for forwarding packets (routing).

Energy for forwarding packet is consumed by only nodes those in final topology FTP. In proposed algorithm every node becomes part of topology with probability P. So energy consumed Ec is given by

$$Ec = Es + (Ef) * P \qquad (6)$$

This results into energy efficient topology control mechanism. Following are major steps in proposed method:

- Every Node broadcast Hello Packet with (ID, Remaining Energy, x coordinate, y coordinate) to Neighbor discovery.
- Every node finds out its distance from neighbor nodes using x and y coordinates received in Hello message.
- Every Node Constructs the Gabriel graph of network locally using distance calculated in previous step and pruning edges with distance more than some constant distance.
- Every Node calculates average remaining energy of all neighbors using collected data during neighbor discovery.
- Every node prunes edges from the topology with nodes having remaining energy less than some factor of average remaining energy producing the reduced topology.
- Above step is useful to save energy of low power nodes.
- Above process is repeated periodically to replace low energy nodes with better energy nodes.

In proposed method, use of Gabriel graph results into low interference as only start and end node will be included in the path. Proposed system will result into equal utilization of energy as Low energy nodes are avoided from being on path and topology is changed periodically. This will result into prolonged network lifetime. But in some cases end to end delay in communication may be little higher as nodes belong to shortest path might be dropped from topology due to lesser energy.

4   Proposed System and Results

Proposed algorithm is evaluated using OMNeT++ simulation building tool. Network area considered for simulation is 500 meter X 500 meter. Unit distance for Gabriel graph is taken as 50 meters. Numbers of nodes in network are taken as 20 and Initial energy of each node is considered as 20000. Fig. 2 and Fig. 3 shows results of simulation. Packets sent in both model is almost same, but simulation results reveal that Energy consumption in proposed model is reduced than existing model. This shows that throughput of system is also improved in proposed system.

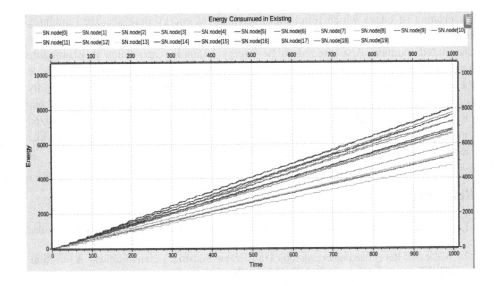

Fig. 2. Energy consumption in SBYoGG

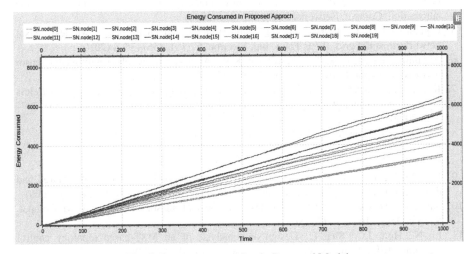

Fig. 3. Energy consumption in Proposed Model

5   Conclusion and Future Scope

From simulation results and mathematical analysis, it is clear that proposed algorithm provides energy efficient topology as energy consumption is in between 3500J to 8200J. Energy requirement of nodes are less than the existing algorithm so throughput of proposed system is improved over existing topology control mechanisms.

By pruning of edges which joins nodes having less energy than average remaining energy in the topology our proposed model provides topology control at local level in distributed manner. Also proved to be energy balancing and helps in prolonging network lifetime as nodes having less energy are keeping apart from the topology so there is very less chance of disconnection within the nodes of topology. So let's only better energy nodes to be the part of topology. In the future study, we can extend our proposed scheme by altering the topology control mechanism, we can use centralized topology control algorithm and at the same time we will use advanced algorithm like Genetic algorithm for getting more accurate result in terms of energy conservation.

References

1. Akyildiz, I.F., Su, W., Sankarasubramaniam, Y., Cayirci, E.: A survey on sensor networks. IEEE Commun. Mag. 40(8), 102–110 (2002)
2. Willig, A.: Recent and emerging topics in wireless industrial communications: A selection. IEEE Trans. Ind. Inform. 4(2), 102–122 (2008)
3. Yoo, S.-E., et al.: Guaranteeing real-time services for industrial wireless sensor networks with IEEE 802.15.4. IEEE Trans. Ind. Electron. 57(11), 3868–3876 (2010)
4. Gungor, V.C., Lambert, F.C.: A survey on communication networks for electric system automation. Comput. Networks 50(7), 877–897 (2006)
5. Gungor, V.C., Lu, B., Hancke, G.P.: Opportunities and challenges of wireless sensor networks in smart grid. IEEE Trans. Ind. Electron. 57(10), 3557–3564 (2010)
6. Akyildiz, I.F., Melodia, T., Chowdhury, K.R.: A survey on wireless multimedia sensor networks. Comput. Networks 51(4), 921–960 (2007)
7. Kouba, A., Severino, R., Alves, M., Tovar, E.: Improving quality-of service in wireless sensor networks by mitigating hidden-node collisions. IEEE Trans. Ind. Inform. 5(3), 299–313 (2009)
8. Adee, S.: IEEE Spectrum (2010),
   http://spectrum.ieee.org/semiconductors/devices/
   wireless-sensors-thatlive-forever
9. Lobello, L., Toscano, E.: An adaptive approach to topology management in large and dense real-time wireless sensor networks. IEEE Trans. Ind. Inform. 5(3), 314–324 (2009)
10. Santi, P.: Topology control in wireless ad hoc and sensor networks. ACM Comput. Surveys 37(2), 164–194 (2005)
11. Li, N., Hou, J.C., Sha, L.: Design and analysis of an MST-based topology control algorithm. IEEE Trans. Wireless Commun. 4(3), 1195–1206 (2005)
12. Li, L., Halpern, J.Y., Bahl, P., Wang, Y.-M., Wattenhofer, R.: A cone-based distributed topology-control algorithm for wireless multi-hop networks. IEEE/ACM Trans. Networking 13(1), 147–159 (2005)
13. Blough, D.M., Leoncini, M., Resta, G., Santi, P.: The K-Neigh protocol for symmetric topology control in ad hoc networks. In: Proc. Int. Symp. Mobile Ad Hoc Networking and Computing (MobiHoc), pp. 141–152 (2003)
14. Santi, P., Blough, D.M.: The critical transmitting range for connectivity in sparse wireless ad hoc networks. IEEE Trans. Mobile Computing 2(1), 25–39 (2003)
15. Kwon, T.J., Gerla, M.: Clustering with power control. In: Proc. IEEE Military Commun. Conf., MILCOM, vol. 2, pp. 1424–1428 (1999)

16. Kawadia, V., Kumar, P.R.: Principles and protocols for power control in wireless ad hoc networks. IEEE J. Sel. Areas Commun. 23(1), 76–88 (2005)
17. Cerpa, A., Estrin, D.: ASCENT: Adaptive self-configuring sensor net-works topologies. IEEE Trans. Mobile Comput. 3(3), 272–285 (2004)
18. Topology control algorithms for wireless sensor networks: A critical survey Alexis apadimitriou and Dimitrios Katsaros and Yannis Manolopoulos

Dynamic Cache Resizing in Flashcache

Amar More and Pramod Ganjewar

MIT Academy of Engineering, Pune, Maharashtra, India
{ahmore,pdganjewar}@comp.maepune.ac.in

Abstract. With an increase in the capacity of storage servers, storage caches have became necessity for better performance. For such storage caches, flash based disks have proved to be one of the best solutions for the better storage performance. With the use of flashcache layer as a part of storage servers it often needs to be expanded or scaled to handle a large work load without affecting the quality of service delivered. Resizing the cache space dynamically results in shutting down the servers for some time or transferring the workload to another server for some time and restarting the servers. It often hits the performance of servers while warming up caches for populating valid data in cache. In this paper, we introduce a solution for dynamic resizing of cache for Facebook Flashcache module, without affecting the performance. Our solution achieves resizing of flash cache on the fly, thus removing the need for restarting the server or transferring the loads to another server. It would also eliminate the need for warming up cache after resizing the cache.

1 Introduction

Growing need for performance and improvements in lifecycle and capacity of SSDs has encouraged many storage vendors to use flash based memory or SSDs as cache for their storage and database servers [2,3]. Such caches deliver significant performance gains for I/O intensive workloads at reasonable cost [1]. Flash cache is one of the best solutions for performance of storage systems provided by major storage solution providers [1]. Integrating such caching mechanisms within operating system can achieve the required throughput for I/O intensive workloads and thereby decreasing the complexity in design. Some storage vendors provide modules for their storage products that can be easily integrated within their systems [1].There are also some open source modules available like Facebook Flashcache [4] and Bcache [5], which can be used in open source operating systems like linux. Flashcache and bcache, both can be used as a block layer cache for linux kernel, but both have significant difference in their operations. Bcache supports multiple backing devices for a single cache device, that can be added or removed dynamically. Bcache is designed according to the performance characteristics of SSDs and makes possible optimal usage of SSDs in its scope, that are used as cache. Facebook Flashcache was originally designed to optimize database I/O performance by acting as block cache and later transformed for general purpose I/Os and other applications as well. Flashcache supports a variety of disks with respect to their I/O latency, to be used as cache. Any portable flash device, SSDs or even some rotational disks with higher speed can be used as cache. All data available on disk can be cached on such devices and made persistent.

In order to meet service level agreements of the clients or meeting specific I/O intensive workloads cache space often needs to be resized dynamically. This results in shutting down of server or transferring the workload to another server for some time while the cache space is resized and restarting the server. When the cache space is increased, newly created cache space is often empty and needs some hours to many days for warming up of cache [6]. Some algorithms were designed to warm up the cache, but it needs to maintain and trace some information like logs of data that is used frequently and some heuristics of data that may be used. This data is then used for warming up the cache that was newly created. In this paper, we present an efficient solution for dynamic resizing of cache based on Facebook Flashcache which eliminates the need for restarting the server.

The rest of the paper is organized as follows: Section two, consists of a brief working of Facebook Flashcache followed by Challenges identified in section three. In section four and section five, we have described our design and evaluation of performance results of our design.

2   Facebooks Flashcache

Flashcache is a device mapper target developed using the Linux Device Mapper. It is a block layer linux kernel module that can used to cache data on a disks with higher speeds than the disks used for secondary storage, allowing it to boost the I/O performance. Flashcache was primarily designed for InnoDB, later transformed to be used as general purpose. In flashcache, cache is divided into uniform sized buckets and I/Os on cache are mapped using a set associative hash. Each bucket consists of data blocks. Metadata is maintained separately for both, block and cache buckets which allows to handle it easily. Cleaning of sets is trigerred on two conditions, firstly when dirty blocks exceed the configured threshhold value and secondly when some blocks are not used from some time i.e. stale blocks.Cache block metadata consists of the state of the block i.e VALID or DIRTY or INVALID.

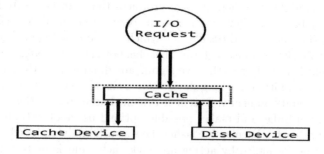

Fig. 1. Overview of Facebook Flashcache

Each I/O arriving at flashcache is divided into block sized requests by device mapper and forwarded to flashcache for mapping. Blocks in a bucket are stored by calculating its target bucket using a hash function. Hash function used for target set mapping can be given as:

$targetset = (dbn/blocksize/setsize)mod(numberof sets)$ After calculating the target bucket of a block, linear probing is used to find the block within that bucket.replacement policy used within bucket is FIFO by default, and can be changed on the fly via sysctl to LRU.

To maintain data persistence metadata is written onto the disk on a scheduled shutdown. In case of unplanned shutdown, only DIRTY blocks persist on cache, and needs warming up of cache only for VALID blocks.

3 Challenges Identified in Facebook Flashcache

In order to resize the cache in flashcache, we need to bring the system offline, resize the cache and restart and reconfigure the flashcache. Though flashcache provides persistence of cache in writeback mode, after resizing warming up of cache degrades the performance of the system. This is because increase in number of buckets in the cache would result in change in mapping of cache blocks to bucket. Following are the issues which are required to be considered while implementing the resizing

3.1 Remapping of Blocks to Their Appropriate Sets While Resizing

Block mapping in flashcache is done through linear probing within the target set which is calculated through a hash function. The hash function is provided with a start sector number as a key to calculate its respective target set. It requires total number of sets in cache. While resizing the cache dynamically, if device is added in the cache the total number of sets present in the cache will change. This difference in total number of sets results in inappropriate calculation of target set, which can ultimately lead to an inconsistent I/O operation. Thus we need to maintain consistent block and target set mapping during and after the resizing process.

3.2 Data Consistency during Resizing in Writeback Mode

In writeback mode of flashcache, data is written only on cache and later on lazily written to the disk in the background, while cleaning the sets. This cleaning is triggered on two conditions, when dirty threshold value of a particular set exceeds its configured threshold value and other way is, when block is not used for a longer period i.e. block lies fallow on cache. So whenever data is written on cache it is not reflected on disk immediately. Metadata update is done only on transition of block from dirty to valid or invalid or vice versa. Here the major challenge is to prevent data loss and incorrect metadata updation and also maintaining the consistency of data on cache, while resizing the cache. In writeback mode most of the data is present only on cache and we cannot bypass the cache even for a single I/O. Another challenge here is to handle the I/Os in an appropriate order while resizing without providing the inconsistent data.

4   Design for Dynamic Resizing

In existing flashcache implementation, only a single cache device is supported. While creating the cache, this cache device is configured and divided into sets and blocks. Arrays are used to keep track of the attributes of each set and block of the cache device. These arrays are created at the time of creation of the cache and every internal algorithm of flashcache depends on these arrays as shown in Figure 2

Dynamic resizing supports multiple devices to be added online. For this purpose we have maintained a list of devices instead of a single cache device. Each device in the list has its own array for sets and blocks which are created at the time of resizing. In order to keep internal algorithms intact, few mapping functions are added to introduce a virtual layer between cache device and the internal algorithms. This virtual layer enables Flashcache to work as if it is working with a single cache device.

Once cache is resized, number of sets in it gets resized which affects the hash function used to map a block of disk to a block of cache. As this change may introduce inconsistency in the data stored, we have implemented a basic framework for resizing. Later, we built an optimal resizing solution on top of the basic framework.

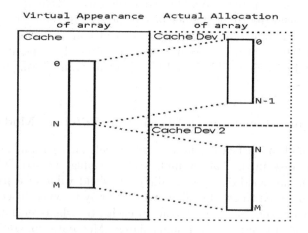

Fig. 2. Array Mapping

4.1 Basic Approach

In this approach, we follow a basic framework for resizing and considering all the complexities. The process begins by configuring the newly added cache device. Once the device is configured, it gets added to the list of cache devices.After adding cache device to the list, all incoming I/O requests are kept on hold. Complete cache is cleaned to avoid inconsistency due to re-mapping. Size of the cache and total number of sets in the cache are updated and I/O requests are released. As the cache is cleaned completely, count of cache miss will be higher on next I/O operations. Performance will slightly degrade unless the cache is refilled again.

4.2 Optimal Resizing Approach

The optimal resizing approach is more complex than the basic approach forresizing but performance is much higher. We have divided the cache sets into three logical categories viz. Trap set, Re-Mapped sets and Disguised sets. Properties for each category is as follows:

1. **Trap set(all blocks are being re-mapped according to their new position after resizing):** All I/Os arriving on this set are holded, until each block in this set is remapped on its original position. At a time only single set will be a trap set.
2. **Re-mapped Sets(all blocks belong to their original position after resizing):** All I/Os arriving on this set will be calculated with new hash function which uses updated number of sets after resizing.
3. **Disguised sets(all blocks in these sets are not in correct position after resizing and need re-mapping):** All I/O arriving on these sets will be mapped by calculating with old hash function.

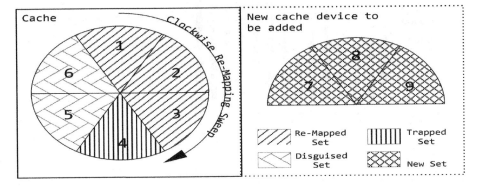

Fig. 3. Set States in Optimal Resizing Approach

As shown in Figure 3, the process for resizing begins by marking a set as trap set, and hold all I/Os arriving on that set only. Visit each block sequentially and apply new hash function on that block, now if that block maps on same set i.e trap set, leave it as it is. If block maps on a different set, it is marked as invalid in the current set and I/O is triggered for that block on its new position after remapping it on a new set. Likewise, we continue this procedure for blocks in trap set. Next, we remove set from trap, i.e. untrap the trapped set and it is marked as Re-Mapped sets and all I/Os arriving at these sets will be now calculated by new hash function. After removing the trap from previous set, next set is visited and it is marked as trap set, and similar procedure is followed. All the I/Os arriving at Re-Mapped set will be calculated using new hash. Until all the sets are remapped, we need to trap the sets and remap each block in it. When all the sets are remapped, we update the superblock and the total number of sets in cache context.

Summarizing the overall process, during resizing we use both the hash functions old hash and new hash, and after completing with resizing and updating the superblock and cache context, we use only new hash function.

5 Evaluation

We have tested Flashcache on a system with Intel dual core processor clocked at 2.6GHz and 4GB RAM. For disk device and cache device we have used loop devices. We have compared the performance of our implementation of flashcache with original flashcache having cache size of 2GB and disk size of 5GB. Our system was having cache size initially 1GB and we resized it dynamically by adding additional 1GB of cache and disk size of 5GB respectively.Following are the test results generated by IOzone benchmark for read /write.

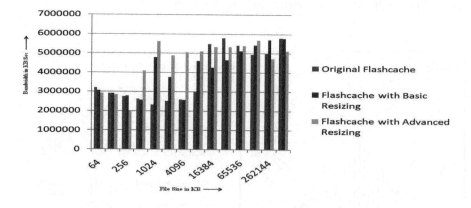

Fig. 4. IOzone Reader Report

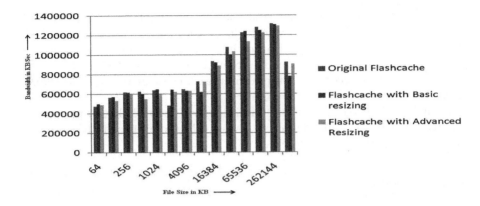

Fig. 5. IOzone Writer Report

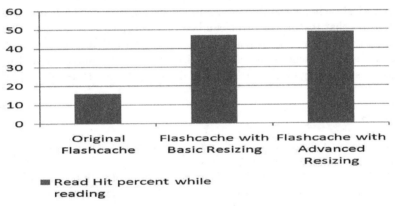

Fig. 6. Flashcache Read/Write Status Report

From the above results of IOzone as shown in Figure 4 and Figure 5, reader and writer, we can observe that our implementation of flashcache does not degrade the normal read write performance. Throughputs of all the variants of flashcache, especially original and advanced, are almost equal in average. Maintaining the actual performance, while introducing resizing was a necessity. Above charts show that the mapping functions in our implementation does not degrade the performance.

Flashcache maintains a count of all the read hits in a particular cache for displaying statistics of a cache. To confirm that our implementation offers a decent read hit percentage even after resizing, we tested it for following scenario.

We created three write back caches using original, basic and advanced flashcache implementations (one for each). For each cache, cache device was of 1GB and disk device was of 5GB. We wrote a 2.4GB file on each cache. Then we resized the caches created using basic and advance implementations to 2GB (added one more 1GB cache device to it). Basic implementation cleaned all blocks in the cache device and advanced implementation re-mapped all blocks. Then we started reading the same file from all three cache devices. Figure 6 shows the read hit percentage obtained from each cache at the time of reading this file.

It could be observed that advanced resizing gives a slightly higher percentage than basic resizing. Original flashcache gives small read hit percent because it is not resized from 1GB to 2GB cache.

6   Conclusion

Flash caches in storage servers prove to be one of the best solutions to boost the I/O performance in an efficient way in terms of cost and energy. However, such caches are often needed to be resized, which needs restarting of server and warming up caches. We have implemented a system for dynamic cache resizing in Facebook Flashcache, without affecting its performance. Cache hits are maintained to ensure that resizing of cache can be done without the need of warming up the cache. In the same way, resizing of the backing disk dynamically can also be useful in a scenario where one cache device needs to be shared by multiple backing devices. Resizing of the backing disk dynamically is left as future work.

References

1. Byan, S., Lentini, J., Madan, A., Pabon, L.: Mercury: Host-side flash caching for the data center. In: MSST, pp. 1–12 (2012)
2. Lee, S., Kim, T., Kim, K., Kim, J.: Lifetime management of flash-based ssds using recovery-aware dynamic throttling. In: Proc. of USENIX FAST 2012 (2012)
3. Oh, Y., Choi, J., Lee, D., Noh, S.H.: Caching less for better performance: balancing cache size and update cost of flash memory cache in hybrid storage systems. In: Proceedings of the 10th USENIX conference on File and Storage Technologies, FAST 2012, Berkeley, CA, USA, p. 25. USENIX Association (2012)
4. Saab, P.: Releasing Flashcache in MySQL at Facebook Blog, http://www.facebook.com/note.php?noteid=388112370932 (accessed April 27, 2011)
5. Stearns, W., Overstreet, K.: Bcache: Caching Beyond Just RAM, http://bcache.evilpiepirate.org/ (accessed July 2, 2012)
6. Zhang, Y., Soundararajan, G., Storer, M.W., Bairavasundaram, L.N., Subbiah, S., Arpaci-Dusseau, A.C., Arpaci-Dusseau, R.H.: Warming up storage-level caches with bonfire. In: Proceedings of the 11th Conference on File and Storage Technologies (FAST 2013), San Jose, California (February 2013)

Heuristic for Context Detection in Time Varying Domains

Sujatha Dandu[1] and B.L. Deekshatulu[2]

[1] Aurora's Technological and Research Institute, Hyderabad, AP, India
sujatha.dandu@gmail.com
[2] Distinguished Fellow, IDRBT, Hyderabad, AP, India
deekshatulu@hotmail.com

Abstract. The primary goal is to find information of co-existing commodities called itemsets in transactional databases. Especially in business to make a proper decision, the knowledge of high support itemset is very important. For example: A business can avoid giving discounts on an item which is more in demand though it is one of the commodities of the same itemset. This phenomenal product's information must be known to data analyst. As we see the change becomes mandatory as the season changes. So it has a great effect on buying habits of customers, not only on the season but also newly introduced merchandise [2, 5, 6]. The primary duty of data analyst is to detect these changes i,e which high-support itemsets withstand the change and which itemset among itself vanishes and which new itemsets emerge. To take this challenge we use a window of the latest market-basket which shows variation in its size time to time. The window grows in the periods of stability, producing an information of the current context. The window reduces in size, once the change is detected. The main objective of this paper is to introduce a new operator for controlling the window size.

Keywords: data stream mining, frequent itemsets, context change.

1 Introduction

The fast expansion of computers in the recent times facilitated individuals and organizations to accumulate and preserve huge amounts of information. The volume of data accumulated daily has been increasing day by day so that conventional statistical and database management systems are becoming inadequate to maintain huge amounts of data [7]. Also traditional approaches failed to estimate the future, since they use simple algorithms for estimating the future which does not give accurate results. So, Data Mining has emerged. Data mining extracts the data in different ways. For instance, if we take a company or business organization, we can predict the future of business in terms of Revenue (or) Employees (or) Customers (or) Orders etc.

An accepted and well researched subfield of data mining is association mining intended for discovery of interesting relations between variables in large databases.

© Springer International Publishing Switzerland 2015   545
S.C. Satapathy et al. (eds.), *Proc. of the 3rd Int. Conf. on Front. of Intell. Comput. (FICTA) 2014*
– *Vol. 1*, Advances in Intelligent Systems and Computing 327, DOI: 10.1007/978-3-319-11933-5_61

The objective of association mining is to look for commonly co-occurring patterns in transactional databases [1]. The main application of association mining is market basket analysis. A market basket is defined as a catalogue of items customers purchase as registered at the checkout desk. Association mining looks for items that are often set up in the similar market baskets. The main aim is to identify groups of items called "itemsets" that frequently co-exist in the same shopping carts. Our research deals with how the itemset supports can vary when customers buying habits change according to the fashion, seasons or newly introduced products. Supermarkets can then gain on or after knowing these changes in the system so that they place connected items on adjoining shelves, promote them in the similar catalogues, or keep away from price cut on more than one allied item [10] [8]. On the other hand, the function of association mining goes well further than market-basket scrutiny and takes in areas like the medical field; the items can possibly be the paperwork in the medical history of a patient. Association mining may be functional to several areas where the data can be articulated in the outline of transactions [9]. In the Internet situation, a transaction may possibly be created by means of links indicating to a Web page, and commonly co-occurring links which then signal connections between Web sites.

Next, we define "context" as a set of circumstances defined by seasons and demographic factors which change in time. When a new block of transactions arrive, the system updates the support values in the current list of itemsets and then decides whether this list has been significantly altered. Major change would indicate that the underlying context might have changed. Special attention has been devoted to the method of performance evaluation.

2   Existing System

Interestingly, based on the existing idea developed by [1] for the context alteration our methodology enumerate the differences between the suggested operators and our newly evaluated operator called the "Summation Operator" .In [3] large item sets come upon all the way through the complete database are marked as global large item sets, and on the other hand, item sets set up only in several parts of the database are considered local large item sets. The model domain comprises of three segments contexts. Every context contains various patterns –concepts–connected with it, and the concept is steady contained by the context. In [4] Contexts modify, either abrupt or steady, are assumed to happen only to a period of definite length. The set of recurrent item sets is updated by subsequently adding up the blocks to the window. In view of the fact that the window is supposed to have only those transactions pertinent to the present context, the system is capable to go again the explanation of the present context at every moment. The system primarily allows a new block and determines the supports of item sets in this block. Afterwards, it makes a decision whether the relations noticed in the new block considerably are at variance from those in the preceding blocks. Such important difference is then regarded as a suggestion of a probable context alteration. A lot will depend on the heuristics that identify the change in the context and the operators that control the window size. For the elimination of the older blocks subsequent to a context change, the following operators were used such as; 1.Harsh operator: If a change is noticed, restore the complete window contents by means of the newest block. 2. Reluctant operator: If a change is noticed, insert the novel block to the window and concern a caution that something suspicious is obtainable on. If the change is established even subsequent to the next block, put back the whole window contents by means of the last two blocks. 3. Opportunistic: If a change is identified remove x% of the blocks and if context change is confirmed even after the nest block replace the x% of the window with the last two blocks.

We suggest: (1) A new operator called the "Summation Operator" to find the detection of context changes in comparison of the existing ones of finding context change and (2) then evaluating the difference between the new and existing approaches using the Jaccard Coefficient and Frequency Distance.

3   Proposed System

The Summation operator compares the current block with both the previous block and next block and if the context change is identified; entire window contents are replaced with the latest block. We call it a summation as it uses the previous, current and the next block while detecting the change. We used two methods namely Jaccard Coefficient and Frequency Distance to evaluate Harsh, Reluctant, Opportunistic and Summation Operator. Jaccard Coefficient

It is popularly used method for measuring the similarity among two sets [11]. It is the ratio between the intersection of the new window and block and Union of new window and block. Let (LB) be the newly arrived block and (LW) be the window for new block, then Jaccard Coefficient is as follows:

$$Jacc(L_w, L_B) = \frac{L_W \cap L_B}{L_W \cup L_B}$$

This coefficient will change after each new block arrival. If the environment is not varying then the context will not change. This value is proportional to the similarity between the two high support itemsets. Once the change is signaled after the new block arrival, the Jacc(LW, LB) is smaller than what it was in the window.

Frequency Distance

Let fb_i be the support for the i-th itemset in new block and fw_i be the support for the i-th itemset in the window. The Frequency Distance [12] calculated as the difference between the new block and the last k blocks by the way of the differences in the supports of the individual

$$dif = \sqrt{\frac{1}{n} \sum_{i=1}^{n} (f_{bi} - f_{wi})^2}$$

4   Results

We used the synthetic data given by IBM generator[13] which models the supermarket domain. We worked on two sets of fimi data as inputs retail.txt and T1014D100K.txt which is taken as two dimensional string of data. The generated transactions contain itemsets created randomly and whose length is also generated randomly.

The Error rate for the operators that make use of flexible window size in the domains is calculated for a list of k itemsets as :

$$error = \sqrt{\frac{1}{k}\sum_{i=1}^{k}(f_{iT} - f_{iR})^2}$$

Where f_{iR} is the real support for the i-th itemset and f_{iT} is the theoretical support. The bar graphs below show that the error rate difference between the Harsh, Reluctant, Oppurtunistic and Summation Operator. The y-axis shows the error rate and the x-axis is the operators.

Error Computation For Jaccard Co-efficient

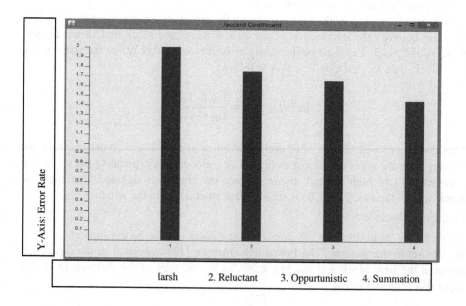

Error Computation for Frequency Distance

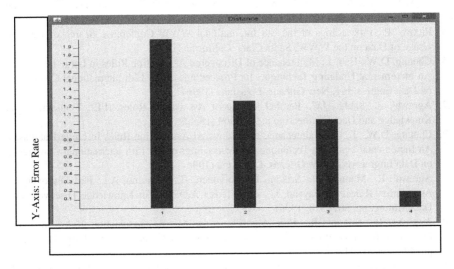

From the graphs it is clear that Summation Operator methodology is more effective than the other three methods.

5   Conclusion and Future Work

Association mining is intended for discovery of regularly co-occurring items within transactional databases. The set of recurrent item sets is updated by subsequent adding up each block to the window. Even though all versions of the program are competent of recognizing the moment of a sudden context alteration with very good correctness, the experiments with steady alteration put forward that methods to differentiate true transform from noise may possibly justify more consideration.

Apart from the search for high-support itemsets, a data analyst dealing with context-sensitive domains may also want to know how much the individual contexts differ from each other, what the frequency of the changes is, and which aspects of the data are most strongly influenced by the changes. We intend to address these questions in our future work.

References

1. Rozyspal, A., Kubat, M.: Association Mining in Time-varying Domain (2003)
2. Raghavan, V.V., Hafez, A.: Dynamic Data Mining. In: Proc. 13th International Conference on Industrial and Engineering Applications of Artificial Intelligence and Expert Systems, IEA/AIE, New Orleans, Louisiana, pp. 220–229 (June 2000)
3. Aumann, Y., Lindell, Y.: A Statistical Theory for Quantitative Association Rules. In: Proceedings of the 5th ACM SIGKDD International Conference on Knowledge Discovery and Data Mining, San Diego, August 15–18, pp. 261–270 (1999)

4. Bayardo, R.J., Agrawal, R.: Mining the Most Interesting Rules. In: Proceedings of the 5th ACM SIGKDD International Conference on Knowledge Discovery and Data Mining, San Diego, California, pp. 145–154 (1999)
5. Pitkow, P.: Proceedings of the 6th International WWW Conference Search of Reliable Usage of Data on the WWW, Santa Clara, California (1997)
6. Cheung, D.W., Han, J.: Maintenance of Discovered Association Rules in Large Databases: An Incremental Updating Technique. In: Proceedings of the 12th International Conference on Data Engineering, New Orleans, Louisiana (1996)
7. Agrawal, R., Shafer, J.C.: Parallel Mining of Association Rules. IEEE Transactions on Knowledge and Data Engineering 8, 962–969 (1996)
8. Cheung, D.W., J., H.: Maintenance of Discovered Association Rules in Large Databases: An Incremental Updating Technique. In: Proceedings of the 12th International Conference on Data Engineering, New Orleans, Louisiana (1996)
9. Agrawal, R., Mannila, H., Srikant, R., Toivonen, H., Verkamo, A.I.: Fast Discovery of Association Rules. In: Fayyad, U., et al. (eds.) Advances in Knowledge Discovery and Data Mining, pp. 307–328 (1996)
10. Agrawal, R., Srikant, R.: Fast Algorithms for Mining Association Rules in Large Databases. In: Proceedings of the 20th International Conference on Very Large Databases, pp. 478–499 (1994)
11. Jain, A.K., Dubes, R.C.: Algorithms for clustering Data. Prentice Hall, Englewood Cliffs (1988)
12. Jobson, J.D.: Applied Multivariate Data Analysis. Categorical and Multivariate Methods, vol. II. Springer Verlag,
13. IBM Generated fimi datasets, http://fimi.ua.ac.be/data/'

Word-Level Script Identification from Handwritten Multi-script Documents

Pawan Kumar Singh[*], Arafat Mondal, Showmik Bhowmik,
Ram Sarkar, and Mita Nasipuri

Department of Computer Science and Engineering,
Jadavpur University, Kolkata, India
pawansingh.ju@gmail.com

Abstract. In this paper, a robust word-level handwritten script identification technique has been proposed. A combination of shape based and texture based features are used to identify the script of the handwritten word images written in any of five scripts *namely, Bangla, Devnagari, Malayalam, Telugu* and *Roman*. An 87-element feature set is designed to evaluate the present script recognition technique. The technique has been tested on 3000 handwritten words in which each script contributes about 600 words. Based on the identification accuracies of multiple classifiers, Multi Layer Perceptron (MLP) has been chosen as the best classifier for the present work. For 5-fold cross validation and epoch size of 500, MLP classifier produces the best recognition accuracy of 91.79% which is quite impressive considering the shape variations of the said scripts.

Keywords: Script identification, Handwritten *Indic* scripts, Texture based feature, Shape based feature, Multiple Classifiers.

1 Introduction

India is a multi-lingual country where people reside at different sections use different languages/scripts. Each script has its own characteristics which is very different from other scripts. Therefore, in this multilingual environment, to develop a successful Optical Character Recognition (OCR) system for any script, separation or identification of different scripts beforehand is utmost important because it is perhaps impossible to design a single recognizer which can identify a variety of scripts/languages. Script identification facilitates many important applications such as sorting the document images, selecting appropriate script specific OCR system and searching digitized archives of document images containing a particular script, etc.

Resemblances among the character set of different scripts are more feasible for handwritten documents rather than for the printed ones. Cultural differences, individual differences, and even differences in the way people write at different times, enlarge the inventory of possible word shapes seen in handwritten documents. Also,

---

[*] Corresponding author.

problems like ruling lines, word fragmentation due to low contrast, noise, skewness, etc. are common in handwritten documents. In general, the visual appearances of the script vary from word to word, and not from character to character. Therefore, the identification of the scripts at word level are more preferable than at character, text line or page-level.

In the context of *Indic* script identification, most of the published methodologies [1-5] have been discussed about printed text documents. A few number of research works [6-9] are available on handwritten text words. Despite these research contributions, it can be noticed that most of researchers have addressed only bilingual or trilingual scripts. But, in a multilingual country like India, this is a pure limitation considering usage of large number of scripts. So, in Indian context, a script recognition system should include more number of *Indic* scripts. This is the primary motivation behind the development of word-level script identification technique for five *Indic* scripts *namely, Bangla, Devnagari, Malayalam, Telugu* along with *Roman* script. We have considered *Roman* script as their use is frequently seen in advertisements, cinemas, and text messaging nowadays.

2   Design of Feature Set

In the present work, different shape and texture based feature are extracted from the word images written in said 5 different scripts.

2.1   Shape Based Features

The one-dimensional function, derived from shape boundary coordinates, is often called shape signature [10-11]. This function usually captures the perceptual feature of the shape. Some of the commonly used shape signatures used in the present work are described below.

Complex Coordinates
A complex coordinates function is simply the complex number generated from the coordinates of boundary points, $P_n(x(n), y(n))$, $n \in [1; N]$. It is expressed as:

$$z(n) = [x(n) - g_x] + i[y(n) - g_y] \tag{1}$$

where, (g_x, g_y) is the centroid of the shape. Summation of complex coordinates values of each pixel on the contour is given by Eqn. 2.

$$sum_{z(n)} = \sum_{i=1}^{n} z(i) \tag{2}$$

where, n is the number of contour pixels. Normalized forms of real and imaginary parts of $sum_{z(n)}$ are taken as feature values F1 and F2 respectively.

Centroid Distance Function

The centroid distance function $r(n)$ is expressed by the distance of the boundary points from the centroid (g_x, g_y) of a shape.

$$r(n) = [(x(n) - g_x)^2 + (y(n) - g_y)^2]^{\frac{1}{2}} \tag{3}$$

Summation of $r(n)$ values of the corresponding word image is taken as feature value F3. Due to the subtraction of centroid from the boundary coordinates of the position of the shape, both complex coordinates and centroid distance representation are invariant to translation.

Tangent Angle

The tangent angle function at a point $P_n(x(n), y(n))$ is defined by a tangential direction of a contour which can be written as:

$$\theta(n) = \theta_n = tan^{-1} \frac{y(n) - y(n - w)}{x(n) - x(n - w)} \tag{4}$$

Since, every contour is a digital curve; w is a small window used to calculate $\theta(n)$ more accurately. Summation of positive values of $\theta(n)$ is taken as feature value F4.

Slope Angle

Slope angle [12] between two points $P_1(x_1, y_1)$ and $P_2(x_2, y_2)$ is calculated by Eqn. (5).

$$Slope = tan^{-1} \frac{y_2 - y_1}{x_2 - x_1} \tag{5}$$

The slope angle is calculated at a point $P_n(x(n), y(n))$ inside a small window w; with midpoint P_n. Slope is calculated at point P_n with its neighboring pixels and its summation is taken as slope at pixel P_n. Summation of slopes of each contour pixel is considered as feature value F5.

Area Function

When the boundary points change along the shape boundary, area of the triangles formed by two successive boundary points and the centroid also change. Suppose $S(n)$ be the area between the successive boundary points P_n, P_{n+1} and centroid G. Area of triangle is calculated using Eqn.6.

$$Area = \sqrt{s * (s - s_1) * (s - s_2) * (s - s_3)} \tag{6}$$

where, s is the semi-perimeter and s_1, s_2 and s_3 are the length of three sides of a triangle. Summation of these areas is considered as feature value F6.

Triangle Area Representation (TAR)

The TAR feature is computed from the area of the triangles formed by the points on the shape boundary. For each three consecutive points $P_{n-t_s}(x_{n-t_s}, y_{n-t_s})$, $P_n(x_n, y_n)$ and $P_{n+t_s}(x_{n+t_s}, y_{n+t_s})$, where $n\epsilon[1, N]$ and $t_s \in [1, N/2 - 1]$ and N is even, the signed area of the triangle formed by these points is given by:

$$TAR(n, t_s) = \frac{1}{2} \begin{vmatrix} x_{n-t_s} & y_{n-t_s} & 1 \\ x_n & y_n & 1 \\ x_{n+t_s} & y_{n+t_s} & 1 \end{vmatrix} \tag{7}$$

Summation of these TAR values of a particular word image is taken as feature value F7.

2.2   Texture Based Features

In general, shapes of the different word images written in a particular script generally differ from word images written in other script. This gives different scripts distinctively different visual appearances. Texture could be defined in simple form as "repetitive occurrence of the same pattern". For the present work, we have used one of the popular texture based features named as Histograms of Oriented Gradients which is described below.

Histograms of Oriented Gradients (HOG)

HOG descriptor [13] counts occurrences of gradient orientation in localized portions of an image. The essential thought behind the HOG descriptors is that local object appearance within an image can be described by the distribution of image gradients or edge directions.

An image gradient is a directional change in the intensity or color in an image. Gradients of a pixel $p(x, y)$ along X-axis and Y-axis are computed using Eqns.(8-9) (for illustration, see Fig. 1) respectively.

$$Gradient_x(x, y) = p(x, y + 1) - p(x, y) \tag{8}$$

$$Gradient_y(x, y) = p(x + 1, y) - p(x, y) \tag{9}$$

Magnitude and direction of a pixel $p(x, y)$ (for illustration, see Fig. 2) are calculated using Eqns. (10-11).

$$Magnitude_{p(x,y)} = \sqrt{Gradient_x(x, y)^2 + Gradient_y(x, y)^2} \tag{10}$$

$$Direction_{p(x,y)} = \tan^{-1} \frac{Gradient_y(x, y)}{Gradient_x(x, y)} \tag{11}$$

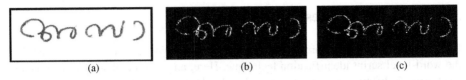

(a)   (b)   (c)

Fig. 1. (a) Sample handwritten *Malayalam* word image, and its gradient along (b) X-axis, (c) Y-axis

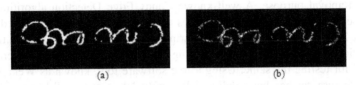

(a)   (b)

Fig. 2. Illustration of (a) Magnitude, and (b) Direction of the word image shown in Fig.1(a)

After the computation of values of the magnitude and direction, each pixel is pigeonholed in certain category according to its direction. The image window is initially divided into 10 small spatial regions, called "cells", and for each cell, a local 1-D histogram of the directions of gradient or edge orientations of the pixels of the cell is accumulated. The entire range of direction (0^0 to 360^0) is then divided into 8 evenly spaced sectors where each sector represents an orientation bin in the histogram. For avoiding the distortion due to presence of noise in the word image, the contribution of the pixel is calculated by sharing the magnitude of each pixel with its 2 nearest bins. Let β and $2\pi/\beta$ denote the angles of the two closest bins for a particular pixel [13]. Then, the contribution of the pixel to the closest bins can be written as $Magnitude_{p(x,y)} \times \left[1 - \frac{T\beta}{2\pi}\right]$ and $Magnitude_{p(x,y)} \times \frac{T\beta}{2\pi}$ respectively. This is illustrated in Fig. 3.

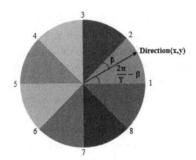

Fig. 3. Illustration of 8 orientation bins for computation of HOG features

After applying the 8 orientation bins scheme on 10 distinct cells gives a total of 80 feature values (F8 to F87).

Hence, altogether a set of 87 (7+80) features has been designed for the present script identification work.

3   Experimental Results and Discussion

For the evaluation of the current work, a total of 3000 words have been collected for the word-level script identification technique. Here, an equal number of word images written in 5 different scripts *namely, Bangla, Devnagari, Malayalam, Telugu* and *Roman* have been considered. The original gray tone word images are binarized using Otsu's global thresholding approach [14]. After that, Gaussian filter [15] is used for the noise removal purpose. A well-known Canny Edge Detection algorithm [15] is then applied for smoothing the outer edges of contour of the binarized word images. A total of 2000 words (400 words per script) have been used for training the script identification system whereas the remaining 1000 words (200 words per script) have been used for testing the same. Using a free software tool, known as Weka [16], the designed feature set has been individually applied to seven well-known classifiers *namely*, Naïve Bayes, Bayes Net, MLP, Support Vector Machine (SVM), Random Forest, Bagging and MultiClass Classifier. The success rates of the said classifiers and their corresponding scores achieved at 95% confidence level are shown in Table 1.

Table 1. Recognition accuracies of the proposed script identification technique using seven well-known classifiers

	Classifiers						
	Naïve Bayes	Bayes Net	MLP	SVM	Random Forest	Bagging	MultiClass Classifier
Success Rate (%)	73.86	78.73	**89.29**	86.6	84.5	80.46	83.17
95% confidence score (%)	83.7	85.63	**95.78**	92.70	90.03	89.1	90.06

Though MLP classifier (as evident form Table 1) outperforms other classifiers, with proper tuning it could produce better results. For this purpose, we have used 3-fold, 5-fold and 7-fold cross validation schemes with different number of epochs of MLP classifiers (see Table 2). From the Table 2, it is observed that for 5-fold cross validation with epoch size 500, the MLP produces the best identification accuracy of 91.79%. The confusion matrix obtained for this best case on the test dataset is shown in Table 3.

Observing the misclassified word images, it can be said that the possible reasons for this are: (1) small words (containing 2-3 characters) which produces less discriminating feature values, (2) presence of skewness in some of the word images (see Fig.4 (a)) and (3) structural similarity in some of the characters of different scripts. Also, discontinuities of Matra in certain words of *Bangla* and *Devnagari* script sometimes appear as *Roman* script words. On the other hand, existence of Matra like structure (found usually at the upper part of most of the characters) in *Roman* script misclassifies them as *Bangla* or *Devnagari* script. The reason for misclassification of *Malayalam* and *Telugu* scripts are mainly due to existence of abrupt spaces in between characters of a single word (see Fig.4 (b-c)).

Table 2. Recognition accuracies of the present script identification technique for different folds and different number of epochs of MLP classifier (the best performance is shaded in grey)

| | Success Rate of MLP classifier (%) | | |
| | #-Fold | | |
Number of Epochs	3-fold	5-fold	7-fold
500	90.12	**91.79**	89.57
1000	88.76	91.23	89.3
1500	90.56	89.63	87.67

Table 3. Confusion matrix produced for the best case of MLP classifier

Scripts	Bangla	Devnagari	Malayalam	Telugu	Roman
Bangla	553	7	15	13	12
Devnagari	5	585	2	1	7
Malayalam	10	4	564	16	6
Telugu	58	3	9	525	5
Roman	33	25	8	7	527

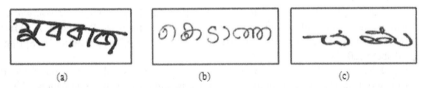

(a)   (b)   (c)

Fig. 4. Sample handwritten word images written in *Bangla, Malayalam* and *Telugu* scripts misclassified due to: (a) presence of skewness and (b-c) presence of abrupt spaces as *Roman, Telugu* and *Bangla* scripts respectively

4   Conclusion

Script identification from handwritten text image is an open document analysis problem. It is necessary to identify the script of handwritten text words correctly before feeding them to corresponding OCR engine. Research in the field of script identification aims at conceiving and establishing an automatic system which has the ability to discriminate a particular script from other scripts. In this paper, we proposed a robust word-level script identification technique for some handwritten *Indic* script documents along with *Roman* script using a combination of shape based and texture based features. This work is first of its kind to the best of our knowledge as far as the number of scripts is concerned. As the texture based features used in this technique show discerning power, in future, this technique could be applicable for recognizing other scripts in any multi-script environment. In future, we also plan to improve the accuracy of the system by minimizing the script dependency of the features.

References

1. Chanda, S., Pal, S., Pal, U.: Word-wise Sinhala, Tamil and English Script Identification Using Gaussian Kernel SVM. IEEE (2008)
2. Chanda, S., Pal, S., Franke, K., Pal, U.: Two-stage Approach for Word-wise Script Identification. In: Proc. of 10th International Conference on Document Analysis and Recognition (ICDAR), pp. 926–930 (2009)
3. Patil, S.B., Subbareddy, N.V.: Neural network based system for script identification in Indian documents. Sadhana 27(pt.1), 83–97 (2002)
4. Pati, P.B., Ramakrishnan, A.G.: Word level multi-script identification. Pattern Recognition Letters 29, 1218–1229 (2008)
5. Dhanya, D., Ramakrishnan, A.G., Pati, P.B.: Script identification in printed bilingual documents. Sadhana 27(pt.1), 73–82 (2002)
6. Roy, K., Majumder, K.: Trilingual Script Separation of Handwritten Postal Document. In: Proc. of 6th Indian Conference on Computer Vision, Graphics & Image Processing, pp. 693–700 (2008)
7. Sarkar, R., Das, N., Basu, S., Kundu, M., Nasipuri, M., Basu, D.K.: Word level script Identification from *Bangla and Devnagari* Handwritten texts mixed with Roman scripts. Journal of Computing 2(2), 103–108 (2010)
8. Singh, P.K., Sarkar, R., Das, N., Basu, S., Nasipuri, M.: Identification of *Devnagari* and *Roman* scripts from multi-script handwritten documents. In: Maji, P., Ghosh, A., Murty, M.N., Ghosh, K., Pal, S.K. (eds.) PReMI 2013. LNCS, vol. 8251, pp. 509–514. Springer, Heidelberg (2013)
9. Singh, P.K., Sarkar, R., Das, N., Basu, S., Nasipuri, M.: Statistical comparison of classifiers for script identification from multi-script handwritten documents. International Journal of Applied Pattern Recognititon 1(2), 152–172 (2014)
10. Zhang, D., Lu, G.: A comparative study of fourier descriptors for shape representation and retrieval. In: Proc. of 5th Asian Conference on Computer Vision (2002)
11. Kauppinen, H., Seppanen, T., Pietikainen, M.: An experimental comparison of autoregressive and fourier-based descriptors in 2-D shape classification. IEEE Transactions on Pattern Analysis and Machine Intelligence 17(2), 201–207 (1995)
12. Mingqiang, Y., Kidiyo, K., Joseph, R.: A survey of shape feature extraction techniques. In: Yin, P.-Y. (ed.) Pattern Recognition, pp. 43–90 (2008)
13. Dalal, N., Triggs, B.: Histograms of Oriented Gradients for Human Detection, http://lear.inrialpes.fr
14. Ostu, N.: A thresholding selection method from gray-level histogram. IEEE Transaction on Systems Man Cybernet. SMC-8, 62–66 (1978)
15. Gonzalez, R.C., Woods, R.E.: Digital Image Processing, vol. I. Prentice-Hall, India (1992)
16. http://www.cs.waikato.ac.nz/ml/weka/documentation.html

A Filtering Based Android Malware Detection System for Google PlayStore

Jithin Thomas Andoor

TIFAC-Core in Cyber Security
Amrita VishwaVidyapeetham
andooranmailbox@gmail.com

Abstract. The paper proposes an advanced filter based android malware detection platform that can be implemented as an add-on to the Google-playstore – the official provider of android applications worldwide. The mechanism uses the signature based, behavioral based technique and the advanced sandboxing technique for detection. It also uses the application rating and provider reputation into account, so as to filter out the input given to the system, this mechanism if implemented efficiently, in long run can be a very effective method to detect the malware when an application is published on the application stores.

Keywords: Android, malware, Signature, Behavioral.

1 Introduction

Malware is a problem that exists since the first generation of personal computers. Initially the attackers mainly concentrated on the operating systems. But through the introduction of Smartphone with complex and multitasking operating system on it, the scenario has drastically changed. Today complete personal information and critical data are handled by the Smartphone. Due to the same reason, most of the attackers find the Smartphone based malware attack more profitable.

Since the first launch of the android OS, it has changed a lot in its structure and popularity. When first introduced, it was an out of the box mobile operating system when compared to the market leaders at that time. The support for java applications to a very great extent with the inbuilt Dalvik virtual machine (DVM) increased its popularity. It has a lot of effective malware prevention mechanism which are inherited from the Linux operating system. But when the attackers changed their motive from mere operating system based attack to fetch personal and critical data, these methods were not effective as before. There is also a fact that even though android is releasing updates with lot of security updates, only about 4% of the total android phones currently being used have updates rolled out to them, availing the rest of the phones vulnerable to known attacks. Hence we see the importance of implementing a server side cloud based malware detection system rather than hoping for the users to be updated. Works is done in the field of android malware detection systems from late 2010. But the main issue with the research on this field is the limitation that the operating system itself has, along with the power and hardware incapability. Many of the

© Springer International Publishing Switzerland 2015
S.C. Satapathy et al. (eds.), *Proc. of the 3rd Int. Conf. on Front. of Intell. Comput. (FICTA) 2014*
– *Vol. 1*, Advances in Intelligent Systems and Computing 327, DOI: 10.1007/978-3-319-11933-5_63

researchers today mainly focuses on one type of malware detection mechanisms, i.e., behavioral or signature base [4][5]. They are implemented on the cloud or the host. But the issue of using both individually is, if a behavioral based analysis is not done the detection of malware with new signatures is a difficult process. Rather if signature based system is not used iterating behavioral based system continuously will lead to unwanted process cost all the time. The Idea proposed in this paper is a combination of both Behavioral and Signature based analysis along with the reputation based scheme for filtering the input.

The rest of this paper is organized as follows.Section 2 gives more insight to the field related to android malware detection and the related works that is being used. Section 3 describe the framework and Section 4 conclusion is done

2   Related Works

The securing model for Android application [1] is a Behavioral based system. The system introduced a concept called AMDA by which automated malware detection is done. Different type of application apk from varies websites including the Android market were collected.for the downloading of check applications; a web-crawler tool was employed. The applications are forwarded to the VirusTotal Malware Verification System (VMS) [6] to classify so that it can be used in later modules. The major drawback of this system was its incapability of classifying malware into completely different classifications aside from simply classifying it as malware. Moreover this method was not fully developed as a client server interactive system rather it behaved as a system that gives an output for an input.

Detecting repackaged applications in third party Android marketplaces [2] is an application source code analysis based mechanism. The purpose of the work is not really meant for the Google Playstore rather it is for the third party marketplace which doesn't have the provision of validating the application and its developer using signature verification. The detection is being done by crosschecking the Androidmainfest.xml file of the suspicious application and the original application downloaded from the Android official market. It is always encouraged to not to widespread the security mechanism to a number of market, since most of them are being managed by private parties. For the same reason sticking on to the official market always remove the above mentioned issue.

AASandboxing[8] is the mechanism that is implemented to enhance the capability of the traditional antivirus software's for the android. Here android kernel is customized and added a new module that will pass on all the system calls to the framework. These system calls were used by the architecture to make a detailed behavioral analysis. The framework has done a static analysis using the signatures of the malwares detected using the behavioral analysis. The logs that obtained were converted to an expression and were used to analyze the behavior of the application. The major issue with this type of architecture was it needed a proper filtering mechanism on the top layer so as to filter out the trusted systems.

Multi-Level Anomaly Detector for Android Malware (MADAM)[10] is a behavioural based system implemented in the mobile platform itself. The system follows a very simple architecture where it extracts 13 features of an application from time to time to detect its behaviour. MADAM can self-adapt to new behaviours by including new elements in the training set learnt at run-time. The system was also able to detect unwanted outgoing SMS stealthily sent by Android malicious applications. But the issue is that the system in long run will reduce the performance rate of the phone since the application starts to expand itself by increasing the datasets and classification elements. Moreover it is not acceptable to run the application on the mobile all the time since it requires a considerable computation speed and power.

3   Framework for the Android malware Detection System

3.1   System Arhitecture

The system is completely divided into 3 scheme namely reputation, signature and behavioral. The output is to obtain the classified result .i.e., classifying the application to white lists and blacklists. This list is being used by that host to detect the malware. The malware detection and classification is being done completely on the cloud. Each time the white list and black list are updated or a new application is being installed on the host the hosts send its whole application package names to the servers. The server compares it with the white and black lists and informs the area that needs necessary remedies.

In reputation based mechanism during first phase, application key is verified and if the provider is reputed (depending on star and feedback) the application is being allowed to pass-through the malware detection mechanism. The next phase is where the application reputation is taken into account, here the success of that application in the Google play store along with its popularity is taken into account, depending on this the application is allowed to bypass the detection mechanism.

In the signature based mechanism the permissions are used along with the basic signatures of the known malware algorithms. Apriori algorithm is used to find deviation of permission required for a specific class of application. If an application taken satisfies the required threshold, it is allowed to bypass, else it's moved to the next stage. The final phase of this classification is the Behavioral analysis. The behavioral analysis has a signature checking on kernel level and application level along with the sandboxing technique.

The advantage of this kind of architecture is that it enables us to utilize the pros of both the Signature and Behavioral scheme. Once an application is found to be a malware, its behavioral characteristics can be extracted and used as a signature for the malware of same type, thereby reducing the running cost of behavioral analysis. Since this is a Cloud based system with all the hosts as client, malware detection from one client is enough to notify all the remaining clients. More over since the detection is being done on the Cloud the device in- capabilities need not be taken into account. The major problem that we may face at the initial stage is to classify all the application in the Google market. To solve this the process first place all the application in the white list and as the client sends reports to the server it take each .apk files and pass it to the malware detection system.

The system is broadly classified into 5 levels along with a database set for the white listed and blacklisted applications. The training set data that is to be fed for the classification algorithm is given in Fig.1.

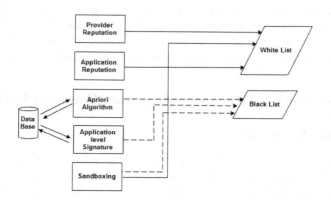

Fig. 1. System Architecture

The first 2 levels are basically for filtering mechanism, which is to filter out the input to the system using the reputation of the application provider or the application provider itself. In the 3rd level, the permissions of the application are extracted and apriori algorithm is applied to find the support and confidence of the given set of permission. If the value is above a threshold, applications are made to pass through this module to the next level. On the other hand those applications failing to attain this value are added to the blacklist. The behavioral analysis scheme is used to detect the malware failed to detect in the level. The application .apk file is being decompressed and the signatures are extracted. These signatures are then compared with the data set of the well-known malware algorithms in the android platform. This enables us to detect the malware following the scheme "old wine in new bottle". The final module in the detection is the sandboxing mechanism. This method enables us to see through the impact of the application at the kernel level and its effect on the operating system itself. Thus preparing the training data set is a difficult job since the efficiency of the whole mechanism depends on them. Methods are incorporated in this mechanism for self learning of new signatures.

3.2   Design and Implementation

In the design and implementation phase the whole system is divided into five modules namely the provider reputation, app reputation, permission set analysis, behavioral analysis, and sandboxing.

Provider Reputation: Normally Google play-store categorizes the application developers as top developers and others hence we can use this categorization of Google play-store to decide whether an application has to be trusted or not.

Fig. 2. Provider Reputation and App Reputation

Fig.2 clearly shows the application Whatsapp messenger can be allowed to be placed in the white list since the Whatsapp Inc is a top developer for Google. To get the information about the provider we rely upon the Google play-store api. The Google normally don't share the application details to external applications for security reasons. For the same reason there is no official api for the Google play-store. Retrieving the application details is a must for our module, for the same reason we can use an unofficial Android-market-api which is capable of doing this work [3].

App Reputation: The next stage is where the Application reputation is taken into account. The most successful applications in the Google play-store are made to by-pass the malware detection mechanism.

In fig.2 mechanism uses the same method as the above module to complete its work. The advantage of removing these kinds of applications is that it clearly focus on the true threat in the Google play-store. A threshold barrier is set for both the number of people installed the application and the rating of the application. If it crosses both the barrier the application is assumed to be safe due to its success in the market

Permission Set Analysis: Every class of applications has a set of permissions that they require to complete its task. The Permission-Based Android Malware Detection [4] is a type where the permissions are extracted and machine learning is done to classify it as whether a malware or not. The process comprises of selecting a collection of trusted applications of each class and applying apriori algorithm to it, there by generating support and confidence value for each permission combination. These values are used to classify whether an application is a malware or not. If the application crosses the threshold it is send to next level of the malware detection else it is blacklisted.

Apriori Algorithm for permission set

Step 1: Select a set of applications of a class
Step 2: Log their permissions as inputs
Step 3: Select all distinct permissions as candidates
Step 4: Take the count for each candidates and apply the threshold value
Step 5: Now take the distinct combinations of candidates and update candidate table
Step 6: Repeat step 4
Step 7: If candidate table contain only one value stop the process

Behavioural Analysis: Fig.4 shows the process that was completed with Random Forest algorithm to make the classification. The training sets were obtained by taking an apk file and checking it with Virustotal [6]. This procedure was done to obtain the well know datasets of different malwares for the android platform. In Behavioural Analysis the extraction of the api calls in the program code and their sequence is done. The sequence by which the api calls are done is much essential because we can utilize a dataflow analysis algorithm [7].

Random forest Algorithm (A variant of bagging)

Step 1: Select ntree, the number of trees to grow, and mtry, a number no larger than   no of variables
Step 2: For i = 1 to ntree:
Step 3: Draw a bootstrap sample from the data. Call those not in the bootstrap sample   the "out-of-bag" data.
Step 4: Grow a "random" tree, where at each node, the best split is chosen among mtry randomly selected variables. The tree is grown to maximum size and not pruned back.
Step 5: Use the tree to predict out-of-bag data.
Step 6: In the end, use the predictions on out-of-bag data to form majority votes.

Prediction of test data is done by majority votes from predictions from the ensemble.

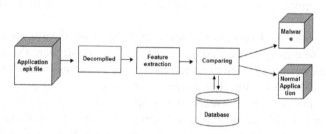

Fig. 3. Behavioral Analysis process

Sandboxing: The sandboxing technique comprised of logging all the system calls in the kernel level. This enables us to monitor the system activities at the lowest level i.e. the system call level. It is essential because Schmidt et al. [9] showed that it is possible to place piggybag Linux malware into Android systems. A loadable kernel module (LKM) was placed in the Android emulator environment [8]. With that all the System calls were intercepted. The android monkey testing tool was used to generate a stream of user input and obtain the system calls log of the application. The obtained log was converted to a vector so that it would be easy for the analysis part.

The method introduced here tries to solve the malware detection at various levels and always try to reduce the input to the mechanism by globally accepted methodologies. The introduction of the reputation based filtering out mechanism not only reduce the input to the system but enable the system to spend more time on the real threat without spending time on applications that has been trusted globally.

3.3   Results and Discussions

The different modules in the proposed architecture were previously implemented and tested. The selection of various Behavioural analysis mechanisms and sandboxing techniques were selected in to the system due to its simplicity and effectiveness. The Modules used in the Signature analysis and reputation based mechanism were needed to be subjected to testing.

The reputation based module was implemented using the Approximate string matching algorithm that is done on the application page on the google playstore. The mechanism showed perfect response since no classification or other type of complex algorithms are involved here. The system looked for "top Developer" string in the applications page and if found classified it as trusted. But this cannot be a trusted method to do the procedure. Google api are to used to extract these informations so that chances of tampering the data can removed.

The Signature analysis was using Apriori Algorithm to classify the data. The System was not highly precise to classify the data but it was effective to perform a good filtering. Table 1 shows the result analysis of Signature analysis. Here FPR is False positive rate, TPR is true positive rate and ACC is Overall Accuracy.

Table 1. Performance Matrix

TPR	.933
FPR	.40
ACC	.85

4   Conclusion

This paper introduced a mechanism that tried to incorporate all the possible type of malware detection mechanism into a single framework at the same time always tried to be focused on real threat. The advantage of this paper is that it is meant to be implemented with the existing architecture of google playstore. There by removing the clients who install the application from the duty to check whether the application is a malware or not. As future work, different ways by which optimised dataset can obtained is to be done.

References

1. Kesavan, M., Sharma, L.R.: Securing Models for Android Market-Places. International Journal of Advance Research 1(7) (July 2013), ISSN 2320-9194
2. Zhou, W., Zhou, Y., Jiang, X., Ning, P.: Detecting repackaged applica-tions in third party Android marketplaces. In: CODASPY 2012, San Antonio, Texas, USA, February 7–9 (2012)
3. Android-market-api, (Online accessed on January 15, 2014), https://code.google.com/p/android-market-api/
4. Aung, Z., Zaw, W.: Permission bases Android Malware Detection. International Journal of Scientific & Technology Research 2(3) (March 2013)
5. Pamuk, O., Jin, H.: Behavioural analysis model and decision tree classification of malware for android. International Journal of Advance Research 1(7) (July 2013) ISSN 2320-9194.
6. VirusTotal (Online accessed on January 15, 2014), https://www.virustotal.com/
7. Enck, W., Octeau, D., McDaniel, P., Chaudhuri, S.: A Study of Android Appli-cation Security. In: Proceedings of the 20th USENIX Security Symposium, USENIX Security
8. Bläsing, T., Batyuk, L., Schmidt, A.-D., Camtepe, S.A., Albayrak, S.: An Android Appli-cation Sandbox system for suspicious software detection. In: 2010 5th International Conference on Malicious and Unwanted Software (MALWARE), October 19-20, pp. 55–62 (2010)
9. Schmidt, A.-D., Schmidt, H.-G., Batyuk, L., Clausen, J.H., Camtepe, S.A., Al-bayrak, S., Yildizli, C.: Smartphone malware evolution revisited: Android next target? In: Proceedings of the 4th IEEE International Conference on Malicious and Unwanted Software (Malware 2009), pp. 1–7. IEEE (2009)
10. Dini, G., Martinelli, F., Saracino, A., Sgandurra, D.: MADAM: A multi-level anomaly de-tector for android malware. In: Kotenko, I., Skormin, V. (eds.) MMM-ACNS 2012. LNCS, vol. 7531, pp. 240–253. Springer, Heidelberg (2012)

Construction of Automated Concept Map of Learning Using Hashing Technique

Anal Acharya[1] and Devadatta Sinha[2]

[1] Computer Science Deparment, St. Xavier's College, Kolkata, India
anal_acharya@yahoo.com
[2] Computer Science and Engineering Deparment,
University of Calcutta, Kolkata, India

Abstract. It has been observed that the growth of communication technologies has led to increased use of computer networks in dissemination of education. An important application area in this perspective where there has been a lot of research is Intelligent Tutoring Systems (ITS). ITS aid the process of learning and evaluation of attainments without human intervention. However ITS are unable to pin point the exact area in a lesson plan where the student is deficient in. In this context, several researchers have used concept maps to perform this identification. However it is time consuming for the educators to construct a concept map of learning manually. Several data mining algorithms have thus been used by the researchers to generate association rules which are used for automated concept map construction. This study proposes automated construction of concept maps using Direct Hashing and Pruning algorithm. The proposed approach was tested with a set of students enrolled in an introductory Java course in some undergraduate colleges in Kolkata and was found to diagnose their learning problems satisfactorily.

Keywords: DHP Algorithm, Concept Maps, Association rules, Learning problems, E-Learning, Remedial Learning Plan.

1 Introduction

In recent times the most important innovation that has changed the face of educational technology is web-based education. This has been possible due to tremendous advancement in the field of computer networks and communication technology. Simultaneously there has been lot of progress in the field of E-Learning as well due to the fact that learners can learn independently of any specific location and platform. One of the particular areas of e-learning that has attracted a lot of researchers is Intelligent Tutoring Systems (ITS). ITS aid the process of learning and evaluation of attainments without human intervention [14]. The first major contribution in this area has been the work of Johnson [12] in which he constructed a authoring environment for building ITS for technical training for IBM-AT class of computers. Another contribution has been the work of Vasandani [13] where he built an ITS to organize system knowledge and operational information to enhance operator performance. The

© Springer International Publishing Switzerland 2015
S.C. Satapathy et al. (eds.), *Proc. of the 3rd Int. Conf. on Front. of Intell. Comput. (FICTA) 2014*
– *Vol. 1*, Advances in Intelligent Systems and Computing 327, DOI: 10.1007/978-3-319-11933-5_64

introduction of mobile devices has added a new dimension to the field of E-Learning. Mobility supplemented to E-Learning gave birth to a new field of research and application domain called Mobile Learning (M-Learning). There has been a lot of work in ITS in M- Learning environment as well. In 2005, Virvou et al [9] implemented a mobile authoring tool which they called Mobile Author. Once the tutoring system is created it can be used by the learners to access learning objects as well as the tests. Around the same time Kazi [10] proposed Voca Test which is an ITS for vocabulary learning using M-Learning approach.

However as identified by Chen-Hsiun Lee et al[6], learner evaluations conducted via online tests in a ITS do not provide a complete picture of student's learning as they show only test scores. They do not help the learner identify the exact area where he is deficient. For this purpose we wish to develop a method for automatic construction of concept maps using Direct Hashing and Pruning (DHP) [5] Algorithm. DHP is used to generate association rules between the concepts, which are in turn used to generate the concept map of learning. These concept maps are used to identify the concepts the student is deficient in. These concepts are called the learning problems of a student.

Concept maps were first used by Hwang [3,4] to generate learning guidance for students in science courses. In his work he created a Concept Effect Relationship (CER) builder which was used for automated construction of concept maps. He tested the system with Physics, Mathematics and Natural Science courses and proved statistically that CER indeed provides better learning diagnosis than traditional ITS. Lee et al[6] in their work has used the Apriori algorithm to generate concept maps which has been used to generate learning guidance. The system built by them is called Intelligent Concept Diagnostic System (ICDS). ICDS generates the Remedial Instruction Path (RIP) for providing proper learning guidance. Tseng et al [16] has proposed a Two-Phase Concept Map Construction (TP-CMC) algorithm. Phase 1 is used to mine association rules from input data whereas phase 2 uses these association rules for creating concept maps. They also developed a prototype system of TP-CMC and used real testing records of students in a junior school to evaluate the results. The experimental results showed that TP-CMC approach works satisfactorily. An agent based system was developed by Ching-Ming Chen [17] to generate learning guidance. This system again used the Apriori algorithm to generate association rules. The system developed was named Personalized E-Learning System (PELS). It was used for diagnosing common misconceptions for a course on Fractions. An advanced form of concept maps called pathfinder networks have been used by Chen [8] to develop personalized diagnosis and remedial learning system (PDRLS). The system was tested with 145 students enrolled in introductory JAVA programming language courses at Central Taiwan Technology University. The experimental results indicate that those students who used PDRLS obtained better learning results. Kohonen's self-organizing map algorithm has been used by Hagiwara[15] to generate Self-organizing Concept Maps (SOCOMs). Computer simulation results done by him have shown the effectiveness of the proposed SOCOM.

From the above discussion it is clear that various algorithms have been used by the researchers for generation of association rules to provide learning guidance. These algorithm works by constructing a candidate set of large itemsets based on some heuristics and then discovering the subsets that contain these large itemsets [5]. Thus the performance of the heuristic to construct the large itemset is critical to the performance of the algorithm. Further research [17] have shown that initial size of the candidate set especially for large 2-itemsets is the key performance improvement issue. A hash-based technique can be used to reduce the size of the candidate itemsets. Thus in this work we have used the Direct Hashing and Pruning (DHP) algorithm proposed in [5] to generate large itemsets and consequently the association rules between the concepts.

The development of concept maps[1] can be traced to the theory of Meaningful Learning proposed by David Ausubel [2,11] in 1963. In meaningful learning the learner is able to relate the new knowledge to the relevant concept he already knows. This psychological foundation led to the development of Concept Maps[1] by Joseph D Novak in Cornell University in 1972. Since then concept Map has been used by a lot of researchers to diagnose the learning problems of students. In brief, let C1 and C2 be two concepts. If learning concept C1 is a prerequisite to learning concept C2 then the rule C1 → C2 exists [4] as shown in Fig 1. Similarly learning concept C2 is prerequisite to learning C3. Thus if a student failed to learn C3 it is due to the failure of mastery over the concepts C1and C2. Associated with the rule C1→C2 there is a confidence level w, which states that if the student fails to understand C1, then the probability for him failing to understand C2 is w [19].

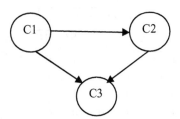

Fig. 1. A Concept Map of learning

The organization of the paper is as follows. The next section discusses generation of concept map of learning using DHP algorithm in details. We then illustrate how this concept map can be used to provide learning guidance to students. Finally, the proposed approach was tested with a set of students enrolled in a introductory Java course in some undergraduate colleges in Kolkata and was found to diagnose their learning barrier satisfactorily.

2   Proposed Approach

Our approach generates association rules between the concepts using DHP Algorithm proposed by Park et al[5]. As discussed earlier this algorithm has several advantages over other data mining algorithms that has been used by several researchers [3,6,8,16,17,18]. Our algorithm has four steps which are discussed below with the help of an example:

(i) Accumulation of Student Results

Let us suppose that there are 4 students S1,S2,S3,S4 who appear at a test containing 5 test items Q1,Q2,Q3,Q4,Q5. Their answer summary is shown in Table 1. We note that 1 denotes a wrong answer whereas 0 denotes a correct answer. This table is called Answer Sheet Summary Table (ASST) [3] and stores the collection of students answer for a test. In order to apply DHP algorithm on ASST we convert it into a table displaying wrong answer consolidated student wise (Table 2).

(ii) Determining the Appropriate Hashing Method and Using DHP to Compute the Association Rules between Test Items

We now apply the DHP Algorithm for efficient large item set generation. Let Ki denote the candidate set of large i-1 item sets. Thus K1={{Q1}, {Q2},{Q3},{Q4},{Q5}}. Computing the count of each of these test items from table 2 and assuming min support as 2 we get the set L1 as {{Q1},{Q2},{Q3},{Q5}}. Next, the 2-itemset of the test items is generated. These are shown student wise in Table 3.

Table 1. ASST for a set of students

| | Students | | | | |
Question	S1	S2	S3	S4	Total
Q1	1	0	1	0	2
Q2	0	1	1	1	3
Q3	1	1	1	0	3
Q4	1	0	0	0	1
Q5	0	1	1	1	3

Table 2. Student wise wrong answer

Student	Questions wrongly answered
S1	Q1,Q3,Q4
S2	Q2,Q3,Q5
S3	Q1,Q2,Q3,Q5
S4	Q2,Q5

Table 3. Student wise 2-itemsets generated from L1

Student	2-itemsets
S1	{Q1,Q3},{Q1,Q4},{Q3,Q4}
S2	{Q2,Q3},{Q2,Q5},{Q3,Q5}
S3	{Q1,Q2},{Q1,Q3},{Q1,Q5},{Q2,Q3},{Q2,Q5}{Q3,Q5}
S4	{Q2,Q5}

For each student, after occurrences of all the 1-subsets are counted, all the 2-subsets of this test item set are generated and hashed into the hash table. There are several types of hash functions that may be used here. We investigate the utility of three types of hash functions for generation of K2. Firstly, Modulo Division method may be used for generating bucket address. If the hash function generates bucket i, the count of bucket i is increased by one. Secondly, Shift Fold method partitions the key value into several parts of equal length and all parts are added to generate the hash address. Thirdly, the Mid square method generates hash address by squaring the key and then extracting the middle digits. Our investigation reveals that the hash table obtained by Modulo Division method is most compact with minimal wastage of space. We use the hash function

$$h(x,y)=((\text{order of }x)\text{X}10+(\text{order of }y))\text{mod } 7 \qquad (1)$$

This yields the hash table shown in Table 4.

Table 4. Hash table generated by modulo division method

Bucket Address	Bucket Count	Bucket Contents
0	3	{Q1,Q4},{Q3,Q5},{Q3,Q5}
1	1	{Q1,Q5}
2	2	{Q2,Q3},{Q2,Q3}
3	0	NULL
4	3	{Q2,Q5},{Q2,Q5},{Q2,Q5}
5	1	{Q1,Q2}
6	3	{Q1,Q3},{Q3,Q4},{Q1,Q3

From the set L1 we construct the 2-item set by computing L1XL1. We also find the bucket count of each of these item sets. Again assuming min support=2 we get the set K2 as {{Q1,Q3},{Q2,Q3},{Q2,Q5},{Q3,Q5}}. Combining K1 and K2, we get the large itemset table along with their support shown in Table 5.

Table 5. Large itemsets generated so far

Item Set	Support	Item Set	Support
{Q1}	2	{Q1,Q3}	2
{Q2}	3	{Q2,Q3}	2
{Q3}	3	{Q2,Q5}	3
{Q5}	3	{Q3,Q5}	2

Corresponding to each of the 2-itemsets we deduce the association rules. The selection of appropriate association rules is done by computing the confidence corresponding to each rule. The confidence level of the test item association rule Q1→Q2 is w implies that if the student answers the question Q1 wrongly, then the probability for him to answer the question Q2 wrongly is w. It is computed using the formula

$$\text{Conf }(Q1\rightarrow Q3) = P(Q1/Q3)=\text{support}(Q1UQ3)/\text{Support}(Q1)=100\% \qquad (2)$$

Similarly,

Conf (Q3→Q1)= 66%, Conf (Q2→Q3)= 66%, Conf (Q3→Q2)= 66%,
Conf (Q2→Q5)= 100%, Conf (Q5→Q2)= 100%, Conf (Q3→Q5)= 66%,
Conf (Q5→Q3)= 66%.

Assuming a minimum threshold of 60% all the rules is chosen. This method is described in Algorithm 1.

Algorithm 1: Association Rule Construction between Test Items
Input: ASST
Output: Association Rules between Questions

Step 1: Construct {K1} containing the original test item set {Qi}
Step 2: Min support=x
  For each Ki
  2.1 if count(Ki)>x
  Copy Ki to Lj
Step 3: Compute the 2-itemsets of the questions from ASST
Step 4: for all buckets set bucket count to zero
  for each 2-itemsets generated in step 3
  4.1 choose suitable hash function
  4.2 compute bucket address using this hash function
  4.3 increment appropriate bucket count
Step 5: Compute L1XL1 along with the corresponding bucket count
  For all item sets in L1XL1
  5.1 if bucket count>x
    Store it in {K2}
Step 6: Construct association rules from {K2}
  For each association rule
  6.1 compute confidence using formula (2)
  6.2 if confidence>M
  Copy it in {A}
Step 7: Return {A}

(iii) Deriving Association Rules between Concepts

Our aim now is to deduce the association rules between the concepts and their respective weights. For this we need to define the Test Item Relationship Table (TIRT) [3,4]. TIRT stores the level of correspondence between Test item Qi and Concept Cj. These are represented in fractions and hence the sum total of these values for a concept corresponding to a test item is 1. A typical TIRT is represented in Table 7. Lee et al[6] has proposed a method for computation of relative weights between the concepts for Apriori algorithm from confidence between test items. It can be adapted for DHP algorithm also. As an example,

$$Q1 \rightarrow Q2 \rightarrow W_{C1C2} = C1 \rightarrow C3 = Conf(Q1 \rightarrow Q3) * R_{Q1C1} * R_{Q3C3} = 1*0.75*0.6 = 0.45. \quad (3)$$

Thus the relative weight between the concepts C1 and C3 is 0.45. It also indicates that the concept C1 should be learnt before concept C3. Also if a student fails to learn C1 then the probability that he fails to learn C3 is 0.45. The entire set of association rules between the concepts and their relative weights are deduced in Table 6.

(iv) Construction of Concept Map of Learning

Based on the above association rules the preliminary concept map of learning can be constructed. However, on having a look at Table 6 we find that the cycle C3→C2→C5→C3 exists. It also shows that C2 is a prerequisite for learning C3 and vice versa which is ambiguous. In this cycle the edge C2→C3 contains the minimum weight. Thus we remove this edge to obtain the final concept map shown in Fig 2.

Table 6. Generation of association rules between concepts and their relative weights

Association rule between Test Items	Corresponding Association rule between Concepts	Formula used for computation of relative weight	Value of relative weight
Q1→Q3	C1→C4	Conf(Q1→Q3)*R_{Q1C1}*R_{Q3C4}	0.30
Q3→Q5	C2→C5	Conf(Q3→Q5)*R_{Q2C5}*R_{Q5C5}	0.19
Q2→Q3	C2→C3	Conf(Q2→Q3)*R_{Q2C2}*R_{Q2C5}	0.16
Q3→Q2	C3→C2	Conf(Q3→Q2)*R_{Q3C3}*R_{Q2C3}	0.18
Q5→Q3	C4→C3	Conf(Q5→Q3)*R_{Q5C4}*R_{Q3C3}	0.14
Q1→Q3	C5→C3	Conf(Q1→Q3)*R_{Q1C5}*R_{Q3C3}	0.15

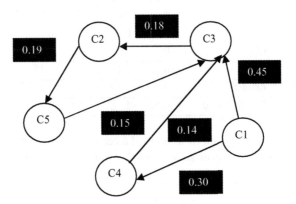

Fig. 2. Final Concept Map

The entire process is represented by the following algorithm:

Algorithm 2: Generate Concept Map
Input: Association rules between test items, TIRT
Output: Concept Map of learning
Step 1: Call Association Rule Construction between Questions Algorithm.
Step 2: For each association rule generated in step 1
    2.1 Compute the association rules between concepts using formula (3) and store them in {R}
    2.2 Compute the corresponding weight
Step 3: for each association rule generated in step 2
    3.1 Check for cycles in R using cycle detection algorithm
    3.2 If a cycle exists remove the rule with lowest weight from R
Step 4: Return R

3   Use of the Above Method to Perform Remedial Learning

In this section we illustrate with the help of a suitable example how the concept map of learning generated using DHP algorithm can be used for diagnosing student learning problems. We assume that a particular student answered the test items Q3 and Q4 wrongly. Using this information and TIRT we intend to identify the concept the student is deficient in. This is illustrated in the Table 7.

Table 7. TIRT used for generating student learning barrier

	C1	C2	C3	C4	C5
Q1	0.75	0	0	0	0.25
Q2	0	0.45	0	0	0.55
Q3	0	0	0.6	0.4	0
Q4	0	0.5	0.5	0	0
Q5	0	0	0	0.35	0.65
S	0.75	0.95	1.10	0.75	1.35
Error(i)	0	0.5	1.10	0.40	0
Error(Ci)	0	0.52	1	0.53	0

In Table 7, S denotes the sum of weights of a certain concept. Error(i) is the sum of weights corresponding to questions answered wrongly by the student. Error(Ci) is computed using the formula

$$Error(Ci)=Error(i)/S \qquad (4)$$

This value is termed as error ratio. If the value of error ratio is greater than 0.5, it means that more than 50% of the concept have been learnt poorly. We thus assume a threshold error of 0.6, i.e. if error ratio corresponding to a concept is greater than this value then the student is deemed to have failed to learn this concept. From the above table we find that corresponding to concept C3 the value of error ratio is greater than 0.6. Thus the student failed to learn the concept C3 properly. From the concept map shown in Fig 2 we find that the path C1→C3 and C4→C3 exists, which indicates that the student failed to learn the concept C3 as he had failed to master the concepts C1 and C4. We call this path as Remedial Learning Path (RLP). Thus after generating the concept map we may apply the connected components algorithm on this graph to find out which of the components contain an edge to C3.

```
.......Calculating the confidences of itemsets.......

ItemSet                    CalCulation              Percentage

Confidence (Q1->Q2)          = 2/3                  =  66%

Confidence (Q2->Q1)          = 2/4                  =  50%

Confidence (Q1->Q3)          = 2/3                  =  66%

Confidence (Q3->Q1)          = 2/4                  =  50%

Confidence (Q2->Q3)          = 2/4                  =  50%

Confidence (Q3->Q2)          = 2/4                  =  50%

Confidence (Q1->Q8)          = 2/3                  =  66%

Confidence (Q8->Q1)          = 2/3                  =  66%

Confidence (Q2->Q5)          = 3/4                  =  75%

Confidence (Q5->Q2)          = 3/3                  = 100%

Confidence (Q3->Q6)          = 2/4                  =  50%

Confidence (Q6->Q3)          = 2/2                  = 100%

Confidence (Q3->Q8)          = 3/4                  =  75%

Confidence (Q8->Q3)          = 3/3                  = 100%

Confidence (Q3->Q5)          = 2/4                  =  50%

Confidence (Q5->Q3)          = 2/3                  =  66%

Confidence (Q2->Q8)          = 2/4                  =  50%

Confidence (Q8->Q2)          = 2/3                  =  66%

Confidence (Q5->Q8)          = 2/3                  =  66%

Confidence (Q8->Q5)          = 2/3                  =  66%
```

Fig. 3. Snap shot of the association rules between Test Items and their Confidences generated by a Java Program

4   Experiments

An experiment was conducted to understand how the proposed diagnostic mechanism could be used to generate concept map of learning and identify student learning problems. The course 'Introduction to Java Programming' was offered to students majoring in Computer science in some undergraduate colleges in Kolkata. The entire course curriculum was divided into 12 concepts and corresponding to these concepts 10 test items were designed. The experiment was conducted on 8 students chosen randomly from a set of 60 students in a class. The concepts were identified as follows:

C1=Variables and Data types, C2=Operators, C3=Library Functions, C4=Loops, C5=Branches, C6=Arrays, C7=Functions, C8=Constructors, C9=Classes and Objects, C10=Exception Handling, C11=Inheritance, C12=File Handling.

Hwang [20] has proposed a method for computation of TIRT. We however, have constructed a TIRT based on general understanding of subject concepts. A *Java Program* was written to generate a set of association rules and their confidences between Questions based on Algorithm 1. A snap shot of this output is shown in Fig 3. Using these association rules and TIRT, association rules between the concepts are generated along with the corresponding weights. After removing the rules with lowest weights which create cycles we get the following association rules which can be used to create the final concept map of learning shown in Fig 4: C1→C2, C2→C4, C2→C5, C3→C4, C3→C5, C4→C6, C5→C6, C6→C9,C7→C9, C8→C11, C9→C11,C10→C12, C11→C12. We assume that a particular student failed to answer the test item Q2 correctly. Using the method specified in section 3, we compute that the student failed to learn the concept C4 properly. From the concept map of learning generated above, we immediately conclude that this is perhaps due to the lack of mastery in concepts C2 and C3.

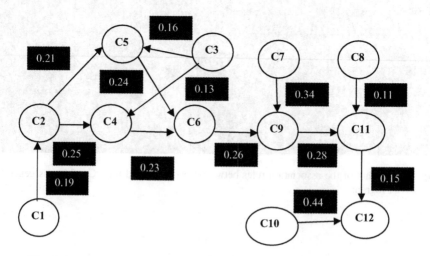

Fig. 4. Concept Map of learning for Java course generated by DHP Algorithm

5   Conclusion

In this work we have proposed a method for development of automated concept map of learning using DHP Algorithm. The proposed method was tested with a set of students enrolled in an introductory JAVA course and was found to identify their learning barriers satisfactorily. This work is actually a part of the work where architecture for Intelligent Diagnostic and Remedial Learning System (IDRLS) is being developed for identification of concepts the students are weak in and suggest remedial lesson plan for these. The system will consist of three modules derived from Ausubel's theory[2] of meaningful learning. One of these modules will generate automated concept map using the method proposed in this work. A prototype version of this system is proposed to be implemented in M-Learning environment using Android Emulator [7,21]. Also a survey should be done on the students to estimate the usefulness of the system so developed. Thus the major research objective of this work has been twofold: Firstly, to develop a method to generate an automated concept map of learning using an efficient heuristic and more significantly, to verify using suitable simulation that this method can indeed identify student learning problems.

References

1. Novak, J.D., Canas, A.J.: Theoretical Origins of Concept Map, How to construct them and their use in Education. Reflecting Education 3(1), 29–42 (2007)
2. Pendidican, F.: Learning Theories, Mathematika dan IInam Alam, Universitas Pendidikan Indonesia
3. Hwang, G.: A conceptual map model for developing Intelligent Tutoring Systems. Computers and Education 40, 217–235 (2003)
4. Hwang, G.: A computer assisted approach to diagnosing student learning problems in science courses. Journal of Information Science and Engineering 5, 229–248 (2003)
5. Park, J.S., Chen, M.S., Philip, S.Y.: An effective Hash based Algorithm for mining Association Rules
6. Lee, C., Lee, G., Leu, Y.: Application of automatically constructed concept maps of learning to conceptual diagnosis of e-learning. Expert Systems with Applications 36, 1675–1684 (2009)
7. Pocatilu, P.: Developing mobile learning applications for Android using web services. Informatica Economica 14(3) (2010)
8. Chen, G., Chan, C., Wang, C.: Ubiquitous learning website: Scaffold learners by mobile devices with information-aware techniques. Computers and Education 50, 77–90 (2008)
9. Virvou, M., Alepis, E.: Mobile Education features in authoring tools for personalized tutoring. Computers and Education 44, 53–68 (2005)
10. Kazi, S.: Voca Test: An intelligent Tutoring System for vocabulary learning using M-Learning Approach
11. Thompson, T.: The learning theories of David P Ausubel: The importance of meaningful and reception learning
12. Johnson, W.B., Neste, L.O., Duncan, P.C.: An authoring environment for intelligent tutoring systems. In: IEEE International Conference on Systems, Man and Cybernetics, pp. 761–765 (1989)

13. Vasandani, V., Govindaraj, T., Mitchell, C.M.: An intelligent tutor for diagnostic problem solving in complex dynamic systems. In: IEEE International Conference on Systems, Man and Cybernetics, vol. 2, pp. 772–777 (1998)

14. Psotka, J., Mutter, S.A.: Intelligent Tutoring Systems: Lessons Learned. Lawrence Erlbaum Associates (1998)

15. Hagiwara, M.: Self-organizing Concept Maps

16. Tseng, S.S.: A new approach for constructing the concept map. Computers and Education 38, 123–141 (2005)

17. Chen, C.C.: Mining learner profile utilizing association rule for web-based learning diagnosis. Expert Systems with applications

18. Haan, J., Kamber, M.: Data Mining: Concepts and Techniques, 3rd edn. Elsevier (2011)

19. Hsiao, C.L., Huang, G.J.: An interactive concept relationship construction assisted system for learning diagnosis (2001)

20. Hwang, G.J.: A Test-Sheet-Generating Algorithm for Multiple Assessment Requirements. IEEE Transactions on Education 46(3) (2003)

21. Acharya, A., Sinha, D.: A Concept Map Approach to Supporting Diagnostic and Remedial Learning Activities. In: Advanced Computing, Networking and Informatics, vol. 1, pp. 557–565. Springer International Publishing (2014)

Automated Localization of Optic Disk, Detection of Microaneurysms and Extraction of Blood Vessels to Bypass Angiography

Patwari Manjiri[1], Manza Ramesh[2], Rajput Yogesh[2],
Saswade Manoj[3], and Deshpande Neha[4]

[1] IMSIT, Vivekanand College, Aurangabad
[2] Department of CS & IT, Dr. BAMU, Aurangabad
[3] Saswade Eye Clinic, Aurangabad
[4] Guruprasad Netra Rugnalaya, Aurangabad
{manjiri.patwari,manzaramesh}@gmail.com,
yogeshrajput128@gmail.com

Abstract. Diabetic Retinopathy is considered as a root cause of vision loss for diabetic patients. For Diabetic patients, regular check-up and screening is required. At times lesions are not visible through fundus image, Dr. Recommends angiography. However Angiography is not advisable in certain conditions like if patient is of very old age, if patient is a pregnant woman, if patient is a child, if patient has some critical disease or if patient has undergone some major surgery. In this paper we propose a system Automated Diabetic Retinopathy Detection System (ADRDS) through which fundus image will be processed in such a way that it will have the similar quality to that of angiogram where lesions are clearly visible. It will also identify the Optic Disk (OD) and extract blood vessels because pattern of these blood vessels near optic disc region plays an important role in diagnosis for eye disease. We have passed 100 images in the system collected from Dr. Manoj Saswade and Dr. Neha Deshpande and got true positive rate of 100%, false positive rate of 3%, and accuracy score is 0.9902.

Keywords: DR, OD, Lesions.

1 Introduction

Diabetic retinopathy, an eye disorder caused by diabetes, is the primary cause of blindness in America and over 99% of cases in India. India and China currently account for over 90 million diabetic patients and are on the verge of an explosion of diabetic populations [1]. Over time, diabetes can damage the heart, blood vessels, eyes, kidneys, and nerves. According to WHO 347 million people worldwide have diabetes. WHO projects that, diabetes deaths will increase by two thirds between 2008 and 2030. This may result in an unprecedented number of persons becoming blind unless diabetic retinopathy can be detected early [2].

For Diabetic patient regular eye check-up and screening is required [3]. Fundus image is taken to view the abnormalities. At times lesions are not visible through

© Springer International Publishing Switzerland 2015
S.C. Satapathy et al. (eds.), *Proc. of the 3rd Int. Conf. on Front. of Intell. Comput. (FICTA) 2014*
– *Vol. 1*, Advances in Intelligent Systems and Computing 327, DOI: 10.1007/978-3-319-11933-5_65

fundus image, Dr. Recommends angiography. However Angiography is not advisable in certain conditions like if patient is of very old age, if patient is a child, if patient is a pregnant woman, if patient is suffering from hypertension ,stroke, or if patient has undergone some major surgery. Sometimes due to dye even a clinically healthy person will have side effects such as Dizziness or faintness, Dry mouth or increased salivation, Hives, Increased heart rate, Metallic taste in mouth, Nausea and Sneezing [4].

To overcome this problem we have come up with a computer aided system (Automated Diabetic Retinopathy Detection System), to detect DR lesions [5] at early stage, to preprocess retinal image, to detect optic disk and to extract blood vessels so that angiography can be avoided. We have applied image processing techniques using MATLAB 2012a. We have performed these operations on 100 images collected from Dr. Manoj Saswade, Dr. Neha Deshpande and got true positive rate of 100%, false positive rate of 3%, and accuracy score is 0.9902.

2   Methodology

Computer assisted diagnosis for various diseases is very common now a days and medical imaging plays a vital role in such diagnosis. Image processing techniques can help in detecting, lesions at early stage, localization of optic disc and extractions of blood vessels to bypass Angiography. The proposed System "Automated Diabetic Retinopathy Detection System" has five stages(fig 1). In first stage preprocessing is done to remove the background noise from input fundus image. Mask of input image is obtained using thresholding technique in the second stage. In the third stage Optic disc is localized using Speeded Up Robust Features techniques. Blood vessels are extracted in the fourth stage using 2-D median filters. In the last stage lesions are detected with the help of intensity transformation function.

2.1   Stage I (Preprocessing)

The Preprocessing is done to remove noise (pixel whose color is distorted) from background and to enhance the image [6]. In the first stage of preprocessing green channel is taken out, because green channel shows high intensity as compare to red and blue. For enhancement of green channel, histogram equalization is used. Mathematical formula for finding green channel is as follows

$$g = \frac{G}{(R + G + B)} \tag{1}$$

Where g is a Green channel and R, G and B are Red, Green and Blue respectively. In the green channel all minute details of image can be viewed. Using red channels only boundary is visible, and in blue channel image shows lots of noise. Due to these reasons green channel is used in the proposed system [7]. For finding histogram equalization of an image following formula is used.

$$h(v) = \text{round} \left(\frac{\text{cdf}(v) - \text{cdf}_{min}}{(M \times N) - \text{cdf}_{min}} \times (L - 1) \right) \tag{2}$$

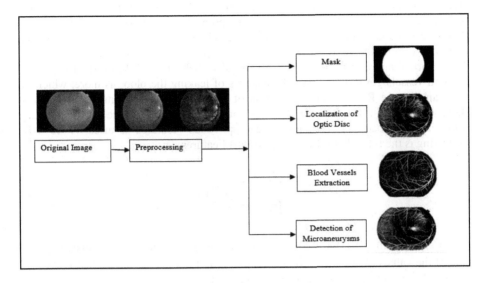

Fig. 1. Flow chart for "Automated Diabetic Retinopathy Detection System"

Here cdf_min is the minimum value of the cumulative distribution function, M × N gives the image's number of pixels and L is the number of grey levels. In histogram equalization frequencies of image are spread out, thus image gets enhanced [6].

2.2   Stage II (Extraction of Mask)

Mask detection is necessary because it displays the exact structure of boundaries of an image. If boundaries are not proper then the image can be discarded to avoid further processing[7]. To extract the mask of fundus image red channel is considered because it is used to detect boundaries. Once red channel is taken out, thresholding[6] operation is performed.

For finding threshold function following formula is used

$$T = \frac{1}{2(m1 + m2)} \tag{3}$$

Where m1 & m2 are the Intensity Values. Threshold is the type of segmentation where required object is extracted from the background[6].

2.3   Stage III (Optic Disk Localization)

A ganglion cell is a cell found in a ganglion, which is a biological tissue mass, most commonly a mass of nerve cell bodies. The optic nerve head in a normal human eye carries from 1 to 1.2 million neurons from the eye towards the brain. The optic disc or optic nerve head is the location where ganglion cell axons exit the eye to form the optic nerve. The optic disc is also the entry point for the major blood vessels that supply the retina. Patten of these blood vessels near optic disc region plays an important role in diagnosis for eye disease. An Ophthalmologist checks this region to detect normal and abnormal vessels. Abnormal vessels are called as tortuous vessels. If blood vessels get tortuous then the chances of leaking the blood is more, which in turn can damage Retina. For localization of Optic Disc green channel is used as it shows high intensity for the pixel values compared to red and blue. After applying some image processing techniques seed-up robust features is used [7 to 10]. Following is the formula for Speed Up Robust Features

$$I_{\Sigma}(x, y) = \sum_{i=0}^{i \leq x} \sum_{j=0}^{j \leq y} I(x, y) \qquad (4)$$

Given an input image I and a point (x; y) the integral image I_{Σ} is calculated by the sum of the values between the point and the origin. Figure 2 shows original images in column A and localized optic disk images in column B and C.

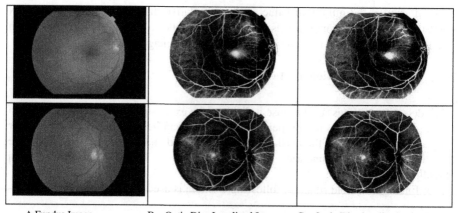

A Fundus Image   B – Optic Disc Localized Image   C – Optic Disc localized color Image

Fig. 2. Optic Disc Localization

2.4   Stage IV (Extraction of Blood Vessels)

For extraction of blood vessels, Image processing operations are performed on green channel image. Figure 3 shows original images in column A and Extracted blood vessels of images are shown in column B and C. Histogram equalization function is used for enhancing the green channel image followed by 2-D median filter[11 to 14]. Morphological structuring element is applied for highlighting the blood vessels of the retina.

$$I_{dilated}(i,j) = \max_{f(n,m)=true} I(i+n,j+m) \tag{5}$$

$$I_{eroded}(i,j) = \min_{f(n,m)=true} I(i+n,j+m) \tag{6}$$

Morphological open function is used for thickening the retinal

$$A \circ B = (A \ominus B) \oplus B \tag{7}$$

Where A ∘B is morphological opening, $\ominus$ is Erosion and $\oplus$ is Dilation.

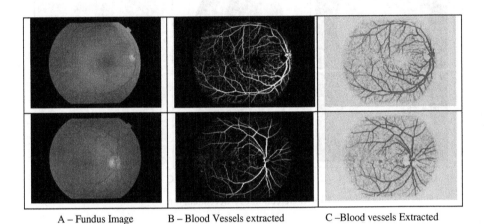

A – Fundus Image   B – Blood Vessels extracted   C –Blood vessels Extracted

Fig. 3. Blood Vessels Extraction

To remove noise 2D median filter is used.

$$y[m,n] = median\{x[i,j], (i,j) \in \omega\} \tag{8}$$

Where ω Represents a neighborhood centered around location (m, n) in the image. In the last Threshold function is used for extracting the retinal blood vessels.

$$T = \frac{1}{2(m1+m2)} \tag{9}$$

Where m1 & m2 are the Intensity Values.

After using image processing techniques in the end intensity-transformation functions is used. Following is the formula for Intensity Transformation Function

$$s = T(r) \tag{10}$$

Where T is Transformation and r is Intensity

2.5   Stage V (Detection of Lesions (Microaneurysm))

Microaneurysms are the first clinically detected lesions. It is Tiny swelling in the wall of a blood vessel. It appears in the retinal capillaries as a small, round, red spot. Usually they are located in the inner nuclear layer of the retina[5]. To begin with the detection of microaneurism, Green channel is taken out, which is followed by compliment function, and histogram equalization function. In figure 4 original images are shown in column A and Microaneurysm detected images are shown in column B and C.

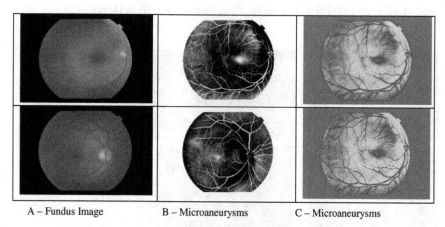

A – Fundus Image   B – Microaneurysms   C – Microaneurysms

Fig. 4. Detection of Microaneurysms

3   Result

We have applied this algorithm on 120 images. For result analysis Receiver Operating Characteristic (ROC) curve[18] is used. This algorithm achieves a true positive rate of 100%, false positive rate of 0%, and accuracy score of 0.9902. For detection of blood vessels this algorithm achieves 0.9937 accuracy (Table1), for Optic disc detection 0.9932(Table2) and for microaneurysms 0.9962(Table 3) compared to other algorithms. We have developed GUI in matlab(Figure 5) for blood vessel extraction, detection of microaneurisms and Optic disc.

Table 1. Comparison of optic disc

	AUC
Rangaraj et al.	0.8890
Usman et al.	0.9632
Chandan et al.	0.9100
Ahmed et al.	0.9500
Proposed	**0.9932**

Table 2. Comparison of blood vessels

	AUC
Chaudhuri et al.	0.9103
Jiang et al.	0.9327
Staal et al.	0.9520
Soares et al.	0.9614
Proposed	**0.9937**

Table 3. Comparison of microaneurysms

	AUC
Yuji et al.	0.6700
Sujith et al.	0.9444
Proposed	**0.9962**

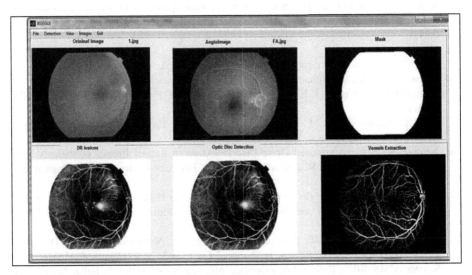

Fig. 5. Graphical User Interface

4   Conclusion

In the developed system "Automated Diabetic Retinopathy Detection System (ADRDS)" we have used Image processing techniques through which fundus image is processed in such a way that it has the similar quality to that of angiogram where lesions are clearly visible. System is also able to identify the Optic Disc (OD) and it can extract blood vessels. Pattern of these blood vessels near optic disc region plays an important role in diagnosis for eye disease. We have passed 120 images collected from Dr. Manoj Saswade, Dr. Neha Deshpande to this system, and we have got true positive rate of 100%, false positive rate of 3%, and accuracy score is 0.9902.

Acknowledgement. We are thankful to University Grant Commission (UGC) for providing us a financial support for the Major Research Project entitled "Development of Color Image Segmentation and Filtering Techniques for Early Detection of Diabetic Retinopathy" F. No.: 41 – 651/2012 (SR) also we are thankful to DST for providing us a financial support for the major research project entitled "Development of multi resolution analysis techniques for early detection of non-proliferative diabetic retinopathy without using angiography" F.No. SERB/F/2294/2013-14.

References

1. Silberman, N., et al.: Case for Automated Detection of Diabetic Retinopathy. In: Association for the Advancement of Artificial Intelligence (2010), http://www.aaai.org
2. (ROC Curve), http://www.who.int/mediacentre/factsheets/fs312/en/
3. Patwari, M.B., Manza, R.R., Saswade, M., Deshpande, N.: A Critical Review of Expert Systems for Detection and Diagnosis of Diabetic Retinopathy. Ciit International Journal of Fuzzy Systems (February 2012), DOI: FS022012001 ISSN 0974-9721, 0974-9608 (IF 0.441)
4. (Angiography), http://www.webmd.com/eye-health/eye-angiogram?page=2
5. Kanski, J.J.: Clinical Ophthalmology: A Systematic Approach, 3rd edn.
6. Gonzalez, R.C., Woods, R.E.: Digital Image processing. Pearson Education, New Delhi (2002)
7. Zhu, X., Rangayyan, R.M., Ells, A.L.: Digital Image Processing for Ophthalmology: Detection of the Optic Nerve Head
8. Aquino, A., et al.: Automated Optic Disc Detection in Retinal Images of Patients with Diabetic Retinopathy and Risk of Macular Edema. International Journal of Biological and Life Sciences 8(2) (2012)
9. Rangayyan, R.M., et al.: Detection of the Optic Nerve Head in Fundus Images of the Retina with Gabor Filters and Phase Portrait Analysis. Journal of Digital Imaging 23(4), 438–453 (2010)
10. Rajput, Y.M., Manza, R.R., Patwari, M.B., Deshpande, N.: RetinalOpticDisc Detection Using Speed Up Robust Features. In: National Conference on Computer & Management Science (CMS-2013), RadhaiMahavidyalaya, Auarngabad- 431003(MS) India, April 25-26 (2013)
11. Jiméneza, S., Alemanya, P., et al.: Automatic detection of vessels in color fundus images. Sociedad Española de Oftalmología (2009)
12. Reza, A.W., et al.: Diabetic Retinopathy: A Quadtree Based Blood Vessel Detection Algorithm Using RGB Components in Fundus Images. Received: Published by Elsevier España, s.larchsocespoftalmol 85(3), 103–109 (2010), Media, LLC 2007
13. Vijayachitra, S., et al.: Analysis Of Diabetic Retinopathy Images Using Blood Vessel Extraction. International Journal of Advanced Engineering Research and Studies E-ISSN2249–8974 IJAERS/Vol. I/ Issue II/January-March, 2012/89-91 Research Article (2012)
14. Rajput, Y.M., Manza, R.R., Patwari, M.B., Deshpande, N.: Third National Conference on Advances in Computing(NCAC-2013). In: Third National Conference on Advances in Computing(NCAC-2013), March 5–6, North Maharashtra University, Jalgaon (2013)

15. Akram, U.M., et al.: Automated Detection of Dark and Bright Lesions in Retinal Images for Early Detection of Diabetic Retinopathy. In: Received: 1 August 2011 / Accepted: 25 October 2011 / Published online: 17. Springer Science+Business Media, LLC (November 2011)

16. Poddar, S., et al.: Quantitative Clinical Marker Extraction from Colour Fundus Images for Non-Proliferative Diabetic Retinopathy Grading. In: International Conference on Image Information Processing (2011)

17. (ICIIP 2011) (ICIIP 2011) 978-1-61284-861-7/11/$26.00 ©2011 IEEE

18. Singh, N., et al.: Automated Early Detection of Diabetic Retinopathy Using Image Analysis Techniques. International Journal of Computer Applications 8(2), 975–8887 (2010)

19. (Blood Vessels Diameter), http://www.vassarstats.net/roc1.html

20. Patwari, M.B., Manza, R.R., Rajput, Y.M., Saswade, M., Deshpande, N.K.: Review on Detection and Classification of Diabetic Retinopathy Lesions Using Image Processing Techniques (IJERT) 2(10) (October 2013) ISSN: 2278-0181

21. Patwari, M.B., Manza, R.R., Rajput, Y.M., Deshpande, N.K., Saswade, M.: Extraction of the Retinal Blood Vessels and Detection of the Bifurcation Points. IJCA (September 18, 2013) ISBN : 973-93-80877-61-7

22. Hatanaka, Y.: Automated microaneurysm detection method based on double-ring filter and feature analysis in retinal fundus images. 978-1-4673-2051-1/12 IEEE

23. SujithKumar, S.B., et al.: Automatic Detection of Diabetic Retinopathy in Non- dilated RGB Retinal Fundus Images. International Journal of Computer Applications 47(19), 888–975 (2012)

24. Reza, A.W., et al.: Diagnosis of Diabetic Retinopathy: Automatic Extraction of Optic Disc and Exudates from Retinal Images using Marker-controlled Watershed Transformation. J. Med. Syst., doi:10.1007/s10916-009-9426-y

Hard Aperiodic Scheduling in Fault Tolerant Cruise System – Comparative Evaluation with Dual Redundancy

Annam Swetha[1,*], V. Radhamani Pillay[1], Sasikumar Punnekkat[2],
and Santanu Dasgupta[3]

[1] Amrita Vishwa Vidyapeetham University, India
[2] Malardalen University, Sweden
[3] Mohandas College of Engineering & Technology, India
{swethaannam4,sdasgupta100}@gmail.com,
vr_pillay@cb.amrita.edu,
sasikumar.punnekkat@mdh.se

Abstract. A scheduling algorithm for hard real-time systems with aperiodic task arrivals and fault tolerance can basically meet the needs of applications in the automotive or avionics domain. In these applications, weight, size and power requirements are crucial. Any resource augmentation technique to satisfy this and ensuring safe functionality under faults can bring in a paradigm shift in the design. This paper is based on a strategy for fault tolerance with task level criticality on dual processor system. An application with parallelizable task set has been used to advantage for resource augmentation under fault free condition. A processor failure (permanent fault) leads to safe recovery mechanism with graceful degradation. This paper deals with fault tolerant periodic task scheduling with arrivals of hard aperiodic events. An algorithm for aperiodic scheduling with admission control plan for hard and soft aperiodic tasks is developed and implemented on LPC2148 processors for the cruise system. A comparison is made with a traditional dual redundant system with appropriate performance metrics for evaluation.

Keywords: Hard real-time systems, Aperiodic scheduling, Dual redundancy, Resource augmentation, Speedup factor, Cruise system.

1 Introduction

Hard real-time systems are very complex and have to operate in dynamic environment where external events and user commands occur aperiodically. Hence they include both periodic, aperiodic tasks and require an integrated scheduling approach that can meet the timing requirements of the system. In addition to this, such systems need to be dependent even in the presence of permanent fault. Various fault tolerant strategies are built into such systems to ensure fail safe conditions during processor failures, redundancy being one of the methods. Under mixed task scheduling with aperiodic tasks, fault tolerance becomes more complex and precautions should be taken to

---

* Corresponding author.

© Springer International Publishing Switzerland 2015
S.C. Satapathy et al. (eds.), *Proc. of the 3rd Int. Conf. on Front. of Intell. Comput. (FICTA) 2014*
– *Vol. 1*, Advances in Intelligent Systems and Computing 327, DOI: 10.1007/978-3-319-11933-5_66

ensure that critical deadlines are met, guaranteeing the minimal safe functionality. Traditional working model of a hard real-time system consist of redundant units which continuously monitor the system (hot standby) and get activated under the occurrence of any fault.

This paper is based on redundancy with task level criticality and resource augmentation on cruise system [17], [18]. A framework is designed and implemented for the admission control and scheduling of aperiodic tasks with fault tolerance. The Traditional Dual Redundant(TDR) scheme and Resource augmented Fault Tolerant(RaFT) scheme, is applied to a case study of cruise control system using an experimental setup of LPC2148 processors and its performance is evaluated by suitable metrics.

Rest of the paper is organized as follows Section 2 gives literature survey in the field of aperiodic scheduling and fault tolerance. Section 3 elaborates the approach with algorithm design and Section 4 describes hardware implementation. Evaluation of the system performance and discussion is presented in Section 5. Conclusion of this work is given in Section 6.

2   Literature Survey

2.1   Mixed Task Scheduling

Aperiodic task scheduling built on static scheduling algorithms like RM and DM are presented by Lehoczky [1]. In the work dealt by Sprunt [2] the Sporadic Server (SS) algorithm is designed to guarantee deadlines for hard aperiodic tasks and provide good responsiveness for soft-deadline aperiodic tasks. Earlier the server based aperiodic scheduling introduced, where a periodic server with highest priority is embedded into static scheduling strategies to serve the aperiodic tasks, leads to more overheads in context switching because of the shorter time period assigned to the server and thus not suitable for aperiodic tasks with low arrival rates [3]. Further aperiodic scheduling with deadlines on multiprocessors was considered by Andersson et al. with constraints on system utilization in both global and partitioned scheduling [4], [5]. Baruah et al. [6] considered multiprocessor implementation of Total Bandwidth Server (TBS) along with dynamic priority scheduling algorithms. This provided a schedulability test which indicates the guaranteed real-time performance of the system. Manimaran [7] proposed an emergency algorithm for combined scheduling of periodic and aperiodic tasks without any comprise on the schedulability of hard periodic tasks. The concept of dropping already scheduled soft tasks to accommodate the hard aperiodic tasks has been introduced and is evaluated to give better performance compared to iterative server rate adjustment algorithms. A survey paper by Davis et al, [8] covers a detailed review of hard real-time scheduling algorithms and schedulability analysis techniques for homogenous multiprocessor systems.

2.2   Fault Tolerance

A. Avizienis [9] integrated fault tolerance with the error detection and fault diagnosis. Fault tolerance in multiprocessor systems with quick recovery techniques is proposed by Krishna [10] and dynamic fault tolerant scheduling analysis is done by Manimaran [11]. A scheduling strategy that integrated the timeliness and criticality to fault tolerance was proposed by Mahmud Pathan [12]. A new paradigm for fault tolerance in hard real time systems for on board computers with effective resource management is given in [13], [14], [15]. The hardware implementation of this fault tolerant strategy has been done in [16] for a dual processor system. Implementation of this approach and a comparison with traditional dual redundant system for a case study of cruise control system is given in [17], [18]. The resource augmentation or speedup factor given by Kalyanasundaram [19] provides an efficient performance metric for comparing the scheduling algorithms with optimal algorithms.

3   Scheduling Strategy

3.1   Assumptions

The priority level of hard aperiodic event is greater than that of hard periodic task in the system. A minimum one execution time unit and with single aperiodic event arrival without bursts are considered. Worst Case Execution Time (WCET) of periodic task include all time overheads and communication delays.

3.2   System Model

The system model consists of a processor which functions as a master node Real-Time Executive Manager (*RTEM*) with two identical processors (P_1, P_2). A mixed task set-hard and soft (critical and non-critical) periodic tasks $\tau_p = \{\tau_{p1}, \tau_{p2}, ... \tau_{pi}\}$, with worst case execution time C_{pi}, time period T_{pi}, Deadline D_{pi} and Utilization $U_{pi}=C_{pi}/T_{pi}$ and aperiodic tasks $\tau_{ap} = \{\tau_{ap1}, \tau_{ap2}, ... \tau_{apj}\}$.

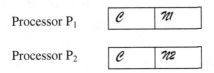

Fig. 1. Task allocation in RaFT scheme

Task allocation and scheduling is done in a Traditional Dual Redundant (TDR) scheme with hot standby. A watchdog timer with self detecting mechanism for faults, detects the permanent fault in any processor. Based on the work in [13], Resource augmented Fault Tolerant (RaFT) scheme, task allocation and scheduling in P_1 and P_2 are done by **RTEM** shown in Fig. 1 where critical($\mathcal{C}$) tasks are duplicated in the processors and non-critical ($\mathcal{N}1$. $\mathcal{N}2$) tasks are shared between processors. Under normal mode (fault free mode), each processor executes a static table driven scheduling, sends periodic health check signals and updates **RTEM**. Absence of health signal (permanent fault) from any processor switches the system to fault mode where the previously allocated non-critical tasks of fault processor are now scheduled in functional processor by **RTEM**, maintaining operational functionality of the system. Any arrival of aperiodic event (hard and soft), an admission control algorithm is executed by **RTEM** to find a feasible Window Of Execution (WOE). A check is made for criticality of aperiodic event and instantaneously schedules the hard aperiodic task in both processors. This assumes significance in the simultaneous occurrence of fault and aperiodic arrival. Soft aperiodic tasks are scheduled in feasible WOE based on its utilization factor U_{apj},

$$\text{where } U_{apj} = \frac{(\sum_{k=1}^{i} C_{pk} + C_{apj})}{\text{Hyper period}}.$$

Admission Control Algorithm (**Al-AdCt**)

Input: User/environment initiated aperiodic event τ_{apj}
1: Check criticality (hard) of aperiodic event
2: **If** (Critical)
3:      Schedule aperiodic event in both processors
4: **Else**
5:      Check for feasible WOE in any processor
6:      **If** $(U_{apj} \leq 1 - \sum_{k=1}^{i} U_{pk})$
7:          Schedule aperiodic event in feasible processor
8:      **Else**
9:          Drop non-critical tasks and schedule the aperiodic event
10:     **End If**
11: **End If**

This algorithm is evaluated with performance metrics, Process Execution Time ($\mathcal{PET}$), Effective Utilization ($\mathcal{EU}$), Deadline Miss Ratio ($\mathcal{DMR}$) and average Response Time ($\mathcal{RT}$). $\mathcal{PET}$ gives total execution time of the process and determines the speedup factor ($\mathcal{PET}_{TDR}$) / ($\mathcal{PET}_{RaFT}$) [17]. $\mathcal{EU}$ is a normalized utilization of processors during execution process [23]. $\mathcal{DMR}$ is the ratio of periodic tasks that have missed deadlines to total number of periodic tasks [24]. $\mathcal{RT}$ is average response time of all soft aperiodic tasks that have been scheduled on the processors [2].

4   Implementation

The RaFT and TDR schemes are implemented in hardware with ARM LPC2148 processors taking the task set of a cruise system with a workload of 75% [21]. Experimental setup (Fig. 2) includes 2 identical processors and a master node which acts as **RTEM.**

UART communication is set between the processors with a baud rate of 9600Bd. Inputs to the system are sensing tasks, control tasks and actuating tasks with time attributes. A performance evaluation board (Fig. 3) indicates the current task executed in each processor by the LED. An LCD displays the real-time scheduling in TDR, RaFT in Fig. 4(a) (b) respectively.

Table 1. Task Set – Cruise System

Task	Time Attributes (time units)		
	C_i	T_i	D_i
τ_1	3	30	15
τ_2	2	20	10
τ_3	2	60	10
τ_4	3	30	15
τ_5	2	60	15
τ_6	10	60	55
τ_7	5	60	30
τ_8	10	60	15

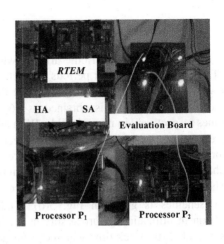

Fig. 2. Experimental Setup

A green LED indicates health status of processor. A simulation of fault injection in a processor is done using the RED switch on evaluation board. To enter into a fault mode the RED switch can be pressed for processor P_1 or P_2 (Fig. 3) and corresponding real-time display is shown in Fig. 4(c),(d). Injection of hard (HA) and soft (SA) aperiodic task are done using switches embedded on the master node (Fig.2). On pressing the HA switch, *RTEM* schedules the hard aperiodic task in both processors as shown in Fig 4(f). Performance metrics are evaluated over five hyper periods for one operational run with the same instantiation of aperiodic arrival. A soft aperiodic task is similarly scheduled with injection of SA during the idle time of processor P_2 (Fig. 4(e)).

Fig. 3. Performance Evaluation Board

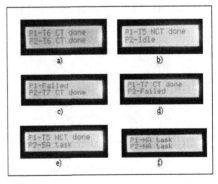

Fig. 4. Real-time display in LCD

CT- Critical Task, NCT- Non Critical Task

Periodic: a) TDR, b) RaFT, c) P_1 failure, d) P_2 failure

Aperiodic: e) Soft (SA) task, f) Hard (HA) task

5   Results and Discussion

For varying number of aperiodic arrivals with varying execution times in TDR and RaFT schemes over 5 hyper periods and with 40% critical task load, the performance metrics are given in Fig. 5. In Fig. 5(a), the Process Execution Time ($\mathcal{PET}$) with only periodic task scheduling is obtained as 56 and 48 time units for TDR and RaFT schemes respectively in normal mode. With aperiodic task arrivals, $\mathcal{PET}$ increases due to preemption cost and aperiodic task execution time. It is found that the speedup factor is 1.2 and there is a minimum 12% reduction in RaFT compared to TDR. This emphasizes the resource augmentation given by the improvement of $\mathcal{PET}$ in RaFT.

The Effective Utilization ($\mathcal{EU}$) of the processors under normal mode for TDR and RaFT schemes is given in Fig. 5(b). Under periodic scheduling, there is 14% reduction in utilization of processors in RaFT compared to TDR and illustrates the presence of extra slack time which can further be used for scheduling any aperiodic events and optional tasks. The effect of different aperiodic execution time on $\mathcal{EU}$ in both schemes is seen with a minimum differnce of 10%. Under fault mode, performance of RaFT equalizes with TDR assuring the safe functionality of system.

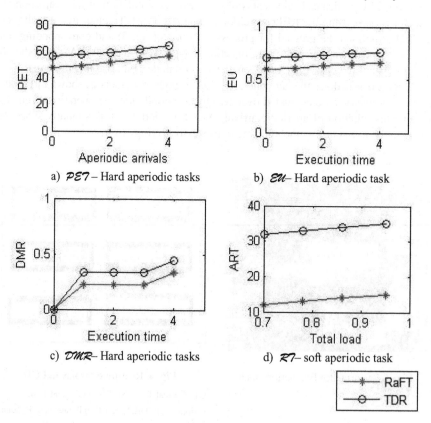

a) $\mathcal{PET}$ – Hard aperiodic tasks   b) $\mathcal{EU}$– Hard aperiodic task

c) $\mathcal{DMR}$– Hard aperiodic tasks   d) $\mathcal{RT}$– soft aperiodic task

Fig. 5. Performance metrics

As execution time of aperiodic event increases, $\mathcal{DMR}$ shows an increases seen in Fig. 5(c). There is a reduction in $\mathcal{DMR}$ for RaFT compared to TDR due to parallelization of tasks in processors. Average Response Time ($\mathcal{RT}$) of soft aperiodic tasks for varying loads of aperiodic task is given in Fig. 5(d), and compared to TDR, there is a miminum reduction of 20 time units emphasizing efficient resource utilisation and performance.

6   Conclusion

In this paper, a prototype of dual fault tolerant system with aperiodic arrival events being scheduled using ARM LPC2148 processors is presented. The evaluation of this resource augmentation scheme for a cruise system over a typical dual redundant scheme with performance metrics illustrates the enhancement in process execution by a speedup factor and provides an additional slack for scheduling the arriving aperiodic tasks and possible optional tasks. Admission control algorithm reduces average response time by efficiently utilizing the additional slack and reducing deadline miss ratio. The results conclude the gain that can be achieved in terms of additional computing resources and improved scheduling with fault tolerant constraints. Further this framework can be applied to any complex applications like avionics, missile guidance and extended with m-processor redundancy. The efficient utilization of resources and higher performance capability can prove to be beneficial in systems where the reduction in weight and volume can mean more/larger payloads. These are desirable factors driving the research goals of fault tolerant scheduling in safety-critical applications.

References

1. Lehoczky, T., Sha, L., Strosnider, J.: Enhanced aperiodic responsiveness in hard real-time environments. In: Proceedings of IEEE Real-Time Systems Symposium, pp. 261–270 (1987)
2. Sprunt, B.: Aperiodic Task Scheduling for Real-Time Systems. Ph.D. Dissertation, Department of Electrical and Computer Engineering. Carnegie Mellon University (1990)
3. Andersson, B., Abdelzaher, T., Jonsson, J.: Global Priority-Driven Aperiodic Scheduling on Multiprocessors. In: Proceedings of the IEEE International Parallel and Distributed Processing Symposium (2003)
4. Tein-Hsiang, Tarng, W.: Scheduling periodic and aperiodic tasks in hard real-time computing systems. In: Proceedings of the 1991 ACM SIGMETRICS Conference on Measurement and Modeling of Computer Systems (1991)
5. Andersson, B., Abdelzaher, T., Jonsson, J.: Partitioned Aperiodic Scheduling on Multiprocessors. In: Proceedings of the IEEE International Parallel and Distributed Processing Symposium (2003)
6. Baruah, S.K., Lipari, G.: A Multiprocessor Implementation of the Total Bandwidth Server. In: Proceedings of the IEEE International Parallel and Distributed Processing Symposium, pp. 40–47 (2004)

7. Duwairi, B., Manimaran, G.: Combined Scheduling of Hard and Soft Real-Time tasks in Multiprocessor systems. In: Pinkston, T.M., Prasanna, V.K. (eds.) HiPC 2003. LNCS (LNAI), vol. 2913, pp. 279–289. Springer, Heidelberg (2003)
8. Davis, R.I., Burns, A.: A Survey of Hard Real-Time Scheduling for Multiprocessor Systems. ACM Computing Surveys 43(4) (October 2011)
9. Avizienis, A.: Design of Fault-Tolerant Computers. Fall Joint Computer Conference 1967 (1967)
10. Krishna, C.M., Shin, K.G.: On scheduling tasks with a quick recovery from failure. In: Proceeding of Fault-tolerant Computing Symposium, pp. 234–239 (1985)
11. Manimaran, G.: Fault tolerant dynamic schedule for multiprocessor real time systems and its analysis. IEEE Transactions on Parallel and Distributed Systems (1998)
12. Pathan, M.: Three Aspects of Real-Time Multiprocessor Scheduling: Timeliness, Fault Tolerance, Mixed Criticality. Ph.D Dissertation, Göteborg, Sweden (2012)
13. Pillay, R., Punnekkat, S., Dasgupta, S.: An improved redundancy scheme for optimal utilization of onboard Computers. In: IEEE INDICON 2009, India (2009)
14. Pillay, R., Chandran, S.K., Punnekkat, S.: Optimizing resources in real-time scheduling for fault tolerant processors. In: IEEE, International Conference on Parallel, Distributed and Grid Computing (PDGC 2010), Solan, India (October 2010)
15. Chandran, S.K., Pillay, R., Dobrin, R., Punnekkat, S.: Efficient scheduling with adaptive fault tolerance in heterogeneous multiprocessor systems. In: International Conference on Computer and Electrical Engineering (ICCEE), China (November 2010)
16. Abraham, S., Sivraj, P., Pillay, R.: Hardware Implementation of an Improved Resource Management Scheme for Fault Tolerant Scheduling of a Multiprocessor System. International Journal of Computer Applications (0975-8887) 27(2) (August 2011)
17. Annam, S., Radhamani, P.V., Punnekkat, S.: Design, Analysis and Implementation of Improved Adaptive Fault Tolerant Model for Cruise Control Multiprocessor System. International Journal of Computer Applications (0975-8887) 86(15) (January 2014)
18. Swetha, A., Pillay, R., Dasgupta, S., Chandran, S.K.: A Real-Time Performance Monitoring Tool for Dual Redundant and Resource Augmented Framework of Cruise Control System. International Journal of Computer Applications (0975 – 8887) 92(14) (April 2014)
19. Kalyanasundaram, B., Pruhs, K.: Speed is as powerful as clairvoyance. In: Proceedings of the Symposium on Foundations of Computer Science, pp. 214–221 (1995)
20. Anand, A., Rajendra, Y., Narayan, S., Pillay, R.: Modelling, Implementation and Testing of an Effective Fault Tolerant Multiprocessor Real-Time System. In: IEEE International Conference on Parallel, Distributed and Grid Computing (PDGC 2012), India (2012)
21. Staines, A.S.: Modeling and Analysis of Real Time Control Systems: A Cruise Control System Case Study. Recent Advances in Technologies. Intech (2009)
22. Knight, J.C.: Safety Critical Systems: Challenges and Directions, Department of Computer Science. In: Proceedings of the 24th International Conference on Software Engineering. University of Virginia (2002)
23. Gurulingesh, R., Sharma, N., Ramamritham, K., Malewar, S.: Efficient Real-Time Support for Automotive Applications: A Case Study. In: Proceedings RTCSA 2006 12th IEEE International Conference on Embedded and Real-Time Computing Systems and Applications, pp. 335–341 (2006)
24. Lu, C., Stankovic, J.A., Abdelzaher, T.F.: Performance Specifications and Metrics for Adaptive Real-Time Systems. In: Proceedings Real-Time Systems Symposium (2000)

Detection of Differentially Expressed Genes in Wild Type HIV-1 Vpr and Two HIV-1 Mutant Vprs

Bandana Barman[1,*] and Anirban Mukhopadhyay[2]

[1] Department of Electronics and Communication Engineering,
Kalyani Government Engineering College,
Kalyani, Nadia, West Bengal, India
[2] Department of Computer Science and Engineering,
University of Kalyani, Kalyani,
Nadia, West Bengal, India
bandanabarman@gmail.com,
anirban@klyuniv.ac.in

Abstract. Identification of differentially expressed genes between two sample groups are important to find which genes are increased in expression (up-regulated) or decreased in expression (down-regulated). We have identified differentially expressed genes between wild type HIV-1 Vpr and two HIV-1 mutant Vprs separately by using statistical t-test and false discovery rate. We also compute q-value of test to measure minimum FDR which occurs. We have found 1524 number of differentially expressed genes between wild type HIV-1 vpr and HIV-1 mutant vpr, R80A. Again we found 1525 diffrential genes between wild type HIV-1 vpr and HIV-1 mutant vpr, F72A/R73A. From these two differentially expressed gene sets we get 941 number of down-regulated genes for both sets and rest genes are found as up-regulated genes.

Keywords: Microarray data, p-value, false discovery data, q-value, normalization, permutation test, differential gene expression.

1 Introduction

Differential expression is assessed by finding ratios of expression levels of different samples and variability of these ratios is not constant. A gene is considered to be differentially expressed between two sample groups, if the gene shows both statistical and biological significance. Microarray gene expression data are introduced in the nineties. From that period so many methods were developed to analyze microarray and also to detect differentially expressed genes [10,4]. Diverse microarray samples are better choice to understand biological phenomena. In the present days a lot of number microarray experiments are being develped to analyze multivariate microarray data [2]. To work with microarray data, the

---

* Corresponding author.

S.C. Satapathy et al. (eds.), *Proc. of the 3rd Int. Conf. on Front. of Intell. Comput. (FICTA) 2014*
– *Vol. 1*, Advances in Intelligent Systems and Computing 327, DOI: 10.1007/978-3-319-11933-5_67

quality of datasets are very much essential for analysis but it is not a easy task. The quality verification or assessment may be done before or after data processing. The researcher proves that microarray data quality control improves the detection of differentially expressed genes [12] and also the useful analysis with the data. Statistical methods like to compare two Poisson means (rates) for detection of differentially expressed genes for RNA-seq data [7] are introduced.

Non-parametric methods such as nonparametric t-test, Wilcoxon rank sum test and heuristic idealized discriminator method are used to find diffentialy expressed genes [14]. Researchers introduced statistical methods to identify diferentially expressed genes in replicated cDNA microarray experiments [8]. There normalized data is used. The univariate testing problem for each identified diferentially expressed genes, correction is made by adjusted pvalue for multiple testing. The principal component analysis (PCA) space and classification of genes between two conditions are done to find differentially expressed genes [13].

In differentially expressed genes, messages are extensively processed and modified prior to translation [1]. To identify differentially expressed genes (DEG), the main key is to understand wrong or fixed under conditions (cancer, stress etc.). On the otherhand DEG may be considered as featuresfor a classifier and also serve as starting point for a model. To detect differential gene expression in cross-species hybridization experiments at gene and probe Level [6] masking procedure is used. A cross-species data set using masking procedure, gene-set analysis are studied. Gene Set Enrichment Analysis (GSEA) and Test of Test Statistics (ToTS) were also investigated.

Transcriptome analysis of DEG relevant to variegation in Peach Flowers [5] is done previously. They identified an informative list of candidate genes associated with variegation in peach flowers by digital expression analysis that offered opportunity to uncover genetic mechanisms of flower color variegation. Using Bayesian model selection approach, DEG in Microarrays were identified [11]. There researchers identified DEG based on a high-dimensional model selection technique and Bayesian ANOVA for microarrays (BAM), that striks a balance between false rejections and false nonrejection.

We have identified differentially expressed genes from wild type HIV-1 Vpr and two mutant HIV-1 Vprs. In our algorithm we used statistical t-test and estimate false discovery rate within the sample groups. We extract the DEG after we plot the volcano plots. The entire algorithm has been discussed in the method section.

2   Material and Methods

To create the global picture of cellular function, Gene expression profiling is a measurement of expression or activity of thousands of genes at once. The DNA microarray technology measures relative activity of genes which are previously identified as target genes. The next generation sequencing is the RNA-Seq. In our approach we used expression profiling by array type data.

We applied algorithm on microarray samples of inducible HIV-1 Vpr protein on cellular gene expression. We get the data set from publicly available government website, http://www.ncbi.nlm.nih.gov.in/geodata/GSE2296. The cell lines express wild type (WT) HIV-1 Vpr and two mutant Vprs, R80A and F72A/R73A. Doxycycline-induced and uninduced and Flp-In TREx 293/Vpr cells were collected as microarray data with 0, 1, 2, 4, 6, 8, 12, 16 and 24 hours post induction (hpi) and the total number of RNA present in the cell are 21,794. In the samples, three 293 cell-derivative cell line, wild type HIV-1 Vpr and two mutant Vprs, F72A/R73A and R80A were constructed using commercially available invitrogen Flp-In TREx 293 cells. The mutant HIV-1 Vprs are defective in the Vpr cell cycle arrest phenotype. The cell lines were characterized for their ability to express Vpr at the RNA and protein levels. A large scale host cell gene expression profiling study using microarray technologies is conducted there to study whether certain host cell genes showed alterations in expression following induced expression of Vpr. In the microarray sample data, purified cDNA was processed for coupling reaction with Cy 5 or Cy 3-ester dye, respectively. To determine whether the differential expression of certain genes was associated with Vpr-induced cell cycle arrest, a comparison between gene expression pattern of wild type Vpr expressing cells and pattern of Vpr mutant F72A/R73A- or R80A-Vpr expressing cells is done.

We applied algorithm on this sample HIV-1 microarray data after normalization. The normalization is done by scaling the values in each column of microarray data, by dividing by the mean column intensity and it results a matrix of normalized microarray data. We used the software Matlab(R2013a) to implement algorithm.

There are several statistical methods used for identification of differentially expressed genes. Those are Student's t-test, T-test, Linear regression model, Nonparametric test (Wilcoxon, or rank-sums test), SAM etc. We have used T-test. It is a statistical hypothesis test. In the statistics, the hypothesis testing plays an important role, and also in statistical inference. The mathematics behind of t-test as follows:

The size of our used samples are equal. If two sample sizes (the number of participants (n) of each group) are equal then it is assumed that sample distributions have same variance.

$$d = \frac{\overline{X_1} - \overline{X_2}}{S_{X_1 X_2} \sqrt{\frac{2}{n}}} \tag{1}$$

where,

$$S_{X_1 X_2} = \sqrt{\frac{1}{2}(S_{X_1}^2 + S_{X_2}^2)} \tag{2}$$

$S_{X_1 X_2}$ is the grand standard deviation. 1 = sample group one, 2 = sample group two. $S_{X_1}^2$ and $S_{X_2}^2$ are the unbiased estimators of the variances of the two samples.

For significance testing, the degrees of freedom for this test is $2n - 2$ where n is the number of participants in each group.

The t-score is defined by:

$$t = \frac{\bar{x} - \mu}{\frac{s}{\sqrt{n}}} \tag{3}$$

where $\bar{x}$ is the sample mean, μ is the population mean, s is the standard deviation of the sample, and n is the sample size. We estimate the result of t-test by false discovery rate. The false discovery rate (FDR) control is a statistical method. It is used in multiple hypothesis testing to correct for multiple comparisons. In a list of findings, FDR procedures are used to control the expected proportion of incorrectly rejected null hypotheses (false discoveries).

The false discovery rate (FDR) is given by:

$$FDR = E\frac{V}{R} \tag{4}$$

The FDR is kept below a threshold value i.e. quantile value (q-value). Where V is the number of false positives or false discoveries.

R is the number of rejected null hypothesis or discoveries
E is estimation.

The q-value of an individual hypothesis test is the minimum FDR at which the test may be called significant.

Our Proposed Algorithm Is as Follows
Input1: *Data matrix object of wild type HIV-1 Vpr (WT) and mutant HIV-1 Vprs R80A;*
dmosubset. Total number of genes present in the sample is 21794.
Input2: *Data matrix object of wild type HIV-1 Vpr (WT) and mutant HIV-1 Vprs F72;*
dmosubsetA. Total number of genes in the sample is 21794.
Step1: *Normalize the both data matrices. We normalized the data matrices to get mean and variance of each gene 0 and 1 rapectively; it results dmosetN, and dmosetAN respectively.*
Step2: *Filtration of normalized datamatrices i.e. dmosubsetN, dmosubsetAN Firstly, filter out genes with very low absolute expression values.*
Secondly, filter out genes with a small variance across samples.
Lastly, we calculate the number genes in the datamatrices.
After filtration the total number of genes are 19615.
Step3: *Extract the data of WT samples and R80A samples from dmosetN (Pheno1 and Pheno2 respectively).*
Extract the WT samples and data of F72A/R73A sample from dmosetAN (Pheno3 and Pheno4 respectively).
Step4: *Standard statistical t-test is done to detect significant changes between the measurement of variables in two groups (Pheno1, Pheno2) and (Pheno3, Pheno4)*
Step5: *Plot normal quantile plot of t-scores and histograms of t-scores and p-values of the t-tests for both variables.*

Step6: *Perform a permutation t-test to compute p-values by permuting the columns of the gene expression data matrix [9]*
Step7: *Set the p-value cutoff of 0.05; The number of genes which have statistical significance at the p-value cutoff are determined.*
Step8: *The false discovery Rate and quantile values (q-values) for each test are estimated.*
Step9: *The number of genes which have q-values less than cutoff value are determined.*
Step10: *Now FDR adjusted p-values are estimated by using the Benjamini-Hochberg (BH) procedure [3]*
Step11:*Plot the -log10 of p-values against the biological effect in a volcano plot*
Step12: *Export the Differentially expressed genes from volcano plot UI*

Output1: *We got 1524 differentially expresed genes from sample group Pheno1 and Pheno2.*
Output2: *We got 1525 differentially expressed genes from sample group Pheno3 and Pheno4.*

After coding of algorithm, the result we got is shown in result section.

3   Results

After filtering the sample groups and performing t-test between Pheno1 and Pheno2, also Pheno3 and Pheno4 (i.e. Wild type HIV-1 Vpr and mutant HIV-1 Vprs, R80A and F72A/R73A) we plotted the sample quantile plot vs theoretical quantile plots fig1(a), fig2(a) respectively.

We also plotted histogram of t-test (i.e.t-score and p-value), as histogram is a graphical representation of the distribution of data. It is an estimate of the probability distribution of a continuous variable, fig1(b), fig2(b) respectively.

In t-score quantile plot, the diagonal line represents sample quantile being equal to the theoretical quantile. Data points of genes which are considered to be differentially expressed lie farther away from this line. Specifically, data points with t-scores within $1 - (\frac{1}{2n})$ and $(\frac{1}{2n})$ display with a circles, where n is the total number of genes.

In fig3(a) and fig4(a) we plot estimated false discovery rate for results Pheno1 and Pheno2, Pheno3 and Pheno4 repectively. The parameter lambda, λ is tunning parameter. It is used to estimate a priori probability of null hypothesis, and its value is > 0 and < 1 and we used polynomial method to choose lambda.

Fig3(b) and fig4(b) shows volcano plot of differentially expressed genes we got from HIV-1 Vprs. Cutoff p-value is to define data points which are statistically significant. It is displayed as a horizontal line on the plot. The volcano plot shows $-\log10$ (p-value) vs $\log2$ (ratio) scatter plot of genes. Genes which are both statistically significant (above the p-value line) and differentially expressed (outside of the fold changes lines) are shown in these plots.

In summary the results are shown in the table1.

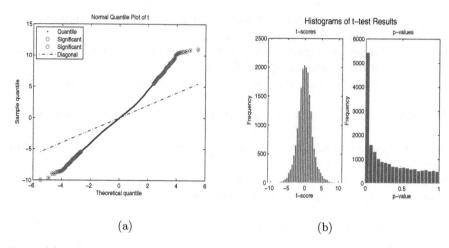

(a)   (b)

Fig. 1. (a) Plot of sample quantile vs theoretical quantile between Wild type HIV-1 vpr and mutant HIV-1 Vpr, R80A; (b) Histogram plot of ttest result of sample used in fig1(a)

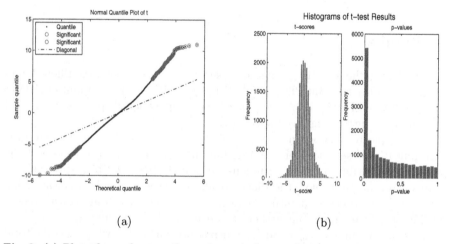

(a)   (b)

Fig. 2. (a) Plot of sample quantile vs theoretical quantile between Wild type HIV-1 vpr and mutant HIV-1 Vpr, F72A/R73A; (b) Histogram plot of ttest result of sample used in fig2(a)

Table 1. Differentially expressed genes we got from wild type HIV-1 Vpr and HIV-1 mutant Vprs

Sample Groups	Differentially expressed genes	Up-regulated	Down-regulated
WT and R80A	1524	583	941
WT and F72A/R73A	1525	584	941

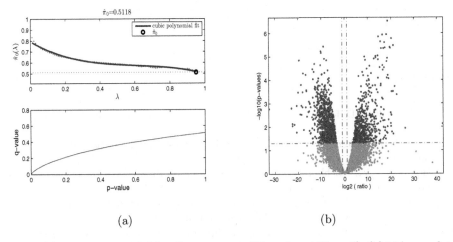

(a)   (b)

Fig. 3. (a) Estimation with false discovery rate (Pheno1 and Pheno2); (b) Volcano plot of differentially expressed genes in Pheno1 and Pheno2

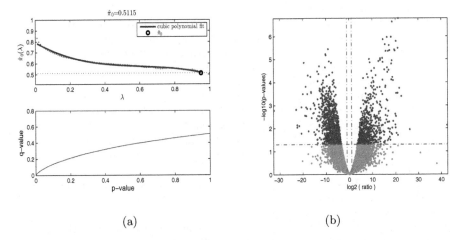

(a)   (b)

Fig. 4. (a) Estimation with false discovery rate (Pheno3 and Pheno4); (b) Volcano plot of differentially expressed genes in Pheno3 and Pheno4

4   Conclusion

Differentially expressed genes shows the variations of samples in different conditions. Our sample data are HIV-1 Vprs data in two variations; one is wild type cell line expression vpr and other two are mutant vprs. Using t-test, we have found 1524 and 1525 number of DEG from two groups, among them down-regulated DEG are 941 in numbers and rest genes are up-regulated. Statistical hypothesis tests are quite appropriate to compare means under relaxed conditions.

From the identified genes we can developed differential network within coexpressed genes. This network may found the metabolic pathway of a particular disease.

References

1. Biology 4361. Differential gene expression. Technical report, Developmental Biology (2008)
2. Bean, G.J., Ideker, T.: Differential analysis of high-throughput quantitative genetic interaction data. Genome Biology (2012)
3. Benjamini, Y., Hochberg, Y.: Controlling the false discovery rate: a practical and powerful approach to multiple testing. J. Royal Stat. Soc. 57, 289–300 (1995)
4. Buechler, S.: Differential expression analysis of microarray data. Technical report, Department of Mathematics, 276B, Hurley Hall (2007)
5. Chen, Y., Mao, Y., Liu, H., et al.: Transcriptome analysis of differentially expressed genes relevant to variegation in peach flowers. PLoS ONE 9(3), e90842 (2014)
6. Chen, Y., Wu, R., Felton, J., et al.: A method to detect differential gene expression in cross-species hybridization experiments at gene and probe level. Biomedical Informatics Insights 3, 1–10 (2010)
7. Chen, Z., Liu, J., Tony Ng, H.K., et al.: Statistical methods on detecting differentially expressed genes for RNA-seq data. BMC Systems Biology 5(3), 1–9 (2011)
8. Dudoit, S., Yang, Y.H., et al.: Statistical methods for identifying diferentially expressed genes in replicated cdna microarray experiments. Technical report, Department of Biochemistry, Stanford University, Stanford University School of Medicine, Beckman Center, B400 Stanford, CA (2000)
9. Dudoit, S., Shaffer, J.P., Boldrick, J.C.: Multiple hypothesis testing in microarray experiment. Statistical Science 18, 71–103 (2003)
10. Dudoit, S., Yang, Y.H., Callow, M.J., et al.: Statistical methods for identifying differentially expressed genes in replicates cDNA microarray experiments. Statistica Sinica 12 (2002)
11. Ishwaran, H., Rao, J.S.: Detecting differentially expressed genes in microarrays using bayesian model selection. Journal of the American Statistical Association 98(462), 438–455 (2003)
12. Kauffmann, A., Huber, W.: Microarray data quality control improves the detection of differentially expressed genes. Genomics 95 (2010)
13. Ospina, Kleine, L.: Identification of differentially expressed genes in microarray data in a principal component space. Springer Plus 2, 60 (2013)
14. Troyanskaya, O.G., Garber, M.E., et al.: Nonparametric methods for identifying differentially expressed genes in microarray data. Bioinformatics 18(11) (2002)

Performance Enhancement
for Audio-Video Proxy Server

Soumen Kanrar and Niranjan Kumar Mandal

Vidyasagar University, CS Department, WB, Midnapore-02, India
kanrar@rahul.ac.in,
Niranjankumarmandl54@gmail.com

Abstract. Bandwidth optimization for real time traffic transmission is a challenging issue for the next generation network. The self control of the traffic rate to the web cache is based on the relative data present in the proxy server or to import the data from the data center node. The smoothness of traffic transmission highly depends on the self configuration of the cache memory of the web proxy server by the optimized page replacement procedure. The web proxy cache memory is finite in size so the enhancement of the system performance depends upon the self optimization of the traffic rate. This paper address the effect of the zipf like distribution parameters to the input traffic request and the effect of the parameters to measure of the bandwidth requirement for the desire Audio/Video data file. This paper presents the result for bandwidth optimization and traffic control depends on the size of the Audio/Video file and the active duration of the transmission stream.

Keywords: Cache Memory, Zipf Distribution, Page Replacement, Traffic Control, Bandwidth Optimization, Self Configuration.

1 Introduction

The cost-effective system development for the smoothly video stream transmission over the wired connection to maintain the grade of service depends on the load to the web cache server. The architecture of video on demand system used web cache to protect the collected data center nods to overcome various types of unwanted attack. The massive traffic load initially submitted to the web cache to fetch the desire video in full range or for the clips or taller/clips of any audio /video. Researchers concentrate over the huge growth of the Internet user and that bring enormous attention to explore the traffic control and bandwidth optimization for the real-time video streaming over the decades. A good number of works considers at the network traffic handle. The hit ratio of a web cache grows to log-like fashion as a function on the cache size [2],[3],[7],[8],[10]. The equivalent capacity in the network and its application to bandwidth allocation in high speed networks for the grade of service [4].The probability based admission control for the incoming traffic rate for the tree based, and graph based network architecture[12]. The problem of bandwidth allocation in the high speed network has been addressed in [4],[5],[6] in various ways

© Springer International Publishing Switzerland 2015
S.C. Satapathy et al. (eds.), *Proc. of the 3rd Int. Conf. on Front. of Intell. Comput. (FICTA) 2014*
– *Vol. 1*, Advances in Intelligent Systems and Computing 327, DOI: 10.1007/978-3-319-11933-5_68

by restrict to bounded ranges of the connection parameters. The performance of multirate request to VoD system from the number of expanding client clusters; optimize by multicusting of the request at the proxy server [11] to the data storage servers. The true traffic control is done at the web proxy server for the limited buffer size of the cache memory. The paper is organized as brief description of the problem and literature survey in the introduction (1). Reaming section describes about the analytic description of the problem (2) and the session initiative Page Replacement with Aggregate Bandwidth algorithm. The section (3) presents the simulation results. The last section (4) describes the about the conclusion remarks further improvement and reference.

2   Analytic Description of the Model

The web proxy traces for the random requests from the infinite heterogeneous population of client. Let A be the given arrival of a page request and B the arrival request for page i. The pages be ranked in order of their popularity (i.e. by the hit count of the page).where page i is the ith most popular page so the page are marked as $1, 1/2, 1/3, 1/4, \ldots 1/i, 1/(1+i)$, $\ldots\ldots$ Now two mutually independent exclusive events occurs A and B, $p(A) = \dfrac{1}{\sum\limits_{i=1}^{N}\dfrac{1}{i}}$ and $p(B) = \dfrac{1}{i}$, so $p(AB) = \dfrac{1}{i} \cdot \dfrac{1}{\sum\limits_{i=1}^{N}\dfrac{1}{i}}$,

where N is a finite natural number. So for the large size of population of submitted request and heterogeneous nature of the request pattern be better approximated [1]

$$p(A) \approx \frac{1}{\sum\limits_{i=1}^{N}\dfrac{1}{i^{\alpha}}} \quad \text{and} \quad p(B) \approx \frac{1}{i^{\alpha}}$$

So, $p_N(i) = p(AB) \approx \dfrac{1}{i^{\alpha}} \cdot \dfrac{1}{\sum\limits_{i=1}^{N}\dfrac{1}{i^{\alpha}}}$. This is the probability for the requested

page i, where $i \in I^{>0}$, assumed that the finite stream of requests R satisfied and the request $(R+1)^{th}$ is made for the ith page, and the page i is not present in the cache memory of the web proxy server . So the distribution obtained (or approximated) by the allowed the repetition permutation $\approx (1 - p_N(i))^R \approx f(i)$ where $1 \le R \le N$, so the corresponding hit rate (miss) express by the hit on demand as

$$H_{demand} = \sum_{i=1}^{N} p_N(i) f(i) \,,$$ Since the cache size is limited to say (C) and

the request comes from the infinite size heterogeneous population .i.e. $R \rightarrow \infty$, so hit

miss be updated as $\lim_{R \rightarrow \infty} \sum_{i=1}^{C} p_N(i)\{1 - p_N(i)\}^R$, clearly $|1 - p_N(i)| \prec 1$ so

the series is convergent. So H_{demand} is equivalent to $\sum_{i=1}^{C} p_N(i)$. If the required

file is not present in the cache, two things be happened simultaneously imported the file from data center nodes connected to web proxy. Update the cache list the page, which is not in used by the dynamic session initiative page replacement algorithm [13]. Figure -1 represent the work flow of the system to the response the submitted request stream.

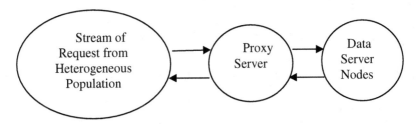

Fig. 1. Traffic flows in the Video on Demand System

So for the hit miss of ith file or page, considering b_i is the required bandwidth

to import the ith file from data center node. Now, Bandwidth $_{demand\ (i)}$ $\approx$ b_i, here, $b_i = s_i.t_i$ in ideal condition. Where S_i the size of the requested is file and t_i is the time duration of the activity of channel.

So, the aggregate bandwidth required on demand approximated

by $\approx \sum_{i=1}^{\infty} H_{demand(i)}.(s_i.t_i) = \sum_{i=1}^{\infty}\sum_{j=1}^{C} p_N(j)(s_i.t_i)$ $= \sum_{i=1}^{\infty}\{\sum_{j=1}^{C} p_N(j)\}(s_i.t_i)$ by

lemma (2.1) we get,

$$= k \sum_{i=1}^{\infty} \alpha.(C^{1-\alpha}).(s_i.t_i) \tag{A}$$

here, k is a threshold value related to packet loss, $0 \prec k \prec 1$ and C is the finite cache memory size, α is the Zipf –like parameter.

2.1   Lemma

For any real or complex value, S the expanding series $\zeta(s) = \sum_{n=1}^{\infty} n^{-s}$ satisfy

$$\zeta(s) \leq |s| n^{1-S}$$

Proof: let s be any complex variable then $s = \alpha + i\beta$, $n^{-s} = \exp(\log\{n^{-s}\})$

$= \exp(-s \log n) = \exp\{(-\alpha - i\beta)\log n\} = [\exp\{(-\alpha - i\beta)\}]^{\log n}$

$$\qquad\qquad\qquad\qquad\qquad \text{by rule } \exp(ab) = [\exp(a)]^b$$

$= [\exp\{(-\alpha) + (-i\beta)\}]^{\log n} = [\exp(-\alpha)]^{\log n} \, [\exp(-i\beta)]^{\log n} = [e^{-\alpha}]^{\log(n)} [e^{-i\beta}]^{\log(n)}$

$= [e^{\log(n)^{-\alpha}}] [e^{-i\beta}]^{\log n} = n^{-\alpha} e^{-i\beta \log n}$

$\Rightarrow n^{-s} = n^{-\alpha} e^{-i\beta \log n}$

So, $|n^{-s}| = |n^{-\alpha} e^{-i\beta \log n}| \leq |n^{-\alpha}| |e^{-i\beta \log n}| \leq n^{-\alpha}$

$$\text{as } \alpha > 0 \quad \text{and } |e^{-i\beta \log n}| \to 1$$

$\zeta(s)$ is analytic in the complex plane except a simple pole at

$$s = 1. \text{Now}, \zeta(s) \frac{1}{s-1} = \sum_{n=1}^{\infty} [n^{-s} - \int_{n}^{n+1} x^{-s} dx]$$

$$= \sum_{n=1}^{\infty} \int_{n}^{n+1} (n^{-s} - x^{-s}) dx$$

$$\text{for, } x \in [n, n+1], \ (n \geq 1) \text{ and } \alpha \succ 0$$

$$\text{We get,} \quad |n^{-s} - x^{-s}| \leq |s \int_{x}^{n} y^{-1-s} dy| \leq |s| n^{1-\alpha}$$

$$|n^{-s} - x^{-s}| \leq |s \int_{x}^{n} y^{-1-s} dy| \leq |s| n^{1-\alpha}$$

$$\Rightarrow \zeta(s) - \frac{1}{s-1} \leq \sum_{n=1}^{\infty} |s| n^{1-\alpha} \text{ So for real, it is } \sum_{n=1}^{\infty} |\alpha| n^{1-\alpha}$$

Session Initiative Page Replacement with Aggregate Bandwidth Algorithm

Var:

i : integer

α, κ, s_i, t_i, C, aggregate_bandwidth of real

List: array [1,2.., n] of string // List contain available video stream name in the system

Hit_count: array [1, 2.., n] of real

Buffer: array [1, 2..., m] of string // m> n

Initialization:

List ← Store the name of the video file name// Preliminary initialization manually

Begin

While (buffer ← index! =Null)

    {

    Read the stream of request to buffer for a session

    }

While (buffer ← index! = Null)

    {

        If (input string match   with any member of the List)

            {

        Hit_count of the file ++

        Stream transfer to the client

            }

        else // Hit Miss

        {

        Step-1: Search the least hit count in Hit_count

        // s_i, t_i is the size of the new video stream and the channel

        // occupancy time, C is the Cache Size, κ is a real threshold

        // α is varying   Zipf parameter

        Step-2: Find   aggregate_bandwidth $\leftarrow k \sum_{i=1}^{\infty} \alpha.(C^{1-\alpha}).(s_i.t_i)$

        Step-3: Replace with new file name entry

        Step-4: Store the file to proxy server from remote storage server

        Step-5: Transfer the video to the client

        }

    }// End while

    Flash buffer

End

End Algorithm

3   Simulation Result and Discussion

In this section, we describe the overall system response on the basis of different values of zipf parameters. The heterogeneous population size of the client request is 100^3 for 10 milliseconds of time and the size as the file varies from (1 to 15 kb), the channel access time varies within (1 to 10 millisecond).

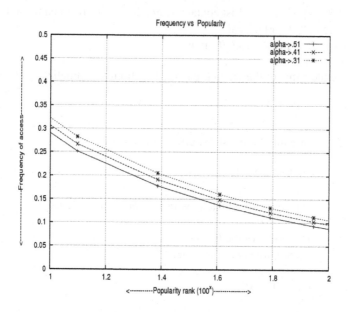

Fig. 2. Traffic Estimation Against Popularity

Figure 2 and 3 present   the frequency of the user choice form the large scale of heterogeneous population of viewers. The traffic traces pattern done at the web proxy server. The Zipf   parameter varies over the range (.98, .75, .64, .51, .41, .31) the proxy server has limited cache memory size. The popularity ranks in the scale 100^x, at x=0, the top rank is 1. According to the plotted simulation figure 2 and figure 3 has a long tail that indicates   the popularity decreases implies the frequency of accessing of the video decreases. The population size of the submitted request is 100^3.

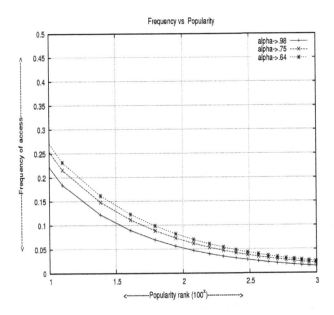

Fig. 3. Traffic Estimation Against Popularity

The simulation shows that the hit misses increases when the value of α increases. The higher value of α indicates the hot spot situation in the proxy server for a particular session of the request stream, where the submitted request for limited files is very high. The figure 2 and 3 indicates the file with rank (1 to $100^{1.1}$) and files with rank (1 to $100^{1.05}$) appears as hot spot situation. The request pattern approximated by different values of α. Figure 4 and figure 5 presents the bandwidth requirement for the video transfer from data center nodes to the heterogeneous client. The data center nodes are distributed, all the requested video streaming through a proxy server. The simulation result in the figure 4 considers Zipf parameters as (.98 and .64). Whereas for the figure 5, it is (.41 and .31). The high bandwidth consumes for the top 100 video. For the simulation purpose, we consider the packet loss and inter packet delay threshold bounded within (0,1) according to the expression (A). Figure 4 and figure 5 shows the comparative scenario for different values of α. Figure 4 indicates bandwidth requirement higher for the lower values of α. The bandwidth required for the file with rank $100^{.475}$ are 9.9 Bits/sec for the α value .98 and 8.9 Bits/sec for the α value .64. The simulation figure shows bandwidth requirement decrease according to the α value decrease. In the video on demand system major request submitted for the high rank video only. The major bandwidth required for the rank (1 to $100^{.75}$) audio/video stream.

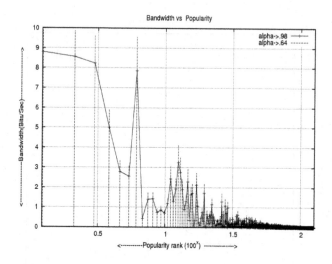

Fig. 4. Bandwidth Estimation Against Popularity

The heterogeneous population size of the client is 100^3 and the size as the file varies from (1 to 15 kb) the channel access time varies within (1 to 10 millisecond). Figure 4 and figure 5 has a long tail that indicated the bandwidth requirement coming less as the popularity goes decrease.

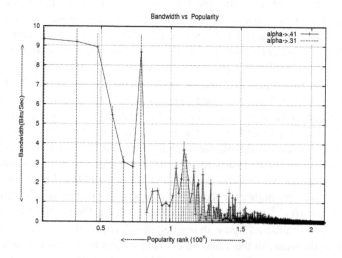

Fig. 5. Bandwidth Estimation Against Popularity

4   Consultation and Remarks

In this paper, we have shown traffic request pattern submitted at the proxy server that matches with the analytic model of the system. We have considered the ideal situation for request response scenario. The hot spot space indicates for a limited number of files, and the required bandwidth estimated for the hit miss and hit rate also. The delay and packet jitter estimation be added to the expression (A) in contrast with the real-world scenario for the further improvement of the paper. The distributed data center node and the routing of the packet to optimize the least hop count between the web proxy and the data center node will present the video streaming more realistic. The new types of algorithms needed to enhance the overall system performance.

References

1. Breslau, L., Pei, C., Li, F., Phillips, G.: Web Caching and Zipf -like distribution: evidence and implications. In: Proceedings of the IEEE INFOCOM 1999, vol. 1, pp. 126–134 (1999) ISSN 0743-166X, doi:10.1109/INFCOM.1999.749260
2. Almeida, V., Bestavros, A., Crovella, M., de Oliveira, A.: Characterizing reference locality in the WWW. In: Proceedings of the Fourth International Conference on on Parallel and Distributed Information Systems, Miami Beach, Florida, USA, December 18-20, pp. 92–107 (1996)
3. Williams, S., Abrams, M., et al.: Removal policies in network caches for www documents. In: ACM SIGCOM (August 1996),
   http://www.acm.org/sigcom/sigcomm96/program.html
4. Guerin, R., Ahmadi, H., Naghshineh, M.: Equivalent Capacity and its Application to Bandwidth allocation in high –speed networks. IEEE Journal of Selected Areas in Communication 9(7), 968–981 (1991)
5. Gill, P., Arlitt, M., et al.: Youtube Traffic characterization, a view from edge. In: Proceeding 7th ACM SIGCOMM Conference on Internet Measurement, IMC 2007, pp. 15–28 (2007), 10.1145/1298306.1298310
6. Wong, A.K.Y.: Web cache Replacement policies: A pragmatic approach. IEEE Network Magazine 20(1), 28–34 (2006), doi:10.1109/MNET.2006.1580916
7. Balamash, A., Krunz, M.: An Overview of Web Caching Replacement Algorithms. IEEE Commun. Surveys & Tutorials 6(2), 44–56 (2004), doi:10.1109/COMST.2004.5342239
8. Shi, L., Gu, Z.-M., Wei, L., Shi, Y.: Quantitative Analysis of Zipf's Law on Web Cache. In: Pan, Y., Chen, D.-X., Guo, M., Cao, J., Dongarra, J. (eds.) ISPA 2005. LNCS, vol. 3758, pp. 845–852. Springer, Heidelberg (2005)
9. Hasslinger, G., Hohlfeld, O.: Efficiency of caches for content distribution on the Internet. In: 2010 22nd International IEEE Teletraffic Congress (ITC), Amsterdam, pp. 1–8 (September 2010), doi:10.1109/ITC.2010.5608730
10. Kanrar, S.: Analysis and Implementation of the Large Scale Video-on-Demand System. arXiv:1202.5094
11. Kanrar, S.: Performance of distributed video on demand system for multirate traffic. In: ReTIS 2011, Kolkata, India, pp. 52–56 (December 2011), 10.1109/ReTIS.2011.6146839
12. Kanrar, S.: Efficient Traffic Control of VoD System. IJCNC 3(5) (September 2011), doi:10.5121/ijcnc.2011.3507
13. Kanrar, S., Mandal, K.N.: Dynamic Page replacement at the Cache Memory for the Video on demand Server. In: Advanced Computing, Networking and Informatics, vol. 2, pp. 461–469. Springer (2014), doi:10.1007/978-3-319-07350-7_51

Design and Implementation of Home Automation System Using Facial Expressions

P.C. Soumya

Computer Science and Engineering, Amrita Vishwa Vidyapeetham, Coimbatore
soumya123v@gmail.com

Abstract. Home automation system means centralized control of lighting, electrical appliances; locking of doors, alarm systems etc. for improving energy efficiency, security, and comfort. Assistive domotics is a home automation system specially developed for elderly and physically challenged persons for their convenience and comport who would otherwise require additional care. In this paper home automation is done using facial expressions which include angry, sad, neutral, fear and smiling. This system is designed for situations were speech control systems and gesture control systems fail.

Keywords: Home Automation, Active Appearance Model (AAM), Relative Sub-Image Based Model (RSB), k-Nearest Neighbors (kNN), Hidden Markov Model (HMM).

1   Introduction

Technology is growing at a rapid rate in our day to day life. Nowadays the introduction of home automation system is playing an important role to make the living easier and comfortable than before. This project implements a home automation system that can be controlled using facial expressions. This system will be beneficial for disabled who have gesture and speech impairments enabling the disabled to be independent of the caregivers.

1.1   Existing Home Automation Systems

There are many automation systems available nowadays. Speech controlled home automation systems are those were controlling of the appliances takes place by the recognition of speech of the user. Speech of the user might be affected when subjected to cold. Hence this system is not reliable. Gesture based home automation systems, were controlling of appliances occurs by the recognition of gestures given by the user as input. Disadvantage of this system is that real time delays causing less accuracy. Bluetooth based home automation system, consisting of a primary controller and a number of Bluetooth sub-controllers [3]. All the home devices will be connected separately to local Bluetooth sub-controller. They communicate with each other via wired connections. A Bluetooth module should be present and is shared by many home devices. This reduces the amount of physical wiring but increases access

© Springer International Publishing Switzerland 2015
S.C. Satapathy et al. (eds.), *Proc. of the 3rd Int. Conf. on Front. of Intell. Comput. (FICTA) 2014*
– *Vol. 1*, Advances in Intelligent Systems and Computing 327, DOI: 10.1007/978-3-319-11933-5_69

delay since a single Bluetooth module is shared between many home devices. Thus the system should be reliable, real time delays should be minimum, less expensive and less access delay.

1.2   Proposed Home Automation System

Here, home automation is controlled using facial expression of the user. Five expressions are focused which include neutral, smiling, sad, fear and angry. Feature extraction is performed using Active Appearance Model (AAM) and Relative Sub-image Model (RSB) in order to compare the performance evaluation between the two. Classification is done Support Vector Machine (SVM). The recognized output is connected to each appliance physically using a wired connection. Once the image is classified the corresponding special character is send through the serial cable to the microcontroller. The microcontroller then performs the respective action.

2   Related Works

Various feature extraction techniques have been implemented such as AAM and RSB. In AAM, key landmarks also known as the fiducial points are labeled on each example. This constitutes the training set. Fiducial points are used to mark the main features. The sets are aligned into a common coordinate frame and represented each by a vector. Then we wrap each example image so that it will match with the mean image. We normalize the example images using scaling and offset to minimize the effects of global lighting variations. To get a linear model principle component analysis (PCA) is performed. More details are discussed in [5].

In RSB, automatic cropping is done to get the maximum face area. This image is divided into many Sub-images and PCA in applied in each Sub-image so as to get local as well as global information. Feature extraction using RSB is fast and accurate whereas AAM is accurate but performs at a slower pace.

As we know there are lots of classifiers in machine learning. One of them is Support Vector Machine (SVM). SVM plays a very important role in pattern classification, for the problem of regression and for learning a ranking function which is by maximizing the margin between various classes. Initially, SVM was implemented for two class problems. In proposed system Multivariate SVM is used where more than two class problems is solved.

Another one is k – Nearest Neighbours algorithm (kNN).The output is a class membership. Classification of objects is performed by a majority vote of its neighbours. Whenever we have a new point to classify, we find its K- nearest neighbours from the training data (vectors in multidimensional feature space each with a class label). Storing of the feature vectors and class labels takes place during the training phase of the algorithm and in classification phase, a query or a test point which is an unlabelled vector is classified by assigning the label which is most frequent and nearest to that query point. The distance is calculated using Euclidean distance, Minkowski distance or Mahalanobis distance. Hidden Markov Model (HMM) is a method applied for pattern recognition such as speech, hand writing, gesture and facial recognition etc. Here the system being modelled is assumed to be a Markov process with unobserved (hidden) states. In [2], the output from kNN is given to HMM in the second classification stage. The HMM reclassifies each input unit by taking into account not only the output of its paired first-stage classifier but also its classification of the previous units. In this way, each unit is classified on the basis of both its feature vector and its temporal relationship with previous units [2].

3   System Architecture

This section describes a compact, flexible and a low cost home automation system.

The software part constitutes image acquisition, noise removal, edge detection, feature extraction, PCA (comparison between traditional and modular PCA algorithm) and classifier. The decision value is the function or output of each recognized expression which is send to the hardware interface.

To ensure that the reproduction of your illustrations is of a reasonable quality, we advise against the use of shading. The contrast should be as pronounced as possible.

If screenshots are necessary, please make sure that you are happy with the print quality before you send the files.

Wired connection is established between the hardware components which consist of a two DC motors, small speaker, LED and GSM (Global System for Mobile Communications) modem (Fig. 1 shows the system architecture).

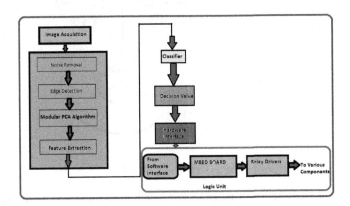

Fig. 1. System Architecture

3.1   Image Acquisition

Image is converted to YCbCr image to identify mouth and eye region. It is observed that Cb value is high around eye region and Cr value is high around mouth region. Then it is converted to binary image.

3.2   Noise Removal

After converting to binary image, few pixels which may not be skin pixels are detected as skin pixels. These pixels are noise. To remove them, a 5 x 5 mask with a low pass filter is implemented. Every pixel of the block is set to white if the number of white pixels is more than half of total pixels in that block; else this 5 × 5 block is transformed to a complete black block [1].

3.3   Edge Detection

For edge detection we use Sobel Edge Detector. Two 3× 3 kernels (h_x, h_y) are convolved with the original binary image and approximations of the derivatives are calculated i.e., one for edges in horizontal direction and other for edges in vertical direction. G_x and G_y are the two images which at each point contain the vertical and horizontal derivative approximations [1].

3.4   Feature Extraction

Here a comparison is done between two feature extraction algorithms i.e., AAM and RSB. In AAM, key landmarks also known as the fiducial points are labeled on each example. This constitutes the training set. Fiducial points are used to mark the main features. The sets are aligned into a common coordinate frame and represented each by a vector. Fiducial points are shown in Fig. 2.

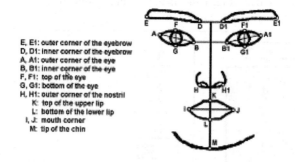

E, E1: outer corner of the eyebrow
D, D1: inner corner of the eyebrow
A, A1: outer corner of the eye
B, B1: inner corner of the eye
F, F1: top of the eye
G, G1: bottom of the eye
H, H1: outer corner of the nostril
K: top of the upper lip
L: bottom of the lower lip
I, J: mouth corner
M: tip of the chin

Fig. 2. Fiducial Points [6]

Then we wrap each example image so that it will match with the mean image. We normalise the example images using scaling and offset to minimize the effects of global lighting variations. To get a linear model principle component analysis (PCA) is performed. More details are discussed in [5]. In RSB, automatic cropping is done to get the maximum face area. This image is divided into many sub-images and PCA in applied in each sub-image so as to get local as well as global information (Modular PCA). RSB features are calculated for each combination [4].

In the proposed system, expressions with **occlusions** (usage of glasses or eye masks) can also be recognized under almost all lighting conditions. The dataset consists of four users each consisting of 40 images (20 with glasses and 20 without glasses) thus making a total of 160 images. The training set for each user consists of 40 images and the test set consists of images which are taken from the web camera constituting 40 images for each user with and without glasses.

Fig. 3. Sample Dataset

3.5   Classifier

Multivariate SVM is used where more than two class problems is solved and works well for unseen examples. Five class means are considered as five nodes and are separated as far as possible. Maximization of margin means more generalization ability. Hyper planes are separated far apart until similar data points are met [4].

3.6   Hardware Implementation

Live image of the expression is taken automatically using a camera. The output from the matlab (recognition of angry, sad, neutral, fear and smiling) is send to each appliance physically using a wired connection. Once the image is classified the corresponding special character is send through the serial cable to the microcontroller. The microcontroller then performs the respective action. Actions performed are given in Table 1.

Table 1. Table showing the events

Expressions	On	Off
Angry	Light	Light
Sad	Fan	Fan
Neutral	Alarm	Alarm
Smile	Door Open	Door Close
Fear	Send Message	Send Message

The switching on and off of the appliances taken place when the same expression is shown simultaneously.

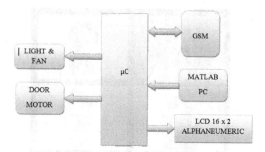

Fig. 4. Block Diagram

Special characters will be sent from the matlab to the microcontroller (PIC16F877A). The program in the PIC is run and appliances are controlled when showing different expressions (Fig. 4). The circuit diagram is given in below (Fig. 5).

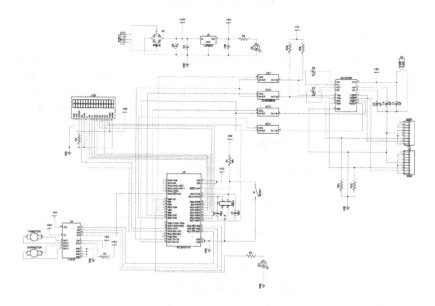

Fig. 5. Circuit Diagram

4   Experimental Analysis

Performance was measured by weighted accuracy (WA) and unweighted accuracy (UA). For binary classification, WA and UA are defined by equation 1 based on the number of true positive (tp), false positive (fp), true negative (tn), and false negative (fn) obtained [2].

$$\mathbf{WA} = tp + tn/ (tp + tn + fp + fn) \,.$$
$$\mathbf{UA} = (tp/ (tp + fn) + tn/ (fp + tn))/2 \,. \tag{1}$$

The proposed system is tested in various lighting conditions and the accuracy for each expression is listed in the Table 2.

Table 2. Performance analysis

Expressions	Accuracy
Angry	95.8 %
Sad	93.0 %
Neutral	98.9 %
Smile	98.2 %
Fear	98.5 %

It is observed that the performance of AAM with occlusion is less as compared with the RSB algorithm with occlusion. AAM is showing only 85 % accuracy whereas RSB algorithm is showing 97 % accuracy.

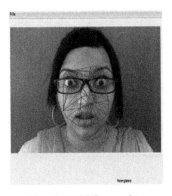

Fig. 6. AAM Screenshot   **Fig. 7.** RSB Screenshot

5   Conclusion

The system is tested under all lighting conditions and it is observed that RSB algorithm is better than AAM for expression recognition. The hardware implemented is simple and cost efficient and is reliable.

References

1. Mittal, R., Srivastava, P., George, A., Mukherjee, A.: Autonomous Robot Control Using Facial Expressions. International Journal of Computer Theory and Engineering 4 (2012)
2. Meng, H., Bianchi-Berthouze, N.: Affective State Level Recognition in Naturalistic Facial and Vocal Expressions. IEEE (2013)

3. Gill, K., Yang, S.-H., Yao, F., Lu, X.: A ZigBee-Based Home Automation System. IEEE (2009)
4. Kudiri, K.M., Md Said, A., Yunus Nayan, M.: Emotion Detection Using Sub-image Based Features through Human Facial Expressions. Computer and Information Science Department, Universiti Teknologi PETRONAS, Bandar Seri Iskandar, 31750 Perak, Malaysia (2012)
5. Cootes, T.F., Edwards, G.J., Taylor, C.J.: Active Appearance Model, Wolfson Image Analysis Unit, Department of Medical Biophysics, University of Manchester, M13 9PT, U.K (1998)
6. Pantic, M., Rothkrantz, L.J.M.: Facial Action Recognition for Facial Expression Analysis From Static Face Images. IEEE (2004)
7. Viola, P., Jones, M.: Robust Real-time Face Detection, Cambridge, MA 02 142 (April 2010)
8. Li, P., Phung, S.L., Bouzerdom, A., Tivive, F.H.C.: Automatic Recognition of Smiling and Neutral Facial Expressions. School of Electrical, Computer and Telecommunication Engineering, University of Wollongong, Wollongong, NSW 2522, Australia (2010)

Semantic Web Services Discovery Using Logic Based Method

Simhadri Srujana, V. Sreenivas Raju, and M. Kranthi Kiran

Department of Computer Science and Engineering,
Anil Neerukonda Institute of Technology and Sciences,
Sangivalasa, Bheemunipatnam [M], Visakhapatnam
srujana_simhadri@yahoo.com,
{vsraju.cse,mkranthikiran.cse}@anits.edu.in

Abstract. In the era of SOA, web services are gaining importance in various fields and e-commerce is one of the promising domains. Many methods are available for discovering web services based on syntax, but lot of irrelevant services are also obtained in this method. To overcome the problem of syntax based method various semantic based discovery methods are existing as in literature survey. In this paper we are proposing scalable semantic web service discovery method without compromising the expressive power with reasoning using prolog and graph database like Allegro graph.

Keywords: SOA, Web Services, Semantic Web Services, Scalable, Allegro Graph.

1   Introduction

Web Services[2] are defined as the services which are provided, published, found and used in web. Web services are web applications components(applications which run on web are called web applications). Before understanding web services we must know the basics of XML. XML is Extensible Markup Language which is designed to store and transport the data.

Importance of web services[2,4] are such as, web applications differ from the normal client server applications, as in client server application not all the clients (users) can access server application. But the web applications, could be accessed from any part of the globe. The standards which are used by web applications for data transferring and communication are HTTP and HTML. Most importantly the reusability, time saving and cost saving are given by these web services. Web services are platform independent as well as hardware, location & also programming language independent and are well known to their interoperability.

Pitfalls of web services[3] are people using internet knew that, the websites provided in internet are not fully available (Availability). The general services are created for customers, where one among them may need some extra feature which is unavailable and should go for another choice/search (Matching requirements). A web service creator should avoid changing of methods or parameters that customers need, as it may cause an interruption in the customers programs. So the only thing can be done is to create new methods and add, but not modify or change(Immutable Interfaces). Most of the services use HTTP, where it is not a reliable protocol doesn't give any guaranteed delivery(Generated Execution). For sending private information to web services, Secure Socket Layer(SSL) program is used. Here the problem is that when large data transfers are done, SSL doesn't work(Security and Privacy).

Before discussing about semantic web services, let us have a glance of what is semantic web? Semantic web[5] is decipherable extension of the world wide web. The information on the websites are used to be presentation oriented but soon changed into content oriented representation only because of the semantic web.

The semantic web service[5] are linguistic extensions of web services and procedure oriented extensions of semantic web. The discovery, composition, invocation and interoperation of services are been backed by meaning oriented description which was been programming interface oriented description of web services. Presently there are two dominant resources, Web Services Modelling Ontology (WSMO) and Web Ontology Language for Services(OWL-S)[8].

2   Semantic Web Services Frameworks

The frameworks of Semantic Web Services are shown in the paragraphs below.

OWL-S (Ontology Web Language for Services): The properties and capabilities of web services in Ontology Web Language[10] can be described by this upper ontology for aiding the automation of web service discovery, execution , composition and interoperation. OWL-S defines an upper ontology. The upper ontology of services is mainly divided into three parts, Service Profile, Service Grounding, Service Model. Service Profile gives all the information which is needed by the user to purely discover a service whereas the service grounding and service model together provides information how to make use of it.

WSMO(Web Service Modelling Ontology): To expedite the automation of discovery, combining and invoking electronic services on the web, WSMO[11] provides a theoretical framework and well-formed web services. For semantic web services related aspects, the WSMO is a meta-model. Meta-Object Facility is a framework for indicating , composing, admonishing   technology neutral meta-models. Web Service Modelling Framework[9] processes and boosts its conception through formal language. To provide a reference architecture and implementation for effective discovery, selection, mediation, invocation and inter-operation of semantic web service based on WSMO specifications. WSMO characterised four top level elements that explain semantic web services, which are Goals, Ontologies, Mediators and web services.

GOALS-As described in the web service capability, the user need aspect along with the requested functionality as opposed to provided functionality.

ONTOLOGIES-It provide the shared common terminology by concept and their relationships along with axioms and instances.

MEDIATORS-Its main aim is to beat the interoperability on data, protocol and process level, where on data level the mismatches between different terminologies are shown. Goals or web services are on process level and web services communications are on protocol level.

Four types of mediators[6],

- ggmediators-provides link to two goals
- oomediators-provides interoperability between two ontologies
-wgmediators-provides link between services and goals.
-wwmediators-mediate between two services

WEB SERVICES-Access to services is been provided by the computational entity. The description contains capabilities, interfaces and internal working. From three different viewpoints, the web services are described in WSMO, they are non-functional properties, Functionality, Behaviour.

WSDL-S(Web Service Description Language-Semantics):Compared to WSDL[12], WSDL-S[14] is light-weighted approach to augment web services with semantics. The semantic models which are related to WSDL-S are controlled outside the WSDL document by WSDL extensibility elements. The inputs, outputs and the operations are annotated by WSDL-S. Along with it, WSDL-S also provides mechanism to specify and annotate preconditions and effects of web services. The automation of the process of service discovery is implemented by combining the preconditions, effects with semantic annotation of inputs and outputs. WSDL-S use the category extension attribute on interface element for publishing web service in UDDI(registry). Some key design principles for WSDL-S are ,services provided by the web standards are fast, Web Services Description Language-Semantics is independent of language, WSDL-S allows multiple annotations, Web Services Description Language-Semantics allows upgrade of deployed WSDL documents to include semantics.

SAWSDL(Semantic Annotations for WSDL):It is itself defined as adding up of semantic annotations to the WSDL components, SAWSDL [13] is not a language but provides mechanism to refer the concepts from the models of semantic. The annotations like preconditions and effects are not considered in SAWSDL. Key design principles for SAWSDL are, Linguistic annotations for web services is completely based on the extensibility framework of WSDL, SAWSDL is needed for both mapping languages and semantic representation, SAWSDL gives semantic annotations for both discovery and invoking web services, SAWSDL annotates WSDL interfaces and operations with categorization information that can be used to publish a web service in a registry, SAWSDL considers bottom-up approach to model web services concepts without considering WSDL.

3   Semantic Web Services Tasks

Semantic Web Services are branched into three different tasks namely Discovery, Selection, Execution.

3.1   Discovery

In general, the web service discovery is used to find web services that elate particular requirements. When grounded in a accurate linguistic formalism, the process becomes a semantic service discovery. Discovery[1] process is nothing but the matchmaking. Services are described by IOPEs- Input, Output, Pre-conditions and effects and the non-functional properties. The process of discovery involves, Firstly the service request is initiated, then the matching process in the repository is done using the matchmaking engine. Next after the matching is done, the results which are obtained are done for negotiation of services which is invoked to communicate with the providers to access information dynamically. Among all the services, the selection of the services are done as per the preferences of the client. Finally the discovery is done by conjuring the services which are selected.

Discovery implementation undergoes three different steps.

- Prefiltering -Process of keyword matching using algorithm.
- Logical matching-IOPEs of the goal and web services.
- Preparing result –List of matching services are taken into consideration with Quality of Services data based on past execution.

3.2   Selection

Selection means ranking up the services which are been discovered accordingly that which is based on Quality of Service (QoS).

3.3   Execution

After the selection, execution[7]is the final task which is done in the process of web services. For a service execution, it should consist of all the necessary actions that need to be taken under consideration during the runtime to get one or more services in an arranged picture. The actions could altogether be initiation, control and validation of service invocations. The most important assets are generated termination and reliability, a consistent state before and after execution in the presence of failure.

4   Related Work

As this paper presents the semantic web service discovery, let us see the various discovery methods. Many existing approaches for web service discovery are mentioned below.

Two-Phase Semantic Web Service Discovery Method[1] was proposed by Laszlo Kovacs, Andras Micisik, Peter Pallinger in which the traditional information retrieval based pre-filter step and a logic based steps combines and uses a technique based on Prolog-Style unification and is used for implementation of INFRAWEBS. The approach finds matching of intersection type and the other two possibilities which are provided are comparison and ranking.

Falak Nawz, Kamram Qadir and H. Farrooq Ahmad proposed push model[15] for web service discovery in which before the discovery implementation, the service notification is provided by the service clients. The descriptions are matched by

OWL-S which is an ontology language in which semantic based web service are used. By the scores assigned for each published web service ranking is given using concept matching. Phases are of two distinct types, subscription module and notification module.after the ranking is done the best matched web service is selected by IOPEs to OWL-S description stored in registry. Matching can be as exact, plug-in, subsume, enclosure, unknown and fail.There is also a limitation that adds overhead in developing and maintaining new components in system architecture.

J.Zhou, T.Zhang, H.Meng, L.Xiao, G.Chen and D.Li proposed web service discovery based on keyword clustering and concept expansion [16].Based on the Pareto principle, the calculation of the similar words in ontology and the similar words are therefore used for semantic reasoning. In the approach Bipartite graphs are used to find the matching degree between the provided services and the users request. The algorithm named Kuhn-Munkres is described to compute the optimal matching of bipartite graph.

Y.TSAI, San-Yih,HWANG and Y.TANG proposed an approach which consists of [17] information about service providers,service and operation descriptions by providers and users , tags and categories and also Quality of Service (QoS) attributes. By using the ontology, the similarities between input/output of query and web service are calculated.If an input is given , the message part is mapped to concept present in the ontology. This test for similarity is performed at three levels. Finally the overall similarities between the given input and the services available are calculated as weighted sum of similarities of all associated attributes.

Guo Wen-Yue,Qu Hai-Cheng and Chen Hang proposed three layers dividing them by using filters to minimise the search area. The approach [18] has been applied on intelligent automotive manufacturing system. The three different matching layers are matching by service category , matching by service functionality and matching by Quality of Service . Using ServiceProfile documents for matching the semantic web service discovery is based on the OWL-S. From all the layers degrees, the service matching degree is calculated. Finally the list of services which meet the best are presented to the clients.

5   Proposed Work

Scalability has become the key issue in semantics-based approaches as and when number of services grow in the repository with time. We are proposing a method in which scalability issue of semantic web services can be resolved without compromising with the expressive power.

In the recent years graph databases like Neo4J and Allegro graph has been gaining popularity in semantic web technologies. We are proposing the following steps for improving the scalability issue.

Step 1: Convert the Service Descriptions written in RDF/OWL into subject, predicate, object and store them as Triple store. We are proposing Allegro graph database to store the triples.

Step 2: Discover Services using reasoning through Prolog based interface from Allegro graph.

Step 3: Display results to the user.

Allegro graph Database can store million of triples and also it has RDFS++ Reasoner which can reason predicates correctly. The benchmark results can be viewed at http://franz.com/agraph/allegrograph3.3/agraph3.3_bench_lubm.lhtml#lubm8000.

6   Conclusion

Based on the various techniques discussed in the above literature we have identified that scalability is one the challenge in Semantic web Service discovery. As we know that Logic based methods has proved efficient in reasoning or discovering the hidden facts from the service descriptions, we are proposing to store the RDF/OWL Service descriptions in the Allegrograph and perform the reasoning using Prolog. We need to still experiment on various Semantic web service descriptions and verify the results in detail.

References

1. Kovács, L., Micsik, A., Pallinger, P.: Two-phase Semantic Web Service Discovery Method for Finding Intersection Matches using Logic Programming. In: SemWS. CEUR Workshop Proceedings, vol. 316. CEUR-WS.org (2006)
2. Web Services Glossary,
   http://www.w3schools.com/webservices/ws_intro.asp
3. Pitfalls of Web Services,
   http://www.pearsonhighered.com/samplechapter/0672325152.pdf
4. Web Services Benefits,
   http://www.altova.com/whitepapers/webservices.pdf
5. Varga, L.Z., Hajnal, A.: Semantic Web Services Description Based on Web Services Description. Position paper for the W3C Workshop on Frameworks for Semantics in Web Services, Computer and Automation Research Institute (2005)
6. Mohebbi, K., Ibrahim, S., Idris, N.B.: Contemporary Semantic Web Service Frameworks:An Overview and Comparisons. International Journal on Web Service Computing (IJWSC) 3(3) (2012)
7. Semantic Web Service Execution,
   http://iscte.pt/~luis/papers/CASCOM_ch12_service_execution.pdf
8. Martin, D., et al.: Bringing Semantics to Web Services: The OWL-S Approach. In: Cardoso, J., Sheth, A.P. (eds.) SWSWPC 2004. LNCS, vol. 3387, pp. 26–42. Springer, Heidelberg (2005)
9. Fensel, D., Bussler, C.: The Web Service Modeling Framework WSMF. Electronic Commerce: Research and Applications 1, 113–137 (2002)
10. Martin, D., et al.: OWL-S: Semantic Markup for Web Services. W3C member submission (December 6, 2004)
11. de Bruijn, J., et al.: Web Service Modeling Ontology (WSMO). W3C member submission, DERI (2005)
12. Chinnici, R., et al.: Web Services Description Language (WSDL) Version 2.0 Part 1: Core Language. W3C recommendation (2007)
13. Farrelland, J., Lausen, H.: Semantic Annotations for WSDL and XML Schema (March 3, 2007)

14. Akkiraju, R., et al.: Web Service Semantics - WSDL-S, W3C member submission, Version 1.0 (2005)
15. Nawaz, F., Qadir, K., Farooq Ahmad, H.: SEMREG-Pro: A Semantic based Registry for Proactive Web Service Discovery using Publish Subscribe Model. In: Fourth International Conference on Semantics, Knowledge and Grid. IEEE Xplore (2008)
16. Zhou, J., Zhang, T., Meng, H., Xiao, L., Chen, G., Li, D.: Web Service Discovery based on Keyword Clustering and Ontology. In: IEEE International Conference, pp. 844–848 (2008)
17. Tsai, Y.-H., San-Yih, H., Tang, Y.: A Hybrid Approach to Automatic Web Services Discovery. In: International Joint Conference on Service Sciences. IEEEXplore (2011)
18. Guo, W.-Y., Qu, H.-C., Chen, H.: Semantic web service discovery algorithm and its application on the intelligent automotive manufacturing system. In: International Conference on Information Management and Engineering. IEEEXplore (2010)

Analysis of Estimation Techniques for Feature Based Navigation of Unmanned Air Vehicles

Shubha Bhat[*], Vindhya P. Malagi, D.R. Ramesh Babu, Krishnan Rangarajan, and K.A. Ramakrishna

Computer Vision Lab. Department of Computer Science and Engineering,
Dayananda Sagar College of Engineering
Bangalore 560 078, India
{shubha22,vindhyapmalagi,bobrammysore,
krishnanr1234,karamakrishna}@gmail.com

Abstract. This paper investigates different linear and non-linear state estimation techniques applicable in the scenario of unmanned air vehicle platform with an image sensor that attempts to navigate in a known urban terrain. Feature detection and matching steps are carried out using DTCWT (Dual tree Complex Wavelet Transform) descriptor. The state parameters of the vehicle are subsequently estimated using the measurements from the image sensor. Due to the non-linear nature of the problem, non-linear filters such as particle filters are often used. Various other methods have been proposed in the literature to the problem ranging from linear Kalman filter to non-linear probabilistic based techniques. This paper therefore attempts to study different state estimation methods and understand two main approaches in particular the Kalman and particle filter for the problem.

Keywords: Unmanned Air Vehicle, Dual Tree Complex Wavelet Transform, Kalman filter, particle filter.

1   Introduction

Unmanned Air Vehicles (UAV) are deployed for a number of defense applications ranging from target recognition, monitoring areas of interest, to border and port surveillance. Consequently, it becomes inevitable that the vehicle navigates in any type of terrain with a reactive mission and accurate outcome in the form of precise target detection or even controlled autonomous navigation. The unmanned air vehicle navigates in a known urban terrain with camera as the sensor that aids in navigation.

Urban terrains are characterized by rich geometric structures with salient features such as sharp corners and prominent edges in contrast to hilly terrains. However they also pose new problems for the autonomous navigation of the vehicle. The UAV has the capability to rotate along all the three planes (X, Y and Z planes; roll, pitch and yaw angles respectively), shift or translate their position and fly high or low presenting new challenges on the captured images which may tend to get distorted.

---

[*] Corresponding author.

© Springer International Publishing Switzerland 2015
S.C. Satapathy et al. (eds.), *Proc. of the 3rd Int. Conf. on Front. of Intell. Comput. (FICTA) 2014*
– *Vol. 1*, Advances in Intelligent Systems and Computing 327, DOI: 10.1007/978-3-319-11933-5_71

For instance, a square may appear as a parallelogram (skewing effect) from a certain viewing angle. In this context it is therefore essential to detect features which are invariant to translation, rotation and scale. A detailed analysis of different feature detectors and descriptors is carried out in [1] motivating to choose Dual-tree Complex Wavelet Transform (DTCWT) as the feature detector and descriptor. DTCWT introduced by Kingsbury [2] provides better directional selectivity, provides information on local phase extraction, is perfectly reconstructable and has limited redundancy than discrete wavelets or even its Fourier counterparts. Moreover with the changing environmental illumination conditions, the feature descriptor has to be robust to localize the features successfully and match accurately with its consecutive frames. This correspondence is further used to estimate the location of the UAV [15].

This paper is organized as follows: in Section 2, the evolution of the estimation theory is discussed and this sets the context for the selection of two main state estimation techniques namely the Kalman filter and the particle filter. Section 3 briefly describes the two chosen state estimation techniques. The associated results and discussed in section 4, followed by the conclusion in Section 5.

2   Evolution of Estimation Theory

This section describes the evolution of the estimation theory to compute the unmanned air vehicle's state parameters. Optimal features are extracted from the image stream using one of the robust approaches as mentioned in the previous section, which are then identified, matched and tracked in the subsequent image frames of the video. The correspondence set so obtained serve as an input to estimate the state of the vehicle.

Vehicle egomotion is usually represented in the form of a state vector comprising a number of parameters namely the location of the vehicle in terms of position (linear motion x, y, z), orientation (roll, pitch, yaw), the altitude of the vehicle from the ground, camera field of view, the velocities, angular rates and the error terms (measured and process noises). Due to the uncertainty and dynamic nature of the system and the process, a probabilistic approach is normally chosen and a Bayesian framework best fits. The literature basically focuses on three major areas of probabilistic based approaches to estimate the states and the parameters: stochastic filtering theory, Bayesian filtering theory and the Monte Carlo methods. The filtering refers to filtering out the unwanted samples or data from the given space.

Stochastic filter theory was pioneered in the early 1940s by Norbert Wiener and Andrey N Kolmogorov which finally converged towards Kalman filters [3]. Many filtering techniques have evolved over the years to help in the process of state estimation and subsequent path planning and Bayesian estimation theory formed the base of all such filters [4].

Over the years Kalman filtering technique and its variants have been followed strongly for the process state estimation. The Kalman Filter is one of the most widely used methods for tracking and estimation due to its simplicity, optimality, tractability and robustness [5].

Today a number of variants of Kalman filter exist applicable to linear Gaussian systems. However Kalman filters suffer from the limitation that they are not capable of handling non-linear and non-Gaussian systems. On the other hand, numerous non-linear filtering methods such as the Extended Kalman filter (EKF), Unscented Kalman filter (UKF), Gaussian quadrature Kalman filter (QKF), Mixture Kalman filter (MKF), Adaptive Kalman filter etc. and their variants have been developed to overcome its limitation to an extent. However there are limitations of the above said methods also. For instance, EKF fail if the system exhibit substantial non-linearity and if the state and the measurement noise are significantly non-Gaussian. Also, due to the presence of Jacobian matrix, the EKF becomes computationally expensive. Several approaches were developed to overcome the limitation by approximating the posterior distribution in a pre-specified function basis expansion. For e.g., in a Gaussian Sum Filter, posterior distribution is approximated by a mixture of Gaussians. Further developments to EKF were the UKF and the QKF which are based on the so-called sigma points and the Gauss-Hermite quadrature integration techniques respectively. These techniques are less computation intensive unlike EKF. Coming to the limitations of UKF and QKF, they are valid only in cases were the posterior distribution can be closely approximated by a Gaussian distribution and hence cannot be applied to all multimodal non-linear state space scenarios.

Bayesian theory particularly the Bayesian inference based approaches [4] (Hidden Markov model (HMM) and Dynamic Bayesian Networks (DBN)), were applied in the field of statistical decision, detection and estimation, pattern recognition and machine learning. In the literature it is evident that Bayesian theory was explored in the area of filtering, recursive Bayesian estimation and moving further in the field of sequential parameter estimation [6] also called by the name Bayesian learning. Several approximation methods were explored (e.g. point mass) in the same Bayesian filtering framework.

Monte Carlo theory [7] initially originated way back in the 18th century, but the modern formulation began only during 1940s. Many Monte Carlo techniques such as the Bayesian Bootstrap, Markov Chain Monte Carlo, Hybrid Monte Carlo and Quasi Monte Carlo, Sequential Monte Carlo (SMC) etc. have been explored in the field of filtering theory. Basically, it uses a statistical sampling and estimation technique that allows on-line estimation of states and parameters with a reasonable expense of computational cost. Along with the Monte Carlo approximation methods (Importance Sampling, Rejection Sampling, Sequential Importance Sampling, Sampling Importance Resampling and Stratified Sampling) also started gaining prominence due to the fact that the estimation accuracy was independent of dimensionality of the state space. However several limitations also were seen w.r.t maintaining consistency, unbiasedness, efficiency, robustness and minimal variance principle.

During the 1960s some of the methods such as the Sequential Monte Carlo started gaining importance due to its on-line state estimation feature which was applied in the field of vehicle state and parameter estimation [8][9]. This resulted in the formulation of the Particle filter concept. This was applied to non-linear, non-Gaussian and non-stationary situations. Several variants of the particle filter were also developed in the 1990s and early 2000s and particularly Rao-Blackwellization [10], a marginalization technique was developed. The main feature of this method was the improvement of the inference efficiency and consequently reducing the variance which were some of the limitations in the Monte Carlo approximation methods.

To summarize, within the sequential state estimation framework, Kalman filters best fits in a linear-Quadratic-Gaussian model, particle filter rooted deeply in Bayesian statistics while the Monte Carlo techniques are emerging as powerful solution candidates in tackling real-time problems within non-linear and non-Gaussian models.

3   KALMAN Filters and Its Variants

The Kalman-Filter (KF) considers linear systems with Gaussian noise. The filtering process begins with the prediction of a noisy estimate of the system state according to a state transition model. Measurements of the system are predicted as if the true system state is equal to the state prediction. The measurement prediction is compared to an actual, noisy system measurement and the state estimate is updated according to a weighted measurement residual. The weight of the measurement residual represents the confidence of the measurement with respect to the model-based state estimate [3].

It updates the unknown parameters in space, stored in the so called state vector, in an iterative manner at every time step. The complete information about the map and the trajectory is stored in the state vector and their covariance matrix. Needless parameters can be eliminated by discarding them from the covariance matrix and from the state vector. The accuracy of the Kalman-Filter depends on the accurate parameter initialization. However, the application of the KF to nonlinear and non-Gaussian systems may give inaccurate results [11].

3.1   Extended KALMAN Filters

As Kalman Filter can be applied for Linear Gaussian systems only, a new approach called the Extended Kalman Filter (EKF) was made use of, which was suitable for non-linear and non- Gaussian type of systems [11]. EKF simply linearize all nonlinear models using nonlinear transformation so that it retains the simplicity of the traditional linear Kalman Filter. Although the EKF (in its many forms)is a widely used filtering strategy, over thirty years of experience with it has led to a general consensus within the tracking and control community that it is difficult to implement, difficult to tune, and only reliable for systems which are almost linear on the time scale of the update intervals. However, linearization is the primary source of error in EKF.

3.2   Unscented KALMAN Filters

The EKF does not retain the optimality of the Kalman filter because the linearization step does not properly handle the statistics of the problem. This inaccuracy is because the random variables no longer have normal distributions after undergoing a nonlinear transformation. The Unscented Kalman Filter [12][17] employs the Unscented Transform (UT) to more accurately estimate the statistics of a random variable that is passed through a nonlinear transformation. The UT is built on the intuition that it is easier to approximate a probability distribution than it is to approximate an arbitrary nonlinear function or transformation. A central and vital operation performed in the

Kalman Filter is the propagation of a Gaussian random variable (GRV) through the system dynamics. In the EKF, the state distribution is approximated by a GRV, which is then propagated analytically through the first-order linearization of the nonlinear system. This can introduce large errors in the true posterior mean and covariance of the transformed GRV, which may lead to sub-optimal performance and sometimes divergence of the filter. The UKF addresses this problem by using a deterministic sampling approach. The state distribution is again approximated by a GRV, but is now represented using a minimal set of carefully chosen sample points. These sample points completely capture the true mean and covariance of the GRV, and when propagated through the *true* nonlinear system, captures the posterior mean and covariance accurately to the 3rd order (Taylor series expansion) for *any* nonlinearity. The EKF, in contrast, only achieves first-order accuracy. The UT calculates the mean and covariance of a variable undergoing a nonlinear transformation by propagating a set of points (called sigma points) [12] through the nonlinear function and estimating the new statistics of the distribution. The sigma points are chosen according to a deterministic algorithm so that they accurately approximate the true mean and covariance of the untransformed distribution. However UKF approach to state estimation is computationally expensive.

3.3   Square Root Unscented KALMAN Filters

The UKF approach to state estimation has been shown to produce superior results to the EKF for nonlinear systems. The UKF, however, is a computationally expensive algorithm in comparison to the EKF. This expense is because the sigma point calculation,

The Square-Root Unscented Kalman Filter (SR-UKF) [12] is an efficient variant of the UKF in which the matrix square root required to generate the sigma points is eliminated by propagating the square root of the covariance matrix in lieu of the full covariance.

4   MONTE CARLO Filtering (PARTICLE Filter)

The Kalman filter and its nonlinear variants (EKF, UKF, SR-UKF) operate under the assumption that the probability distribution of the state estimate is Gaussian, or able to be parameterized by a mean and covariance. The particle filter is a Monte Carlo (MC) method [13][16] for state estimation which represents the probability distribution as a set of weighted random samples. This formulation allows the estimation process to handle non-Gaussian and multi-modal [14] probability density functions. Like the Kalman filtering approaches, the particle filter operates in a recursive fashion. The state is predicted according to a process model and, subsequently, updated by comparing sensor measurements to measurement predictions calculated with a measurement model. However, while the Kalman filter propagates a normal distribution through the nonlinear models, the particle filter [18] propagates a set of weighted samples called particles.

5   Results and Discussion

In the following section the estimation of the position of the vehicle along X, Y and Z axis using the basic Kalman filter is simulated and plotted as shown in Fig 1, Fig 2 and Fig 3 respectively. Input to the simulator are the aerial synthetic video sequences generated by the tool, from which features are extracted and matched between successive frames. The transformation between the successive images are calculated to yield the vehicle parameters in X, Y and Z directions. The Kalman filter refines and resamples these parameters to increase the accuracy.

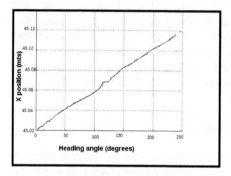

Fig. 1. UAV position along X-axis

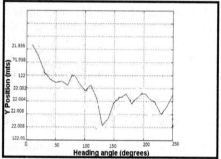

Fig. 2. UAV position along Y-Axis

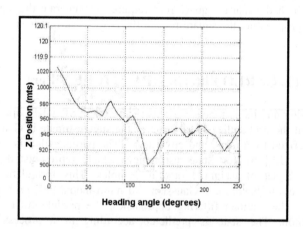

Fig. 3. UAV position results using Kalman filter along Z-Axis

Below shown table illustrates the UAV position estimates along X, Y and Z-Axis using basic Kalman Filter.

Table 1. UAV position estimates for varying heading angles using Kalman filter

X axis	Y axis	Z axis (altitude)	Heading angle (Θ)
50	22.04	930	150
45.01	22.04	960	200
45.04	22.002	970	50

The following section illustrates the estimation of the position of the vehicle along X, Y and Z axis using Particle filter. The video sequences are generated from a simulation tool containing a terrain with two waypoints, where waypoints are represented using latitude, longitude and altitude as follows. Waypoint1 is (-522.490994, -808.256392, 300) and waypoint2 is (-515.918875, -802.092870, 300) with a heading angle of 48 degrees. The sample images extracted from the video are shown in Fig 4 and the estimated, measured and true paths of the vehicle are plotted as shown in Fig 5, Fig 6 and Fig 7 respectively.

Fig. 4. Aerial image frames generated from the Simulator

The following results depict the state parameters along with corresponding errors along X direction.

Table 2. State Parameters and Corresponding Errors Along X Direction

ESTIMATED	MEASURED	TRUE	MEASURED ERROR	ESTIMATED ERROR
-522.505442	-522.490994	-522.490994	0.61707	0.617149
-522.486545	-522.483291	-522.142319	0.9	0.944078
-522.475543	-522.475463	-521.758346	-0.103627	-0.191049
-521.465465	-521.554433	-519.440364	1.172353	1.172353
-519.564432	-519.565675	-518.429247	1.410609	1.128508
-517.993423	-517.996476	-517.795679	0.714686	0.612663
-522.490994	-515.905451	-515.918875	1.363907	1.001695

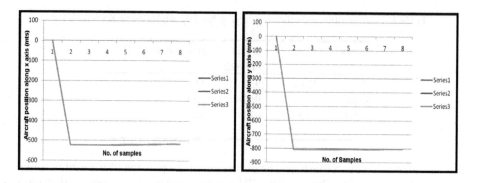

Fig. 5. Estimated, measured and true paths of aircraft along X and Y direction

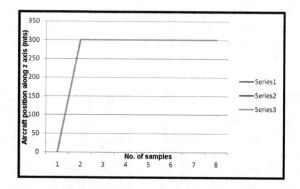

Fig. 6. Estimated, measured and true paths of aircraft along Z-axis

In Fig 5 and Fig 6, X-axis is the number of samples and y axis is the values of an aircraft. Series 1 shows estimated path, series 2 shows measured path and series 3 shows true path for the results obtained in Table 2, Table 3 and Table 4 respectively. From the above figure it can be shown that the measured and estimated values are very close to the true values obtained from visual simulation tool.

Table 3. State parameters and corresponding errors along Y direction

ESTIMATED	MEASURED	TRUE	MEASURED ERROR	ESTIMATED ERROR
-808.873541	-808.873462	-808.256392	0.61707	0.617149
-808.873541	-808.829463	-807.929463	0.9	0.944078
-807.378244	-807.465666	-807.569293	-0.103627	-0.191049
-806.567777	-806.567777	-805.395424	1.172353	1.172353
-805.575676	-805.857777	-804.447168	1.410609	1.128508
-804.465654	-804.567677	-803.852991	0.714686	0.612663
-803.094565	-803.456777	-802.092870	1.363907	1.001695

Table 4. State parameters and corresponding errors along Z direction

ESTIMATED	MEASURED	TRUE	MEASURED ERROR	ESTIMATED ERROR
300	300	300	0	0
300	300	300	0	0
299.923154	300	300	0	0.076846
300	300	300	0	0
300	300	300	0	0
300	300	300	0	0
299.892134	300	300	0	0.107866

From the above experiments it is observed that to estimate the non-linear, non-Gaussian state parameters of the vehicle, particle filter performs the best and gives better results than Kalman filter.

6   Conclusion and Future Work

In this paper, the existing estimation techniques beginning from the early 1960s until the 2000s ranging from stochastic filtering theory to Bayesian filitering theory and finally to Monte Carlo methods are reviewed. In particular, the SMC controlled by Markov Chain Monte Carlo and other batch based inference methods are discussed. Further this paper includes analysis of two main filtering techniques, the Kalman and the Particle filter. Through experiments it is found that particle filter based technique fits well to estimate non-linear non-Gaussian state parameters of unmanned air vehicles.

References

1. Bhat, S., Malagi, V.P., Ramesh Babu, D.R., Ramakrishna, K.A.: Analysis and Classification of Feature Extraction Techniques: A Study. International Journal of Engineering Research and Development 2(12), 07–12 (2012)
2. Kingsbury, N.: Multiscale Keypoint Detection Using the Dual-Tree Complex Wavelet Transform. In: ICIP 2006, pp. 1625–1628 (2006)
3. Welch, G., Bishop, G.: An Introduction to Kalman Filter, Course 8. In: SIGGRAPH (2001)
4. Gordon, N.J., Salmond, D.J., Smith, A.F.M.: Novel approach to nonlinear/non-Gaussian Bayesian state estimation. IEEE Proceedings-F 140(2) (April 1993)
5. Funk, N.: A Study of Kalman Filter applied to Visual Tracking. Tecjnical report for Project CMPUT 652, University of Alberta (December 2003)
6. Doucet, A.: On sequential Simulation-Based Methods for Bayesian Filtering, Technical report CUED/F-INFENG/TR.310. University of Cambridge (1998)
7. Cappe, O., Godsill, S.J., Moulines, E.: An overview of existing methods and recent advances in sequential Monte Carlo. In: Collaboration with Telecom Paris and Signal Processing and Communication Lab, University of Cambridge, UK

8. Fox, D., Burgard, W., Dellaert, F., Thrun, S.: Monte Carlo Localization: Efficient Position Estimation for Mobile Robots. In: AAAI 1999 Proceedings (1999), http://www.aaai.org
9. Thrun, S., Fox, D., Burgrand, W., Dellaert, F.: Robust Monte Carlo Localization for Mobile Robots. Artificial Intelligence 128, 99–141 (2001)
10. de Freitas, N.: Rao-Blackwellised Particle Filtering for Fault Diagnosis. IEEE Proceedings, 0-7803-7231-X/01, Dept of Computer Science, University of British Columbia (2002)
11. Chen, Z.: Bayesian Filtering: From Kalman Filters to Particle Filters and Beyond. Manuscript, Communication Research Laboratory, McMaster University, Ontario, Canada (2003)
12. Wan, E.A., van der Merwe, R.: The Unscented Kalman Filter for Nonlinear Estimation. Technical report, Oregon Graduate Institute of Science and Technology, Oregon (2000) (in preparation)
13. Storvik, G.: Particle filters for state-Space Models With the Presence of Unknown Static Parameters. IEEE Transactions on Signal Processing 50(2) (February 2002)
14. Vermaak, J., Doucet, A., Perez, P.: Maintaining Multi-Modality through Mixture tracking. Engineering dept. Cambridge University and Microsoft Research, Cambridge (2002)
15. Fox, D., Hightower, J., Liao, L., Schulz, D., Borriello, G.: Bayesian Filtering for Location Estimation, Published by the IEEE CS and IEEE ComSoc, 1536-1268/03, Published in the Journal Pervasive Computing (July-September 2003), http://computer.org/pervasive
16. Seliger, D.: State Estimation: Particle Filter, technical report, Programming and Software technique (June 2012)
17. Ahn, K.W., Chan, K.S.: Approximate conditional least squares estimation of a nonlinear state-space model via an unscented Kalman filter. Computational Statistics and Data Analysis 69, 243–254 (2014)
18. Gyorgy, K., Kelemen, A., David, L.: Unscented Kalman Filters and Particle Filter Methods for Non-linear State Estimation. In: Proceedings of 7th International Conference Interdisciplinarity in Engineering, INTER-ENG 2013 (2013)

Divide-and-Conquer Computational Approach to Principal Component Analysis

Vijayakumar Kadappa[1] and Atul Negi[2]

[1] Dept. of Computer Applications, BMS College of Engineering, Bangalore, India
vijaykumar.mca@bmsce.ac.in, kadappakumar@gmail.com
[2] School of CIS, University of Hyderabad, Hyderabad, India
atulcs@uohyd.ernet.in, atul.negi@gmail.com

Abstract. Divide-and-Conquer (DC) paradigm is one of the classical approaches for designing algorithms. Principal Component Analysis (PCA) is a widely used technique for dimensionality reduction. The existing block based PCA methods do not fully comply with a formal DC approach because (i) they may discard some of the features, due to partitioning, which may affect recognition; (ii) they do not use recursive algorithm, which is used by DC methods in general to provide natural and elegant solutions. In this paper, we apply DC approach to design a novel algorithm that computes principal components more efficiently and with dimensionality reduction competitive to PCA. Our empirical results on palmprint and face datasets demonstrate the superiority of the proposed approach in terms of recognition and computational complexity as compared to classical PCA and block-based SubXPCA methods. We also demonstrate the improved gross performance of the proposed approach over the block-based SubPCA in terms of dimensionality reduction, computational time, and recognition.

Keywords: Pattern Recognition, Dimensionality Reduction, Principal Component Analysis, Block based PCA, Divide-and-Conquer approach.

1 Introduction

Divide-and-Conquer (DC) paradigm is one of the popular techniques for designing computer algorithms. The DC technique is based on a simple and natural idea: *Divide the given problem into small-enough subproblems, Solve these subproblems independently, then Combine these solutions to arrive at a final solution.* The DC method is successfully applied in applications such as anomaly detection [1], switching linear systems [2], image segmentation [3].

Principal Component Analysis (PCA) is a well-established technique for dimensionality reduction and is used in numerous applications such as retweet detection [4], face recognition [5] and speaker recognition [6] to name a few. PCA, an optimal linear dimensionality reduction method, finds Principal Components (PC) by computing linear combinations of original variables. Some of the issues with classical PCA are its *high computational complexity* with high dimensional data and its *inability to extract local region based variations.*

© Springer International Publishing Switzerland 2015
S.C. Satapathy et al. (eds.), *Proc. of the 3rd Int. Conf. on Front. of Intell. Comput. (FICTA) 2014*
– *Vol. 1*, Advances in Intelligent Systems and Computing 327, DOI: 10.1007/978-3-319-11933-5_72

To counter the issues of classical PCA, many block PCA methods were proposed; they *divide the given pattern into sub-patterns and then extract features locally from these sub-patterns*. Block PCA methods are used in applications such as change detection [7], automated target detection [8], face recognition [9][10]. These block based methods do not fit into the formal framework of DC approach because (i) they may discard a few features to satisfy some constraints such as *equally-sized sub-patterns*; discarding features may affect the recognition rate, (ii) they do not follow recursive formulation of the algorithm; DC methods, in general, use recursive algorithms to provide natural and elegant solutions [11], and (iii) they show lower dimensionality reduction as compared to classical PCA since they ignore inter-sub-pattern correlations. We follow the definition of a formal DC approach as presented in a well-known book [11].

To overcome the drawbacks of block PCA methods, SubXPCA [12] was proposed; it extracts global PCs using local PCs obtained from sub-patterns. However, SubXPCA needs more computational time with the increased number of local PCs and may discard a few features while partitioning the patterns. To resolve the issues of SubXPCA, PCA, and other block PCA methods, we develop a novel DC approach to compute PCs with the near or similar variance as that of PCA. The proposed method shows increased computational efficiency and better or competitive recognition over SubXPCA and PCA; it also exhibits improved overall performance over SubPCA [10], a block PCA method.

The rest of the paper is organized as follows. In section 2, we present the proposed method in detail. The experimental results on palmprint and face recognition are shown in section 3. Finally we conclude in section 4.

2   Divide-and-Conquer Based PCA (DCPCA)

We begin here by presenting our approach in a more detailed and formal manner. Let $[\mathbf{X}]_{N \times d} = [\mathbf{X}_1 \mathbf{X}_2 \ldots \mathbf{X}_N]^T$ be set of N training patterns with dimensionality d; $[\mathbf{Z}]_{N \times w} = [\mathbf{Z}_1 \mathbf{Z}_2 \ldots, \mathbf{Z}_N]^T$ be the set of finally reduced patterns with w PCs corresponding to $[\mathbf{X}]_{N \times d}$; t be the threshold used for partitioning.

The algorithm of DCPCA is presented here (Fig. 1). The algorithm is called initially by using $[\mathbf{Z}, w] \leftarrow$ DCPCA$(1, d, \mathbf{X}, N)$. The algorithm accepts $[\mathbf{X}]_{N \times d}$ as the input and produces $[\mathbf{Z}]_{N \times w}$ as the output.

Algorithm DCPCA$(l, h, \mathbf{X}, N)$ {
    1. $d_1 = h - l + 1$
    2. if $(d_1 > t)$//divide into subproblems
        2.1. $d_2 = \lfloor \frac{l+h-1}{2} \rfloor$
        2.2. $[\mathbf{Y}^1, r_1] \leftarrow$ DCPCA$(l, d_2, \mathbf{X}^1, N)$ //SubProblem-1
        2.3. $[\mathbf{Y}^2, r_2] \leftarrow$ DCPCA$(d_2 + 1, h, \mathbf{X}^2, N)$ //SubProblem-2
        2.4. $[\mathbf{Z}, w] \leftarrow$ Combine-PCs$(r_1, r_2, \mathbf{Y}^1, \mathbf{Y}^2, N)$
        2.5. return$(\mathbf{Z}, w)$
    3. else
        3.1. $[\mathbf{Z}, w] \leftarrow$ Compute-PCs$(l, h, \mathbf{X}, N)$
        3.2. return$(\mathbf{Z}, w)$ }

Algorithm Combine-PCs $(r_1, r_2, \mathbf{Y}^1, \mathbf{Y}^2, N)$ {
[The algorithm combines the local PCs of partitions, $\mathbf{Y}^1{}_{N \times r_1}$ and $\mathbf{Y}^2{}_{N \times r_2}$
into $\mathbf{Y}_{N \times (r_1+r_2)}$, then computes PCs, $\mathbf{Z}_{N \times w}$ from $\mathbf{Y}_{N \times (r_1+r_2)}$]
1. $r_3 = r_1 + r_2$
//[Steps 2-3 concatenate $\mathbf{Y}^1$ and $\mathbf{Y}^2$]
2. for $(i = 1, 2, \ldots r_1)$ $\mathbf{Y}[i] = \mathbf{Y}^1[i]$
3. for $(i = r_1 + 1, r_1 + 2, \ldots r_3)$ $\mathbf{Y}[i] = \mathbf{Y}^2[i - r_1]$
//[Combining $\mathbf{Y}^1$ and $\mathbf{Y}^2$ based on inter-sub-pattern correlations]
4. $([\mathbf{Z}, w] \leftarrow$ Compute-PCs $(1, r_3, \mathbf{Y}, N)$
5. return $(\mathbf{Z}, w)$ }

Algorithm Compute-PCs $(l, h, \mathbf{X}, N)$ {
[The algorithm takes input as $[\mathbf{X}]_{N \times (h-l+1)}$, computes w PCs, $[\mathbf{Z}]_{N \times w}$.]
1. $u_1 = h - l + 1$
2. $[\mathbf{M}]_{u_1 \times u_1} \leftarrow$ Compute-CovMatrix$(l, h, \mathbf{X}, N)$ //$\mathbf{M}$ is the covariance
matrix //[Computes eigenvectors, $\mathbf{E}$ and eigenvalues, $\mathbf{\Lambda}$ from $[\mathbf{M}]_{u_1 \times u_1}$]
3. $[\mathbf{E}, \mathbf{\Lambda}] \leftarrow$ Compute-EigVectors-EigValues$(u_1, \mathbf{M})$
4. $w \leftarrow$ Select-Number-of-PCs$(u_1, \mathbf{\Lambda})$
5. $[\mathbf{Z}]_{N \times w} = [\mathbf{X}]_{N \times u_1}[\mathbf{E}]_{u_1 \times w}$ //[Reduces the dimensionality to w]
6. return $(\mathbf{Z}, w)$ }

How to Choose the Threshold, t? In the Algorithm DCPCA(), we can choose
the threshold (t) such that (i) the subproblems fit into memory; (ii) $d > \frac{N}{10}(= t)$
for better classification [13]; (iii) t is chosen specific to an application.
The DCPCA is illustrated in Fig. 1 with parameters, $d = 4u$, $t = u$ & $r_1 = r_2$.

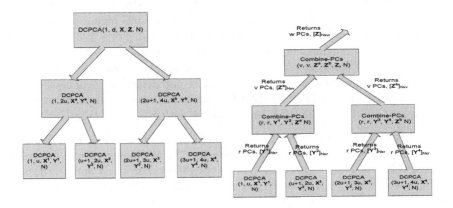

(a) Recursive calls of DCPCA()

(b) Recursive calls of Combine-PCs()

Fig. 1. Visualizing DCPCA method

3   Experimental Results and Discussion

In this section, we compare DCPCA with other methods empirically based on the criteria: Recognition rate, Computational time and Summarization of variance in PCs.

Data Sets Used: (i) *Palmprint data set.* We chose 498 images (about 20 images per subject) from first 25 subjects of PolyU palmprint database [14]. Each image is of dimension, 284×384 (PGM format). (ii) *UMIST face data set* [15] consists of gray scale images of 20 persons amounting to 565 images in total. Each image is of dimension, 112×92 (PGM format). We used 10 images per subject for training and the rest of them for testing in both the cases.

Experimental Setup: We conducted experiments by using PCA [16] on 5 different training and test data sets of PolyU Palmprint [14] and UMIST face data [15]. The training and test data sets are generated randomly. For palmprint data, we used $t = 13632, 6816, 3408$, which yields $8, 16, 32$ partitions respectively; we have chosen $10, 4, 6$ local PCs (r) respectively from these partitions for SubXPCA and DCPCA. First 5 local PCs are not considered from each of the partitions. For face data set, we used $t = 2576, k = 4$ and local PCs (viz. $2, 3, 5, 10$) for DCPCA and SubXPCA.

The local PCs are those extracted from a partition of sub-patterns (sub-problems). The classification is done based on Nearest Neighbour rule. The average of 5 classification results and computational time are shown in Figs. 2-5.

3.1   Experiments Based on Varying Number of PCs and Partitions

For Palmprint Data: Both SubXPCA and DCPCA show improved recognition upto 2%, 3% higher rates as compared to PCA and SubPCA respectively for $t = 13632$ (i.e. $k = 8), r = 10$ (Fig. 2(a)). For $t = 6816$ (i.e. $k = 16), r = 4$, DCPCA shows improved recognition upto 3%, 1%, 1.6% higher rates against PCA, SubXPCA, and SubPCA respectively (Fig. 2(b)). For $t = 3408$ (i.e. $k = 32), k = 6$, DCPCA shows better or competitive recognition against PCA and SubPCA (upto 1% higher rate)(Fig. 2(c)). For all these cases, SubPCA takes more PCs to show good recognition which is not desirable in dimensionality reduction point of view (Fig. 2). DCPCA shows decreased computational time as compared to PCA, SubXPCA, and SubPCA with $k = 16, 32$ (Figs. 3(a)-3(c)).

For UMIST Face Data: For $t = 2576$ (i.e. $k = 4$) and $r = 2, 3$, both DCPCA and SubXPCA show superior recognition upto 7%, 5% higher rate as compared to SubPCA, and PCA respectively (Figs. 4(a)-4(b)). For $k = 4$ and $r = 5$, both DCPCA and SubXPCA show better recognition upto 1% higher rate against PCA and by 7% higher rate against SubPCA (at 4 PCs) (Fig. 4(c)); SubPCA shows improved recognition upto 3% higher rate as compared to DCPCA and SubXPCA at 8 & 12 PCs (Fig. 4(c)). For $k = 4$ and $r = 10$, DCPCA, SubXPCA, and PCA are competitive and shows better recognition at 4 PCs against SubPCA by 6%; however, SubPCA shows improved recognition upto 3% higher rate at 8, 12, & 16 PCs (Fig. 4(d)) as compared to DCPCA and SubXPCA. For $k = 4$

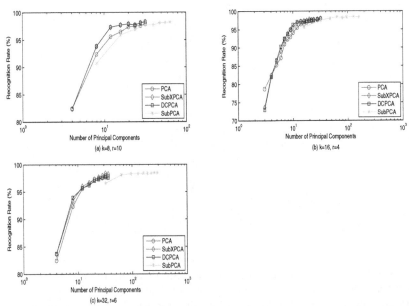

Fig. 2. *Recognition rate (Palmprint data).* (a) Both SubXPCA and DCPCA show improved recognition rates. (b) DCPCA shows much better recognition rate against other methods. (c) DCPCA shows better or competitive recognition against PCA and SubPCA. SubPCA shows good recognition but with higher dimensionality.

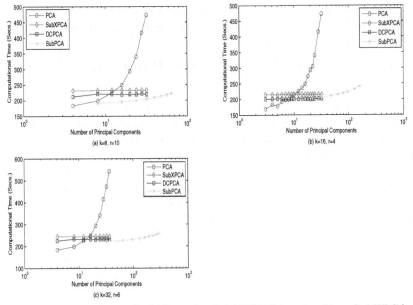

Fig. 3. *Computational time (Palmprint data).* (a) DCPCA outperforms SubXPCA and PCA methods. However, SubPCA shows much decreased computational time. (b)-(c) DCPCA shows better computational efficiency over other methods.

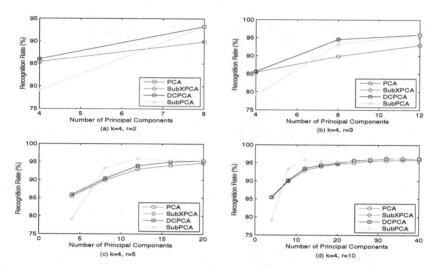

Fig. 4. *Recognition rate (UMIST face data; $t = 2576, k = 4$).* (a)-(b): Both DCPCA and SubXPCA show improved recognition over SubPCA and PCA. (c) Both DCPCA and SubXPCA show better recognition against PCA and also SubPCA (at 4 PCs); (d) DCPCA, SubXPCA, and PCA are competitive and show better recognition against SubPCA at 8 PCs. However SubPCA shows improved recognition at 8, & 12 PCs.

and $r = 2, 3, 5$, DCPCA shows decreased computational time in most of the cases as compared to PCA, SubXPCA and SubPCA (Figs.5(a)-5(c)); for $r = 10$, both SubPCA and DCPCA show better computational time as compared to SubXPCA and PCA (Fig. 5(d)). Since the local variance captured is varied with different values of k, r, the performance of the proposed method is also varied.

3.2 Why Does DCPCA Appear Better Than PCA, SubXPCA, and SubPCA?

DCPCA versus PCA. DCPCA approximates PCA method in terms of dimensionality reduction with increased number of local PCs (Fig. 6). DCPCA is scalable to high-dimensional data because of its divide-and-conquer principle and shows decreased computational time (Figs. 3 & 5). DCPCA extracts both local and global variations which results in improved recognition as compared to PCA (Figs. 2 & 4). PCA extracts only global features which is the reason why it may not perform well if the local variations are prominent in the patterns. We note that PCA becomes a special case of DCPCA if $t = d$.

DCPCA versus SubXPCA. DCPCA partitions the patterns into sub-patterns without discarding any features such that $u \leq t$. In contrast to DCPCA, SubXPCA may discard features if the pattern size is not a multiple of sub-pattern size because it divides the patterns into equally-sized sub-patterns. The features discarded by SubXPCA may be crucial for recognition. DCPCA computes smaller covariance matrices and consequently takes lower computational time as compared to SubXPCA (Figs. 3 & 5). DCPCA also shows improved recognition as

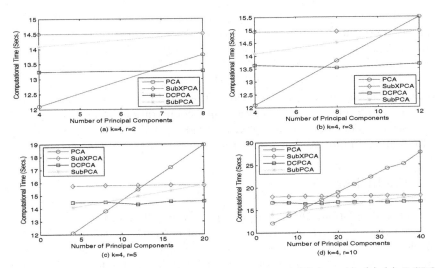

Fig. 5. *Computational Time (UMIST face data; t = 2576, k = 4).* (a)-(c) DCPCA shows decreased computational time over other methods. (d) Both SubPCA and DCPCA show better computational time as compared to SubXPCA and PCA.

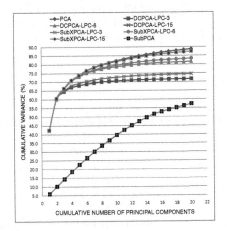

Fig. 6. *Summarization of variance in PCs (Palmprint data; t = 6816, k = 16).* Both DCPCA and SubXPCA show increased variance in PCs with the increased number of local PCs (3, 6, & 15) and move closer to PCA. However, SubPCA shows decreased variance in its PCs, resulting in lower dimensionality reduction.

compared to SubXPCA (Fig. 2(b)) and it is competitive with SubXPCA in other cases. From Fig. 6, it is evident that DCPCA and SubXPCA are competitive in terms of dimensionality reduction.

DCPCA versus SubPCA. DCPCA shows much superior variance summarized in its PCs (i.e. higher dimensionality reduction) as compared to SubPCA (Fig. 6). It is also to be noted that variance of DCPCA moves closer to PCA with increased number of local PCs (Fig. 6). In terms of recognition, DCPCA shows improved performance over SubPCA (Figs. 2(a)-(c), 4(a)-(d)). Also DCPCA shows decreased computational time against SubPCA (Figs. 3(b)-(c), 5(a)-(c)). However SubPCA also shows better recognition rate in some cases (Figs. 4(c)-(d)) and decreased computational time (Figs. 3(a), 5(d)).

From the above discussion, it is evident that DCPCA method performs much better in terms of overall recognition and computational time as compared to classical PCA, block based SubXPCA and SubPCA methods. In addition, DCPCA shows improved dimensionality reduction over SubPCA method.

4   Conclusion

We proposed a formal Divide-and-Conquer based PCA method which shows improved recognition rate and computational efficiency as compared to PCA and block based SubXPCA methods. PCA becomes a special case of DCPCA when threshold is equal to pattern size. DCPCA method is faster and showed better recognition rates as compared to classical PCA and SubXPCA methods. DCPCA shows improved gross performance over SubPCA (a block PCA method) in terms of dimensionality reduction, recognition, and time. In this paper we applied DCPCA for high dimensional palmprint and face data and it can be used for reducing other high dimensional data as well. The DCPCA method can be adapted to compute PCs in parallel environments such as Grid.

References

1. Liu, J., et al.: A divide and conquer approach to anomaly detection, localization and diagnosis. Mechanical Systems and Signal Processing 23(8), 2488–2499 (2009)
2. Li, R., et al.: Divide, conquer and coordinate: Globally coordinated switching linear dynamical system. IEEE Trans. PAMI 34(4), 654–669 (2012)
3. Appia, V., et al.: Localized PCA based curve evolution: A divide and conquer approach. In: IEEE Int. Conf. on Computer Vision, pp. 1981–1986 (2011)
4. Morchid, M., et al.: Feature selection using PCA for massive retweet detection. Pattern Recognition Letters 49, 33–39 (2014)
5. Luan, X., et al.: Extracting sparse error of robust PCA for face recognition in the presence of varying illumination and occlusion. Pat. Reco. 47, 495–508 (2014)
6. Jeong, Y.: Adaptation of HMM mean parameters using two-dimensional PCA with constraint on speaker weight. IEEE Electronics Letters 50(7), 550–552 (2014)
7. Prinet, Q.B., et al.: Multi-block PCA method for image change detection. In: IEEE 12th Int. Conf. on Image Analysis and Processing, pp. 385–390 (2003)
8. Rizvi, S.A., Saadawi, T.N., Nasrabadi, N.M.: A clutter rejection technique for FLIR imagery using region based PCA. Pattern Recognition 33, 1931–1933 (2000)
9. Gottumukkal, R., Asari, V.K.: An improved face recognition technique based on modular PCA approach. Pattern Recognition Letters 25, 429–436 (2004)
10. Chen, S., Zhu, Y.: Subpattern-based principal component analysis. Pattern Recognition 37, 1081–1083 (2004)
11. Horowitz, E., Sahni, S., Rajasekharan, S.: Fundamentals of Computer Algorithms. Galgotia Publications Pvt. Ltd., 5, Ansari Road, Daryaganj New Delhi (2003)

12. VijayaKumar, K., Negi, A.: SubXPCA and a generalized feature partitioning approach to PCA. Pattern Recognition 41(4), 1398–1409 (2008)
13. Jain, A.K., Duin, R.P.W., Mao, J.: Statistical pattern recognition: A review. IEEE Transactions on Pattern Analysis and Machine Intelligence 22(1), 4–37 (2000)
14. http://www.comp.polyu.edu.hk/~biometrics/
15. http://www.sheffield.ac.uk/eee/research/iel/research/face
16. Guangming Lu, D.Z., Wang, K.: Palmprint recognition using eigenpalms features. Pattern Recognition Letters 24, 1463–1467 (2003)

An Efficient Mining of Biomedical Data from Hypertext Documents via NLP

Ravi Shankar Shukla[*], Kamendra Singh Yadav, Syed Tarif Abbas Rizvi,
and Faisal Haseen

Department of Computer Science and Engineering,
Invertis University, Bareilly, Uttar Pradesh, India
{ravipraful,kamendrasinghyadav,tarifabbas786110,
faisal.haseen31}@gmail.com

Abstract. As we all know that searching over the internet for the information and then filtering out the correct and right information is exhaustive and time consuming task. Even in the most popular and successful search engine Google searching the right information can expand across many web pages before encountering the web page with the right information. A more efficient and more efficient technique was developed to overcome this problem which made use of biomedical words in the documents for filtering the information for the searching purposes. This technique shows much more positive and efficient results. Here, in this paper we increase efficiency and effectiveness through the use of NLP. This technique shows much more improved results

Keywords: Biomedical data, NLP, Ontology, Unified Medical Language System, Pagerank.

1 Introduction

As all of us is familiar with the fact that finding anything over the internet we uses various search engines name by Google, Ask, Bing just to name the few . We know that internet is the interconnection of huge databases across many platforms and these databases contains vast amount of information and data but the difficult and critical part is that to find the right information from such vast databases.

Whenever we required searching anything over the internet most of us uses the query word and type it on the Google and usually opens the very first tab that appears on the searched web page but many times this technique fails because the first link is not the correct link in many cases. So, from this point we deploys the technique of hit and trial and we open every link thereafter in order to find out the correct result and. though many times we didn't find the desired even after going through numerous sites and web pages.

Firstly to understand the methodology of how a search is performed we take the example of the most famous search engine Google and see how it results a search

---

[*] Corresponding author.

© Springer International Publishing Switzerland 2015
S.C. Satapathy et al. (eds.), *Proc. of the 3rd Int. Conf. on Front. of Intell. Comput. (FICTA) 2014*
– *Vol. 1*, Advances in Intelligent Systems and Computing 327, DOI: 10.1007/978-3-319-11933-5_73

query from the vast database. Google search engine employs the Pagerank [2] and Hypertext Induced Topis Search [3] techniques to allocate rank and retrieve searches. Google takes the keyword query [1] to search related documents and then it ranks them on the basis of hyperlinks either pointing inward or outward to the documents. Though the technique stated is effective but it suffers from the fact that many times ranking documents on the basis of hyperlinks does not return the right document.

Many websites in order to increase the ranking web pages uses the fake links and many a times we counter web pages which have many advertisements link which unnecessarily increases our time. Another drawback is that most users only browse first twenty to forty result pages while discarding the rest of the pages. The documents which are discarded may contain some relevant information related to the searched topic.

To overcome these stated problems in this paper the retrieved documents will be processed and stored in the using sorting algorithm and then it is classified into the particular domain. The proposed system classifies the retrieved document in the particular domain in the biomedical system. The proposed system enhances the work using the classification of the retrieved documents.

2   Literature Review

Ontology can be referred to a glossary or a simple dictionary but the only difference is that ontology offers more refined detail and more simplified structure that can be processed by the computer. Ontology is the collection of axioms, concepts and various relationships that define the domain of interest.

Ontology finds its great importance in medical informatics research [5]. Ontology can be used in many biomedical applications. Tan et al. [13] explains the importance of text mining for the field of biomedicine demands a great knowledge of domain. All most all of the information can be taken from medical ontology. Our paper describes a structure for the selection of most current ontology for a specified text mining application. Shatkey et al. [12] analyses the field involved in unstructured analysis of the text then classifies ongoing work in biomedical literature mining in accordance with these fields. The numbers of biomedical database articles are increasing exponentially day by day. To search meaningful information many systems are deployed to search the right information automatically. One such system was developed by Conan [10] who utilises the various database XML files.

Another paper was published by Mukherjea et a.l [9] put forward a technique to find biomedical terms and relations from vast World Wide Web database. It first put forward a query to web search engine with lexicon syntantatic patterns to extract relevant information. The knowledge derived out from the web search engines can be used to augment the ontology and knowledge databases and made a semantic web for various life sciences. Various kinds of relations among biomedical entities can be finding out by the technique. Though we know that web is full of information of interlinking databases. In order to make our task bit easier to work with. Bunescu R.Mooney et al. [2] proposed supervised machine learning method in the biomedical area with great success but it suffers from the data sparseness because of the fast that no much of biomedical data is available.

The mechanism which is used by Google to rank web pages consist of two algorithm, Pagerank [2] and HITS [3]. The value of pagerank is calculated on basis of number of web pages that are pointing to it.

In the paper published by Sazia *et al.* [1] they use the process of introducing the biomedical terms for the purpose of ranking the documents instead of simple keywords and this technique shows that hit rate is much improved. The technique is effective but it suffers from many web pages contains the fake links.

3 Methodology

Our methodology includes five steps which are:

3.1 Collecting the Documents

The web documents collected from various search engines that are available, as html documents. The html documents are then converted into text documents for further processing.

3.2 Pre-processing of Collected Documents via NLP

The first step of document pre-processing is to reduce the dimensionality by discarding the irrelevant words from the documents. The collected documents are pre-processed to perform the effective document search and to improve the efficiency of the system. It does not need to search the documents/read the words that are not related to the query. To achieve removal of stop words is done via NLP (Natural Language Processing). The goal of document pre-processing is to optimize the performance of next step:-step3 and the irrelevant text on the web document such as prepositions, conjunction, verb, noun, pronoun etc. are categorized as stop words. The web documents contain large quantity of stop words which have no meaning for the research purposes. Finally, the stop words are identified and removed from the document, which is now simplified.

3.3 Extracting the Biomedical Terms

UMLS, the Biomedical Ontology is used to extract the biomedical terms from the pre-processed documents. To extract the biomedical terms from the documents   text mining and data mining techniques are used, since this is where the actual information extraction happens. The words from the text document are compared to the words in the UMLS. If the word is found in the UMLS then it is counted as biomedical words and saved in the document else removed from the document.

3.4 Assigning the Frequency to the Document

Assigning the frequency means that counting the number of biomedical terms in the documents and on the basis the importance of web pages is calculated and is assigned to the page saved. Document which is having more biomedical terms gets the is given the first position and so on.

3.5   Classification of Retrieved Documents

In this step the documents are classified in two categories which are immunology and other category. The documents which are having some biomedical terms are classified into the immunology and the other documents which do not contain any biomedical words are classified into the other category. The benefit of this process is that results are much better when compared to the existing process which is used for classification.

The block diagram of the entire five steps is shown in figure 1.

The online version of the volume will be available in LNCS Online. Members of institutes subscribing to the Lecture Notes in Computer Science series have access to all the pdfs of all the online publications. Non-subscribers can only read as far as the abstracts. If they try to go beyond this point, they are automatically asked, whether they would like to order the pdf, and are given instructions as to how to do so.

Please note that, if your email address is given in your paper, it will also be included in the meta data of the online version.

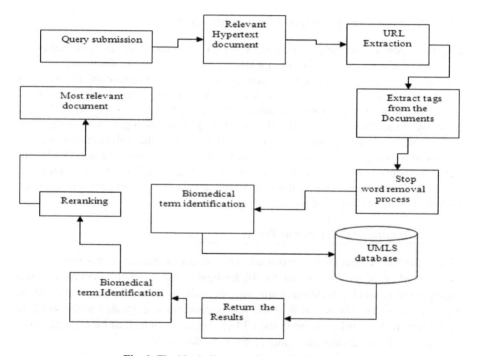

Fig. 1. The block diagram of the entire five steps

4   Experiment

This experiment follows the methodology proposed above for identifying the biomedical word form a multilingual collection of web documents using NLP.

We have collected 100 documents for pre-processing and after pre-processing we found 31 documents are most relevant for further steps. These 31 documents have been ranked in an ascending order based on the number of biomedical term in each document. These 31 documents are classified on the basis of biomedical term, into two categories: Immunology and Other-category. When biomedical term antibody, gene, and cd are encountered, then document is classified into immunology else into other-category.

filename	domain
D:\project_2014\MiningBioMedicalDatafromHypertextDocuments\coding\Preprocess\1.htm	othercategory
D:\project_2014\MiningBioMedicalDatafromHypertextDocuments\coding\Preprocess\3.htm	immunology
D:\project_2014\MiningBioMedicalDatafromHypertextDocuments\coding\Preprocess\1.htm	immunology
D:\project_2014\MiningBioMedicalDatafromHypertextDocuments\coding\Preprocess\3.htm	immunology
D:\project_2014\MiningBioMedicalDatafromHypertextDocuments\coding\Preprocess\1.htm	immunology
D:\project_2014\MiningBioMedicalDatafromHypertextDocuments\coding\Preprocess\3.htm	immunology
D:\project_2014\MiningBioMedicalDatafromHypertextDocuments\coding\Preprocess\CD44 from adhesion molecules to signa.htm	immunology
D:\project_2014\MiningBioMedicalDatafromHypertextDocuments\coding\Preprocess\CD44 from adhesion molecules to signa.htm	immunology
D:\project_2014\MiningBioMedicalDatafromHypertextDocuments\coding\Preprocess\3.htm	immunology
D:\project_2014\MiningBioMedicalDatafromHypertextDocuments\coding\Preprocess\CD44 from adhesion molecules to signa.htm	immunology
D:\project_2014\MiningBioMedicalDatafromHypertextDocuments\coding\Preprocess\101 ways to save money.htm	immunology
D:\project_2014\MiningBioMedicalDatafromHypertextDocuments\coding\Preprocess\101 ways to save money.htm	immunology
D:\project_2014\MiningBioMedicalDatafromHypertextDocuments\coding\Preprocess\2008 Olympics in Beijing, China.htm	othercategory
D:\project_2014\MiningBioMedicalDatafromHypertextDocuments\coding\Preprocess\CD44 activation and associated primary.htm	othercategory
D:\project_2014\MiningBioMedicalDatafromHypertextDocuments\coding\Preprocess\2008 Olympics in Beijing, China.htm	othercategory
D:\project_2014\MiningBioMedicalDatafromHypertextDocuments\coding\Preprocess\CD44 activation and associated primary.htm	othercategory
D:\project_2014\MiningBioMedicalDatafromHypertextDocuments\coding\Preprocess\3 Lions Sport.htm	immunology
D:\project_2014\MiningBioMedicalDatafromHypertextDocuments\coding\Preprocess\3 Lions Sport.htm	othercategory
D:\project_2014\MiningBioMedicalDatafromHypertextDocuments\coding\Preprocess\CD44 and its role in inflammati...htm	othercategory
D:\project_2014\MiningBioMedicalDatafromHypertextDocuments\coding\Preprocess\CD44 and its role in inflammati...htm	immunology
D:\project_2014\MiningBioMedicalDatafromHypertextDocuments\coding\Preprocess\2008 Olympics in Beijing, China.htm	immunology
D:\project_2014\MiningBioMedicalDatafromHypertextDocuments\coding\Preprocess\3 Lions Sport.htm	othercategory

Fig. 2. The classifications of documents in two categories

Let us take the example of ranking a document in the Google with that of our biomedical search engine. We have taken three keyword namely antibody, gene and cd, we search these keywords in Google and then we save the document found. Now we applied our process on these documents and then we classify them on the basis of biomedical words found and rank them accordingly in ascending order. We then compare our rank with that of the rank given to them by the Google. There are two columns in our table named 'Google search 1' and 'Google search 2' because as we know that Google continuously re-rank document both online and offline in every second and hence we see the difference in the ranking of documents given by the Google at two different instance.

Table 1. Comparison of Ranks

S. no	Document Name	Thesis Rank	Google rank 1	Google rank 2	Category
1	CD44 from adhesion molecules to	1	17	36	Immunology
2	1.htm	2	23	21	Other-category
3	CD44 and its role in inflammation .htm	3	9	52	immunology
4	CD44 activation and associated	4	19	31	immunology
5	3 Lions Sport. htm	5	21	11	immunology
6	3.htm	6	54	37	Other-category
7	2008 Olympics in Beijing, China.	7	32	71	other-category
8	Cutting edge an inducible sialida	8	15	29	Immunology
9	Cytokine ability of human monoc	11	42	56	immunology
10	Hyaluronan binding functio	12	23	39	immunology
11	Induction of interactions between CD44 and hyalur .htm	14	41	13	immunology

5   Performance Evaluation

The benefit of the classification of retrieved documents into two categories is that it gives the better performance in searching the biomedical words. We draw the graph by taking the number of documents on x-axis and classification of retrieved document on y-axis.

The comparison between the performance of existing system and the proposed system is shown in the graph given below.

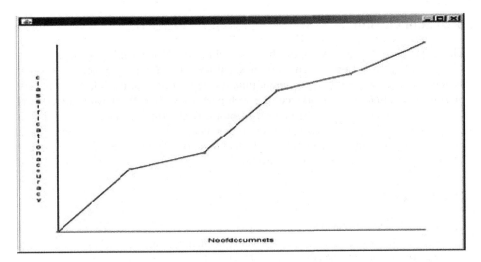

Fig. 3. The classification accuracy of the proposed system

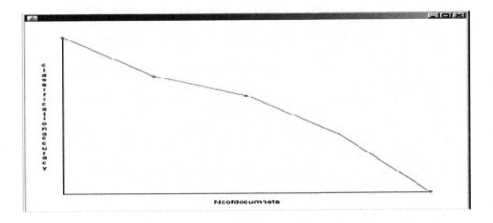

Fig. 4. The classification accuracy of the existing system

Comparing to the existing system the classification accuracy of the proposed system is high. Here the biomedical document is handled and classified using the biomedical terms. But the existing system does not consider the classification of retrieved result; so that the accuracy of the existing one is low compare our proposed one.

For example, in a million word corpus (collection of texts) I have laying around, *the* alone is 6.8% of the total number of tokens. The twelve most frequent tokens (see below) is 25.6% of the total number of tokens. Considering that you'll often want to be dealing with huge amounts of text, this can free up a tremendous amount of resources! To filter out those most-frequent tokens, many NLP applications use a list of these words. Anything on the list is removed from the stream of input tokens before any further processing is done.

6   Conclusion

This paper proposed a new approach for searching the biomedical terms using NLP. This process proves to be much more efficient and effective as compared to present system. Every web page has a different importance for different users hence it is very difficult to evaluate the importance of a web page for a user. We have evaluated the rank of web pages based on the number of biomedical terms it contains. But it might be possible that page having lower biomedical terms has richer biomedical information. Though this method is effective but for the future aspect we can use biomedical images for more refined results.

References

1. Rahman, R.M., Salahuddin, S.: Comparative study of classification techniques on biomedical data from hypertext documents. International Journal of Knowledge Engineering and Soft Data Paradigms 4(1), 21–41 (2013)
2. Lawrence, Brin, S., Motwani, R., Winograd, T.: The Page Rank Citation Ranking: Bringing Order to the Web. Technical Report. Stanford InfoLab (1999)
3. Kleinberg, J.M.: Authoritative sources in a hyperlinked environment. J. ACM 46(5), 604–632 (1999), http://www.cs.cornell.edu/home/kleinber/auth.ps
4. Musen, M.A.: Medical Informatics: Searching for Underlying Components. Methods Inf. Med. 41(1), 12–19 (2002)
5. Tan, H., Lambrix, P.: Selecting an ontology for biomedical text mining. In: Workshop on BioNLP (2009)
6. Shatkay, H., Feldman, R.: Mining the Biomedical Literature in the Genomic Era: An Overview. Journal of Computational Biology (JCB) 10(6), 821–856 (2003)
7. Malik, R., Siebes, A.: CONAN: An Integrative System for Biomedical Literature Mining. In: Bento, C., Cardoso, A., Dias, G. (eds.) EPIA 2005. LNCS (LNAI), vol. 3808, pp. 248–259. Springer, Heidelberg (2005)
8. Mukherjea, S., Sahay, S.: Discovering Biomedical Relations Utilizing the World-Wide Web. In: Pacific Symposium on Biocomputing, pp. 164–175 (2006)
9. UMLS® Knowledge Source Server (July 7, 2010), https://uts.nlm.nih.gov/home.html
   Frequent Itemset Mining Implementations Repository, http://fimi.cs.helsinki.fi
10. Yi, J., Nasukawa, T., Bunescu, R., Niblack, W.: Sentiment analyzer: Extracting sentiments about a given topic using natural language processing techniques. In: IEEE Intl. Conf. on Data Mining (ICDM), pp. 427–434 (2003)

Passive Copy Move Forgery Detection Using SURF, HOG and SIFT Features

Ramesh Chand Pandey, Rishabh Agrawal, Sanjay Kumar Singh,
and K.K. Shukla

Department of Computer Science &Engineering
Indian Institute of Technology, BHU
Varanasi-221005, India
{rameshcse19,rg1995007}@gmail.com,
{sks.cse,kkshuka.cse}@iitbhu.ac.in

Abstract. Copy-Move in an image might be done to duplicate something or to hide an undesirable region. So in this paper we propose a novel method to detect copy-move forgery detection (CMFD) using Speed-Up Robust Features (SURF), Histogram Oriented Gradient (HOG) and Scale Invariant Features Transform (SIFT), image features. SIFT and SURF image features are immune to various transformations like rotation, scaling, translation etc., so SIFT and SURF image features help in detecting Copy-Move regions more accurately in compared to other image features. We have compared our method for different features and SIFT features show better results among them. For enhancement of performance and complete localization to Copy Move region, a hybrid SURF-HOG and SIFT-HOG features are considered for CMFD. We are getting commendable results for CMFD using hybrid features.

Keywords: Copy-Move forgery, CMFD, SIFT, SURF, HOG, hybrid features.

1 Introduction

The advancements in image processing has made image editing very easy which makes their authenticity uncertain. In court of law, where images are presented as basic evidence, its verification plays a crucial role as images can be edited to change its meaning and hence influence the judgment. Many prominent personalities of film industry are also persecuted by image tampering. So, it is important to prove the authenticity of the image and bring the world towards truth.

In the past, digital media used to be authenticated using digital watermarking [4],[5] technique. Multimedia Forensics [1],[2],[3] aims at devising methods which test the authenticity of a digital asset in the absence of watermarks. These methods are defined as "passive" because they can conclude results based on the digital asset itself and allow the user to determine whether the asset is authentic [6],[7] or which was the acquisition device used [8],[9]. Focusing on the task of acquisition device identification requires the study of two main aspects: the first is to understand which kind of device generated a digital image (e.g. a scanner, a digital camera or is a

© Springer International Publishing Switzerland 2015
S.C. Satapathy et al. (eds.), *Proc. of the 3rd Int. Conf. on Front. of Intell. Comput. (FICTA) 2014*
Vol. 1, Advances in Intelligent Systems and Computing 327, DOI: 10.1007/978-3-319-11933-5_74

computer graphics product) [10],[11] and the second is to determine which specific camera or scanner (by recognizing model and brand) acquired that specific content [8],[9]. Good copy-move forgery detection should be robust to some types of transformation, such as rotation and scaling. Most existing methods do not deal with all these manipulations or are often computationally prohibitive. For e.g. the method in [13] is not able to detect scaling or rotation transformation, whereas with the method [14] only small variations in rotation and scaling are identifiable as reported in [15]. Authors in [18] proposed a SIFT based forensic method for Copy Move and transformation recovery.

In this paper this issue of image forgery using a copy-move attack is investigated. We are considering different features of image like SURF, HOG and SIFT for CMFD. We have also considered hybrid features for CMFD. The proposed method is able to detect a cloned region within an image and evaluate the authenticity of the image.

The paper is divided into seven sections. The second section describes the use of SURF features for CMFD, third section describes the use of HOG features for CMFD and the fourth section describes the use of SIFT features for CMFD. Our method is explained in detail in the fifth section. The experimental results are shown in the sixth section. Future work and conclusions are drawn in the seventh section.

2   SURF Features in Copy-Move Forgery Detection

Speed up Robust features (SURF) [16], is a novel scale- and rotation-invariant interest point detector and descriptor. The detector is based on Hessian Matrix and uses very basic approximation. It relies on integral images to reduce computational time. The descriptor describes a distribution of Haar-wavelet responses within the interest point neighborhood. Integral images are exploited for speed. Only 64 dimensions are used, which reduce the time for feature computation and matching and simultaneously increasing the robustness. Indexing step is based on the sign of the Laplacian, which increases the matching speed and the robustness of the descriptor.

2.1   Procedure for CMFD

i)   Calculate SURF features for an image.
ii) Determine the Euclidian distance between every pair of SURF keypoints.
iii) Determine the best matches based on a threshold applied to the minimum Euclidian distance.
iv) Make the best match keypoints as the center of their respective clusters.
v)   Match both clusters as a whole using a threshold for the Euclidian distance and cluster size.
vi) If both clusters have a minimum of two points, display the clusters.
vii) Decide whether the image is forged or authentic.

3   HOG Features in Copy-Move Forgery Detection

To extract Histogram oriented gradients (HOG) descriptor, first horizontal and vertical gradients are calculated with no smoothing, then gradient orientations and magnitude are computed [17]. For a 64×128 image, the image is divided in 16×16 sized blocks with 50% overlap. Each block is divided in cells of size 8×8 each. Gradient magnitudes are quantized in 9 bins. Then all the histograms for all the blocks in a window are concatenated to compute the descriptor which is 3780 dimension vector.

3.1   Procedure for CMFD

i)   Input the image for which the authenticity is to be determined.
ii) Resize the image and divide it in blocks of 64x128.
iii) Compute HOG descriptor for each block.
iv) Determine the best matches based on a threshold applied to the min Euclidian distance.

4   SIFT Features in Copy-Move Forgery Detection

Scale Invariant Features Transform (SIFT) [12] presents a method for extracting distinctive invariant features from images that can be used to perform reliable matching between different views of an object or scene. The features are invariant to image scale and rotation, and are shown to provide robust matching across a substantial range of affine distortion, change in 3D viewpoint, addition of noise, and change in illumination.

4.1   Procedure for CMFD

i)   Calculate SIFT features for an image.
ii)   Determine the Euclidian distance between every pair of SIFT keypoints.
iii) Determine the best matches based on a threshold applied to the min Euclidian distance.
iv)   Make the best match keypoints as the center of their respective clusters.
v)   Match both clusters as a whole using a threshold for the Euclidian distance and cluster size.
vi)   If both clusters have a minimum of two points, display the clusters.
vii) Decide whether the image is forged or authentic.

5   The Proposed Method

The proposed method works in two steps. The first step consists of feature extraction and key point matching, the second step is devoted to keypoint clustering and forgery detection.

5.1   Features Extraction and Keypoints Matching

Given a test image calculate the keypoints $x = \{x1, x2, x3, ... xn\}$ where n is the no of keypoints with the corresponding descriptors $f = \{f1, f2, ... fn\}$. For each keypoint xi, calculate the Euclidian distances $d = \{d1, d2, ... dn - 1\}$ for (n-1) other neighboring keypoints and decide the best match based on the ratio of the nearest (shortest distance) and second nearest neighbor (next shortest distance) using a threshold value. Store the best matches as $m = \{m1, m2, \ldots\ldots mn\}$ for n matched keypoints. Sort the n matched keypoints in the order of their increasing Euclidean distance. Apply a scale factor to the minimum distance and use this as a threshold to find the best matches among the keypoints, i.e. if their distance is less than or equal to the threshold.

$$Threshold = Scale\ Factor * min\ distance \tag{1}$$

5.2   Keypoint Clustering and Forgery Detection

The best matches are used as the centers of their respective clusters to find the cluster points using a threshold radius for each cluster using $(x - xi)^2 + (y - yi)^2 \leq r^2$.: Here, r is the radius of the cluster and (xi, yi) is a point inside the cluster. Check all the good matches for their respective cluster and eliminate the outliers. If the cluster still has more than or equal to two keypoints then it is a forged image otherwise it is authentic. Flowchart for the proposed method is in Fig.1.

6   Experimental Results

In this section we evaluate results for each of the proposed methods and determine the accuracy for each of them. For evaluating the results, a dataset of images from MICC – F220 is taken [18]. The results for different scale factors and cluster radius for each of SURF, HOG and SIFT image features are taken. For the images, a *Scale Factor* of 3 and a *Cluster Size* of 30 was selected. For larger sized images, *Cluster Size* can vary between 50 and 100. We define accuracy and precision proposed method as given below.

$$True\ Matches = No.\ of\ true\ positive\ matches + No.\ of\ true\ negative\ matches \tag{2}$$

$$False\ Matches = No.\ of\ false\ postive\ matches + No.\ of\ false\ negative\ matches \tag{3}$$

$$Matches = True\ Matches + False\ Matches \tag{4}$$

$$Final\ Matches = No.\ of\ true\ positive + No.\ of\ false\ positive \tag{5}$$

$$Accuracy = \frac{No.\ of\ true\ positive\ matches + No.\ of\ true\ negative\ matches}{No.\ of\ Matches} \tag{6}$$

$$Precision = \frac{No.\ of\ true\ positive\ matches}{No.\ of\ Final\ Matches} \tag{7}$$

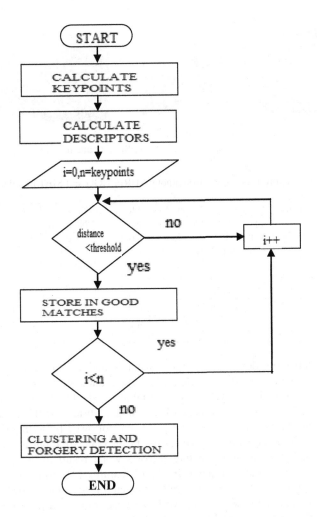

Fig. 1. Flowchart depicting the proposed method

Fig. 2. CMFD using SURF image features

Fig. 3. CMFD using SIFT image features

Table 1. A comparison among SURF, HOG and SIFT based CMFD

CMFD using different image features	Mean No Of Matched Keypoints	Mean No Of final Matched Keypoints	Mean No of True positive Matches	Mean No of false negative matches	Accuracy	Precision
SURF	12.6	6.2	6.1	0	48.4%	94.6%
HOG	10.4	2.1	0.9	5.7	0.8%	42.8%
SIFT	24.2	22.2	22.2	0	91.7%	100%

we are only considering the forged matches and not the authentic matches, true negatives matches are 100% correct., i.e., every authentic image region is detected as authentic. A graphical representation for table 1 and table 2 is given in figure 4. Based on the results we can say that CMFD using Hybrid or SIFT image features give results with the better accuracy and precision, followed by SURF image feature then HOG. SURF image features give results faster than SIFT but the accuracy is less in comparison to SIFT. HOG feature provide worst result in comparison, SURF and SIFT. To improve the accuracy using HOG features a Hybrid features of SIFT,HOG or SURF ,HOG is considered where the detected keypoints by SIFT or SURF CMFD are used to calculate their respective HOG descriptors and increase the capability of hybrid HOG features for CMFD. These hybrid features can detect the *copied regions completely* instead of just detecting the *keypoints*. In table 2 we have shown results of CMFD using hybrid features.

Table 2. A comparison between hybrid features for CMFD

Hybrid Features	Copied-Region Matched	Precision	Accuracy
SURF and HOG	92.2%	96.3%	95.5%
SIFT and HOG	96.8%	100%	98.5%

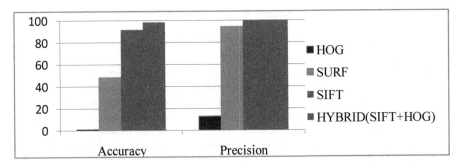

Fig. 4. A comparison among HOG, SURF, SIFT, and HYBRID based CMFD

7   Conclusions

A methodology to detect Copy-Move Forgery has been proposed to support image forgery detection. The results were recorded using three different image features namely SURF, HOG and SIFT among which SIFT provided best results in the form of accuracy and precision. By applying same method on different features we have shown that how one feature provides better results in comparison to others. After considering hybrid features (SURF-HOG or SIFT-HOG), we are getting better result for CMFD in comparison to SIFT or SURF or when HOG is used alone. The advantage of using hybrid features is this, that it will provide copy-move area completely instead keypoint as shown in the SURF and SIFT -CMFD. The CMFD using SURF is very fast in comparison to SIFT or hybrid features. So all the features what we have taken have some advantage and some disadvantages. If we want commendable precision and accuracy in CMFD then we should select SIFT or Hybrid features or if we want fast CMFD in term of time then we should select SURF or if we want better localization of Copy Move region then we should select Hybrid features .Future work will be mainly dedicated to investigating how to improve the detection phase for multiple cloned objects with respect to the cloned image patch and for patches with highly uniform texture where salient keypoints are not recovered by SIFT-like technique.

References

1. Lyu, S., Farid, H.: How realistic is photorealistic? IEEE Transactions on Signal Processing 53(2), 845–850 (2005)
2. Farid, H.: Photo fakery and forensics. Advances in Computers 77, 1–55 (2009)
3. Redi, J.A., Taktak, W., Dugelay, J.L.: Digital image forensics, a booklet for beginners. Multimedia Tools and Applications 51(1), 133–162 (2011)
4. Cox, I.J., Miller, M.L., Bloom, J.A.: Digital watermarking. Morgan Kaufmann, San Francisco (2002)
5. Barni, M., Bartolini, F.: Watermarking Systems Engineering: Enabling Digital Assets Security and Other Applications. Marcel Dekker (2004)

6. Farid, H.: A survey of image forgery detection. IEEE Signal Processing Magazine 2(26), 16–25 (2009)
7. Popescu, A., Farid, H.: Statistical tools for digital forensics. In: Proc. of Int.'l Workshop on Information Hiding, Toronto, Canada (2005)
8. Swaminathan, A., Wu, M., Liu, K.: Digital image forensics via intrinsic fingerprints. IEEE Transactions on Information Forensics and Security 3(1), 101–117 (2008)
9. Chen, M., Fridrich, J., Goljan, M., Lukas, J.: Determining image origin and integrity using sensor noise. IEEE Transactions on Information Forensics and Security 3(1), 74–90 (2008)
10. Khanna, N., Chiu, G.T.C., Allebach, J.P., Delp, E.J.: Forensic techniques for classifying scanner, computer generated and digital camera images. In: Proc. of IEEE ICASSP, Las Vegas, USA (2008)
11. Caldelli, R., Amerini, I., Picchioni, F.: A DFT-based analysis to discern between camera and scanned images. International Journal of Digital Crime and Forensics 2(1), 21–29 (2010)
12. Lowe, D.G.: Distinctive image features from scale-invariant keypoints. Int.'l Journal of Computer Vision 60(2), 91–110 (2004)
13. Popescu, A., Farid, H.: Exposing digital forgeries by detecting duplicated image regions, Dartmouth College, Computer Science, Tech. Rep. TR2004-515 (2004)
14. Bayram, S., Sencar, H.T., Memon, N.: An efficient and robust method for detecting copy-move forgery. In: Proc. of IEEE ICASSP, Washington, DC, USA (2009)
15. Bayram, S., Sencar, H.T., Memon, N.: A survey of copy-move forgery detection techniques. In: Proc. of IEEE Western New York Image Processing Workshop, Rochester, NY (2008)
16. Bay, H., Tuytelaars, T., Gool, L.-V.: SURF: Speeded Up Robust Features. In: Proc. of ETH Zurich (2006)
17. Dalal, N., Triggs, B.: Histograms of Oriented Gradients for Human Detection. In: Proc. of INRIA Rhone-Alps, 655 avenue de l'Europe, Montbonnot 38334, France (2005)
18. Amerini, I., Ballan, L., Caldelli, R., Bimbo, A.D.: A SIFT-based forensic method for copy-move attack detection and transformation recovery. In: Proc. of IEEE Transactions on Information Forensics and Security (2011)

A Bit-Level Block Cipher Diffusion Analysis Test - BLDAT

Dipanjan Bhowmik[1], Avijit Datta[2], and Sharad Sinha[1]

[1] Dept. of Computer Science and Application, University of North Bengal
[2] Dept. of Computer Application, Siliguri Institute of Technology, Siliguri

Abstract. The paper describes a scheme aimed at measuring the diffusion characteristic of a block cipher. Cryptographic strength of a cipher is directly proportional to the extent to which diffusion is achieved by the underlying cipher. The scheme used to measure this characteristic of a cipher and the test results are analyzed to come to conclusion. Potentially, the scheme described in the paper can be added as part of the already existing, varied test suit to act as a distinguisher based on the diffusion characteristic of the underlying cipher.

Keywords: Diffusion, Random Sequence, Bit-level Block Cipher, Randomness.

1 Introduction

Many test suits in [1] have been designed to test the extent of randomness approximated by a block cipher. In [2] most of these tests measure the degree of randomness of change at block level by changing a bit in the original block. However, while operating at block level, a situation may arise where the ith bit of the block changes with a very high frequency whereas some other j^{th} bit hardly changes. This gives a false implication that all the bits are changing with a probability of 0.5.

In this paper a scheme is being proposed which is not significantly different in nature with some of the existing test sets but is rather different in terms of how it is implemented to measure the diffusion characteristic of the concerned cipher.

The scheme named "Bit-level Block Cipher Diffusion Analysis Test (BLDAT)" is aimed at how vulnerable the underlying block cipher is with regards to a particular bit.

Like most test in this field, the proposed scheme also treats the underlying block cipher as a black box and the results are based on solely the input and output patterns provided to and generated by the cipher, respectively.

1.1 Terminology

Truly Random Sequence
An n bit sequence is a truly random sequence if each bit is independent to every other bit in the sequence.

© Springer International Publishing Switzerland 2015
S.C. Satapathy et al. (eds.), *Proc. of the 3rd Int. Conf. on Front. of Intell. Comput. (FICTA) 2014*
– *Vol. 1*, Advances in Intelligent Systems and Computing 327, DOI: 10.1007/978-3-319-11933-5_75

Informally, it can be stated that the probability of regeneration of a truly random sequence is very low, though we cannot guarantee the non regeneration of such a sequence.

Relatively Random Sequence

Two n bit sequence are relatively random if number of bit by bit successful matches between the two sequences is n/2.

2   Bit-Level Diffusion Analysis Test (BLDAT)

The scheme uses a randomly selected n bit block of plain text (say P), which is then encrypted using the underlying cipher to produce the corresponding cipher block (say C). Then, a matrix of size $n \times n$ is produced, where each row of the matrix is say P_i a new plain text block in itself derived from the original block by flipping the bit at the ith position i.e. $P_i[i] = P \oplus e_i$, where e_i is a zero vector containing 1 at i^{th} position.

$$\begin{pmatrix} p[0,0] & \cdots & p[0,n-1] \\ \vdots & \ddots & \vdots \\ p[n-1,0] & \cdots & p[n-1,n-1] \end{pmatrix}$$

Next, each row of the P_i matrix is fed as input to the underlying cipher to produce the corresponding cipher text, which is stored as the ith row of the C_i matrix of size $n \times n$.

$$C_i = \begin{pmatrix} c[0,0] & \cdots & c[0,n-1] \\ \vdots & \ddots & \vdots \\ c[n-1,0] & \cdots & c[n-1,n-1] \end{pmatrix}$$

i.e. $C_i[i] = E(P_i[i])$ where E() denotes encryption using the underlying block cipher.

At this point, the scheme kicks in to produce another matrix (say X) of size $n \times n$, where i^{th} row of the matrix is obtained by bit wise addition (modulo 2 additions) of C_i vector with C vector [3] , $X[i] = C_i[i] \oplus C$

$$X = \begin{pmatrix} x[0,0] & \cdots & x[0,n-1] \\ \vdots & \ddots & \vdots \\ x[n-1,0] & \cdots & x[n-1,n-1] \end{pmatrix}$$

The scheme further produces the diffusion-factor by scanning each column of the X matrix. The algorithm of the proposed scheme is given as:

2.1   Algorithm

Step-1: Randomly select a binary string of n bits (P).

Step-2: Encrypt the plaintext with the concerned encryption algorithm to generate the corresponding cipher text (C).

Step-3: Encrypt P_i's with the particular encryption algorithm to generate C_i's, where $P_i = P \oplus e_i$ and e_i is a string of zeros with the ith bit 1 and $E(P_i) = C_i$.

Step-4: $X_i = C_i \oplus C$ is stored in the ith row of the matrix of size $n \times n$ where n, the number of bits is in the original plaintext.

Step-5: Find the number of 1s in each column (j).

As it is evident from the algorithm, there are n bits in the block and at every instance we had changed only one bit, as a result there are n blocks such that $H(P, P_i[i] = 1)$, where H denotes the Hamming Distance. The test finds the number of times a particular bit has changed when each of the n newly generated blocks are encrypted using the underlying cipher with respect to the cipher text of the original block. Ideally each bit should change n/2 times, if a particular bit has changed with very high or very low frequency, it might motivate an attack.

The number of times a particular bit has changed is referred to as the vulnerability factor of the bit. An extremely low or extremely high vulnerability factor associated with a particular bit may act as a motivation for attackers to exploit this idea.

The time taken to construct the X matrix is $O(n^2)$ and the time taken to determine the bit-vulnerability factors is also $O(n^2)$. So, the time complexity of the algorithm is $O(n^2)$.

3   Experiment

A single block cipher has been used for applying the scheme. The cipher text generated by the block cipher is subjected to the proposed test analysis.

3.1   Describing the Test Cipher

The analysis of the scheme is done using a simple Test Cipher. The Test Cipher is a simple substitution – permutation network which takes 8-bit block as input and produces 8-bit cipher block. At first, as found in [4], the 8 bit block is bit wise XOR-ed with an 8-bit key and then passed through two 4-bit S-boxes (for simplicity the two S-boxes are considered to be identical). The outputs of the S-boxes are permuted to generate the cipher text. The block diagram of the test cipher is depicted by fig. 1.

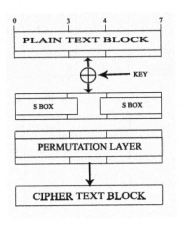

Fig. 1. Test Cipher

3.2   Objective

The objective is to determine the secret key as is the case with other cryptanalysis techniques.

3.3   Assumptions

It is assumed that the plain text block is known, the corresponding cipher text block is also known and the results of the bit-level diffusion test are available. Using exhaustive key search method, the correct key can be determined with 28 trials (in the worst case). The sub-goal would be to reduce the number of trials using the results of BLDAT. If it is observed from the results of the BLDAT that a particular bit (say j^{th} bit) of the cipher block seldom changes i.e. if the ith bit of the plain text is 1/0, and remains 1/0 in the cipher text at j^{th} position with a very high degree of probability, then it will be easy to identify the particular key bit. Linear cryptanalysis and differential cryptanalysis are well known for mapping an input plain text bit to an output cipher text bit. If the observed plain text and cipher text are dissimilar, then it is clearly due to the key bit which got XOR-ed with the ith plain text bit and will be a 1with a very degree of probability. And if both the observed bits are the same, then it implies that the key bit has not affected the plain text bit which in turn implies that the bit is a 0 with a very high degree of probability.

3.4   Experimental Results of BLDAT

Two well known ciphers, namely Data Encryption Standard and Rijndael (Advanced Encryption Standard) are put to test, and the results obtained are analyzed in[5]. Say δ is deviation where $\delta = 0$ is the ideal case. Table 1 and Table 2 lists the results of BLDAT on DES and Rijndael Cipher (AES) respectively.

Table 1. Experimental result on DES

Key	No. of Deviations			
	$\delta = 0$	$\delta = 4$	$\delta = 8$	$\delta > 8$
Sparse Key	5	40	16	3
Moderately Dense Key	5	47	9	3
Dense Key	5	43	13	3
Random Key	7	34	21	2

Table 2. Experimental result on AES

Key	No. of Deviations			
	$\delta = 0$	$\delta = 8$	$\delta = 16$	$\delta > 16$
Sparse Key	8	89	13	1
Moderately Dense Key	11	99	18	0
Dense Key	5	102	20	1
Random Key	6	97	25	0

4   Analysis of BLDAT

The results obtained from both DES and AES are analyzed using the established statistical tools to finally draw the conclusion. The Chi-square (χ^2) test has been used to determine the goodness of fit between theoretical and experimental data. The observed values and expected values are to be tested here.

4.1   Chi-Square (χ^2) Test on Experimental Result of DES

In the experiment of BLDAT for DES, with 64 bit plaintext block and 56 bit key, the observed bit changes in cipher text block for every bit change in plaintext given in Table 3.

Table 3. Observed bit changes in cipher text using DES

v[0]=29	v[8]=27	v[16]=30	v[24]=38	v[32]=32	v[40]=37	v[48]=37	v[56]=24
v[1]=38	v[9]=37	v[17]=36	v[25]=33	v[33]=32	v[41]=28	v[49]=30	v[57]=29
v[2]=31	v[10]=31	v[18]=30	v[26]=38	v[34]=30	v[42]=25	v[50]=33	v[58]=35
v[3]=35	v[11]=32	v[19]=28	v[27]=34	v[35]=33	v[43]=32	v[51]=24	v[59]=26
v[4]=26	v[12]=33	v[20]=24	v[28]=28	v[36]=31	v[44]=31	v[52]=24	v[60]=23
v[5]=28	v[13]=29	v[21]=34	v[29]=31	v[37]=34	v[45]=33	v[53]=40	v[61]=27
v[6]=28	v[14]=38	v[22]=39	v[30]=30	v[38]=36	v[46]=31	v[54]=35	v[62]=30
v[7]=26	v[15]=32	v[23]=32	v[31]=31	v[39]=32	v[47]=42	v[55]=31	v[63]=33

The change of cipher text block is ideally n/2 where n is the block size and is estimated 32. Table 4 is constructed to calculate the Chi-square distribution and goodness of fit for Chi-square.

Table 4. DES Observed values with corresponds to estimated value with their occurrence

(T)	(O)	(E)	O-E	(O-E)^2	Y=(O-E)^2/E
3	29	32	-3	9	0.28125
4	38	32	6	36	1.125
8	31	32	-1	1	0.03125
3	35	32	3	9	0.28125
3	26	32	-6	36	1.125
5	28	32	-4	16	0.5
2	27	32	-5	25	0.78125
3	37	32	5	25	0.78125
6	33	32	1	1	0.03125
7	32	32	0	0	0
6	30	32	-2	4	0.125
2	36	32	4	16	0.5
4	24	32	-8	64	2
3	34	32	2	4	0.125
1	39	32	7	49	1.53125
1	25	32	-7	49	1.53125
1	42	32	10	100	3.125
1	40	32	8	64	2
1	23	32	-9	81	2.53125

Where T is occurrences of observed value, O is observed value, E is estimated value. Using these values the chi-square is calculated with the consideration of occurrence of values and the calculated chi-square value of the experimental data is:

$$\chi^2 = \sum \frac{(O - E)^2}{E}$$
$$\chi^2 = 37.25$$

and fig. 2 is the graphical representation of observed value, estimated value and calculated value:

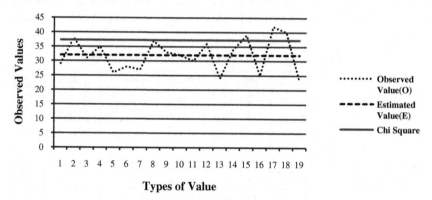

Fig. 2. Graph for Chi-Square of DES

In the experimental result (Table 4) it is visible that there exist 19 different sets of value (v) i.e. number of time changes of every bit in ciphertext while changing every bit of plaintext for at most once. Therefore the Degree of Freedom (df) is:

$$df = v - 1 = 19 - 1 = 18$$

From the Table 4 it is now easy to calculate chi-square value for df = 18 α = .005:

$$GF_{18,.005} = \chi^2_{18,.005} = 34.71875$$

Moreover, chi-square value for df = 18 and α = 5% is:

$$GF_{18,5\%} = \chi^2_{18,5\%} = 31.5$$

From the chi-square distribution table [6], is found that $\chi^2_{18,.005}$ is 37.156 and $GF_{18,.005} < 37.156$.

So, according to [7] it may be concluded that either (i) this model is valid but that a statistically improbable excursion of χ^2 has occurred, (ii) too conservatively, overestimated the values of α or (iii) data is 'too good to be true'.

4.2   Chi-Square (χ^2) Test on Experimental Result of AES

In the experiment of BLDAT for AES, for 128 bit plaintext block with 56 bit key and observed bit changes in cipher text block for every bit change in plaintext. The Chi-square (χ^2) test has been used. The change of cipher text block is ideally n/2 where n is the block size and is estimated 64. The calculated chi-square value of the experimental data of AES is:

$$\chi^2 = \sum \frac{(O - E)^2}{E}$$

$$\chi^2 = 73.609375$$

Fig. 3 is the graphical representation of observed value, estimated value and calculated value.

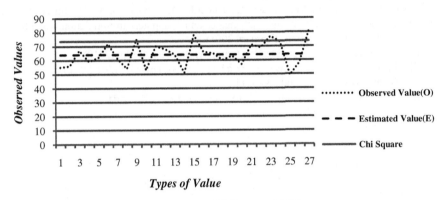

Fig. 3. Graph for Chi-Square of AES

In the experimental it is visible that there exist 27 different sets of value (v) i.e. number of time changes of every bit in cipher text while changing every bit of plaintext for at most once. Therefore the Degree of Freedom (df) is:

$$df = v - 1 = 27 - 1 = 26$$

Now it is easy to calculate chi-square value for df = 18 α = .005:

$$GF_{26,.005} = \chi^2_{26,.005} = 69.09375$$

Moreover, chi-square value for df = 26 and α = 5% is:

$$GF_{26,5\%} = \chi^2_{26,5\%} = 63.8046875$$

From the chi-square distribution table [6], is found that $\chi^2_{26,.005}$ is 48.290 and $GF_{26,.005} > 48.290$.

So, according to [7] it may be concluded that either (i) this model is valid one but that a statistically improbable excursion of χ^2 has occurred or (ii) that this model is poorly chosen that an unacceptable large value of χ^2 has resulted. The theory of chi-square test relies on the assumption that chi-square is the sum of the squares of random normal derivatives, that is, that each x_i is normally distributed its mean value μ_i.

5   Conclusion

Even if the Hamming Distance of the plain text block and the cipher text block is ideal i.e. n/2 where n is the block size, the key space can be reduced using a scheme such as the proposed BLDAT.

References

1. Toz, D., Doğanaksoy, A., Turun, M.S.: Statistical Analysis of Block Ciphers. In: Ulusal Kriptologi Sempozyumu, Ankara, Turkey, pp. 56–66 (2005)
2. Soto, J., Bassham, L.: Randomness Testing of the Advanced Encryption Standard Finalist Candidates. In: Computer Security Division. National Institute of Standards and Technology (2000)
3. Katos, V.: A Randomness Test for Block Ciphers. In: Applied Mathematics and Computation, vol. 162, pp. 29–35. Elsevier Publication (2005)
4. Castro, J.C.H., Sieria, J.M., Seznec, A., Izquierdo, A., Ribagorda, A.: The Strict Avalanche Criterion Randomness Test. In: Mathematics and Computers in Simulation, vol. 68(2005), pp. 1–7. Elsevier Publication (February 2005)
5. Paar, C., Pelzl, J.: Understanding Cryptography. Springer, Berlin (2010)
6. Chi Square Tests, ch.10, http://uregina.ca/~gingrich/ch10.pdf
7. Chi-Square: Testing for Goodness of Fit, http://www.physics.ucsc.edu/~drip/133/ch4.pdf

Curved Videotext Detection and Extraction: LU-Decomposition and Maximal H-Transform Based Method

P. Sudir[1] and M. Ravishankar[2]

[1] Dept. of E&C, SJCIT, Chikkaballapur, India
[2] Dept. of IS, DSCE, Bangalore, India
{sudirhappy,ravishankarmcn}@gmail.com

Abstract. Proposed approach explores a new framework for Curved video text detection and extraction by utilizing fast texture descriptor LU-transform and Maximal H-Transform for the text detection. Connected Component filtering method and Nearest Neighborhood (NN) constraint is utilized for false positive elimination. B-Spline curve is fitted to centroid of each character and each oriented character is vertically aligned by finding angle that aligns the normal vector to the vertical axis. The aligned text string is fed to OCR for recognition. Experiments on various curved video data and Hua's horizontal video text dataset shows that proposed method is efficacious and robust in detecting and recognizing multioriented videotext.

Keywords: Curve Video text, LU Decomposition, Maximal H-Transform.

1 Introduction

Due to tremendous advancement in Video technology, copious amount of video information available calls for the need for Semantic Video analysis and management. Since Video text is very vital information content, some of the Video processing techniques focus on text data embedded in videos. Video text detection and extraction is a vital task for applications such as Video Indexing and Content based video retrieval. Videotexts are classified into two types, first is Superimposed Text which is added as a part of editing process and the second is Scene text which appears naturally in a Scene captured by a camera. Despite the fact that many methods exist, text detection and extraction is still a challenging task because of unconstrained colors, sizes, and alignments of the characters. Custom Text detection and extraction methods fall into three subgroups: Connected Component (CC), Edge based and Texture-based. CC-based methods [1][2] group small CCs to larger ones based on color and contrast until all text regions are identified in the image thus limiting itself to low contrast and Non-Uniform Color text. Edge based approach [3],[4],[5] requires text to have a better contrast to the background in order to detect the edges hence they exhibit poor result in case of complex backgrounds. Texture based methods [6],[7] utilize various local texture descriptors, such as Gabor Filter [8] and Wavelet Filters [9].

© Springer International Publishing Switzerland 2015
S.C. Satapathy et al. (eds.), *Proc. of the 3rd Int. Conf. on Front. of Intell. Comput. (FICTA) 2014*
– *Vol. 1*, Advances in Intelligent Systems and Computing 327, DOI: 10.1007/978-3-319-11933-5_76

Despite varying degrees of success these methods fail to provide solutions in the case of Multi Oriented Video Texts. In this regard Shivakumara et al.[10] proposed a method for detecting arbitrarily oriented text based on Morphological approach which fails in case of complex background. Yong et al.[11] proposed a corner and skeleton based method for arbitrarily oriented text which is quiet robust only when the corners are detected effectively and fails in case of low resolution Videos. Shivakumara et al. [12] proposed an arbitrarily oriented scene text detection method which extracts text lines of any orientation based on Gradient Vector Flow (GVF) and Neighbor component grouping which does not give good accuracy for less spaced text lines. Shivakumara et al. [13] proposed a curved text detection scheme which uses enhancement criterion and quad tree approach. Although efficient in detection proposed method fails to eliminate false positives.

Clearly all the above methods though efficient in detection fail to produce an OCR ready input. Hence in this paper, we propose an efficient text detection and extraction scheme for Curved Videotexts. Vital contributions of the proposed method are as follows: 1) A LU-Transform and Maximal H-Transform based curved text detection method 2) A Character alignment correction method based on normal vector calculation for generating OCR friendly curve texts for recognition.

2   Proposed Approach

Proffered method works in Three steps: 1) Probable Text region detection by LU decomposition and Maximal H-Transform based true text pixels detection. 2) Size and nearest neighborhood based false positive elimination. 3) Estimation and correction of orientation of Individual characters.

2.1   Text Region Detection by LU-Decomposition

Inspired by the work explained in [14] successful texture response generation of text regions using LU decomposition was attained by the proposed method. In numerical analysis, any matrix can be factored as the product of a lower triangular matrix (L) and an upper triangular matrix (U). All elements above the diagonal are zero in L matrix and all the elements below the diagonal are zero in U matrix. For example, for a 3-by-3 matrix A, its LU decomposition looks like this:

$$\begin{bmatrix} a_{11} & a_{12} & a_{13} \\ a_{21} & a_{22} & a_{23} \\ a_{31} & a_{31} & a_{33} \end{bmatrix} = \begin{bmatrix} l_{11} & 0 & 0 \\ l_{21} & l_{22} & 0 \\ l_{31} & l_{31} & l_{33} \end{bmatrix} \begin{bmatrix} u_{11} & u_{12} & u_{13} \\ 0 & u_{22} & u_{23} \\ 0 & 0 & u_{33} \end{bmatrix} \qquad (1)$$

$$[L] \qquad\qquad [U]$$

$$\begin{bmatrix} 4 & 3 \\ 6 & 3 \end{bmatrix} = \begin{bmatrix} 1 & 0 \\ 1.5 & 1 \end{bmatrix} \begin{bmatrix} 4 & 3 \\ 0 & -1.5 \end{bmatrix} \qquad (2)$$

Our Experiments on various video frames showed that diagonal of U fires in image areas of rough texture and successfully captures the text information. The actual texture response of text region is calculated as the maximum value of all diagonal values of the U matrix.

$$\Omega_p(w) = \max(|u_{kk}|) \tag{3}$$

where u_{kk} are the diagonal elements of U, w is the window size(4 in our experiment). Homogeneous NonText regions exhibit low magnitude Ω because of their linear dependence and low value of rank r. Conversely text regions exhibit large Ω magnitude regions because of high value of r.

2.2   Text Localization Method

Further an extended maxima transform [15] is used to find the peaks which are n intensity values higher than the background in probable text regions. It is a robust peak finder which depends only on contrast. It performs first H-maxima transform followed by recognition of regional maxima. H-maxima transform is based on morphological reconstruction, repeated dilations of image followed by masking. The H-maxima transform is given by Koh et al. [15]:

$$Hmax_h(f) = R_f(f-h) \tag{4}$$

$R_f(f-h)$ is the morphological reconstruction by dilation of image f with respect to (f-h). This morphological operation ignores regions whose intensity values with respect to their neighbors is smaller than a threshold level H. All regional maximas are identified by an extended maxima operation. Regional maxima are those whose external boundary pixels have a lower value and which are connected components of pixels with a constant intensity value. The extended maxima operation is defined by:

$$EMAX_h(f) = RMAX(Hmax_h(f)) \tag{5}$$

It removes local peaks which are lower than h intensity values from the background. Based on the analysis done during our experiments, h is set to 40. The curve text localized by using extended maxima transform is shown in Fig.1 (c).

2.3   False Positive Removal Method

False Positive elimination is executed based on size of CC, height of CC and distance between CCs. Since it is a known fact that all the characters in a circular text will be of almost of same height. Height of a CC of circular text is chosen as maximum among length of CC and width of CC. In our experiment we ignore CCs whose size is less than 200 and height less than 20, height greater than 70 and nearest neighbor CC centroid distance greater than 45.The result of false positive elimination is shown in Fig.1 (f).

2.4   Text Extraction and Recognition Procedure

In view of inability of Commercial OCR's in recognizing curved text we propose an efficient method to align videotext laid out in a variety of layouts such as arc, wave or their combination with linearly-skewed text lines. Restoration of the inter-character gaps is taken care so that the rectified text string can be 'meaningfully' recognized. Proposed method is the first of its kind which renders curved video text a linear form so that it can be recognized using off-the-shelf OCRs. An 8-connectivity based component labeling is performed on the binary image to obtain a set of 'n' disjoint components $\{CC_j\}$, j = 1, 2, ...n. For each identified CC, the centroid of the constituent CC is identified and represented as follows:

$$C_j^k = \left\{ \left(C_{x,j}^k, C_{y,j}^k \right) \right\}, j = 1,2,....N^k \tag{6}$$

Where N^k is the number of CCs. These centroid points serve as the control points for fitting B-Spline curves. The ith point in the resulting curve is denoted by:

$$p_i^k = \{x_i^k, y_i^k\} \tag{7}$$

The normal vector at the ith point of the curve is then computed using the following relation

$$n_i^k = \begin{bmatrix} \cos\left(\dfrac{\pi}{2}\right) & -\sin\left(\dfrac{\pi}{2}\right) \\ \sin\left(\dfrac{\pi}{2}\right) & \cos\left(\dfrac{\pi}{2}\right) \end{bmatrix} \times \frac{1}{2}\left(\frac{\left(p_i^k - p_{i-1}^k\right)}{\left\| p_i^k - p_{i-1}^k \right\|_2} + \frac{\left(p_{i+1}^k - p_i^k\right)}{\left\| p_{i+1}^k - p_i^k \right\|_2} \right) \tag{8}$$

Here $\|.\|_2$ denotes the L_2 norm.

The normal vectors thus obtained give fairly good estimate of the orientation of the character. The characters are rotated such that the normal vector is aligned vertically. The required angle of rotation for any character is computed as:

$$\theta_i^k = (90 - \angle\, n_i^k), i=1,2,...,N^k \tag{9}$$

Thus each character is individually rotated and stacked from left to right such that the horizontal spacing between the characters is proportional to the corresponding inter-character centroid distances. The characters are recognized using Tesseract OCR [19]. Sample results are shown in Fig.2. Due to zigzag pattern of the Centroid of the CCs 'small' vertical misalignments still exist.

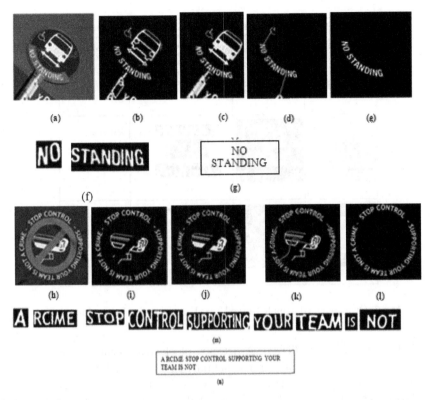

Fig. 1. Curved Text Detection and Recognition result (a,h) Video Frame (b,i) Text region detected using LU-Decomposition (c,j) Output after Maximal H- Transform (d,k) Centroids connected based on Distance calculation (e,l) Text detected after Distance constraint (f,m) Orientation Corrected CC's (g,n) OCR output

3   Experiments Results

Experiments were carried out on our own 20 mpeg-1 videos collected each with 320x240 pixel resolution due to lack of benchmarked Curved video text dataset. It consists of 92 words and 670 characters. Also a small dataset of 45 video frames, which is available, publicly [10] was chosen to evaluate the performance of the method. Four methods were chosen for comparison of which first is a recently developed method by Shivakumara et al.[13] which address curved text detection in video through an enhancement criterion and use of quad tree. Second is Laplacian method [10] which works well for non-horizontal Video text by combining Laplacian and Fourier. Third is Bayesian method [21] which explores Bayesian classifier and gradient. Fourth is a method for arbitrary text detection by Sharma et al. [16] which proposes gradient directions and two stage grouping. Following performance criterion has been defined for horizontal, vertical and multioriented videotext (except curved text).

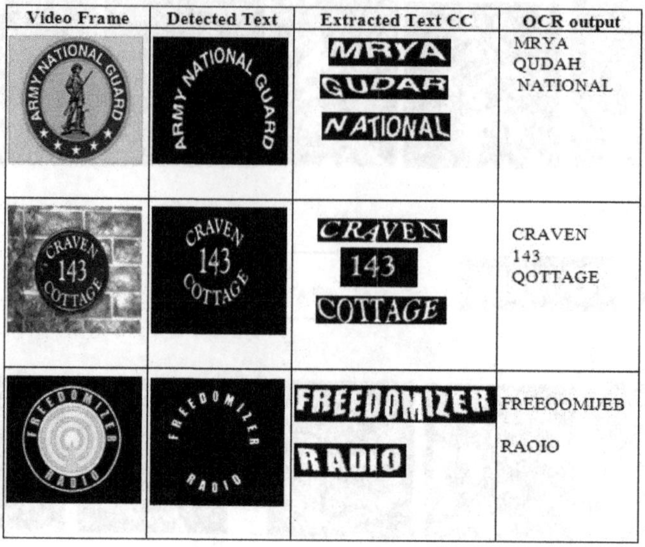

Video Frame	Detected Text	Extracted Text CC	OCR output
		MRYA GUDAR NATIONAL	MRYA QUDAH NATIONAL
		CRAVEN 143 COTTAGE	CRAVEN 143 QOTTAGE
		FREEDOMIZER RADIO	FREEOOMIJEB RAOIO

Fig. 2. Own collected Sample Curved Text Extraction and Recognition results

- Truly Detected Block (TDB): A detected block that contains at least one true character. Thus, a TDB may or may not fully enclose a text line.
- Falsely Detected Block (FDB): A detected block that does not contain text.
- Text Block with Missing Data (MDB): A detected block that misses more than 20% of the characters of a text line (MDB is a subset of TDB).

The performance measures are defined as follows.

$$1)\ \ Recall(R) = \frac{TDB}{ATB} \quad 2)\ \ Precision(P) = \frac{TDB}{TDB + FDB} \quad 3)\ \ F - measure(F) = \frac{2PR}{P + R}$$

Average Processing Time (APT) per frame is also considered as a measure to the time complexity of the proposed method.

3.1   Performances on Hua's Data

Text detection results on sample video frames from Hua's dataset are shown in Fig.3. Proposed method achieves results at par with the state of the art as indicated in Table 1 and is robust to the orientation, perspective, color, and lighting of the text object. It detected most text objects successfully.

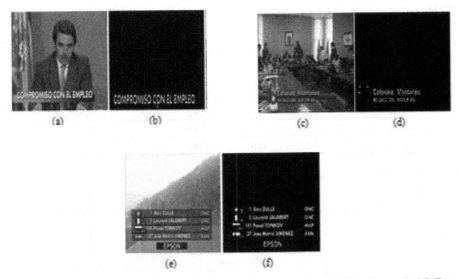

Fig. 3. Text Detection and Localization result on Hua's dataset (a,c,e)-Video Frame (b,d,f) Text Detection result

3.2 Performance Analysis on Curved Data

Sample results of the proposed method for curved text extraction versus the Quad tree based method are shown in Fig.4. The proposed method outperforms the existing methods in terms of recall, precision-measure and misdetection rate as illustrated in Table 2. The main reason for poor results of the existing methods is that the existing methods are developed for horizontal and non-horizontal text detection but not for curved video text detection. Table 3 shows the recognition accuracy obtained on the curved dataset. The Accuracy is fairly good and there is still scope for improvement in this regard.

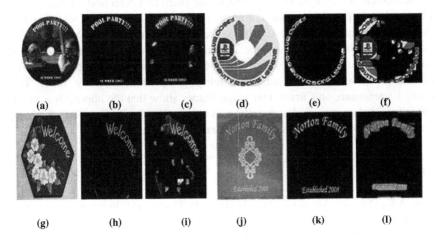

Fig. 4. Text Detection result (a,d,g,j) Input Frames,(b,e,h,k) Proposed method result,(c,f,i,l) Quad tree based method

Table 1. Comparison of proposed method with Existing Detection Methods for Hua's database

Methods	R	P	F	MDR	APT(secs)
Proposed	**0.87**	**0.9**	**0.9**	**0.06**	**1.1**
Quad tree	0.82	0.88	0.84	0.06	2.0
Sharma	0.88	0.77	0.82	0.32	9.0
Bayesian	0.87	0.85	0.85	0.18	5.6
Laplacian	0.93	0.81	0.87	0.07	11.7

Table 2. Performance of the Proposed and Existing Detection Methods on Curved Data

Methods	R	P	F	MDR	APT(secs)
Proposed	**0.82**	**0.87**	**0.84**	**0.20**	**0.71**
Quad tree	0.80	0.83	0.81	0.25	2.3
Sharma	0.73	0.88	0.79	0.28	10.3
Bayesian	0.59	0.52	0.55	0.27	12.1
Laplacian	0.55	0.68	0.60	0.42	9.9

Table 3. Character and Word Recognition Performance by Proposed Recognition Method

Dataset	Accuracy	
	Character	Word
Own Curved Data	82	85

4   Conclusion

Proposed Method unifies a framework for Curved video text detection, extraction and recognition. In this paper we propose a novel method for curved video text detection which exploits the intrinsic characteristic of the text by using the LU-Decomposition and Maximal H-Transform for effective Text Detection and localization. Orientation of each character are found and corrected by B-Spline curve fitting and Normal vector calculation. Proposed method generates fairly good OCR friendly input Text segments. Experiments on Curved text video frames show that the approach is viable and effective in detecting and recognition in comparison to state of Art methods. Improvement of Recognition rate will be further focused.

References

1. Jain, A.K., Yu, B.: Automatic Text Location in Images and Video Frames. Pattern Recognition 31(12), 2055–2076 (1998)
2. Lienhart, R., Stuber, F.: Automatic Text Recognition in DigitalVideos. In: Proceedings of SPIE Image and VideoProcessing IV 2666, pp. 180–188 (1996)

3. Agnihotri, L., Dimitrova, N.: Text Detection for Video Analysis. In: IEEE Workshop on CBAIVL, pp. 109–113 (1999)
4. Agnihotri, L., Dimitrova, N., Soletic, M.: Multi-layered Videotext Extraction Method. In: IEEE International Conference on Multimedia and Expo (ICME), Lausanne (Switzerland), August 26-29 (2002)
5. Hua, X.-S., Chen, X.-R., et al.: Automatic Location of Text in Video Frames. In: Intl. Workshop on Multimedia Information Retrieval In Conjunction with ACM Multimedia (2001)
6. Kim, H.-K., et al.: Efficient Automatic Text Location Method and Content-based Indexing and Structuring of Video Database. Journal of Visual Communication and Image Representation 7(4), 336–344 (1996)
7. Li, H., Doermann, D., Kia, O.: Automatic Text Detection and Tracking in Digital Video. IEEE Trans. on Image processing. 9(1), 147–155 (2000)
8. Jain, A.K., Bhatarcharjee, S.: Text Segmentation using Gabor Filters for automatic document processing. Machine Vision and Application 5, 169–184 (1992)
9. Li, H., Doermann, D., Kia, O.: Automatic Text Detection and Tracking in Digital Video, Univ. of Maryland, College Park, Tech.Reps. LAMP-TR-028, CAR-TR-900 (1998)
10. Shivakumara, P., Phan, T.Q., Tan, C.L.: A Laplacian Approach to Multi-Oriented Text Detection in Video. IEEE Trans. on PAMI, 412–419 (2011)
11. Zhang, Y., Lai, J.: Arbitrarily Oriented Text Detection Using Geodesic Distances Between Corners and Skeletons. In: 1st International Conference on Pattern Recognition (ICPR), Tsukuba, Japan (2012)
12. Shivakumara, P., Phan, T.Q., Lu, S., Tan, C.L.: Gradient Vector Flow and Grouping based Method for Arbitrarily Oriented Scene text Detection in Video Images. IEEE Transactions on Circuits and Systems for Video Technology 23(10) (2013)
13. Shivakumara, P., Basavaraju, H.T., Guru, D.S., Tan, C.L.: Detection of Curved Text in Video: Quad Tree based Method. In: ICDAR (2013)
14. Targhi, A.T., Hayman, E., Eklundh, J.O.: Real-time texture detection using the LU-Transform. In: CIMCV (2006)
15. Koh, H.K., Shen, W., Shuter, B., Kassim, A.A.: Segmentation of kidney cortex in MRI studies: a constrained morphological 3D h-maxima transform approach. International Journal of Medical Engineering and Informatics 1(3), 330–341 (2009)
16. Sharma, N., Shivakumara, P., Pal, U., Bluemenstein, M., Tan, C.L.: A New Method for Arbitrarily-Oriented Text Detection in Video. In: Proc. DAS, pp. 74–78 (2012)
17. Yao, C., Bai, X., Liu, W., Ma, Y., Tu, Z.: Detecting Texts of Arbitrary Orientations in Natural Images. In: CVPR (2012)
18. Roy, S., Roy, P.P., Shivakumara, P., Pal, U., Tan, C.L.: HMM based Multi Oriented Text Recognition in Natural Scene Image. In: The 2nd Asian Conference on Pattern Recognition, ACPR 2013 (2013)
19. http://code.google.com/p/tesseract-ocr/
20. Hua, X.S., Wenyin, L., Zhang, H.J.: An Automatic Performance Evaluation Protocol for Video Text Detection Algorithms. IEEE Trans. on CSVT, 498–507 (2004)
21. Shivakumara, P., Sreedhar, R.P., Phan, T.Q., Lu, S., Tan, C.L.: Multi-Oriented Video Scene Text Detection through Bayesian Classification and Boundary Growing. IEEE Trans. on CSVT, 1227–1235 (2012)

i-Door: Intelligent Door Based on Smart Phone

Kumar Shashank,[1] M.T. Gopalakrishna[1], and M.C. Hanumantharaju[2]

[1] Dept. of ISE, DSCE
{zoom2shashank,gopalmtm}@gmail.com
[2] Dept. of ECE, BMSIT
mchanumantharaju@gmail.com

Abstract. Face recognition system have been widely developed. The machine vision system becomes an interest of many researchers in various fields of science. It provides the most important characteristic of natural interaction that is personalization. Automatic face recognition is a challenging problem, since human faces have a complex pattern. This paper presents a method for recognition of frontal human faces on gray scale images. The system is developed so that user can access the room just stand in front of the webcam. The webcam will send the image captured to the computer for recognition. In the proposed method, Discrete Cosine Transform (DCT) is used to extract the facial feature of an image the distance between the of the test image and train I and Euclidean Classifier is used to for the selection of best match between test image and trained image that has already been stored in database. When match occurred, computer will send signal to the microcontroller to open the lock through UART, else computer will send the unrecognized image to the owner's mobile. When owner wants to open the door for the visitor then, using his mobile owner will send signal to the microcontroller to unlock the door. This paper will develop an intelligent door based on smart phone that can be implemented in real-life applications.

Keywords: Discrete Consine Tranform, Euclidean Distance Classifier, Face Recognition, Feature Extraction.

1 Introduction

A face recognition system is a computer application that can be used for automatic identification and verification of human face image from a video source. One of the ways to do this is by comparing the features of selected test image with trained facial image database. The human face plays an important role in identification by means of biometric signature. The most difficulty in developing a face recognizer is that human face can change in timely manner or it can also be affected by wearing different face accessories like glasses, earring, piercings and makeup. A person can change his/her hair from long/short and can wear beard or mustache. It also depends on many facial expressions like smiling or sad. These facial changes make the recognition task very difficult even human beings do some recognition mistakes from time to time. In this paper we are going to

© Springer International Publishing Switzerland 2015
S.C. Satapathy et al. (eds.), *Proc. of the 3rd Int. Conf. on Front. of Intell. Comput. (FICTA) 2014*
– *Vol. 1,* Advances in Intelligent Systems and Computing 327, DOI: 10.1007/978-3-319-11933-5_77

implement the discrete cosine transform (DCT) algorithm for face recognition. The face recognition commonly includes feature extraction, face reduction and recognition or classification. Feature extraction is to find the most representative description of the faces, which is used to distinguish the faces. Face reduction is to not only decompose and compress the original features but also not destroy the most important information. Euclidean distance classifier is used to compares the relevant feature of test image with the feature of trained image which is stored in the database. When test image matches with the train image then computer will send signal to the microcontroller through UART to open the lock, else computer will send unrecognized to the owner's mobile.so that owner will know the unknown person's image. Now, if owner wants to open the lock, owner can signal to the microcontroller to open the lock or remain unlocked. Through the proposed method, now owner have total control of their home even when he/she is not present in the home. Outlay of the rest of paper is as follows; In section 2, literature survey is discussed and in section 3, proposed method is presented , section 4 presents the experimental results and conclusion is stated in section 5.

2   Review of Literature

Ayushi Gupta et al. [1] proposed system for face recognition using Principal Component Analysis (PCA) method where every image in the training set can be represented as a linear combination of weighted eigen vectors called as Eigen faces. These eigen vectors are obtained from covariance matrix of a training image set called as basis function. The weights are found out after selecting a set of most relevant Eigenfaces. Recognition is performed by projecting a new image (test image) onto the subspace spanned by the eigenfaces and then classification is done by distance measure methods such as Euclidean distance. Recognition rate saturates after a certain amount of increase in the Eigen value. Recognition rate decreased in noisy images.

Taewan Kim et.al [2] proposes a smart system using both face recognition and sound localization techniques to identify foreign faces through a door phone in a more efficient and accurate way. In this a visitor's voice is used to localize the proper location of the speaker. The location information is then used to adjust the door phone camera position. To locate the facial position of the visitor 4 microphones are positioned in a cross configuration. The system was implemented by integrating both sound localization and face detection/recognition schemes on a single FPGA to reduce power consumption and system size.Therefore, the smart door phone system is a robust way to provide safety to homes.

M. Meenakshi [3] presents the design, implementation and validation of a Digital Signal Processor (DSP)-based Prototype facial recognition and verification system. This system is organized to capture an image sequence, find facial features in the images, and recognize and verify a person. In this images captured using a Webcam, compares it to a stored database using methods of Principal Component Analysis (PCA) and Discrete Cosine Transform (DCT). The algorithms were validated in real-time using different test images. Finally automatic opening and closing of the door is considered and proved that both algorithms are capable of identifying someone reaching the door.

O.T. Arulogun et al. [4] proposed a security system which assures that value will not be taken. It employs two of the emerging artificial intelligence technologies: Facial Recognition and Artificial Neural Networks for developing a secure keyless door where authentications of authorized faces are the only guarantee for entry. The system has the capability to increase the number of users that have access to the secured environment and to remove users which depends upon the security system administrator.

Mayank Agarwal et al. [5] proposed a methodology for face recognition based on information theory approach of coding and decoding the face image. This method is based on two stages, Feature extraction using principle component analysis and recognition using the feed forward back propagation Neural Network. The proposed technique is analyzed by varying the number of Eigen faces used for feature extraction and the maximum recognition rate obtained by using whole dataset.

In this paper, we propose a face recognition algorithm based on discrete cosine transform and Euclidean distance algorithm. After the face recognition computer will send signal to the micro controller to open the door, if face is unrecognized it will send the unrecognized image to the owners mobile, if owner wants the visitor to access the door then owner can send signal to the micro controller to open it. The important contribution of the proposed method is the efficient face recognition with more privilege to the owner by which he will be having the authority to open the open the door even though if he is not in the home.

3   Proposed Method

In this paper, method is proposed which recognizes image of a person to access door. In order to access the door, the person should stand at the front door facing the webcam which is planted on the door. In the beginning, a person needs to register his face to the database so that the system has all the information of an authorized person. For the real time recognition, webcam will capture the person image and detect only the face region so that the system can have better face image. The door equipped with lock, will be unlocked after receiving the unlock signal from the microcontroller. Here we use image processing techniques for face detection and recognition microcontroller part to lock or unlock the door.

Fig .1 shows the overall flow diagram of the system i.e when a person comes in front of the door, the webcam will acquire the person image i.e test image and send it to the MATLAB for further processing. Now MATLAB face recognition algorithm starts calculating the features of a test image and compares the features with the trained image that is already stored in the database. If it is matching then it will send signal to the microcontroller to open the door. If the test image is not matching then it will send the test image to the owner mobile. If owner wishes to allow that person to enter then owner can send signal to the microcontroller to open the door.

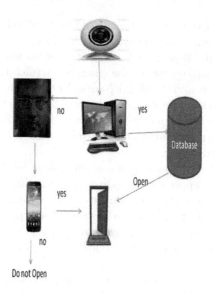

Fig. 1. Block Diagram of the Proposed System

In this paper, the algorithm used is discrete cosine transform (DCT) which is the simplest way in represent information content of face Image. DCT is the most widely used in the image processing applications for feature extraction. The DCT of an image basically consists of three frequency components namely low, middle, high each containing some detail and information in an image. The low frequency generally contains the average intensity of an image which is the most intended in Face Recognition systems. The main features of the DCT which make it attractive for facial recognition are: 1. Efficient computation, due to its relationship to the Fourier transforms, and 2. Data compression (energy compaction) property for highly correlated images, commonly used in image/video compression.

The definition of DCT for an N x N image is:

$$F(u,v) = \alpha(u)\alpha(v)\Sigma_{x=0}^{N-1}\Sigma_{y=0}^{N-1}f(x,y)cos\left[(2x+1)u\pi/2N\right]cos\left[(2y+1)v\pi/2N\right]$$
(1)

where

$$\alpha(u),\alpha(v) = \left\{\sqrt{\frac{1}{N}}...u,v=0; \sqrt{\frac{2}{N}}...u,v\neq0;\right\}$$
(2)

with u, v, x, y = 0, 2, 3,.. N-1, where x and y are spatial coordinates in the sample domain while u, v are coordinates in the transform domain. f(x, y) is the pixel intensity in row x and column y. F(u, v) is the DCT coefficient in row u and column v of the DCT matrix.

4   Feature Extraction

To obtain the feature vector which represents the face, DCT must be computed. The process of feature extraction is to extract the most relevant information to be used to distinguish a person, i.e. the information that is common for a person and is different for the other persons. Fig .2 shows the Face Recognition System using DCT.To perform the feature extraction, the approach is used to select the lowest DCT frequency coefficients. This approach is fast and simple, because it neither evaluates the DCT coefficient of the image, nor performs calculations or any kind of comparison.

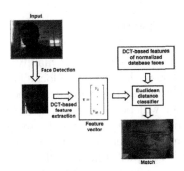

Fig. 2. Face Recognition Process using DCT

Fig.3 shows the properties of the DCT coefficients in N x N blocks with the zigzag pattern in order to process the DCT coefficients. Although the total energy remains the same in the N x N blocks, the energy distribution changes with most energy being compacted to the low-frequency coefficients. Thus, it is essentially the low frequency components, which play a key role in Face Recognition system. Within each DCT block of the image, the pixels differ from a low variation to a high variation in a zig-zag pattern, which is known as the raster scan order and it is basis for selecting the number of features in the selected blocks. The raster scan order which is the basis for selection of number of coefficients within the block size of 8*8 and that consist the arrangement of DCT coefficients as a matrix as shown in Fig.3.

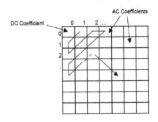

Fig. 3. Block Feature of DCT Coefficients and their Selection in Zig-Zag Pattern

5 Euclidean Classifier

Euclidean classifier is used to find the best match between the train and test images. It is calculated as follows:

$$D = \sqrt{\Sigma_{i=1}^{N}(p_i - q_i)^2} \tag{3}$$

pi and qi are the coordinates of p and q in the N dimensional space, corresponding to the train and test images. Minimum distance will tell the maximum correlation.

6 Hardware and Software

The hardware development of this project is mainly based on microcontroller PIC16F877 that controls the push button and lock. The webcam that is planted on the door will be connected to the computer. The communication between laptop and microcontroller will be established by UART. Microcontroller will control lock, buzzer and LED. Fig. 4 shows the overall representation of the hardware implementation. If the person wants to come into the room he needs to press the push button. The push button will send signal through UART to signal the MATLAB in the computer so that image acquisition process will start and the process of recognition can be done. After the recognition is done, MATLAB will send signal to the microcontroller. If the image is known person, the microcontroller will open the lock.

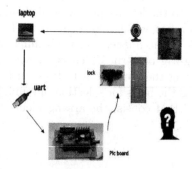

Fig. 4. Basic Idea of Door Access System based on Face Recognition

If the person image is unknown then computer will send an image of the unknown person to the owner mobile. So that owner of the house will know who is standing outside his home. Now if the owner allows the person to enter his house, owner has to press the pass-button from his mobile. After pressing the pass- button MATLAB will send signal to microcontroller to open the door.

7 Experimental Results

Fig.5 shows the hardware interfacing with the computer. The most important thing in face recognition system is to differentiate between two people. The test can be done by taking image of multiple people so that accuracy of an algorithm can be calculated. The train image of different people is saved to folder called train database. The database folder will have several image of each person. Fig. 6 shows the result when an image is recognized by the system. When an Image is not recognized then Fig 7 shows the result of unrecognized image. For the real implementation, it needs to perform well. The challenge for the face recognition is that the variation in light, environment background and face expression. The background and the illumination must be same for both test and trained image.

Fig. 5. Hardware Interface with Computer

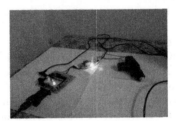

Fig. 6. Result of Recognized Image

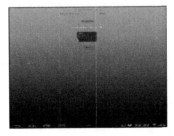

Fig. 7. Mobile Interface for Unrecognized Image

8 Conclusion

In this paper, we propose a novel approach for face recognition, based on the discrete cosine transform (DCT) and Euclidean distance classifier. The experimental result confirms its usefulness and robustness. DCT has played a key role in efficient feature extraction and also used to reduce the number of features being obtained. For the security applications DCT algorithm is reliable and suitable. As such, continuous works have been carried out in order to achieve satisfactory results for security system. In our future work we will focus on the elimination/reduction of shadowing effect as well as working with more efficient hardware.

References

1. Gupta, A., Sharma, E., Sachan, N., Tiwari, N.: Door Lock System Face Recognition Using MATLAB. International Journal of Scientific Research in Computer Science and Engineering 1(3) (June 2013)
2. Kim, T., Park, H., Hong, S.H., Chung, Y.: Integrated System of Face Recognition and Sound Localization for a Smart Door Phone. IEEE Transactions on consumer Electronics 59(3) (August 2013)
3. Meenakshi, M.: Real-Time Facial Recognition System-Design, Implementation and Validation. Journal of Signal Processing Theory and Applications 1, 1–18 (2013)
4. Arulogun, T., Omidiora, E.O., Olaniyi, O.M., Ipadeola, A.A.: Development of Security system Using Facial Recognition. The Pacific Journal of Science and Technology 9 (November 2008)
5. Manikantan, K., Govindarajan, V., SasiKiran, V.V.S., Ramachandra, S.: Face Recognition Using Block-Based DCT Feature Extraction. Journal of Advanced Computer Science and Technology 1(4) (2012)
6. Kaur, S., Virk, R.S.: DCT Based Fast Face Recognition Using PCA and ANN. International Journal of Advanced Research in Computer and Communication Engineering 2(5) (May 2013)
7. Sharif, M., Mohsin, S., Javed, M.Y., Ali, M.A.: Single Image Face Recognition Using Laplacian Of Gaussian and Discrete Cosine Transform. The Internation Arab Journal of Information Technology 9(6) (November 2012)

Optimized Heap Sort Technique (OHS) to Enhance the Performance of the Heap Sort by Using Two-Swap Method

S. Sumathi[1], A.M. Prasad[1], and V. Suma[2]

[1] Dayananda Sagar College of Engineering,
Karnataka, India
{sumathi5990,prasaddsce}@gmail.com
[2] Reaserch and Industry Incubation Center (RIIC)
Dayanada Sagar Institution,
Karnataka, India
sumavdsce@gmail.com

Abstract. In the field of computer science and mathematics, Sorting is arranging elements in a certain order, there are many sorting techniques are existing such as Bubble sort, Heap sort, Merge sort, Insertion sort etc. In this paper we are concentrating on heap sort, heap sort works by first organizing the data to be sorted into a special type of binary tree called a heap. The heap itself has, by definition the largest value or the smallest value at the top of the tree based on max heap or min heap. Heap sort method sorts a single element at a time. In the existing heap sort method for every construction of heap only a single root element will be sorted, so it slowdowns the performance of heap sort technique. In order to optimize the performance of heap sort we are proposing the optimized heap sort technique (OHS) by using Two-swap method. The Two-swap method sorts two-elements at a time for each heap construction, and it reduces 30-50% of the time complexity of heap sort in the construction of heap as well as in sorting of array elements.

Keywords: Heap, Complete Binary tree, Two-Swap method.

1   Introduction

Sorting is a process of rearranging a list of numbers in an array into increasing or decreasing order. There are various techniques are available to perform sorting i.e., Bubble sort, Selection sort, Insertion sort, Merge sort, Quick sort, Shell sort, Heap sort etc [1][2][3][4][5].

Heap sort is an algorithm that sorts an array of numbers in order, that being formatted first as a heap. A heap stores data in a complete binary tree such that all the children of a parent are either greater than or less than itself based on Max-heap or Min-heap respectively. Binary tree is a tree data structure where each node has at most two children, referred to as left child and the right child namely [6][7][8][9].

© Springer International Publishing Switzerland 2015
S.C. Satapathy et al. (eds.), *Proc. of the 3rd Int. Conf. on Front. of Intell. Comput. (FICTA) 2014*
– *Vol. 1*, Advances in Intelligent Systems and Computing 327, DOI: 10.1007/978-3-319-11933-5_78

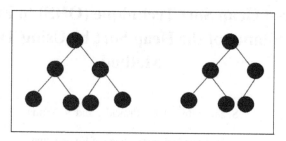

Fig. 1. Complete Binary Tree

Fig 1 shows complete binary tree. A binary tree T with N levels is complete if all levels except possibly the last are completely full, and the last level as its all nodes to the left side.

Working of Heap Sort: In an array 'A' of N elements, heap sort performs following function

1) Construction of heap
2) Sorting of an array element for each heap construction

1) Construction of heap: The binary heap data structure is an array that can be viewed as a complete binary tree. Each node of the binary tree corresponds to an element of the array. The array is completely filled on all levels except possibly lowest. Consider an un-sorted array 'A' which contains of elements [34, 62, 72, 84, 92, 10, and 7] it can be represented in complete binary tree form as follows:

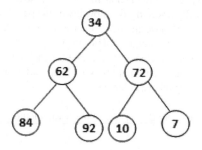

Fig. 2. Un-sorted Complete Binary Tree

Heap Properties

1) It must be a complete binary tree.
2) Child nodes of each Parent node must be either "greater than or equal" or "lesser than or equal" to its parent node i.e. $A[parent(i)] \geq A[i]$ or $A[parent(i)] \leq A[i]$.

There are two-kinds of heap

1) Max-heap
2) Min-heap

Max-heap: Max-heap is a complete binary tree in which the value in each parent node is greater than or equal to the value in the children of that node $A[parent(i)] \geq A[i]$.Construction of Max-heap for an array 'A'. Consider which contains the following elements. A= [34, 62, 72, 84, 92, 10, and 7]

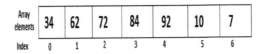

Fig. 3. Array Representation of Un-sorted Array A[ ] Elements

Algorithm 1: Max-heap construction
1. **Input:** array 'A' of 'n' elements i.e. $A[n]$
2. **Output:** Max-heap of $A[n]$
3. **for** $i = 1\ to\ n$ **do**
4. $c = i$
5.   **do**
6. $root = (c - 1)/2$
7.   **if** $A[root] < A[c]$ **then**
8.   **swap** ($A[root]$, $A[c]$)
9.   **end if**
10. $c = root$
11.   **while** (c!=0)
12. **end for**
13. **return** Max-heap $A[n]$

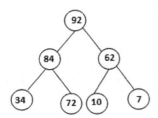

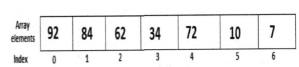

Fig. 4. Max-Heap of an Array   **Fig. 5.** Array Representation of Max-heap of an Array A[ ]
A[ ]   Elements

Min-heap: Min-heap is a complete binary tree in which the value in each parent node is lesser than or equal to the value in the children of that node $A[parent(i)] \leq A[i]$.Construction of Min-heap for an array 'A'. Consider which contains the following elements. A= [34, 62, 72, 84, 92, 10, and 7].

Algorithm 2: Min-heap construction
1. **Input:** array 'A' of 'n' elements i.e. $A[n]$
2. **Output:** Min-heap of $A[n]$
3. **for** $i = 1\ to\ n$ **do**
4. $c = i$
5.   **do**
6. $root = (c - 1)/2$
7.   **if** $A[root] > A[c]$ **then**
8.   **swap** ($A[root]$, $A[c]$)
9.   **end if**
10. $c = root$
11.   **while** (c!=0)
12. **end for**
13. **return** Min-heap $A[n]$

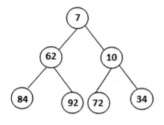

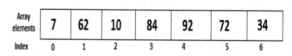

Fig. 6. Max-heap of an Array   **Fig. 7.** Array Representation of Min-heap of an Array A[ ]
A[ ]   Elements

2. Sorting of an Array Element: After construction of heap, sorting of an array elements takes place. Sorting will be performed based on whether an array to be sorted in an ascending order or descending orders. Max-heap will be used as an input when the array elements to be sorted in an ascending order. Min-heap will be used in order to sort elements in a descending order [10] [11].

Algorithm 3: Heap-sort
1. **Input:** Max-heap of $A[n]$
2. **Output:** Sorted array $A[n]$
3. **for** $j = n - 1\ to\ 0$ **do**
4.   **Swap** ($A[0]$, $A[j]$)
5. $root = 0$
6.   **do**
7. $c = 2 * root + 1$
8.   **if** $((A[c] < A[c + 1])\ \&\&\ c < j - 1$ **then**
9.   increment c
10.   **end if**
11.   **if** $((A[c] < A[c + 1])\ \&\&\ c < j)$ **then**

12.   **Swap** ($A[root]$, $A[c]$)
13.   **end if**
14. $root = c$
15.   **while** $(c < j)$
16. **end for**
17. **return** Sorted array $A[n]$

Algorithm 3 shows heap-sort performing on an array A[ ] of 'n' elements where it takes input as either Max-heap or Min-heap of an array $A[n]$ and sorts the elements using Algorithm 3 and returns output as sorted array $A[n]$.

2   Proposed Scheme

In this paper we are proposing an Optimized Heap Sort technique (OHS) in order to improve the performance of the heap sort.OHS technique improves the performance by sorting two-elements at a time after the construction of heap. The process of sorting two elements at a time is called Two-swap method because swapping will be performed twice on heap array.

Working of OHS Technique

1) Construction of heap
2) Sorting of two-elements at a time in an array for each construction of heap.

1. **Construction of heap:** Heap will be constructed as similar as in the heap sort. The algorithm 1 will be used to construct Max-heap of an array. And the algorithm 2 will be used to construct Min-heap of an array.

2. **Sorting of array elements using Two-swap method**

   Two-Swap method: when we construct an heap of any array, consider if it's a Max-heap then root will contain largest number and the second large number will be placed in any one of the following array index that may be either *index* 2 *or index* 3 .In two-swap method we find out the second large number along with the first in each heap construction and we arrange both the large numbers in an appropriate place while sorting. In this way we are reducing time complexity of heap sort by 2.

Algorithm 4: OHS method to sort array in an ascending order
1. **Input:** Max-heap of $A[ ]$ and number of elements 'n'.
2. **Output:** Sorted array $A[n]$
3. $j = n - 1$
4. **if** $(n \geq 5)$ **then**
5.   **Swap** ($A[0]$, $A[j]$)
6.   **if** $(A[1] > A[2])$ **then**
7.   **Swap** ($A[1]$, $A[j - 1]$)
8.   **else**
9.   **Swap** ($A[2]$, $A[j - 1]$)

10.   **end if**
11.   **Call** Max-heap ($A[\,]$, $n - 2$)
12. **else**
13.   **Swap** ($A[0]$, $A[j]$)
14.   **Call** Max-heap ($A[\,]$, $n - 1$)
15. **end if**
16. **return** sorted array $A[n]$

Algorithm 4 shows an OHS method to sort array elements in an ascending order. In the above algorithm we are sending constructed Max-heap of an array A[ ] and number of elements 'n' as inputs. Two-elements are sorted at a time for each construction of heap. Provides sorted array of a[n] in ascending order.

Algorithm 5: OHS method to sort array in a descending order
1.   **Input:** Min-heap of $A[\,]$ and number of elements 'n'.
2.   **Output:** Sorted array $A[n]$
3. $j = n - 1$
4.   **if** $(n \geq 5)$ **then**
5.   **Swap** ($A[0]$, $A[j]$)
6.   **if** $(A[1] < A[2])$ **then**
7.   **Swap** ($A[1]$, $A[j - 1]$)
8.   **else**
9.   **Swap** ($A[2]$, $A[j - 1]$)
10.   **end if**
11.   **Call** Min-heap ($A[\,]$, $n - 2$)
12. **else**
13.   **Swap** ($A[0]$, $A[j]$)
14.   **Call** Min-heap ($A[\,]$, $n - 1$)
15. **end if**
16. **return** sorted array $A[n]$

Algorithm 5 shows an OHS method to sort array elements in a descending order. It takes input as a Min-heap of A[ ] and number of elements and provides sorted array of A[n] in descending order.

3   Performance Evolution

Heap Sort

1)   Construction of heap: Construction of heap whether it may be Max-heap or Min-heap of 'N' elements takes O (n) time where O(n) is the time complexity of heap sort.

2)   Sorting an array element: After the construction of heap, the root element will be deleted and it will be inserted into the nth position in an array. This will be performed using swapping function, Then again for n-1 the heap will be constructed again. It takes O (log n) time.

The time complexity of heap sort [12].

$$O(n) * O(\log n) = O(n \log n)$$

Optimized Heap Sort (OHS) Scheme

1) **Construction of heap:**   In OHS scheme first time the heap will be constructed for 'n' elements. Each time two Max-elements will be deleted from the heap, again the heap will be reconstructed. So the time complexity to construct heap will be O(n/2).

2) **Sorting of array elements:** In OHS scheme after construction of heap, consider if it's a Max-heap then root element will be deleted and placed in an 'nth' position of an array. Then it compares left child (root) and right child (root) the max-value will be deleted and placed in an (n-1)th   position.

The time complexity of OHS scheme:

$$O\left(\frac{n}{2}\right) * O\left(\frac{\log n}{2}\right) = O(\frac{n \log n}{2})$$

Table 1 shows the practical result of Heap Sort and OHS scheme for Random input data which is varied from 100-2000. To obtain the Table1 readings some of hardware and software requirements have been considered. Those are listed in the Table 2.

Table 1. The Sorting Time of Heap Sort and OHS Scheme for Random Input Data

Sorting Algorithms	Number of input data items				
	100	200	300	400	500
Heap sort	0.0000	0.0000	0.0000	0.0000	0.0000
OHS scheme	0.0000	0.0000	0.0000	0.0000	0.0000
	600	700	800	900	1000
Heap sort	0.54945	0.54945	0.109890	0.109890	0.164835
OHS scheme	0.0000	0.54945	0.54945	0.109890	0.109890
	1100	1200	1300	1400	1500
Heap sort	0.164835	0.219780	0.219780	0.274725	0.274725
OHS scheme	0.164835	0.164835	0.164835	0.219780	0.274725
	1600	1700	1800	1900	2000
Heap sort	0.329670	0.389670	-	-	-
OHS scheme	0.274725	0.329670	0.329670	0.384615	0.384615

Table 2. Hardware and Software Requirements

Hardware and Software Requirements	
Operating System	Windows XP
Version	2002
Processor	Intel ® core ™2 Duo CPU E7500 @ 2.93GHz
RAM	1.96GB
Application Software	Turbo C

4   Conclusion

As we know Sorting is a process of arranging elements in a certain order. Where heap sort is one of the sorting technique which takes O(n log n) time to sort elements in an ascending or descending order. In this paper we are proposing new sorting scheme OHS in order to improve the performance of heap sort. The OHS is a modified heap sort, where two elements are sorted at a time after the construction of heap and also the construction of heap will also reduced by half the percent. The time complexity of OHS scheme will be O((n log n)/2). We produced practical results of heap sort and OHS scheme in terms of number of inputs versus time complexity.

References

[1] Cormen, T.H., Leiserson, C.E., Rivest, R.L., Stein, C.: Introduction to Algorithms, 2nd edn., ch. 7. MIT Press, Cambridge (2001)
[2] Knuth, D.E.: The Art of Computer Programming. Sorting and Searching, vol. 3. Addison Wesley, Reading (1998)
[3] Floyd, R.W.: Algorithm 245: Treesort 3. Communications of ACM 7(4), 701 (1964)
[4] Cormen, et al.: Introduction to Algorithms, Chap. 6
[5] Hoare, C.A.R.: Quicksort. Computer Journal 5(1), 10–15
[6] Williams, J.W.J.: Algorithm 232: HEAPSORT. Communications of ACM 7(4), 347–348 (1964)
[7] Wegner, I.: The Worst Case Complexity of McDiarmid and Reed's Variant of BOTTOM-UP HEAP SORT. Information and Computation 97(1), 86–96 (1992)
[8] Carlsson, S.: A variant of HEAPSORT with almost optimal number of comparisons. Information Processing Letters 24, 247–250 (1987)
[9] Gonnet, G.H., Munro, J.I.: Heaps on Heaps. SIAM Journal on Computing 15(6), 964–971 (1986)
[10] McDiarmid, C.J.H., Reed, B.A.: Building Heaps Fast. Journal of Algorithms 10(3), 352–365 (1989)
[11] Dutton, R.D.: Weak Heap Sort. BIT 33(3), 372–381 (1993)
[12] Carlsson, S., Chen, J.: The Complexity of Heaps. In: The Third Annual ACM SIAM Symposium on Discrete Algorithms, pp. 393–402. SIAM, Philadelphia (1992)

An Approach to Utilize FMEA for Autonomous Vehicles to Forecast Decision Outcome

Samitha Khaiyum, Bishwajit Pal, and Y.S. Kumaraswamy

MCA (VTU) Department,
Dayananda Sagar College of Engineering,
Bangalore-78, India
{samitha.athif,bishwajit.pal}@gmail.com,
yskldswamy2@yahoo.com

Abstract. Every autonomous vehicle has an analytic framework which monitors the decision making of the vehicle to keep it safe. By tweaking the FMEA (Failure Mode Effect Analysis) framework and applying this to the decision system will make significant increase in the quality of the decisions, especially in series of decision and its overall outcome. This will avoid collisions and better quality of decision.the proposed methodology uses this approach to identify the risks associated with the best alternative selected. The FMEA requires to be running at real time. It has to keep its previous experiences in hand to do quick/split time decision making. This paper considers a case study of FMEA framework applied to autonomous driving vehicles to support decision making. It shows a significant increase in the performance in the execution of FMEA over GPU. It also brings out a comparison of CUDA to TPL and sequential execution.

Keywords: Autonomous vehicle, FMEA, CUDA, TPL, Sequential execution.

1   Introduction

Majority of technological advances in the automotive industry were evolutionary. Now, the automotive industry finds a significant change that will now have a huge impact on the way we perceive mobility through the autonomous vehicle [11]. The concept of autonomous driving has given rise to many opportunities but at the same time it has raised lot of concerns. However, study has proven many benefits of these vehicles in terms of improved fuel economy, enhanced safety and reduced traffic congestion. Studies have proven to yield a number of clear safety enhancements and reduction of casualties along with cost [7]. This paper discusses the various aspects to modify the FMEA framework to support the decision making system in autonomous vehicle. It also discusses the various ways to execute this framework at higher speed equivalent to the graphic processor and supporting and evaluating the decisions taken by an autonomous vehicle. There is also a comparison drawn of the result with the traditional sequential execution and Task parallel Libraries (TPL).

© Springer International Publishing Switzerland 2015   701
S.C. Satapathy et al. (eds.), *Proc. of the 3rd Int. Conf. on Front. of Intell. Comput. (FICTA) 2014*
– *Vol. 1*, Advances in Intelligent Systems and Computing 327, DOI: 10.1007/978-3-319-11933-5_79

2   Literature Survey

The GPU pipelining holds promises for executing complex or lengthy arithmetical model, without degrading camera tracking and reconstruction. With reference [2] to the "Real time depth camera 3D reconstruction" the author has successfully demonstrated the real time pipelining on GPU devices to attain a real time 3D reconstruction of the environment. We propose to use a similar technique to generate a Hybrid FMEA structure in real time to aid the Decision making system of an autonomous vehicle. A very similar work [1] on Real-Time GPU programming is carried out demonstrating the real time capability of GPU to render dense reconstruction. Authors in [9] state that autonomous driving vehicles work on large scale data collection of driver, vehicle and environment information in the real world. Safety critical events are currently identified using kinematic triggers, for instance searching for deceleration below a certain threshold signifying harsh braking. Due to the low sensitivity and specificity of this filtering procedure, manual review of video data is currently necessary to decide whether the events identified by the triggers are actually safety critical. There are three inter-related spaces involved in fault diagnosis: fault space, observation space and diagnosis space. Each space contains the totality of all possibilities in their respective areas. Fault space contains all faults that can occur on the vehicle, observation space contains all observations from the point of view of the vehicle and diagnosis space contains all possible diagnoses available by using the observation space.[6] Authors in [8] through their research show that usage of autonomous driven vehicles significantly increases the highway capacity by making use sensors and having a vehicle to vehicle communication.

3   Failure Types

The challenge in RT system is to constantly interact with the real world entities. This must be continuously improved regarding safety engineering activities and the built prototypes should fit into the real world traffic conditions. Failures in RT system can be categorized as represented below. These failures mainly concentrate in the requirement elicitation, design, development, testing phases along with the maintenance phase. Project failures can be categorized as the technical and process failures, which deal with technical issues such as planning, defining requirements, design, code and tests, eliminating redundancy, rework, schedule, etc. Secondly, the business and organizational failures include approvals from management, cost overruns, contract issues, and change in business strategies and so on. Lastly, the resource failures are failures occurring from third party vendors in terms of tools, technological support, training, manpower and so on. However, the Product Failures leads to failures of the system or project which can affect the users and they directly impact the quality of the product. They can be categorized as quality failures affect the functionality, reliability, usability, performance, etc. and Harmful failures which cause physical harm to life, logical harm to data or money, etc. [10].

4   FMEA in Autonomous Vehicles

The external environment whenever changes, the algorithm will be constantly be facing new threats. Hence a FMEA is required which will be pooling all the results and its detection algorithm for its effectiveness in various situations. In an autonomous platform the usage of FMEA is very likely as the environment is always changing and the algorithm\s used for driving the vehicle is more challenged. An FMEA can give a potential prediction model to help the autonomous vehicle make quality decision and rule out risky decisions. For example it can keep many paths planning algorithm, and execute all at the same time for a given condition to pick the best or the robust one, or use multiple fuzzy matching algorithm and take the average, to produce an optimal option. Multiple algorithms are used in parallel for the same situation. This will yield different ways of achieving the same thing. The FMEA is required to take all the alternatives and run its detection and severity calculation and present the result for a good choice. Table 1 shows the different index levels for factors like severity, occurrence, detection and the scales that determine the relative risk and then provide a mechanism to prioritize work. The impact or relative risk is measured using three different scales namely severity to measure consequence of failure if it occurs, occurrence which gives the frequency of failure and detection which is the ability to measure the potential failure before the consequences are observed.

Table 1. Index levels of Severity, Occurrence and Detection

Ratings	Severity level	Occurrence level	Detection level
7-10	Blocker	Probable	Delayed
4-6	Medium	Remote	Slow
1-3	Cosmetic	Improbable	Early

Table 2. Matrix upon merging probability and severity

	Improbable	Remote	Probable
Cosmetic	Pick Any	Pick Average	Pick with all Consideration
Medium	Pick Average	Pick with all Consideration	Pick with all Consideration
Blocker	Pick with all Consideration	Pick with all Consideration	Pick with all Consideration

Table 2 shows a matrix obtained upon merging factors of severity and occurrence. This helps the algorithms used by autonomous vehicles to pick from based on the result generated.

In a normal driving condition, a "simple" impact, with a "Not likely" occurrence will result in selecting result generated by any ("Pick any") algorithm. It means the vehicle can rely on any algorithm for a "Simple impact" and "not likely" kind of decision. Similarly for a simple impact and certain failure, it has to consider all the algorithms (executing in parallel) and select the best fit result.[4] When the vehicle is in motion, consideration must be given to the relationship of the target of evaluation (TOE) to form a checklist where each interaction between the vehicle and environment has to be considered in turn as to whether a property of, or a failure in, the TOE may produce a hazard at the boundary of the moving vehicle [3]. For the final result emitted by FMEA we propose to product a Δt (where Δt is the critical time failure barrier of the system) variable with the final result of the FMEA. This result will then be mapped to an existing system decision criticality chart. The Δt is converted to a more meaningful variable and linked to the FMEA table, as time till the analysis is valid. SEV-Severity is given a qualitative scale (1-10), 1 having the Least effect and 10 being the Most Hazardous, OCC-Occurrence is given a qualitative scale (1-10), 1 being the Almost Never and 10 being the Almost Certain and DET-Detection is given a qualitative scale (1-10), 1 being the Almost Certain and 10 being the Almost Impossible.

Table 3. FMEA table for autonomous driven vehicles

Key Process Step or Input	What is the Process Step or Input? Environment Limits/ Co-ordinate.	Cam generated movable area. Generated 2d path. {12,33}, {33,45}
	What is the Process Step or Input? Path Planning algorithm generated Point Sequence.	Path Planning generated 2d path.{12,33}, {33,45}
Potential Failure Mode	In what ways can the Process Step or Input fail?	Lines intersect in the given range. This will be found by the algorithm for line intersection.
Potential Failure Effects	What is the impact on the Key Output Variables once it fails (customer or internal requirements)?	Decision on: 1. Direct Collision. 2. Close Brushing/ Partial. 3. Minimal Get through. 4. Optimistic clearance 5. Widely Cleared.
SEV	How Severe is the effect to the customer?	0 to 10 scale.
Potential Causes	What causes the Key Input to go wrong?	1. Camera fault. 2. Algorithm Fault. 3. Algorithm Accuracy.
OCC	How often does cause or FM occur?	0 to 10 scale.
Current Controls	What are the existing controls and procedures that prevent either the Cause or the Failure Mode?	1. Algorithms to recalculate Path Planned. 2. Fall back to next proposed Path.
DET	How well can you detect the Cause or the Failure Mode?	0 to 10 scale.
Actions Recommended	What are the actions for reducing the occurrence of the cause, or improving detection?	Algorithms to cross check its validity.
Responsibility	Who is Responsible for the recommended action?	Controller Name
Actions Taken	Note the actions taken. Include dates of completion.	Include time within which the action to be taken

The vehicle is in constant movement, so are the obstacle points which are constantly updated in the System planned path list. The severity will be detected be mathematical rules and some will be derived based on the experience made by the rover. The detection is limited to the rover's sensor capability due to its sensing range and refresh rate.

5   System Architecture. (Autonomous Vehicle Brief)

The AV (Autonomous Vehicle) uses a classic model of layered architecture where some of the layers and components are distributed over different machines. The AV has a high speed (1-4 microsecond response time) UDP Multicast network within the master and the subsystem. The actuator and motor driver system manages the operations of the drive system (Figure 1). It monitors the overheating, speed, deceleration, acceleration and many other tasks for the drive system. The task generator module works closely with the main program to generate the sequence of task for other components to follow up. The process FMEA monitors and analyses its decision making capabilities and keeps its records updated. The Process FMEA will forecast the criticality of the decisions to the main program. The main program can then update or change its perception to the task scheduler.

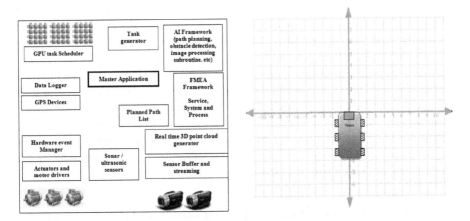

Fig. 1. System architecture of autonomous vehicle   **Fig. 2.** Coordinate system for AV

Figure 2 shows the Coordinate System, which represents the x,y co-ordinate system which the AV uses. Its origin is from its camera (0, 0). All values falling on the left will have its x coordinate value which is negative and values falling behind the vehicle will have the Y coordinate values in the negative range.

5.1   FMEA Application to AV

The 3D vision system: The front view of the path/road seen by the 3D cameras in the AV is as below in Figure 3.

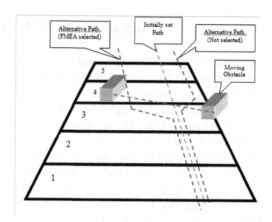

Fig. 3. Front view of the path/ road as seen by the camera in the AV

This region is divided into many horizontal and vertical sections. The sections near the camera/vehicle are more critical than the ones farther. In the below example the AV is moving on a planned path marked as "Initially set path". Normally the AV will continue to move in that direction until it encounters an obstacle. Assuming there are multiple obstacle avoidance algorithm running, if there is a moving obstacle in front of the AV, then there may be multiple alternative path to avoid the obstacle. In a system without the FMEA a standard calculation will be made taking the speed and external conditions to calculate the new path. But it may not consider a concrete severity index to choose the final path. On the other hand the AV considering the FMEA model will pick a path with lesser severity index.

6   Case Study (RT-FMEA Implementation on GPU Processor)

The AV will be executing tasks at the speed, nearly equivalent to human response (200 microseconds). We have chosen the NVIDIA provided framework named CUDA (Compute Unified Device Architecture) to implement this solution. We have also done comparative study of the same implementation on single or multiple CPU machines. The implementation was first written on a .net program without any parallel processing support. Followed we converted the code to run on parallel CPU. This was done using the .NET 4.5 TPL libraries. In this model we did not program the parallel computing load as the .net Runtime is designed to take care of the loading and execution. The final model was to implement this on GPU hardware.

Table 4. Algorithms under .net, .net and TPL and CUDA implementation

Case study with .NET implementation	Case Study with .NET and TPL (Task Parallel Library).	Case study with CUDA library
Start.	Start.	Start.
Initialize the proposed list of Path with Starting and Ending point. Starting point: (X1, Y1) and ending Point (X2, Y2).	Initialize the proposed list of Path with Starting and Ending point. Starting point: (X1, Y1) and ending Point (X2, Y2).	Initialize the proposed list of Path with Starting and Ending point. Starting point: (X1, Y1) and ending Point (X2, Y2).
Initialize the list of environment lines with Starting and Ending point. Starting point: (eX1, eY1) and ending Point (eX2, eY2).	Initialize in Parallel (Using TPL) the list of environment lines with Starting and Ending point. Starting point: (eX1, eY1) and ending Point (eX2, eY2).	Initialize the list of environment lines with Starting and Ending point. Starting point: (eX1, eY1) and ending Point (eX2, eY2).
Calculate Slope for all the point in the Proposed List of path and list of Environment lines using the formula $Slope = (Y2 - Y1) / (X2 - X1)$.	Calculate in Parallel (Using TPL) Slope for all the point in the Proposed List of path and list of Environment lines using the formula $Slope = (Y2 - Y1) / (X2 - X1)$.	Load the List of Paths and List of Environment Lines to the GPU device Memory.
Calculate B1 for all the point in the Proposed List of path and list of Environment lines using the formula $B1 = Y1 - (Slope * X1)$.	Calculate in Parallel (Using TPL) B1 for all the point in the Proposed List of path and list of Environment lines using the formula $B1 = Y1 - (Slope * X1)$.	Calculate in Parallel (Using CUDA) Slope for all the point in the Proposed List of path and list of Environment lines using the formula $Slope = (Y2 - Y1) / (X2 - X1)$.
Calculate B2 for all the point in the Proposed List of path and list of Environment lines using the formula $B2 = Y2 - (Slope * X2)$.	Calculate in Parallel (Using TPL) B2 for all the point in the Proposed List of path and list of Environment lines using the formula $B2 = Y2 - (Slope * X2)$.	Calculate in Parallel (Using CUDA), B1 for all the point in the Proposed List of path and list of Environment lines using the formula $B1 = Y1 - (Slope * X1)$.
Compare B1 and B2. If B1 is not equal to B2, go to step 8.	Compare in Parallel (Using TPL), B1 and B2. If B1 is not equal to B2, go to step 8.	Calculate in Parallel (Using CUDA), B2 for all the point in the Proposed List of path and list of Environment lines using the formula $B2 = Y2 - (Slope * X2)$.
Iterate through all the List of Path and compare each one of these to the List of Environment Lines, using the formula $Xintsct = ((B2 - eB2) / (M1 - eM2)$.	Iterate in Parallel (Using TPL) through all the List of Path and compare each one of these to the List of Environment Lines, using the formula $Xintsct = ((B2 - eB2) / (M1 - eM2)$.	Compare in Parallel (Using CUDA), B1 and B2. If B1 is not equal to B2, go to step 8.
Calculate $Yintsct = (Slope * Xintsct) + B$.	Calculate $Yintsct = (Slope * Xintsct) + B$.	Iterate in Parallel (Using CUDA) through all the List of Path and compare each one of these to the List of Environment Lines, using the formula $Xintsct = ((B2 - eB2) / (M1 - eM2)$.
If the intersection line falls within the two lines, mark that line as colliding to an obstacle.	If the intersection line falls within the two lines, mark that line as colliding to an obstacle.	Calculate in Parallel (Using CUDA) $Yintsct = (Slope * Xintsct) + B$
Error: Log Error to the system Log and move to the next iteration. Go back to the same step from where the control came to this step.	Error: Log Error to the system Log and move to the next iteration. Go back to the same step from where the control came to this step.	If the intersection line falls within the two lines, mark that line as colliding to an obstacle.
Stop.	Synchronize all thread (i.e. wait for all parallel thread to finish).	Error: Log Error to the system Log and move to the next iteration. Go back to the same step from where the control came to this step.
	Stop.	Synchronize all thread (i.e. wait for all CUDA blocks to finish).
		Copy results from Device memory to CPU memory. And Clear the GPU Device Memory.
		Stop.

Table 4. (*continued*)

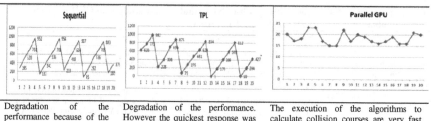

| Degradation of the performance because of the garbage collection. Conclusively sequential execution doesn't have a reliable time frame for completing the execution | Degradation of the performance. However the quickest response was better than the sequential run. The same garbage collection issue was observed while executing 100,000 random paths. | The execution of the algorithms to calculate collision courses are very fast and are very much under a microsecond. |

6.1   Comparison of CUDA to TPL and Sequential Execution

The execution on CUDA is very fast compared to TPL and sequential run. The above tests were executed on a NVIDIA GT 9800 Force 4 graphic card clocked at 1.5 GHz and 512 DDR 3 RAM, containing 112 parallel CUDA Blocks. In spite of having clock speed lesser than the CPU (nearly lesser than 1.1 GHz), the GPU wins by its raw 112 processors working in parallel to calculate the line intersections. The time taken to execute is also very reliable as compared to TPL and sequential run. The execution time is also very consistent. On the other hand the CPU has shown a large variation in execution time because of garbage collection. The memory allocation / deallocation in .net are managed by the .NET runtime whereas in CUDA there are dedicated statements to free memory and reuse the same for the next iterations. This makes CUDA more fast and efficient. From the above results it's clear that the GPU is almost 50 times faster than the CPU when it comes to parallel computation (especially arithmetic intensive steps). It's a very ideal framework to achieve real time analysis and Generating the FMEA table for the AV on a continuous pipeline.

6.2   Conclusion

This paper brings out the application of FMEA for autonomous vehicles to forecast decision outcomes. The results show significant increase in the performance in the execution of FMEA over GPU. This model is most likely to be used on the AV platform. Because this model can also be distributed over another machine the FMEA analysis is executed without interrupting the main program execution. With the response time of less than a microsecond the entire analysis table will be available to the main program to pool the right decision in real time. With this speed ($< 1ms$) it is also achieving the real time analysis of path generated by different algorithms. The main program can now rely on multiple path generating algorithms to achieve optimized decisions on which path to follow and also keep the AV from harms away.

References

1. Izadi, S., Kim, D., Hilliges, O., Molyneaux, D., Newcombe, R., Kohli, P., Shotton, J., Hodges, S., Freeman, D., Davison, A., Fitzgibbon, A.: KinectFusion: Real-Time Dense Surface Mapping and Tracking. UIST (2011)
2. Newcombe, R.A., Davison, A.J., Izadi, S., Kohli, P., Hilliges, O., Shotton, J., Molyneaux, D., Hodges, S., Kim, D., Fitzgibbon, A.: KinectFusion: Real-time 3D Reconstruction and Interaction Using a Moving Depth Camera, at Microsoft Research, Published by ACM Symposium on User Interface Software and Technology (2011)
3. Jesty, P.H., Hobley, K.M., Evans, R., Ian: Safety Analysis of Vehicle - Based Systems (2010)
4. Walker, M., Papadopoulos, Y., Parker, D., Lonn, H., Torngren, M., Chen, D., Johannson, R., Sandberg, A.: Semi-Automatic FMEA Supporting Complex Systems with Combinations and Sequences of Failures by at SAE World Congress & Exhibition (2012)
5. Recommended Failure Modes and Effects Analysis (FMEA) Practices for Non-Automobile Applications (2011)
6. Hamilton, K., Lane, D., Taylor, N., Brown, K.: Fault Diagnosis on Autonomous Robotic Vehicles with recovery: An Integrated Heterogeneous-Knowledge Approach. In: IEEE International Conference on Robotics & Automation (2011)
7. Avery, M., et al.: Autonomous Braking Systems and Their Potential Effect on Whiplash Injury Reduction. In: ESV 2009, Paper Number 09-0328 (2009)
8. Tientrakool, P.: Highway Capacity Benefits from Using Vehicle-to-Vehicle Communication and Sensors for Collision Avoidance. In: Vehicular Technology Conference VTC Fall, September 5-8. IEEE (2011)
9. Malta, L., LjungAust, M., Faber, F., Metz, B., Saint Pierre, G., Benmimoun, M., Schäfer, R.: Final results: Impacts on traffic safety, EuroFOT (2012)
10. Khaiyum, S., Kumaraswamy, Y.S., Karibasappa, K.: Classification of failures in real time embedded software projects. In: International Conference on Systemics, Cybernetics and Informatics Proceedings (2014) ISSN 0973-4864, SP-1.3
11. Silberg, G., Wallace, R.: Self driving cars: The next revolution (2012), https://www.kpmg.com/US/en/IssuesAndInsights/ArticlesPublications/Documents/self-driving-cars-next-revolution.pdf

Image Clustering Based on Different Length Particle Swarm Optimization (DPSO)

Somnath Mukhopadhyay[1], Pragati Mandal[2], Tandra Pal[2],
and Jyotsna Kumar Mandal[3]

[1] Department of Computer Science and Engineering,
Aryabhatta Institute of Engineering and Management, Durgapur, India-713148
som.cse@live.com
[2] Department of Computer Science and Engineering,
National Institute of Technology, Durgapur, India-713209
{pragatimandal.bu,tandra.pal}@gmail.com
[3] Department of Computer Science and Engineering,
University of Kalyani, Kalyani, India-741235
jkm.cse@gmail.com

Abstract. Partitioning image pixels into several homogeneous regions is treated as the problem of clustering the pixels in the image matrix. This paper proposes an image clustering algorithm based on different length particle swarm optimization algorithm. Three evaluation criteria are used for the computation of the fitness of the particles of PSO based clustering algorithm. A novel Euclidean distance function is proposed based on the spatial and coordinate level distances of two image pixels towards measuring the similarity/dissimilarity. Different length particles are encoded in the PSO to minimize the user interaction with the program hence the execution time. PSO with different length particles automatically finds the number of cluster centers in the intensity space.

The performance of the proposed algorithm is demonstrated by clustering different standard digital images. Results are compared with some well known existing algorithms.

Keywords: Crisp clustering, digital image, Euclidean distance, image clustering, mean square error, quantization error, different length particle swarm optimization.

1 Introduction

Clustering is a process of grouping a set of samples or data so that they are similar within each group. The groups are called clusters [3]. Clustering techniques are used in many applications, such as image processing, object recognition, data mining, machine learning, etc. The clustering algorithms try to minimize or maximize certain objective functions.

A popular partitioning clustering algorithm is K-means[15]. This algorithm clusters the samples based on *Euclidean distance* as similarity/dissimilarity measure. The algorithm is suitable for large data set and easy to implement. In any

© Springer International Publishing Switzerland 2015   711
S.C. Satapathy et al. (eds.), *Proc. of the 3rd Int. Conf. on Front. of Intell. Comput. (FICTA) 2014*
– *Vol. 1*, Advances in Intelligent Systems and Computing 327, DOI: 10.1007/978-3-319-11933-5_80

fixed length clustering algorithm e.g., K-means algorithm, the clustering is obtained by iteratively minimizing a fitness function that is dependent on the distance of the pixels to the cluster centers. However, the K-means algorithm, like most of the existing clustering algorithms, assume a priori knowledge of the number of clusters, K, while in many practical situations, this information cannot be determined in advance. It is also sensitive to the selection of the initial cluster centers and may converge to the local optima. Finding an optimal number of clusters is usually a challenging task and several researchers have used various combinatorial optimization methods to solve the problem. Some other fixed length image clustering algorithms [9,8,16] exist in the literature. Various approaches [6,14,13,5,7] toward the image clustering based on variable length chromosome genetic algorithm and variable length particle swarm optimization have also been proposed in the recent past years for clustering. Some cluster validity indices [12,11] are also proposed for fuzzy and crisp clustering.

In this paper, a new approach termed as *Different length Particle Swarm Optimization (DPSO)* is proposed for image clustering. In Particle Swarm Optimization [4,1], the solution parameters are encoded in the form of strings called particles. A collection of such strings are called a swarm. Initially a random population of swarm is created, which represents random different points in the search space. An objective or fitness is associated with each string that represents the degree of goodness of the solution encoded in the particles. In the proposed DPSO algorithm, a swarm of particles of different lengths, automatically determines the number of clusters and simultaneously clusters the data set with minimal user interference. By using a novel fitness function which contains three evaluation criteria such as intra cluster distance, inter cluster distance and an error minimizer function. A novel weighted *Euclidean* distance function is used as distance function between two pixels in the image matrix. The algorithm terminates when the *(gBest)* converges to optimal solution or it meets a finite number of iterations. Since the actual number of clusters is considered to be unknown, the string of different particles in the same swarm population are allowed to contain different number of clusters. As a consequence, the different particles have different lengths having different number of cluster centers.

Rest of the paper is organized as follows. The standard PSO algorithm is described in section 2. Section 3 described the proposed PSO of different length particles for image clustering. Experimental results and discussions are provided in section 4. Finally we conclude in section 5.

2   Particle Swarm Optimization

Particle Swarm Optimization (PSO) is a population based stochastic optimization technique modeled on the social behavior of bird flocks [4]. It maintains a population of particles, where each particle represents a potential solution of the optimization problem. Each particle is assigned a velocity. The particles then flow through the problem space. The aim of PSO is to find the particle position that results the best fitness value. A fitness function is associated with a given optimization problem, which gives a fitness value for each particle. Each particle keeps track of the following information in the problem space: x_i, the current position of the particle; v_i, the current velocity of the particle; and y_i, the personal best position of the particle which is the best position that it has achieved so far. This position yields the best fitness value for that particle. The fitness value of this position is called *pBest*. There is another parameter in PSO, called global best *(gBest)*. For *(gBest)*, the best particle is determined from the entire swarm. The best value obtained so far by any particle in the population is *(gBest)*. The PSO changes the velocity and position of each particle at each time step so that it moves toward its personal best and global best locations, using (1) and (2) respectively. The process is repeated for maximum iterations or sufficient good fitness value.

$$v_p(i+1) = h(i) * v_p(i) + \Psi_p * r_p * (x_{pB}(i) - x_p(i)) + \Psi_g * r_g * ((x_{gB}(i) - x_p(i)) \quad (1)$$

$$x_p(i+1) = x_p(i) + v_p(i+1) \quad (2)$$

In those equations, Ψ_p and Ψ_g are the positive learning factors (or acceleration constants). r_p and r_g are random numbers in $[0, 1]$. i is the generation number in $[1, I_{MAX}]$. I_{MAX} is the maximum number of iterations. $h(i) \in [0, 1]$ is the inertia factor. $f_{pB}(i)$ and $f_{gB}(i)$ are the *(pBest)* value and *(gBest)* values at i^{th} generation, respectively. $x_{pB}(i)$ and $x_{gB}(i)$ are respectively the personal and global best positions of pth particle at i^{th} generation.

3   Different Length PSO (DPSO) Based Clustering

In this paper, a novel particle swarm optimization algorithm with different length particles is proposed. In this algorithm, a swarm of particles of different length, automatically determines the number of clusters and simultaneously clusters the data set with minimal user interference. It starts with random partitions of the image, encoded in each particle of the swarm.

The fitness function proposed by Omran and Salman [10,9] has been associated in the proposed clustering. The proposed fitness function defined in (7), contains three evaluation criteria such as intra cluster distance measure, inter cluster distance and the quantization error minimization function. These criteria are defined respectively in (4), (5) and (6). We consider the same weight of all these three criteria to the fitness of the corresponding particle. The weighted *Euclidean* distance function [9], given in (3), is used to compute the distance between two pixels in the image matrix, which is used for computing intra cluster distance measure, inter cluster distance and the quantization error. Vector containing the X-coordinate (say x), Y-coordinate (say y) and pixel intensity value (say z) are used to obtain the Euclidean distance between two pixels. The difference between two pixels termed as gray level distance as well as their spatial distance is calculated in the proposed distance function. The proposed weighted

Euclidean distance function between the i-th pixel and j-th pixel in the image matrix is given in (3).

$$d(x, y, z) = \sqrt{w_1((x_i - x_j)^2 + (y_i - y_j)^2) + w_2(z_i - z_j)^2} \qquad (3)$$

In the Euclidean distance function, different weights are assigned for two different distances, the spatial and gray level distances. The spatial distance has been assigned with less weight over the gray level distance. The intensity value similarity/dissimilarity of any two pixels is given with more weight over the positional similarity/dissimilarity of the pixels. The difference in intensity values between the pixels are multiplied by a weight factor $w_2=0.5$, whereas coordinate level difference is multiplied by $w_1=0.1$.

Different length PSO (DPSO) is proposed for image clustering, where a solution gives a set of cluster centers. Let $Z = (z_1, z_2, z_3, ..., z_{N_p})$ be the digital image with n number of pixels. The DPSO maintains a swarm of particles, where each particle represents a potential solution to the clustering problem and each particle encodes partition of the image Z. DPSO tries to find the number of clusters, N_c. The DPSO based image clustering method has various parameters. N_p, N_c, z_p, m_j, C_j and $|C_j|$, which are respectively the number of image pixels to be clustered, N_c number of clusters, p-th pixel of the image, mean or center of cluster j, set of pixels in cluster j and the number of pixels in cluster j. Each particle can be represented by $\{m_{i1}, ...,m_{ij}, ..., m_{iN_c}\}$, where m_{ij} refers to the j-th cluster center vector of the i-th particle. In this algorithm, particles have different lengths since the number of clusters is unknown. The particles are initialized with random number of cluster centers in the range $[K_{min}, K_{max}]$, where K_{min} is usually assigned to 2 and K_{max} describes the maximum particle length, which represents the maximum possible number of clusters. K_{max} depends on the size and type of image. DPSO, the proposed algorithm for image clustering using different length particle swarm optimization is presented in Algorithm 1.

The intra-cluster distances of all the clusters are measured and the maximum one among all the clusters is selected in d_{max} which is defined in (4), where Z is a partition matrix representing the assignment of pixels to clusters of particle i. A smaller value of d_{max} means that the clusters are more compact.

$$d_{max}(Z, x_i) = \max_{j=1 \ to \ N_c} \left\{ \sum_{\forall z_p \in C_{ij}} d(z_p, m_{ij}) / |C_{ij}| \right\} \qquad (4)$$

Inter-cluster separation distances for all clusters are measured and the minimum distance between any two clusters is calculated using (5). A large value of d_{min} means that the clusters are well separated.

$$d_{min}(x_i) = \min_{\forall j1, j2, j1 \neq j2} \{d(m_{ij_1}, m_{ij_2})\} \qquad (5)$$

The quantization error function [2,9] is proposed in the clustering of image pixels which calculates the average distance of the pixels of a cluster to its cluster centers, followed by the average distances of all clusters and hence calculates new average. The problem of Esmin et al. [2] is that any cluster with one pixel would

---

Algorithm 1. DPSO Algorithm

Input: Gray Scale Image Matrix
Output: Partition Matrix
1: **begin**
2: Initialize the maximum number of cluster centers K_{max} and all the constant parameters
3: Initialize each particle with K randomly selected cluster centers
4: Initialize each particle x_i with the $pBest_i$ and also the *(gBest)*
5: **while** gen $< I_{max}$ **do**   ▷ I_{max} is the maximum number iterations
6: **for** i=1 to NOP **do**   ▷ Number of particles
7:    **for** x=1 to rows **do**   ▷ Number of rows of the image matrix
8:       **for** y=1 to cols **do**   ▷ Number of columns of the image matrix
9:          Let (x,y) be the coordinate of the pth pixel
10:         Find Euclidean distance between p^{th} pixel and all centers of ith particle
11:         Assign p^{th} pixel to j^{th} centers of ith particle
12:      **end for**
13:   **end for**
14:   Compute Intra cluster distance of ith particle using (4)
15:   Compute Inter cluster distance of ith particle using (5)
16:   Compute Quantization error of ith particle using (6)
17:   Compute the fitness value of ith particle using (7), which uses (4), (5) and (6)
18:   Update *(pBest)* position $x_{pB}(i)$ and *(pBest)* value $f_{pB}(i)$ of ith particle
19: **end for**
20: Update *(gBest)* from all the particles in the swarm
21: Update velocity and then position of particles using (1) and (2)
22: Update inertia weight
23: **end while**
24: **end**

---

affect the final result with another cluster containing many pixels. Suppose for i^{th} particle in a cluster which has only one pixel and very close to the center and there is another cluster that has many pixels which are not so close to the center. The problem has been resolved by assigning less weight to the cluster containing only one pixel than with cluster having many pixels. The weighted quantization error function is given in (6), where N_0 is the total number of data vectors to be clustered. The fitness function is constructed by intra-cluster distance d_{max}, inter-cluster distance d_{min} along with the quantization error Q_e function. The fitness function used to minimize $f(x_i, Z)$ [16] which is given in (7). Here z_{max} is the maximum intensity value of the digital images which is 255 for 8-bit gray scale images. In the optimization function, equal weights are assigned to the three distance functions. The fitness function is given to the PSO based optimization technique and which minimizes the value of f in each generation to make the noisy image well clustered.

$$Q_e = \{ \sum_{\forall j=1 \ to \ N_c} [(\sum_{\forall z_p \in c_{ij}} d(z_p, m_{ij}) / \ |C_{ij}|.(N_0/ \ |C_{ij}|)] \} \tag{6}$$

$$f(x_i, Z) = d_{max}(z, x_i) + (z_{max} - d_{min}(x_i)) + Q_e \qquad (7)$$

4   Results and Discussion

The proposed algorithm is tested by using three standard gray scale images: *lena*, *pepper* and *airplane*. The performance of the proposed algorithm is measured by three evaluation metrics, *intra cluster distance, inter cluster distance* and *quantization error*. The performance of the algorithm is compared with three existing algorithm *K-means, Man et al.* [16] and *FPSO* [9] algorithms. For comparison purpose, the following parameter values are used for the algorithms:

- Gray scale image resolution = 512 × 512
- Number of particles (NOP) = 20
- Maximum number of clusters = 20
- Number of iterations = 50

Number of iterations for Man *et al.*, FPSO and proposed DPSO algorithms are set to 50. For K-means, the number of iterations will be 50 × number of particles, because in each iteration the fitness of 20 particles are computed in PSO based clustering algorithms. The inertia factor is set to 1 initially and decreased linearly with the number of iterations. Both the acceleration constants are set to 2. For DPSO algorithm, minimum (K_{min}) and maximum (K_{max}) number of clusters are set to 2 and 20 respectively. For K-means algorithm, only Mean Square Error (MSE) is used as fitness function. For Man *et al.*, FPSO and DPSO algorithm, the same fitness function is used for evaluation which is given in (7).

Table 1. Clustering results using Intra distance (d_{max}), Quantization Error (Q_e), Inter Distance (d_{min}) and Fitness Value

Algorithm/Image	d_{max}	Q_e	d_{min}	Fitness Value
(1.) K-Means				
Pepper	13.57	10.50	32.03	247.04
Lena	9.85	8.70	29.03	244.52
Airplane	16.30	10.12	20.57	260.85
(2.) Man et al.				
Pepper	12.11	9.87	40.20	236.78
Lena	9.35	8.49	34.72	238.12
Airplane	11.63	10.42	40.60	236.45
(3.) FPSO				
Pepper	11.82	9.50	41.60	234.72
Lena	9.11	8.23	35.70	236.64
Airplane	11.22	10.04	41.56	234.70
(4.) **DPSO**				
Pepper	10.68	9.00	42.49	232.19
Lena	8.44	7.79	36.41	234.82
Airplane	10.21	9.47	42.11	232.57

Table 1 shows the intra cluster distance, inter cluster distance, weighted quantization error and fitness value by the existing and proposed DPSO algorithms. For K-means algorithm, all the three performance metrics, intra cluster distance, inter cluster distance, weighted quantization error are worse than those of all other algorithms. FPSO is better than Man *et al.* and K-means. Proposed DPSO is best with respect to all those measures. The last column of the table shows the fitness values of the algorithms. Less the value, more is fitness. We can see from the table that the DPSO outperforms all the other algorithms in all respect.

5 Conclusion

This paper proposed a novel image clustering algorithm based on the different length particle swarm optimization (DPSO) algorithm. DPSO uses a novel fitness function with three evaluation criterion, the results show that DPSO performs better than three other existing algorithms. The weighted Euclidean function measures the similarity and/or dissimilarity among the pixels. Using the Euclidean distance and weighted quantization error functions as fitness criteria, DPSO outperforms the K-Means algorithm, Man *et al.* and FPSO algorithms with a wide margin. The algorithm can minimize the user intervention during the program run as it can find the number of cluster centers automatically in a specified range. A limitation of the proposed work is that the number of cluster centers assigned initially for a particle remains fixed. This limitation is resolved to some extent by using a large number of particles to increase the diversity regarding the number of cluster centers.

In future, cluster validity indices may be used in the fitness function of the algorithm for clustering. Another attempt may be to use *Multi-objective evolutionary algorithm (MOEA)* to get a a set of optimal solutions where three criteria used in proposed DPSO will be used as objectives. To overcome the limitations of the fixed length particles, variable length PSO based clustering may be considered the limitations of this work by merging or splitting the particles.

Acknowledgments. Authors expressed deep sense of gratitude towards the Department of Computer Science and Engineering, University of Kalyani and National Institute of Technology, Durgapur and the PURSE Project (DST, Govt. of India) of the university where the computational resources are used for the work.

References

1. Eberhart, R., Kennedy, J.: A new optimizer using particle swarm theory. In: Proceedings of the Sixth International Symposium on Micro Machine and Human Science, MHS 1995, pp. 39–43. IEEE (October 1995)

2. Esmin, A.A.A., Pereira, D.L., de Arajo, F.P.A.: Study of different approach to clustering data by using particle swarm optimization algorithm. In: Proceedings of the IEEE World Congress on Evolutionary Computation (CEC 2008), Hong Kong, China, pp. 1817–1822 (June 2008)
3. Gose, E., Johnsonbough, R., Jost, S.: Pattern recognition and image analysis. Prentice-Hall (1996)
4. Eberhart, J.K., Particle, R.C.: swarm optimization. In: IEEE International Conference on Neural Network, Perth, Australia, pp. 1942–1948 (1995)
5. López, J., Lanzarini, L., De Giusti, A.: VarMOPSO: Multi-objective particle swarm optimization with variable population size. In: Kuri-Morales, A., Simari, G.R. (eds.) IBERAMIA 2010. LNCS, vol. 6433, pp. 60–69. Springer, Heidelberg (2010)
6. Katari, V., Ch, S., Satapathy, R., Ieee, M., Murthy, J., Reddy, P.P.: Hybridized improved genetic algorithm with variable length chromosome for image clustering abstract. International Journal of Computer Science and Network Security 7(11), 1121–1131 (2007)
7. Maulik, U., Bandyopadhyay, S.: Fuzzy partitioning using a real-coded variable-length genetic algorithm for pixel classification. IEEE Transactions on Geoscience and Remote Sensing 41(5), 1075–1081 (2003)
8. Mukhopadhyay, S., Mandal, J.K.: Adaptive Median Filtering based on Unsupervised Classification of Pixels. In: Handbook of Research on Computational Intelligence for Engineering, Science and Business. IGI Global, 701 E. Chocolate Ave, Hershey, PA 17033, USA (2013)
9. Mukhopadhyay, S., Mandal, J.K.: Denoising of digital images through pso based pixel classification. Central European Journal of Computer Science 3(4), 158–172 (2013)
10. Omran, M., Enge;brecht, A., Salman, A.: Particle swarm optimization method for image clustering. International Journal of Pattern Recognition and Artificial Intelligence 19, 297–322 (2005)
11. Pakhira, M.K., Bandyopadhyay, S., Maulik, U.: Validity index for crisp and fuzzy clusters. Pattern Recognition 37(3), 487–501 (2004),
http://www.sciencedirect.com/science/article/pii/S0031320303002838
12. Wong, M.T., He, X., Yeh, W.-C.: Image clustering using particle swarm optimization. In: 2011 IEEE Congress on Evolutionary Computation (CEC), pp. 262–268 (June 2011)
13. Qiu, M., Liu, L., Ding, H., Dong, J., Wang, W.: A new hybrid variable-length ga and pso algorithm in continuous facility location problem with capacity and service level constraints. In: IEEE/INFORMS International Conference on Service Operations, Logistics and Informatics, SOLI 2009, pp. 546–551 (July 2009)
14. Srikanth, R., George, R., Warsi, N., Prabhu, D., Petry, F., Buckles, B.: A variable-length genetic algorithm for clustering and classification. Pattern Recognition Letters 16(8), 789–800 (1995),
http://www.sciencedirect.com/science/article/pii/016786559500043G,
Genetic Algorithms
15. Tan, P., Steinbach, M., Kumar, V.: Introduction to data mining. Pearson Education (2006)
16. Wong, M.T., He, X., Yeh, W.-C.: Image clustering using particle swarm optimization. In: 2011 IEEE Congress on Evolutionary Computation (CEC), pp. 262–268 (June 2011)

Design of PI Controller for a Non-linear Spherical Tank System Using Enhanced Bacterial Foraging Algorithm

G. Sivagurunathan[1] and K. Saravanan[2]

[1] Department of Electronics and Instrumentation Engg., St. Joseph's College of Engineering, Chennai – 600 119, Tamilnadu, India
[2] Department of Chemical Engineering, Kongu Engineering College, Perundurai, Erode, Tamilnadu, India- 638 052
gsgnathan@gmail.com, rumisivaesh@yahoo.com

Abstract. This paper presents a Bacterial Foraging Algorithm (BFA) based controller parameter selection procedure for a nonlinear spherical tank system. In this work Enhanced BFA (EBFA) is adopted to find the K_p and K_i value for the PI controller. Minimization of Integral Absolute Error (IAE), peak overshoot (M_p) and settling time (t_s) are chosen as the objective function. The performance of the proposed method is validated with a comparison of Particle Swarm Optimization (PSO). This study confirms that, the proposed method offers satisfactory performance in terms of time domain and error performance criteria and also verified with simulation and real time results.

Keywords: Non-linear system, spherical tank, PI controller, EBFA, PSO.

1 Introduction

Most of the process industries, nonlinear process are common in nature. Selection of controller parameters is essential in all the process industries. Design of controllers for the linear and stable processes are quite simple compared to the nonlinear and unstable processes [2]. In order to get the optimal controller parameters, the conventional PID tuning methods need more numerical calculations and a suitable mathematical model [3]. To overcome the above mentioned problem, the research on auto tuning of PID controller using intelligent soft computing approaches such as Neural Network (NN), Genetic Algorithm (GA), PSO and BFA are used [9], [10]. The advantage of PSO algorithm is that, it is an auto tuning method, it does not require detailed mathematical description of the process and finds the optimal PID controller parameters based on the performance index provided for algorithm convergence.. Although PSO has the characteristics of fast convergence, good robustness, easy implementation, and has been successfully applied in many areas, it has the shortcomings of premature convergence and low searching accuracy. In order to overcome the above mentioned limitations, many researchers have attempted to improve the PSO algorithm. The BFA has more flexibility and high rate of convergence to overcome the limitations of PSO.

In this work, the controller parameters are estimated using PSO and BFA and the simulations works are carried out. The comparison between PSO and BFA were made

© Springer International Publishing Switzerland 2015
S.C. Satapathy et al. (eds.), *Proc. of the 3rd Int. Conf. on Front. of Intell. Comput. (FICTA) 2014*
– *Vol. 1*, Advances in Intelligent Systems and Computing 327, DOI: 10.1007/978-3-319-11933-5_81

in terms of setpoint tracking and load disturbance rejection. Further BFA based controller is implemented in real time spherical tank level process.

The remainder of this article is organized as follows. The real time experimental setup is presented in section 2. The mathematical model of the level process is described in section 3. Section 4 describes PSO, BFA-tuning method and proposals for defining the fitness function. Section 5 presents simulated results and real time implementation. Conclusion of the present work is given in section 6.

2   Experimental Setup

Figure 1 shows the real time experimental setup of a spherical tank. The system consists of a spherical tank, a water reservoir, pump, rotameter, a differential pressure transmitter, an electro pneumatic converter (I/P converter), a pneumatic control valve with positioner, an interfacing module (DAQ) and a Personal Computer (PC). The differential pressure transmitter output is interfaced with computer using data acquisition RS-232 port of the PC. The programs written in script code using MATLAB software is then linked via the interface module. The operation of the experimental setup is discussed in detail [6].

3   System Identification

This section describes the modeling procedure for the spherical tank system.

3.1   Mathematical Model of Spherical Tank System

The development of mathematical model for a nonlinear spherical tank system is considered in this work. The nonlinear dynamics of a spherical tank system is described by the first order differential equation

$$\frac{dV}{dt} = F_{in} - F_{out} \tag{1}$$

where V represents volume of tank, F_{in}, F_{out} are the inflow and outflow rate respectively.

$$V = \frac{4}{3}\pi h^3 \tag{2}$$

where h is the total height of the tank in cm.

Applying the steady state values and solving equations (1) and (2), for linearizing the non-linearity in the spherical tank;

$$\frac{H(s)}{Q(s)} = \frac{R_t}{\tau s + 1} \tag{3}$$

Where $R_t = \dfrac{2h_s}{F_{out}}$, $\tau = 4\pi R_t h_s$

h_s - Height of the tank at steady state.

3.2   Block Box Modeling

A general first order process with dead time is represented by

$$y(s) = \frac{k_p e^{-t_{ds}}}{\tau s + 1} u(s) \tag{4}$$

The output response to a step change input

$$y(t) = 0 \text{ for } t < t_d \tag{5}$$

$$y(t) = k_p \Delta u\{1 - \exp(-(t - t_d)/\tau)\} \text{ for } t \geq t_d \tag{6}$$

The procedure to develop a model for various operating regions has been discussed in detail [6]. The developed model is subjected to the framed controller and tested in a real time environment. The process dynamics are analyzed in four operating points so as to obtain their corresponding suitable model. The obtained model parameters of four operating points are shown in Table 1.

Fig. 1. Experimental setup of spherical tank

Table 1. Calculated values of k, τ and τ_d for different operating regions

Operating point (cm)	Model parameters		
	k_p	$\tau(s)$	$\tau_d(s)$
11	2.275	157.5	77.5
20	3.42	486	94
30	4.5	465	85
40	4.52	547.5	57.5

4   Methodology

This section discuss about the heuristic algorithms considered in this work.

4.1   Particle Swarm Optimization

The traditional PSO algorithm was initially developed by Kennedy and Eberhart in 1995 [9].

The PSO algorithm considered in this study is given below.

Mathematically the search operation is described by the following equations;

$$V_{i,D}^{t+1} = W\left[V_{i,D}^t + C_1.R_1.(P_{i,D}^t - S_{i,D}^t) + C_2.R_2.(G_{i,D}^t - S_{i,D}^t)\right] \tag{7}$$

$$W = W_{max} - \frac{W_{max} - W_{min}}{iter_{max}} iter \tag{8}$$

$$S_{i,D}^{t+1} = S_{i,D}^{t} + V_{i,D}^{t+1}$$
(9)

where W = inertia weight; $V_{i,D}^{t}$ = current velocity of the particle; $S_{i,D}^{t}$= current position of the particle; R_1, R_2 are the random numbers in the range 0-1; C_1, C_2 are the cognitive and global learning rate respectively, W_{max} - Maximum iteration number, W_{min} – Minimum iteration number, iter – current iteration, $iter_{max}$ – maximum iteration, $V_{i,D}^{t+1}$ = updated velocity; $S_{i,D}^{t+1}$ - updated position; $G_{i,D}^{t}$ - global best position(gbest); $P_{i,D}^{t}$ - local best position(pbest).

In order to design an optimal controller the following algorithm parameters are considered; dimension of search space is two (i.e., K_p, K_i), number of swarm and bird step is considered as 20, the assigned value of cognitive parameter C_1 is 0.7 and global search parameter is C_2 is 0.3, the inertia weight "W" is set as 0.6.

4.2   Brief Overview of Algorithms in the Study

EBFA: The classical BFA was initially proposed by Passino in 2002 [7]. This offers benefits, such as good computational efficiency, easy implementation and stable convergence, but also comes with a major drawback when compared to other heuristic methods: the large number of algorithm parameters to be assigned. Recently, an empirical procedure was discussed in the literature to assign BFA parameters which require only two values, such as population size and search dimension [1][3].

The initial algorithm parameters are assigned as follows:

Number of bacteria = N; $N_c = \dfrac{N}{2}$; $N_s = N_{re} \approx \dfrac{N}{3}$; $N_{ed} \approx \dfrac{N}{4}$; $N_r = \dfrac{N}{2}$; $P_{ed} =$

$\left(\dfrac{N_{ed}}{N + N_r} \right)$; $d_{attr} = W_{attr} = \dfrac{N_s}{N}$;   and $h_{rep} = W_{rep} = \dfrac{N_c}{N}$

where N_c = number of chemotactic steps, N_s = swim length during the search, N_{re} = number of reproduction steps, N_{ed} = number of elimination – dispersal events, N_r = number of bacterial reproduction, P_{ed} = probability of the bacterial elimination, d_{attr} = depth of attraction, W_{attr} = width of attraction, h_{rep} = height of repellant and W_{rep} = width of repellant signal. In this work, the EBFA parameters are assigned as discussed in [1].

4.3   Objective Function

The overall performance of of PSO algorithm depends on Objective Function (OF), which monitors the optimization search [2]. In this work, OF is chosen as a minimization problem. In the literature, there exist a number of weighted sum-based objective functions [5], considered the following with three parameters such as peak over shoot- M_p, Integral Absolute Error (IAE), and Settling time- t_s as follows;

$$J(p)_{min} = w_1 * M_p + w_2 * IAE + w_3 * t_s$$
(10)

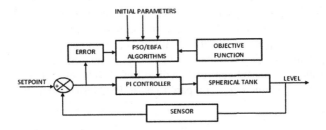

Fig. 2. Block diagram of PSO/BFO based controller

The block diagram of PSO and BFO based controller design procedure considered in this work is depicted in figure 2. The PSO/ BFO algorithm continuously adjust K_p, K_i until J(p) is minimized.

5   Results and Discussions

This section presents the developed PSO and EBFA based PI controllers are applied in simulation and real time environment of level process of spherical tank system using the model of the spherical tank at four operating regions. Tuning of PSO and EBFA based controller is attempted and the obtained optimal controller parameters are presented in table 2. The selected optimal PI (K_p, K_I) parameters will minimize the IAE, peak overshoot and settling time when the process is in steady state. To validate the EBFA tuned PI controller, the performance is compared with basic PSO based PI controller.

5.1   Servo Response

After selecting the optimal controller values, the proposed controller settings are applied in simulation mode to study the controller performance on the spherical tank with different operating regions. The simulation is also carried out with basic PSO controllers and the results are compared. The simulated responses of the two controllers for the two operating points (11cm and 30cm) are shown in figures 3 and 4.

Table 2. PI tuning values for different operating points

Setpoint	Method	K_p	K_I
11 cm	PSO	0.6131	0.0029
	EBFA	0.5652	0.0029
20 cm	PSO	0.7608	0.0016
	EBFA	0.7650	0.0015
30 cm	PSO	0.7201	0.0016
	EBFA	0.6857	0.0015
40 cm	PSO	1.0159	0.0018
	EBFA	1.0488	0.0019

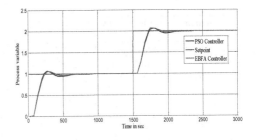

Fig. 3. Servo responses of PSO and EBFA based PI controller at operating point of 11 cm

Figures 3 and 4 shows the setpoint tracking performance of the model (11 cm and 30cm) for multiple set points. The performances of controllers are analyzed by considering rise time, peak time, settling time and percentage peak overshoot and the performance evaluation is presented in table 3-5. From the figure 3 and 4   it is observed that the BFO based controller will follow the changes of set point with less overshoot and settling time when compare to basic PSO based controller

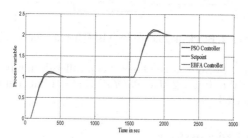

Fig. 4. Servo responses of PSO and EBFA based PI controller at operating point of 30 cm

Table 3. Comparison of time domain specification for PSO and EBFA based PI controller

Setpoint	Method	Rise time (s)	Peak time (s)	Settling time (s)	Peak overshoot (%)
11 cm	PSO	223	270	1320	6.2
	EBFA	240	300	1170	5.0
20 cm	PSO	350	450	1332	5.0
	EBFA	350	440	1172	3.0
30 cm	PSO	250	335	1800	12.5
	EBFA	270	350	815	8.7
40 cm	PSO	227	295	750	3.7
	EBFA	220	280	600	3.7

Table 4. Performance indices comparison

Setpoint	Method	ISE	IAE
11 cm	PSO	118.1	161.2
	EBFA	120.7	159.1
20 cm	PSO	158.2	206.6
	EBFA	158	204.3
30 cm	PSO	135.1	181.4
	EBFA	137	179.9
40 cm	PSO	98.61	128.3
	EBFA	97.62	126.3

Table 5. Performance indices comparison (servo response)

Setpoint	Method	ISE	IAE
11 cm	PSO	240.2	325.3
	EBFA	245.3	323.4
20 cm	PSO	320.6	423.2
	EBFA	320.6	416.1
30 cm	PSO	274.2	373.3
	EBFA	278.2	369.7
40 cm	PSO	200	263.1
	EBFA	197.6	257.6

5.2   Real Time Implementation

The performance of EBFA tuned PI controller is validated in real time on a nonlinear spherical tank system. The hardware details of the considered experimental setup are given in section 2. The reference tracking performance of the system for multiple set points for the operating region of 11cm is shown in Fig 5.

Fig. 5 depicts the setpoint tracking performance of EBFA based PI controller for a set point of 11cm (22% for a tank diameter of 50cm). Initially, the reference tracking is studied with a single reference input. Later, 10% change is added with the initial set point at 1200 sec and the set point is increased from 11cm to 16 cm. From the response it is noted that the controller track the set point without any overshoot, when we apply 10% changes of set point (5cm) the controller follow the changes of set point with small overshoot.

5.3   Regulatory Response

After estimated the tuning parameters, the developed controller is tested in simulation and real time environment with 20% load changes at the operating point of 11cm. The responses of simulated and real time are recorded in figure 6 and 7.

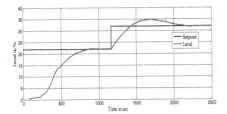

Fig. 5. Real time servo response of EBFA based PI controller at the operating point of 11 cm

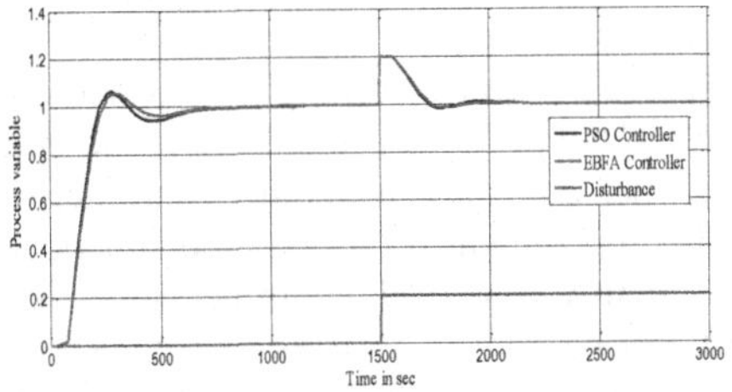

Fig. 6. Regulatory response of EBFA based PI controller at the operating point of 11 cm with 20% disturbance

Table 6. Performance indices comparison of PSO and EBFA based PI controller (regulatory response)

Setpoint	Method	ISE	IAE
11cm	PSO	123.1	193.8
	EBFA	125.9	192.5
20cm	PSO	165.4	251.2
	EBFA	165.1	247.7
30cm	PSO	141.0	221.6
	EBFA	143.0	219.6
40cm	PSO	103.2	158.2
	EBFA	101.8	155.5

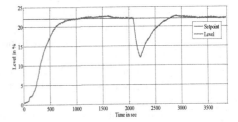

Fig. 7. Real time regulatory response of EBFA based PI controller at the operating point of 11 cm with 20% disturbance

Figure 6 shows that for a sudden load change applied at 1500 sec, the EBFA based PI controller returns to the set point with negligible overshoot when compared PSO based PI controller. The EBFA tuned PI based controller is capable to compensate for the load changes considerably better than PSO tuned PI controller. From table 8 it is also noted that IAE values are considerably low but ISE values are high when compared with other controller. Further, the proposed EBFA based PI controller implemented in real time liquid level process at the operating stage of 11cm with 20% load change. The recorded response is shown in figure 7. From the figure it is observed that the designed controller rejects the given disturbances within 1350 sec.

6   Conclusion

A PI controller using the EBFA algorithm was developed with IAE, peak overshoot and settling time as the objective function to improve the transient and steady state response of nonlinear spherical tank system. The estimated controller parameters are tested in simulation environment to study its performance. From the simulation results it is noted that, the time domain specifications such as peak over shoot and settling time of EBFA based PI controller are improved when compared with basic PSO based PI controller. Further, the performance indices of the optimized controllers are analyzed, IAE value is less when compared with other controller. The performance of EBFA based PI controller is then implemented on real time spherical tank process. The real time responses show that EBFA tuned PI controller gives smooth response for set point tracking and disturbance rejection. It is concluded that the proposed technique design optimal controller parameter gains are found satisfactory performance in terms of setpoint tracking and disturbance rejection operations in all the operating regions.

References

[1] Rajinikanth, V., Latha, K.: Controller Parameter Optimization for Nonlinear Systems Using Enhanced Bacteria Foraging Algorithm. Applied Computational Intelligence and Soft Computing 2012, Article ID 214264, 12 pages (2012)
[2] Rajinikanth, V., Latha, K.: Setpoint weighted PID controller tuning for unstable system using heuristic algorithm. Archives of Control Science 22(4), 481–505 (2012a)
[3] Rajinikanth, V., Latha, K.: I-PD Controller Tuning for Unstable System Using Bacterial Foraging Algorithm: A Study Based on Various Error Criterion. Applied Computational Intelligence and Soft Computing 2012, Article ID 329389, 10 pages (2012b)
[4] Latha, K., Rajinikanth, V., Surekha, P.M.: PSO-Based PID Controller Design for a Class of Stable and Unstable Systems. ISRN Artificial Intelligence 2013, Article ID 543607 (2013)
[5] Liu, G.P., Yang, J.B., Whidborne, J.F.: Whidborne, Multiobjective Optimization and Control. Printice Hall, NewDelhi (2008)
[6] Sivagurunathan, G., Saravanan, K.: Evolutionary Algorithms based Controller Optimization for a Real Time Spherical Tank System. Australian Journal of Basic and Applied Sciences 8(3), 244–254 (2014)
[7] Passino, K.M.: Biomimicry of bacterial foraging for distributed optimization and control. IEEE Control Systems Magazine 22(3), 52–67 (2002)

[8] Kennedy, J., Eberhart, R.C.: Particle swarm optimization. In: Proceedings of IEEE international Conference on Neural Networks, pp. 1942–1948 (1995), doi:10.1109/ ICNN.1995.488968.

[9] Kotteeswaran, R., Sivakumar, L.: Optimal Tuning of Decentralized PI Controller of Nonlinear Multivariable Process Using Archival Based Multiobjective Particle Swarm Optimization. Modelling and Simulation in Engineering 2014, Article ID 504706 (2014)

[10] Soni, Y.K., Bhatt, R.: BF-PSO optimized PID Controller design using ISE, IAE, IATE and MSE error criteria. International Journal of Advanced Research in Computer Engineering & Technology 2(7), 2333–2336 (2013)

Diffusion and Encryption of Digital Image Using Genetic Algorithm

Subhajit Das[1], Satyendra Nath Mandal[2], and Nabin Ghoshal[3]

[1] Nayagram Bani Bidyapith, Nayagram, India
Subhajit.batom@gmail.com
[2] Kalyani Govt. Engg. College, Dept. of Information Tech. Kalyani, Nadia, India
satyen_kgec@rediffmail.com
[3] University of Kalyani, Dept. of Engg. and Tech. Studies, Nadia, India
nabin_ghoshal@yahoo.co.in

Abstract. The security level of any symmetric key algorithm is directly proportional to its execution time. The algorithm will be more secured if the number of iteration is increase. Soft computing methods are trying to reduce time complexity without compromise the security level. In this paper, an image has been encrypted by number of steps. At first, a key set has been developed based on 16 arbitrary characters and a large number. The image has been diffused in the next step. The key set and diffused image have been computed by genetic algorithm. Here, ring crossover and order changing mutation operator is used in genetic algorithm. The encrypted image is constructed based on logical operation between the diffused image and key. The effectiveness of the algorithm has been tested by number of statistical tests. Finally, a comparative study has been made between our proposed algorithm and some other algorithms. It has been observed that the proposed algorithm has given better result.

Keywords: Symmetric key, prime number, Genetic algorithm, Ring crossover Order Changing Mutation and Statistical Test.

1   Introduction

Cryptography is the technique which is used to convert readable information into unreadable form. Now a days many branches of science is used to increase the security and decrease the computation time of encryption algorithm Recently, soft computing techniques are used widely to produce a new encryption method with its powerful features. A. Kumar et.al. [3] have described encryption technique by using crossover operator and pseudorandom sequence generated by NLFFSR (Non- Linear Feed Forward Shift Register). Pseudorandom sequence decided the crossover point for which fully encrypted data is achieved. A. Kumar et.al. [4] have further extended this work using mutation after encryption. Finally, encrypted data has been hidden inside the steno-image. Aarti Soni and Suyash Agrawal [5] have used a genetic algorithm based symmetric key encryption algorithm. They have generated secret key with the help of pseudo random number generator (PRNG). For encryption technique

© Springer International Publishing Switzerland 2015
S.C. Satapathy et al. (eds.), *Proc. of the 3rd Int. Conf. on Front. of Intell. Comput. (FICTA) 2014
– Vol. 1*, Advances in Intelligent Systems and Computing 327, DOI: 10.1007/978-3-319-11933-5_82

they have used AES algorithm Rasul Enayatifar and Abdul Hanan Abdullah [6] have proposed a new method based on a hybrid model consisting of a genetic algorithm and chaotic function. They have applied their proposed algorithm to an image. At first, a number of encrypted images have been made by chaotic function. The new encrypted images have been taken as initial population. Genetic algorithm has been used to optimize the encrypted image as much as possible. Best cipher image has been chosen based on highest entropy and lowest correlation coefficient between adjacent pixels. Rasul Enayatifar et al. [8] have used a chaos based image encryption technique using hybrid genetic algorithm and DNA sequence. They have used DNA and logistic map functions to create the number of initial DNA masks. The crossover has been used to affect the masks of DNA.

From above journals, it has been observed that the authors have used genetic algorithm either key generation or encrypt the images. In this paper, genetic algorithm has been used to generate key and image encryption in both cases. The numbers of tests have been applied to find the security level of this algorithm. It has been observed that the proposed algorithm produced better result compare to others.

2   Ring Crossover

Genetic Algorithms are adaptive search procedures which are footed on Charles Darwin theory of the survival of the fittest [13]. It consists of four parts, determine the chromosome structure, define initial population, perform crossover and mutation and repeat last two steps until optimal solution obtained [7]. There are many types of crossover namely single point, two point, intermediate, heuristic and ring crossover [2]. Steps of ring cross over is furnished in fig 1.

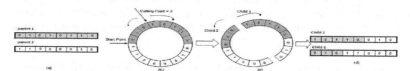

Fig. 1. Ring Crossover

3   Order Changing Mutation

In genetic algorithm mutation alters one or more gene values in a chromosome from its initial state. In order changing mutation two specified location of chromosome is selected and their values are altered [7]. The steps of order changing mutation have been furnished in fig 2.

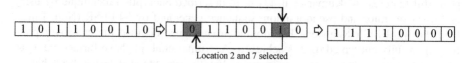

Location 2 and 7 selected

Fig. 2. Order chaining mutation

4   Proposed Algorithm

The algorithm is divided into five parts key generation, image diffusion, encryption, decryption and plain image reconstruction.

4.1   Key Generation

Input: 16 random characters and a large integer number.
Output: A set of Effective key value.
Method:

> Step 1. To compute remainder set
>> Set 1.1. To take integer number (n)
>> Set 1.2. To compute prime numbers from2 to n and store in reverse.
>> Set 1.3. To divide all the numbers by 16 and store their remainder in $\mathbb{Q}$.
>
> Step 2. To separate 16 random characters into two half and each half is treated as initial parent chromosome.
>
> Step 3: To compute key set
>> Step 3.1. To convert each half (as gene)into its equivalent ASCII value and to make key set empty.
>> Step 3.2. Repeat step 3.3 to 3.4 until all elements of remainder set are taken
>> Step 3.3. To apply ring crossover between two genes and the cross over points are elements taken from remainder set $\mathbb{Q}$ in a sequential manner.
>> Step 3.4. To apply order changing mutation by select 1st and 8th gene of each chromosome.
>> Step 3.5. To store two new children into key set and they are taken as next parent.

4.2   Image Diffusion

Input: - A grey scale image, set of remainders $\mathbb{Q}$.
Output: - Diffuse Image.
Method :
Step 1. To take grey scale image into matrix $\mathbb{M}$

$$\mathbb{M} = \{a_{ij}\}_{m \times n}$$

Step 2. To represent each pixel into its eight bits binary as

$$a_{ij} = a_8^{ij} a_7^{ik} \ldots \ldots a_1^{ik} \text{ Where} a_1^{ij}, \ldots, a_1^{ik}, \ldots = 0 \text{ or } 1.$$

Step 3. To apply ring crossover between each a_{ij} and a_{ij+1} and store it back into same place. Elements of $\mathbb{Q}$ taken sequentially act as a crossover point of each crossover operation.

$$a'_{ij}, a'_{ij+1} = a_{ij} \otimes a_{ij+1} \mid \otimes \text{ represent crossover.}$$

Step 4. To apply order changing mutation in each a'_{ij} by selecting first and last bit.

$$a''_{ij} = a'_{ij1} \ a'_{ij7} \ a'_{ij6} \ a'_{ij5} a'_{ij4} \ a'_{ij3} \ a'_{ij2} \ a'_{ij8}$$

4.3   Image Encryption

Input: - Diffused image, set of key value $\mathbb{K}$.
Output: - Encrypted Image.
Method: At this point logical XOR operation has been performed between each a_{ij} and each element of $\mathbb{K}$ starting from left corner of the plain image, to obtain an encrypted image. If total number of a_{ij} is grater then the number of elements $\mathbb{K}$ then $\mathbb{K}$ taken as circularly.

4.4   Image Decryption

Input: An Encrypted image, set of key value $\mathbb{K}$
Output: Diffused image.
Method: To perform bitwise XOR between elements of set $\mathbb{K}$ and grey value of pixels starting from left corner of the encrypted image. If the number of pixels > number of elements of $\mathbb{K}$ then the elements of the $\mathbb{K}$ is to be taken circularly.

4.5   Image Reconstruction

Input: Diffused image set of remainders $\mathbb{Q}$
Output: Plain Image.
Method:
Step 1. To convert each pixel a''_{ij} of defused image into its equivalent 8 bit binary value.
Steps 2. To perform order changing mutation at LSB and MSB of each a''_{ij} and obtained a'_{ij}.
Step 3. To perform ring crossover between each a'_{ij} and a'_{ij+1}, where each of the element of $\mathbb{Q}$ taken one by one in a circular manner, subtract it from 16 and resultant denotes the crossover point of each crossover.

5   Result and Analysis

Experimental analysis of the new algorithm is presented here with several images shown in Fig. 3. The key has been taken as "tA#SANTR@!W^*RET98765".

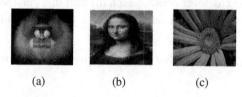

(a)   (b)   (c)

Fig. 3. Input images a) Babun b) Monalisa c) Flower

5.1   Exhaustive Key Search

In this paper total128 bits are needed to represent 16 characters and 17 bits are required to represent a 5 digit decimal number. Hence, an exhaustive key search is required 2^{145} times attempt to get exact key. If a desktop computer can try 17 billion key in an hour then it require 14,93,46,69,95,03,08,300 days.

5.2   Histogram Analysis

The histograms of encrypted images are furnished in Fig.4. The frequencies of pixels are almost uniform in encrypted images. It is difficult to recover original images from decrypted images.

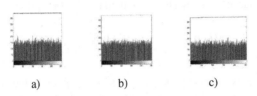

a)   b)   c)

Fig. 4. Histogram of encrypted a) Babun. b) Monalisa c) Flower

5.3   Correlation of Two Adjacent Pixels

In this paper, to test the correlation of pixels (vertical, horizontal, diagonal), randomly 3000 pairs of adjacent pixels have been selected both from plain image and encrypted image. Calculate the correlation coefficients of pixels according the following formula [9] as follows

$$E(x) = \frac{1}{N} \sum_{i=1}^{N} x_i \tag{1}$$

$$D(x) = \frac{1}{N} \sum_{i=1}^{N} (x_i - E(x))2 \tag{2}$$

$$COV(x, y) = \frac{1}{N} \sum_{i=1}^{N} (x_i - E(x))(y_i - E(y)) \tag{3}$$

$$r = \frac{COV(x, y)}{\sqrt{D(x)}\sqrt{D(y)}} \tag{4}$$

Where x and y are grey-scale values of two adjacent pixels in the image. The correlation coefficients of horizontal, vertical and diagonal pixels of input images and its corresponding cipher images are shown in Table 1. These correlation analysis is proved that the proposed encryption algorithm satisfy zero co-correlation.

Table 1. Correlation coefficient of two images

image name	plain image			ciphered image		
	horizontal	vertical	Diagonal	horizontal	Vertical	Diagonal
Babun	0.9834	0.9824	0.9742	0.0085	-0.0051	-0.0023
Monalisa	0.9792	0.9809	0.966	0.0035	0.0012	0.0029
Flower	0.9587	0.9544	0.9287	0.0016	0.0124	0.0015

5.4   Information Entropy Analysis

Information entropy is thought to be one of the most important features of randomness. Information entropy H(m) is calculated by the following formula[10]:

$$H(m) = \sum p(m_i) \log_2 \frac{1}{p(m_i)} \tag{5}$$

Where $p(m_i)$ represents the probability of the symbol (pixel value) m_i. Theoretically, a true random system should generate 2^8 symbols with equal probability, Table no 2 shows the entropy of the plain image and ciphered image.

Table 2. Entropy of different plain and encrypted image

Image name	Plain image	Ciphered image
Babun	7.4812	7.9933
Monalisa	7.4069	7.995
Flower	7.2549	7.9929

From the above table, it is cleared that the proposed scheme has hidden information randomly, and information leakage in encryption process is negligible.

5.5   Peak Signal-to-Noise Ratio (PSNR)

Peak signal-to-noise ratio can be used to evaluate the performance of an encryption scheme. PSNR reflects the encryption quality. It is a measurement which indicates the changes in pixel values between plain image and the cipher image [11]. PSNR value can be calculated using Eqns (6).

$$PSNR = 10 \times log_{10} \left[\frac{M \times N \times 255^2}{\sum_{i=0}^{M-1} \sum_{i=0}^{N-1} ((P(i,j) - C(i,j))^2} \right] \tag{6}$$

Where M is the width and N is the height of digital image. P(i; j) is pixel value of the plaintext image at location (i; j) and C(i; j) is pixel value of the cipher image of location (i,j). The lower value of PSNR represents better encryption quality. PSNR values of input and its corresponding cipher image have been furnished in table 3.

Table 3. PSNR of different plain and encrypted image

Image name	PSNR
Babun	7.8657
Monalisa	8.0425
Flower	7.9971

6   Comparative Study

Performance of our proposed algorithm has been compared between ref [12], Haar wavelet transform and Modified fast haar wavelet transform. The comparative study has been made between two images of size (128 × 128) in table 4.

Table 4. Comparisons with other algorithms

Image		Our algorithm	Ref.[12]	M.FHWT	HWT
	PSNR	7.9991	46.896	50.093	46.759
	Correlation coefficient	0.0129	-0.0164	-0.0082	.0063
	Entropy	7.9929	7.9896	7.9897	7.9888
	PSNR	7.9844	46.9425	51.237	69.7082
	Correlation coefficient	.0093	-0.00027	0.003	-0.00173
	Entropy	7.9948	7.9882	7.9872	7.9891

7   Conclusion

In this paper, image has been encrypted using genetic algorithm based on ring crossover. The ring crossover has been used in image diffusion and secure key. A logical XOR operation has been used between them to produce cipher image. The different experimental tests have been given satisfactory result. The algorithm is also given better result compare to other algorithms.

References

1. Al-Husainy, M.A.F.: Image encryption using genetic algorithm. Information Technology Journal 5(3), 516–519 (2006)
2. Kaya, Y., Uyar, M., Tekin, R.: A Novel Crossover Operator for Genetic Algorithms: Ring Crossover (2011)
3. Kumar, A., Rajpal, N.: Application of Genetic Algorithm in the Field of Steganography. Journal of Information Technology 2(1), 12–15 (2004)
4. Kumar, A., Rajpal, N., Tayal, A.: New Signal Security System for Multimedia Data Transmission Using Genetic Algorithms. In: NCC 2005, pp. 579–583 (January 2005)

5. Soni, A., Agrawal, S.: Using Genetic Algorithm for Symmetric keyGeneration in Image Encryption. International Journal of Advanced Research in Computer Engineering & Technology (IJARCET) 1(10), 137–140 (2012)
6. Rasul Enayatifar, A.H.A.: Image Security via Genetic Algorithm. In: International Conference on Computer and Software Modeling IPCSIT IACSIT, pp. 198–203. IACSIT Press, Singapore (2011)
7. Gupta, R.K.: Genetic Algorithms-An overview. IMPULSE 1, 30–39 (2006)
8. Enayatifar, R., Abdullah, A.H., FauziIsnin, I.: Chaos-based image encryption using a hybrid genetic algorithm and a DNA sequence. Optics and Lasers in Engineering 56, 83–93 (2014)
9. Wang, X., Luan, D.: A novel image encryption algorithm using chaos and reversible cellular automata. Commun. Nonlinear Sci. Numer. Simulat. 18, 3075–3085 (2013)
10. Cao, G., Hu, K., Zhang, Y., Zhou, J., Zhang, X.: Chaotic Image Encryption Based on Running-Key Related to Plaintext. Scientific World Journal 2014, Article ID 490179, 9 pages (2014)
11. El-lskandarani, M., Darwish, S., Abuguba, S.: A robust and secure scheme for image transmission over wireless channels in Security Technology. In: 42nd Annual IEEE International Carnahan Conference on ICCST 2008, pp. 51–55. IEEE (2008)
12. Sethi, N., Vijay, S.: Comparative Image Encryption Method Analysis Using New Transformed - Mapped Technique. In: Conference on Advances in Communication and Control Systems (CAC2S 2013), pp. 46–50 (2013)
13. Bhasin, H., Bhatia, S.: Use of Genetic Algorithms for Finding Roots of Algebraic Equations. IJCSIT 2(4), 1693–1696

Character Recognition Using Teaching Learning Based Optimization

Stuti Sahoo, S. Balarama Murty, and K. Murali Krishna

Anil Neerukonda Institute of Technology and Sciences, Visakhapatnam, India
stutisahoo@gmail.com, balaramamurty.ece@anits.edu.in,
mkasi71@yahoo.com

Abstract. Character recognition is an image analysis method, where handwritten images are given as input to a system and then the job of the system is to recognize them on the basis of information available about them. Pattern recognition capability of human beings cannot be imitated, however up to a certain extent it can be achieved by the use of neural network. In this paper an attempt is made to recognize English characters by the use of back propagation algorithm (BPA) as well as Teaching Learning Based Optimization (TLBO). TLBO is a recent algorithm used to solve many real world problems, which is inspired by practical environment of a class room, whereas Back Propagation algorithm is a generalization of Least Mean Square algorithm.

Keywords: Handwritten character recognition, Teaching-learning-based optimization algorithm, Back Propagation algorithm, Neural Networks.

1 Introduction

Though human beings posses a superior pattern recognition capability, but to make their jobs easier they search for an artificial machine. Character recognition system was originated from such a requirement. These systems introduced non keyboard computer system, automatic mailing classification system and automatic reading machine [1]. If one knows the basic style of a character he/she can understand as well as write any language easily by the use of character recognition system. On the other hand it can also be useful for visually impaired people as well as recognizing poor quality documents [3]. Character recognition systems may either be optical character recognition system (offline) or intelligent character recognition system (online). Offline system recognizes hand written, typed or printed text whereas, online system first observes the direction of the motion while writing a character and recognizes in real time [7]. Recognition of a character goes through a certain number of phases. In first phase it collects the unknown input in the form of hand written text, and then transforms the input to a suitable image, which is called as preprocessing method. After getting the image it uses feature extraction method to extract character feature from that image. Finally with the help of character recognition method that character information is kept in the form of bits and the best fitness is being displayed [8].

Various evolutionary algorithms have been used for character recognition. That may be Genetic algorithm (GA), Particle Swarm Optimization (PSO) or Differential

© Springer International Publishing Switzerland 2015
S.C. Satapathy et al. (eds.), *Proc. of the 3rd Int. Conf. on Front. of Intell. Comput. (FICTA) 2014*
– *Vol. 1*, Advances in Intelligent Systems and Computing 327, DOI: 10.1007/978-3-319-11933-5_83

Evolution (DE). GA is based on survival of the fittest in natural selection process [6], whereas PSO is inspired by the natural behavior of schools of fish or flocks of birds [1] and DE performs mutation based on the distribution of the solution in the current population. Before applying any algorithm, a pool of images has to be made and converted to graph. Now for getting more accuracy these graphs have to be intermixed to generate a number of styles. Various algorithms can be used for recognizing a character. An algorithm may consider either convergence speed or quality of the result as prime factor. Sometimes it may consider computational time as the first priority or it may focus on accuracy neglecting other parameters [2].

1.1   Back Propagation Algorithm

This algorithm consists of two phases- forward phase and backward phase. In forward phase the activations are propagated from input to output layer, whereas in backward phase the actual value and the predicted value are propagated backward in order to modify the weight as well as bias [11]. So, it is also called as feed forward network. The following figure shows the architecture of back propagation algorithm. Initially a number of images will be given input as training set and then testing set will be fed to check the accuracy. System needs to be given diverse training data so the difference between unknown data and training data will be less.

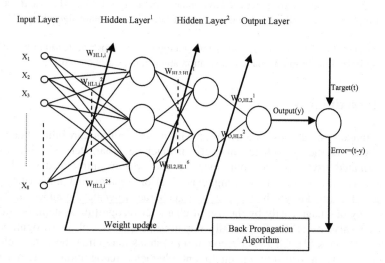

Fig. 1. Back Propagation Neural Network

1.1.1   Derivation
Since BPA uses the gradient descent method, one needs to calculate the derivative of the squared error function with respect to the weights of the network. The squared error function is:

$$E = \frac{1}{2}(t-y)^2 \tag{1}$$

Where

E= the squared error, t = target output and y= actual output of the output neuron
Activation function φ is needed to introduce nonlinearity into the network.

$$y = \varphi(net), \text{ where } net = \sum_{i=1}^{n} w_i x_i$$

w_i = the ith weight and x_i = the ith input value to the neuron

Calculating the partial derivative of the error with respect to a weight,

$$\frac{\partial E}{\partial w_i} = \frac{dE}{dy}\frac{dy}{dnet}\frac{\partial net}{\partial w_i} \tag{2}$$

The partial derivative of the sum with respect to a weight w_i is the just the corresponding input x_i and the partial derivative of the sum with respect to an input value x_i is just the weight w_i :

$$\frac{\partial net}{\partial w_i} = x_i \text{ and } \frac{\partial net}{\partial x_i} = w_i \Rightarrow \frac{dy}{dnet} = \frac{d}{dnet}\varphi$$

If activation function is logistic function: $y = \dfrac{1}{1+e^{-z}}$,

which has a derivative: $\dfrac{dy}{dz} = y(1-y)$

Finally, the derivative of the error E with respect to the output y is:

$$\frac{dE}{dy} = \frac{d}{dy}\frac{1}{2}(t-y)^2 \Rightarrow \frac{dE}{dy} = y - t$$

Putting it all together:

$$\frac{\partial E}{\partial w_i} = \frac{dE}{dy}\frac{dy}{dnet}\frac{\partial net}{\partial w_i} \tag{3}$$

So, $\dfrac{\partial E}{\partial w_i} = (y-t)(1-y)x_i$

To update the weight w_i using gradient descent, one must choose a learning rate α. The change in weight after learning would be the product of the learning rate and the gradient:

$$\Delta w_i = -\alpha\frac{\partial E}{\partial w_i} \tag{4}$$

1.2   Teaching Learning Based Optimization

A recent algorithm which is influenced by practical environment of a classroom is Teaching Learning Based Optimization (TLBO), which is proposed by R.V.Rao et al in 2011 [10]. It is a global optimization method, which is being used over a continuous space. When a teacher teaches in a classroom all the knowledge transformed by him/her is not acquired by the students. Normally the knowledge acquired by a student depends on both his/her teachers and colleagues [5]. In practice, a same teacher may not be able to deliver equal amount of knowledge to all students. So successful teaching can be distributed under Gaussian law [4]. Rarely a few students will be able to receive all the knowledge delivered by the teacher and there is a possibility that some of the students may not acquire anything. A group of learners, in search of optimal solution is considered as a population [5]. A classical school learning process can be divided into two phases

- Teacher phase
- Learner phase

In teacher phase learners learn through teacher, who is considered as the most experienced person in a society and he always wants to deliver maximum knowledge to as many students as possible. As the amount of knowledge acquired by a student reflects in results, teacher tries to increase the mean result of classroom [9].Whereas in learner phase in order to acquire maximum knowledge students interact among themselves. Although the knowledge acquired by both students is not same after a discussion, always knowledge flows from a student having more knowledge to the other one [12-13].

1.2.1 TLBO Algorithm

TLBO algorithm is a population based method, which tries to reach the optimal solution. Here different subjects offered to the learners are taken as design variables and learner's result is equivalent to fitness [9]. The flowchart for TLBO algorithm has been given in figure (2). By explaining the figure the TLBO method can be implemented.

Initialization. Population is the number of learners in class (P), Design variable is the number of subjects offered to them (D) and epoch is number of iteration
Population is initialized randomly with P rows and D columns.

$$Population = \begin{pmatrix} X_{11} & X_{12} & X_{13} & \cdots\cdots & X_{1D} \\ X_{21} & X_{22} & X_{23} & \cdots\cdots & X_{2D} \\ \vdots & \vdots & \vdots & \ddots & \vdots \\ X_{P1} & X_{P2} & X_{P3} & \cdots\cdots & X_{PD} \end{pmatrix}$$

Teacher Phase. Calculate the mean of population column wise, which gives the mean of the corresponding subject as M = [m_1, m_2, m_3,...........m_D]. The learner with minimum objective function value is selected as teacher ($X_{teacher}$) for the respective iteration. The teacher phase tries to shift the mean towards the best solution i.e. the teacher. The difference between new mean and old mean is expressed as:

$$Difference\_mean = r_i (M_{new} - T_F * M_{old}) \tag{5}$$

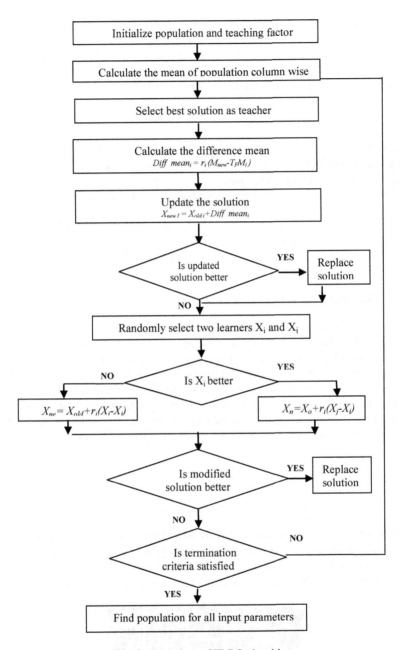

Fig. 2. Flow chart of TLBO algorithm

Here r_i represents uniformly distributed random value in the range of (0, 1) and T_F is the teaching factor, which decides the value of the mean to be changed.

$$T_F = round\ [1 + rand\ (0,\ 1)\ \{2\text{-}1\}] \qquad (6)$$

T_F is not a parameter of TLBO. This value is decided randomly by the help of equation (6). If the value of T_F is between 1 and 2 algorithm performs better [2]. If T_F is 1 there is no increase in knowledge level and if T_F is 2 complete transfer of knowledge occurs. To update the new set of modified learners Difference_mean has to be added with the old learner set at iteration i,

$$X_{new,i} = X_{old,i} + Difference\_mean_i \qquad (7)$$

Learner Phase. When two different learners interact between themselves knowledge flows from the better learner. So the knowledge acquired by a particular learner can be improved after a mutual interaction.

At a particular iteration consider any two learners X_i and X_j, where $i \neq j$.

$$X_{new,i} = X_{old,i} + r_i(X_i - X_j) \quad if \quad f(X_i) < f(X_j)$$
$$else$$
$$X_{old,i} + r_i(X_j - X_i) \quad if \quad f(X_j) < f(X_i) \qquad (8)$$

2   Database and Experiment

Database: In our experiment scanned input of handwritten English character is taken as database. The database consists of 5 characters (A, N, I, T and S) and in order to check how the algorithm works various types of characters from the above database were taken into consideration. Image of each character was taken in individual files. MATLAB software is being used to extract the required information from those files. A neural network having two hidden layers is being selected. The first hidden layer consists of three neurons and the second layer consists of two neurons. A sample set of English characters considered for experiment are shown as figure (3).

Fig. 3. Sample set of handwritten characters

Experiment: First image is being read by MATLAB, after reading it converts the image to gray scale image. And then obtained image is converted to binary image and edge detection is done. After detecting the edges output is shown in figure (4). After normalization process the size of each character is taken as a form of 8*8 matrixes.

Fig. 4. Some sample images after normalization

That matrix is taken as input for both BPA and TLBO separately, so simultaneously 8 inputs will be fed to the neural network. A total of 32 weights were initialized randomly. Like-from input to first hidden layer 24 weights, from first hidden layer to second hidden layer 6 weights and from second hidden layer to output neuron 2 weights. Once the output value is obtained error is calculated as the difference between target and output. That error value decided modification of weight values for further iterations. Mean square error (MSE) is calculated for comparing BPA and TLBO algorithm. MSE is calculated as the average of the squares of the 'errors', i.e. the difference between the estimator and what is estimated.

$$\frac{1}{epoch} \sum_{i=1}^{epoch} (estimater-estimated\_value)^2 \qquad (9)$$

3   Results and Discussions

Both BPA and TLBO algorithms were compared by calculating their respective MSE. It is observed that MSE keeps reducing as the number of iteration increases. Here 500 epochs have been taken for experiment. Figure (5) shows the plot between the mean square error and epochs. At an interval of every 50 epochs performance of the algorithms is plotted. It is being observed that at first iteration MSE for BPA is 1.8 and TLBO is 1respectively. When the weight values were modified for further iterations, it came to 0.45 for BPA and 0.23 for TLBO after 500 iterations. Since back propagation uses the gradient descent method, the derivative of the squared error function needs to be calculated with respect to the weights of the network. So, there is a possibility that the solution may trap in local minima, from where the derivation of the squared error will become zero. So TLBO can be considered as more reliable than BPA.

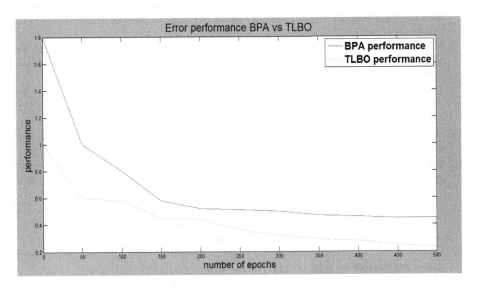

Fig. 5. Performance comparison between BPA and TLBO

4   Conclusion

In this paper an attempt is made for recognizing English handwritten characters using both BPA and TLBO algorithm. By implementing both algorithms for a single problem it is observed that TLBO shows better performance in terms of root mean square error. TLBO finds the global solution in relatively short computational time, whereas BPA may be trapped in local minima and the computational time is comparatively more. As TLBO has no parameters it can be used easily for complex real world problems. Also it uses very small population size and gives better performance in less convergence time.

References

1. Abed, M.A., Alasadi, H.A.A.: Simplifying Handwritten Characters Recognition Using a Particle Swarm Optimization Approach. European Academic Research I(5) (August 2013)
2. Satapathy, S.C., Naik, A., Parvathi, K.: Weighted Teaching-Learning-Based Optimization for Global Function Optimization. Applied Mathematics 4, 429–439 (2013)
3. Pal, U., Chaudhuri, B.B.: Indian script character recognition: a survey. Pattern Recognition 37, 1887–1899 (2004)
4. Wang, K.-L., Wang, H.-B., Yu, L.-X., Ma, X.-Y., Xue, Y.-S.: Toward Teaching-Learning-Based Optimization Algorithm for Dealing with Real-Parameter Optimization Problems. In: Proceedings of the 2nd International Conference on Computer Science and Electronics Engineering, ICCSEE 2013 (2013)
5. Rao, R.V., Kalyankar, V.D.: Parameters optimization of advanced machining processes using TLBO algorithm. In: EPPM, Singapore, September 20-21 (2011)
6. Kala, R., Vazirani, H., Shukla, A., Tiwari, R.: Offline Handwriting Recognition using Genetic Algorithm. IJCSI International Journal of Computer Science Issues 2(1) (March 2010)
7. Saraf, V., Rao, D.S.: Devnagari Script Character Recognition Using Genetic Algorithm for Get Better Efficiency. International Journal of Soft Computing and Engineering (IJSCE) 2(4) (August 2013) ISSN: 2231-2307
8. Pornpanomchai, C., Wongsawangtham, V., Jeungudomporn, S., Chatsumpun, N.: Thai Handwritten Character Recognition by Genetic Algorithm (THCRGA). IACSIT International Journal of Engineering and Technology 3(2) (April 2011)
9. Satapathy, S.C., Naik, A., Parvathi, K.: A teaching learning based optimization based on orthogonal design for solving global optimization problems. Springer Plus 2, 130 (2013)
10. Rao, R.V., Savsani, V.J., Vakharia, D.P.: Teaching-learning-based optimization: A novel method for constrained mechanical design optimization problems. Computer-Aided Design 43(3), 303–315 (2011)
11. Lagudu, S., Sarma, C.V.: Hand Writing Recognition Using Hybrid Particle Swarm Optimization & Back Propagation Algorithm. International Journal of Application or Innovation in Engineering & Management (IJAIEM)
12. Črepinšek, M., Liu, S.-H., Mernik, L.: A note on teaching–learning-based optimization algorithm. Information Sciences 212, 79–93 (2012)
13. Satapathy, S.C., Naik, A.: Modified Teaching–Learning-Based Optimization algorithm for global numerical optimization —A comparative study. Swarm and Evolutionary Computatio 16, 28–37 (2014)

A Novel FLANN with a Hybrid PSO and GA Based Gradient Descent Learning for Classification

Bighnaraj Naik, Janmenjoy Nayak, and H.S. Behera

Department of Computer Science & Engineering
Veer Surendra Sai University of Technology, Burla, Odisha, India
mailtobnaik@gmail.com

Abstract. In this paper, it is an attempt to design a PSO & GA based FLANN model (PSO-GA-FLANN) for classification with a hybrid Gradient Descent Learning (GDL). The PSO, GA and the gradient descent search are used iteratively to adjust the parameters of FLANN until the error is less than the required value. Accuracy and convergence of PSO-GA-FLANN is investigated and compared with FLANN, GA-based FLANN and PSO-based FLANN. These models have been implemented and results are statistically analyzed using ANOVA test in order to get significant result. To obtain generalized performance, the proposed method has been tested under 5-fold cross validation.

Keywords: Classification, Functional Link Artificial Neural Network (FLANN), Gradient Descent Learning, Particle Swarm Optimization, Genetic Algorithm.

1 Introduction

Zhang et al [1] realized that neural networks models are alternative to various conventional classification methods which are based on statistics. Classical neural networks are unable to automatically decide the optimal model of prediction or classification. In last few years, to overcome the limitations of conventional ANNs, some researchers have focused on higher order neural network (HONN) models [2][3]. In this paper, an attempt has been made to design and implementation of HONN with fast, stable and accurate gradient descent learning based on PSO and GA. Prior to this, a FLANN model for prediction of financial indices and an another Chebyshev neural network model with Chebyshev polynomials functional expansion have been presented by J. C. Patra et al. [4]. B.B. Misra et al. [5] has developed a classification method using FLANN with least complex architecture which is efficient in terms of ability of handling linearly non-separable classes by increasing dimension of input space. S. Dehuri et al. [6] have proposed a genetic algorithm (GA) based hybrid functional link artificial neural network for optimal input feature selection by using functionally expanded selected features which is suitable for non-linearity nature of classification problems. A PSO and back propagation learning based

© Springer International Publishing Switzerland 2015
S.C. Satapathy et al. (eds.), *Proc. of the 3rd Int. Conf. on Front. of Intell. Comput. (FICTA) 2014*
– *Vol. 1*, Advances in Intelligent Systems and Computing 327, DOI: 10.1007/978-3-319-11933-5_84

FLANN is proposed by S. Dehuri et al. [7] . An efficient FLANN for predicting stock price of US stocks has been suggested by J. C. Patra et al. [8] which is compared with (MLP)-based model through several experiments. Prediction of the causing genes in gene diseases by FLANN model is proposed by J. Sun et al. [9] and compared with Multi-layer Perceptron (MLP) and Support Vector Machines (SVM). To predict the stock market indices, a Functional Link Neural Fuzzy (FLNF) Model is developed by S. Chakravarty et al. [10] . Classification of online Indian customer behavior using FLANN is achieved by R. Majhi et.al. [11] and they claim that their method is better than the other statistical approaches by using discriminant analysis. Genetic algorithm (GA) based FLANN model is proposed by S. C. Nayak et al. [12] for forecasting of stock exchange rates. Various models of FLANNs like Power FLANN, Laguerre FLANN , Legendre FLANN and Chebyshev FLANN have been proposed by D.K. Bebarta et.al. [13] for forecasting stock price index and the results were compared by using various performance measure like standard deviation error , squared error etc. PSO based FLANN, GA based FLANN and Differential Evolution (DE) based FLANN for classification task in data mining are proposed by Faissal MILI et al. [14] and are compared and tested with various expansion functions. S. Mishra et al [15] have used FLANN classification model based on Bat inspired optimization and compared with FLANN and hybrid PSO-FLANN model. R. Mahapatra et al. [16] suggested Chebyshev FLANN classifier with various dimension reduction strategy is for cancer classification. MLP, FLANN and PSO-FLANN classification model are used and tested by S Mishra et al. [17] for classification of biomedical data. S Dehuri et.al. [18] have proposed an improved PSO based FLANN classifier and compared with previously available alternatives. In this paper, a FLANN model with hybrid Gradient descent learning based on PSO and GA for classification has been proposed and compared with previously available alternatives. The remaining part of the paper is organized as follows: Preliminaries in section 2, Proposed Method in section 3, Experimental Results and Analysis in section 4, Cross Validation in section 5, Statistical analysis in section 6, Conclusion in section 7.

2   Preliminaries

2.1   Functional Link Artificial Neural Network

Functional Link Neural Network (FLANN, Fig. 1) [19,20] is a class of Higher Order Neural Networks that utilize higher combination of its inputs and is much more modest than MLP since it has a single-layer network compared to the MLP but still is able to handle a non-linear separable classification task.

If x is a dataset with data in a matrix of order m x n then functionally expanded values can be generated by using euation-1.

$$\varphi(x_i(j)) = \{ x_i(j), \cos \Pi x_i(j) , \sin \Pi x_i(j),$$

$$\cos 2\Pi x_i(j) , \sin 2\Pi x_i(j) \dots \cos n\Pi x_i(j) , \sin n\Pi x_i(j) \} \qquad (1)$$

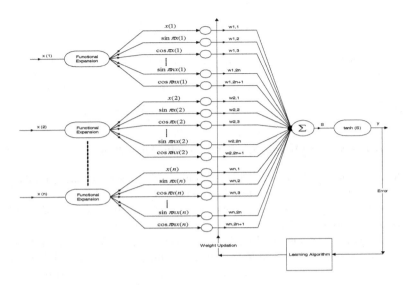

Fig. 1. Functional Link Artificial Neural Network

for i=1,2….m and j=1,2…n, Where m and n are number of input pattern and number of input values of each input pattern respectively except class level (Probably last column of dataset x).

$$\varphi = \{\{\varphi(x_1(1)), \varphi(x_1(2)) \dots \varphi(x_1(n))\}^T$$
$$, \{\varphi(x_2(1)), \varphi(x_2(2)) \dots \varphi(x_2(n))\}^T \dots$$
$$\dots \{\varphi(x_m(1)), \varphi(x_m(2)) \dots \varphi(x_m(n))\}^T\}$$

W_i is the weight vector Initialized randomly for a single input value of a input pattern as $W_i = \{w_{i,1}, w_{i,2}, \dots w_{i,2n+1}\}$, Where i=1,2….n. Hence for the set of input value of a single pattern weight vector is $W = \{W_1, W_2 \dots W_m\}^T$. Then, net output of FLANN network is obtained as follows.

$Y = \tanh(S) = \{\tanh(s_1), \tanh(s_2) \dots \tanh(s_m)\} = \{y_1, y_2 \dots y_m\}$,

Where S is calculated as $S = \varphi \times W = \{s_1, s_2 \dots s_m^T\}$.

2.2 Gradient Descent Learning

Error of kth input pattern is generated as $e(k) = Y(k) - t(k)$ which is used to compute error term in euation-2.

$$\delta(k) = \left(\frac{1 - y_k^2}{2}\right) \times e(k) \tag{2}$$

If $\varphi = (\varphi_1, \varphi_2 \dots \varphi_L)$, $e = (e_1, e_2 \dots e_L)$ and $\delta = (\delta_1, \delta_2 \dots \delta_L)$ are vector which represent set of functional expansion, set of error and set of error tern respectively then weight factor of w $'\Delta W'$ can be computed by using equation-3.

$$\Delta W_q = \left(\frac{\sum_{i=1}^{L} 2 \times \mu \times \varphi_i \times \delta_i}{L} \right) \tag{3}$$

Where $= 1,2 \dots L$, $q = 1,2 \dots L \times (2n + 1)$ and L is the number of input pattern. Weight updating is done as $w_{new} = w + \Delta W$ where $w = (w_1, w_2 \dots w_{L \times (2n+1)})$ and $\Delta W = (\Delta W_1, \Delta W_2 \dots \Delta W_{L \times (2n+1)})$.

2.3   Particle Swarm Optimization and Genetic Algorithm

Particle swarm optimization (PSO) [21] [22] is a widely used stochastic based search algorithm and it is able to search global optimized solution. Like other population based optimization methods the particle swarm optimization starts with randomly initialized population for individuals and it works on the social behavior of particle to find out the global best solution by adjusting each individual's positions with respect to global best position of particle of the entire population (Society). Each individual is adjusting by altering the velocity according to its own experience and by observing the experience of the other particles in search space by use of equation - 4 and equation - 5. Equation-4 is responsible for social and cognition behavior of particles respectively where c1 and c2 are the constants in between 0 to 2 and rand(1) is random function which generates random number in between 0 to 1.

$$V_i(t + 1) = V_i(t + 1) + c_1 * rand(1) * (lbest_i - X_i) + c_2 * rand(1) * (gbest_i - X_i) \tag{4}$$

$$X_i(t + 1) = X_i(t) + V_i(t + 1) \tag{5}$$

Genetic algorithm (GA) [23] is a computational model of machine learning inspired by evolution. The development of GA has now reached a stage of maturity, due to the effort made in the last decade by academics and engineers all over the world. They are less vulnerable to getting 'stuck' at local optima than gradient search methods. The pioneering work is contributed by J.H. Holland for various scientific and engineering applications. GA is inspired by the mechanism of natural selection, a biological process in which stronger individuals are likely be the winners in a competing environment. Fatnesses (goodness) of the chromosomes are used for solving the problem and in each cycle of genetic operation (known as evolving process) a successive generation is created from the chromosomes with respect to the current population. To facilitate the GA evolution cycle, an appropriate selection procedure and two major fundamental operators called crossover and mutation are required to create a population of solutions in order to find the optimal solution (chromosome).

3   Proposed Method

In this section, a FLANN model with a hybrid gradient descent learning scheme based on PSO and GA have been presented. Basic problem solving strategy of PSO and GA are used to design a fast and accurate learning method (Algorithm-1).

Initially, population P is initialized with pre-defined number of weight-set, where weight-set is a set of weights of an instance of FLANN at a particular time. In first iteration, initial population considered to be local best (lbest). **'Fitfromtrain'** algorithm gradient descent learning which is used to evaluate fitness of all weight-sets (individuals) based on root mean square error (equation-7) of FLANN. The functionally expended data ($\boldsymbol{\varphi}$) is the input to FLANN model and error are calculated for each pattern by comparing with target 't'. During calculation of fitness of weight-sets, network is trained with gradient descent learning which is demonstrated in Algorithm–2.

Table 1. Formula Used In This Section

Classification Accuracy	If cm is confusion matrix of order accuracy of classification is computed as $$Accuracy = \frac{\sum_{i=1}^{n}\sum_{\substack{j=1, \\ i==j}}^{m} cm_{i,j}}{\sum_{i=1}^{n}\sum_{j=1}^{m} cm_{i,j}} \times 100$$	(6)
Root Mean Square Error	Root Mean Square Error (RMSE) of predicted output values $\hat{y}_i$ of a target variable y is computed for n different predictions as follows: $$RMSE = \sqrt{\frac{\sum_{i=1}^{n}(y_i - \hat{y}_i)^2}{n}}$$	(7)
Min-Max Normalization	$$v' = \frac{v - min_A}{max_A - min_A}(new\_max_A - new\_min_A) + new\_min_A$$ Min-Max normalization maps values of dataset v to v' in the range $[new\_max_A \text{ to } new\_min_A]$ of an attribute A.	(8)

Algorithm – 1 PSO-GA-GDL-FLANN for Classification

INPUT: Dataset, Target vector 't' Functional Expansion 'φ', Gradient Descent Learning parameter 'μ', Population 'P'.
OUTPUT: Weight Sets with minimum RMSE.
Iter = 0;
1. **while** (1)
2.   Compute local best from population P based of fitness of individual weight-set of the population P.
3.   Iter = Iter + 1;
4.   **if** (iter == 1)
5.   lbest = P;
6.   **else**
7.   Calculate fitness of all weight-sets by using algorithm-2 in the population. Select local best weight-sets
8.   'lbest' is generated by comparing fitness of current and previous weight-sets
9.   **end**
10.   Evaluate fitness of each weight-set (individuals) in population P based on RMSE.
11.   Select global best weight set from population based on fitness of all individuals (weight-sets) by using fitness vector F.
12.   Create mating pool by replacing weak individuals (having minimum fitness value) with   global best weight set.
13.   Perform two point crossover in mating pool 'MatingPool' and replace all the weigh set of population P with weight sets of mating pool.
14.   Compute new velocity Vnew of all weight-set (individuals) by using equation – 4.
15.   Update positions of all weight-set (individuals) by using equation – 5.
16.   P = P + Vnew;
17.   **If** population is having 95 % similar weight sets (individuals) or if maximum number of iteration is   reached,
18.   **Then** stop the iteration.
19.   **End**
20. **end**

Algorithm – 2 Fitfromtrain Procedure

1. **function** F=fitfromtrain (φ, w, t, μ)
2.   S = φ X w
3.   Y = tanh(S);
4.   If $\varphi = (\varphi_1, \varphi_2 \dots \varphi_L)$, e = $(e_1, e_2 \dots e_L)$ and $\delta = (\delta_1, \delta_2 \dots \delta_L)$ are vector which represent set of functional expansion, set of error and set of error tern respectively then weight factor of w 'ΔW' is computed as follow $\Delta W_q = \left(\frac{\sum_{i=1}^{L} 2\times\mu\times\varphi_i\times\delta_i}{L}\right)$.
5.   compute error term $\delta(k) = \left(\frac{1-y_k^2}{2}\right) \times e(k)$, for k=1,2...L where L is the number of pattern.
6.   e = t - y;
7.   Compute root mean square error (RMSE) by using equition-7 from target value and output.
8.   F=1/RMSE, where F is fitness of the network instance of FLANN model.
9. **end**

4   Experimental Results and Analysis

In this section, the comparative study on the efficiency of our proposed method has been presented. Benchmark datasets (table-2) from UCI machine learning repository [24] have been used for classification and the result of proposed PSO-GA-FLANN model is compared with FLANN, GA-FLANN based on Genetic Algorithm and PSO-FLANN based on Particle Swarm Optimization. Datasets information is presented in table-2. Datasets have been normalized and scaled in the interval -1 to +1 using Min-Max normalization (Eq. 8, table-1) before training and testing is made. Classification accuracy (Eq. 6, table-1) of models has been calculated in terms of number of classified patterns are listed in table-3. Number of epochs required for convergence of RSME (Eq. 7, table -1) is noted for each datasets during training as well as testing phase has been presented from Fig.2 to Fig. 9.

4.1   Parameter Setting

During simulation, c1 and c2 constants of PSO has been set to 2, learning parameter of gradient descent method 'μ' is set to 0.13 and two point crossover is used throughout the experiment. We obtained the value of 'μ' by testing the models in the range 0 to 3. During functional expansion, each input of input pattern is expanded to 11 number of functionally expanded inputs by setting n=5. (As in FLANN, 2n+1 number of functionally expanded input corresponds to a single input of input pattern).

Table 2. Data Set Information

Dataset	Number of Pattern	Number of Features (excluding class label)	Number of classes	Number of Pattern in class-1	Number of Pattern in class-2	Number of Pattern in class-3
Monk 2	256	06	02	121	135	-
Hayesroth	160	04	03	65	64	31
Heart	256	13	02	142	114	-
New Thyroid	215	05	03	150	35	30
Iris	150	04	03	50	50	50
Pima	768	08	02	500	268	-
Wine	178	13	03	71	59	48
Bupa	345	06	02	145	200	-

Table 3. Performance Comparison in Terms Of Accuracy

Dataset	Accuracy of Classification in Average							
	FLANN		GA-FLANN		PSO-FLANN		PSO-GA-FLANN	
	Train	Test	Train	Test	Train	Test	Train	Test
Monk 2	93.82813	92.04303	96.54689	93.19913	97.453134	95.46585	97.867196	96.3112
Hayesroth	90.35938	82.3125	91.062504	83.5625	91.265628	83.9375	91.46875	84.875
Heart	88.962966	78.48149	89.407408	79.07408	89.777762	79.85185	88.91667	78.2963
New Thyroid	93.918596	76.55813	94.197648	77.53487	94.3023	78.79069	94.36044	79.20929
Iris	96.84712	97.36815	97.12973	98.16639	97.352249	98.65	97.847622	99.428836
Pima	78.416	78.7608	78.64	78.80	80.126048	79.47073	80.64	80.58045
Wine	92.76	93.18595	94.36842	95.53644	97.762	95.6274	98.58207	97.370274
Bupa	72.16	72.76	74.3208	75.5	76.3842	76.75	77.1883	78.95

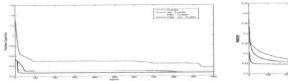

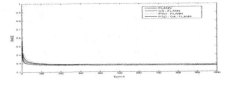

Fig. 2. Performance of models on a sample of monk2.dat dataset

Fig. 6. Performance of models on a sample of iris.dat dataset

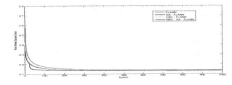

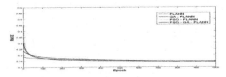

Fig. 3. Performance of models on a sample of hayesroth.dat dataset

Fig. 7. Performance of models on a sample of pima.dat dataset

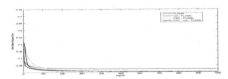

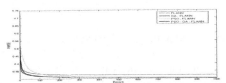

Fig. 4. Performance of models on a sample of heart.dat dataset

Fig . 8. Performance of models on a sample of wine.dat dataset

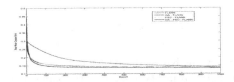

Fig. 5. Performance of models on a sample of newthyroid.dat dataset

Fig. 9. Performance of models on a sample of bupa.dat dataset

5   Cross Validation

Cross-Validation [25] is a statistical method to estimate performance of the learned model from data which compare learning algorithms by dividing data into two segments: training data & testing data used to train and evaluate the model respectively. In k-fold cross-validation [26] the data is partitioned into k equally or nearly equally sized segments on which training and validation are performed such that, in each iteration different fold of the data is used for training and validation. The proposed PSO-GA-FLANN model has been implemented using MATLAB 9.0 and the data (Table-4) are being prepared using 5-fold validation processed by KEEL Data-Mining Software Tool [27]. Like monk-2, all the dataset has been prepared for test in same way. Accuracy of training and testing of models on monk-2 dataset under 5-folds cross validation is shown in table-5.

Table 4. Datasets In 5-Fold For Cross Validation

Dataset	Data files	Number of Pattern	Task	No. of Pattern inclass-1	No. of Pattern in class-2
	monk-2-5-1tra.dat	256	Training	105	151
	monk-2-5-1tst.dat	87	Testing	41	46
	monk-2-5-2tra.dat	256	Training	108	148
	monk-2-5-2tst.dat	87	Testing	41	46
	monk-2-5-3tra.dat	256	Training	111	145
Monk 2	monk-2-5-3tst.dat	86	Testing	41	45
	monk-2-5-4tra.dat	256	Training	110	146
	monk-2-5-4tst.dat	86	Testing	41	45
	monk-2-5-5tra.dat	256	Training	108	148
	monk-2-5-5tst.dat	86	Testing	40	46

Table 5. Accuracy Of Classification In Cross Validation

Datasets		Task	Accuracy of Classification in %			
Data	5 - Folds		FLANN	GA-FLANN	PSO-FLANN	PSO-GA-FLANN
	monk-2-5-1tra.dat	Training	92.03125	94.14065	96.36718	97.81249
	monk-2-5-1tst.dat	Testing	91.95403	93.10344	96.23697	97.7011
	monk-2-5-2tra.dat	Training	90.62501	94.60938	96.13281	98.24218
	monk-2-5-2tst.dat	Testing	89.54023	91.26433	93.56322	95.977
Monk-2	monk-2-5-3tra.dat	Training	93.39845	96.4844	97.65626	96.32814
	monk-2-5-3tst.dat	Testing	89.18602	90.93026	94.186	95.34885
	monk-2-5-4tra.dat	Training	95.23436	98.8281	97.89062	98.12502
	monk-2-5-4tst.dat	Testing	95.34881	94.186	95.81396	95.58136
	monk-2-5-5tra.dat	Training	97.8516	98.67192	99.2188	98.82815
	monk-2-5-5tst.dat	Testing	94.18605	96.51163	97.5291	96.9477

6   Statistical Analysis

All the four methods have been executed for 5 number of run on each datasets and during each run, RMSE are being generated for up to 1000 epochs. The simulated results of proposed method are analyzed under ANOVA by using SPSS-16.0 statistical tool to prove the result statistically significant. The test has been carried out using one way ANOVA in Duncan multiple test range with 95 % confidence interval, 0.05 significant level and linear polynomial contrast. The snap shot of the result is listed below (Fig. 10). In descriptive table of Fig. 10, row-1, row-2, row-3 and row-4 corresponds to performance of FLANN, GA-FLANN, PSO-FLANN and Hybrid PSO-GA FLANN respectively. The result of one way ANOVA test shows that, the proposed method is statistically significant and having minimum standard error.

Descriptives

Sample	N	Mean	Std. Deviation	Std. Error	95% Confidence Interval for Mean		Minimum	Maximum
					Lower Bound	Upper Bound		
1	400	87.0580	8.96739	.44837	86.1766	87.9395	65.12	100.00
2	400	88.0731	8.52111	.42606	87.2355	88.9107	65.12	100.00
3	400	88.8556	8.31227	.41561	88.0385	89.6727	67.44	100.00
4	400	89.3220	8.04954	.40248	88.5308	90.1133	67.44	100.00
Total	1600	88.3272	8.50471	.21262	87.9102	88.7442	65.12	100.00

ANOVA

Sample			Sum of Squares	df	Mean Square	F	Sig.
Between Groups	(Combined)		1177.891	3	392.564	5.473	.001
	Linear Term	Contrast	1147.449	1	1147.449	15.997	.000
		Deviation	30.242	2	15.121	.211	.810
Within Groups			114478.066	1596	71.728		
Total			115655.758	1599			

Fig. 10. One Way ANOVA Test

7   Conclusion

In this paper, a FLANN classification model together with a hybrid PSO-GA based Gradient Decent learning scheme has been proposed, which facilitates detecting class level of patterns over real data efficiently. The proposed PSO-GA-FLANN model can be computed with a low cost due to less complex architecture and still enables us to prune out accurate class label. The experimental analysis shows that the proposed method performs relatively better than other models and it confer promising results on various training and testing datasets under 5-fold cross validation. Analysis of the result of simulation under ANOVA provides evidence of the statistical correctness of proposed method. The future work may comprise with the integration of other competitive optimization techniques with other means of higher order neural network in diversified applications of data mining.

Acknowledgments. This work is supported by Department of Science & Technology (DST), Ministry of Science & Technology, New Delhi, Govt. of India, under grants No. DST/INSPIRE Fellowship/2013/585.

References

1. Zhang, G.P.: Neural networks for classification: a survey. IEEE Transactions on Systems Man and Cybernetics. Part C: Applications and Reviews 30(4), 451–462 (2000)
2. Redding, N., et al.: Constructive high-order network algorithm that is polynomial time. Neural Networks 6, 997–1010 (1993)
3. Goel, A., et al.: Modified Functional Link Artificial Neural Network. International Journal of Electrical and Computer Engineering 1(1), 22–30 (2006)
4. Patra, J.C., et al.: Financial Prediction of Major Indices using Computational Efficient Artificial Neural Networks. In: International Joint Conference on Neural Networks, Canada, July 16-21, pp. 2114–2120. IEEE (2006)
5. Mishra, B.B., Dehuri, S.: Functional Link Artificial Neural Network for Classification Task in Data Mining. Journal of Computer Science 3(12), 948–955 (2007)
6. Dehuri, S., Mishra, B.B., Cho, S.-B.: Genetic Feature Selection for Optimal Functional Link Artificial Neural Network in Classification. In: Fyfe, C., Kim, D., Lee, S.-Y., Yin, H. (eds.) IDEAL 2008. LNCS, vol. 5326, pp. 156–163. Springer, Heidelberg (2008)

7. Dehuri, S., Cho, S.: A comprehensive survey on functional link neural networks and an adaptive PSO–BP learning for CFLNN. Springer, London (2009), doi:10.1007/s00521-009-0288-5

8. Patra, J.C., et al.: Computationally Efficient FLANN-based Intelligent Stock Price Prediction System. In: Proceedings of International Joint Conference on Neural Networks, Atlanta, Georgia, USA, June 14-19, pp. 2431–2438. IEEE (2009)

9. Sun, J., et al.: Functional Link Artificial Neural Network-based Disease Gene Prediction. In: Proceedings of International Joint Conference on Neural Networks, Atlanta, Georgia, USA, June 14-19, pp. 3003–3010. IEEE (2009)

10. Chakravarty, S., Dash, P.K.: Forecasting Stock Market Indices Using Hybrid Network. In: World Congress on Nature & Biologically Inspired Computing, pp. 1225–1230. IEEE (2009)

11. Majhi, R., et al.: Classification of Consumer Behavior using Functional Link Artificial Neural Network. In: IEEE International Conference on Advances in Computer Engineering, pp. 323–325 (2010)

12. Nayak, S.C., et al.: Index Prediction with Neuro-Genetic Hybrid Network: A Comparative Analysis of Performance. In: IEEE International Conference on Computing, Communication and Applications (ICCCA), pp. 1–6 (2012)

13. Bebarta, D.K., et al.: Forecasting and Classification of Indian Stocks Using Different Polynomial Functional Link Artificial Neural Networks (2012) 978-1-4673-2272-0/12/Crown

14. Mili, F., Hamdi, H.: A comparative study of expansion functions for evolutionary hybrid functional link artificial neural networks for data mining and classification, pp. 1–8. IEEE (2013)

15. Mishra, S., et al.: A New Meta-heuristic Bat Inspired Classification Approach for Microarray Data. C3IT, Procedia Technology 4, 802–806 (2012)

16. Mahapatra, R.: Reduced feature based efficient cancer classification using single layer neural network. In: 2nd International Conference on Communication, Computing & Security, Procedia Technology, vol. 6, pp. 180–187 (2012)

17. Mishra, S.: An enhanced classifier fusion model for classifying biomedical data. Int. J. Computational Vision and Robotics 3(1/2), 129–137 (2012)

18. Dehuri, S.: An improved swarm optimized functional link artificial neural network (ISO-FLANN) for classification. The Journal of Systems and Software, 1333–1345 (2012)

19. Pao, Y.H.: Adaptive pattern recognition and neural networks. Addison-Wesley Pub. (1989)

20. Pao, Y.H., Takefuji, Y.: Functional-link net computing: theory, system architecture, and functionalities. Computer 25, 76–79 (1992)

21. Kennedy, J., Eberhart, R.: Swarm Intelligence Morgan Kaufmann, 3rd edn. Academic Press, New Delhi (2001)

22. Kennedy, J., Eberhart, R.C.: Particle swarm optimization. In: IEEE International Conference on Neural Networks, Perth, Australia, pp. 1942–1948 (1995)

23. Holland, J.H.: Adaption in Natural and Artificial Systems. MIT Press, Cambridge (1975)

24. Bache, K., Lichman, M.: UCI Machine Learning Repository. University of California, School of Information and Computer Science, Irvine,
http://archive.ics.uci.edu/ml

25. Larson, S.: The shrinkage of the coefficient of multiple correlation. J. Educat. Psychol. 22, 45–55 (1931)

26. Mosteller, F., Turkey, J.W.: Data analysis, including statistics. In: Handbook of Social Psychology. Addison-Wesley, Reading (1968)

27. Alcalá-Fdez, J., et al.: KEEL Data-Mining Software Tool: Data Set Repository, Integration of Algorithms and Experimental Analysis Framework. Journal of Multiple-Valued Logic and Soft Computing 17(2-3), 255–287 (2011)

A New Ecologically Inspired Algorithm for Mobile Robot Navigation

Prases K. Mohanty, Sandeep Kumar, and Dayal R. Parhi

Robotics Laboratory,
National Institute of Technology, Rourkela
Odisha, India
{pkmohanty30,sandeepkumarintegrates}@gmail.com,
dayalparhi@yahoo.com

Abstract. The present paper describes a new optimal route planning for mobile robot navigation based on the invasive weed optimization (IWO) algorithm. This nature inspired meta-heuristic algorithm is based on the colonizing property of weeds. A new objective function has been framed between the robot to position of the goal and obstacles, which satisfied both obstacle avoidance and target seeking behavior of robot present in the environment. Depending upon the objective function value of each weed in the colony the robot that avoids obstacles and moves towards the goal. The mobile robot shows robust performance in various complex environments and local minima situation. Finally, the effectiveness of the developed path planning algorithm has been analyzed in various scenarios populated variety of static obstacles.

Keywords: Invasive weed optimization, Mobile robot, motion planning, Obstacle avoidance.

1 Introduction

Recent times mobile robots having lots of applications in a wide variety of areas such as automatic driving, exploration, surveillance, guidance for the blind and disabled, exploration of dangerous or hostile terrains, transportation, and collecting topographical data in unknown environments, etc. The most important issue of an autonomous mobile robot is to control independently in an unknown or partially known environment. So the primary aim of the robot is to develop a path planning system, which consists of the planning and implementation of collision free motion within the unknown terrain. The sensor based motion planning approaches uses either global or local path planning depending upon the surrounding environment. Global path planning or deliberative approaches requires the environment to be completely known and the terrain should be static; on other side local path planning means the environment is completely or partially unknown for the mobile robot. Many exertions have been paid in the past to improve various robot navigation techniques.

Several different navigation techniques and approaches for intelligent robotics have been discussed by many researchers. Since motion planning problems for mobile

© Springer International Publishing Switzerland 2015
S.C. Satapathy et al. (eds.), *Proc. of the 3rd Int. Conf. on Front. of Intell. Comput. (FICTA) 2014*
– *Vol. 1*, Advances in Intelligent Systems and Computing 327, DOI: 10.1007/978-3-319-11933-5_85

robots have high computational intricacies due to a set of several constraints, various heuristic based approaches can be proposed as the substitute techniques. Nature inspired meta-heuristics algorithm in general provides efficient or feasible solutions to optimization problems [1-2]. Many researchers consider a system with complete information about the environment [3-4]. Due to the complexity and uncertainty of the path planning problem, classical path planning techniques, such as visibility Graph [5], voronoi diagrams [6], grids [7], cell decomposition [8] and artificial potential method [9] are not suitable for online implementations. The use of the above algorithms for path finding for mobile robot requires more time and the finding of this path will not completely feasible for real-time movement. In recent times most popular bio-inspired algorithms such as genetic algorithm (GA) [10-11], ant colony optimization (ACO) [12-13], cuckoo search [14] and particle swarm optimization (PSO) [15-16] have been implemented for path planning of mobile robot.

This paper introduces a new variant of an Invasive weed optimization algorithm for solving the navigation problem of mobile robot. Invasive weed optimization is a metaheuristic search algorithm developed by Mehrabian and Lucas in 2006 [17], which was inspired by the colonizing property of weeds. It is potentially more generic to implement to a wider class of optimization problem and also it has proved that the Invasive weed optimization algorithm satisfy the global convergence necessities and thus has guaranteed global convergence properties. Finally, a simulation study over a set of challenging terrains is presented to verify the effectiveness of the proposed algorithm.

2   An Overview of Invasive Weed Optimization Algorithm

The Invasive Weed Optimization algorithm was firstly introduced by Mehrabian and Lucas in 2006 [17]. This meta-heuristic optimization (IWO) is a population based stochastic algorithm that mimics the colonizing behavior of weeds. The basic behavior of a weed is that it grows its population entirely or predominantly in any specified geographical area which can be significantly large or small. The IWO has some special characteristics in comparison with other numerical search algorithms like reproduction, spatial, dispersal, and competitive exclusion.

The process of this developed algorithm is summarized in the following steps [17].

I. Initializing a Population

A finite number of weeds are being distributed randomly in the feasible search space. Each weed in the feasible search space representing a trial solution for the optimization problem.

II. Reproduction

Each weed in the population is allowed to produce seeds depending on its own as well as colony's lowest and highest fitness, such that the number of seeds produced by each weed increases linearly from minimum possible seed for a weed with minimum fitness to the maximum number of seeds for a weed with maximum fitness.

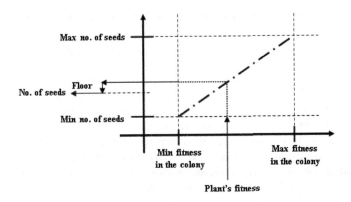

Fig. 1. Seed production procedure in a colony of weeds

III. Spatial Dispersal

The produced seeds in the previous step are being randomly distributed over d-dimensional problem space by a normal distribution with mean equal to zero but varying variance. This stage ensures that the generated seeds will be abode near to the parent seed, leading to a local search around each weed. However, the standard deviation (SD), 'σ' of the random function will be decreased from a previously defined initial value, $\sigma_{initial}$ to a final value, σ_{final}, in every step (generation). In the simulation, a non-linear alteration has been shown satisfactory performance, which is given in the following equation (1)

$$\sigma_{iter} = \left(\frac{iter_{max} - iter}{iter_{max}} \right)^{n} \left(\sigma_{initial} - \sigma_{final} \right) + \sigma_{final} \tag{1}$$

where $iter_{max}$ is the maximum number of iterations, σ_{iter} is the SD at the present time for a particular iteration and 'n' is the non-linear modulation index.

This stage implies that the probability of dropping a seed in a distant area decreases non-linearly at each time step which results in grouping fitter plants and the elimination of inappropriate plants.

IV. Competitive Exclusion

If a plant leaves no offspring, then it would go extinct, otherwise they would take over the world. Thus, there is a need of some kind of competition between plants for limiting the maximum number of plants in a colony. After completing some iterations, the number of plants in a colony will reach its maximum by fast reproduction, however, it is expected that the fitter plants have reproduced more than undesirable plants. After reaching the maximum number of plants in a colony, p_{max}, a mechanism for eliminating the plants with poor fitness in the generation activates. From thereon, only the fitter plants, among the existing ones and the reproduced ones, are taken in the colony and steps 1-4 are repeated until the maximum number of iterations has been reached, i.e. the colony population size is fixed thereon to p_{max}.

3   Architecture of Robot Path Planning with IWO Algorithm

In this section, a new meta-heuristic optimization algorithm inspired from the ecological process of weeds colonization and distribution has been implemented in the local path planning of a mobile robot in a static environment. Path planning layout for mobile robot is one of the most fundamental issues in the robotic research field. The main objective of this research paper is determining an optimal safe layout for a mobile robot in an unknown or partially known environment containing a variety of static obstacles. If a robot moves in an unstructured environment with unknown obstacles, it is very essential to detect and avoid obstacles as the robot moves towards the target. Based on the sensory information about location of the target and obstacles, we have implemented the IWO algorithm to solve the above stated problem. Firstly, we have transformed the navigation problem into a minimization one and formulated a new objective function based on the location of the target and obstacles present in the environment. Then we have implemented the IWO algorithm to solve the above optimization problem. During this process of visualization, the locations of the fittest or the best weeds in the colony in each iterative are selected and the robot reaches these locations in sequence. When the robot does not detect any obstacles on its target path, it will move directly towards its destination. Then it is not necessary to implement any intelligence computing technique to steer the robot within its environment.

Formulation of the Objective Function Using the IWO Algorithm

1.   Obstacle Avoidance Behavior

The obstacle avoidance behavior is used to avoid hitting with obstacles (such as walls) present in the environment. The position of the fittest weed in the colony (according to objective function values) must have kept up maximum safe distance from the nearest obstacle. The Euclidean distance between the weed and nearest obstacle is determined by the following expression, in terms of an the objective function as follows[18]:

$$ob_{(OBS-W)} = \sqrt{\left(x_{OBS} - x_{weed_n}\right)^2 + \left(y_{OBS} - y_{weed_n}\right)^2} \tag{2}$$

Note: The nearest obstacle to the robot can be calculated by the following expression:

$$d_{ROB-OBS} = \sqrt{\left(x_{OBS_n} - x_r\right)^2 + \left(y_{OBS_n} - y_r\right)^2}$$

where, x_{weed} and y_{weed} are the weed's position at x and y coordinates. $ob_{(OBS-W)}$ is the maximum safe distance from the weed position to obstacles. x_r and y_r are the robot's position at x and y coordinates. x_{OBS} and y_{OBS} are the centre coordinates of the obstacles.

2. Target Seeking Behavior

In its simplest form, the move to the target behavior can express the desire to move to a specific robot's position. The position of the fittest weed in the colony (according to objective function values) must have kept up the minimum distance from the goal. The Euclidean distance between the weed and goal is calculated by the following expression in terms of an the objective function as follows:

$$ob_{(TAR-W)} = \sqrt{\left(x_{t\,\arg\,et} - x_{weed_n}\right)^2 + \left(y_{t\,\arg\,et} - y_{weed_n}\right)^2} \tag{3}$$

where, x_{target} and y_{target} are the target position at x and y coordinates. $ob_{(TAR-W)}$ is the minimum distance from the weed position to the robot.

Based on the above two criteria the objective function of each weed of the whole path planning, optimization problem can be expressed as follows

$$objective\ function\left(f_i\right) = \alpha.\frac{1}{\min\limits_{OBS_n \in OBS_d} \left\|ob_{(OBS_d - W)}\right\|} + \beta.\left\|ob_{(TAR-W)}\right\| \tag{4}$$

Here we assumed that the 'n' number of obstacles are present in the environment and represent them as OBS_1, OBS_2, OBS_3.....OBS_n, their coordinates are $(x_{OBS1},\ y_{OBS1})$, $(x_{OBS2},\ y_{OBS2})$, $(x_{OBS3},\ y_{OBS3})$...... $(x_{OBSn},\ y_{OBSn})$. Due to the certain threshold of the robot, sensor, in each move it can recognize the number of obstacles present in the environment, and the number of obstacles being recognized by the robot sensor in some step is denoted as $OBS_d \in [OBS_1,\ OBS_2,\ OBS_3.....OBS_n]$. It can be observed from the objective function that when the position of the weed is nearer to the goal, the cost of $\left\|ob_{(TAR-W)}\right\|$ will be minimum and when the position of the weed is away from the obstacles, the cost of $\min\limits_{OBS_n \in OBS_d} \left\|ob_{(OBS-W)}\right\|$ will be maximum. So from the above discussions, we have concluded that the path planning problem for mobile robot solved by the IWO algorithm is a minimization one.

The objective of IWO algorithm, is to minimize the objective function (f_i). When a robot falls in an obstacle sensing range, obstacle avoidance behavior based on IWO algorithm is activated to find the best position for the weed. From equation no. (4) it can be clearly seen that the weed having minimum objective function value can be treated as best weed in the colony and the corresponding weed is maintaining the maximum safe distance from the nearest obstacle and minimum distance from the target. Then the robot will move towards the best weed position and this process will continue until the robot reaches its target.

From the objective function equation, it can be clearly understood that the weight parameters or controlling parameters α and β are specifying the relative importance of achieving obstacle avoidance behavior and reaching the target. Proper selection of controlling parameters may result in faster convergence of the objective function and elevation of local minimum point. In this proposed algorithm, we have chosen the controlling parameters by trial and error methods.

Steps of IWO Algorithm for Mobile Robot Navigation

Step-1: Initialize the start and goal position of the robot.

Step-2: Navigate the robot towards the goal until it is stuck by an obstacle.

Step-3: When the target path is obstructed by an obstacles, implement the IWO algorithm.

Step-4: Generate the initial population of weeds, each representing one trial solution of the proposed optimization problem.

Step-5: Calculate the objective function value of each weed and then find the global or best weed in the colony by equation no.4.

Step-6: When the population size is more than population maximum, eliminate weeds with lower objective function values.

Step-7: Navigate the robot towards the best weed position.

Step-8: Repeat the steps (2-7) until the robot avoids obstacles and reaches its goal.

4   Simulation Results and Discussion

In this section the simulation experiments are performed in two dimensional path planning under partially or totally unknown environments. The simulation space is 40cm by 40cm rectangular environment and safe range from robot to obstacle is 3cm. The simulations are conducted using MATLAB R2008 processing under Windows XP. The best parametric values of proposed algorithm have been selected after performing a series of simulation experiments (shown in Fig. 2) on partially or totally unknown environments.

Details of parameter values selection in IWO algorithm for Mobile Robot Navigation are given in Table-1.

Table 1. Parameters used in IWO algorithm

Symbol	Quantity	Value
N_o	Number of initial populations	20
$iter_{max}$	Maximum number of iterations	300
P_{max}	Maximum number of plant populations	30
S_{max}	Maximum number of seeds	10
S_{min}	Minimum number of seeds	0
n	Non-linear modular index	3
$\sigma_{Initial}$	Initial value of standard deviation	4
σ_{Final}	Final value of standard deviation	0.01
α	Controlling parameter-1	1
β	Controlling parameter-2	1×10^{-6}

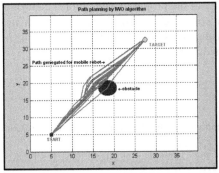

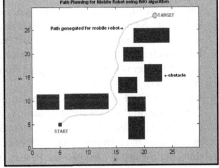

Fig. 2. Paths generated for a single robot by varying the parameters of IWO algorithm

Fig. 3. Single robot avoiding a Wall using IWO algorithm

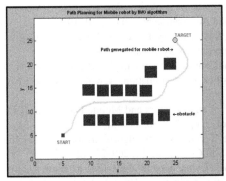

Fig. 4. Single robot escaping from a corridor using IWO algorithm

Fig. 5. Single robot navigating in a maze scenario using IWO algorithm

Fig. 2 shows the paths generated for a single robot by varying parameters in proposed algorithm. Fig (3-5) illustrates the simulation results for the different environmental scenarios. From the simulation experiment, it can be clearly seen that using the proposed path planning algorithm the robot has reached the target in an efficient manner without colliding with obstacles present in the environment.

5   Conclusions and Future Work

In this article a new meta-heuristic algorithm based on the colonizing property of weeds has been introduced for mobile robot navigation in a partially or completely unknown environment populated by variety of static obstacles. This nature inspired navigational algorithm is implemented to achieve a versatile and robust behavior based local motion planning. It has been observed that the proposed algorithm is capable of avoiding collisions among the obstacles and effectively controlling the mobile robot moving from the start point to the desired goal point with optimum/shortest path length. The authenticity of the proposed navigational controller has been verified and proven by simulation results using MATLAB. In future, real time implementation is to be carried out using robot and multiple robots are to be taken into consideration instead of a single mobile robot.

References

1. Back, T., Fogel, D.B., Michalewicz, Z.: Handbook of Evolutionary computation. Oxford University Press, Oxford (1997)
2. Raidl, G.: Hybrid Evolutionary Algorithms for Combinatorial Algorithms. Habilitation Thesis, Vienna University of Technology (2002)
3. Latombe, J.C.: Robot Motion Planning. Kluwer Academic Publishers, New York (1990)
4. Canny, J.E.: The Complexity of Robot Motion Planning. MIT Press, Cambridge (1988)
5. Mitchell, J.S.B.: An algorithm approach to some problems in terrain navigation. Artificial Intelligence 37(1-3), 171–201 (1988)
6. Takahashi, O., Schilling, R.J.: Motion planning in a plane using generalized Voronoi diagrams. IEEE Transactions on Robotics and Automation 5(2), 143–150 (1989)
7. Weigl, M., Siemiaatkkowska, B., Sikorski, K.A., Borkowski, A.: Grid- based mapping for autonomous mobile robot. Robotics and Autonomous Systems 11(1), 13–21 (1993)
8. Lingelbach, F.: Path planning using probabilistic cell decomposition. In: Proceedings of the IEEE International Conference on Robotics and Automation, New Orleans, La, USA, pp. 467–472 (2004)
9. Khatib, O.: Real time Obstacle Avoidance for manipulators and Mobile Robots. In: IEEE Conference on Robotics and Automation, vol. 2, pp. 505–505 (1985)
10. Castillo, O., Trujillo, L., Melin, P.: Multiple objective genetic algorithms for path-planning optimization in autonomous mobile robots. Soft Computing 11, 269–279 (2007)
11. Ram, A., Arkin, R., Boone, G., Pearce, M.: Using genetic algorithms to learn reactive control parameters for autonomous robotic navigation. Journal Adaptive Behavior archive 2(3), 277–305 (1994)
12. Chen, X., Kong, Y., Fang, X., Wu, Q.: A fast two-stage ACO algorithm for robotic path planning. Neural Computing and Applications 22(2), 313–319 (2013)
13. Cen, Y., Song, C., Xie, N., Wang, L.: Path Planning Method for Mobile Robot Based on Ant Colony Optimization Algorithm. In: 3rd IEEE Conference on Industrial Electronics and Applications (ICIEA), Singapore, June 3-5, pp. 289–301 (2008)
14. Mohanty, P.K., Parhi, D.R.: Cuckoo Search Algorithm for the Mobile Robot Navigation. In: Panigrahi, B.K., Suganthan, P.N., Das, S., Dash, S.S. (eds.) SEMCCO 2013, Part I. LNCS, vol. 8297, pp. 527–536. Springer, Heidelberg (2013)
15. Zhang, Y., Gong, D.W., Zhang, J.H.: Robot path planning in uncertain environment using multi-objective particle swarm optimization. Neurocomputing 103, 172–185 (2013)
16. Tang, Q., Eberhard, P.: A PSO-based algorithm designed for a swarm of mobile robots. Structural and Multidisciplinary Optimization 44, 483–498 (2011)
17. Mehrabian, A.R., Lucas, C.: A novel numerical optimization algorithm inspired from weed colonization. Ecological Informatics 1, 355–366 (2006)
18. Mohanty, P.K., Parhi, D.R.K.: A new efficient optimal path planner for mobile robot based on invasive weed optimization algorithm. Frontiers of Mechanical Engineering (article in press, 2014)

Design of Classifier for Detection of Diabetes Mellitus Using Genetic Programming

Madhavi Pradhan[1] and G.R. Bamnote[2]

[1] Department of Computer Engineering, AISSMS College of Engineering,
University of Pune, Pune, Maharashtra, India
[2] Department of Computer Science and Engineering, PRMIT&R, SSGBAU, Amravati,
Maharashtra, India
{madhavipradhan,grbamnote}@rediffmail.com

Abstract. Diabetes Mellitus is the one of the most serious health challenges. During the last 20 years the total number of diabetes patients has risen from 30 million to 230 million. It is a major health problem worldwide. So there is need of predictive model for early and accurate detection of diabetes. Diabetes disease diagnosis with proper interpretation of the diabetes data is an important classification problem. This research work proposes a Classifier for detection of Diabetes using Genetic programming (GP). Classification expression evaluation is used for creation of classifier. Reduced function pool of just arithmetic operators is used which allows for lesser validation and leniency during crossover and mutation.

Keywords: Classifier, Diabetes Mellitus, Genetic Programming, Predictive Model.

1   Introduction

Diabetes is a chronic disease wherein human body does not produce or properly uses insulin. Diabetes can go undetected for many years during which it can cause irreparable damage to the human body. Early, easy and cost effective detection of diabetes is a need for many developed and developing countries. Many classification algorithms have been used for classification of diabetes data. Use of GP for classification of diabetes data has been limited and much research can be conducted in this field. M. W. Aslam and A. K. Nandi [1] proposed a method which uses a variation of genetic programming (GP) called GP with comparative partner selection (CPS) for designing of classifier for detection of diabetes. It consists of two stages, first one being the generation of a single feature from available features using GP from the training data. The second stage consists of using the test data for checking of this classifier. The results of GP with CPS approach can change drastically by experimenting with the crossover probability, mutation probability or by selection of different methods of population replacement and selection. The highlight of this approach was the increased accuracy by using CPS method for selection of the partner during crossover. The system proposed in this paper was able to achieve the 78.5

© Springer International Publishing Switzerland 2015   763
S.C. Satapathy et al. (eds.), *Proc. of the 3rd Int. Conf. on Front. of Intell. Comput. (FICTA) 2014*
– *Vol. 1*, Advances in Intelligent Systems and Computing 327, DOI: 10.1007/978-3-319-11933-5_86

±2.2% accuracy. D. P. Muni, N. R. Pal and J. Das [2] proposed a GP approach to design classifiers. It needs only a single GP run to evolve an optimal classifier for a multiclass problem. For an n-class problem, a multi-tree classifier consists of trees, where each tree represents a classifier for a particular class. Unlike other approaches, paper took an integrated view where GP evolves considering performance over all classes at a time. This paper also proposes algorithms for various operations like crossover, mutation and others for a generalized n-class classification problem. It also demonstrates the effectiveness of concept of unfitness of the individual solutions. To obtain a better classifier they have proposed a new scheme for OR-ing two individuals. They have used a heuristic rule-based scheme followed by a weight-based scheme to resolve conflicting situations. The heuristic rules model typical situations where the classifier indicates ambiguous decisions and try to resolve them. The weight-based scheme assigns a higher weight to a tree which is less responsible for making mistakes. P. Day and A. K. Nandi [3] introduce a new approach to evaluate each individual's strengths and weaknesses and represent it in the form of binary strings. Supported by this approach the author introduced novel population evaluations, which demonstrated the promising ability of distinguishing between runs that converged at effective solutions and those that did not. This ability is based on the assumption that an effective solution should have a good overall fitness value and perform equally well with all training cases. If individuals are not capable of doing this, then it is important to preserve enough genetic fragments in the population to solve all of the training cases equally for long enough so that these can be recomposed later. CPS method tries to reduce the population wide weaknesses while trying to promote the strengths. This method leads to additional overheads because of the CPS procedure but the increased accuracy outweighs the additional overheads incurred. Although this method does guarantee an optimal solution but it does increase the probability of getting an optimal solutions. J. K. Kishore et al [4] used genetic programming for multiclass problem. A number of classification techniques have been proposed before for classification of diabetes dataset.[5] But GP has certain advantages over others that make it favourable for classification of large dataset. GP is not affected by increase in number of parameters. Another advantage is that GP is domain independent so very less knowledge is required regarding the domain. GP is relatively new technique and thus there are many areas yet unexplored. So, we have decided to design a classifier using Genetic Programming.

2   Proposed Classifier

The block diagram of the proposed classifier is given in Fig 1. We have use Pima Indian diabetes [6] data set available at the UCI Repository of Machine Learning Databases, which is owned by National Institute of Diabetes, Digestive and Kidney Disease. This Data set has eight attributes and one output variable which has either a value '1', for diabetic, or '0', for non-diabetic. The eight attributes are No. of times pregnant, Plasma Glucose concentration a 2 hours oral glucose tolerance test, Diastolic Blood Pressure (mmHg), Triceps Skin fold thickness (mm), Hour serum insulin (mu U/ml), Body Mass Index, Diabetes Pedigree Function, Age. There are 268(34.9%) cases for class '1' and 500(65.1%) cases for class '0'. The inconsistent and missing attributes entries are deleted from the dataset. Then it is divided into two parts training and testing dataset. Training dataset is used to train the classifier using GP, which gives us the best tree forming the classifier.

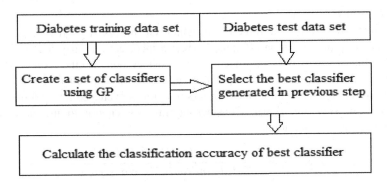

Fig. 1. Block diagram of proposed Classifier

2.1   Genetic Programming

Classification using GP has many and varied implementation possibilities. Without loss of generality these implementations can be divided into 3 categories namely Algorithm Evolution, Rule Evolution and Classifier Expression. We have used classifier expression evolution for creation of the classifier. These expressions use the attributes of data as variables and serve as a discriminating function between classes. The output of this function can be used for distinguishing between the classes [7]. Diabetes data classification being a binary classification problem, a threshold of positive and negative numbers can serve as a natural boundary between two classes.

2.1.1 Function Pool
Function pool represents the set of functions that are used as the intermediate nodes in the tree. Thus, the function pool forms the operators in the classifier expression. We have used a reduced function pool consisting of only arithmetic operators i.e. {+, -, *, /}. The advantage of using just arithmetic operators is that all operators are binary and hence it allows for lesser validation and leniency during crossover and mutation operations.

2.1.2 Initialisation
Like all other evolutionary approaches the initial population in GP is also randomly generated. The improvement in performance and accuracy of GP is dependent on variable sized representations which can be achieved with proper initialization of the first population. Initialization plays an important role in success of an evolutionary algorithm. A poor initial population can cause a good algorithm to get stuck in local optima. Whereas a good initialization can make most of the algorithms work sufficiently well. Three of the most popular initialization techniques are Full method, Grow method and Ramped half-and-half method.

In order to get maximum diversity and thus increase the probability of getting the optimal solution we have used the ramped half and half method. The population size is fixed at 100. Out of these 100 trees, 50 trees have been constructed using the full method and remaining 50 have been constructed using grow method. The tree depth has been fixed at 5 for both methods i.e. depth 5 contains only the leaf nodes.

2.1.3 Fitness Function

Fitness function governs the direction in which GP moves, a good fitness function may lead to optimal solution efficiently while a incorrect one may lead GP astray. Another metric that can be taken into consideration is the complexity. A mathematically complex and computation intensive fitness function will require significant time for calculation and as fitness function is evaluated for each individual, this time can slow GP down. Considering this, we choose to use accuracy to determine fitness of individuals.

$$Fitness = \frac{Correctly\ classified\ individuals}{Total\ individuals\ classified}$$

Although, this fitness function is simple and computation time is less but it was observed that more than one individuals were evaluated to have same highest fitness value in the last generation. Thus to distinguish between these individuals, having highest fitness, another fitness function was employed. Aim of this fitness function was to increase the distance between these classes while decreasing the distance between the points within each class [8]

$$Fitness = \left[\frac{|m_1 - m_2|}{\sqrt{\sigma_1^2 + \sigma_2^2}}\right]^{-1}$$

where, m1, m2 are the means of the two classes
σ1, σ2 are the standard deviations of the two classes.
This fitness function tries to increase the distance between the means of the two classes while minimizing the variance of the same classes. This fitness function has been optimized to give fitness values between 0 and 1. Here, fitness value of 0 represents best solution while 1 represents the worst solution.

2.1.4 Selection of Individuals

GP follows Darwinian principle of survival of fittest. It dictates that better individuals have better chance of passing their genetic material to the next generation. Selection of better individuals may help in achieving optimal solution efficiently. Different selection methods have been devised but the most commonly used are Roulette wheel selection, ranked roulette wheel selection and tournament selection.

Roulette wheel selection is a classic fitness-proportional selection. The probability of selection of each tree depends on the fitness value of the tree, as a consequence even the worst tree may also get selected. [9]

$$p_i = \frac{f_i}{\sum_{k=1}^{n} f_k}$$

where, pi is the probability of selection of ith individual, f_i is the fitness value of the ith individual and n is population size.

Ranked roulette wheel selection is an enhancement over roulette wheel selection where the probability of selection depends on fitness rank of individual relative to the entire population. It introduces a uniform scaling across all the individuals and is not affected by super-fit trees. [9]

$$Rank(Pos) = 2 - SP + \left[2 * (SP - 1) * \frac{(Pos - 1)}{(n - 1)} \right]$$

Where, SP is the selective pressure.
For linear mapping value is between $2.0 \geq SP \geq 1.0$,
n is the Population size and Pos is the Position of the individual in population.
Least fit individual has Pos = 1 and fittest individual has Pos = n

As the name suggests, Tournament selection is a competition conducted among a set of individuals that are selected at random. The number of individuals in the set is called the tournament size. A larger tournament size leads to higher selection pressure but may also slow down GP on account of increased number of comparisons required. It was observed that tournament selection with tournament size of 10 for a population size of 100 gave best results without slowing down GP. Also, we can see that roulette wheel and ranked roulette wheel require sorting and summation operations which can be time consuming.

2.1.5 Genetic Operators
Genetic operators operate on the selected individuals and the offspring created in this process form the new generation. They are responsible for continuity of GP. The genetic operators used in this approach are;

Crossover
It is a binary operator, it takes two parents and creates two off-springs. Crossover is performed in hope that offspring will be better than the parents as crossover causes swapping of the genetic material of the parents. It leads to exploitation of a specific region in the search space.We have selected two parents by using the tournament selection. A random crossover point is selected, in both of the parents, between 0 - 100. The sub trees rooted below these selected crossover points are then swapped to get offspring. By selecting crossover between 0 - 100 we are probabilistically restricting the depth of the trees.

Mutation
Mutation is a unary operator. A random mutation point is selected and the sub-tree rooted below that point is then replaced by a randomly generated tree. This operation causes addition of new genetic material and thus prevents GP from stagnating a local optimal solution. We have performed mutation on the offspring generated by crossover process.

2.1.6 Replacement
Replacement strategies dictate how the newly generated offspring are included in the population. Two of the most popular strategies are Steady State Update and Generational Update. We have fixed the population size 100. The generational update

strategy has been used as it becomes easy to keep track of the number of generations taking place. Two best individuals of current population are transferred to next generation by the elitism strategy. The remaining 98 offspring are created by crossover and mutation operations.

2.1.7 Termination Condition

The commonly used termination criteria are completion of a given number of generations, success in finding a solution of desired fitness or completion of specific time period (after the start of algorithm). Initially, the termination condition was completion of 500 generations. But due to the probabilistic nature of GP, these 500 generations may be completed within 10 seconds or may require more than 30 minutes depending on the size of the trees. Such unpredictable behavior not being appropriate, the termination criteria was changed to finding a solution of 0.9 and above fitness value. But restricting the fitness value indirectly represented restricting of the accuracy. As restriction of accuracy was undesirable, the termination criterion was later on decided to be 5 minutes. The proposed system runs for 5 minutes and the best solution from the last generation is considered as the optimal classifier.

3   The Experimental Results

The system is developed in JAVA (Netbeans 7.0) and Pima Indian diabeties dataset is used for training and testing.

Table 1. Characteristics of the GP system used

Sl. No.	Parameter	Value
1	Function pool	{+, -, *, /}
2	Terminal set	Eight attributes of the diabetes data set.
3	Population size	100
4	Initialization method	Ramped half and half
5	Maximum initial tree depth	5
6	Crossover probability	0.8
7	Mutation probability	0.4
8	Selection method	Tournament selection with tournament size 10.
9	Replacement strategy	Generational update with elitism.

With the characteristics mentioned in Table 1, the classifier was generated 40 times. The accuracy, sensitivity and specificity were calculated as shown in Table 2 above. We were able to obtain a classification accuracy of 77.54 ± 3.83% with a maximum accuracy of 89%.

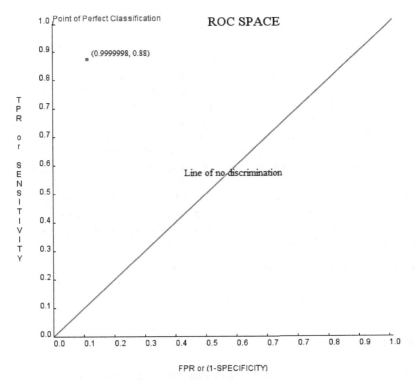

Fig. 2. Receiver Operating Characteristics (ROC) plot

Table 2. Confusion Matrix

		Actual Value	
		Diabetic	Non-diabetic
Test Outcome	Diabetic	44	6
	Non-diabetic	5	45
	Sensitivity	= 0.88	
	Specificity	= 0.9	

For holistic interpretation of the classifier, ROC of the classifier was also plotted as shown in Fig 2 above. ROC is a plot of false positive rate i.e. (1- Specificity) versus the true positive rate i.e. Sensitivity. The 45° line in ROC represents the line of no discrimination which is equivalent to a random guess. The region above the line of no discrimination represents better classifier while region below it represents worse classifiers. The point (0,1) on ROC is point of perfect classification.

4   Conclusion

In this paper, we have presented an approach for the binary classification problem using Genetic Programming and Pima Indian Diabetes Data Set for training and testing. It can be seen that we were able to achieve comparable accuracy using a simplified function pool. Various selection methods were implemented and the one that gave optimal result i.e. tournament selection was selected. The proposed classifier lies above the line of no Discrimination and thus it is a better classifier. During implementation, one of the points that was observed is that increase in tree size was enormous and it sometimes hampered the algorithm as trees become more complex to traverse. It would be interesting to analyze the effects of restricting tree growth beyond a specific depth and document its effect on overall accuracy. Increase in the computation power and the time required for the generation of classifier, will help in improving accuracy. It can be useful in detection of diabetes at low cost in rural areas where the costlier tests like HbA1C tests may not be available. It can be used as a second opinion to inexperienced physician.

Acknowledgement. This work was supported in part by University of Pune Research Grant Scheme fund from Board of College and University Development (BCUD) Ref. No. # O.S.D./ BCUD / 330.

References

[1] Aslam, M.W., Nandi, A.K.: Detection of Diabetes Using Genetic Programming. In: 18th European Signal Processing Conference (EUSIPCO 2010), Aalborg, Denmark, pp. 1184–1188 (2010)

[2] Muni, D.P., Pal, N.R., Das, J.: A Novel Approach to Design Classifiers Using Genetic Programming. IEEE Trans. on Evolutionary Computation 8(2), 183–196 (2004)

[3] Day, P., Nandi, A.K.: Binary String Fitness Characterization and Comparative Partner Selection in Genetic Programming. IEEE Trans. on Evolutionary Computation 12(6), 724–755 (2008)

[4] Kishore, J.K., Patnaik, L.M., Mani, V., Agawal, V.K.: Application of genetic programming for multi-category pattern classification. IEEE Trans. on Evolutionary Computation 4, 242–258 (2000)

[5] Zhang, B., Vijaya Kumar, B.V.K.: Detecting Diabetes Mellitus and Non proliferative Diabetic retinopathy using Tongue color, Texture and Geometry features. IEEE Transaction on Biomedical Engineering 61(2), 491–501 (2014)

[6] UCI repository of machine learning Databases, Pima Indian Diabetes Dataset, http://archive.ics.uci.edu/ml/datasets/Pima+Indians+Diabetes

[7] Jabeen, H., Baig, A.R.: Review of Classification Using Genetic Programming. International Journal of Engineering Science and Technology 2(2), 0975–5462 (2010)

[8] Pradhan, M.A., Rahman, A., Acharya, P., Gawade, R., Pateria, A.: Design of classifier for Detection of Diabetes using Genetic Programming. In: International Conference on Computer Science and Information Technology, Pattaya, Thailand, pp. 125–130 (2011)

[9] Razali, N.M., Geraghty, J.: Genetic Algorithm Performance with Different Selection Strategies in Solving TSP. In: Proceedings of the World Congress on Engineering, vol. 2 (2011)

Genetic Algorithm: Paradigm Shift over a Traditional Approach of Timetable Scheduling

Suresh Limkar, Ajinkya Khalwadekar, Akash Tekale,
Manasi Mantri, and Yogesh Chaudhari

Department of Computer Engineering, AISSMS IOIT, Pune, 411001, India

Abstract. The development of education and college expansion and consolidation in the Educational Management System has made the Course Scheduling System complex, and therefore it has become necessary to design one for development, and reuse. A Course Timetabling Problem (CTP) is an NP-hard combinatorial optimization problem which lacks analytical solution methods. During the last two decades several algorithms have been proposed, most of which are based on heuristics like evolutionary computation methods. This paper proposes a solution based on genetic algorithm .Genetic Algorithm (GA) emerges as one automation timetabling method to solve this problem by searching solution in multi-points and the ability to refine and optimizing the existing solution to a better solution. The experimental results show that the proposed GAs are able to produce promising results for the course timetabling problem.

1 Introduction

Organizations like universities, colleges and schools use timetables to schedule classes, courses and lectures, assigning times and places to future events in a way that makes best use of the resources available. Timetabling Problem is an NP-complete problem, which is very difficult to solve by conventional methods and the amount of computation required to find optimal solution increases exponentially with problem size.[19] It is a search for good solutions in a space of possible timetables. The problem of creating a valid timetable involves scheduling classes, lecturers and rooms into a fixed number of periods, in such a way that no teacher, class or room is used more than once per period. Timetabling becomes a problem when the assigning task becomes hard to imply when specific requirements need to be followed. There are a number of hard and soft constraints which need to be satisfied. A few of them may be stated as follows:

- Hard constraints
  1. There are 6 classes a day from Monday to Friday.
  2. All classes must be placed in the timetable.
  3. The same teacher must be not place at the same time.
  4. The same class must be not place at the same time.

© Springer International Publishing Switzerland 2015   771
S.C. Satapathy et al. (eds.), *Proc. of the 3rd Int. Conf. on Front. of Intell. Comput. (FICTA) 2014*
– *Vol. 1*, Advances in Intelligent Systems and Computing 327, DOI: 10.1007/978-3-319-11933-5_87

5. There is the class that opening time is fixed like a seminar.
6. Problem of facilities (whether using a computer room or a projector)
7. Externship day (as for each exclusive teacher takes externship day in a week)
8. The class to affect graduation requirements like a required subject locates for 2-4 periods on timetable.

– Soft constraints

1. The classes in the required subject do not open in the same time as possible.
2. The class of the same teacher does not leave the periods as much as possible
3. On a duty day of the same part-time teacher opens on the same day as possible.

2   Related Work

The last two decades have seen growing interest in AI based Software Engineering technology. Our review of this area shows that there have been only few approaches that provide automated tools for generation of timetables.

Jian Ni in [1] proposed the strings representation for every chromosome along with the various hard and soft constraints that are considered in timetable design.. The new sets of chromosomes are generated from the initial population which is simplified using strings.

Kuldeep Kumar, Sikander, et al[3] used a penalty scheme for calculating the fitness of the new population that is generated. This penalty scheme is based on the principle of minimization of the total cost to optimize the solution. [3] also uses single point technique for crossover of the chromosomes.

The case based approach system in [4] used sequential methods for solving the timetabling problem. Generally, they order the events using domain-specific heuristics and then assign the events sequentially into valid time slots in such a way that no constraints are violated for each timeslot.

W. Legierski [5] proposed constraint based method to solve this problem. The timetabling problem is modeled as a set of variables (events) to which values (resources such as teachers and rooms) have to be assigned in order to satisfy a number of hard and soft constraints.

J.J. Thomas et al.[6] proposed cluster based approach which divides the problem into number of event sets. Each set is defined so that it satisfies all hard constraints. Then, the sets are assigned to real time slots to satisfy the soft constraints as well.

E. Aycan and T. Ayav [7] used Simulated Annealing algorithm which gave results comparable to that of GA. But since SA relies on a random set of permutations it cannot guarantee the true minimum cost. It is often necessary to perform multiple runs to determine whether the cost is the best. Thus the annealing program might need running several time to find the variance in the final cost.

K. H. Zhipeng Lu [8] discussed use of Tabu Search for solving this problem. Though the results show that Tabu search alone can be used to solve the problem, to improve the efficiency and optimality, it has to be hybridized with other algorithms like Perturbation.

Ping Guo et al. [9] solved the timetabling problem by merging genetic algorithm with greedy approach. The greedy approach is used to generate the initial set of population. The crossover technique used to achieve the result is single point crossover which gives less randomness in the new population.

M.H. Husin et al. [10] proved that the timetabling problem cannot be solved with simple genetic algorithm. The high constraints and limited free slots can be satisfied only after modifying the traditional approach of genetic algorithm.

J. Lassig and K. H. Hoffmann [11] focused on the problem of selecting individuals from the current population in genetic algorithms for crossover. It proves that the best performance can be achieved using threshold selection, but it is only of limited use in practice.

R. Watson et al. [12] used GA for allocation of platforms to the trains at a particular railway station. To reduce the execution time and complexity of the algorithm, the fitness function was redesigned by comparing only the clashing chromosomes instead of comparing the entire set of population in the same time window.

B.R.Rajakumar et al.[13] Suggested the use of innovation institution designed by Goldberg. The innovation institution deals with combination of selection and mutation strategy or selection and recombination strategy. These combinations are done in the real world by humans and can improve the results of simple GA.

A clear analysis of previous works infers that attempts have been made for the construction of timetables for schools and colleges. But the methods used are very complex and the results have not always been satisfactory. Automated tools for this process are still not very famous.

Currently, there is no framework for auto-generation of timetable that can be universally used in the educational institutes. Most of the earlier tools were developed considering only a few constraints as hard. This led to exclusion of many important constraints from the created timetable which further caused clashes after the execution of the timetable. Some existing tools require tremendous human interactions while making any changes in the created timetables. They also require a large data set in the beginning as input. Furthermore provisions to update the timetables when needed are not provided.

The different steps of GA like Selection, Crossover, Mutation, etc require different algorithms. The algorithms generally used were chosen based on their simplicity of implementation, which in turn did not provide the required randomness and optimality in an efficient manner.

The significance of our method is to resolve each of these above cited problems. The system takes basic information about the classrooms, courses, teachers and time-slots as an input & generates timetables which can schedule the class room and laboratory lectures while satisfying all the constraints. The GUI provided by the system is user friendly that allows user to modify, print or edit generated timetables.

3 Proposed System

In the previous section, we have reviewed the most recent works. We have proposed a desktop application to ease the process of timetable generation for educational institutes. It provides automation to course timetabling, thereby reducing efforts and time wastage. It is based on Genetic Algorithm.

This approach satisfies all the hard constraints effectively and aims at satisfying as many soft constraints as possible. The aim of our approach is to efficiently apply Genetic Algorithm to achieve a fast and accurate result.

Proposed System is divided into two modules as follows. Module 1 is the Timetable Creation (the process of creating the timetable based on the information provided by the client), whereas Module 2 is Timetable Modification (the process of updating the timetables which are generated by the module 1).

3.1   Timetable Creation

In this, we are going to generate a timetable based on the information provided by the user. This information includes all the details required for satisfying the hard and soft constraints. They can be:

- Teacher-Subject Relationships
- Room-Subject Relationships
- Teacher-Time Relationships, etc.

This information is given as an input to genetic algorithm for further processing and timetable creation.

3.2   Genetic Algorithm

Genetic algorithms (GA) mimic the process of natural selection. GAs maintains many individual solutions in the form of population. Individuals i.e. parents are chosen from the population and are then mated to form a new individual (a child). To introduce diversity into the population children are further mutated. [3] A simple genetic algorithm is as follows:

```
Create a population of creatures.
Evaluate the fitness of each creature.
While the population is not fit enough:
{
Select two population members.
Combine their generic material to create a new creature.
Cause a few random mutations on the new creature.
Evaluate the new creature and place it in the population.
}
```

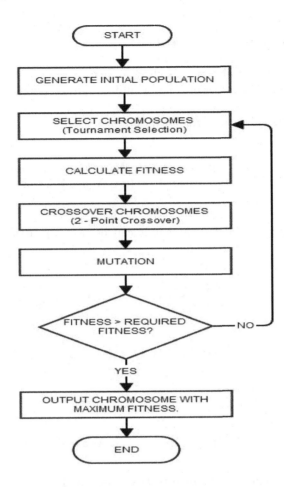

Fig. 1. Flowchart of Genetic Algorithm

3.3 Timetable Modification

Once a timetable is generated, it might need to be altered under conditions like unavailability of the teachers or the rooms. The system, TTGene provides a facility to make these changes while satisfying the constraints and maintaining optimality.

4 System Implementation

4.1 Problem Description

Let us assume that,

X   set of all teachers
Y   set of all subjects

Z   set of all rooms
X_y gives the teacher-subject association
Y_z gives the subject-room association
L_t is the maximum load that a teacher can handle
L_s is the maximum number of times a subject is to be taught

We also define the following decision variables:

$$W_{xyz} = \begin{cases} 1 & \textit{if teacher } x \textit{ teaches subject } y \textit{ in room } z \\ 0 & \textit{otherwise} \quad (x \in X, y \in Y, z \in Z) \end{cases}$$

The system is implemented such that it can be used to satisfy the following equation.

Maximise

$$\sum_x \sum_y \sum_z (X_y + Y_z) \times W_{xyz} \tag{1}$$

subject to

$$1 \leq \sum_x W_{xyz} \leq L_t \qquad y \in Y, z \in Z \tag{2}$$

$$1 \leq \sum_y W_{xyz} \leq L_s \qquad x \in X, z \in Z \tag{3}$$

$$\sum_y W_{xyz} \leq 1 \qquad x \in X, z \in Z \tag{4}$$

The objective function (1) reflects that associated teachers, subjects and rooms are assigned in the timetable.

Equation (2) ensures that each teacher has to teach at least one course, while not allowing the teacher to teach more than the maximum load of courses allowed for each teacher.

Equation (3) ensures that the number of slots assinged for each subject does not exceed the maximum load of the subject, per week.

Equation (4) ensures that each teacher and each room is free when it is assigned a slot in the timetable.

4.2 Problem Definition

The timetable is used by engineering students & faculty. It must schedule lectures and practical sessions. It should not cause any clashes for the students, teachers, classrooms or the time slots. It can be viewed by the students, teachers and anyone else related to the department of computer engineering. It can be created or modified only by the admin(s) after the identity is authenticated.

4.3   Solution

To solve the above mentioned problem, we have implemented the system named "TTGene" using the proposed approach of Genetic Algorithm.

Each timetable is considered as a Chromosome and each cell of the timetable is considered as a Gene. Each gene is made up of following parts :

- Subject
- Teacher
- Room

The genes are combined together to form a timetable. The different Genetic Operations are performed on the genes to get the best solution.

1. Initially, 5000 random timetables are generated to create the random initial population.
2. The fitness of each timetable is calculated based on the equation from section 4.1.
3. Then, tournaments are held between 125 timetables for selecting the parents. The best timetables that are obtained after selection are forwarded to the next step.
4. 2 - point crossover is performed on the selected timetables to improve the fitness.
5. The weaker timetables then undergo mutation procedure. This step might improve or reduce the fitness.

These steps are repeated continuously until the timetable with required fitness is obtained.

The system uses the following tables as sample initial data. Using this data, the steps of Genetic algorithm are executed. The system can be fed with the data, as and when required and timetables are generated.

Table 1. Rooms Table

Id	Name	Lab

Table 2. Subjects Table

Id	Name	Class	Practical	Update	Dayskipped	Lectures per week

Table 3. Teachers Table

Id	Name	No._subject	Subject1	Subject2	Subject3	Update	Dayskipped

5   Comparative Analysis and Results

This section presents the results of a practical performance evaluation of the TTGene system. Although the problem space used for the study is small, it is sufficient enough to compare between the different methods.

The timetabling problem as discussed in [19] is a NP-Complete problem. When solved by manual method and the computations needed increase exponentially with problem size. Hence, optimal solutions cannot be expected from the traditional manual approach. The following table discusses the drawbacks of the approaches based on their performance.

Table 4. Performance evaluation between manual and automated approach

Sr. No.	Factors	Manual Approach	Automated Approach
1	Resource Utilization	Less	More
2	Efficiency	Poor	Higher
3	Human Effort	More	Less
4	Optimality	Less	Better
5	Conflict Resolving Capability	Good	Comparatively Bad

6   Conclusion and Future Scope

In today's global and competitive environment, the use of scarce resources in the best possible way is more important than ever, and time is one of the critical resources for almost all problems. Our proposed method demonstrates the use of Genetic algorithm techniques for generation and modification of timetables to achieve optimality in the generated timetables. Each timetable is evaluated by testing the number of times it breaches each constraint. Thus timetables are evolved with a minimum number of constraint violations. The designed system can generate timetables efficiently without much human interactions and the experiment results show the optimized timetables are more reasonable. The performance of GA is compared with other algorithms and it is observed that GA outperforms the others.

The system currently can be used only to generate the school/college timetables. The improved optimization techniques used in the system TTGene can be further applied to generate University Exam timetables. The same concept can also be used in hospitals where doctors have to be assigned as a resource to the various patients.

References

1. Yang, N.-N., Ni, J.: Genetic algorithm and its application in scheduling system. TELKOMNIKA 11(1), 1934–1939 (2013)

2. Ramik, J., Perzina, R.: Self-learning genetic algorithm for a timetabling problem with fuzzy constraints. International Journal of Innovative Computing, Information and Control 9(11), 4565–4582 (2013)

3. Sharma, R., Mehta, K., Kumar, K., Sikander: Genetic algorithm approach to automate university timetable. International Journal of Technical Research 1(1) (March 2012)

4. Yang, Y., Petrovic, S., Dror, M.: Case-based selection of initialisation heuristics for metaheuristic examination timetabling. Expert Systems with Applications 33(3), 772–785 (2007)

5. Legierski, W.: Search strategy for constraint-based classteacher timetabling. In: Burke, E.K., De Causmaecker, P. (eds.) PATAT 2002. LNCS, vol. 2740, pp. 247–261. Springer, Heidelberg (2003)

6. Belaton, B., Thomas, J.J., Khader, A.T.: A visual analytics framework for the examination timetabling problem. In: Proceedings of the Fifth International Conference on Computer Graphics, Imaging and Visualisation, vol. 1(1), pp. 305–310 (August 2008)

7. Aycan, E., Ayav, T.: Solving the course scheduling problem using simulated annealing. In: IEEE International Advance Computing Conference, IACC 2009, vol. 1(1), pp. 462–466 (2009)

8. Hao, J.-K., Lu, Z.: Adaptive tabu search for course timetabling. European Journal of Operational Research 200(1), 235–244 (2010)

9. Zhu, L., Guo, P., Chen, J.-X.: The design and implementation of timetable system based on genetic algorithm. In: International Conference on Mechatronic Science, Electric Engineering and Computer, vol. 1(1) (August 2011)

10. Chai, S., Sabri, M.F.M., Husin, M.H.: Development of a timetabling software using soft-computing techniques with a case study. In: The 2nd International Conference on Computer and Automation Engineering (ICCAE), vol. 5(1), pp. 394–397 (February 2010)

11. Lassig, J., Hoffmann, K.H.: On the structure of a best possible crossover selection strategy in genetic algorithms. In: Research and Development in Intelligent Systems XXVI, vol. 26, pp. 263–276. Springer (April 2010)

12. Withall, M.S., Jackson, T.W., Phillips, I.W., Brown, S., Clarke, M., Hinde, C.J., Watson, R.: Allocating railway platforms using a genetic algorithm. In: Research and Development in Intelligent Systems XXVI, vol. 26, pp. 421–434. Springer (April 2010)

13. Binu, D., George, A., Rajakumar, B.R.: Genetic algorithm based airlines booking terminal open/close decision system. In: ICACCI 2012, vol. 1(1), ACM (August 2012)

14. Melanie, M.: An Introduction to Genetic Algorithms. First MIT Press paperback edition (1998)

15. Luke, S.: Essentials of Metaheuristics Online version (June 2013)

16. Haupt, R.L., Haupt, S.E.: Practical Genetic Algorithms, 2nd edn. John Wiley & Sons, Inc. (2004)

17. de Oliveira Rech, L., Lung, L.C., Ribeiro, G.O., de Campos, A.J.R.: Generating timetables of work scales for companies using genetic algorithms. In: 2012 XXXVIII Conferencia Latinoamericana En IEEE Informatica (CLEI), vol. 1(1), pp. 1–10 (October 2012)

18. Adachi, Y., Ataka, S.: Study on timetable design for osaka international university by differential evolution. In: The 1st IEEE Global Conference on Consumer Electronics, vol. 1(1), pp. 1–10 (2012)

19. Cooper, T.B., Kingston, J.H.: The complexity of timetable construction problems. University of Sydney, Technical report, vol. 495(1) (February 1995)
20. Beaty, S.J.: Genetic Algorithms versus Tabu Search for Instruction Scheduling. Artificial Neural Nets and Genetic Algorithms 1, 496–501 (1993)
21. Aladağ, C.H., Hocaoglu, G.: A Tabu Search Algorithm to Solve a Course Timetabling Problem. Hacettepe Journal of Mathematics and Statistics 36(1), 53–64 (2007)

A New Particle Swarm Optimizer with Cooperative Coevolution for Large Scale Optimization

Shailendra S. Aote[1], M.M. Raghuwanshi[2], and L.G. Malik[3]

[1] Research Scholar, GHRCE, Nagpur, India
shailendra_aote@rediffmail.com
[2] RGCER, Nagpur, India
m_raghuwanshi@rediffmail.com
[3] CSE, GHRCE, Nagpur, India
lgmalik@rediffmail.com

Abstract. With the increasing demands in solving larger dimensional problems, it is necessary to have efficient algorithm. Efforts were put towards increasing the efficiency of the algorithms. This paper presents a new approach of particle swarm optimization with cooperative coevolution. The proposed technique [NPSO-CC] is built on the success of an early CCPSO2 that employs an effective variable grouping technique random grouping. The technique of moving away out of the local minima is presented in the paper. Instead of using simple velocity update equation, the new velocity update equation is used from where the contribution of worst particle is subtracted. Experimental results show that our algorithm performs better as compared to other promising techniques on most of the functions.

Keywords: article Swarm Optimization, Self adaptive, Co operative co evolution.

1 Introduction

Due to wider application area of evolutionary algorithms, many researchers have found their interest. Many stochastic algorithms suffer the problem of "curse of dimensionality" i.e. it performs well for smaller dimension problem but performance degrades as dimensionality increases. Finding global optima for simple unimodal function to rotated shifted multimodal function has become much more challenging. As most of the problems in real life are equivalent to solving multimodal problem with higher dimension, it is necessary to design the effective optimization algorithm.

Particle swarm optimization(PSO) is global optimization method proposed initially by James Kennedy and Russell Eberhart [1] in 1995. The particle swarm concept originated as a simulation of simplified social system. However it was found that particle swarm model can be used as an optimizer. Suppose the following scenario: a flock of birds are randomly searching food in an area. There is only one piece of food in the area being searched. All the birds are unaware of the location of food. But they know how far the food is in each iteration. So what`s the best strategy to find the

© Springer International Publishing Switzerland 2015
S.C. Satapathy et al. (eds.), *Proc. of the 3rd Int. Conf. on Front. of Intell. Comput. (FICTA) 2014*
– *Vol. 1*, Advances in Intelligent Systems and Computing 327, DOI: 10.1007/978-3-319-11933-5_88

food? The effective one is to follow the bird which is nearest to the food. PSO is a part of evolutionary computing which is mostly used algorithm in almost all kind of optimization problems. Due to its simple equation and involvement of less parameters, it becomes one of the promising algorithm for solving optimization problems. Different PSO variants are proposed from its inception to till date. Efforts were put to solve higher dimensional as well as multidimensional problems. Because of premature convergence of different PSO variants, it is necessary to handle higher dimensional problems. The better approach is to use divide and conquer strategy, which divides total dimensions into number of sub dimensions and handling each of these subcomponents individually. Jong and Potter [2] in 1994 started their efforts towards cooperative coevolution (CCGA) which is extended by Frans van den Bergh and Andries P. Engelbrecht in 2004. They proposed CPSO-Sk [3] and CPSO-Hk techniques which is tested on 30 dimensions only. But how it performs on larger dimension upto 1000 is unanswered.

Recent studies by Z. Yang [4] suggested a new decomposition strategy based on random grouping. Without prior knowledge of the nonseparability of a problem, it was shown that random grouping increases the probability of two interacting variables being allocated to the same subcomponent, thereby making it possible to optimize these interacting variables in the same subcomponent, rather than across different subcomponents. Adaptive weighting scheme is proposed by Zhenyu yang which uses evolutionary computing in cooperative coevolution. To solve the problems of larger dimensions, CCPSO technique [5] was proposed. Author has used random grouping and adaptive weighting scheme in his work. Based on this idea, CCPSO2 model [6] was introduced where a new PSO model using Cauchy and Gaussian distributions was used instead of generalized PSO velocity and position update equation. It uses new adaptive scheme which dynamically determines the subcomponent sizes for random grouping. The performance was tested on CEC 2008 test suit [7] for 1000 and 2000 dimension. Though it performs well, results are not much satisfactory on some functions. To increase the performance, a new approach namely NPSO-CC is proposed.Results taken in this direction are more promising as compared to other techniques proposed. Results are taken for 1000 dimension and performance is evaluated on CEC-2010 test suit [8].

Rest of the paper is organized as follows. Section 2 represents the overview and different variants of PSO. Section 3 represents proposed methodology. Experimental results and analysis is given into Section 4, followed by Conclusion and future direction in section 5.

2   Overview of PSO

Particle Swarm Optimization (PSO) incorporates swarming behaviors observed in flocks of birds, schools of fish, or swarms of bees, and even human social behavior. As an algorithm, the main strength of PSO is its fast convergence, which compares favorably with many global optimization algorithms like Genetic Algorithms (GA), Simulated Annealing (SA) etc. In a particle swarm optimizer, individuals are "evolved" by cooperation and competition among the individuals. Each particle adjusts its flying according to its own flying experience and its companions' flying experience. Each individual is named as a "particle" which, in fact, represents a potential solution to a problem. Each particle is treated as a point in a D dimensional space. The ith particle is represented as

$$X_i = (X_{i1}, X_{i2}, \ldots \ldots \ldots, X_{iD})$$

The best previous position (the position giving the best fitness value) of any particle is recorded and represented as

$$P_i = (P_{i1}, P_{i2}, \ldots \ldots, P_{iD})$$

The index of the best particle among all the particles in the population is represented by the symbol g. The rate of the position change (velocity) for particle i is represented as

$$V_i = (V_{i1}, V_{i2}, \ldots \ldots, V_{iD})$$

$$V_{id} = V_{id} + c_1 * \text{rand}(\ )*(P_{id} - X_{id} + c_2 * \text{rand}(\ )*(P_{gd} - X_{id}) \qquad \text{(a)}$$

The Particle are manipulated according to equation

$$X_{id} = X_{id} + V_{id} \qquad \text{(b)}$$

Where, rand () is a random function which generates value in the range [0, 1]. The second part of the equation (a) is the "cognition' part, which represents the private thinking of the particle itself. The third part is the "social" part, which represents the collaboration among the particles. The equation (a) is used to calculate the particle's new velocity according to its previous velocity and the distances of its current position from its own best experience (position) and the group's best experience. Then the particle flies toward a new position according to equation (b). The performance of each particle is measured according to a predefined fitness function, which is related to the problem to be solved.

The above PSO equation is a basic equation, which is continuously updated in previous years. The inertia weight w is brought into the equation (a) as shown in equation (c) by Yuhui Shi in 1998[9]. This w plays the role of balancing the global search and local search. It can be a positive constant or even a positive linear or nonlinear function of time.

$$V_{id} = w * V_{id} + c_1 * \text{rand}(\ )*(P_{id} - X_{id} + c_2 * \text{rand}(\ )*(P_{gd} - X_{id}) \qquad \text{(c)}$$

Its value is taken as 0.7 in initial days which is further considered as a dynamic for better exploration initially and then fine exploitation. To remove the problem of stagnation, various PSO variants like GCPSO, MPSO, OPSO, QPSO, HPSO, RegPSO were proposed which were tested upto 30 dimension problems. But solution for higher dimensional problems upto 1000 were unanswered. Now a days cooperative co evolution method is used in different evolutionary algorithms like genetic algorithm(GA), differential evolution(DE) and PSO etc. Zhenyu Yang proposed a new cooperative coevolution framework that is capable of optimizing large scale nonseparable problems by using differential evolution. A random grouping scheme and adaptive weighting were introduced in problem decomposition and coevolution. Proposed DECC-G[4] was very effective and efficient in tackling large optimization problems with dimensions up to 1000. Jia Zhao proposed TSEM-PSO which divides the swarm into two subswarms, where first swarm follows basic PSO equation and second follows other equation proposed by author and then particles of subswarm exchanges information between them.

Efforts were also made by combining the features of PSO & DE in cooperative coevolution. PSO has the tendency to distribute the best personal positions of the swarm particles near to the vicinity of problem's optima and to efficiently guide the evolution and enhance the convergence, the DE algorithm is used without destroying the search capabilities of the algorithm. CCPSO performs integrated random grouping and adaptive weighting scheme which scales up the PSO on high dimensional non separable problems. Based on the success of CCPSO2, Xiaodong Li proposed CCPSO2 which adopts a new PSO position update rule that relies on Cauchy and Gaussian distributions to sample new points in the search space, and a scheme to dynamically determine the coevolving subcomponent sizes of the variables. By considering promising results of these cooperative strategies, NPSO-CC algorithm is proposed in this paper.

3   Proposed Method

NPSO-CC is an extended version of CCPSO2 where some features are replaced by new methodology. Here we perform random grouping of variables at each iteration.

3.1   More Frequent Random Grouping

Here we randomly divide the D-dimensional space into n subcomponents of fixed size. Number of particles remains same in this process. The advantage of dividing it randomly that the probability of placing two interacting variables into the same subcomponent becomes higher, over an increasing number of iterations. The probability of assigning two interacting variables xi and xj into a single subcomponent for at least k cycles is [4]

$$Pk = \sum_{r=k}^{N} \binom{N}{r} \left(\frac{1}{m}\right)^k \left(1 - \frac{1}{m}\right)^{N-r}$$

where N is the total number of cycles and m is the number of subcomponents. If we run the algorithm for 50 iterations, there are 50 executions of random groupings will occur. The probability of optimizing the two variables in the same subcomponent for at least one iteration is

$$P(x \geq 1) = p(1) + p(2) + \ldots + p(50)$$

$$= 1 - p(0)$$

$$= 1 - \binom{50}{0} (0.1)0 (1 - 0.5)50$$

$$= 0.9948$$

Where x is number of observed successes over 50 runs. It shows relatively high probability of interacting variables into same subcomponent.

3.2   Velocity Update Equation

When particles move in the search space, it updates its velocity based on its personal as well as swarm experience. The worst particle in each iteration is also contributing to find the new velocity. The contribution is subtracted from the original equation. The new velocity update equation is i.e. equation (c) is modified as,

$$Vid=w*Vid+c1*rand()*(Pid-Xid)+c2*rand()*(Pgd-Xid)-c2*rand()*(Pwd-Xid)   (d)$$

Where, Pwd = Worst particle in the swarm

d= Number of dimensions

The main idea behind subtracting the contribution of worst particle is as given. After considering the performance of all particles together, worst particle actually degrades the performance. Subtracting performance plays very vital role in increasing the performance.

3.3   A New Technique to Deal with Stagnation

Though the above strategies are used to enhance the performance of PSO, the problem of stagnation is still persisting.

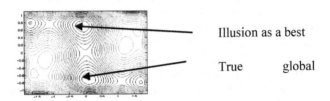

Illusion as a best

True   global

Fig. 1. Illusion as a best

Fig. 1 shows the nature of multimodal function. Due to large number of uphills and downhills, it is difficult to identify true global optimum. Here we proposed simple technique to deal with this. If for 50 numbers of iterations, the fitness value remains constant, those particles will be initialized in another territory. To find new region the radius is calculated by dividing the range by some constant. Particles must be initiated outside the circle of radius rad to carry out the further search. The steps of the algorithm are presented as follows.

D : Dimension   N : Number of particles
START
1. Divide the swarm into subswarm along s dimension where D = No of group *s
2. For each group
    Find Velocity by subtracting the contribution of worst particle in subswarm
    Update position for each subswarm
    Evaluate the subswarm

   Divide the subswarm into two groups along number of particles based on fitness
      First half particle remain unchanged
      For remaining half particle
            pos = rand * pos
      end
3.   Merge the subswarm along dimensions
4.   Evaluate the swarm
5.   If for 50 iterations fitness unchanged
         Find radius , rad = range / constant;
         Initialize the swarm at different position outside the circle of radius, rad
6.   Repeat the procedure from step 1.
END

When we divide the subswarm into two groups along number of particles, first part remains unchanged. For remaining half of the particles, we multiply the position by random value to bring randomness in its position. This is because these particles does not perform well on previous position. So it is better to change the position randomly. The reason for taking number of iteration as 50 is as explained. We have performed the experiment for 10, 50, 100, 200, 500 iterations. It is found that there are 90 % chances of changing the solution after 10 iterations, so the value 10 is not suitable. We also checked the change in solution after 100, 200, 500 iterations, it is found that there are less than 2 % chances of changing the values. So instead of waiting for large number of iterations, where we cannot get better value, it is better to change the territory after 50 iterations. When we want to update the position after 50 iterations, it is necessary to shift that particle in another region. The region is nothing but the circle of radius, rad. If for example bounds are set in between -100 to 100, the value of range becomes 200. If we divide this value by 40, we get radius as 5. Then move the particle outside the circle of radius 5 in the search space. The value of constant varies from 40 to 80, as these values produces better results. If we set the value below 40, radius increases, which will lead to set the particle beyond bigger circle. That time it will have a lower space to explore. As value increases beyond 80, particle may have smaller territory to explore.

4   Experimental Results and Analysis

Above algorithm is evaluated on CEC-2010 & CEC-2008 test suit, as these are the mostly used for comparison on single objective optimization problems. Experiment is run on Matlab 7.9, windows i3 processor with 4 GB RAM. The values of w,c1,c2 are taken as 0.72, 1.49,1.49 respectively. Algorithm is evaluated for 100000 iterations in 25 runs. Its mean, median and standard deviation values are recorded for comparison. NPSO-CC is compared with ith SPSO-2011[12] & DECC-G on CEC-2010 test suit. CEC-10 suit consist of 20 fitness function where first 3 functions are separable, last 2 are fully non separable and others are group nonseparable.

Table 1. Comparison among NPSO-CC, SPSO-11, DECC-G on CEC-2010

FN	NPSO-CC	SPSO-2011	DECC-G	Sign
f1	3.2734e+007(1.34e+007)	1.98e+011(2.24e+010)	**3.2734e-018(1.985e-019)**	(- , -)
f2	**2.1e-001(4.865e-002)**	1.81e+004(1.77e+002)	3.1e+002(9.7742)	(+ , +)
f3	**9.3832e-003(2.94e-004)**	2.12e+001(8.94e-002)	3.9126(2.011e-001)	(+ , +)
f4	9.3892e+012(1.214e+012	1.17e+015(3.45e+014)	4.8493e+012(**2.534e+011**	(~ , -)
f5	**2.6067e+005(2.017e+003**	6.43e+008(5.23e+007)	7.9827e+007(8.183e+006	(+ , +)
f6	**5.3631e+006(7.812e+005**	1.95e+007(8.234e+006)	9.3324e+006(**3.528e+004**	(+ , -)
f7	**1.2616e+007(9.126e+005**	3.84e+011(1.914e+011)	2.2621e+011(2.2621e+011	(+ , +)
f8	**6.3027e+007(4.214e+006**	3.33e+016(6.1148+014)	1.9072e+012(4.0532e+01	(+ , +)
f9	**2.3796e+008(1.027e+008**	2.19e+011(7.093e+010)	4.6198e+010(3.1127e+00	(+ , ~)
f10	**936.96(1.1283)**	1.82e+004(3.75e+002)	2.28e+004(9.734e+002)	(+ , +)
f11	**27.5232(8.6129)**	2.32e+002(1.0075e+002	935.28(6.217)	(+ , ~)
f12	**27811.6(92.211)**	1.19e+007(8.524e+005)	1.2891e+006(8.741e+005	(+ , +)
f13	**9526.58(9377.22)**	1.03e+012(4.941e+011)	5.88e+006(1.925e+006)	(+ , +)
f14	**6.0891e+008(1.124e+007**	2.36e+011(1.163e+011)	1.1254e+010(8.454e+008	(+ , +)
f15	**2.8621(9.428e-001)**	1.82e+004(3.329e+002)	267.98(28.911)	(+ , +)
f16	**3.1172(2.107e-001)**	4.22e+002(3.715e+001)	511.72(128.135)	(+ , +)
f17	**9822.8(8219.24)**	4.01e+007(8.6412e+006	9.0127e+006(1.7635e+00	(+ , +)
f18	**4.2341e+008(8.984e+007**	2.27e+012(1.884e+012)	2.3981e+012(2.1934e+01	(+ , +)
f19	8.3920e+008(8.3920e+008)	7.83e+007(2.214e+005)	**4.3328e+005(1.283e+005**	(- , -)
f20	1.2001e+012(**3.261e+011**	2.17e+012(2.17e+012)	**1.0824e+012(**1.0824e+012	(- , +)
Total	30	Nil	07	--------

Values outside the brackets shows the mean performance, whereas values inside the brackets shows standard deviation. '+' sign denotes better performance of NPSO-CC, '-' sign denotes better performance of some other algorithm, '~' sign denotes equal performance of NPSO-CC with other algorithm.

Above table shows the performance of NPSO-CC, SPSO-11, DECC-G algorithms. T-test is used to compare the performance of these algorithms on both test suits. From the table it is found that NPSO-CC performs well   for 30 values whereas DECC-G performs well for 7 values. SPSO -2011 does not perform well for any function. Proposed algorithm & DECC-G performs statistically equals on f4 for its mean values and on f9, f11 for its standard deviation. It is also found that proposed algorithm does not perform well on fully non separable functions. This is because proposed algorithm divides the dimensions into sub dimensions and because of interdependencies among dimensions, there may be chance of presence of most interdependent components in different subcomponents. Simple cooperative coevolution is not suitable for fully non separable functions.

Figure 2 is a graphical comparison of these algorithms on their mean values. Algorithm is also evaluated on CEC-2008, where it considers six functions for performance evaluation. Following table shows the comparison of proposed algorithm with CCPSO2 and DMS-PSO.

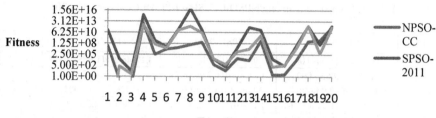

Fig. 2. Comparison of above three algorithms on their mean values

Table 2. Comparison between NPSO-CC, CCPSO2, DMS-PSO

FUN	NPSO-CC	CCPSO2	DMS-PSO	Sign
f1	4.3321e-18 (2.981e-19)	5.18E-13 (9.61E-14)	**0.00E+00 (0.00E+00)**	(- , -)
f2	**2.2991e-03 (4.22e-03)**	7.82E+01 (4.25E+01)	9.15E+01 (7.14E-01)	(+ , +)
f3	**1.3217e-01 (8.2391e-03)**	1.33E+03 (2.63E+02)	8.98E+09 (4.39E+08)	(+ , +)
f4	**7.9320e-05 (3.2181e-06)**	1.99E-01 (4.06E-01)	3.84E+03 (1.71E+02)	(+ , +)
f5	3.3876e-05 (3.3876e-05)	1.18E-03 (3.27E-03)	**0.00E+00 (0.00E+00)**	(+ , +)
f6	**5.2139e-13 (1.915e-14)**	1.02E-12 (1.68E-13)	7.76E+00 (8.92E-02)	(- , -)
Total	**08**	Nil	04	---------

-    *Results of CCPSO2 & DMS PSO are taken from [6]*

From the result it is concluded that NPSO-CC outperformed CCPSO2 and DMS-PSO on 4 functions for its mean and standard deviation. For remaining two functions it performs well as compared to CCPSO2 but does not perform well as compared to DMS-PSO. Statistically it performs well on all the values. First two functions are unimodal and remaining four functions are multidimensional in nature. For complex problems with higher dimensions, our algorithm performs well. For multimodal functions, there are always the chances of stagnation, which gets removed by exploiting another region.

5   Conclusion and Future Direction

Multidimensional problems are always difficult to solve. Simple PSO could not find the global optimum and fails to converge it. In this paper, we have proposed new algorithm which forms groups along dimensions so that dimensional space is decreased which results in finding better value. The technique of moving the swarm outside the local minima contributes better results. Test suit of 2008 and 2010 is used to compare performance of the algorithm. Performance is compared with well known algorithms SPSO2011, DMS-PSO, DECC-G & CCPSO (2012) and it is found that this algorithm performs extremely well.

As this algorithm doesn't perform well for fully non separable function, strategies must be designed to deal with these kinds of problems. It is better to hybrid this technique with other intelligent techniques present in the market.

References

1. Eberhart, R., Kennedy, J.: A new optimizer using particle swarm theory. In: Proc. 6th Int. Symp. Micro Mach. Human Sci., pp. 39–43 (October 1995)
2. Potter, M., Jong, K.D.: A cooperative coevolutionary approach to function optimization. In: Proc. 3rd Conf. Parallel Problem Solving Nat., pp. 249–257 (1994)
3. van den Bergh, F., Engelbrecht, A.: A cooperative approach to parnticle swarm optimization. IEEE Trans. Evol. Comput. 8(3), 225–239 (2004)
4. Yang, Z., Tang, K., Yao, X.: Large scale evolutionary optimization using cooperative coevolution. Information sciences 178, 2985–2999 (2008)
5. Li, X., Yao, X.: Tackling high dimensional nonseparable optimization problems by cooperatively coevolving particle swarms. In: Proc. IEEE CEC, pp. 1546–1553 (May 2009)
6. Li, X., Yao, X.: Cooperatively Coevolving Particle Swarms for Large Scale Optimization. IEEE Transactions on Evolutionary Computation 16(2), 210–224 (2012)
7. Tang, K., Yao, X., Suganthan, P., MacNish, C., Chen, Y., Chen, C., Yang, Z.: Benchmark functions for the CEC'2008 special session and competition on large scale global optimization," Nature Inspired Computat. Applicat. Lab., Univ. Sci. Technol. China, Hefei, China, Tech,rep (2007), http://nical.ustc.edu.cn/cec08ss.php
8. Tang, K., Li, X., Suganthan, P., Yang, Z., Weise, T.: Benchmark functions for the CEC'2010 special session and competition on large scale global optimization, Nature Inspired Computat. Applicat. Lab., Univ. Sci. Technol. China, Hefei, China, Tech. Rep (2009), http://nical.ustc.edu.cn/cec10ss.php
9. Shi, Y., Eberhart, R.: A Modified Particle Swarm Optimizer. In: IEEE International Conference on Evolutionary Computation, Anchorage, Alaska, May 4-9, pp. 69–73 (1998)
10. Ji, H., Jie, J., Li, J., Tan, Y.: A Bi-swarm Particle Swarm Optimization with Cooperative Co-evolution. In: International Conference on Computational Aspects of Social Networks, pp. 323–326 (2010)
11. Zhao, J., Li, L., Sun, H., Zhang, X.-W.: A Novel Two Sub-swarms Exchange Particle Swarm Optimization Based on Multi-phases. In: IEEE International Conference on Granular Computing, pp. 626–629 (2010)
12. Clerc, M.: Standard Particle Swarm Optimisation (2006–2011)

Efficient Digital Watermarking
Based on SVD and PSO with Multiple Scaling Factor

Meghdut Roychowdhury[1], Suraday Sarkar[2], Sulagna Laha[3], and Souvik Sarkar[4]

[1] Techno India, Salt Lake, Kolkata
[2] Netaji Subhas Engg. College, Kolkata
[3] Jadavpur University, Kolkata
[4] IBM, Hyderabad
{meghdut.tig,lahasulagna}@gmail.com,
suryadaysarkar@ymail.com, souviksarkar@in.ibm.com

Abstract. Digital Watermarking, that is used for protection against digital piracy, tampering etc, has a lot of uses apart from protection like media forensics, improvement to digital communication quality, accurate audience monitoring, medical applications and so on. We present, here in this work, an invisible robust non-blind Digital Watermarking algorithm to maintain acceptable levels of imperceptibility and robustness at the same time using Singular Value Decomposition (SVD) and Particle Swarm Optimization (PSO) with multiple scaling factor. Application of such scheme provides an avenue for optimization of the strength-of-the-watermark (scaling factor) for attaining high robustness against watermarking attacks while maintaining the visual quality of the cover object. Our algorithm is found to exhibit significant performance improvement over the usage of constant scaling factors in Singular Value Decomposition based Digital Watermarking algorithm.

Keywords: SVD, PSO, scaling factor, robustness, imperceptibility.

1   Introduction

Even though the practise of watermarking has been there for several centuries, the last two decades have seen a positive avalanche in the research for various watermarking techniques. This has been mainly because of the need for protection of intellectual properties on the digital front. But along with the positive benefits there also have been the negative aspects of copyright infringements and piracy. Today, there is no such thing as protection of personal data items on a moral basis - the protection has to be enforced. Digital watermarking comes into play in the prevention of unauthorized copying of intellectual property and even detection of tampering with the digital objects. With the growth of high speed computer networks and wide availability of digital consumer devices, access to digital information and electronics transaction became easier. This leads towards the problem of illegal copying and redistribution of information.

Visual Information hiding technique becomes a potential solution to this intimidation by means of Intellectual Property Right protection, authentication and

© Springer International Publishing Switzerland 2015   791
S.C. Satapathy et al. (eds.), *Proc. of the 3rd Int. Conf. on Front. of Intell. Comput. (FICTA) 2014*
– *Vol. 1*, Advances in Intelligent Systems and Computing 327, DOI: 10.1007/978-3-319-11933-5_89

integrity verification. Visual Information hiding aims at ensuring a privacy property without damaging the information during an electronic transaction. Mainly, the research on Digital Watermarking is focussed on two arenas: spatial and transform domain. Spatial domain watermarks are easy to incorporate within the cover object, have less computational requirement and tend to preserve the perceptual quality of the cover object. However, the spatial watermarks are also generally easy to be removed or corrupted due to malicious or non-malicious involvements (watermarking attacks). On the other hand transform domain watermarks make use of more computation intensive methods to be embed a watermark within the cover object. However, once into the cover object, they are relatively more difficult to be distorted or removed than their spatial counterparts. However, the perceptual quality of the cover object after taking in the transform domain watermarks are not always up to the mark, thus once again defeating the purpose of invisible digital watermarking. Hence, the research for the perfect invisible digital watermarking algorithm that preserves the perceptual quality of the cover object and at the same time offers tough resistance to the removal or distortion of the watermark embedded in the cover object still going on.

When, a watermark is inserted into a digital cover object, the invisibility and robustness of the watermark depends upon several factors. They are: (i) the algorithm used being for the watermarking technique (ii) the innate nature of the watermark (iii) the strength of the watermark being embedded. Now, a designer might choose an algorithm and a watermark that provides a general good performance. However, simply estimating the strength of the watermark required for providing adequate imperceptibility but at the same time enough robustness to withstand distortions to the cover image is a dicey situation. This is because; different attacks might react differently for a particular strength of the watermark. Thus for the same strength of the watermark, an attack might not be able to disrupt the watermark at all on the other hand some other attack might totally render it indistinguishable. Hence, the concept of multiple watermarking is introduced i.e. watermarking using multiple copies of watermark each of which might have different strengths. The strength of watermark is otherwise known as scaling factor. Thus this kind of watermarking focuses on multiple scaling factors.

2   Theory

In this work we demonstrate an invisible robust non-blind Digital Watermarking algorithm to maintain acceptable levels of imperceptibility and robustness at the same time using Singular Value Decomposition (SVD) and Particle Swarm Optimization (PSO) with multiple scaling factor. Let us now put some effort to brief about SVD, PSO and the proposed algorithm.

2.1   Singular Value Decomposition (SVD)

The Singular Value Decomposition (SVD) is a factorization technique of linear algebra. It is an orthogonal transformation and it also leads to diagonalization. It breaks a matrix of higher order into summation of product of matrices of smaller orders. The method is applicable not only to square matrices but also rectangular matrices, which makes it flexible enough to be applied to a varied no of applications.

Through SVD any matrix M_{mxn} ($M \in R^{mxn}$, $m > n$) having rank r can be represented as the product of three matrices U_{mxm} ($U \in R^{mmn}$), S_{mxn} ($S \in R^{mxn}$) and V_{nxn} ($V \in R^{nxn}$).

$$M = U * S * V^T$$

S-matrix

$S = \text{diag} \ (\sigma_1, \sigma_2, ..., \sigma_n)$ consists of positive values in the decreasing order of magnitude. σ_i is known as the *singular value* of M. The number of non-zero elements in the diagonal of S is equal to the rank of M.

$$\sigma_1 \geq \sigma_2 \geq ..... \geq \sigma_{r-1} \geq \sigma_r > \sigma_{r+1} > ......\sigma_n = 0$$

Also $\sigma = \Lambda^{1/2}$ where Λ represents the *Eigen-value* of both $M^T M$ and MM^T.

$$(M^T M) \ v_i = \lambda_i \ v_{i;} \ \ \sigma = \Lambda^{\frac{1}{2}} \ ; (MM^T) \ u_i = \lambda_i \ u_i$$

V-matrix

V is *orthogonal* matrix i.e. $\quad\quad VV^T = V^T V = I$

$V = [v_1, v_2, ..., v_n]$ consists of *unit eigenvectors* of $M^T M$.   & $(M^T M) \ v_i = \lambda_i \ v_i$

The columns of V are known as *right-singular values* of M   & $MV = US$

U-matrix

U is an *orthogonal* matrix i.e. $UU^T = U^T U = I$

The columns of U are known as *left-singular values* of M & $U^T M = SV^T$

The columns of $U = [u_1, u_2, .., u_m]$ also consist of *unit eigenvectors* of MM^T & (MM^T) $u_i = \lambda_i \ u_i$

2.2   Particle Swarm Optimization (PSO)

From the various works we can observe that the symbiosis of digital watermarking and soft-computing tools is still in its infancy and has been used in the very recent years only. However, the usage of soft-computing tools in digital watermarking is increasing very fast due to the strong optimization and searching capability of soft computing tools. In this thesis we intend to use Particle Swarm Optimization to increase the performance of the algorithm proposed.

Particle Swarm Optimization (PSO) is an Evolutionary Computation technique developed by Kennedy and Eberhart In 1995. The algorithm was inspired by the coordinated, seamless but unpredictable motion of the flight of a flock of birds. The effort was initially being made to model the flight of birds but then the model was seen to act as an optimizer and hence it was refined to ultimately form the Particle Swarm Optimization technique. In this technique, each solution to the problem is referred to as a 'particle'. The particle may be multi-dimensional with each dimension consisting of a different variable to be optimized. At the start of the algorithm a population of particles is initialized. Each particle has a 'position' and a 'velocity' associated with it. Initially, the position and velocity of a particle are chosen randomly. Then the fitness of each particle are evaluated according to some fitness function. The best position a particular particle attains throughout the search for the solution corresponding to the best value of fitness function of the individual particle, is tagged as 'pbest'. Similarly, the best position among all the particles that has been attained throughout the search for the solution corresponding to the best value of fitness function attained globally is tagged as 'gbest'. The PSO mechanism consists of driving i.e. accelerating the position of the particles towards their respective individual pbest solution and towards the global gbest solution. The search for solution to the problem stops when all the particles attain position values i.e. as close as possible to the gbest value. The PSO iterations are driven by the following equations:

$$v_{id} = w * v_{id} + c_1 * \text{rand}() * (p_{id} - x_{id}) + c_2 * \text{Rand}() * (p_{gd} - x_{id})$$

$$x_{id} = x_{id} + v_{id}$$

Here, x_{id}: position of each particle.

v_{id}: velocity of each particle.
p_{id}: the personal best position of the particle.
p_{gd}: the global best position among all the particles.
c_1 and c_2 : constants whose values are usually taken as 2.
w: weight of vid. It is taken in the range (0, 1].
rand() and Rand(): random number generators within the range (0,1].

3   Algorithm

We are now in a position to explain the proposed algorithm.

Steps of the Algorithm

The algorithm can be summarized in the form of the following steps:

1. Decompose the host into its 3 component 2-D matrices: part_1, part_2 and part_3.
2. For each of the parts:
    a. Optimize the scaling factor according to category-1, categoty-2 or category-3 (The three categories must correspond to the three parts).
    b. Embed the watermark using the optimized scaling factor according to the steps in section 5.1.1.A.
    c. Calculate the PSNR
3. Calculate the PSNR according to the equation 5.1.
4. Simulate the attacks given in section 5.1.1.B.
5. Extract the watermarks from the 3 parts of the stego by the steps outlined in section 5.1.1.C.: watermark_ex1, watermark_ex2 and watermark_ex3
6. Calculate the NCC corresponding to each watermark_ex: NCC_1, NCC_2 and NCC_3
7. Chosen extracted_watermark = watermark_ex having the maximum[NCC_1, NCC_2, NCC_3]

4   Results and Discussion

In our present work, we present an invisible robust non-blind Digital Watermarking algorithm to maintain acceptable levels of imperceptibility and robustness at the same time using Singular Value Decomposition (SVD) and Particle Swarm Optimization (PSO) with multiple scaling factor.   In this paper the PSO will be used to optimize a single parameter depending upon two factors i.e. the scaling factor of a watermark will be optimized to maximize the robustness and imperceptibility of the proposed digital watermarking algorithm.   But due to brevity we are citing limited results. However, those results establish suitability of our algorithm.   We have considered here five different pictures and five   attacks to validate our algorithm. Results are given in tabular form. The Table 1 shows that the PSNR for each of the 5 hosts that have been tested have are very satisfactory.   They are all above the acceptable level of PSNR i.e. usually held to be

Table 1. Peak Signal-to-Noise Ratio (PSNR)

Host	PSNR	PSNR_host
Host1	57.76	37.32
Host2	20.62	65.50
Host3	25.58	37.88
Host4	18.35	59.74
Host5	18.38	36.47

Table 2. Normalized Cross Correlation (NCC)

Attacks	Host1	Host2	Host3	Host4	Host5
JPEG(CR=25)	0.9127	0.9976	0.9956	0.9993	0.9965
Gaussian Noise	0.5154	0.8073	0.4694	0.7989	0.8427
Rotation-30°	0.8879	0.9181	0.8283	0.8393	0.8680
Motion	0.5272	0.6241	0.8035	0.7059	0.7316
Averaging	0.7478	0.8964	0.9662	0.9504	0.9338

35dB to 37dB. Thus from the respect of imperceptibility the performance of the proposed algorithm is very satisfactory. Also, it must be mentioned that in spite of the heavy capacity the imperceptibility maintained is laudable. The Table 2 shows that all obtained NCC values for each of the hosts and found to be better than earlier reported values.

5   Conclusion

The algorithm proposed in the thesis has made the use of optimization of the scaling factor. The algorithm proposed in this thesis has a superior performance in terms of imperceptibility and capacity. The imperceptibility and robustness of the SVD based Digital Watermarking algorithm is dependent on the value of the scaling factors. The optimized scaling factors yield a boost-up the performance of the algorithm as compared to the constant scaling factors. The boost-up is mainly in terms of robustness of the algorithm.

References

1. Bianchi, T., Piva, A.: Secure Watermarking for Multimedia Content Protection: A Review of its Benefits and Open Issues. IEEE Signal Processing Magazine 30(2), 87–96 (2013)
2. Tsai, J.-S., Huang, W.-B., Kuo, Y.-H.: On the Selection of Optimal Feature Region Set for Robust Digital Image Watermarking. IEEE Transactions on Image Processing 20(3), 735–743 (2011)
3. Lai, C.-C., Ko, C.-H., Yeh, C.-H.: An adaptive SVD-based watermarking scheme based on genetic algorithm. In: International Conference on Machine Learning and Cybernetics (ICMLC), vol. 4, pp. 1546–1551 (July 2012)

Ant Colony Based Optimization from Infrequent Itemsets

Diti Gupta and Abhishek Singh Chauhan

NIIST, Bhopal, M.P., India

Abstract. Data Mining is the area of research by which we can find relevant patterns from the data set. It is used in several areas. In this paper we are focusing on finding relevant patterns from Positive and Negative Rules. For this we have applied Ant Colony Optimization (ACO) technique on the positive and negative rules. Our algorithm has achieved better global optimum value and chances of finding are improved. So the chances of Positive or relevant rules are more in comparison to the traditional technique. We are also applying the optimization to the negative rules so that there are equal chances for achieving the global optimum. But the negative rules are not qualifying the global optimum value and hence the relevant rules find by our algorithm are verified.

Keywords: Data Mining, Ant Colony Optimization (ACO), Positive Association, Negative Association, Global Optimum Value.

1   Introduction

Mining association rules is an important task. Past transaction data can be analyzed to discover customer purchasing behaviors such that the quality of business decision can be improved. The association rules describe the associations among items in the large database of customer transactions. However, the size of the database can be very large. A large amount of research work has been devoted to this area, and resulted in such techniques as k-anonymity [1], data perturbation [2], [3], [4], [5], and data mining based on [6], [7]. Data Mining and association rules are applied in several application areas including mobile technology [8][9]. It will be associated with several level of subset, superset based criteria of filtering the data [10]. This signifies the research direction in several fields. We can use ARM and data mining application in health care, medical database, classification and combining these techniques with other approach extensively increases the potential behavior and applicability. In [11] author suggests that many of the researchers are generally focused on finding the positive rules only but they not find the negative association rules. But it is also important in analysis of intelligent data. It works in the opposite manner of positive rule finding. But problem with the negative association rule is it uses large space and can take more time to generate the rules as compare to the traditional mining association rule [12][13]. So better optimization technique can find a better solution in the above direction. In this paper we have applied Ant Colony optimization technique for finding the better optimization technique.

© Springer International Publishing Switzerland 2015
S.C. Satapathy et al. (eds.), *Proc. of the 3rd Int. Conf. on Front. of Intell. Comput. (FICTA) 2014*
– *Vol. 1*, Advances in Intelligent Systems and Computing 327, DOI: 10.1007/978-3-319-11933-5_90

2   Literature Review

In 2008, He Jiang et al. [14] suggest that the weighted association rules (WARs) mining are made because importance of the items is different. Negative association rules (NARs) play important roles in decision-making. But according to the authors the misleading rules occur and some rules are uninteresting when discovering positive and negative weighted association rules (PNWARs) simultaneously. So another parameter is added to eliminate the uninteresting rules. They propose the support-confidence framework with a sliding interest measure which can avoid generating misleading rules. An interest measure was defined and added to the mining algorithm for association rules in the model. The interest measure was set according to the demand of users. The experiment demonstrates that the algorithm discovers interesting weighted negative association rules from large database and deletes the contrary rules. In 2009, Yuanyuan Zhao et al. [15] suggest that the Negative association rules become a focus in the field of data mining. Negative association rules are useful in market-basket analysis to identify products that conflict with each other or products that complement each other. The negative association rules often consist in the infrequent items. The experiment proves that the number of the negative association rules from the infrequent items is larger than those from the frequent. In 2011, Weimin Ouyang et al. [16] suggest three limitations of traditional algorithms for mining association rules. Firstly, it cannot concern quantitative attributes; secondly, it finds out frequent itemsets based on the single one user-specified minimum support threshold, which implicitly assumes that all items in the data have similar frequency; thirdly, only the direct association rules are discovered. They propose mining fuzzy association rules to address the first limitation. In this they put forward a discovery algorithm for mining both direct and indirect fuzzy association rules with multiple minimum supports to resolve these three limitations. In 2012, Yihua Zhong et al. [17] suggest that association rule is an important model in data mining. However, traditional association rules are mostly based on the support and confidence metrics, and most algorithms and researches assumed that each attribute in the database is equal. In fact, because the user preference to the item is different, the mining rules using the existing algorithms are not always appropriate to users. By introducing the concept of weighted dual confidence, a new algorithm which can mine effective weighted rules is proposed by the authors. The case studies show that the algorithm can reduce the large number of meaningless association rules and mine interesting negative association rules in real life. In 2012, He Jiang et al. [18] support the technique that allows the users to specify multiple minimum supports to reflect the natures of the itemsets and their varied frequencies in the database. It is very effective for large databases to use algorithm of association rules based on multiple supports. The existing algorithms are mostly mining positive and negative association rules from frequent itemsets. But the negative association rules from infrequent itemsets are ignored. Furthermore, they set different weighted values for items according to the importance of each item. Based on the above three factors, an algorithm for mining weighted negative association rules from infrequent itemsets based on multiple supports(WNAIIMS) is proposed by the author. In 2013, Anjana Gosain et al. [22] suggest that the traditional algorithms for mining association rules are built on binary attributes databases, which has two limitations. Firstly, it cannot concern quantitative attributes; secondly, it treats each item with the same significance although different item may have different significance [23]. Also a binary association rule suffers from sharp boundary problems [24]. According to the authors many real world transactions consist of quantitative attributes. That is why several researchers have been working on generation of association rules for quantitative data. They present different algorithms given by various researches to generate association rules among quantitative data. They have done comparative analysis of different algorithms for association rules based on various parameters. They also suggest a future suggestion that there is the need of a framework of association rules for data warehouse that overcome the problems observed by various authors. In 2013, Luca Cagliero et al. [25] tackle the issue of discovering rare and weighted itemsets, i.e., the Infrequent Weighted Itemset (IWI) mining problem. They proposed two novel quality measures to drive the IWI mining process. Two algorithms that perform IWI and Minimal IWI mining efficiently, driven by the proposed measures, are presented. Experimental results show efficiency and effectiveness of the proposed approach. In 2013, Johannes K. Chiang et al. [26] aims at providing a novel data schema and an algorithm to solve the some drawbacks in conventional mining techniques. Since most of them perform the plain mining based on predefined schemata through the data warehouse as a whole, a re-scan must be done whenever new attributes are added. Secondly, an association rule may be true on a certain granularity but fail on a smaller one and vise verse, they are usually designed specifically to find either frequent or infrequent rules. A forest of concept taxonomies is used as the data structure for representing healthcare associations patterns that consist of concepts picked up from various taxonomies. Then, the mining process is formulated as a combination of finding the large item sets, generating, updating and output the association patterns. Their research presents experimental results regarding efficiency, scalability, information loss, etc. of the proposed approach to prove the advents of the approach.

3   Proposed Work

The figure 1 shows the working process of our algorithm. We are using the famous dataset T10I4D100K.dat form http://fimi.ua.ac.be/data/T10I4D100K.dat.

Then we first consider the T10I4D100K data set. First we consider the initial value of the dataset. Then we find the occurrences of each item set in the dataset. Then we evaluate the support of each item separately. Then based on multiple dynamic minimum support entered by the users we will able to classify the dataset. The dataset is classified in two different categorizes one is positive rules and other is negative rules. Then we set the values as the agents and then we find the individual supports of each agent. Then we apply the optimization technique in our case we have consider ant colony optimization to optimize the initial ants. We have found negative and positive association first then negative and positive set are optimized separately with our algorithm shown below. After optimization we find the global optimum value based on we can compare the classification accuracy which is better in terms of the previous technique. The process is applied both for negative and positive rules. By our algorithm the positive rules are so optimized that the selection chances are higher. After applying the same process in the negative rules so that the chances are equal but the negatives rules are not highly optimized and the chances are still weak. So by our approach we can prove it that the negative and positive classifications are justified ans suitable for use and classification in different area. The algorithm of our approach is shown below.

Algorithm:
Assumptions:

TD: T10I4D100K.dat
R1 and R2 are the relational sets
IR1: Initial set
T_V: Cumulative Value
P_t : Pheromone Trail
E_p : Evaporation Value
O_{AC}: Overall Accuracy

Input:
- WS(ws1,ws2....wsn)
- LS(ls1,ls2....lsn)

Output:
- R1 U R2 –IR1
- AC((R1 U R2) –IR1)

Step 1: Input Set
Step 2: Find the frequency of the individual dataset.
Step 3: $$support = \frac{(X \cup Y).count}{n}$$
Step 4: User Based Multiple support.
Step 5: Positive and Negative association.
Step 6: Separate the List.
Step 7: Optimization will be applied separately.
Step 8: Initialize the Initial values.
Step 9: Check the IR set for the relevancy
For 1 to 5
$T_V = (IR_1 + IR_2 + IR_3 + ... + IR_n)/n$
$P_t = T_v - R_p$
$E_p = \{0.2, 0.4, 0.6, 0.8\}$
If($P_{t1} > P_{tn-1}$)
$P_{t1} = P_{tn-1}$
Step 10: Final R set
For 1 to 8
$T_V = (R_1 + R_2 + R_3 + ... + R_n)/n$
$P_t = T_v - R_p$

$P_t = T_v - R_p$

$E_p = \{0.2, 0.4, 0.6, 0.8\}$

If($P_{t1} > P_{tn-1}$)

$P_{t1} = P_{tn-1}$

Step 11: Overall Accuracy

$O_{AC} = \sum P_{t1} + P_{t2} + P_{t3} + \ldots$

Step 12: Step 9 to 11 will be applied for negative association also.

Step 13: Min and max value are calculated.

Step 14: Match the values.

Step 15: Final Optimized Results.

Step 16: Finish

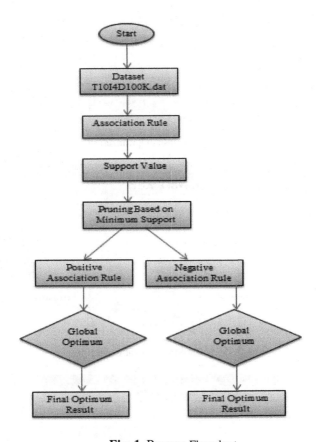

Fig. 1. Process Flowchart

4   Results and Evaluation

In our approach we first select the dataset as specified above. Then we find the frequency and support value. Then based on the minimum support value we classify it in two parts first is for positive association and second is for negative association.

Then we apply the optimization iteration. The limit of iteration is 5. We are considering the minimum support of 20 %. Based on the consideration we achieve five iteration of Tmax and five for Tmin. Based on the support value we are achieving the equal negative and positive value. The qualification rate of the positive association is improved as shown in figure 2, 3, 4.

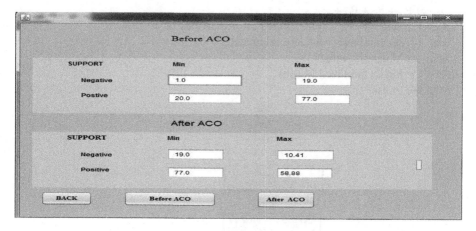

Fig. 2. ACO Based Optimization

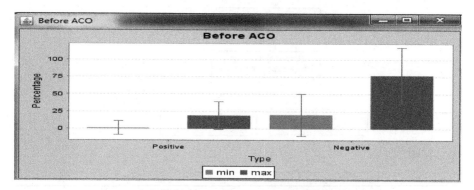

Fig. 3. ACO Based Optimization (Before)

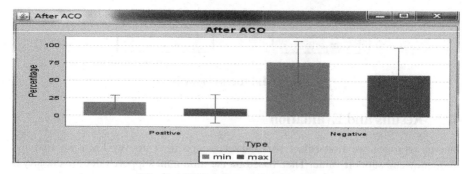

Fig. 4. ACO Based Optimization (After)

5   Conclusions

In this paper we have applied support based association rules on T10I4D100K. Based on our results the classifications as well as the selection chances are increases in the case of positive rules. But the same things will be applied for negative rules. But the chances of the selection are still low. This signifies that our approach classification is better.

References

1. Sweeney, L.: K-anonymity: a model for protecting privacy. International Journal on Uncertainty, Fuzziness and Knowledge-based Systems 10(5), 557–570 (2002)
2. Agrawal, R., Srikand, R.: Privacy preserving data mining. In: Proc. of ACM SIGMOD Conference, pp. 439–450 (2000)
3. Chen, K., Liu. L.: A random rotation perturbation approach to privacy data classification. In: Proc. of IEEE Intl. Conf. on Data Mining (ICDM), pp. 589–592 (2005)
4. Xu, S., Zhang, J., Han, D., Wang, J.: Singular value decomposition based data distortion strategy for privacy distortion. Knowledge and Information System 10(3), 383–397 (2006)
5. Mukherjeee, S., Chen, Z., Gangopadhyay, A.: A privacy-preserving technique for Euclidean distance-based mining algorithms using Fourier related transforms. Journal of VLDB 15(4), 293–315 (2006)
6. Vaidya, J., Clifton, C.: Privacy preserving k-means clustering over vertically partitioned data. In: Prof. of ACM SIGKDD Conference, pp. 206–215 (2003)
7. Vaidya, J., Yu, H., Jiang, X.: Privacy preserving SVM classification. Knowledge and Information Systems 14, 161–178 (2007)
8. Dubey, A.K., Shandilya, S.K.: A novel J2ME service for mining incremental patterns in mobile computing. In: Das, V.V., Vijaykumar, R. (eds.) ICT 2010. Communications in Computer and Information Science, vol. 101, pp. 157–164. Springer, Heidelberg (2010)
9. Dubey, A.K., Kushwaha, G.R., Shrivastava, N.: Heterogeneous data mining environment based on DAM for mobile computing environments. In: Das, V.V., Thomas, G., Lumban Gaol, F. (eds.) AIM 2011. CCIS, vol. 147, pp. 144–149. Springer, Heidelberg (2011)
10. Dubey, A.K., Dubey, A.K., Agarwal, V., Khandagre, Y.: Knowledge Discovery with a Subset-Superset Approach for Mining Heterogeneous Data with Dynamic Support. In: Conseg-2012 (2012)
11. Jain, N., Sharma, V., Malviya, M.: Reduction of Negative and Positive Association Rule Mining and Maintain Superiority of Rule Using Modified Genetic Algorithm. International Journal of Advanced Computer Research (IJACR) 2(4(6)) (December 2012)
12. Khare, P., Gupta, H.: Finding Frequent Pattern with Transaction and Occurrences based on Density Minimum Support Distribution. International Journal of Advanced Computer Research (IJACR) 2(3(5)) (September 2012)
13. Deshpande, L.A., Prasad, R.S.: Efficient Frequent Pattern Mining Techniques of Semi Structured data: a Survey. International Journal of Advanced Computer Research (IJACR) 3(1(8)) (March 2013)
14. Jiang, H., Zhao, Y., Dong, X.: Mining Positive and Negative Weighted Association Rules from Frequent Itemsets Based on Interest, Computational Intelligence and Design. In: International Symposium on Computational Intelligence and Design, ISCID 2008, 17-18 October, vol. 2, pp. 242–245 (2008)

15. Zhao, Y., Jiang, H., Geng, R., Dong, X.: Mining Weighted Negative Association Rules Based on Correlation from Infrequent Items. In: International Conference on Advanced Computer Control,ICACC 2009, January 22-24, pp. 270–273 (2009)
16. Ouyang, W., Huang, Q.: Mining direct and indirect fuzzy association rules with multiple minimum supports in large transaction databases. In: Eighth International Conference on Fuzzy Systems and Knowledge Discovery (FSKD), July 26-28, vol. 2, pp. 947–951 (2011)
17. Zhong, Y., Liao, Y.: Research of Mining Effective and Weighted Association Rules Based on Dual Confidence. In: Fourth International Conference on Computational and Information Sciences (ICCIS), August 17-19, pp. 1228–1231 (2012)
18. Jiang, H., Luan, X., Dong, X.: Mining Weighted Negative Association Rules from Infrequent Itemsets Based on Multiple Supports. In: International Conference on Industrial Control and Electronics Engineering (2012)
19. Gosain, A., Bhugra, M.: A Comprehensive Survey of Association Rules On Quantitative Data in Data Mining. In: IEEE Conference on Information and Communication Technologies (2013)
20. Ouyang, W., Huang, Q.: Mining Direct and Indirect Weighted Fuzzy Association Rules in Large Transaction Databases. In: IEEE Eighth International Conference on Fuzzy Systems and Knowledge Discovery (2011)
21. Lee, K.-M.: Mining Generalized Fuzzy Quantitative Association Rules with Fuzzy Generalization Hierarchies. IEEE (2011)
22. Cagliero, L., Garza, P.: Infrequent Weighted Itemset Mining using Frequent Pattern Growth. IEEE Transactions on Knowledge and Data Engineering (2013)
23. Chiang, J.K., Huang, S.-Y.: Multidimensional Data Mining for Healthcare Service Portfolio Management. IEEE (2013)
24. Rameshkumar, K., Sambath, M., Ravi, S.: Relevant Association Rule Mining from Medical Dataset Using New Irrelevant Rule Elimination Technique. IEEE (2013)

The Fuzzy Robust Graph Coloring Problem

Arindam Dey[1], Rangaballav Pradhan[2], Anita Pal[3], and Tandra Pal[2]

[1]Department of Computer Science and Engineering,
Saroj Mohan Institute of Technology, Hooghly
`arindam84nit@gmail.com`
[2]Department of Computer Science and Engineering,
National Institute of Technology, Durgapur
`{rangaballav.pradhan,tandra.pal}@gmail.com`
[3]Department of Mathematics,
National Institute of Technology, Durgapur
`anita.buie@gmail.com`

Abstract. Fuzzy graph model can represent a complex, imprecise and uncertain problem, where classical graph model may fail. In this paper, we propose a fuzzy graph model to represent the examination scheduling problem of a university and introduce a genetic algorithm based method to find the robust solution of the scheduling problem that remains feasible and optimal or close to optimal for all scenarios of the input data. Fuzzy graph coloring method is used to compute the minimum number of days to schedule the examination. But problem arises if after the examination schedule is published, some students choose new courses in such a way that it makes the schedule invalid. We call this problem as fuzzy robust coloring problem (FRCP). We find the expression for robustness and based on its value, robust solution of the examination schedule is obtained. The concept of fuzzy probability of fuzzy event is used in the expression of robustness, which in turn, is used for fitness evaluation in genetic algorithm. Each chromosome in the genetic algorithm, used for FRCP, represents a coloring function. The validity of the coloring function is checked keeping the number of colors fixed. Fuzzy graphs with different number of nodes are used to show the effectiveness of the proposed method.

Keywords: Fuzzy graph, fuzzy probability, fuzzy event, fuzzy graph coloring, robustness.

1 Introduction

The Graph coloring problem (GCP) are used for a wide variety of applications, such as frequency assignment, time-table scheduling, register allocation, bandwidth allocation, circuit board testing, etc. For graph coloring problem, the decision environment is often characterized by the following facts:

1. Uncertainty is essential to human being at all levels of their interaction with the real world. So, the data are uncertain / inexact.

© Springer International Publishing Switzerland 2015   805
S.C. Satapathy et al. (eds.), *Proc. of the 3rd Int. Conf. on Front. of Intell. Comput. (FICTA) 2014*
– *Vol. 1*, Advances in Intelligent Systems and Computing 327, DOI: 10.1007/978-3-319-11933-5_91

2. Two adjacent vertices, assigned different colors, must remain feasible for all meaningful realizations of the data.

3. Bad optimal solutions, i.e., optimal solutions which become severely infeasible in the case of even relatively small changes in the nominal data, are not uncommon. Those solutions are not robust.

Here, we propose a method for graph coloring that would be able to handle those facts, mentioned above, which basically represents an uncertain environment. In this paper, examination scheduling problem of a university is considered. Here courses are represented by the nodes of a graph and every pair of incompatible courses is connected by an edge. The coloring of this graph provides a feasible schedule of the courses and computes the minimum number of time slots which is equal to the chromatic number of the graph. Problem arises if after the examination schedule is published, some students choose a new course that makes the schedule invalid. This type of uncertainty exists in real world. Uncertainty of data comes from two different sources: randomness and incompleteness. Fuzzy graph model is used when uncertainty is there in the description of the objects or in its relationships or in both.

Let x^* be an optimal solution of an examination scheduling problem. But after scheduling, the input data changes. The change may be caused due to many reasons, e.g., new preliminaries are to be considered, data is updated or there may be some disturbances. In that new situation, the previous optimal solution x^* may not be feasible or if feasible, may not be optimal. A solution is considered as robust if it is still feasible and optimal or nearly optimal if changes of the input data occur within a prescribed uncertainty set. It protects the decision maker against parameter ambiguity and stochastic uncertainty.

There are few works on graph coloring problems with robustness. However, there is no work, as per our knowledge, that has incorporated robustness and fuzzy graph coloring in an integrated manner for scheduling problem. Works on robustness in scheduling problem can be found in [9], [8], [4], [3]. Some works related to fuzzy graph coloring are there in [1], [6], [7], [5]. A work, somewhat similar to us, can be found in [9]. However, the authors have used crisp graph to represent the model. In their robust graph coloring problem (RGCP) at most c colors can be assigned to all the vertices of the graph, where c is the given number of available colors. RGCP imposes a penalty for each pair of nonadjacent vertices if they are equally colored. Accordingly, the objective of the RGCP is to minimize the rigidity level of the coloring, which is defined as the sum of such penalties. They have proposed a genetic algorithm to find the robust solution for the problem.

In this paper, we propose a method to find the robust solution of an examination scheduling problem of a university in uncertain environment. We call this problem as Fuzzy robust coloring problem (FRCP). Probability of fuzzy event is used to bridge the crisp world (probability) and real world (fuzzy). We use a genetic algorithm for FRCP, where each chromosome represents an ordered arrangement of nodes without repetition and is checked its validity keeping the number of colors fixed. Here, we propose a fitness function for the chromosome based on the concept of fuzzy probability of a fuzzy event. Roulette wheel selection, two point crossover and one point mutation are used in the genetic algorithm.

The paper is organized as follows. Section 2 briefly reviews some basic definations associated with fuzzy graph coloring and probablity. In Section 3, we describe the examination scheduling problem by fuzzy graph model. We present an algorithm based on GA for FRCP in Section 4. The results of the proposed approach are presented in Section 5. Finally, we conclude in Section 6.

2   Preliminaries

In this section, we present some definitions related to fuzzy graph and fuzzy probability, which are associated with the proposed model.

Definition 1 [7]. Let V be a finite non empty set. A fuzzy graph G is defined by a pair of functions σ and μ, $G = (\sigma, \mu)$. σ is a fuzzy subset of V and μ is a symmetric fuzzy relation on σ, i.e., $\sigma\colon V \longrightarrow [0, 1]$ and $\mu\colon V \times V \longrightarrow [0, 1]$ such that

$$\mu(u, v) \leq (\sigma(u) \cap \sigma(v)), \forall u, \forall v \in V \tag{1}$$

A fuzzy graph is a generalization of crisp graph.

Definition 2 [6]. A fuzzy graph $G = (\sigma, \mu)$ is a complete fuzzy graph if

$$\mu(x, y) = (\sigma(u) \cap \sigma(v)), \forall u, \forall v \in \sigma \tag{2}$$

Definition 3 [7]. The complement of a fuzzy graph $G = (\sigma, \mu)$ is also a fuzzy graph and it is denoted as $\overline{G} = (\overline{\sigma}, \overline{\mu})$, where $\overline{\sigma} = \sigma$ and

$$\overline{\mu}(u, v) = \sigma(u) \wedge \sigma(v) - \mu(u, v) \tag{3}$$

Definition 4 [10][11]. Let event $\widetilde{A}$ is a fuzzy event or a fuzzy set considered in the space $\Re^n$. All events in the space $\Re^n$ are mutually exclusive and the probability of each event is as follows.

$$P(\widetilde{A}) = \sum_{x \in \widetilde{A}} \mu_{\widetilde{A}}(x)P(x) \tag{4}$$

Here, $\mu_{\widetilde{A}}(x)$ is the membership function of the event $\widetilde{A}$.

Definition 5 [5]. A family $\Gamma = \{\gamma_1, \gamma_2, ..., \gamma_k\}$ of fuzzy subsets on V is called a k- fuzzy coloring of $G = (\sigma, \mu)$ if
(a)$\gamma_1 \cup \gamma_2 \cup ... \cup \gamma_k = \sigma$
(b)$min\{ \gamma_i(u), \gamma_j(u) \mid 1 \leq i \neq j \leq k\} = 0, \forall u \in V$
(c)For every strongly adjacent vertices u, v of G, $min\{ \gamma_i(u), \gamma_i(v)\} = 0$ $(1 \leq i \leq k)$

Fuzzy chromatic number of fuzzy graph G is the least value of k for which the G has k-fuzzy coloring and is denoted by $\chi_F(G)$.

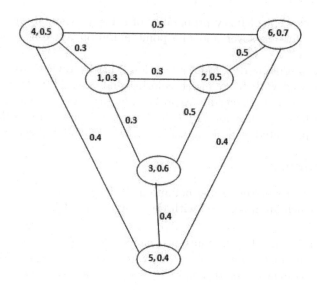

Fig. 1. Fuzzy graph model of an examination schedule

3   The Examination Scheduling Problem

Let us consider an example of examination scheduling problem, where the examination of 6 courses of a particular program at a university has to be scheduled. Two courses sharing at least one student cannot be scheduled on the same day. Taking into account the course incompatibilities, a fuzzy graph $G = (\mu, \sigma)$, shown in Fig 1, is constructed, where a node / vertex represents a course and an edge $\{i, j\}$ is included in the edge set when the courses i and j share at least one student. Here, σ and μ are respectively the membership values of the vertices and edges. Membership value of a node measures the degree to which an event occurs. Let the membership values of the vertices of the fuzzy graph are $\sigma(1)$ =0.3, $\sigma(2) = 0.5$, $\sigma(3) = 0.6$, $\sigma(4) = 0.5$, $\sigma(5)= 0.4$ and $\sigma(6)= 0.7$. We use a complete fuzzy graph to represent the problem and find the membership values of the edges using (2).

In order to minimize the duration of examinations, a minimal coloring problem is considered. From the Definition 5, the fuzzy chromatic number of the fuzzy graph is computed as $\chi_F(G) = 3$ and the following coloring γ_1^0, in this sense, is optimal.

$$\gamma_1^0(u_1) = \begin{cases} 0.3 & i = 1 \\ 0.4 & i = 5 \\ 0, & \text{otherwise} \end{cases} \qquad \gamma_1^0(u_2) = \begin{cases} 0.5 & i = 2 \\ 0.5 & i = 4 \\ 0, & \text{otherwise} \end{cases}$$

$$\gamma_1^0(u_3) = \begin{cases} 0.6 & i = 3 \\ 0.7 & i = 6 \\ 0, & \text{otherwise.} \end{cases}$$

Table 1. Probability and membership values of edges of $\overline{G}$

$\{i,j\} \in \overline{E}$	pr_{ij}	$\mu_{i,j}$
1-5	0.0506	0.3
1-6	0.0127	0.3
2-4	0.4557	0.5
2-5	0.3038	0.4
3-4	0.1519	0.5
3-6	0.0253	0.6

We measure the probability of the fuzzy event [11], [10] and use it as robustness value of the model. The fuzzy probability that a complementary edge will be added to the graph is computed. Using (3), we also compute the membership values of all complementary edges of the fuzzy graph, shown in Table 1. The robustness of the coloring can be measured as the fuzzy probability of such coloring after one random complementary edge has been added to the edge set. We assume that the probability of the complementary edge $\{i,j\}$ be proportional to the number of students registered in courses i and j. Let $n = 50$ be the total number of students who has chosen two out of the six courses and n_i be the number of students registered in course i. In this example, we consider $n_1 = 5$, $n_2 = 30$, $n_3=10$, $n_4=30$, $n_5=20$ and $n_6 = 5$.

Given the coloring γ_1^0, the probability that the coloring remains valid, assuming statistical independence, can be computed as follows.

$$(1 - pr_{15} \cdot \mu_{1,5})\,(1 - pr_{24} \cdot \mu_{2,4})\,(1 - pr_{36} \cdot \mu_{3,6})=0.9477$$

Let us consider the following coloring γ_1^1,

$$\gamma_1^1(u_1) = \begin{cases} 0.3 & i = 1 \\ 0.7 & i = 6 \\ 0, & \text{otherwise} \end{cases} \qquad \gamma_1^1(u_2) = \begin{cases} 0.5 & i = 2 \\ 0.4 & i = 5 \\ 0, & \text{otherwise} \end{cases}$$

$$\gamma_1^1(u_3) = \begin{cases} 0.6 & i = 3 \\ 0.5 & i = 4 \\ 0, & \text{otherwise.} \end{cases}$$

Then the corresponding fuzzy probability is

$$(1 - pr_{16} \cdot \mu_{1,6})\,(1 - pr_{25} \cdot \mu_{2,5})\,(1 - pr_{34} \cdot \mu_{3,4})=0.80866$$

If the number of days to schedule the examination is not 3 as before, say 4, the fuzzy probability increases choosing the following coloring γ_1^2.

$$\gamma_1^2(u_1) = \begin{cases} 0.3 & i = 1 \\ 0.4 & i = 5 \\ 0, & \text{otherwise} \end{cases} \qquad \gamma_1^2(u_2) = \begin{cases} 0.6 & i = 3 \\ 0.7 & i = 6 \\ 0, & \text{otherwise} \end{cases}$$

$$\gamma_1^2(u_3) = \begin{cases} 0.6 & i = 2 \\ 0, & \text{otherwise} \end{cases} \qquad \gamma_1^2(u_4) = \begin{cases} 0.5 & i = 4 \\ 0, & \text{otherwise} \end{cases}$$

The fuzzy probability that this coloring γ_1^2 remains valid is $(1 - pr_{15} \cdot \mu_{1,5})(1 - pr_{36} \cdot \mu_{3,6})$=0.96987, which is more than the previous two colorings.

4   Proposed Genetic Algorithm for FRCP

Genetic Algorithm [2] is a probabilistic heuristic search process based on the concepts of natural genetics. Given an examination scheduling problem, proposed genetic algorithm finds the robust solution for this problem. Below we describe the chromosome encoding, fitness function, coloring function and genetic operators.

4.1   Encoding

Each node in the graph is represented by a unique integer value. A chromosome represents an order of all nodes without repetition. Thus, the chromosome length is same as the number of vertices in the graph. We have to find a coloring function for each chromosome and check its validity keeping the number of colors fixed.

4.2   Fitness Function

The value of the fitness function is a measure of the quality of a solution. More is the fitness, better the solution is. For FRCP, the fitness of ith chromosome, given below in (5), is calculated using the penalty values and the membership values of the complimentary edges. $R(C_i)$ represents the rigidity level of the coloring function C of the i_{th} chromosome. Thus, it is a minimization problem.

$$R(C_i) = \sum_{\{i,j\} \in \overline{E}, C(i)=C(j)} p_{ij} \cdot \mu_{i,j} \tag{5}$$

Here, $\mu_{i,j}$ and p_{ij} are respectively the membership value and the penalty value of the edge (i, j). The penalty values for the complimentary edges and membership values of the vertices are generated randomly with uniform distribution in the interval $[0, 1]$.

4.3   Greedy Algorithm

Given the adjacency matrix of a graph and an optimal order of nodes $optorder$, greedy algorithm, presented below in Algorithm 1, construct a coloring $C{:}\to \{1, 2, ..., c\}$, such that for every edge (i, j), $C(i) \neq C(j)$.

Algorithm 1. Greedy Algorithm

1: **begin**
2: **for** i=1 to n **do**
3:     set color(i) = 0
4: **end for**
5: Set color($optorder(1)$) = 1
6: **for** j=2 to n **do**
7:     Choose a color k >0 for the node $optorder(j)$ that differs from its neighbor's colors
8: **end for**
9: **end**

Let us consider the chromosome 4 1 3 2 6 5, which is an ordered arrangement of the nodes for the fuzzy graph shown in Fig. 1. We take this order as the optimal order. First, node 4 is colored with color 1.The next node in the optimal order is node 1, which is adjacent to node 4, cannot be colored with color 1. So, node 1 is assigned color 2. Similarly, we traverse through the optimal ordering one by one and assign colors to each node so that every node has a color different from the colors of its neighboring nodes. We finally get a coloring C as $C(1) = 2$, $C(2) = 3$, $C(3) = 1$, $C(4) = 1$, $C(5) = 3$, and $C(6) = 2$.

4.4 Genetic Operators

We use Roulette wheel selection. A two point crossover is used in the proposed approach, which is described in Fig. 2. Two parents $P1$ and $P2$ are selected from the mating pool and the following operation is performed to obtain two children $C1$ and $C2$.

Step 1. Choose two random points $cp1$ and $cp2$ such that $cp1$ <$cp2$.
Step 2. Copy the nodes from $P2$ into $C1$ starting from position $cp1$ to the position $cp2$.
Step 3. Traverse the nodes of $P1$ one by one from 1 to n, if a node of $P1$ is not already present in $C1$ then copy it into $C1$ at the first available position from 1 to n.
Step 4. Copy the nodes from $P1$ into $C2$ starting from position $cp1$ to the position $cp2$.
Step 5. Traverse the nodes of $P2$ one by one from 1 to n, if a node of $P2$ is not already present in $C2$ then copy it into $C1$ at the first available position from 1 to n.

For mutation two points or positions are selected randomly from a chromosome and the nodes in those positions are exchanged. It is illustrated in Fig. 3.

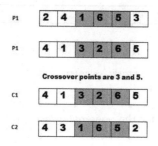

Fig. 2. Illustration of crossover operator

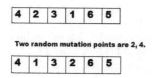

Fig. 3. Illustration of mutation operator

Table 2. Results with proposed GA and binary programming

no. of vertices	no. of colors	Rigidity (binary programming)	Rigidity (proposed)
6	3	0.69892	0.69892
6	4	0.50345	0.50263
8	4	1.4256	1.327433
10	3	1.9991562	1.999152
10	4	1.89543	1.900034
12	4	2.897612	2.897625
12	5	2.56734	2.6274512

5   Implementation and Results

In our experiment, number of generations, population size, crossover probability and mutation probability are respectively 30, 20, 0.6 and 0.1. The results of the proposed approach on a set of graphs with number of nodes 6, 8, 10 and 12 are shown in Table 2. The results for same set of graphs using binary programming with the same penalty and membership values are also shown in Table 2.

Our results are more or less similar to those of binary programming. It shows the effectiveness of the proposed approach on GA. Here, we consider binary programming as we did not find, best of our knowledge, any existing algorithm for robustness on fuzzy graph. The results show that for a particular graph when the number of colors increases the rigidity level of the coloring function decreases. The rigidity level is a measure of penalty when the coloring function becomes invalid on addition of any complimentary edge. The lower is the value of rigidity, the probability of the coloring function to be invalid is less.

6   Conclusions

In this work, we present the fuzzy robust graph coloring model that is simple, flexible, and easy to communicate to decision maker. With this model, we address the examination scheduling problem under uncertainty based on genetic algorithm, which produces robust solutions. Uncertainty in data comes from two different sources: randomness and vagueness. This model handles both type of uncertainty in decision making and bridges fuzzy graph coloring with robust optimization. Our model is capable of solving many real world optimization problems in uncertain environment.

References

1. Blue, M., Bush, B., Puckett, J.: Unified approach to fuzzy graph problems. Fuzzy Sets and Systems 125(3), 355–368 (2002)
2. Goldberg, D.E., et al.: Genetic algorithms in search, optimization, and machine learning, vol. 412. Addison-Wesley, Reading (1989)
3. Hasuike, T.: Robust shortest path problem based on a confidence interval in fuzzy bicriteria decision making. Information Sciences 221, 520–533 (2013)
4. Lim, A., Wang, F.: Robust graph coloring for uncertain supply chain management. In: Proceedings of the 38th Annual Hawaii International Conference on System Sciences, HICSS 2005, p. 81b. IEEE (2005)
5. Muñoz, S., Teresa Ortuño, M., Ramırez, J., Yáñez, J.: Coloring fuzzy graphs. Omega 33(3), 211–221 (2005)
6. Sunitha, M.: Studies on fuzzy graphs. Ph.D. thesis, Cochin University of Science and Technology (2001)
7. Sunitha, M., Vijayakumar, A.: Complement of a fuzzy graph. Indian Journal of Pure and Applied Mathematics 33(9), 1451–1464 (2002)
8. Wang, F., Xu, Z.: Metaheuristics for robust graph coloring. Journal of Heuristics 19(4), 529–548 (2013)
9. Yáñez, J., Ramırez, J.: The robust coloring problem. European Journal of Operational Research 148(3), 546–558 (2003)
10. Zadeh, L.A.: Fuzzy sets versus probability. Proceedings of the IEEE 68(3), 421 (1980)
11. Zadeh, L.A.: Probability measures of fuzzy events. Journal of Mathematical Analysis and Applications 23(2), 421–427 (1968)

A Real Time Fast Non-soft Computing Approach towards Leaf Identification

Pratikshya Mohanty[1,*], Avinash Kranti Pradhan[1,*],
Shreetam Behera[1], and Ajit K. Pasayat[2]

[1] Dept. of ECE, Centurion University of Technology and Management, Jatni, 752050, India
{pm.pratikshya,kranti.avi,shreetam}@gmail.com
[2] Dept. of CSE, Centurion University of Technology and Management, Jatni, 752050, India
akpasayat@gmail.com

Abstract. In an agricultural country like India, majority of population depend on plant produce for their survival. Plants occupy a large portion of our ecosystem. In order to derive different benefits from plants in an optimum manner, one needs to be aware of the properties being possessed by plants. For that purpose, one needs to have proper source carrying significant information about plants and an expert so as to respond to ones queries. However, both these are not available in adequate which drives the need to create automation in the process of recognition of leaves for plant classification. Thus, a novel algorithm has been developed which helps in recognizing different varieties of leaves without human interference. The system uses real time images of leaves and extracts physiological as well as morphological features of the leaves, which are then fed as input to a classifier. The same has been implemented on a Back propagation based neural network classifier and a comparative study has been made. The study shows that the recognition rates of the proposed method are more accurate than that of BPNN and the proposed algorithm is found to be an efficient one.

Keywords: Features, Physiological features, Morphological features, Classification System, Recognition rate, Real time implementation.

1 Introduction

India is an agricultural country. It is the agriculture sector that provides people with food, raw materials and huge employment opportunities. More than 70% of the country's population depends on agriculture directly. Thus with the advancement in technology, there should be significant improvement in the field of agriculture, that will not only enhance yield and productivity but also beneficial to the various stakeholders. Now-a-days the study of leaf identification is in demand because of the availability of huge varieties of plants with very little or almost no information. Moreover, the major problem lies with the shortage of experts and lack of knowledge base. Thus, it

---

* Corresponding author.

© Springer International Publishing Switzerland 2015
S.C. Satapathy et al. (eds.), *Proc. of the 3rd Int. Conf. on Front. of Intell. Comput. (FICTA) 2014*
– *Vol. 1*, Advances in Intelligent Systems and Computing 327, DOI: 10.1007/978-3-319-11933-5_92

has become important to develop an algorithm which will recognize different plant species and in turn, will help to maintain a database of plants, providing significant information about plants.

In general, identification of plants is done on the basis of shape, color and structure of leaves, flowers and fruits which are 3D structures. Instead of analyzing plants on these complex 3D structures, plants can be recognized from simple 2D images of leaves also. However, in leaf recognition, one cannot consider color as a feature because almost all leaves are green in color. Thus in this study, different physiological as well as morphological features are extracted from the digital images of leaves and are fed to a classification system for recognition of plants. In this paper, a robust plant identification system has been proposed which would help not only farmers and agriculturists but also common people with limited botanical expertise.

Several research works have been conducted in the field of agriculture for the application of leaf recognition and plant classification. In [1], a leaf identification system has been developed in which the shape and vein of the leaf has been considered as the basis of identification. The system performance has been found to be 97.19%. A method based on leaf feature extraction has been implemented [2]. This system uses algorithms based on image processing and machine learning techniques.

[3, 4] presents the concept of widely used machine vision which uses machines instead of human eyes in carrying out measurements and judgements. This concept of machine vision has been applied in the field of agriculture since 1970s, mainly for testing of quality and classification of fruits [5,6].Many image processing techniques were employed to extract texture features from captured images in various applications [7,8]. [9] Presents a paper on fruits defect detection by feature extraction. In this algorithm the weights are being calculated for different features. In [10], a novel color thresholding approach has been implemented for the successful demarcation of rotten pointed gourds from mixture containing fresh and rotten ones, considering two color textures i.e. green and yellow. A combination of machine vision and machine intelligence has been used for betel leaf classification. The machine vision segment extracted the different leaf features and the machine intelligence part implemented the Back propagation neural network [11]. A method has been presented for the identification of medicinal plants based on their leaf features considering area and edge [12]. In [13], a physiological based classification system has been developed which identifies leaves based on its basic physiological features. [14] Presents a legume leaf classification method based on the leaf vein analysis.

2   Proposed Methodology

In this study, the leaf identification is done by considering the physiological as well as morphological features of leaves. Here, 13 features have been taken into account. First, the leaf image is acquired and pre-processed to extract its features. The images of the leaves are stored to form a database. When an image of a leaf is taken for testing purpose, the particular leaf is pre-processed and its features are compared with the features of the stored images on a classification system.

2.1   Image Acquisition

The real time image of leaves has been captured using a webcam (Vimicro USB2.0 UVC PC Camera). The images are taken from the top with white background. Five different varieties of leaves are taken. The samples of leaf images are shown in Figure 1.

(a) Betel   (b) Jackfruit   (c) Money plant   (d) Mango   (e) Oleander

Fig. 1. Leaf samples

2.2   Image Pre-processing

Here, the RGB image of the leaf is converted to gray scale image by using the following formula:

Gray scale image= red component * 0.3 + green component * 0.59 + blue component
* 0.11   (1)

The gray scale image is then, converted to black and white. Max filtering is applied for noise reduction and making the image smooth.

2.3   Feature Extraction

40 leaves of each class are taken and 13 physiological and morphological features are extracted as follows:

- Area:
The leaf area is calculated by counting the total number of pixels embedded within the leaf boundary.

- Perimeter:
The perimeter of a leaf is calculated by counting the number of pixels in the edge of the leaf. Sobel operator is used for edge detection.

- Length:
The distance between the two ends of the main vein of leaf is called its length.

- Width:

The leaf width is defined as the distance between the intersection point with length at the centroid and its opposite side on the margin of the leaf.

- Aspect Ratio:

It is the ratio of leaf length to leaf width.

$$Aspect\ ratio = Length\ of\ leaf/\ Width\ of\ leaf \tag{2}$$

- Diameter:

It is the length of a circle having area same as that of the leaf region.

- Form factor:

It is the difference between a leaf and a circle. It can be calculated as follows:

$$form\ factor = (4\Pi * leaf\ area) \div (leaf\ perimeter)^2 \tag{3}$$

- Rectangularity:

It is the similarity between rectangle and leaf area. It can be calculated as follows:

$$rec\tan gularity = (leaf\ length * leaf\ width) \div leaf\ area \tag{4}$$

- Length to perimeter ratio = leaf length/ leaf perimeter $\tag{5}$

- $$Ratio = leaf\ perimeter \div (leaf\ length + leaf\ width) \tag{6}$$

- $$Mean\ X_j = \sum_{i=1}^{n} x_i / n \tag{7}$$

- $$Narrow\ factor = diameter \div leaf\ length. \tag{8}$$

- Perimeter to diameter ratio= leaf perimeter/ leaf diameter. $\tag{9}$

2.4 Database Creation

The images of different leaf samples are acquired by using a webcam which are stored in a database in .jpg format. Each class of leaf consists of forty (40) image samples stored in the database in .jpg format. The database showing leaf samples from different class of leaves is shown in Figure 2.

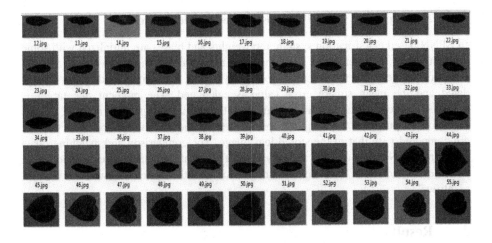

Fig. 2. Leaf Database

2.5   Proposed Classification System

A novel approach has been proposed for identification of different class of leaves. In this study, first a database is created which consists of the images of different samples of leaves. At the time of testing, the images present in the database are pre-processed and their features are extracted and at the same time, the image to be tested undergoes pre-processing so that its features could be extracted. Then the extracted features of stored leaf and tested leaf are compared with a consideration of 2% error. The image whose maximum features match with the features of the stored leaf image is thus identified and the one whose features do not match at all could not be identified. Here, the error of 2% is taken into account because out of 13 features, there may be chance of mismatch of two or three features. However, this error can be reduced by increasing the number of features.

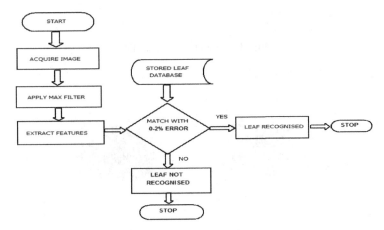

Fig. 3. Proposed Flow Chart

2.6   A Comparison with BPNN Based Classification System

In BPNN based classification system, the weights and biases are initialized with some random values, a target vector is defined and the system is trained by specifying the number of epochs. By the completion of every epoch, the weights and biases are updated and the mean square error value is calculated. The moment, the target is achieved or the number of epochs complete, the training stops. When a leaf is captured for the purpose of testing, it is pre-processed and its features are extracted and are compared with that of the stored leaf image. The comparison and matching takes place in BPNN based classification system. If the features match, the defined target is achieved and hence the leaf is identified or else the system shows "Leaf is not identified".

3   Results

When an image of a leaf sample is acquired for the purpose of testing, first it is pre-processed to find its features. At the same time the image stored in the database is pre-processed to extract its features. The features of both the images are compared. The one, whose features matches maximum with an error of 2%, is identified equivalent to its corresponding leaf variety. The results are shown in the next subsequent figures. A comparative study of recognition rates of the proposed system with that of the BPNN Classification system is shown in Table 1.

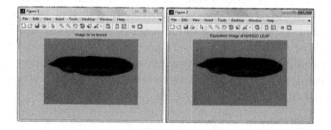

Fig. 4. Mango Leaf is identified

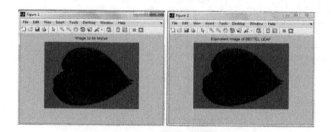

Fig. 5. Betel Leaf is identified

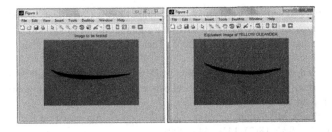

Fig. 6. Oleander Leaf is identified

Table 1. A comparative study between proposed system & BPNN system

Leaf Sample	No. of Leaf Samples	No. of samples for testing	No.of samples identified correctly using BPNN System	No.of samples identified correctly using proposed system	Recognition rate in case of BPNN system	Recognition rate in case of proposed system
Betel	40	20	17	19	85%	95%
Jack-fruit	40	15	13	15	86.67%	100%
Mango	40	15	12	15	80%	100%
Money Plant	40	15	12	14	80%	93.33%
Yellow Oleander	40	17	15	16	88.23%	94.11%

4   Conclusion

This paper introduces a novel approach towards identification of different varieties of leaves implemented on platform MATLAB version 7.0. The proposed method is simple, cheap and easy to implement. Moreover, the time complexity is reduced since there is no need to train the system unlike other machine learning complex algorithms. Also, the database is easy to be maintained. The results are found to be more accurate in case of the proposed algorithm since an error of 2% is assumed. Thus the proposed system is found to be robust and efficient as compared to BPNN based classification system. This work can be extended further to include more number of leaves and features. The same work can be implemented on other classification system and a comparative study can be made.

References

1. Lee, K.B., Hong, K.S.: An Implementation of Leaf Recognition System using Leaf Vein and Shape. International Journal of Bio-Science and Bio-Technology 5(2), 57–66 (2013)

2. Tzionas, P., Papadakis, E.S., Manolakis, D.: Plant Leaves Classification based on morphological features and fuzzy surface selection technique. In: 5th International Conference on Technology and Automation, ICTA 2005, Thessaloniki, Greece, October 15-16, pp. 365–370 (2005)

3. Lee, D.J., Schoenberger, R., Archibald, J.: Development of a machine vision system for Automatic date grading using digital reflective near- Infrared imaging. Journal of Food Engineering 86, 388–398 (2008)

4. Feng, B., Wang, M.: Computer vision classification of fruit based on fractal color. Transactions of the CSAE 18(2), 141–144 (2002)

5. Bao, X., Zhang, R.: Apple grade Identification method based on artificial neural network and image processing. Transactions of the CSAE 20(3), 109–112 (2004)

6. Pearson, T.: Machine vision system for automated detection of stained pistachio nuts. Lebensmittel-Wissenschaft & Technology 29, 203–209 (1996)

7. Weszka, J.S., Rosenfeld, A.: An application of texture analysis to materials inspection. Pattern Recogn. J., 195–199 (1976)

8. Garcia, P., Petrou, M., Kamata, S.: The use of Boolean model for texture analysis of grey images. Computer Vis. Image Understand. 74(3), 227–235 (1999)

9. Patel, H.N., Jain, R.K., Joshi, M.V.: Fruit Detection using Improved Multiple Features based Algorithm. International Journal of Computer Applications 13(2), 1–5 (2011)

10. Mohanty, P., Pradhan, A.K., Behera, S.: A Real time based Image Segmentation Technique to Identify Rotten Pointed Gourds. International Journal of Engineering and Innovative Technology 3(4), 144–148 (2013)

11. Kumar, S.: A novel Neural network based approach for the Classification of Betel Leaves. International Journal of Emerging Trends & Technology in Computer Science 1(2), 10–16 (2012)

12. Kumar, S.: Leaf Colour, Area and Edge features based approach for Identification of Indian Medicinal Plants. International Journal of Computer Science and Engineering 3(3), 436–442 (2012)

13. Mohanty, P., Pradhan, A.K., Behera, S.: A Real time based Physiological Classifier for Leaf Recognition. International Journal of Advanced Computer Research 4(14), 337–345 (2014)

14. Larese, M.G., Namias, R., Craviotto, R.M., Arango, M.R.: Automatic classification of legumes using leaf vein image features. Pattern Recognition 47, 158–168 (2014)

Cost Reduction in Reporting Cell Planning Configuration Using Soft Computing Algorithm

S.R. Parija[1,*], P. Addanki[1], P.K. Sahu[1], and S.S. Singh[2]

[1] Department of Electrical Engineering, NIT Rourkela
{smita.parija,pratima.addanki}@gmail.com,
pksahu@nitrkl.ac.in
[2] KIIT University, School of Electrical Science, BBSR
sudh_69@yahoo.com

Abstract. This paper presents Binary Genetic Algorithm (BGA) is a heuristic, adaptive population based method and which has shown to be a very powerful global search method used for optimization process. Using BGA the objective of this work is used to minimize the location management cost thereby achieve trade-off between location update and paging cost based on reporting cell planning configuration. This BGA algorithm is used to solve location management cost using reporting cell planning problem. With the use of reporting cell location management some cells are designated as reporting cells where mobile station (MS) updates its location upon entering the same coverage. The effectiveness of the technique is tested for collected real data for validation and presented in the paper. The simulation results obtained from this work with reasonable degree of accuracy are very encouraging.

Keywords: Cost function, Location management,Reporting Cell Planning, Genetic Algorithm.

1 Introduction

In recent years mobile subscribers increasing drastically that has resulted in exploitation of mobile network resources, being applied to other newly and renovated applications such as fax services, voice and data transfer [1]. This communication network must support a large number of subscribers and maintaining a good response without losing quality also availability. To keep this quality, it is necessary to consider location management while making the design of network infrastructure.

In the area of location management when a call routed to the user the mobile network has to keep the information of the user in that specific coverage area is known. Even when the mobile station is not active it keeps the mobile network informed of its location. The location management basically management of the network design and configuration.

---

* Corresponding author.

© Springer International Publishing Switzerland 2015
S.C. Satapathy et al. (eds.), *Proc. of the 3rd Int. Conf. on Front. of Intell. Comput. (FICTA) 2014*
– *Vol. 1*, Advances in Intelligent Systems and Computing 327, DOI: 10.1007/978-3-319-11933-5_93

There are different schemes used to solve the location management issue. One is location area scheme and another one is reporting cell scheme. Reporting cell scheme is explained in next section. In location area scheme the location area consist of cluster of cells. Location update occurs when a user moves from one coverage area to another coverage area otherwise no location updating performed by the MS. When a call or SMS arrives to a MS then location area of that MS is need to be searched by the mobile network for which some amount is to be incurred to the user. The cost involved for the location management cost is location update cost and paging cost.

There is a trade-off between this location update cost and paging cost [2, 3]. The optimum location area configurations consistent to the minimized cost function shown below are:

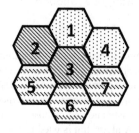

Fig. 1. Optimal LA planning for 7 cell network

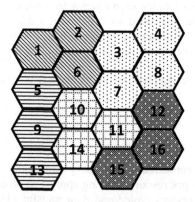

Fig. 2. Optimal LA planning for 16 cell network

So with the objective of reduction of total cost in mobile network using different soft computing techniques to optimize the network for solving various issues. Binary Genetic Algorithm is one of the techniques used to optimize the resources used for minimization of the location cost [4].

This paper is structured as follows: In section 2 explains an overview of reporting cell planning and its involvement in location cost. Section 3 describes Genetic based algorithm is described and its related parameter elaborately explained. In section 4 presents analysis of simulation results. Finally section 5 includes conclusion and future scope.

2   Location Management

Location management is an important issue in cellular network as it includes location update as well as paging. Location update that enables the mobile network to find the current location of the mobile station is static or dynamic within a coverage area. Whenever a mobile station moves to a new coverage area with a new cell, known as location update. Paging is a process when network broadcast the messages to all the mobile terminals in all the possible cells in order to find the mobile to which the call should be routed. In cellular network, notification of the current location involves location update performed by the mobile station and tracing of the mobile station by the mobile network is paging. The cost involved for the user movement and tracing is the location management cost and the main objective of the paper is to minimize the involved cost [5].

Reporting cell planning is another approach to location management was proposed by Bar and Kessler [6] which is characterizing by subset of cell. In cellular network using reporting cell location management cost can be reduced. Reporting Cell Planning (RCP) is one of the approaches to Location Management. In RCP, few cells in the cellular network are assigned as reporting cells and the cells other than reporting cells (RCs) are non-reporting cells. Reporting cells can be adjacent to each other sharing a boundary or scattered in a specified coverage area. Location update is performed whenever a mobile terminal enters a reporting cell. Next location update is done only when the mobile terminal enters or crosses a new reporting cell. During call arrival to a mobile terminal, paging is restricted to the last updated RC where the MT's location update has been last performed and to its neighboring non reporting cells. In simple terms, we can consider this set of reporting cells as a boundary, where all the call routing and call procedures are carried out.

3   Genetic Algorithm

Optimization is a process of finding a best solution for a given problem. RCC is a discrete optimization problem, where the location management cost has to be minimized and the corresponding optimal set of reporting cells are to be determined.

Like Genetic algorithm, other evolutionary algorithm such as binary particle swarm optimization and binary differential evolution algorithms are used to solve RCP problem. Each of these algorithms is discussed in detail below.

Evolutionary Computing is a major research area in the field of artificial intelligence. Evolutionary algorithms use randomness and genetic inspired operations. These algorithms start with an initial potential solution set called as population. Each solution in the population is called as a chromosome or an individual. Each chromosome consists of a set of genes. In accordance with the problem statement, each RCC solution is a chromosome and individual cell is a gene. Major operations involved in evolutionary algorithms are selection, crossover, mutation and competition of the individuals in the population. The general evolutionary process is show in the figure below.

GA is an adaptive heuristic approach able to find good, possibly optimal solutions, to optimization problems with huge state spaces to be searched. It acts as a global probabilistic search method. In this paper, Genetic Algorithm is used to minimize the location management cost and thereby obtain the optimal location area planning configuration [7,8].Procedure of basic Binary Genetic Algorithm is as follows:

1. Generate random N number of RCC solution.
2. Evaluate the total   location management cost of each   RCC solution, which is known as fitness value in terms of GA.
3. Create new RCC solutions by repeating following steps for *n* (number of solutions) times:

> a. Selection – Here RCC solutions are selected using Roulette wheel. The solution is selected on the basis of cost value. Probabilities are assigned to each solution. The solution with lesser cost value has higher probability of being selected.
> b. Crossover – Select two RCC solutions using roulette wheel and combine them to form a new RCC.
> c. Mutation-With a mutation probability, (taken as 0.05) the new solutions at random cell positions are mutated in the binary RCC solution vectors.

4. Updating the RCC – If the cost of the new RCC is better than the previous, replace the previous RCC with the new RCC.
5. Terminate - If the end condition is satisfied i.e. maximum number of iterations, stop and return the best RCC solution vector set in current RCC.
6. Loop – Repeat the process from step 2.

Evolution of genetic algorithm is presented below.

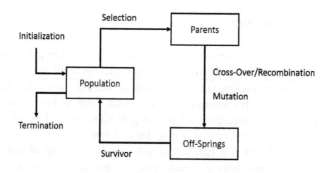

Fig. 3. Evolution of genetic algorithm

3.1   Problem Statement

In RCP location management method, mobile stations need to update their locations whenever they cross a reporting cell. During call arrival, users are located by paging in their last location updated reporting cell as well as its neighboring non reporting cells without crossing another reporting cell. For example, in the reporting cell configuration shown in Fig 4 below, cells 3, 6, 9, 10, 11, 13 and 16 are reporting cells and the others are non-reporting cells. Let us say, an MT's location has been last updated in cell 3. Therefore, when there is a call arrival to that mobile station, paging is done in cells 1, 2, 3, 4, 5, 7, 8 and 12. Location Management cost consists of location update cost and paging cost [9].

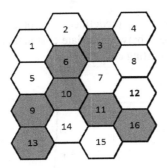

Fig. 4. Reporting cell configuration

Therefore, total cost of a given cellular network is equal to the sum of total number of location updates and total number of paging transactions over a certain period of time.

$$\text{Total Location Cost} = C * N_{LU} + N_P \tag{1}$$

Where, C is a positive constant which represents the cost ratio of location update to paging, and is taken as 10, i.e. C=10, because cost of location update is much higher than paging cost. NLU is the total number of location updates and NP is the total number of paging transactions in the given network. In a given cellular network, each cell i is associated with two weights: movement weight (W_{mi}) and call arrival weight (W_{ci}). W_{mi} represents the frequency (or total number) of movement of a MS into a cell. W_{ci} represents the frequency of call arrivals in a cell. The total number of location updates and paging transactions in a network can be calculated using these two weights as follows:

$$N_{LU} = C * \sum_{i \in s} w_{mi} \tag{2}$$

$$N_P = \sum_{j=0}^{N} w_{ci} * v(j) \tag{3}$$

Where N is the total number of cells in the network, S is the set of reporting cells and v(j) is the vicinity value of cell j. Vicinity value is defined as the maximum number of cells that can be searched if an incoming call is received in cell j. Thus, Location management cost for a reporting cell configuration is obtained as:

$$\text{Total Cost} = C * \sum_{i \in s} w_{mi} + \sum_{j=0}^{N} w_{ci} * v(j) \qquad (4)$$

$$\text{Cost Per call Arrival} = \frac{TotalCost}{\sum_{j=0}^{N} w_{cj}} \qquad (5)$$

Movement and call weights are predefined. The objective is to minimize the cost per call arrival with a trade-off between update cost and paging cost and find the corresponding set of optimal reporting cell set [10].

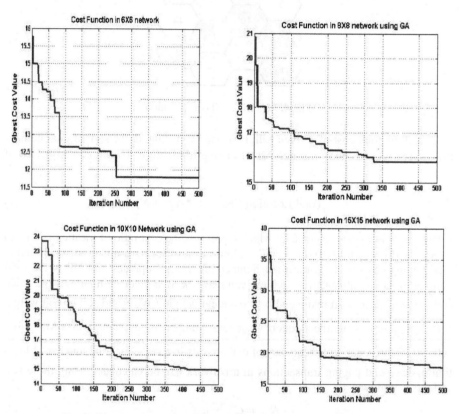

Fig. 5. Simulation result for different cell network (36, 64,100and125)

4   Results and Discussion

Location management cost function behavior using genetic algorithm for different test networks is shown in this section. Equation (5) is optimized using GA for 6X6, 8X8, 10X10, and 15x15 test networks. Each simulation is run for 500 iterations. The minimum location cost value for every iteration is plotted against the iteration number.

Simulation has been carried out in MATLAB for 500 iterations for the collected real data. The GA based RCP planning is verified on a 500 data set with cell of different sizes 36, 64, 100 and 125. The best optimum result for different network is shown in Fig 5. With the help of Pentium Core duo Processor 3.2 GHz CPU, 4 GB based RAM and 360 GB storage capacity based personal computer code is simulated to get the optimum result.

The population number is taken to be 200 in each simulation. Initial population is such that each cell belongs to one location area, which gives N location areas where N is the number of cells in the network. As seen from the above graphs, the total cost is decreasing by a reasonable amount of accuracy for the RCP based location management problem. The convergence of the graph improves with the increase in the number of populations.

Fig 5 is the performance curves for GA based on RCP in cellular network. In this work GA based solution obtained for the system with N=36, 64, 100, and 125 cells and it is observed that there is the minimization of the objective function. Here other parameters were also varied such as size of the network, mutation probability (0.8), number of population, potential solution etc. Roulette wheel cost function is used for minimizing the location area. The minimized cost is presented in Table 1.

Table 1. Cost comparison of Different cells network

Sl No	Network size	Initial cost	Minimized cost
1	36 cell	15	11.75
2	64 cell	20.9	15.9
3	100 cell	23	15
4	125 cell	36	19

5   Conclusion and Future Work

To minimize total location management cost Genetic Algorithm (GA) has been used and it is verified using collected real data. In this work location management cost is reduced by optimizing the network using reporting cell configuration. The future work of the same problem includes hybridization with other optimization techniques such as differential evolution, particle swarm optimization etc. that involves more detailed and in depth comparison with other search algorithms. Also in dynamic location management [11,12] one of the current challenging issues where the number of subscribers is increasing day by day, and beyond from 2G transitions, location management must be efficiently carried out, while allocating appropriate reporting cell planning in real-time, practical and implementable fashion.

References

1. Wong, V., Leung, V.: Location management for next generation personal communication networks. IEEE Network 14(5), 18–24 (2000)

2. Zhang, J.: Location Management in Cellular networks. In: Handbook of Wireless networks and Mobile Computing, pp. 27–49
3. Demestichas, P., Georgantas, N., Tzifa, E., Demesticha, V., Striki, M., Kilanioti, M., Theologou, M.: Computationally efficient algorithms for location area planning in future cellular systems. Computer Communications 23(13), 1263–1280 (2000)
4. Almeida-Luz, S., Vega-Rodriguez, M.A., Gomez-Pulido, J.A., Sanchez-Perez, J.M.: Applying Differential Evolution to the Reporting Cells Problem. In: Proceedings of the International Multiconference on Computer Science and Information Technology, pp. 65–71 (2008)
5. Sidhu, B., Singh, H.: Location management in cellular networks. In: Proc. of World Academy of Science, Engineering and Technology, vol. 21, pp. 314–319 (2007)
6. Subrata, R., Zomaya, A.: Artificial Life Techniques for Reporting Cell Planning in Mobile Computing. In: Proceedings of the International Parallel and Distributed Processing Symposium (IPDPS 2002), pp. 169–187 (2003)
7. Lin, Y.-B., Chlamatac, I.: Wireless and Mobile Network Architecture. John Wiley and Sons, Inc. (2001)
8. Agrawal, D.P., Zeng, Q.-A.: Introduction to Wireless and Mobile Systems. Thomson Brooks/Cole Inc. (2003)
9. Al-Tawil, K., Akrami, A., Youssef, H.: A new authentication protocol for GSM networks. In: Proceedings of the 23rd Annual Conference on Local Computer Networks, LCN 1998, October 11-14, pp. 21–30 (1998)
10. Jie, L., Kameda, H., Keqin, L.: Optimal dynamic location update for PCS networks. In: Proceedings of the 19th IEEE International Conference on Distributed Computing Systems (1998)
11. Vroblefski, M., Brown, E.C.: A grouping genetic algorithm for registration area planning. Omega 34(3), 220–230 (2006)
12. Gondim, P.R.: Genetic algorithms and the location area partitioning problem in cellular networks. In: Proc. of Vehicular Technology Conference, pp. 1835–1838 (1996)

Detection of Moving Object: A Modular Wavelet Approach

Latha Anuj[1], M.T. Gopalakrishna[2], and M.C. Hanumantharaju[3]

[1] Department of Information Science and Engineering,
Dayananda Sagar College of Engineering, Bangalore, India
[2] Department of Computer Science and Engineering,
KS School of Engineering and Management, Bangalore, India
[3] Department of Electronics and Communication Engineering,
BMS Institute of Technology, Bangalore, India
{aplatha123,gopalmtm,mchanumantharaju}@gmail.com

Abstract. In video surveillance, identification is a very significant element for target tracking, activity recognition, traffic monitoring, military etc. The identification process classifies the pixels into either foreground or background and a common approach used to achieve such a classification is background removal. A Novel method is proposed for the moving object detection based on Modular Wavelet approach, where two consecutive image from image sequences are divided into four parts and then, the Wavelet Energy (WE) is applied to each sub image. The sub image in turn has two energy values of WE, namely, the percentage of energy corresponding to the approximation and the detail. Comparing the energy values corresponding to the detail, the moving object is recognized. Since the discrete wavelet transform has a pleasant property that it can divide an image into four different frequency bands without loss of the spatial information and most of the fake motions in the background can be decomposed into the high frequency wavelet sub-band. Proposed method is compared with existing methods and proposed algorithm gives an enhanced performance.

Keywords: Video Surveillance, Background Model, Foreground Detection, Wavelet Transform, Wavelet Energy.

1 Introduction

Moving objects can be viewed as lower level vision tasks to achieve higher level event understanding. In traditional security systems, problems are spotted using the alarm functions of invasion detectors based on surveillance cameras. However, such systems often react to changes in the natural environment, which makes the identification of invasions is difficult task. Manoj S. Nagmode et al. [1] has proposed an algorithm in which Partitioning and Normalized Cross Correlation method is used for the detection and tracking of moving object from the image sequence. Sang Hyun Kim [2] proposed an algorithm in which moving edge extraction is done by using the concept of entropy and cross entropy. The cross

© Springer International Publishing Switzerland 2015
S.C. Satapathy et al. (eds.), *Proc. of the 3rd Int. Conf. on Front. of Intell. Comput. (FICTA) 2014*
– *Vol. 1*, Advances in Intelligent Systems and Computing 327, DOI: 10.1007/978-3-319-11933-5_94

entropy is applied to dynamic scene analysis which provides enhancement of detection for the dynamically changed area. Detecting regions of change in multiple images of the same scene taken at different times is of widespread interest due to a large number of applications in diverse disciplines [3]. Daviest, et. al. [4] has addressed the problem of detection and tracking of small, low contrast objects by using wavelet as well as Kalman filter. Li-Qun Xu [5] has addresses primarily the issue of robustly Detection of multiple objects. In this, first morphological reconstruction is used to remove cast shadows. Yiwei Wang, et. al. [6] uses wavelet decomposition and multi-resolution analysis to detect the moving object and dispersion calculation to track that object which takes more processing time. The main difficulty is to build a system that runs in real time on a standard PC and performs accurate detection of vehicles even under unfavorable illumination and weather conditions. Measures to evaluate quantitatively the performance of video object segmentation and tracking methods without ground-truth (GT) segmentation maps are presented in [7]. In [8] a real-time video tracking and recognition system for video surveillance applications is presented. The system takes a video sequence as the input of a background learning module and gives the output as a statistical background model that describes the static parts of the scene. From the survey, there is a necessity to build a efficient algorithm, which can employ in noisy and variable illumination conditions as well. In this paper, The Modular Wavelet Energy (MWE) is a novel and robust parameter to identify moving objects and is derived using discrete wavelet transformation.

2   Wavelet Transform

Wavelet transform (WT) is a technique for analyzing the time frequency domain that is most sited for non stationary signal. The importance of WT is to capture the localizing features of the signal. A continuous WT (CWT) maps a given function in time domain into two dimensional functions of $'s'$ and $'t'$. The parameter s is called the scale corresponding to frequency in Fourier transform and t is the translation of the wavelet functions. The CWT is defined by

$$CWT(s,t) = \frac{1}{\sqrt{s}} \int S(T) \phi \frac{(T-t)}{S} dt \qquad (1)$$

Where $S(T)$ is the signal and $\phi(T)$ is the basic wavelet and $\phi \frac{(T-t)}{\sqrt{s}}$ is the wavelet basis function. The discreet wavelet transform (DWT) is normally used for short time analysis. The DWT for a signal can be written as

$$DWT(m,n) = \frac{1}{2^m} \sum_{i=1}^{N} S(I)(i)[2^{-m}(i-n)] \qquad (2)$$

For many natural signals, the wavelet transform is a more effective tool than the Fourier transform. The wavelet transform provides a multi-resolution representation using a set of analyzing functions that are dilations and translations of a

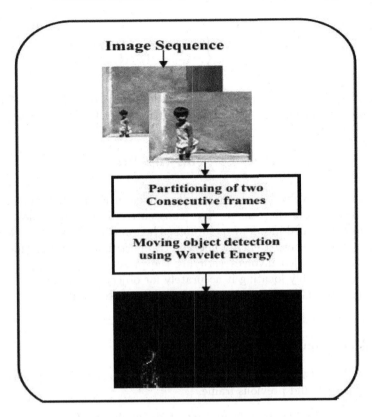

Fig. 1. Block Diagram of the Proposed System

few functions (wavelets). Calculating wavelet coefficients at every possible scale is a fair amount of work, and it generates an awful lot of data. If the scales and positions are chosen based on powers of two, the so-called dyadic scales and positions, then calculating wavelet coefficients are efficient and just as accurate. This is obtained from discrete wavelet transform (DWT). They have advantages over traditional Fourier methods in analyzing physical situations where the signal contains discontinuities and sharp spikes.

2.1   Wavelet Energy

Wavelet Energy is mainly used for measuring variance between two images. It is useful in feature recognition and registration. For basic wavelet selection, second derivative of a Gaussian, Haar, Shannon are the common ones. Among these wavelets, since the Haar wavelet achieves the best performance in our experiments, it is used in this work. The Haar wavelet in Eqn. (2) is described.

$$\phi[2^{-m}(i-n)] = \begin{cases} 1, & 0 \le 2^{-m}(i-n) \le \frac{1}{2} \\ -1, & \frac{1}{2} \le 2^{-m}(i-n) \le 1 \\ 0, & otherwise \end{cases}$$

The outputs of DWT at different scales contain different amounts of moving object and non moving object information. The Wavelet Energy is computed in Eqn.(3).

$$WE = \frac{1}{2^{-\frac{m}{2}}} \sum_{i=1}^{N} S(I)(i)[2^{-m}(i-n)] \tag{3}$$

The Basic steps involved in the process are given in Figure 1. Input image sequence is taken from the camera. Two consecutive frames from the image sequence are partitioned into four quadrants. Then moving object detection takes place after finding wavelet energy between two partitioned frames. Moving Object detection in video involves verifying the presence of an object in image sequence and possibly locating it accurately for object identification. The basic algorithm steps for the detection of moving objects are given below.

– Algorithm for MWE

Input: Read two consecutive frames from the image sequence called as current frame and previous frame.

1. **Step-1**: Divide these frames into four quadrants. In this situation current frame is divided into four parts called as P1, P2, P3 and P4. Similarly, previous frame is divided into four parts called as Q1, Q2, Q3 and Q4.
2. **Step-2** : Now find out the WE of each sub image of current frame and the previous frame. This gives detailed coefficients. After this there are eight detailed coefficients values of WE, called as Ed1, Ed2, Ed3, Ed4, Ed5, Ed6, Ed7 and Ed8.
3. **Step-3**: Now compare the value of WE from these detailed coefficients values.
4. **Step-4**: Suppose the Ed1 is greater than Ed5 value of WE is obtained at the first quadrant, it means that the moving object is present in that quadrant.
5. **Step-5**: Now operate in the first quadrant. Take the difference between the first quadrants of two consecutive frames.
6. **Step-6**: If the second Ed2 value is also greater than Ed6 then it means that the moving object is present in that quadrant. Now, identify the location of second moving object and Detect that object.
7. **Step-7**: Repeat the same procedure for the next frame.

Table 1. TP, FP and FN of different Image Sequences

Input Sequences	TP	FP	FN
S1	74	0	0
S2	75	0	0
S3	50	0	0
S4	50	25	25
S5	100	25	50
S6	200	0	50
S7	34	12	10
S8	32	1	2
S9	46	0	30
S10	55	0	30
S11	42	0	10
S12	46	0	5

Table 2. DR and FAR for Different Image Sequences

Input Sequences	Detection Rate	False Alarm Rate
S1	100.00	0.00
S2	100.00	0.00
S3	100.00	0.00
S4	66.67	33.33
S5	66.67	20.00
S6	80.00	0.00
S7	77.27	26.09
S8	94.12	3.03
S9	60.53	0.00
S10	64.71	0.00
S11	80.77	0.00
S12	90.20	0.00

3 Experimental and Comparative Study

In the proposed method, the performances of the detection scheme are tested with standard and our own collected dataset and are reported in Figure. 2. The performance evaluation for moving object detection is done by using performance metrics. There are two metrics for characterizing the Detection Rate (DR) and the False Alarm Rate (FAR) of the system represented in Eqn.(4) and (5) respectively. The scalars are combined to define the following metrics: TP, FP and FN values for different image sequences are shown in Table 1. And then, the obtained values of TP, FP and FN from Table 1, detection rate false alarm rate obtained, as shown in Table 2. The comparative study of state-of-art methods

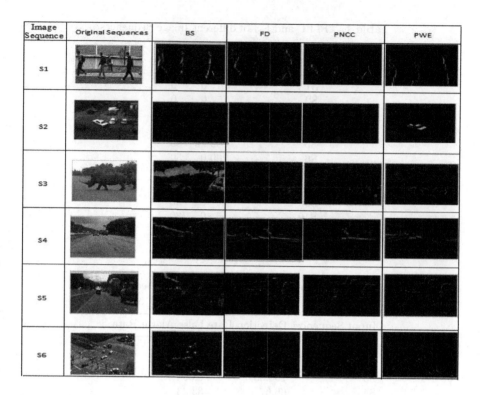

Fig. 2. Detection sequence of multiple objects by All Four Methods(Image Sequence From S1-S6)

likes Frame Difference (FD), Background Subtraction (BS) along with PNCC [1] methods against the proposed method MWE is done. Experimental results of moving object detection of an image sequence are shown in Figure 2. It is observed that the FD method and BS method are sensitive to noise and variation in illumination which gives more number of false foreground pixels and provides a false detection. Therefore, it fails to detect the moving object accurately. However, by the use of MWE method the static objects are detected as background and the moving objects are well detected under varies environmental conditions in addition to that camera movements handled as well. As observed from Figure 3, the FAR has a high value for FD, BS and PNCC methods. Evidently, the MWE method gives much efficient results.

$$DR = \frac{TP}{(TP + FN)} \tag{4}$$

$$FAR = \frac{FP}{(TP + FP)} \tag{5}$$

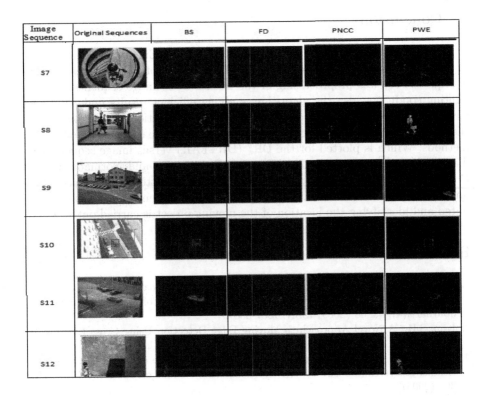

Fig. 3. Detection sequence of multiple objects by All Four Methods(Image Sequence From S7-S12)

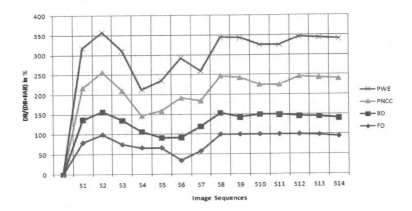

Fig. 4. Graph of DR/(DR + FAR) versus image sequences

4 Conclusions

An algorithm is proposed by Partitioning and Wavelet Energy for the detection of moving object from the image sequence. Advantage of this algorithm is that it requires very less preprocessing of the frames from image sequence. In poor lighting conditions also the algorithm is giving efficient results under different environment. From Fig. 4, it is observed that the Detection Rate is better and the False alarm rate is less for MWE method as compare to FD, BS and PNCC methods. Which is plotted for the DR/ (DR+FAR) versus image sequences indicates the comparative result of FD, BS, Partitioning and Normalized Cross Correlation and PWE method. This algorithm gives better performance as the images are partitioned into different sub images and then wavelet energy computed and identifying the location of the moving object is carried out.

References

1. Nagmode, M.S., Joshi, M.A., Sapkal, A.M.: A Novel approach to Detect and Track Moving Object using Partitioning and Normalized Cross Correlation. ICGST-GVIP Journal 9(4), 49–58 (2009) ISSN: 1687-398X
2. Kim, S.H.: A Novel Approach to Moving Edge Detection Using Cross Entropy. In: GVIP 2005 Conference, December 19-21, pp. 21–24. CICC, Cairo (2005)
3. Radke, R.J., Andra, S., Al-ofahi, O., Roysam, B.: Image change Detection Algorithms: A Systematic Survey. IEEE Transactions on Image Processing 14(3), 294–307 (2005)
4. Daviest, D., Palmert, P.L., Mirmehdit, M.: Detection and Tracking of Very Small low Contrast Objects. In: Proceedings of the 9th British Machine Vision Conference, p. 1 (1998)
5. Xu, L.Q.: Robust Detection and Tracking of Multiple Objects in Cluttered Scenes. In: One Day BMVA symposium at the Royal Statistical Society, London, UK (March 24, 2004)
6. Wang, Y., Van Dyck, R.E., Doherty, J.F.: Tracking moving objects in video sequences. In: Conference on Information Sciences and Systems, vol. 2, pp. 24–29 (March 2000); Paysan, P.: Stereo vision based vehicle classification using support vector Machines. thesis Submitted to the University of Applied Sciences, Fachhochschule Esslingenon (February 28, 2004)
7. Erdem, C.E., Sankur, B., Tekalp, A.M.: Performance Measures for Video Object Segmentation and Tracking. IEEE Transactions on Image Processing 13(7), 937–951 (2004)
8. Tsai, G., Chiang, A., Yang, T., Lai, C.C., Wang, S.-W., Liu, C.-D.: Video Tracking and Recognition System, Multimedia Lab, China
9. Nehme, M.A., Khoury, W., Yameen, B., Al-Alaoui, M.A.: Real Time Color Based Motion Detection and Tracking. In: Proceedings of the 3rd IEEE International Symposium on Signal Processing and Information Technology (ISSPIT 2003), December 14-17, pp. 696–700 (2003)

Image Steganography Using Integer Wavelet Transform Based on Color Space Approach

K. Sathish Shet[1], Nagaveni[1], and A.R. Aswath[2]

[1] Dept. of ECE, JSS Academy of Technical Education, Bangalore, India
[2] Dept. of Telecommunication, Dayananda Sagar College of Engineering, Bangalore, India
{satish.personal,yargolnagaveni02m,aswath.ar}@gmail.com

Abstract. Steganography is one in all the foremost persuasive approaches to mask the presence of hidden data inside a cover object. Images are one of the suitable cover objects for the Steganography. Current trends support, digital image files as the cover file to hide multiple secret images. Steganography presents the clandestine information can be covered in mediums like image, video and audio. This paper provides a LSB image Steganography technique to hide multiple secret images in a cover image which is in a YCbCr color space format using Integer Wavelet Transform (IWT). There is no optical variation between the stego image and the original cover image. The proposed hardware architecture gives very fast, programmable & cost effective hardware solution in the area of Secure Communication.

Keywords: Steganography, Inverse Wavelet Transform, Peak Signal to Noise Ratio, RGB Color Space.

1 Introduction

The exchange of multimedia system data doable by victimization the internet and wireless network. The strength of software package and latest equipments has provided users round the world with the ability to access, exchange and alter the multimedia system objects. There's invariably the chance of exchanging the information over a group of people or strive crack the information that is hidden. Steganography is an art and science of perpetuating the presence of the unrevealed broadcasting data by hiding the information within other information. Solely by the expected receiver it's not possible to decipher the hidden information from the stego image. Differing kinds of cover media are used, very fact that digital images are most commonly used due to that they're widely used over internet. Steganography is known as "covered writing" in Greek.

A Steganography system generally is to satisfy three main demands. They arc invisibility of embedding, reconstruction of original data and high capacity. In a Steganography work, the tactic that is employed for the encrypting the data is not known to unauthorized persons except the sender and deliberate persons. In this paper, digital image is taken as the cover object and two grayscale images are considered as secret data. The stego image is almost similar to the cover image so that

© Springer International Publishing Switzerland 2015
S.C. Satapathy et al. (eds.), *Proc. of the 3rd Int. Conf. on Front. of Intell. Comput. (FICTA) 2014*
– *Vol. 1*, Advances in Intelligent Systems and Computing 327, DOI: 10.1007/978-3-319-11933-5_95

no one can predict the presence of secret data. It can be determined using the parameter called PSNR.

2 Literature Survey

Karim, *et al.*, [1] proposed a new approach based on LSB using a secret key. The secret key encrypts the hidden information and then it is stored in different position of LSB of the image. This provides very good security, but the capacity of secret data storage is less.

In order to overcome the drawback of low capacity, XIE Qing *et al.*, [2] proposed a system where information is hidden in all RGB planes based on HVS (Human Visual System). This degrades the quality of the stego image. Sachdeva et al., [3] proposed a method in which Vector Quantization (VQ) table is utilized to conceal the covert message which increases the capacity along with stego size.

Sankar Roy *et al.*, [5] proposed an improved Steganography approach for hiding text messages within lossless RGB images which will suffer from withstanding the signal processing operations. Minimum deviation of fidelity based data embedding technique has been proposed by J. K. Mandal *et al.*, [6] where two bits per byte have been replaced by choosing the position randomly between LSB and up to four bits towards MSB. A DWT based mostly frequency domain steganographic method, termed as WTSIC is additionally planned by the same authors, [6] wherever secret message/image bit stream are embedded in horizontal, vertical and diagonal parts. Saeed Sarreshtedari *et al.*, [7] proposed a method to achieve a higher quality of the stego image using BPCS (Bit Plane Complexity Segmentation) in the wavelet domain. The capacity of each DWT block is estimated using the BPCS. Anjali Sejul, *et al.*, [8] proposed a method where binary images are taken as secret images. By taking the HSV (Hue, Saturation and Value) standards for cover image, the secret images are hidden. The secret image is inserted into the cover image by cropping the cover image according to the skin tone detection and then applying the DWT. In this method the capacity is too low.

A DWT-based Color Image Steganography Scheme, proposed by LIU Tong and QIU Zheng-ding [9] in which the secret information is hidden into a publicly accessed color image by a quantization related approach, so the moving of the covert information is unnoticed by criminal eavesdropper. Using this, the secret information is implanted in the wavelet area of each chrominance parts, therefore the hiding capacity is incremented. Abbas Cheddad *et al.*, [10] presented a Steganographic system which exploits the YCbCr color space.

Hemalatha S, *et al.*, [11] proposed a novel image steganography technique to hide multiple secret images and keys in the color cover image using Integer Wavelet Transform (IWT).

Maya C and Sabarinath proposed an Optimized FPGA Implementation of LSB replacement steganography using DWT [12]. This work focuses on the image steganography with an image compression using Discrete Wavelet Transform (DWT). Dr. Ahlam Fadhil Mahmood, et al., [13] proposed a steganography hardware approach where Linear Feedback Shift Register (LFSR) method has been used to hide the information in the image.

3   Proposed Work

Digital color images can be described in different color spaces such as RGB (Red Green,Blue),HSV(Hue,Saturation,Value),YUV,YIQ,YCbCr(Luminance/Chrominance) etc. YCbCr is not a complete color space relatively; it is a technique of encoding RGB data. The Y component represents the intensity of the light (luminance). The Cb and Cr components indicate the intensities of the blue and red components relative to the green component.

This color space exploits the properties of the human eye. The eye is more responsive to light intensity variations that can be found out easily by human eye rather than hue changes. The eye is sensitive to the very small modifications in intensity of light component, but the changes which are done for the cb and Cr components are less. Hence, in this paper the cover image is converted into Cb and Cr components. Fig. 1 shows the YCbCr format of the cover image.

Fig. 1. (a) Pepper (b) luminance component, Y of (a), (c) chrominance component Cb of (a), (d) chrominance component Cr of (a)

YCbCr signals are obtained from digital image RGB signals are as follows:

$$Y=0.299R+0.587G+0.114B \tag{1}$$

$$Cb=0.169R-0.331G+0.500B \tag{2}$$

$$Cr=0.500R-0.419G+0.081B \tag{3}$$

In this paper, for the cover image integer wavelet transform technique is used. A wavelet is a small packet of data with finite energy. It is an efficient approach for lossless compression. Here the coefficients are expressed by finite precision values. This wavelet transforms ports integers to integers. The equations that are used for the transformation are as follows. If the original image is A and total number of pixels are MxN, then the IWT coefficients are given by

$$LL_{i,j}=[(A_{2i,2j}+A_{2i+1,2j})/2] \tag{6}$$

$$HL_{i,j}=A_{2i+1,2j}-A_{2i,2j} \tag{7}$$

$$LH_{i,j}=A_{2i,2j+1}-A_{2i,2j} \tag{8}$$

$$HH_{i,j}=A_{2i+1,2j+1}-A_{2i,2j} \tag{9}$$

The inverse transform is given by

$$I_{2i,2j}=LL_{i,j}-\{HL_{i,j}\}/2 \tag{10}$$

$$I_{2i,2j+1}=LL_{i,j}+\{HL_{i,j+1}\}/2 \tag{11}$$

$$I_{2i+1,2j}=I_{2i,2j+1}+LH_{i,j}-HL_{i,j} \tag{12}$$

$$I_{2i+1,2j+1}=I_{2i+1,2j}+HH_{i,j}-LH_{i,j} \tag{13}$$

Where $1\leq i\leq M/2$ and $1\leq j\leq N/2$

The proposed block diagram is as shown in fig 2.

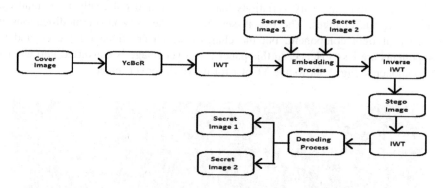

Fig. 2. Proposed Method

The proposed system is divided into two parts, such as encoding and decoding. In the encoding section two secret images are hidden in the cover image components of Cb and Cr using LSB technique. Integer wavelet transform is used here for the compression of cover images. The concept of YCbCr is considered which takes the advantage of the Human Vision System. In the extraction process inverse process of embedding is used in order to get back the original secret images back.

3.1   Embedding Processes

1. Select the cover image of size 256x256.
2. The selected cover image is transformed into YCbCr format .
3. Apply IWT for Cb and Cr components which results in four parts such as LL, LH, HL and HH.
4. These images are used for the embedding of two secret images. The Cb is used to hide one secret image and Cr is used for the other secret image.
5. The secret images are converted into the binary bit stream which is stored as a text file. Least Significant Bit is the lowest bit (right hand side)where data is hidden in a cover image.
6. Apply inverse IWT for both cover images. Combine the Y, Cb and Cr components resulting will be the stego image which is same as that of the cover image.

3.2   Extraction Process

1. The stego image is represented in YCbCr color space.
2. Obtain IWT of Cb and Cr components.
3. Start from the suitable pixel position of stego-image (binary received data). Check every pixel and store the first two bits, i.e., 0th and 1st bits of the components for extracting secret images.
4. Store the text file of resulting binary values of the secret images. This text file is converted back into secret images.

5   Experimental Results

The inputs which are described in the modules are functionally verified using the Matlab. Next the design is transformed into verilog and simulation results are compared with the Matlab results . After the successful completion of simulation, the design is synthesized using Xilinx.

In this work, an image file with ".bmp" format has been chosen as a cover image. The size of the cover image is 256x256. Two gray scale images of size 128x128 are taken as the secret images. The secret images are shown as "football" and "earth." As shown in fig (3(a) and 3(b)), are considered for testing the algorithm. The "football" is hidden in Cb component and "earth" is hidden in the Cr composition of pepper image. The cover image is shown in fig. 4 (a). The image with embedded data is called as stego image which resembles the cover image. Fig. 4 (b) shows the stego image.

Fig. 3. (a) football   (b) earth   **Fig. 4.** Matlab Result (a) cover Image (b)screte Image

The Stegno top module is the module of entire program as shown in fig 5.Table 1 describes the signal details of the top module.

The simulation results for the stegno top module are given below. In this module the two cover images Cb, Cr and two secret images are considered as an input. All these data are of eight bit in nature. The module operates in the presence of positive edge of 'clk' and 'rst'=0. Initially when the 'clk' signal is at the positive edge and 'rst'=1, no data evaluation process taking place as shown in fig 6.

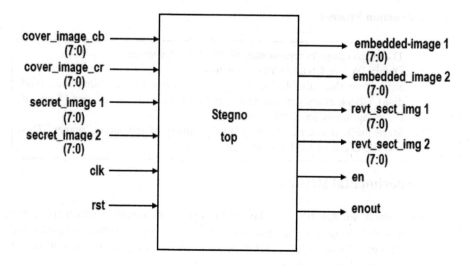

Fig. 5. Stegno Top Module

Table 1. Signal description of Top Level module

Signals	I/O	Description
Clk	Input	Synchronous clock signal of stegno top module
Rst	Input	Reset signal
image_in	Input	64 bit of cover image used to hide secret message
cover_image_cb	Input	8 bit of cover image of Cb part
cover_image_cr	Input	8 bit of cover image of Cr part
secrete_image1	Input	First 8 bit secrete image
secrete_image2	Input	Second 8 bit secrete image
embedded_image_1	output	8 bit of stego image 1
embedded image_2	output	8 bit of stego image 2
revt_sect_img 1	output	8 bit of extracted secret image 1
revt_sect_img 2	output	8 bit of extracted secret image 2
En	output	Enable output signal for embedded images
enout	output	Enable output signal for retrieved images

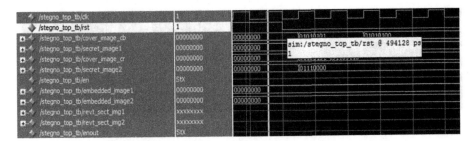

Fig. 6. Simulation Result of Initial Condition When 'rst' = 1

The data reading of input cover images and secret images are shown when 'rst' becomes zero which is shown in fig 7.

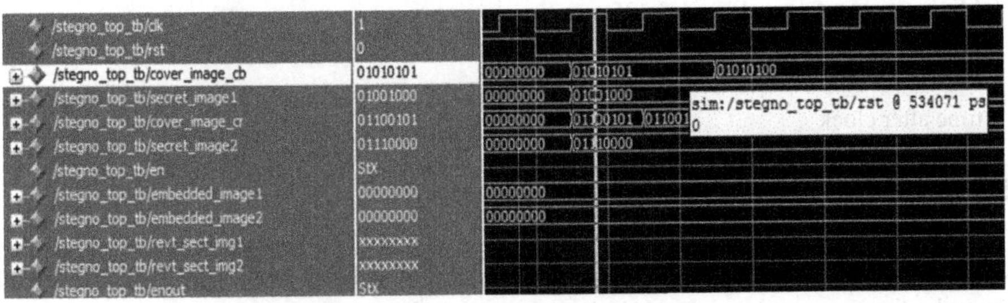

Fig. 7. Simulation result for data reading of Stegno top module

The output of two stego images and two extracted secret images are as shown in fig 8.

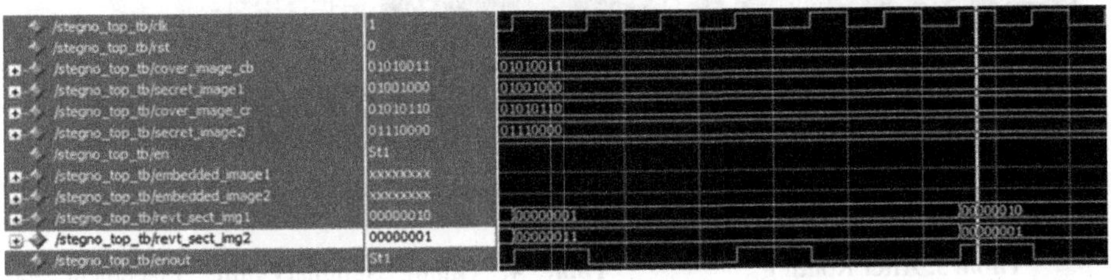

Fig. 8. Simulation results of Stegno Top Module

5.1   Synthesis Results

The generated RTL Schematic is as shown in fig 9. The Steganography using IWT shows delay of 3.157ns.The timing summary and power consumed are given in table 2.

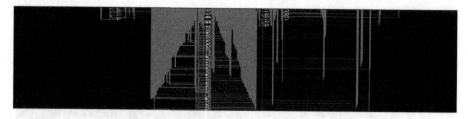

Fig. 9. RTL Schematic

Table 2. Timing summary and Power Consumed

Delay	3.157ns	Supply	Total	Dynamic	Quiescent
Maximum Frequency	316.768 MHz	Power (W)	0.187	0.022	0.165
Minimum input arrival time before clock	3.426ns				
Maximum output required time after clock	5.848ns				

5.2   Placement and Routing

The design is placed and routed to the device 4VLX15SF363-12 in Xilinx after the synthesis process. The resultant cell layout is shown in fig 10. The Table 3 and 4 describe the memory usage and timing summary after routing respectively.

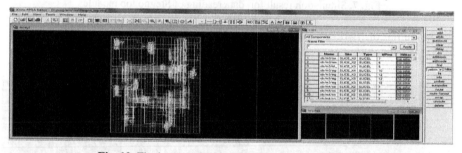

Fig. 10. Final routed Cell layout of the Proposed Design

Table 3. After Routing

Increased Memory	48Mb
Total memory	323Mb
Peak Memory	371Mb

Table 4. Timing Summary after Placement and Routing (PAR)

Set up Time	3.357ns
Hold Time	0.369ns
Total REAL time to PAR completion	27secs
Total CPU time to PAR completion	22secs

6   Conclusion

The experimental results obtained through simulation and synthesis shows that this method provides a hardware solution for hiding two secret images in a single cover image. In this implementation the authors used YCbCr color space concept which takes the advantage of Human Vision System (HVS). The data is encrypted and concealed with a steganography process provides an extra layer of protection and reduces the probability of the hidden data being detected. This method is one of the more secured ways of transmitting confidential information. At the receiver side by using inverse process, both the secret images can be successfully extracted from the cover image. In this paper an effort is made to enhance the overall processing speed when compared to the general purpose processor by implementing a hardware solution. It provides a very fast, programmable and cost effective hardware solution in the area of secure communication. Steganography isn't intended to switch cryptography, but rather to supplement it.

References

1. Masud Karim, S.M., Saifur Rahman, M., Ismail Hossain, M.: A New Approach for LSB Based Image Steganography using Secret Key. In: Proceedings of 14th International Conference on Computer and Information Technology, pp. 286–291. IEEE Conference Publications
2. Xie, Q., Xie, J., Xiao, Y.: A High Capacity Information Hiding Algorithm in Color Image. In: Proceedings of 2nd International Conference on E-Business and Information System Security, pp. 1–4. IEEE Conference Publications (2010)
3. Sachdeva, S., Kumar, A.: Colour Image Steganography Based on Modified Quantization Table. In: Proceedings of Second International Conference on Advanced Computing & Communication Technologies, pp. 309–313. IEEE Conference Publications (2012)
4. Roy, S., Parekh, R.: A Secure Keyless Image Steganography Approach for Lossless RGB Images. In: Proceedings of International Conference on Communication, Computing & Security, pp. 573–576. ACM Publications (2011)
5. Mandal, J.K., Sengupta, M.: Steganographic Technique Based on Minimum Deviation of Fidelity (STMDF). In: Proceedings of Second International Conference on Emerging Applications of Information Technology, pp. 298–301. IEEE Conference Publications (2011)
6. Mandal, J.K., Sengupta, M.: Authentication/Secret Message Transformation Through Wavelet Transform based Sub-band Image Coding (WTSIC). In: Proceedings of International Symposium on Electronic System Design, pp. 225–229. IEEE Conference Publications (2010)
7. Sarreshtedari, S., Ghaemmaghami, S.: High Capacity Image Steganography in Wavelet Domain. In: Proceedings of 2010 7th IEEE Consumer Communications and Networking Conference (CCNC), Las Vegas, Nevada, USA, January 9-12, pp. 1–5. IEEE Conference Publications (2010)
8. Shejul, A.A., Kulkarni, U.L.: A Secure Skin Tone based Steganography (SSTS) using Wavelet Transform. International Journal of Computer Theory and Engineering 3(1), 16–22 (2011)

9. Tong, L., Zheng-ding, Q.: A DWT-based Color Image Steganography Scheme. In: 6th International Conference on signal Processing, vol. 2, pp. 1568–1571. IEEE (2002)
10. Cheddad, A., Condell, J., Curran, K., Mc Kevitt, P.: Skin Tone Based Steganography in Video Files Exploiting the YCbCr Colour Space. In: School of Computing and Intelligent Systems, Londonderry, BT48 7JL, Northern Ireland, United Kingdom 978-1-4244-2571-6/08, pp. 905–908. IEEE (2008)
11. Hemalatha, S., Dinesh Acharya, U., Renuka, A., Kamath, P.R.: A Secure and high capacity image steganography technique. Signal & Image Processing: An International Journal (SIPIJ) 4(1), 83–89 (2013)
12. Maya, C., Sabarinath, G.: An Optimized FPGA Implementation of LSB Replacement Steganography Using DWT. International Journal of Advanced Research in Electrical, Electronics and Instrumentation Engineering (An ISO 3297: 2007 Certified Organization) 2(1), 586–593 (2013)
13. Mahmood, A.F., Kanai, N.A., Mohmmad, S.S.: An FPGA Implementation of Secured Steganography Communication System. Tikrit Journal of Engineering Sciences 19(4), 14–23 (2012)

Author Index

Printed in the United States
By Bookmasters